Advanced Studies in Theoretical and Applied Econometrics

Volume 52

This book series aims at addressing the most important and relevant current issues in theoretical and applied econometrics. It focuses on how the current data revolution has affected econometric modeling, analysis and forecasting, and how applied work benefited from this newly emerging data-rich environment. The series deals with all aspects macro-, micro-, financial-, and econometric methods and related disciplines, like for example, program evaluation or spatial analysis.

The volumes in the series can either be monographs or edited volumes, mainly targeting researchers, policymakers, and graduate students.

This book series is listed in Scopus.

More information about this series at http://www.springer.com/series/5667

Peter Fuleky
Editor

Macroeconomic Forecasting in the Era of Big Data

Theory and Practice

Editor
Peter Fuleky
UHERO and Department of Economics
University of Hawaii at Manoa
Honolulu, HI, USA

ISSN 1570-5811 ISSN 2214-7977 (electronic)
Advanced Studies in Theoretical and Applied Econometrics
ISBN 978-3-030-31152-0 ISBN 978-3-030-31150-6 (eBook)
https://doi.org/10.1007/978-3-030-31150-6

Mathematics Subject Classification (2010): 62-00, 62M10, 62M20, 91B64, 91B82

This Springer imprint is published by the registered company Springer Nature Switzerland AG.
The registered company address is: Gewerbestrasse 11, 6330 Cham, Switzerland

Foreword

The econometrics and statistics professions have made enormous progress on exploiting information in Big Data for a variety of purposes. In macroeconomics, Big Data analysis has a long history. In 1937, NBER researchers Wesley Mitchell and Arthur Burns studied 487 monthly and quarterly economic time series to advise policy makers on the state of the US business cycle. Their refined list of 71 variables formed a system of leading, lagging, and coincident indicators that has had lasting impact on business-cycle analysis and real-time macroeconomic monitoring and forecasting.

My own interest in the subject came 50 years later, in 1987, when the NBER asked Jim Stock and me to revisit the indicators and bring "modern" (circa 1980s) time series methods to bear on the Mitchell–Burns project. Like Mitchell and Burns, we began by collecting data on hundreds of monthly and quarterly time series. For each series, we produced a variety of plots and summary statistics, which we printed and stored in a large blue three-ring binder that we called the "blue book." Ultimately, the project culminated in an 11-variable dynamic factor model that could be used to monitor (nowcast) and forecast the state of the macroeconomy and the probability of a recession. In some respects, this was progress—indeed, it was viewed as state of the art in the late 1980s—but it certainly didn't exploit Big Data.

"Why," Jim asked, "do we have to limit the analysis to such a small set of variables? Can't we somehow use all the variables in the blue book? Aren't we throwing away a lot of information?" Yes, we were throwing away important information, but in 1987 macroeconomists didn't have the necessary Big Data tools to extract it. That has changed, and the contributions in this volume provide a systematic survey of Big Data methods currently used in macroeconomics and economics more generally. I wish I had this volume thirty-some years ago for the blue book project, but I'm certainly glad I have it now for future projects.

Princeton, NJ, USA
August 2019

Mark Watson

Foreword

Preface

The last three decades have seen a surge in data collection. During the same period, statisticians and econometricians have developed numerous techniques to digest the ever-growing amount of data and improve predictions. This volume surveys the adaptation of these methods to macroeconomic forecasting from both the theoretical and the applied perspective. The reader is presented with the current state of the literature and a broad collection of tools for analyzing large macroeconomic datasets. The intended audience includes researchers, professional forecasters, instructors, and students. The volume can be used as a reference and a teaching resource in advanced courses focusing on macroeconomic forecasting.

Each chapter of the book is self-contained with references. The topics are grouped into five main parts. Part I sets the stage by surveying big data types and sources. Part II reviews some of the main approaches for modeling relationships among macroeconomic time series, including dynamic factor models, vector autoregressions, volatility models, and neural networks. Part III showcases dimension reduction techniques yielding parsimonious model specifications. Part IV surveys methods that deal with model and forecast uncertainty. Several techniques described in Parts III and IV originated in the statistical learning literature and have been successfully adopted in econometrics. They are frequently combined to avoid overfitting and to improve forecast accuracy. Part V examines important extensions of the topics covered in the previous three sections.

This volume assumes prior training in econometrics and time series analysis. By filling a niche in forecasting in data-rich environments, it complements more comprehensive texts such as Economic Forecasting by Elliott and Timmermann (2016) and Applied Economic Forecasting Using Time Series Methods by Ghysels and Marcellino (2018). The chapters in this book attempt to balance the depth and breadth of the covered material, and are more survey-like than the papers published in the 2019 *Journal of Econometrics* special issue titled Big Data in Dynamic Predictive Econometric Modeling. Given the rapid evolution of big data analysis, the topics included in this volume have to be somewhat selective, focusing on the time series aspects of macroeconomic forecasting—as implied by the title—rather than on spatial, network, structural, and causal modeling.

I began working on this volume during my sabbatical at the Central European University, where I received lots of valuable advice and encouragement to proceed from László Mátyás. Working on this book has been a great learning experience. I would like to thank the wonderful team of contributors for excellent chapters produced in a timely manner. I appreciate their responsiveness and patience in dealing with my requests. I would also like to thank my colleagues at the University of Hawaii for letting me focus on this project during a busy academic year.

Honolulu, HI, USA
May 2019

Peter Fuleky

Contents

List of Contributors

Mohamed Affan Maldives Monetary Authority, Malé, Republic of Maldives

George Athanasopoulos Department of Econometrics and Business Statistics, Monash University, Caulfield, VIC, Australia

Marco Avella-Medina Department of Statistics, Columbia University, New York, NY, USA

Federico Bassetti Politcenico di Milano, Milan, Italy

Mauro Bernardi Department of Statistical Sciences, University of Padova and Istituto per le Applicazioni del Calcolo "Mauro Picone" - CNR, Roma, Italy

Giovanni Bonaccolto School of Economics and Law, University of Enna Kore, Enna, Italy

Tom Boot Department of Economics, Econometrics & Finance, University of Groningen, Groningen, The Netherlands

Jianfei Cao University of Chicago Booth School of Business, Chicago, IL, USA

Massimiliano Caporin Department of Statistical Sciences, University of Padova, Padova, Italy

Roberto Casarin University Ca' Foscari of Venice, Venice, Italy

Felix Chan School of Economics, Finance and Property, Curtin University, Perth, WA, Australia

Joshua C. C. Chan Purdue University and UTS, West Lafayette, IN, USA

Mingmian Cheng Department of Finance, Lingnan (University) College, Sun Yat-sen University, Guangzhou, China

Jianghao Chu Department of Economics, University of California, Riverside, CA, USA

Thomas R. Cook Federal Reserve Bank of Kansas City, Kansas, MO, USA

Michele Costola SAFE, Goethe University, Frankfurt, Germany

Catherine Doz Paris School of Economics and University Paris 1 Panthéon-Sorbonne, Paris, France

Martin Feldkircher Oesterreichische Nationalbank (OeNB), Vienna, Austria

Peter Fuleky UHERO and Department of Economics, University of Hawaii, Honolulu, HI, USA

Puwasala Gamakumara Department of Econometrics and Business Statistics, Monash University, Clayton, VIC, Australia

Philip ME Garboden Department of Urban and Regional Planning and UHERO, University of Hawaii Manoa, Honolulu, HI, USA

Bettina Grün Department of Applied Statistics, Johannes Kepler University Linz, Linz, Austria

Chris Gu Scheller College of Business, Georgia Institute of Technology, Atlanta, GA, USA

Paul Hofmarcher Salzburg Centre of European Union Studies (SCEUS), Department of Business, Economics and Social Theory, Paris Lodron University of Salzburg, Salzburg, Austria

Florian Huber Salzburg Centre of European Union Studies (SCEUS), University of Salzburg, Salzburg, Austria

Rob J Hyndman Department of Econometrics and Business Statistics, Monash University, Clayton, VIC, Australia

Anders B. Kock University of Oxford, Oxford, UK
Aarhus University and CREATES, Aarhus, Denmark

Tae-Hwy Lee Department of Economics, University of California, Riverside, CA, USA

Runze Li Department of Statistics and The Methodology Center, The Pennsylvania State University, State College, PA, USA

Wanjun Liu Department of Statistics, The Pennsylvania State University, State College, PA, USA

Marcelo C. Medeiros Pontifical Catholic University of Rio de Janeiro, Rio de Janeiro, Brazil

Didier Nibbering Department of Econometrics & Business Statistics, Monash University, Clayton, VIC, Australia

Anastasios Panagiotelis Department of Econometrics and Business Statistics, Monash University, Caulfield, VIC, Australia

Laurent Pauwels Discipline of Business Analytics, University of Sydney, Sydney, NSW, Australia

Michael Pfarrhofer Salzburg Centre of European Union Studies (SCEUS), University of Salzburg, Salzburg, Austria

Jeremy Piger Department of Economics, University of Oregon, Eugene, OR, USA

Francesco Ravazzolo Free University of Bozen-Bolzano, Bolzano, Italy
BI Norwegian Business School, Oslo, Norway

Marco Reale School of Mathematics and Statistics, University of Canterbury, Christchurch, New Zealand

Stephan Smeekes Department of Quantitative Economics, School of Business and Economics, Maastricht University, Maastricht, The Netherlands

Sylvia Soltyk School of Economics, Finance and Property, Curtin University, Perth, WA, Australia

Norman R. Swanson Department of Economics, School of Arts and Sciences, Rutgers University, New Brunswick, NJ, USA

Aman Ullah Department of Economics, University of California, Riverside, CA, USA

Gabriel Vasconcelos Pontifical Catholic University of Rio de Janeiro, Rio de Janeiro, Brazil

Ran Wang Department of Economics, University of California, Riverside, Riverside, CA, USA

Yike Wang Department of Economics, London School of Economics, London, UK

Etienne Wijler Department of Quantitative Economics, School of Business and Economics, Maastricht University, Maastricht, The Netherlands

Chun Yao Department of Economics, School of Arts and Sciences, Rutgers University, New Brunswick, NJ, USA

Part I
Introduction

Chapter 1
Sources and Types of Big Data for Macroeconomic Forecasting

Philip M. E. Garboden

1.1 Understanding What's Big About Big Data

Nearly two decades after the term "Big Data" first appeared in print, there remains little consensus regarding what it means (Lohr, 2012; Shi, 2014). Like many a scientific craze before it, the term Big Data quickly became an omnipresent buzzword applied to anything and everything that needed the gloss of being cutting edge. Big Data was going to solve many of society's most complex problems (Mayer-Schonberger & Cukier, 2013), while simultaneously sowing the seeds of its self-destruction (O'Neil, 2017). Of course, neither of these predictions was accurate; Big Data do have the potential to advance our understanding of the world, but the hard problems remain hard and incrementalism continues to dominate the social sciences. And while critics have expressed legitimate concerns, specifically at the intersection of data and governance, the consequences are by no means as dire as some would make them out to be.

Hyperbole aside, Big Data do have enormous potential to improve the timeliness and accuracy of macroeconomic forecasts. Just a decade ago, policymakers needed to wait for periodic releases of key indicators such as GDP and inflation followed by a series of subsequent corrections. Today, high frequency economic time series allow researchers to produce and adjust their forecasts far more frequently (Baldacci et al. 2016, Bok, Caratelli, Giannone, Sbordone, & Tambalotti 2018, Einav & Levin 2014a, Einav & Levin 2014b, Swanson & Xiong 2018), even in real time (Croushore, 2011). Not only are today's time series updated more frequently, but there are more of them available than ever before (Bok et al., 2018) on a much more

P. M. E. Garboden (✉)
Department of Urban and Regional Planning and UHERO, University of Hawaii Manoa, Honolulu, HI, USA
e-mail: pgarbod@hawaii.edu

P. Fuleky (ed.), *Macroeconomic Forecasting in the Era of Big Data*, Advanced Studies in Theoretical and Applied Econometrics 52,
https://doi.org/10.1007/978-3-030-31150-6_1

heterogeneous set of topics (Einav & Levin, 2014b), often with near population-level coverage (Einav & Levin, 2014a).

In this chapter, we consider the types of Big Data that have proven useful for macroeconomic forecasting. We begin by adjudicating between the various definitions of the term, settling on one we believe is most useful for forecasting. We review what the literature has presented as both the strengths and weaknesses of Big Data for forecasters, highlighting the particular set of skills necessary to utilize non-traditional data resources. This chapter leaves any in-depth discussion of analytic tools and data structures to the rest of the volume, and instead highlights the challenges inherent in the data themselves, their management, cleaning, and maintenance. We then propose a taxonomy of the types of data useful for forecasting, providing substantive examples for each. While we are neither the first nor the last researchers to take on such a categorization, we have structured ours to help readers see the full range of data that can be brought to bear for forecasting in a Big Data world.

1.1.1 How Big is Big?

Not surprisingly, there is no specific threshold after which a dataset can be considered "Big." Many commentators have attempted to describe the qualitative differences that separate Big Data from traditional data sources. Two themes emerge: First, Big Data are generally collected for purposes other than academic research and statistical modeling (Baldacci et al., 2016; Einav & Levin, 2014a). Second, they generally require processing beyond the capabilities of standard statistical software (Taylor, Schroeder, & Meyer 2014, Shi 2014, Hassani & Silva 2015). As much as neither of these assertions is wholly correct, there is some value in both.

Administrative data have been used for forecasting well before the advent of Big Data (Bok et al., 2018). Data from Multiple Listing Services, for example, have played a key role in forecasting future housing demand for decades, well before anyone thought to call it "Big" (Vidger, 1969). Nor are all Big Data administrative data. Data collected for scientific purposes from the Sloan Digital Sky Survey consist of over 175,000 galaxy spectra, prompting researchers to develop novel data management techniques (Yip et al., 2004). Despite numerous exceptions, however, much of the data we now consider "Big" is collected as a part of regular business or governing processes and thus, as we will describe below, present a number of technical challenges related to data management and cleaning.

The second argument, that Big Data are "Big" because of the computational requirements associated with their analysis, is similarly resonant but emerges more from the discipline's origins in computer science than from its uses in macroeconomics. Most of the definitions in this vein focus on whether or not a dataset is "difficult to process and analyze in reasonable time" (Shi, 2014; Tien, 2014) or more flippantly when a dataset is so large that "you can't use STATA"

(Hilger quoted in Taylor et al., 2014). And while these definitions clearly present a moving target (just think about STATA from 10 years ago), they reflect the fact that Big Data are an inherently intradisciplinary endeavor that often requires economists to collaborate with computer scientists to structure the data for analysis.

Rather than focusing on precise definitions, we believe it is more important to distinguish between the types of Big Data, specifically the dimensions by which a dataset can be considered "Big." Many canonical attempts have been made to define these parameters, from "volume, velocity, and variety" (Laney, 2001), to its more recent augmented form "volume, velocity, variety, and veracity" (Shi, 2014), to "tall, fat, and huge" (Buono, Mazzi, Kapetanios, Marcellino, & Papailias, 2017).

But most valuable for forecasting is the idea that time series data are big if they are huge in one or more of the following dimensions: the length of time (days, quarters, years) the data is collected (T), the number of samples per unit time that an observation is made (m),[1] and the number of variables that are collected at this rate (K) for an X matrix of dimensions ($mT \times K$) (Diebold, 2016b). Time series data are thus Big if they are tall (huge T), wide (huge K), dense (huge m), or any combination of those. The value of this approach is that it distinguishes between the types of data that are big because they have been collected for a very long time (US population, for example), those that are big because they are collected very frequently (tick-by-tick stock fluctuations), and those that contain a substantial number of variables (satellite imaging data). From a forecasting perspective, this differentiation provides a common language to determine the strengths of a dataset in forecasting volatility versus trend in both the short and the long term. As Diebold points out, dense data (huge m) are largely uninformative for forecasting long-term trends (for which one wants tall data) (Diebold, 2016c), but can be quite useful for volatility estimation (Diebold, 2016a).

1.1.2 The Challenges of Big Data

Big Data, however defined, present a unique set of challenges to macroeconomists above and beyond the need for new statistical tools. The data sources themselves require a level of technical expertise outside of the traditional methods curriculum. In this section, we outline those challenges.

Undocumented and Changing Data Structures

Most Big Data are not created for the benefit of economic researchers but exists as a byproduct of business or governmental activities. The Internal Revenue Service

[1]The value of m may not be constant for a particular time series and, in some cases such as tick data, may be dependent on the data generating process itself.

keeps tax information on all Americans so that it can collect taxes. Google stores billions of search queries so that it can improve its algorithms and increase advertisement revenue. Electronic medical records ensure continuity of care and accurate billing. None of these systems were designed with academic research in mind. While many businesses have begun to develop APIs and collaborate with academic institutions, these projects exist far outside of these firms' core business model. This has several consequences.

First of all, Big Data generally come to researchers uncleaned, unstructured, and undocumented (Baldacci et al., 2016; Einav & Levin, 2014b; Laney, 2001). The structure of "Big Data" can be quite complex often incorporating spatial and temporal elements into multi-dimensional unbalanced panels (Buono et al., 2017; Matyas, 2017). While traditional survey data include metadata that can help expedite analysis, most business and governmental data are simply exported from proprietary systems which store data in the way most convenient to the users (Bok et al., 2018). Larger more technologically friendly companies have invested resources in making their data more accessible for research. Data from the likes of Zillow, Twitter, and Google, for example, are made available through APIs, online dashboards, and data sharing agreements, generally with accompanying codebooks and documentation of database structure. But not all organizations have the resources necessary to prepare their data in that manner.

Second, traditional longitudinal data sources take care to ensure that the data collected is comparable across time. Data collected in later waves must be identical to or backwards compatible with earlier waves of the survey. For Big Data collected from private sources, the data change as business needs change, resulting in challenges when constructing time series over many years (huge T data). Moreover, much of technology that could realistically allow companies to store vast amounts of data was developed fairly recently, meaning that constructing truly huge T time series data is often incompatible with Big Data. To make matters worse, local government agencies are often required to store data for a set period of time, after which they can (and in our experience often do) destroy the information.

Third, many sources of Big Data rely on the increasing uptake of digital technologies and thus are representative of a changing proportion of the population (Baldacci et al., 2016; Buono et al., 2017). Cellular phone data, for example, have changed over just the last 20 years from including a small non-representative subset of all telecommunications to being nearly universal. Social networks, by comparison, go in and out of popularity over time as new competitors attract younger early adapters out of older systems.

Need for Network Infrastructure and Distributed Computing

A second core challenge for macroeconomists looking to utilize Big Data is that it can rarely be stored and processed on a personal computer (Bryant, Katz, & Lazowska 2008, Einav & Levin 2014a, Einav & Levin 2014b). Instead, it requires access to distributed cluster computing systems connected via high-speed networks.

Ironically, access to these technologies has put industry at an advantage over academia, with researchers' decision to embrace big data coming surprisingly "late to the game" (Bryant et al., 2008).

Fortunately, many campuses now host distributed computing centers to which faculty have access, eliminating a serious obstacle to big data utilization. Unfortunately, usage of these facilities can be expensive and require long-term funding streams if data infrastructure is to be maintained. Moreover, hardware access is necessary but not sufficient. The knowledge to use such equipment falls well outside of the usual Economics training and requires a great deal of specialized knowledge generally housed in computer science and engineering departments. While these challenges have forced Big Data computing to become a rare "interdisciplinary triumph" (Diebold, 2012; Shi, 2014), they also incur costs associated with the translational work necessary to link expertise across disciplines.

Costs and Access Limitations

While some agencies and corporations have enthusiastically partnered with macroeconomists, others have been far more hesitant to open their data up to outside scrutiny. Many corporations know that their data have value not just to academics but for business purposes as well. While many companies will negotiate data sharing agreements for non-commercial uses, some big data are, quite simply, very expensive.

A complementary concern is that because much of Big Data is proprietary, private, or both, organizations have become increasingly hesitant to share their data publicly, particularly when there are costs associated with deidentification. A security breech can not only harm the data provider's reputation (including potential litigation) but can also greatly reduce the amount of data the provider is willing to make available to researchers in the future. In the best case scenario, data owners are requiring increasingly costly security systems to be installed to limit potential harm (Bryant et al., 2008). In the worst, they simply refuse to provide the data.

Data Snooping, Causation, and Big Data Hubris

One of the main challenges with Big Data emerges from its greatest strength: there's a lot of it. Because many of the techniques used to extract patterns from vast datasets are agnostic with respect to theory, the researcher must remain vigilant to avoid overfitting (Baldacci et al., 2016; Hassani & Silva, 2015; Taylor et al., 2014). Although few macroeconomists would conflate the two, there is a tendency for journalists and the general public to interpret a predictive process with a causal one, a concern that is amplified in situations where variables are not pre-selected on theoretical grounds but instead are allowed to emerge algorithmically.

Big Data generation, moreover, rarely aligns the sampling logics, making traditional tests of statistical significance largely inappropriate (Abadie, Athey,

Imbens, & Wooldridge, 2014, 2017). The uncertainty of estimates based on Big Data is more likely to be based on the data generating process, design, or measurement than on issues associated with random draws from a theoretically infinite population.

When ignored, these issues can lead to what some researchers have referred to as "big data hubris" (Baldacci et al., 2016), a sort of overconfidence among researchers that having a vast amount of information can compensate for traditional econometric rigor around issues of selection, endogeneity, and causality.

Perhaps the most common parable of Big Data hubris comes from Google's attempt to build a real-time flu tracker that could help public health professionals respond more quickly (Lazer, Kennedy, King, & Vespignani, 2014). The idea behind the Google Flu Tracker (GFT) was to use search terms such as "do I have the flu?" to determine the spread of the flu much more rapidly than the CDC's traditional tracking, which relied on reports from hospital laboratories. The GFT was released to enormous fanfare and provided a sort of proof-of-concept for how Big Data could benefit society (Mayer-Schonberger & Cukier, 2013). Unfortunately, as time went on, the GFT became less and less effective, predicting nearly double the amount of flu than was actually occurring (Lazer et al., 2014).

The reason for this was fairly simple. Google had literally millions of search terms available in its database but only 1152 data points from the CDC. Because machine learning was utilized naively—without integration with statistical methods and theory—it was easy to identify a set of search terms that more or less perfectly predicted the CDC data. Going forward, however, the algorithm showed itself to have poor out-of-sample validity. According to Lazer et al. (2014), even 3-week-old CDC data do a better job of predicting current flu prevalence than Google's tracker, even after extensive improvements were made to the system.

While there is no reason to give up on what would be a useful public health tool, the shortcomings of GFT illustrate the risks of looking for patterns in data without expertise in economic forecasting and time series analysis. Moreover, Google Search is an evolving platform both in terms of its internal algorithms but also its usage, making it particularly hard to model the data generating function in a way that will be reliable across time.

1.2 Sources of Big Data for Forecasting

Having outlined the challenges of Big Data, we now turn to various forms of Big Data and how they can be useful for macroeconomic forecasting. They are: (1) Financial Market Data; (2) E-Commerce and Credit Cards; (3) Mobile Phones; (4) Search; (5) Social Media Data; (6) Textual Data; (7) Sensors, and The Internet of Things; (8) Transportation Data; (9) Other Administrative Data. There are a number of existing taxonomies of Big Data available in the literature (Baldacci et al., 2016; Bryant et al., 2008; Buono et al., 2017) and we do not claim any superiority to our structure other than a feeling for its inherent logic in the forecasting context. In each section we briefly describe the types of data that fall into each category and present exemplary (or at least interesting) examples of how each type of data has been used in forecasting. While all examples are related to forecasting, we admittedly look outside of macroeconomics for examples of the less common data types. Our goal was to err on the side of inclusion, as many of these examples may have relevance to macroeconomics.

1.2.1 Financial Market Data

Many core economic indicators such as inflation and GDP are released several months after the time period they represent, sometimes followed by a series of corrections. Because forecasts of these measures with higher temporal granularity are enormously valuable to private firms as well as government agencies, the issue of how to forecast (or nowcast[2]) economic indicators has received significant attention.

The sheer number of hourly, daily, weekly and monthly data series that can be applied to such analyses is staggering. Nearly all aspects of financial markets are regularly reported including commodity prices, trades and quotes of both foreign and domestic stocks, derivatives, option transactions, production indexes, housing starts and sales, imports, exports, industry sales, treasury bond rates, jobless claims, and currency exchange rates, just to name a few (Buono et al. 2017, Bańbura, Giannone, Modugno, & Reichlin 2013). The challenge is then how to manage such abundance, particularly when the number of items in each series (mT) is less than the number of series (K), and when the different data series relevant to the forecast are reported at different frequencies (m) and different release lags creating a ragged edge at the end of the trend (Bańbura et al., 2013).

It is well beyond the scope of this chapter to summarize all the examples of how high (or at least higher) frequency financial data have been used to improve macroeconomic forecasts (for some examples see Stock & Watson 2002, Aruoba, Diebold, & Scotti 2009, Giannone, Reichlin, & Small 2008, Angelini, Camba-Mendez, Giannone, Reichlin, & Rünstler 2011, Baumeister, Guérin, & Kilian, 2015, Monteforte & Moretti 2013, Andreou, Ghysels, & Kourtellos 2013, Kim & Swanson 2018, Pan, Wang, Wang, & Yang 2018). But a few examples are worth highlighting.

Modugno (2013) examines the issue of constructing an inflation forecast that can be updated continuously, rather than waiting for monthly releases (as is the case in the USA). Modugno uses daily data on commodity prices from the World Market Price of Raw Materials (RMP), weekly data on energy prices from the Weekly Retail Gasoline and Diesel Prices data (WRGDP) from the Energy Information Administration, monthly data on manufacturing from the Institute for Supply Management (released 2 weeks prior to inflation), and daily financial data from the

[2]We generally favor the use of the word "forecast" throughout the chapter even for predictions of contemporaneous events. In a philosophical sense, we see little difference in predicting a number that is unknown because it has not yet occurred or because it has not yet been observed.

US dollar index, the S&P 500, the Treasury constant maturity rate, and the Treasury-bill rate. He finds that for zero and 1 month horizons the inclusion of these mixed frequency data outperforms standard models but exclusively due to an improvement in the forecasting of energy and raw material prices.

Degiannakis and Filis (2018) attempt to use high-frequency market data to forecast oil prices. Arguing that because oil markets are becoming increasingly financialized, there are benefits in looking beyond market fundamentals to improve forecasts. Their model combines traditional measures on the global business cycle, oil production, oil stocks, and the capacity utilization rate with "ultra-high" frequency tick-by-tick data on exchange rates, stock market indexes, commodities (oil, gold, copper, gas, palladium, silver), and US T-bill rates. They find that for long-term forecasts, the fundamentals remain critical, but the inclusion of highly granular market data improves their short-term estimates in ways robust to various comparisons.

1.2.2 E-commerce and Scanner Data

In order to construct the consumer price index, the census sends fieldworkers out to collect prices on a basket of goods from brick and mortar stores across the country. While this represents a sort of gold standard for data quality, it is not without its limitations. It is both expensive to collect and impossible to monitor in real time (Cavallo & Rigobon, 2016). Nor is it able to address bias related to the substitution of one good for another or to provide details on how quality may shift within the existing basket; factors that are critical for an accurate measure of inflation (Silver & Heravi, 2001).

To fill this gap, economists have begun collecting enormous datasets of prices, relying either on bar-code scanner data (Silver & Heravi 2001, Berardi, Sevestre, & Thébault 2017) or by scraping online listings from e-commerce retailers (Cavallo & Rigobon, 2016; Rigobón, 2015). Perhaps the most famous of these is the MIT Billion Prices Project which, in 2019, was collecting 15 million prices every day from more than 1000 retailers in 60 countries (Project, 2019). The genesis of the project came from Cavallo (2013) and his interest in measuring inflation in Argentina. Cavallo suspected, and would later prove, that the official releases from Argentine officials were masking the true rate of inflation in the county. By scraping 4 years' worth of data from supermarket websites in Argentina (and comparison data in Brazil, Chile, Columbia, and Venezuela), Cavallo was able to show an empirical inflation rate of 20%, compared to official statistics which hovered around 4%. Building on this methodology, the Billion Prices Project began scraping and curating online sales prices from around the world allowing not only for inflation forecasting, but additional empirical research on price setting, stickiness, and so forth. Other similar work has been done using data from Adobe Analytics, which collects sales data from its clients for the production of business metrics (Goolsbee & Klenow, 2018).

While e-commerce has increased its market share substantially (and is continuing to do so), its penetration remains dwarfed by brick and mortar businesses particularly in specific sectors (like groceries). For this reason, researchers have partnered with particular retailers to collect price scanner data in order to construct price indexes (Ivancic, Diewert, & Fox, 2011), more precisely measure inflation (Silver & Heravi, 2001), and to examine the influence of geopolitical events on sales (Pandya & Venkatesan, 2016). While is seems like that the relative ease of collecting online prices, combined with the increasing e-commerce market share will tend to push research into the online space, both forms of price data have enormous potential for forecasting.

1.2.3 Mobile Phones

In advanced economies, nearly 100% of the population uses mobile phones, with the number of cellphone accounts exceeding the number of adults by over 25% (Blondel, Decuyper, & Krings, 2015). Even in the developing world, mobile phones are used by three quarters of the population; in a country such as Cote d'Ivoire, where fewer than 3% of households have access to the web, 83% use cell phones (Mao, Shuai, Ahn, & Bollen, 2015). This has pushed many researchers to consider the value of mobile phone data for economic forecasting, particularly in areas where traditional demographic surveys are expensive or dangerous to conduct (Blondel et al. 2015, Ricciato, Widhalm, Craglia, & Pantisano 2015).

The mobile phone data that is available to researchers is fairly thin, generally consisting solely of the Call Detail Records (CDR) that simply provide a unique identifier of the mobile device, the cell tower to which it connected, and type of connection (data, call, etc.), the time, and duration of the call (Ricciato et al., 2015). Some datasets will also include data on the destination tower or, if the user is moving, the different connecting towers that he or she may utilize during the route. To compensate for this thinness, however, is the data's broad coverage (particularly in areas with a single cellular carrier) and the potential for near real-time data access.

Mobile phone data have been used to forecast demographic trends such as population densities (Deville et al., 2014), poverty (Mao et al. 2015, Smith-Clarke, Mashhadi, & Capra 2014, Blumenstock, Cadamuro, & On 2015), and unemployment (Toole et al., 2015). In general, these papers look at the distribution of the number of cell phone calls and the networks of connections between towers (including density and heterogeneity) to improve forecasting at smaller levels of temporal and spatial granularity than is typically available.[3]

In one of the more ingenious analyses, Toole et al. (2015) looked at patterns of cell phone usage before and after a mass layoff in an undisclosed European

[3]This approach has particular value in times of disaster, upheaval or other unexpected events for which data are necessary for an effective response.

country. Using a Bayesian classification model, they found that after being laid off, individuals significantly reduced their level of communication. They made fewer calls, received fewer calls, and spoke with a smaller set of people in the period after the layoff than they had prior to the layoff (or than a control group in a town that did not experience the employment shock). The researchers used the relationship identified in this analysis to train a macroeconomic model to improve forecasts of regional unemployment, finding significant improvements to their predictions once mobile phone data were included.

Mobile phones and other always-on devices have also allowed economists to increase the frequency with which they survey individuals as they go about their daily routines. MacKerron and Mourta's Mappiness project, for example, utilizes a mobile phone app to continually ping users about their level of happiness throughout the day resulting in over 3.5 million data points produced by tens of thousands of British citizens (MacKerron & Mourato, 2010). While this particular project has yet to be employed for forecasting, it's clear that this sort of real-time attitudinal data could contribute to the prediction of multiple macroeconomic time series.

1.2.4 Search Data

Once the purview of marketing departments, the use of search data has increased in forecasting, driven by the availability of tools such as Google Trends (and, more recently, Google Correlate). These tools report a standardized measure of search volume for particular user-identified queries on a scale from 0, meaning no searches, to 100, representing the highest search volume for a particular topic during the time window. Smaller areas of spatial aggregation, namely states, are also available. While keywords with a search volume below a particular threshold are redacted for confidentiality reasons, Google Trends provides otherwise unavailable insight into a population's interest in a particular topic or desire for particular information.

It is hardly surprising given the nature of this data, that it would be used for forecasting. Although attempts have not been wholly successful (Lazer et al., 2014), many researchers have attempted to use search terms for disease symptoms to track epidemics in real time (Ginsberg et al., 2009; Yuan et al., 2013). Similar approaches have used search engine trend data to forecast hotel room demand (Pan, Chenguang Wu, & Song, 2012), commercial real estate values (Alexander Dietzel, Braun, & Schäfers, 2014), movie openings, video game sales, and song popularity (Goel, Hofman, Lahaie, Pennock, & Watts, 2010). The various studies using search data have indicated the challenges of making such predictions. Goel et al. (2010), for example, look at a number of different media and note significant differences in search data's ability to predict sale performance. For "non-sequel games," search proved a critical early indicator of buzz and thus first month sales. But this pattern did not hold for sequel games for which prequel sales were a much stronger predictor, or movies for which the inclusion of search in the forecasting models provided very little improvement over traditional forecasting models.

One of the places where search data have proved most valuable is in the early prediction of joblessness and unemployment (Choi & Varian, 2009, 2012; D'Amuri & Marcucci, 2017; Smith, 2016; Tefft, 2011). D'Amuri and Marcucci (2017), for example, use search volume for the word "jobs" to forecast the US monthly unemployment rate finding it to significantly outperform traditional models for a wide range of out-of-sample time periods (particularly during the Great Recession). In an interesting attempt at model falsification, the authors use Google Correlate to find the search term with the highest temporal correlation with "jobs" but that is substantively unrelated to employment (in this case, the term was "dos," referring to the operating system or the Department of State). The authors found that despite its strong in-sample correlation, "dos" performed poorly in out-of-sample forecasting. This combined with several other robustness checks, provides evidence that unlike early attempts with Google Flu Trends, online searchers looking for job openings are a sustainable predictor of unemployment rates.

1.2.5 Social Network Data

Since the early days of the internet, social networks have produced vast quantities of data, much of it in real time. As the use of platforms such as Facebook and Twitter have become nearly ubiquitous, an increasing number of economists have looked for ways to harness these data streams for forecasting. Theoretically, if the data from these networks can be collected and processed at sufficient speeds, forecasters may be able to gather information that has not yet been incorporated into the prices of stock or the betting markets, presenting opportunities for arbitrage (Giles 2010, Arias, Arratia, & Xuriguera 2013).

Generally speaking, the information provided by social media can be of two distinct kinds. First, it may provide actual information based on eyewitness accounts, rumors, or whisper campaigns that has not yet been reported by mainstream news sources (Williams & Reade, 2016). More commonly, social media data are thought to contain early signals of the sentiments or emotional states of specific populations which is predictive of their future investment behaviors (Mittal & Goel, 2012). Whether exogenous or based on unmeasured fundamentals, such sentiments can drive market behavior and thus may be useful data to incorporate into models for forecasting.

The most common social network used for forecasting is Twitter, likely because it lacks the privacy restrictions of closed networks such as Facebook and because of Twitter's support for the data needs of academic researchers. Each day, approximately 275 million active Twitter users draft 500 million tweets often live tweeting events such as sporting events, concerts, or political rallies in close to real time.

Several papers have examined how these data streams can be leveraged for online betting, amounting to a test of whether incorporating social media data can improve outcome forecasting more accurately than the traditional models used by odds makers. For example, Brown, Rambaccussing, Reade, and Rossi (2018) used

13.8 million tweets responding to events in UK Premier League soccer[4] matches. Coding each tweet using a microblogging dictionary as either a positive or negative reaction to a particular team's performance, the researchers found that social media contained significant information that had not yet been incorporated into betting prices on the real-time wagering site Betfair.com. Similar research has looked at Twitter's value in predicting box office revenue for films (Arias et al., 2013; Asur & Huberman, 2010) and the results of democratic elections (Williams & Reade, 2016).

Not surprisingly, a larger literature examines the ability of social networks to forecast stock prices, presenting itself as challenges to the efficient market hypothesis (Bollen, Mao, & Zeng 2011, Mittal & Goel 2012). Bollen et al. (2011), in a much cited paper, used Twitter data to collect what they term "collective mood states" operationally defined along six dimensions (calm, alert, sure, vital, kind, and happy). The researchers then used a self-organizing fuzzy neural network model to examine the non-linear association between these sentiments and the Dow Jones Industrial Average (DJIA). They found that the inclusion of some mood states (calm in particular) greatly improved predictions for the DJIA, suggesting that public sentiment was not fully incorporated into stock prices in real time. Similar work with similar findings has been done using social networks more directly targeting potential stock market investors such as stock message boards (Antweiler & Frank, 2004), Seeking Alpha (Chen, De, Hu, & Hwang, 2014) and The Motley Fool's CAPS system, which crowdsources individual opinions on stock movements (Avery, Chevalier, & Zeckhauser, 2015).

1.2.6 Text and Media Data

If social media is the most popular textual data used for forecasting, it is far from the only one. Text mining is becoming an increasingly popular technique to identify trends in both sentiment and uncertainty (Nassirtoussi, Aghabozorgi, Wah, & Ngo 2014, Bholat, Hansen, Santos, & Schonhardt-Bailey 2015). The most popular text data used in such analyses come from online newspapers, particularly business related newspapers like the *Wall Street Journal* or the *Financial Times* (Alanyali, Moat, & Preis 2013, Schumaker & Chen 2009, Thorsrud 2018, Baker, Bloom, & Davis 2016). But other sources are used as well such as minutes from the Fed's Federal Open Market Committee (FOMC) (Ericsson, 2016, 2017) and Wikipedia (Moat et al. 2013, Mestyán, Yasseri, & Kertész 2013).

In one of the field's seminal works, Baker et al. (2016) develop an index of economic policy uncertainty (EPU) which counts the number of articles using one or more terms from each of the following three groups: (1) "economic" or "economy," (2) "uncertainty," or "uncertain," and (3) "Congress," "deficit," "Federal Reserve," "legislation," "regulation," or "White House." For the last two decades in the USA,

[4]Football.

the researchers constructed this measure using the top 10 daily newspapers, with other sources used internationally and further back in time. While this approach will assuredly not capture all articles that suggest policy uncertainty, the index strongly correlates with existing measures of policy uncertainty and can improve economic forecasts. When the index of uncertainty increases, investment, output and employment all decline.

A wholly different use of textual data is employed by Moat et al. (2013). Rather than focus on the content of news sources, their research considers how investors seek information prior to trading decisions. Collecting a count of views and edits to Wikipedia pages on particular DJIA firms, the authors find a correlation between Wikipedia usage and movements in particular stocks. Fortunately or unfortunately (depending on your view), they did not find similar associations with more generic Wikipedia pages listed on the *General Economic Concepts* page such as "modern portfolio theory" or "comparative advantage."

1.2.7 Sensors, and the Internet of Things

The Internet of Things, like many technology trends, is more often discussed than used. Nevertheless, there is no doubt that the technology for ubiquitous sensing is decreasing dramatically in cost with potentially profound implications for forecasting. There are currently few examples of sensor data being used for economic forecasting, and that which does exist comes from satellite images, infrared networks, or sophisticated weather sensing hardware (rather than toasters and air conditioners).

One of the primary uses of sensor data is to collect satellite information on land use to predict economic development and GDP (Park, Jeon, Kim, & Choi 2011, Keola, Andersson, & Hall 2015, Seto & Kaufmann 2003). Keola et al. (2015), for example, use satellite imagery from the Defense Meteorological Satellite Program (DMSP) to estimate the level of ambient nighttime light and NASA's Moderate Resolution Imaging Spectroradiometer (MODIS) to determine whether non-urbanized areas are forests or agricultural land. The authors find that these two measures combined can be useful for forecasting economic growth. This technology, the authors argue, is particularly valuable in areas where traditional survey and administrative measures are not yet reliably available, specifically the developing world.

In a much more spatially limited implementation, Howard and Hoff (2013) use a network of passive infrared sensors to collect data on building occupancy. Applying a modified Bayesian combined forecasting approach, the authors are able to forecast building occupancy up to 60 min into the future. While seemingly inconsequential, the authors argue that this 1-hour-ahead forecast has the potential to dramatically reduce energy consumption as smart heating and cooling systems will be able to pre-cool or pre-heat rooms only when occupancy is forecast.

Looking to the future, an increasing number of electronics will soon be embedded with some form of low-cost computer capable of communicating remotely either to end users or to producers (Fleisch, 2010; Keola et al., 2015). Because this technology is in its infancy, no economic forecasts have incorporated data collected from these micro-computers (Buono et al., 2017). It is not hard, however, to imagine how such data could function in a forecasting context. Sensors embedded in consumer goods will be able to sense the proximity of other sensors, the functioning of the goods themselves, security threats to those goods, and user behavior (Fleisch, 2010). Data that suggest high levels of product obsolescence could be used to forecast future market demand. Data on user behavior may some day serve as a proxy for sentiments, fashion, and even health, all of which could theoretically forecast markets in a way similar to social network data.

1.2.8 Transportation Data

Over the last few years, there has been an enormous increase in both the quantity and quality of transportation data, whether through GPS enabled mobile phones, street level sensors, or image data. This has led to a sort of renaissance of forecasting within the transportation planning field. To date, the bulk of this work has been detached from the work of macroeconomists as it has primarily focused on predicting traffic congestion (Xia, Wang, Li, Li, & Zhang 2016, Yao & Shen 2017, Lv, Duan, Kang, Li, & Wang 2015, Polson & Sokolov 2017), electric vehicle charging demand (Arias & Bae, 2016), or transportation-related crime and accidents (Kouziokas, 2017).

One paper that suggests a locus of interaction between macroeconomists and transportation forecasters comes from Madhavi et al. (2017). This paper incorporates transportation data into models of electricity load forecasting and finds that, in addition to traditional weather variables, the volume of traffic into and out of a particular area can be highly predictive of energy demands.

While little work exists, there are a number of plausible uses of transportation data to forecast large-scale trends. Of course, transportation inefficiencies are likely predictive of gasoline prices. The pattern of peak travel demand could be indicative of changes of joblessness, sector growth or decline, and other labor force variables. And finally, the movement of vehicles throughout a metropolitan area could be used to construct a forecast of both residential and commercial real estate demand within particular metropolitan areas.

1.2.9 Other Administrative Data

The reliability and consistency of some government, nonprofit, and trade association data has made it difficult to construct reliable time series, particularly at a national level. In the presence of undocumented changes in the data collection process (either because of regulation or technology), it becomes difficult to disentangle the data generating process from changes in data coverage and quality. One exception to this trend is use of data on housing sales, collected through the National Association of Realtors' Multiple Listing Service (MLS) in the USA (for examples outside of the USA, see Baltagi & Bresson, 2017). These data, which track all property sales for which a real estate agent was involved, have been invaluable to those attempting to forecast housing market trends (Chen, Ong, Zheng, & Hsu 2017, Park & Bae 2015).

Park and Bae (2015), for example, use a variety of machine learning algorithms to forecast trends in housing prices in a single county in Fairfax, Virgina. Using the MLS data, they obtained 15,135 records of sold properties in the county, each of which contained a rich set of property-level attributes. In similar work in the international context, Chen and colleagues use a Support Vector Machine approach using administrative sales records from Taipei City. In both cases, the Big Data forecasting approach improved over previous methods.

One place where big administrative data have significant potential is in the area of local and national budget forecasting. At the national level, the approach to revenue and expenditure forecasting has followed traditional methods conducted by up to six agencies (The Council of Economic Advisers, The Office of Management and Budget, The Federal Reserve Board, The Congressional Budget Office, The Social Security Administration, and the Bureau of Economic Analysis) (Williams & Calabrese, 2016). While some of these agencies may be including Big Data into their forecasts, there is a dearth of literature on the subject (Ghysels & Ozkan, 2015) and federal budget forecasts have traditionally shown poor out-of-year performance (Williams & Calabrese, 2016). It is almost certain that the vast majority of local governments are not doing much beyond linear interpolations, which an abundance of literature suggests is conservative with respect to revenue (see Williams & Calabrese, 2016, for review).

1.2.10 Other Potential Data Sources

We have attempted to provide a fairly comprehensive list of the sources of Big Data useful for forecasting. In this, however, we were limited to what data *have* been tried rather than what data *could* be tried. In this final section, we propose several data sources that have not yet bubbled to the surface in the forecasting literature often because they have particular challenges related to curation or confidentiality.

The most noticeable gap relates to healthcare. Electronic medical records, while extremely sensitive, represent enormous amounts of data about millions of patients. Some of this, such as that collected by Medicare and Medicaid billing, is stored by the government, but other medical data rest with private providers. While social media may be responsive to rapidly spreading epidemics, electronic medical records would be useful in predicting healthcare utilization well into the future. In the future, the steady increase of "wearables" such as Apple watches and FitBits has the potential to provide real-time information on health and wellness, potentially even enabling researchers to measure stress and anxiety, both of which the literature suggests are predictive of market behavior.

Second, while the use of textual data has increased, audio and video data has been largely ignored by the forecasting literature. Online content is now produced and consumed via innumerable media ranging from YouTube videos, to podcasts, to animated gifs. As speech recognition software improves along with our ability to extract information from images and video recordings, it is likely that some of this data may prove useful for forecasting.

Finally, as noted above, there appears to be a significant under-utilization of administrative data, particularly that which is collected at the local level but relevant nationally. The challenges here have more to do with federalism and local control than they do with data science and econometrics. Data that are primarily collected by state and local governments are often inaccessible and extremely messy. This limits, for example, our ability to use education or court data to forecast national trends; the amount of local autonomy related to the collection and curation of this data presents a simply insurmountable obstacle, at least for now.

1.3 Conclusion

This chapter has outlined the various types of Big Data that can be applied to macroeconomic forecasting. By comparison to other econometric approaches, the field is still relatively new. Like all young fields, both the challenges and opportunities are not fully understood. By and large, the myriad publications outlined here have taken the approach of adding one or more big datasets into existing forecasting approaches and comparing the out-of-sample performance of the new forecast to traditional models. While essential, this approach is challenged by the simple fact that models that fail to improve performance have a much steeper hill to climb for publication (or even public dissemination). And yet knowing what does not improve a model can be very valuable information. Moreover, it remains unclear how approaches combining multiple Big Data sources will perform versus those that select one or the other. Over a dozen papers attempt to use Big Data to forecast unemployment, all using clever and novel sources of information. But are those sources redundant to one another or would combining multiple sources of data produce even better forecasts? Beyond simply bringing more data to the forecasting table, it is these questions that will drive research going forward.

References

Abadie, A., Athey, S., Imbens, G.W., & Wooldridge, J. M. (2014). *Finite population causal standard errors*. Cambridge: National Bureau of Economic Research.

Abadie, A., Athey, S., Imbens, G. W., & Wooldridge, J. M. (2017). *Samplingbased vs. design-based uncertainty in regression analysis*. arXiv Preprint :1706.01778.

Alanyali, M., Moat, H. S., & Preis, T. (2013). Quantifying the relationship between financial news and the stock market. *Scientific Reports, 3*, 3578.

Alexander Dietzel, M., Braun, N., & Schäfers, W. (2014). Sentiment-based commercial real estate forecasting with Google search volume data. *Journal of Property Investment & Finance, 32*(6), 540–569.

Andreou, E., Ghysels, E., & Kourtellos, A. (2013). Should macroeconomic forecasters use daily financial data and how? *Journal of Business & Economic Statistics, 31*(2), 240–251.

Angelini, E., Camba-Mendez, G., Giannone, D., Reichlin, L., & Rünstler, G. (2011). Short-term forecasts of Euro area GDP growth. *The Econometrics Journal, 14*(1), C25–C44.

Antweiler, W., & Frank, M. Z. (2004). Is all that talk just noise? the information content of internet stock message boards. *The Journal of Finance, 59*(3), 1259–1294.

Arias, M. B., & Bae, S. (2016). Electric vehicle charging demand forecasting model based on big data technologies. *Applied Energy, 183*, 327–339.

Arias, M., Arratia, A., & Xuriguera, R. (2013). Forecasting with Twitter data. *ACM Transactions on Intelligent Systems and Technology (TIST), 5*(1), 8.

Aruoba, S. B., Diebold, F. X., & Scotti, C. (2009). Real-time measurement of business conditions. *Journal of Business & Economic Statistics, 27*(4), 417–427.

Asur, S., & Huberman, B. A. (2010). Predicting the future with social media. In *Proceedings of the 2010 IEEE/WIC/ACM international conference on web intelligence and intelligent agent technology-volume 01* (pp. 492–499). Silver Spring: IEEE Computer Society.

Avery, C. N., Chevalier, J. A., & Zeckhauser, R. J. (2015). The CAPS prediction system and stock market returns. *Review of Finance, 20*(4), 1363–1381.

Baker, S. R., Bloom, N., & Davis, S. J. (2016). Measuring economic policy uncertainty. *The Quarterly Journal of Economics, 131*(4), 1593–1636.

Baldacci, E., Buono, D., Kapetanios, G., Krische, S., Marcellino, M., Mazzi, G., & Papailias, F. (2016). Big data and macroeconomic nowcasting: From data access to modelling. Luxembourg: Publications Office of the European Union.

Baltagi, B. H., & Bresson, G. (2017). Modelling housing using multi-dimensional panel data. In L. Matyas (Ed.), *The econometrics of multi-dimensional panels* (pp. 349–376). Berlin: Springer.

Bańbura, M., Giannone, D., Modugno, M., & Reichlin, L. (2013). Now-casting and the real-time data flow. In *Handbook of economic forecasting* (Vol. 2, pp. 195–237). Amsterdam: Elsevier.

Baumeister, C., Guérin, P., & Kilian, L. (2015). Do high-frequency financial data help forecast oil prices? the MIDAS touch at work. *International Journal of Forecasting, 31*(2), 238–252.

Berardi, N., Sevestre, P., & Thébault, J. (2017). The determinants of consumer price dispersion: Evidence from french supermarkets. In L. Matyas (Ed.), *The econometrics of multi-dimensional panels* (pp. 427–449). Berlin: Springer.

Bholat, D., Hansen, S., Santos, P., & Schonhardt-Bailey, C. (2015). *Text mining for central banks*. Available at SSRN 2624811.

Blondel, V. D., Decuyper, A., & Krings, G. (2015). A survey of results on mobile phone datasets analysis. *EPJ data science, 4*(1), 10.

Blumenstock, J., Cadamuro, G., & On, R. (2015). Predicting poverty and wealth from mobile phone metadata. *Science, 350*(6264), 1073–1076.

Bok, B., Caratelli, D., Giannone, D., Sbordone, A. M., & Tambalotti, A. (2018). Macroeconomic nowcasting and forecasting with big data. *Annual Review of Economics, 10*(0), 615–643.

Bollen, J., Mao, H., & Zeng, X. (2011). Twitter mood predicts the stock market. *Journal of computational science, 2*(1), 1–8.

Brown, A., Rambaccussing, D., Reade, J. J., & Rossi, G. (2018). Forecasting with social media: Evidence from tweets on soccer matches. *Economic Inquiry, 56*(3), 1748–1763.

Bryant, R., Katz, R., & Lazowska, E. (2008). *Big-data computing: Creating revolutionary breakthroughs in commerce, science, and society*. Washington: Computing Community Consortium.

Buono, D., Mazzi, G. L., Kapetanios, G., Marcellino, M., & Papailias, F. (2017). Big data types for macroeconomic nowcasting. *Eurostat Review on National Accounts and Macroeconomic Indicators, 1*(2017), 93–145.

Cavallo, A. (2013). Online and official price indexes: Measuring Argentina's inflation. *Journal of Monetary Economics, 60*(2), 152–165.

Cavallo, A.,& Rigobon, R. (2016). The billion prices project: Using online prices for measurement and research. *Journal of Economic Perspectives, 30*(2), 151–78.

Chen, H., De, P., Hu, Y. J., & Hwang, B.-H. (2014). Wisdom of crowds: The value of stock opinions transmitted through social media. *The Review of Financial Studies, 27*(5), 1367–1403.

Chen, J.-H., Ong, C. F., Zheng, L., & Hsu, S.-C. (2017). Forecasting spatial dynamics of the housing market using support vector machine. *International Journal of Strategic Property Management, 21*(3), 273–283.

Choi, H., & Varian, H. (2009). Predicting initial claims for unemployment benefits. *Google Inc*, 1–5.

Choi, H., & Varian, H. (2012). Predicting the present with Google trends. *Economic Record, 88*(S1), 2–9.

Croushore, D. (2011). Frontiers of real-time data analysis. *Journal of Economic :iterature, 49*(1), 72–100.

D'Amuri, F., & Marcucci, J. (2017). The predictive power of Google searches in forecasting US unemployment. *International Journal of Forecasting, 33*(4), 801–816.

Degiannakis, S., & Filis, G. (2018). Forecasting oil prices: High-frequency financial data are indeed useful. *Energy Economics, 76*, 388–402.

Deville, P., Linard, C., Martin, S., Gilbert, M., Stevens, F. R., Gaughan, A. E., et al. (2014). Dynamic population mapping using mobile phone data. *Proceedings of the National Academy of Sciences, 111*(45), 15888–15893.

Diebold, F. X. (2012). On the origin(s) and development of the term 'big data' (September 21, 2012). PIER working paper No. 12-037. Available at SSRN: https://ssrn.com/abstract=2152421orhttp://dx.doi.org/10.2139/ssrn.2152421

Diebold, F. X. (2016a). Big data for volatility vs. trend. https://fxdiebold.blogspot.com. Accessed: 2019 March 21.

Diebold, F. X. (2016b). Big data: Tall, wide, and dense. https://fxdiebold.blogspot.com. Accessed: 2019 March 21.

Diebold, F. X. (2016c). Dense data for long memory. https://fxdiebold.blogspot.com. Accessed: 2019 March 21.

Einav, L., & Levin, J. (2014a). Economics in the age of big data. *Science, 346*(6210), 1243089.

Einav, L., & Levin, J. (2014b). The data revolution and economic analysis. *Innovation Policy and the Economy, 14*(1), 1–24.

Ericsson, N. R. (2016). Eliciting GDP forecasts from the FOMC's minutes around the financial crisis. *International Journal of Forecasting, 32*(2), 571–583.

Ericsson, N. R. (2017). Predicting Fed forecasts. *Journal of Reviews on Global Economics, 6*, 175–180.

Fleisch, E. (2010). What is the internet of things? an economic perspective. *Economics, Management & Financial Markets, 5*(2), 125–157.

Ghysels, E., & Ozkan,N. (2015). Real-time forecasting of the US federal government budget: A simple mixed frequency data regression approach. *International Journal of Forecasting, 31*(4), 1009–1020.

Giannone, D., Reichlin, L., & Small, D. (2008). Nowcasting: The real-time informational content of macroeconomic data. *Journal of Monetary Economics, 55*(4), 665–676.

Giles, J. (2010). Blogs and tweets could predict the future. *New Scientist, 206*(2765), 20–21.

Ginsberg, J., Mohebbi, M. H., Patel, R. S., Brammer, L., Smolinski, M. S., & Brilliant, L. (2009). Detecting influenza epidemics using search engine query data. *Nature, 457*(7232), 1012.
Goel, S., Hofman, J. M., Lahaie, S., Pennock, D. M., & Watts, D. J. (2010). Predicting consumer behavior with web search. *Proceedings of the National academy of sciences, 107*(41), 17486–17490.
Goolsbee, A. D., & Klenow, P. J. (2018). Internet rising, prices falling: Measuring inflation in a world of e-commerce. In *Aea papers and proceedings* (Vol. 108, pp. 488–92).
Hassani, H., & Silva, E. S. (2015). Forecasting with big data: A review. *Annals of Data Science, 2*(1), 5–19.
Howard, J., & Hoff,W. (2013). Forecasting building occupancy using sensor network data. In *Proceedings of the 2nd international workshop on big data, streams and heterogeneous source mining: Algorithms, systems, programming models and applications* (pp. 87–94). Ney York: ACM.
Ivancic, L., Diewert,W. E., & Fox, K. J. (2011). Scanner data, time aggregation and the construction of price indexes. *Journal of Econometrics, 161*(1), 24–35.
Keola, S., Andersson, M., & Hall, O. (2015). Monitoring economic development from space: Using nighttime light and land cover data to measure economic growth. *World Development, 66*, 322–334.
Kim, H. H., & Swanson, N. R. (2018). Methods for backcasting, nowcasting and forecasting using factor-MIDAS: With an application to Korean GDP. *Journal of Forecasting, 37*(3), 281–302.
Kouziokas, G. N. (2017). The application of artificial intelligence in public administration for forecasting high crime risk transportation areas in urban environment. *Transportation Research Procedia, 24*, 467–473.
Laney, D. (2001). 3d data management: Controlling data volume, velocity and variety. *META Group Research Note, 6*(70), 1.
Lazer, D., Kennedy, R., King, G., & Vespignani, A. (2014). The parable of Google Flu: Traps in big data analysis. *Science, 343*(6176), 1203–1205.
Lohr, S. (2012). How big data became so big. *New York Times, 11*, BU3.
Lv, Y., Duan, Y., Kang, W., Li, Z., & Wang, F.-Y. (2015). Traffic flow prediction with big data: A deep learning approach. *IEEE Transactions on Intelligent Transportation Systems, 16*(2), 865–873.
MacKerron, G., & Mourato, S. (2010). LSE's mappiness project may help us track the national mood: But how much should we consider happiness in deciding public policy? *British Politics and Policy at LSE.*
Madhavi, K. L., Cordova, J., Ulak, M. B., Ohlsen, M., Ozguven, E. E., Arghandeh, R., & Kocatepe, A. (2017). Advanced electricity load forecasting combining electricity and transportation network. In *2017 North American power symposium (NAPS)* (pp. 1–6). Piscataway: IEEE.
Mao, H., Shuai, X., Ahn, Y.-Y., & Bollen, J. (2015). Quantifying socio-economic indicators in developing countries from mobile phone communication data: Applications to Côte d'Ivoire. *EPJ Data Science, 4*(1), 15.
Matyas, L. (2017). *The econometrics of multi-dimensional panels*. Berlin: Springer.
Mayer-Schonberger, V., & Cukier, K. (2013). Big data: *A revolution that will transform how we live, work, think.* London: Taylor & Francis.
Mestyán, M., Yasseri, T., & Kertész, J. (2013). Early prediction of movie box office success based on Wikipedia activity big data. *PloS one, 8*(8), e71226.
Mittal, A., & Goel, A. (2012). *Stock prediction using Twitter sentiment analysis*.
Moat, H. S., Curme, C., Avakian, A., Kenett, D. Y., Stanley, H. E., & Preis, T. (2013). Quantifying Wikipedia usage patterns before stock market moves. *Scientific reports, 3*, 1801.
Modugno, M. (2013). Now-casting inflation using high frequency data. *International Journal of Forecasting, 29*(4), 664–675.
Monteforte, L., & Moretti, G. (2013). Real-time forecasts of inflation: The role of financial variables. *Journal of Forecasting, 32*(1), 51–61.
Nassirtoussi, A. K.,Aghabozorgi, S.,Wah, T.Y., & Ngo, D. C. L. (2014). Text mining for market prediction: A systematic review. *Expert Systems with Applications, 41*(16), 7653–7670.

O'Neil, C. (2017). *Weapons of math destruction: How big data increases inequality and threatens democracy*. New York: Broadway Books.
Pan, B., Chenguang Wu, D., & Song, H. (2012). Forecasting hotel room demand using search engine data. *Journal of Hospitality and Tourism Technology, 3*(3), 196–210.
Pan, Z., Wang, Q., Wang, Y., & Yang, L. (2018). Forecasting US real GDP using oil prices: A time-varying parameter MIDAS model. *Energy Economics, 72*, 177–187.
Pandya, S. S., & Venkatesan, R. (2016). French roast: Consumer response to international conflict—evidence from supermarket scanner data. *Review of Economics and Statistics, 98*(1), 42–56.
Park, B., & Bae, J. K. (2015). Using machine learning algorithms for housing price prediction: The case of Fairfax County, Virginia housing data. *Expert Systems with Applications, 42*(6), 2928–2934.
Park, S., Jeon, S., Kim, S., & Choi, C. (2011). Prediction and comparison of urban growth by land suitability index mapping using GIS and RS in South Korea. *Landscape and Urban Planning, 99*(2), 104–114.
Polson, N. G., & Sokolov, V. O. (2017). Deep learning for short-term traffic flow prediction. *Transportation Research Part C: Emerging Technologies, 79*, 1–17. 10.1080/07350015.2018.1506344
Project, B. P. (2019). Billion prices projectwebsite. http://www.thebillionpricesproject.com. Accessed: 2019 March 21.
Ricciato, F., Widhalm, P., Craglia, M., & Pantisano, F. (2015). *Estimating population density distribution from network-based mobile phone data*. Luxembourg: Publications Office of the European Union.
Rigobón, R. (2015). Presidential address: Macroeconomics and online prices. *Economia, 15*(2), 199–213.
Schumaker, R. P., & Chen, H. (2009). Textual analysis of stock market prediction using breaking financial news: The AZFin text system. *ACM Transactions on Information Systems, 27*(2), 12.
Seto, K. C., & Kaufmann, R. K. (2003). Modeling the drivers of urban land use change in the Pearl River Delta, China: Integrating remote sensing with socioeconomic data. *Land Economics, 79*(1), 106–121.
Shi, Y. (2014). Big data: History, current status, and challenges going forward. *Bridge, 44*(4), 6–11.
Silver, M., & Heravi, S. (2001). Scanner data and the measurement of inflation. *The Economic Journal, 111*(472), 383–404.
Smith, P. (2016). Google's MIDAS touch: Predicting UK unemployment with internet search data. *Journal of Forecasting, 35*(3), 263–284.
Smith-Clarke, C., Mashhadi, A., & Capra, L. (2014). Poverty on the cheap: Estimating poverty maps using aggregated mobile communication networks. In *Proceedings of the sigchi conference on human factors in computing systems* (pp. 511–520). New York: ACM.
Stock, J. H., & Watson, M.W. (2002). Forecasting using principal components from a large number of predictors. *Journal of the American statistical association, 97*(460), 1167–1179.
Swanson, N. R., & Xiong, W. (2018). Big data analytics in economics: What have we learned so far, and where should we go from here? *Canadian Journal of Economics, 51*(3), 695–746.
Taylor, L., Schroeder, R., & Meyer, E. (2014). Emerging practices and perspectives on big data analysis in economics: Bigger and better or more of the same? *Big Data & Society, 1*(2), 2053951714536877.
Tefft, N. (2011). Insights on unemployment, unemployment insurance, and mental health. *Journal of Health Economics, 30*(2), 258–264.
Thorsrud, L. A. (2018). Words are the new numbers: A newsy coincident index of the business cycle. *Journal of Business & Economic Statistics*, 1–17. https://doi.org/10.1080/07350015.2018.1506344
Tien, J. (2014). Overview of big data, a US perspective. *The Bridge–Linking Engineering and Society, 44*(4), 13–17.

Toole, J. L., Lin, Y.-R., Muehlegger, E., Shoag, D., González, M. C., & Lazer, D. (2015). Tracking employment shocks using mobile phone data. *Journal of The Royal Society Interface, 12*(107), 20150185.

Vidger, L. P. (1969). Analysis of price behavior in san francisco housing markets: The historical pattern (1958–67) and projections (1968–75). *The Annals of Regional Science, 3*(1), 143–155.

Williams, D. W., & Calabrese, T. D. (2016). The status of budget forecasting. *Journal of Public and Nonprofit Affairs, 2*(2), 127–160.

Williams, L. V., & Reade, J. J. (2016). Prediction markets, social media and information efficiency. *Kyklos, 69*(3), 518–556.

Xia, D.,Wang, B., Li, H., Li, Y., & Zhang, Z. (2016). A distributed spatial–temporal weighted model on MapReduce for short-term traffic flow forecasting. *Neurocomputing*, 179, 246–263.

Yao, S.-N., & Shen, Y.-C. (2017). Functional data analysis of daily curves in traffic: Transportation forecasting in the real-time. In *2017 computing conference* (pp. 1394–1397). Piscataway: IEEE.

Yip, C.-W., Connolly, A., Szalay, A., Budavári, T., SubbaRao, M., Frieman, J., et al. (2004). Distributions of galaxy spectral types in the sloan digital sky survey. *The Astronomical Journal, 128*(2), 585.

Yuan, Q., Nsoesie, E. O., Lv, B., Peng, G., Chunara, R., & Brownstein, J. S. (2013). Monitoring influenza epidemics in china with search query from baidu. *PloS one, 8*(5), e64323.

Part II
Capturing Dynamic Relationships

Chapter 2
Dynamic Factor Models

Catherine Doz and Peter Fuleky

2.1 Introduction

Factor analysis is a dimension reduction technique summarizing the sources of variation among variables. The method was introduced in the psychology literature by Spearman (1904), who used an unobserved variable, or factor, to describe the cognitive abilities of an individual. Although originally developed for independently distributed random vectors, the method was extended by Geweke (1977) to capture the comovements in economic time series. The idea that the comovement of macroeconomic series can be linked to the business cycle has been put forward by Burns and Mitchell (1946): "a cycle consists of expansions occurring at about the same time in many economic activities, followed by similarly general recessions, contractions, and revivals which merge into the expansion phase of the next cycle; this sequence of changes is recurrent but not periodic." Early applications of dynamic factor models (DFMs) to macroeconomic data, by Sargent and Sims (1977) and Stock and Watson (1989, 1991, 1993), suggested that a few latent factors can account for much of the dynamic behavior of major economic aggregates. In particular, a dynamic single-factor model can be used to summarize a vector of macroeconomic indicators, and the factor can be seen as an index of economic conditions describing the business cycle. In these studies, the number of time periods in the dataset exceeded the number of variables, and identification of the factors required relatively strict assumptions. While increments in the time dimension were

C. Doz
Paris School of Economics and University Paris 1 Panthéon-Sorbonne, Paris, France
e-mail: catherine.doz@univ-paris1.fr

P. Fuleky (✉)
UHERO and Department of Economics, University of Hawaii, Honolulu, HI, USA
e-mail: fuleky@hawaii.edu

P. Fuleky (ed.), *Macroeconomic Forecasting in the Era of Big Data*,
Advanced Studies in Theoretical and Applied Econometrics 52,
https://doi.org/10.1007/978-3-030-31150-6_2

limited by the passage of time, the availability of a large number of macroeconomic and financial indicators provided an opportunity to expand the dataset in the cross-sectional dimension and to work with somewhat relaxed assumptions. Chamberlain (1983) and Chamberlain and Rothschild (1983) applied factor models to wide panels of financial data and paved the way for further development of dynamic factor models in macroeconometrics. Indeed, since the 2000s dynamic factor models have been used extensively to analyze large macroeconomic datasets, sometimes containing hundreds of series with hundreds of observations on each. They have proved useful for synthesizing information from variables observed at different frequencies, estimation of the latent business cycle, nowcasting and forecasting, and estimation of recession probabilities and turning points. As Diebold (2003) pointed out, although DFMs "don't analyze *really* Big Data, they certainly represent a movement of macroeconometrics in that direction," and this movement has proved to be very fruitful.

Several very good surveys have been written on dynamic factor models, including Bai and Ng (2008b); Bai and Wang (2016); Barigozzi (2018); Barhoumi, Darné, and Ferrara (2014); Breitung and Eickmeier (2006); Lütkepohl (2014); Stock and Watson (2006, 2011, 2016). Yet we felt that the readers of this volume would appreciate a chapter covering the evolution of dynamic factor models, their estimation strategies, forecasting approaches, and several extensions to the basic framework. Since both small- and large-dimensional models have advanced over time, we review the progress achieved under both frameworks. In Sect. 2.2 we describe the distinguishing characteristics of exact and approximate factor models. In Sects. 2.3 and 2.4, we review estimating procedures proposed in the time domain and the frequency domain, respectively. Section 2.5 presents approaches for determining the number of factors. Section 2.6 surveys issues associated with forecasting, and Sect. 2.8 reviews the handling of structural breaks in dynamic factor models.

2.2 From Exact to Approximate Factor Models

In a factor model, the correlations among n variables, $\boldsymbol{x}_1 \dots \boldsymbol{x}_n$, for which T observations are available, are assumed to be entirely due to a few, $r < n$, latent unobservable variables, called factors. The link between the observable variables and the factors is assumed to be linear. Thus, each observation x_{it} can be decomposed as

$$x_{it} = \mu_i + \boldsymbol{\lambda}_i' \boldsymbol{f}_t + e_{it},$$

where μ_i is the mean of $\boldsymbol{x}_i$, $\boldsymbol{\lambda}_i$ is an $r \times 1$ vector, and e_{it} and $\boldsymbol{f}_t$ are two uncorrelated processes. Thus, for $i = 1 \dots n$ and $t = 1 \dots T$, x_{it} is decomposed into the sum of two mutually orthogonal unobserved components: the common component, $\chi_{it} = \boldsymbol{\lambda}_i' \boldsymbol{f}_t$, and the idiosyncratic component, $\xi_{it} = \mu_i + e_{it}$. While the factors drive the correlation between $\boldsymbol{x}_i$ and $\boldsymbol{x}_j$, $j \neq i$, the idiosyncratic component arises from features that are specific to an individual $\boldsymbol{x}_i$ variable. Further assumptions placed on the two unobserved components result in factor models of different types.

2.2.1 Exact Factor Models

The exact factor model was introduced by Spearman (1904). The model assumes that the idiosyncratic components are not correlated at any leads and lags so that ξ_{it} and ξ_{ks} are mutually orthogonal for all $k \neq i$ and any s and t, and consequently all correlation among the observable variables is driven by the factors. The model can be written as

$$x_{it} = \mu_i + \sum_{j=1}^{r} \lambda_{ij} f_{jt} + e_{it}, \quad i = 1 \ldots n,\ t = 1 \ldots T, \tag{2.1}$$

where f_{jt} and e_{it} are orthogonal white noises for any i and j; e_{it} and e_{kt} are orthogonal for $k \neq i$; and λ_{ij} is the loading of the jth factor on the ith variable. Equation (2.1) can also be written as

$$x_{it} = \mu_i + \boldsymbol{\lambda}_i' \boldsymbol{f}_t + e_{it}, \quad i = 1 \ldots n,\ t = 1 \ldots T,$$

with $\boldsymbol{\lambda}_i' = (\lambda_{i1} \ldots \lambda_{ir})$ and $\boldsymbol{f}_t = (f_{1t} \ldots f_{rt})'$. The common component is $\chi_{it} = \sum_{j=1}^{r} \lambda_{ij} f_{jt}$, and the idiosyncratic component is $\xi_{it} = \mu_i + e_{it}$.

In matrix notation, with $\boldsymbol{x}_t = (x_{1t} \ldots x_{nt})'$, $\boldsymbol{f}_t = (f_{1t} \ldots f_{rt})'$, and $\boldsymbol{e}_t = (e_{1t} \ldots e_{nt})'$, the model can be written as

$$\boldsymbol{x}_t = \boldsymbol{\mu} + \boldsymbol{\Lambda} \boldsymbol{f}_t + \boldsymbol{e}_t, \tag{2.2}$$

where $\boldsymbol{\Lambda} = (\boldsymbol{\lambda}_1 \ldots \boldsymbol{\lambda}_n)'$ is a $n \times r$ matrix of full column rank (otherwise fewer factors would suffice), and the covariance matrix of e_t is diagonal since the idiosyncratic terms are assumed to be uncorrelated. Observations available for $t = 1, \ldots T$ can be stacked, and Eq. (2.2) can be rewritten as

$$\boldsymbol{X} = \iota_T \otimes \boldsymbol{\mu}' + \boldsymbol{F}\boldsymbol{\Lambda}' + \boldsymbol{E},$$

where $\boldsymbol{X} = (\boldsymbol{x}_1 \ldots \boldsymbol{x}_T)'$ is a $T \times n$ matrix, $\boldsymbol{F} = (\boldsymbol{f}_1 \ldots \boldsymbol{f}_T)'$ a $T \times r$ matrix, $\boldsymbol{E} = (\boldsymbol{e}_1 \ldots \boldsymbol{e}_T)'$ is a $T \times n$ matrix, and ι_T is a $T \times 1$ vector with components equal to 1.

While the core assumption behind factor models is that the two processes $\boldsymbol{f}_t$ and $\boldsymbol{e}_t$ are orthogonal to each other, the exact static factor model further assumes that the idiosyncratic components are also orthogonal to each other, so that any correlation between the observable variables is solely due to the common factors.

Both orthogonality assumptions are necessary to ensure the identifiability of the model (see Anderson, 1984; Bartholomew, 1987). Note that $\boldsymbol{f}_t$ is defined only up to a premultiplication by an invertible matrix $\boldsymbol{Q}$ since $\boldsymbol{f}_t$ can be replaced by $\boldsymbol{Q}\boldsymbol{f}_t$ whenever $\boldsymbol{\Lambda}$ is replaced by $\boldsymbol{\Lambda}\boldsymbol{Q}^{-1}$. This means that only the respective spaces spanned by $\boldsymbol{f}_t$ and by $\boldsymbol{\Lambda}$ are uniquely defined. This so-called indeterminacy problem of the factors must be taken into account at the estimation stage.

Factor models have been introduced in economics by Geweke (1977), Sargent and Sims (1977), and Engle and Watson (1981). These authors generalized the model above to capture dynamics in the data. Using the same notation as in Eq. (2.2), the dynamic exact factor model can be written as

$$\boldsymbol{x}_t = \boldsymbol{\mu} + \boldsymbol{\Lambda}_0\boldsymbol{f}_t + \boldsymbol{\Lambda}_1\boldsymbol{f}_{t-1} + \cdots + \boldsymbol{\Lambda}_s\boldsymbol{f}_{t-s} + \boldsymbol{e}_t,$$

or more compactly

$$\boldsymbol{x}_t = \boldsymbol{\mu} + \boldsymbol{\Lambda}(L)\boldsymbol{f}_t + \boldsymbol{e}_t, \tag{2.3}$$

where L is the lag operator.

In this model, $\boldsymbol{f}_t$ and $\boldsymbol{e}_t$ are no longer assumed to be white noises, but are instead allowed to be autocorrelated dynamic processes evolving according to $\boldsymbol{f}_t = \boldsymbol{\Theta}(L)\boldsymbol{u}_t$ and $\boldsymbol{e}_t = \boldsymbol{\rho}(L)\boldsymbol{\varepsilon}_t$, where the q and n dimensional vectors $\boldsymbol{u}_t$ and $\boldsymbol{\varepsilon}_t$, respectively, contain iid errors. The dimension of $\boldsymbol{f}_t$ is also q, which is therefore referred to as the number of *dynamic factors*. In most of the dynamic factor models literature, $\boldsymbol{f}_t$ and $\boldsymbol{e}_t$ are generally assumed to be stationary processes, so if necessary, the observable variables are pre-processed to be stationary. (In this chapter, we only consider stationary variables; non-stationary models will be discussed in Chap. 17.)

The model admits a *static representation*

$$\boldsymbol{x}_t = \boldsymbol{\mu} + \boldsymbol{\Lambda}\boldsymbol{F}_t + \boldsymbol{e}_t, \tag{2.4}$$

with $\boldsymbol{F}_t = (\boldsymbol{f}'_t, \boldsymbol{f}'_{t-1}, \ldots, \boldsymbol{f}'_{t-s})'$, an $r = q(1+s)$ dimensional vector of *static factors*, and $\boldsymbol{\Lambda} = (\boldsymbol{\Lambda}_0, \boldsymbol{\Lambda}_1, \ldots, \boldsymbol{\Lambda}_s)$, a $n \times r$ matrix of loading coefficients. The dynamic representation in (2.2.1) and (2.3) captures the dependence of the observed variables on the lags of the factors explicitly, while the static representation in (2.4) embeds those dynamics implicitly. The two forms lead to different estimation methods to be discussed below.

Exact dynamic factor models assume that the cross-sectional dimension of the dataset, $n < T$, is finite, and they are usually used with small, $n \ll T$, samples. There are two reasons for these models to be passed over in the $n \to \infty$ case. First, maximum likelihood estimation, the typical method of choice, requires specifying a full parametric model and imposes a practical limitation on the number of parameters that can be estimated as $n \to \infty$. Second, as $n \to \infty$, some of the unrealistic assumptions imposed on exact factor models can be relaxed, and the approximate factor model framework, discussed in the next section, can be used instead.

2.2.2 *Approximate Factor Models*

As noted above, exact factor models rely on a very strict assumption of no cross-correlation between the idiosyncratic components. In two seminal papers Chamberlain (1983) and Chamberlain and Rothschild (1983) introduced approximate factor models by relaxing this assumption. They allowed the idiosyncratic components to be mildly cross-correlated and provided a set of conditions ensuring that approximate factor models were asymptotically identified as $n \to \infty$.

Let $\boldsymbol{x}_t^n = (x_{1t}, \ldots, x_{nt})'$ denote the vector containing the tth observation of the first n variables as $n \to \infty$, and let $\boldsymbol{\Sigma}_n = \text{cov}(\boldsymbol{x}_t^n)$ be the covariance matrix of $\boldsymbol{x}_t$. Denoting by $\lambda_1(\boldsymbol{A}) \geq \lambda_2(\boldsymbol{A}) \geq \cdots \geq \lambda_n(\boldsymbol{A})$ the ordered eigenvalues of any symmetric matrix $\boldsymbol{A}$ with size $(n \times n)$, the assumptions underlying approximate factor models are the following:

(CR1) $\text{Sup}_n \lambda_r(\boldsymbol{\Sigma}_n) = \infty$ (the r largest eigenvalues of $\boldsymbol{\Sigma}_n$ are diverging)
(CR2) $\text{Sup}_n \lambda_{r+1}(\boldsymbol{\Sigma}_n) < \infty$ (the remaining eigenvalues of $\boldsymbol{\Sigma}_n$ are bounded)
(CR3) $\text{Inf}_n \lambda_n(\boldsymbol{\Sigma}_n) > 0$ ($\boldsymbol{\Sigma}_n$ does not approach singularity)

The authors show that under assumptions (CR1)–(CR3), there exists a unique decomposition $\boldsymbol{\Sigma}_n = \boldsymbol{\Lambda}_n \boldsymbol{\Lambda}_n' + \boldsymbol{\Psi}_n$, where $\boldsymbol{\Lambda}_n$ is a sequence of nested $n \times r$ matrices with rank r and $\lambda_i(\boldsymbol{\Lambda}_n \boldsymbol{\Lambda}_n') \to \infty$, $\forall i = 1 \ldots r$, and $\lambda_1(\boldsymbol{\Psi}_n) < \infty$. Alternatively, $\boldsymbol{x}_t$ can be decomposed using a pair of mutually orthogonal random vector processes $\boldsymbol{f}_t$ and $\boldsymbol{e}_t^n$

$$\boldsymbol{x}_t^n = \boldsymbol{\mu}_n + \boldsymbol{\Lambda}_n \boldsymbol{f}_t + \boldsymbol{e}_t^n,$$

with $\text{cov}(\boldsymbol{f}_t) = \boldsymbol{I}_r$ and $\text{cov}(\boldsymbol{e}_t^n) = \boldsymbol{\Psi}_n$, where the r common factors are pervasive, in the sense that the number of variables affected by each factor grows with n, and the idiosyncratic terms may be mildly correlated with bounded covariances.

Although originally developed in the finance literature (see also Connor & Korajczyk, 1986, 1988, 1993), the approximate static factor model made its way into macroeconometrics in the early 2000s (see for example Bai, 2003; Bai & Ng, 2002; Stock & Watson, 2002a,b). These papers use assumptions that are analogous to (CR1)–(CR3) for the covariance matrices of the factor loadings and the idiosyncratic terms, but they add complementary assumptions (which vary across authors for technical reasons) to accommodate the fact that the data under study are autocorrelated time series. The models are mainly used with data that is stationary or preprocessed to be stationary, but they have also been used in a non-stationary framework (see Bai, 2004; Bai & Ng, 2004). The analysis of non-stationary data will be addressed in detail in Chap. 17.

Similarly to exact dynamic factor models, approximate dynamic factor models also rely on an equation linking the observable series to the factors and their lags, but here the idiosyncratic terms can be mildly cross-correlated, and the number of series is assumed to tend to infinity. The model has a dynamic

$$\boldsymbol{x}_t^n = \boldsymbol{\mu}_n + \boldsymbol{\Lambda}_n(L) \boldsymbol{f}_t + \boldsymbol{e}_t^n$$

and a static representation

$$x_t^n = \mu_n + \Lambda_n F_t + e_t^n$$

equivalent to (2.3) and (2.4), respectively.

By centering the observed series, μ_n can be set equal to zero. For the sake of simplicity, in the rest of the chapter we will assume that the variables are centered, and we also drop the sub- and superscript n in Λ_n, x_t^n, and e_t^n. We will always assume that in exact factor models the number of series under study is finite, while in approximate factor models $n \to \infty$, and a set of assumptions that is analogous to (CR1)–(CR3)—but suited to the general framework of stationary autocorrelated processes—is satisfied.

2.3 Estimation in the Time Domain

2.3.1 *Maximum Likelihood Estimation of Small Factor Models*

The static exact factor model has generally been estimated by maximum likelihood under the assumption that $(f_t)_{t \in \mathbb{Z}}$ and $(e_t)_{t \in \mathbb{Z}}$ are two orthogonal iid Gaussian processes. Unique identification of the model requires that we impose some restrictions on the model. One of these originates from the definition of the exact factor model: the idiosyncratic components are set to be mutually orthogonal processes with a diagonal variance matrix. A second restriction sets the variance of the factors to be the identity matrix, $Var(f_t) = I_r$. While the estimator does not have a closed form analytical solution, for small n the number of parameters is small, and estimates can be obtained through any numerical optimization procedure. Two specific methods have been proposed for this problem: the so-called zig-zag routine, an algorithm which solves the first order conditions (see for instance Anderson, 1984; Lawley & Maxwell, 1971; Magnus & Neudecker, 2019) and the Jöreskog (1967) approach, which relies on the maximization of the concentrated likelihood using a Fletcher–Powell algorithm. Both approaches impose an additional identifying restriction on Λ. The maximum likelihood estimators $\hat{\Lambda}$, $\hat{\Psi}$ of the model parameters Λ, Ψ are $\sqrt{T}$ consistent and asymptotically Gaussian (see Anderson & Rubin, 1956). Under standard stationarity assumptions, these asymptotic results are still valid even if the true distribution of f_t or e_t is not Gaussian: in this case $\hat{\Lambda}$ and $\hat{\Psi}$ are QML estimators of Λ and Ψ.

Various formulas have been proposed to estimate the factors, given the parameter estimates $\hat{\Lambda}$, $\hat{\Psi}$. Commonly used ones include

- $\widehat{f}_t = \widehat{\Lambda}'(\widehat{\Lambda}\widehat{\Lambda}' + \widehat{\Psi})^{-1} x_t$, the conditional expectation of f_t given x_t and the estimated values of the parameters, which, after elementary calculations, can also be written as $\widehat{f}_t = (I_r + \widehat{\Lambda}'\widehat{\Psi}^{-1}\widehat{\Lambda})^{-1}\widehat{\Lambda}'\widehat{\Psi}^{-1} x_t$, and

- $\widehat{f}_t = (\widehat{\Lambda}'\widehat{\Psi}^{-1}\widehat{\Lambda})^{-1}\widehat{\Lambda}'\widehat{\Psi}^{-1}x_t$ which is the FGLS estimator of f_t, given the estimated loadings.

Additional details about these two estimators are provided by Anderson (1984). Since the formulas are equivalent up to an invertible matrix, the spaces spanned by the estimated factors are identical across these methods.

A dynamic exact factor model can also be estimated by maximum likelihood under the assumption of Gaussian $(f_t', e_t')_{t\in\mathbb{Z}}$.[1] In this case, the factors are assumed to follow vector autoregressive processes, and the model can be cast in state-space form. To make things more precise, let us consider a factor model where the vector of factors follows a VAR(p) process and enters the equation for x_t with s lags

$$x_t = \Lambda_0 f_t + \Lambda_1 f_{t-1} + \cdots + \Lambda_s f_{t-s} + e_t, \tag{2.5}$$

$$f_t = \Phi_1 f_t + \cdots + \Phi_p f_{t-p} + u_t, \tag{2.6}$$

where the VAR(p) coefficient matrices, Φ, capture the dynamics of the factors. A commonly used identification restriction sets the variance of the innovations to the identity matrix, $\text{cov}(u_t) = I_r$, and additional identifying restrictions are imposed on the factor loadings. The state-space representation is very flexible and can accommodate different cases as shown below:

- If $s \geq p-1$ and if e_t is a white noise, the measurement equation is $x_t = \Lambda F_t + e_t$ with $\Lambda = (\Lambda_0\ \Lambda_1 \ldots \Lambda_s)$ and $F_t = (f_t' f_{t-1}' \cdots f_{t-s}')'$ (static representation of the dynamic model), and the state equation is

$$\begin{bmatrix} f_t \\ f_{t-1} \\ \vdots \\ f_{t-p+1} \\ \vdots \\ f_{t-s} \end{bmatrix} = \begin{bmatrix} \Phi_1 \ldots \Phi_p & O \ldots O \\ I_q\ O \ldots\ldots\ldots & O \\ O\ I_q\ O \ldots\ldots & O \\ \vdots\ \ddots\ \ddots\ \ddots & \vdots \\ \vdots\ \ \ \ddots\ \ddots\ \ddots & \vdots \\ O \ldots\ldots O & I_q\ O \end{bmatrix} \begin{bmatrix} f_{t-1} \\ f_{t-2} \\ \vdots \\ f_{t-p} \\ \vdots \\ f_{t-s-1} \end{bmatrix} + \begin{bmatrix} I_q \\ O \\ \vdots \\ O \end{bmatrix} u_t$$

- If $s < p-1$ and if e_t is a white noise, the measurement equation is $x_t = \Lambda F_t + e_t$ with $\Lambda = (\Lambda_0\ \Lambda_1 \ldots \Lambda_s\ O \ldots O)$ and $F_t = (f_t' f_{t-1}' \cdots f_{t-p+1}')'$, and the state equation is

$$\begin{bmatrix} f_t \\ f_{t-1} \\ \vdots \\ f_{t-p+1} \end{bmatrix} = \begin{bmatrix} \Phi_1\ \Phi_2 \ldots \Phi_p \\ I_q\ O \ldots O \\ \vdots\ \ddots\ \ddots\ \vdots \\ O \ldots I_q\ O \end{bmatrix} \begin{bmatrix} f_{t-1} \\ f_{t-2} \\ \vdots \\ f_{t-p} \end{bmatrix} + \begin{bmatrix} I_q \\ O \\ \vdots \\ O \end{bmatrix} u_t \tag{2.7}$$

[1]When (f_t) and (e_t) are not Gaussian, the procedure gives QML estimators.

- The state-space representation can also accommodate the case where each of the idiosyncratic components is itself an autoregressive process. For instance, if $s = p = 2$ and e_{it} follows a second order autoregressive process with $e_{it} = d_{i1}e_{it-1} + d_{i2}e_{it-2} + \varepsilon_{it}$ for $i = 1 \ldots n$, then the measurement equation can be written as $\boldsymbol{x}_t = \boldsymbol{\Lambda}\boldsymbol{\alpha}_t$ with $\boldsymbol{\Lambda} = (\boldsymbol{\Lambda}_0\ \boldsymbol{\Lambda}_1\ \boldsymbol{\Lambda}_2\ \boldsymbol{I}_n\ \mathbf{O})$ and $\boldsymbol{\alpha}_t = (\boldsymbol{f}_t'\ \boldsymbol{f}_{t-1}'\ \boldsymbol{f}_{t-2}'\ \boldsymbol{e}_t'\ \boldsymbol{e}_{t-1}')'$, and the state equation is

$$\begin{bmatrix} \boldsymbol{f}_t \\ \boldsymbol{f}_{t-1} \\ \boldsymbol{f}_{t-2} \\ \boldsymbol{e}_t \\ \boldsymbol{e}_{t-1} \end{bmatrix} = \begin{bmatrix} \boldsymbol{\Phi}_1 & \boldsymbol{\Phi}_2 & \mathbf{O} & \mathbf{O} & \mathbf{O} \\ \boldsymbol{I}_q & \mathbf{O} & \mathbf{O} & \mathbf{O} & \mathbf{O} \\ \mathbf{O} & \boldsymbol{I}_q & \mathbf{O} & \mathbf{O} & \mathbf{O} \\ \mathbf{O} & \mathbf{O} & \mathbf{O} & \boldsymbol{D}_1 & \boldsymbol{D}_2 \\ \mathbf{O} & \mathbf{O} & \mathbf{O} & \boldsymbol{I}_n & \mathbf{O} \end{bmatrix} \begin{bmatrix} \boldsymbol{f}_{t-1} \\ \boldsymbol{f}_{t-2} \\ \boldsymbol{f}_{t-3} \\ \boldsymbol{e}_{t-1} \\ \boldsymbol{e}_{t-2} \end{bmatrix} + \begin{bmatrix} \boldsymbol{I}_q & \mathbf{O} \\ \mathbf{O} & \mathbf{O} \\ \mathbf{O} & \mathbf{O} \\ \mathbf{O} & \boldsymbol{I}_n \\ \mathbf{O} & \mathbf{O} \end{bmatrix} \begin{bmatrix} \boldsymbol{u}_t \\ \boldsymbol{\varepsilon}_t \end{bmatrix}, \tag{2.8}$$

  with $\boldsymbol{D}_j = \operatorname{diag}(d_{1j} \ldots d_{nj})$ for $j = 1, 2$ and $\boldsymbol{\varepsilon}_t = (\varepsilon_{1t} \ldots \varepsilon_{nt})'$. This model, with $n = 4$ and $r = 1$, was used by Stock and Watson (1991) to build a coincident index for the US economy.

One computational challenge is keeping the dimension of the state vector small when n is large. The inclusion of the idiosyncratic component in $\boldsymbol{\alpha}_t$ implies that the dimension of system matrices in the state equation (2.8) grows with n. Indeed, this formulation requires an element for each lag of each idiosyncratic component. To limit the computational cost, an alternative approach applies the filter $\boldsymbol{I}_n - \boldsymbol{D}_1 L - \boldsymbol{D}_2 L^2$ to the measurement equation. This pre-whitening is intended to control for serial correlation in the idiosyncratic terms, so that they do not need to be included in the state equation. The transformed measurement equation takes the form $\boldsymbol{x}_t = \boldsymbol{c}_t + \widetilde{\boldsymbol{\Lambda}}\boldsymbol{\alpha}_t + \boldsymbol{\varepsilon}_t$, for $t \geq 2$, with $\boldsymbol{c}_t = \boldsymbol{D}_1\boldsymbol{x}_{t-1} + \boldsymbol{D}_2\boldsymbol{x}_{t-2}$, $\widetilde{\boldsymbol{\Lambda}} = \left(\widetilde{\boldsymbol{\Lambda}}_0 \ \ldots \ \widetilde{\boldsymbol{\Lambda}}_4\right)$ and $\boldsymbol{\alpha}_t = (\boldsymbol{f}_t\ \boldsymbol{f}_{t-1}\ \cdots\ \boldsymbol{f}_{t-4})'$, since

$$\begin{aligned}(\boldsymbol{I}_n - \boldsymbol{D}_1 L - \boldsymbol{D}_2 L^2)\boldsymbol{x}_t =& \boldsymbol{\Lambda}_0\boldsymbol{f}_t + (\boldsymbol{\Lambda}_1 - \boldsymbol{D}_1\boldsymbol{\Lambda}_0)\boldsymbol{f}_{t-1} + \\ & (\boldsymbol{\Lambda}_2 - \boldsymbol{D}_1\boldsymbol{\Lambda}_1 - \boldsymbol{D}_2\boldsymbol{\Lambda}_0)\boldsymbol{f}_{t-2} - \\ & (\boldsymbol{D}_1\boldsymbol{\Lambda}_2 + \boldsymbol{D}_2\boldsymbol{\Lambda}_1)\boldsymbol{f}_{t-3} - (\boldsymbol{D}_2\boldsymbol{\Lambda}_2)\boldsymbol{f}_{t-4} + \boldsymbol{\varepsilon}_t\end{aligned} \tag{2.9}$$

The introduction of lags of $\boldsymbol{x}_t$ in the measurement equation does not cause further complications; they can be incorporated in the Kalman filter since they are known at time t. The associated state equation is straightforward and the dimension of $\boldsymbol{\alpha}_t$ is smaller than in (2.8).

Once the model is written in state-space form, the Gaussian likelihood can be computed using the Kalman filter for any value of the parameters (see for instance Harvey, 1989), and the likelihood can be maximized by any numerical optimization procedure over the parameter space. Watson and Engle (1983) proposed to use a score algorithm, or the EM algorithm, or a combination of both, but any other numerical procedure can be used when n is small. With the parameter estimates, $\widehat{\boldsymbol{\theta}}$, in hand, the Kalman smoother provides an approximation of $\boldsymbol{f}_t$ using information from all observations $\widehat{\boldsymbol{f}}_{t|T} = E(\boldsymbol{f}_t|\boldsymbol{x}_1,\ldots,\boldsymbol{x}_T,\widehat{\boldsymbol{\theta}})$. Asymptotic consistency and normality of the parameter estimators and the factors follow from general results concerning maximum likelihood estimation with the Kalman filter.

To further improve the computational efficiency of estimation, Jungbacker and Koopman (2014) reduced a high-dimensional dynamic factor model to a low-dimensional state space model. Using a suitable transformation of the measurement equation, they partitioned the observation vector into two mutually uncorrelated subvectors, with only one of them depending on the unobserved state. The transformation can be summarized by

$$\boldsymbol{x}_t^* = \boldsymbol{A}\boldsymbol{x}_t \quad \text{with} \quad \boldsymbol{A} = \begin{bmatrix} \boldsymbol{A}^L \\ \boldsymbol{A}^H \end{bmatrix} \quad \text{and} \quad \boldsymbol{x}_t^* = \begin{bmatrix} \boldsymbol{x}_t^L \\ \boldsymbol{x}_t^H \end{bmatrix},$$

where the model for $\boldsymbol{x}_t^*$ can be written as

$$\boldsymbol{x}_t^L = \boldsymbol{A}^L \boldsymbol{\Lambda} \boldsymbol{F}_t + \boldsymbol{e}_t^L \quad \text{and} \quad \boldsymbol{x}_t^H = \boldsymbol{e}_t^H,$$

with $\boldsymbol{e}_t^L = \boldsymbol{A}^L \boldsymbol{e}_t$ and $\boldsymbol{e}_t^H = \boldsymbol{A}^H \boldsymbol{e}_t$. Consequently, the $\boldsymbol{x}_t^H$ subvector is not required for signal extraction and the Kalman filter can be applied to a lower dimensional collapsed model, leading to substantial computational savings. This approach can be combined with controlling for idiosyncratic dynamics in the measurement equation, as described above in (2.9).

2.3.2 Principal Component Analysis of Large Approximate Factor Models

Chamberlain and Rothschild (1983) suggested to use principal component analysis (PCA) to estimate the approximate static factor model, and Stock and Watson (2002a,b) and Bai and Ng (2002) popularized this approach in macro-econometrics. PCA will be explored in greater detail in Chap. 8, but we state the main results here.

Considering centered data and assuming that the number of factors, r, is known, PCA allows to simultaneously estimate the factors and their loadings by solving the least squares problem

$$\min_{\boldsymbol{\Lambda},\boldsymbol{F}} \frac{1}{nT} \sum_{i=1}^{n} \sum_{t=1}^{T} \left(x_{it} - \lambda_i' \boldsymbol{f}_t\right)^2 = \min_{\boldsymbol{\Lambda},\boldsymbol{F}} \frac{1}{nT} \sum_{t=1}^{T} (\boldsymbol{x}_t - \boldsymbol{\Lambda} \boldsymbol{f}_t)'(\boldsymbol{x}_t - \boldsymbol{\Lambda} \boldsymbol{f}_t). \tag{2.10}$$

Due to the aforementioned rotational indeterminacy of the factors and their loadings, the parameter estimates have to be constrained to get a unique solution. Generally, one of the following two normalization conditions is imposed

$$\frac{1}{T}\sum_{t=1}^{T}\widehat{\boldsymbol{f}}_t\widehat{\boldsymbol{f}}_t' = \boldsymbol{I}_r \quad \text{or} \quad \frac{\widehat{\boldsymbol{\Lambda}}'\widehat{\boldsymbol{\Lambda}}}{n} = \boldsymbol{I}_r.$$

Using the first normalization and concentrating out $\boldsymbol{\Lambda}$ gives an estimated factor matrix, $\widehat{\boldsymbol{F}}$, which is T times the eigenvectors corresponding to the r largest eigenvalues of the $T \times T$ matrix $\boldsymbol{X}\boldsymbol{X}'$. Given $\widehat{\boldsymbol{F}}$, $\widehat{\boldsymbol{\Lambda}} = (\widehat{\boldsymbol{F}}'\widehat{\boldsymbol{F}})^{-1}\widehat{\boldsymbol{F}}'\boldsymbol{X} = \widehat{\boldsymbol{F}}'\boldsymbol{X}/T$ is the corresponding matrix of factor loadings. The solution to the minimization problem above is not unique, even though the sum of squared residuals is unique. Another solution is given by $\widetilde{\boldsymbol{\Lambda}}$ constructed as n times the eigenvectors corresponding to the r largest eigenvalues of the $n \times n$ matrix $\boldsymbol{X}'\boldsymbol{X}$. Using the second normalization here implies $\widetilde{\boldsymbol{F}} = \boldsymbol{X}\widetilde{\boldsymbol{\Lambda}}/n$. Bai and Ng (2002) indicated that the latter approach is computationally less costly when $T > n$, while the former is less demanding when $T < n$. In both cases, the idiosyncratic components are estimated by $\widehat{\boldsymbol{e}}_t = \boldsymbol{x}_t - \widehat{\boldsymbol{\Lambda}}\widehat{\boldsymbol{f}}_t$, and their covariance is estimated by the empirical covariance matrix of $\widehat{\boldsymbol{e}}_t$. Since PCA is not scale invariant, many authors (for example Stock & Watson, 2002a,b) center and standardize the series, generally measured in different units, and as a result PCA is applied to the sample correlation matrix in this case.

Stock and Watson (2002a,b) and Bai and Ng (2002) proved the consistency of these estimators, and Bai (2003) obtained their asymptotic distribution under a stronger set of assumptions. We refer the reader to those papers for the details, but let us note that these authors replace assumption (CR1) by the following stronger assumption[2]

$$\lim_{n\to\infty}\frac{\boldsymbol{\Lambda}'\boldsymbol{\Lambda}}{n} = \boldsymbol{\Sigma}_{\boldsymbol{\Lambda}}$$

and replace assumptions (CR2) and (CR3), which were designed for white noise data, by analogous assumptions taking autocorrelation in the factors and idiosyncratic terms into account. Under these assumptions, the factors and the loadings are proved to be consistent, up to an invertible matrix $\boldsymbol{H}$, converging at rate $\boldsymbol{\delta}_{nT} = 1/\min(\sqrt{n}, \sqrt{T})\boldsymbol{\iota}_r$, so that

$$\widehat{\boldsymbol{f}}_t - \boldsymbol{H}\boldsymbol{f}_t = O_P(\boldsymbol{\delta}_{nT}) \text{ and } \widehat{\boldsymbol{\lambda}}_i - \boldsymbol{H}^{-1}\boldsymbol{\lambda}_i = O_P(\boldsymbol{\delta}_{nT}),\ \forall\, i = 1 \ldots n.$$

[2]A weaker assumption can also be used: $0 < \underline{c} \leq \liminf_{n\to\infty} \lambda_r(\frac{\boldsymbol{\Lambda}'\boldsymbol{\Lambda}}{n}) < \limsup_{n\to\infty} \lambda_1(\frac{\boldsymbol{\Lambda}'\boldsymbol{\Lambda}}{n}) \leq \overline{c} < \infty$ (see Doz, Giannone, & Reichlin, 2011).

Under a more stringent set of assumptions, Bai (2003) also obtains the following asymptotic distribution results:

- If $\sqrt{n}/T \to 0$ then, for each t: $\sqrt{n}(\widehat{\boldsymbol{f}}_t - \boldsymbol{H}'\boldsymbol{f}_t) \underset{d}{\to} \mathcal{N}(\mathbf{0}, \boldsymbol{\Omega}_t)$ where $\boldsymbol{\Omega}_t$ is known.
- If $\sqrt{T}/n \to 0$ then, for each i: $\sqrt{T}(\widehat{\boldsymbol{\lambda}}_i - \boldsymbol{H}^{-1}\boldsymbol{\lambda}_i) \xrightarrow{d} \mathcal{N}(\mathbf{0}, \boldsymbol{W}_i)$ where $\boldsymbol{W}_i$ is known.

2.3.3 Generalized Principal Component Analysis of Large Approximate Factor Models

Generalized principal components estimation mimics generalized least squares to deal with a nonspherical variance matrix of the idiosyncratic components. Although the method had been used earlier, Choi (2012) was the one who proved that efficiency gains can be achieved by including a weighting matrix in the minimization

$$\min_{\boldsymbol{\Lambda},\boldsymbol{F}} \frac{1}{nT} \sum_{t=1}^{T} (\boldsymbol{x}_t - \boldsymbol{\Lambda}\boldsymbol{f}_t)' \boldsymbol{\Psi}^{-1} (\boldsymbol{x}_t - \boldsymbol{\Lambda}\boldsymbol{f}_t).$$

The feasible generalized principal component estimator replaces the unknown $\boldsymbol{\Psi}$ by an estimator $\widehat{\boldsymbol{\Psi}}$. The diagonal elements of $\boldsymbol{\Psi}$ can be estimated by the variances of the individual idiosyncratic terms (see Boivin & Ng, 2006; Jones, 2001). Bai and Liao (2013) derive the conditions under which the estimated $\widehat{\boldsymbol{\Psi}}$ can be treated as known. The weighted minimization problem above relies on the assumption of independent idiosyncratic shocks, which may be too restrictive in practice. Stock and Watson (2005) applied a diagonal filter $\boldsymbol{D}(L)$ to the idiosyncratic terms to deal with serial correlation, so the problem becomes

$$\min_{\boldsymbol{D}(L),\boldsymbol{\Lambda},\widetilde{\boldsymbol{F}}} \frac{1}{T} \sum_{t=1}^{T} (\boldsymbol{D}(L)\boldsymbol{x}_t - \boldsymbol{\Lambda}\widetilde{\boldsymbol{f}}_t)' \widetilde{\boldsymbol{\Psi}}^{-1} (\boldsymbol{D}(L)\boldsymbol{x}_t - \boldsymbol{\Lambda}\widetilde{\boldsymbol{f}}_t),$$

where $\widetilde{\boldsymbol{F}} = (\widetilde{\boldsymbol{f}}_1 \dots \widetilde{\boldsymbol{f}}_T)'$ with $\widetilde{\boldsymbol{f}}_t = \boldsymbol{D}(L)\boldsymbol{f}_t$ and $\widetilde{\boldsymbol{\Psi}} = \mathrm{E}[\widetilde{\boldsymbol{e}}_t\widetilde{\boldsymbol{e}}_t']$ with $\widetilde{\boldsymbol{e}}_t = \boldsymbol{D}(L)\boldsymbol{e}_t$. Estimation of $\boldsymbol{D}(L)$ and $\widetilde{\boldsymbol{F}}$ can be done sequentially, iterating to convergence. Breitung and Tenhofen (2011) propose a similar two-step estimation procedure that allows for heteroskedastic and serially correlated idiosyncratic terms.

Even though PCA was first used to estimate approximate factor models where the factors and the idiosyncratic terms were iid white noise, most asymptotic results carried through to the case when the factors and the idiosyncratic terms were stationary autocorrelated processes. Consequently, the method has been widely used as a building block in approximate dynamic factor model estimation, but it required extensions since PCA on its own does not capture dynamics. Next we review several approaches that attempt to incorporate dynamic behavior in large scale approximate factor models.

2.3.4 Two-Step and Quasi-Maximum Likelihood Estimation of Large Approximate Factor Models

Doz et al. (2011) proposed a two-step estimator that takes into account the dynamics of the factors. They assume that the factors in the approximate dynamic factor model $\boldsymbol{x}_t = \boldsymbol{\Lambda} \boldsymbol{f}_t + \boldsymbol{e}_t$ follow a vector autoregressive process, $\boldsymbol{f}_t = \boldsymbol{\Phi}_1 \boldsymbol{f}_{t-1} + \cdots + \boldsymbol{\Phi}_p \boldsymbol{f}_{t-p} + \boldsymbol{u}_t$, and they allow the idiosyncratic terms to be autocorrelated but do not specify their dynamics. As illustrated in Sect. 2.3.1, this model can easily be cast in state-space form. The estimation procedure is the following:

Step 1 Preliminary estimators of the loadings, $\widehat{\boldsymbol{\Lambda}}$, and factors, $\widehat{\boldsymbol{f}}_t$, are obtained by principal component analysis. The idiosyncratic terms are estimated by $\widehat{e}_{it} = x_{it} - \widehat{\boldsymbol{\lambda}}_i' \widehat{\boldsymbol{f}}_t$, and their variance is estimated by the associated empirical variance $\widehat{\psi}_{ii}$. The estimated factors, $\widehat{\boldsymbol{f}}_t$, are used in a vector-autoregressive model to obtain the estimates $\widehat{\boldsymbol{\Phi}}_j, j = 1 \ldots p$.

Step 2 The model is cast in state-space form as in (2.7), with the variance of the common shocks set to the identity matrix, $\text{cov}(\boldsymbol{u}_t) = \boldsymbol{I}_r$, and $\text{cov}(\boldsymbol{e}_t)$ defined as $\boldsymbol{\Psi} = \text{diag}(\psi_{11} \ldots \psi_{nn})$. Using the parameter estimates $\widehat{\boldsymbol{\Lambda}}, \widehat{\boldsymbol{\Psi}}, \widehat{\boldsymbol{\Phi}}_j, j = 1 \ldots p$ obtained in the first step, one run of the Kalman smoother is then applied to the data. It produces a new estimate of the factor, $\widehat{\boldsymbol{f}}_{t|T} = \text{E}(\boldsymbol{f}_t | \boldsymbol{x}_1, \ldots, \boldsymbol{x}_T, \widehat{\boldsymbol{\theta}})$, where $\widehat{\boldsymbol{\theta}}$ is a vector containing the first step estimates of all parameters. It is important to notice that, in this second step, the idiosyncratic terms are misspecified, since they are taken as mutually orthogonal white noises.

Doz et al. (2011) prove that, under their assumptions, the parameter estimates, $\widehat{\boldsymbol{\theta}}$, are consistent, converging at rate $\left(\min(\sqrt{n}, \sqrt{T})\right)^{-1}$. They also prove that, when the Kalman smoother is run with those parameter estimates instead of the true parameters, the resulting two step-estimator of $\boldsymbol{f}_t$ is also $\min(\sqrt{n}, \sqrt{T})$ consistent.

Doz, Giannone, and Reichlin (2012) proposed to estimate a large scale approximate dynamic factor model by quasi-maximum likelihood (QML). In line with Doz et al. (2011), the quasi-likelihood is based on the assumption of mutually orthogonal *iid* Gaussian idiosyncratic terms (so that the model is treated as if it were an exact factor model, even though it is not), and a Gaussian VAR model for the factors. The corresponding log-likelihood can be obtained from the Kalman filter for given values of the parameters, and they use an EM algorithm to compute the maximum likelihood estimator. The EM algorithm, proposed by Dempster, Laird, and Rubin (1977) and first implemented for dynamic factor models by Watson and Engle (1983), alternates an expectation step relying on a pass of the Kalman smoother for the current parameter values and a maximization step relying on multivariate regressions. The application of the algorithm is tantamount to successive applications of the two-step approach. The calculations are feasible even when n is large, since in each iteration of the algorithm, the current estimate of the ith loading, $\boldsymbol{\lambda}_i$, is obtained by ordinary least squares regression of x_{it} on the current estimate of the factors. The authors prove that, under a standard set of assumptions, the estimated factors are mean square consistent. Their results remain valid even if the processes are not Gaussian, or if the idiosyncratic terms are not iid, or not mutually orthogonal, as long as they are only weakly cross-correlated. Reis and Watson (2010) apply this approach to a model with serially correlated idiosyncratic terms. Jungbacker and Koopman (2014) also use the EM algorithm to estimate a dynamic factor model with autocorrelated idiosyncratic terms, but instead of extending the state vector as in Eq. (2.8), they transform the measurement equation as described in Sect. 2.3.5.

Bai and Li (2012) study QML estimation in the more restricted case where the quasi-likelihood is associated with the static exact factor model. They obtain consistency and asymptotic normality of the estimated loadings and factors under a set of appropriate assumptions. Bai and Li (2016) incorporate these estimators into the two-step approach put forward by Doz et al. (2011) and obtain similar asymptotic results. They also follow Jungbacker and Koopman (2014) to handle the case where the idiosyncratic terms are autoregressive processes. Bai and Liao (2016) extend the approach of Bai and Li (2012) to the case where the idiosyncratic covariance matrix is sparse, instead of being diagonal, and propose a penalized maximum likelihood estimator.

2.3.5 *Estimation of Large Approximate Factor Models with Missing Data*

Observations may be missing from the analyzed dataset for several reasons. At the beginning of the sample, certain time series might be available from an earlier start date than others. At the end of the sample, the dates of final observations may differ depending on the release lag of each data series. Finally, observations may be missing within the sample since different series in the dataset may be sampled at different frequencies, for example, monthly and quarterly. DFM estimation techniques assume that the observations are missing at random, so there is no endogenous sample selection. Missing data are handled differently in principal components and state space applications.[3]

The least squares estimator of principal components in a balanced panel given in Eq. (2.10) needs to be modified when some of the nT observations are missing. Stock and Watson (2002b) showed that estimates of $\boldsymbol{F}$ and $\boldsymbol{\Lambda}$ can be obtained

[3]Mixed data sampling (MIDAS) regression models, proposed by Ghysels, Santa-Clara, and Valkanov (2004), represent an alternative way of dealing with missing data.

numerically by

$$\min_{\mathbf{\Lambda},\boldsymbol{F}} \frac{1}{nT} \sum_{i=1}^{n} \sum_{t=1}^{T} S_{it} \left(x_{it} - \boldsymbol{\lambda}_i' \boldsymbol{f}_t \right)^2$$

where $S_{it} = 1$ if an observation on x_{it} is available and $S_{it} = 0$ otherwise. The objective function can be minimized by iterations alternating the optimization with respect to $\mathbf{\Lambda}$ given $\boldsymbol{F}$ and then $\boldsymbol{F}$ given $\mathbf{\Lambda}$; each step in the minimization has a closed form expression. Starting values can be obtained, for example, by principal component estimation using a subset of the series for which there are no missing observations. Alternatively, Stock and Watson (2002b) provide an EM algorithm for handling missing observations.

Step 1 Fill in initial guesses for missing values to obtain a balanced dataset. Estimate the factors and loadings in this balanced dataset by principal components analysis.
Step 2 The values in the place of a missing observations for each variable are updated by the expectation of x_{it} conditional on the observations, and the factors and loadings from the previous iteration.
Step 3 With the updated balanced dataset in hand, reestimate the factors and loadings by principal component analysis. Iterate step 2 and 3 until convergence.

The algorithm provides both estimates of the factors and estimates of the missing values in the time series.

The state space framework has been adapted to missing data by either allowing the measurement equation to vary depending on what data are available at a given time (see Harvey, 1989, section 6.3.7) or by including a proxy value for the missing observation while adjusting the model so that the Kalman filter places no weight on the missing observation (see Giannone, Reichlin, & Small, 2008).

When the dataset contains missing values, the formulation in Eq. (2.9) is not feasible since the lagged values on the right-hand side of the measurement equation are not available in some periods. Jungbacker, Koopman, and van der Wel (2011) addressed the problem by keeping track of periods with missing observations and augmenting the state vector with the idiosyncratic shocks in those periods. This implies that the system matrices and the dimension of the state vector are time-varying. Yet, the model can still be collapsed by transforming the measurement equation and partitioning the observation vector as in Jungbacker and Koopman (2014) and removing from the model the subvector that does not depend on the state. Under several simplifying assumptions, Pinheiro, Rua, and Dias (2013) developed an analytically and computationally less demanding algorithm for the special case of jagged edges, or observations missing at the end of the sample due to varying publication lags across series.

Since the Kalman filter and smoother can easily accommodate missing data, the two-step method of Doz et al. (2011) is also well-suited to handle unbalanced panel datasets. In particular, it also allows to overcome the jagged edge data problem.

This feature of the two-step method has been exploited in predicting low frequency macroeconomic releases for the current period, also known as nowcasting (see Sect. 2.6). For instance, Giannone et al. (2008) used this method to incorporate the real-time informational content of monthly macroeconomic data releases into current-quarter GDP forecasts. Similarly, Banbura and Rünstler (2011) used the two-step framework to compute the impact of monthly predictors on quarterly GDP forecasts. They extended the method by first computing the weights associated with individual monthly observations in the estimates of the state vector using an algorithm by Koopman and Harvey (2003), which then allowed them to compute the contribution of each variable to the GDP forecast.

In the maximization step of the EM algorithm, the calculation of moments involving data was not feasible when some observations were missing, and therefore the original algorithm required a modification to handle an incomplete dataset. Shumway and Stoffer (1982) allowed for missing data but assumed that the factor loading coefficients were known. More recently, Banbura and Modugno (2014) adapted the EM algorithm to a general pattern of missing data by using a selection matrix to carry out the maximization step with available data points. The basic idea behind their approach is to write the likelihood as if the data were complete and to adapt the Kalman filter and smoother to the pattern of missing data in the E-step of the EM algorithm, where the missing data are replaced by their best linear predictions given the information set. They also extend their approach to the case where the idiosyncratic terms are univariate AR processes. Finally they provide a statistical decomposition, which allows one to inspect how the arrival of new data affects the forecast of the variable of interest.

Modeling mixed frequencies via the state space approach makes it possible to associate the missing observations with particular dates and to differentiate between stock variables and flow variables. The state space model typically contains an aggregator that averages the high-frequency observations over one low-frequency period for stocks, and sums them for flows (see Aruoba, Diebold, & Scotti, 2009). However, summing flows is only appropriate when the variables are in levels; for growth rates it is a mere approximation, weakening the forecasting performance of the model (see Fuleky & Bonham, 2015). As pointed out by Mariano and Murasawa (2003) and Proietti and Moauro (2006), appropriate aggregation of flow variables that enter the model in log-differences requires a non-linear state space model.

2.4 Estimation in the Frequency Domain

Classical principal component analysis, described in Sect. 2.3.2, estimates the space spanned by the factors non-parametrically only from the cross-sectional variation in the data. The two-step approach, discussed in Sect. 2.3.4, augments principal components estimation with a parametric state space model to capture the dynamics of the factors. Frequency-domain estimation combines some features of the previous two approaches: it relies on non-parametric methods that exploit variation both over time and over the cross section of variables.

Traditional static principal component analysis focuses on contemporaneous cross-sectional correlation and overlooks serial dependence. It approximates the (contemporaneous) covariance matrix of $\boldsymbol{x}_t$ by a reduced rank covariance matrix. While the correlation of two processes may be negligible contemporaneously, it could be high at leads or lags. Discarding this information could result in loss of predictive capacity. Dynamic principal component analysis overcomes this shortcoming by relying on spectral densities. The $n \times n$ spectral density matrix of a second order stationary process, $\boldsymbol{x}_t$, for frequency $\omega \in [-\pi, \pi]$ is defined as

$$\boldsymbol{\Sigma}_{\boldsymbol{x}}(\omega) = \frac{1}{2\pi} \sum_{k=-\infty}^{\infty} e^{-ik\omega} \boldsymbol{\Gamma}_{\boldsymbol{x}}(k),$$

where $\boldsymbol{\Gamma}_{\boldsymbol{x}}(k) = \mathrm{E}[\boldsymbol{x}_t \boldsymbol{x}'_{t-k}]$. In analogy to their static counterpart, dynamic principal components approximate the spectrum of $\boldsymbol{x}_t$ by a reduced rank spectral density matrix. Static principal component analysis was generalized to the frequency domain by Brillinger (1981). His algorithm relied on a consistent estimate of $\boldsymbol{x}_t$'s spectral density, $\widehat{\boldsymbol{\Sigma}}_{\boldsymbol{x}}(\omega)$, at frequency ω. The eigenvectors corresponding to the r largest eigenvalues of $\widehat{\boldsymbol{\Sigma}}_{\boldsymbol{x}}(\omega)$, a Hermitian matrix, are then transformed by the inverse Fourier transform to obtain the dynamic principal components.

The method was popularized in econometrics by Forni, Hallin, Lippi, and Reichlin (2000). Their generalized dynamic factor model encompasses as a special case the approximate factor model of Chamberlain (1983) and Chamberlain and Rothschild (1983), which allows for correlated idiosyncratic components but is static. And it generalizes the factor model of Sargent and Sims (1977) and Geweke (1977), which is dynamic but assumes orthogonal idiosyncratic components. The method relies on the assumption of an infinite cross section to identify and consistently estimate the common and idiosyncratic components. The common component is a projection of the data on the space spanned by all leads and lags of the first r dynamic principal components, and the orthogonal residuals from this projection are the idiosyncratic components.

Forni et al. (2000) and Favero, Marcellino, and Neglia (2005) propose the following procedure for estimating the dynamic principal components and common components.

Step 1 For a sample $\boldsymbol{x}_1 \dots \boldsymbol{x}_T$ of size T, estimate the spectral density matrix of $\boldsymbol{x}_t$ by

$$\widehat{\boldsymbol{\Sigma}}_{\boldsymbol{x}}(\omega_h) = \frac{1}{2\pi} \sum_{k=-M}^{M} w_k e^{-ik\omega_h} \widehat{\boldsymbol{\Gamma}}_{\boldsymbol{x}}(k), \quad \omega_h = 2\pi h/(2M+1), \quad h = -M \dots M,$$

where $w_k = 1 - |k|/(M+1)$ are the weights of the Bartlett window of width M and $\widehat{\boldsymbol{\Gamma}}_{\boldsymbol{x}}(k) = (T-k)^{-1} \sum_{t=k+1}^{T} (\boldsymbol{x}_t - \overline{\boldsymbol{x}})(\boldsymbol{x}_{t-k} - \overline{\boldsymbol{x}})'$ is the sample covariance matrix of $\boldsymbol{x}_t$ for lag k. For consistent estimation of $\boldsymbol{\Sigma}_{\boldsymbol{x}}(\omega)$ the window width has to be chosen such that $M \to \infty$ and $M/T \to 0$ as $T \to \infty$. Forni et al. (2000) note that a choice of $M = \frac{2}{3}T^{1/3}$ worked well in simulations.

Step 2 For $h = -M \dots M$, compute the eigenvectors $\boldsymbol{\lambda}_1(\omega_h) \dots \boldsymbol{\lambda}_r(\omega_h)$ corresponding to the r largest eigenvalues of $\widehat{\boldsymbol{\Sigma}}_{\boldsymbol{x}}(\omega_h)$. The choice of r is guided by a heuristic inspection of eigenvalues. Note that, for $M = 0$, $\boldsymbol{\lambda}_j(\omega_h)$ is simply the jth eigenvector of the (estimated) variance-covariance matrix of $\boldsymbol{x}_t$: the dynamic principal components then reduce to the static principal components.

Step 3 Define $\boldsymbol{\lambda}_j(L)$ as the two-sided filter

$$\boldsymbol{\lambda}_j(L) = \sum_{k=-M}^{M} \boldsymbol{\lambda}_{jk} L^k, \quad k = -M \dots M,$$

where

$$\boldsymbol{\lambda}_{jk} = \frac{1}{2M+1} \sum_{h=-M}^{M} \boldsymbol{\lambda}_j(\omega_h) e^{-ik\omega_h},$$

The first r dynamic principal components of $\boldsymbol{x}_t$ are $\widehat{\boldsymbol{f}}_{jt} = \boldsymbol{\lambda}_j(L)'\boldsymbol{x}_t$, $j = 1 \dots r$, which can be collected in the vector $\widehat{\boldsymbol{f}}_t = (\widehat{\boldsymbol{f}}_{1t} \cdots \widehat{\boldsymbol{f}}_{rt})'$.

Step 4 Run an OLS regression of $\boldsymbol{x}_t$ on present, past, and future dynamic principal components

$$\boldsymbol{x}_t = \boldsymbol{\Lambda}_{-q} \widehat{\boldsymbol{f}}_{t+q} + \dots + \boldsymbol{\Lambda}_p \widehat{\boldsymbol{f}}_{t-p},$$

and estimate the common component as the fitted value

$$\widehat{\boldsymbol{\chi}}_t = \widehat{\boldsymbol{\Lambda}}_{-q} \widehat{\boldsymbol{f}}_{t+q} + \dots + \widehat{\boldsymbol{\Lambda}}_p \widehat{\boldsymbol{f}}_{t-p},$$

where $\widehat{\boldsymbol{\Lambda}}_l, l = -q \dots p$ are the OLS estimators, and the leads q and lags p used in the regression can be chosen by model selection criteria. The idiosyncratic component is the residual, $\widehat{\xi}_{it} = x_{it} - \widehat{\chi}_{it}$.

Although this method efficiently estimates the common component, its reliance on two-sided filtering rendered it unsuitable for forecasting. Forni, Hallin, Lippi, and Reichlin (2005) extended the frequency domain approach developed in their earlier paper to forecasting in two steps.

Step 1 In the first step, they estimated the covariances of the common and idiosyncratic components using the inverse Fourier transforms

$$\widehat{\boldsymbol{\Gamma}}_{\chi}(k) = \frac{1}{2M+1} \sum_{k=-M}^{M} e^{ik\omega_h} \widehat{\boldsymbol{\Sigma}}_{\chi}(\omega_h) \quad \text{and}$$

$$\widehat{\boldsymbol{\Gamma}}_{\xi}(k) = \frac{1}{2M+1} \sum_{k=-M}^{M} e^{ik\omega_h} \widehat{\boldsymbol{\Sigma}}_{\xi}(\omega_h)$$

where $\widehat{\boldsymbol{\Sigma}}_{\chi}(\omega_h)$ and $\widehat{\boldsymbol{\Sigma}}_{\xi}(\omega_h) = \widehat{\boldsymbol{\Sigma}}_{x}(\omega_h) - \widehat{\boldsymbol{\Sigma}}_{\chi}(\omega_h)$ are the spectral density matrices corresponding to the common and idiosyncratic components, respectively.

Step 2 In the second step, they estimated the factor space using a linear combination of the $\boldsymbol{x}$'s, $\boldsymbol{\lambda}'\boldsymbol{x}_t = \boldsymbol{\lambda}'\boldsymbol{\chi}_t + \boldsymbol{\lambda}'\boldsymbol{\xi}_t$. Specifically, they compute r independent linear combinations $\widehat{\boldsymbol{f}}_{jt} = \widehat{\boldsymbol{\lambda}}_j' \boldsymbol{x}_t$, where the weights $\widehat{\boldsymbol{\lambda}}_j$ maximize the variance of $\boldsymbol{\lambda}'\boldsymbol{\chi}_t$ and are defined recursively

$$\widehat{\boldsymbol{\lambda}}_j = \arg\max_{\boldsymbol{\lambda}\in\mathbb{R}^n} \boldsymbol{\lambda}'\widehat{\boldsymbol{\Gamma}}_{\chi}(0)\boldsymbol{\lambda} \quad \text{s.t.} \quad \boldsymbol{\lambda}'\widehat{\boldsymbol{\Gamma}}_{\xi}(0)\boldsymbol{\lambda} = 1 \quad \text{and} \quad \boldsymbol{\lambda}'\widehat{\boldsymbol{\Gamma}}_{\xi}(0)\widehat{\boldsymbol{\lambda}}_m = 0,$$

for $j = 1 \dots r$ and $1 \le m \le j-1$ (only the first constraint applies for $j = 1$). The solutions of this problem, $\widehat{\boldsymbol{\lambda}}_j$, are the generalized eigenvectors associated with the generalized eigenvalues $\widehat{\nu}_j$ of the matrices, $\widehat{\boldsymbol{\Gamma}}_{\chi}(0)$ and $\widehat{\boldsymbol{\Gamma}}_{\xi}(0)$,

$$\widehat{\boldsymbol{\lambda}}_j'\widehat{\boldsymbol{\Gamma}}_{\chi}(0) = \widehat{\nu}_j \widehat{\boldsymbol{\lambda}}_j' \widehat{\boldsymbol{\Gamma}}_{\xi}(0), \quad j = 1 \dots n,$$

with the normalization constraint $\widehat{\boldsymbol{\lambda}}_j'\widehat{\boldsymbol{\Gamma}}_{\xi}(0)\widehat{\boldsymbol{\lambda}}_j = 1$ and $\widehat{\boldsymbol{\lambda}}_i'\widehat{\boldsymbol{\Gamma}}_{\xi}(0)\widehat{\boldsymbol{\lambda}}_j = 0$ for $i \neq j$. The linear combinations $\widehat{\boldsymbol{f}}_{jt} = \widehat{\boldsymbol{\lambda}}_j'\boldsymbol{x}_t, j = 1 \dots n$ are the generalized principal components of $\boldsymbol{x}_t$ relative to the couple $(\widehat{\boldsymbol{\Gamma}}_{\chi}(0), \widehat{\boldsymbol{\Gamma}}_{\xi}(0))$. Defining $\widehat{\boldsymbol{\Lambda}} = (\widehat{\boldsymbol{\lambda}}_1 \dots \widehat{\boldsymbol{\lambda}}_r)$, the space spanned by the common factors is estimated by the first r generalized principal components of the $\boldsymbol{x}$'s: $\widehat{\boldsymbol{\Lambda}}'\boldsymbol{x}_t = \widehat{\boldsymbol{\lambda}}_1'\boldsymbol{x}_t \dots \widehat{\boldsymbol{\lambda}}_r'\boldsymbol{x}_t$. The forecast of the common component depends on the covariance between χ_{iT+h} and $\widehat{\boldsymbol{\Lambda}}'\boldsymbol{x}_T$. Observing that this covariance equals the covariance between χ_{T+h} and $\widehat{\boldsymbol{\Lambda}}'\boldsymbol{\chi}_T$, the forecast of the common component can be obtained from the projection

$$\widehat{\boldsymbol{\chi}}_{T+h|T} = \widehat{\boldsymbol{\Gamma}}_{\chi}(h)\widehat{\boldsymbol{\Lambda}}(\widehat{\boldsymbol{\Lambda}}'\widehat{\boldsymbol{\Gamma}}_{\chi}(0)\widehat{\boldsymbol{\Lambda}})^{-1}\widehat{\boldsymbol{\Lambda}}'\boldsymbol{x}_T.$$

Since this two-step forecasting method relied on lags but not leads, it avoided the end-of-sample problems caused by two-sided filtering in the authors' earlier study. A simulation study by the authors suggested that this procedure improves upon the forecasting performance of Stock and Watson (2002a,b) static principal components method for data generating processes with heterogeneous dynamics and heterogeneous variance ratio of the common and idiosyncratic components. In line with most of the earlier literature, Forni, Giannone, Lippi, and Reichlin (2009) continued to assume that the space spanned by the common components at any time t has a finite dimension r as n tends to infinity, allowing a static representation of the model. By identifying and estimating cross-sectionally pervasive shocks and their dynamic effect on macroeconomic variables, they showed that dynamic factor models are suitable for structural modeling.

The finite-dimension assumption for the common component rules out certain factor-loading patterns. In the model $x_{it} = a_i(1 - b_i L)^{-1} u_t + \xi_{it}$, where u_t is a scalar white noise and the coefficients b_i are drawn from a uniform distribution (-0.9, 0.9), the space spanned by the common components χ_{it}, $i \in \mathbb{N}$ is infinite dimensional. Forni and Lippi (2011) relaxed the finite-dimensional assumption and proposed a one-sided estimator for the general dynamic factor model of Forni et al. (2000). Forni, Hallin, Lippi, and Zaffaroni (2015, 2017) continued to allow the common components—driven by a finite number of common shocks—to span an infinite-dimensional space and investigated the model's one-sided representations and asymptotic properties. Forni, Giovannelli, Lippi, and Soccorsi (2018) evaluated the model's pseudo real-time forecasting performance for US macroeconomic variables and found that it compares favorably to finite dimensional methods during the Great Moderation. The dynamic relationship between the variables and the factors in this model is more general than in models assuming a finite common component space, but, as pointed out by the authors, its estimation is rather complex.

Hallin and Lippi (2013) give a general presentation of the methodological foundations of dynamic factor models. Fiorentini, Galesi, and Sentana (2018) introduced a frequency domain version of the EM algorithm for dynamic factor models with latent autoregressive and moving average processes. In this paper the authors focused on an exact factor model with a single common factor, and left approximate factor models with multiple factors for future research. But they extended the basic EM algorithm with an iterated indirect inference procedure based on a sequence of simple auxiliary OLS regressions to speed up computation for models with moving average components.

Although carried out in the time domain, Peña and Yohai's (2016) generalized dynamic principal components model mimics two important features of Forni et al.'s (2000) generalized dynamic factor model: it allows for both a dynamic representation of the common component and nonorthogonal idiosyncratic components. Their procedure chooses the number of common components to achieve a desired degree of accuracy in a mean squared error sense in the reconstruction of the original series. The estimation iterates two steps: the first is a least squares estimator of the loading coefficients assuming the factors are known, and the second is updating the factor estimate based on the estimated coefficients. Since the authors do not place restrictions on the principal components, their method can be applied to potentially nonstationary time series data. Although the proposed method does well for data reconstruction, it is not suited for forecasting, because—as in Forni et al. (2000)—it uses both leads and lags to reconstruct the series.

2.5 Estimating the Number of Factors

Full specification of the DFM requires selecting the number of common factors. The number of static factors can be determined by information criteria that use penalized objective functions or by analyzing the distribution of eigenvalues. Bai and Ng (2002) used information criteria of the form $IC(r) = \ln V_r(\widehat{\mathbf{\Lambda}}, \widehat{\boldsymbol{F}}) + rg(n, T)$, where $V_r(\widehat{\mathbf{\Lambda}}, \widehat{\boldsymbol{F}})$ is the least squares objective function (2.10) evaluated with the principal components estimators when r factors are considered, and $g(n, T)$ is a penalty function that satisfies two conditions: $g(n, T) \to 0$ and $\min[n^{1/2}, T^{1/2}] \cdot g(n, T) \to \infty$, as $n, T \to \infty$. The estimator for the number of factors is $\widehat{r}_{IC} = \min_{0 \leq r \leq rmax} IC(r)$, where $rmax$ is the upper bound of the true number of factors. The authors show that the estimator is consistent without restrictions between n and T, and the results hold under heteroskedasticity in both the time and cross-sectional dimension, as well as under weak serial and cross-sectional correlation. Li, Li, and Shi (2017) develop a method to estimate the number of factors when the number of factors is allowed to increase as the cross-section and time dimensions increase. This is useful since new factors may emerge as changes in the economic environment trigger structural breaks.

Ahn and Horenstein (2013) and Onatski (2009, 2010) take a different approach by comparing adjacent eigenvalues of the spectral density matrix at a given frequency or of the covariance matrix of the data. The basic idea behind this approach is that the first r eigenvalues will be unbounded, while the remaining values will be bounded. Therefore the ratio of subsequent eigenvalues is maximized at the location of the largest relative cliff in a scree plot (a plot of the ordered eigenvalues against the rank of those eigenvalues). These authors also present alternative statistics using the difference, the ratio of changes, and the growth rates of subsequent eigenvalues.

Estimation of the number of dynamic factors usually requires several steps. As illustrated in Sect. 2.2.1, the number of dynamic factors, q, will in general be lower than the number of static factors $r = q(1 + s)$, and therefore the spectrum of the common component will have a reduced rank with only q nonzero eigenvalues. Based on this result, Hallin and Liska (2007) propose a frequency-domain procedure which uses an information criterion to estimate the rank of the spectral density matrix of the data. Bai and Ng (2007) take a different approach by first estimating the number of static factors and then applying a VAR(p) model to the estimated factors to obtain the residuals. They use the eigenvalues of the residual covariance matrix to estimate the rank q of the covariance of the dynamic (or primitive) shocks. In a similar spirit, Amengual and Watson (2007) first project the observed variables $\boldsymbol{x}_t$ onto p lags of consistently estimated r static principal components $\boldsymbol{f}_t \dots \boldsymbol{f}_{t-p}$ to obtain serially uncorrelated residuals $\widehat{\boldsymbol{u}_t} = \boldsymbol{x}_t - \sum_{i=1}^{p} \widehat{\mathbf{\Pi}}_i \widehat{\boldsymbol{f}}_{t-i}$. These residuals then have a static factor representation with q factors. Applying the Bai and Ng (2002) information criterion to the sample variance matrix of these residuals yields a consistent estimate of the number of dynamic factors, q.

2.6 Forecasting with Large Dynamic Factor Models

One of the most important uses of dynamic factor models is forecasting. Both small scale (i.e., n small) and large scale dynamic factor models have been used to this end. Very often, the forecasted variable is published with a delay: for instance, Euro-area GDP is published 6 weeks after the end of the corresponding quarter. The long delay in data release implies that different types of forecasts can be considered. GDP predictions for quarter Q, made prior to that quarter, in $Q-1, Q-2, \ldots$ for instance, are considered to be "true" out of sample forecasts. But estimates for quarter Q can also be made during that same quarter using high frequency data released within quarter Q. These are called "nowcasts." Finally, since GDP for quarter Q is not known until a few weeks into quarter $Q+1$, forecasters keep estimating it during this time interval. These estimates are considered to be "backcasts." As we have seen, dynamic factor models are very flexible and can easily handle all these predictions.

Forecasts using large dimensional factor models were first introduced in the literature by Stock and Watson (2002a). The method, also denoted diffusion index forecasts, consists of estimating the factors $\boldsymbol{f}_t$ by principal component analysis, and then using those estimates in a regression estimated by ordinary least squares

$$y_{t+h} = \boldsymbol{\beta}'_f \widehat{\boldsymbol{f}}_t + \boldsymbol{\beta}'_w \boldsymbol{w}_t + \varepsilon_{t+h},$$

where y_t is the variable of interest, $\widehat{f_t}$ is the vector of the estimated factors, and $\boldsymbol{w}_t$ is a vector of observable predictors (typically lags of y_t). The direct forecast for time $T+h$ is then computed as

$$y_{T+h|T} = \widehat{\boldsymbol{\beta}}'_f \widehat{\boldsymbol{f}}_T + \widehat{\boldsymbol{\beta}}'_w \boldsymbol{w}_T.$$

The authors prove that, under the assumptions they used to ensure the consistency of the principal component estimates,

$$\text{plim}_{n\to\infty} \left[(\widehat{\boldsymbol{\beta}}'_f \widehat{\boldsymbol{f}}_T + \widehat{\boldsymbol{\beta}}'_w \boldsymbol{w}_T) - (\boldsymbol{\beta}'_f \boldsymbol{f}_T + \boldsymbol{\beta}'_w \boldsymbol{w}_T) \right] = 0,$$

so that the forecast is asymptotically equivalent to what it would have been if the factors had been observed. Furthermore, under a stronger set of assumptions Bai and Ng (2006) show that the forecast error is asymptotically Gaussian, with known variance, so that forecast intervals can be computed. Stock and Watson (2002b) considered a more general model to capture the dynamic relationship between the variables and the factors

$$y_{t+h} = \alpha_h + \boldsymbol{\beta}_h(L) \widehat{\boldsymbol{f}}_t + \gamma_h(L) y_t + \varepsilon_{t+h},$$

with forecast equation

$$\widehat{y}_{T+h|T} = \widehat{\alpha}_h + \widehat{\boldsymbol{\beta}}_h(L)\widehat{\boldsymbol{f}}_T + \widehat{\gamma}_h(L)y_T. \tag{2.11}$$

They find that this model produces better forecasts for some variables and forecast horizons, but the improvements are not systematic.

While the diffusion index methodology relied on a distributed lag model, the approach taken by Doz et al. (2011) and Doz et al. (2012) captured factor dynamics explicitly. The estimates of a vector-autoregressive model for $\widehat{\boldsymbol{f}}_t$ can be used to recursively forecast $\widehat{\boldsymbol{f}}_{T+h|T}$ at time T. A mean square consistent forecast for period $T+h$ can then be obtained by $y_{T+h|T} = \widehat{\boldsymbol{\Lambda}}\widehat{\boldsymbol{f}}_{T+h|T}$. This approach has been used by Giannone et al. (2008) and by many others. However Stock and Watson (2002b) point out that the diffusion index and two step approaches are equivalent since the recursive forecast of the factor in the latter implies that $\widehat{\boldsymbol{f}}_{T+h|T}$ is a function of $\widehat{\boldsymbol{f}}_T$ and its lags, as in (2.11).

2.6.1 *Targeting Predictors and Other Forecasting Refinements*

In the diffusion index and two-step forecasting methodology, the factors are first estimated from a large number of predictors, $(x_{1t} \ldots x_{nt})$, by the method of principal components, and then used in a linear forecasting equation for y_{t+h}. Although the method can parsimoniously summarize information from a large number of predictors and incorporate it into the forecast, the estimated factors do not take into account the predictive power of x_{it} for y_{t+h}. Boivin and Ng (2006) pointed out that expanding the sample size simply by adding data without regard to its quality or usefulness does not necessarily improve forecasts. Bai and Ng (2008a) suggested to target predictors based on their information content about y. They used hard and soft thresholding to determine which variables the factors are to be extracted from and thereby reduce the influence of uninformative predictors. Under hard thresholding, a pretest procedure is used to decide whether a predictor should be kept or not. Under soft thresholding, the predictor ordering and selection is carried out using the least angle regression (LARS) algorithm developed by Efron, Hastie, Johnstone, and Tibshirani (2004).

Kelly and Pruitt (2015) proposed a three-pass regression filter with the ability to identify the subset of factors useful for forecasting a given target variable while discarding those that are target irrelevant but may be pervasive among predictors. The proposed procedure uses the covariances between the variables in the dataset and the proxies for the relevant latent factors, the proxies being observable variables either theoretically motivated or automatically generated.

Pass 1 The first pass captures the relationship between the predictors, $\boldsymbol{X}$, and $m \ll \min(n, T)$ factor proxies, $\boldsymbol{Z}$, by running a separate time series regression of each predictor, $\boldsymbol{x}_i$, on the proxies, $x_{it} = \alpha_i + \boldsymbol{z}_t'\boldsymbol{\beta}_i + \varepsilon_{it}$, for $i = 1 \ldots n$, and retaining $\widehat{\boldsymbol{\beta}}_i$.

Pass 2 The second pass consists of T separate cross section regressions of the predictors, $\boldsymbol{x}_t$, on the coefficients estimated in the first pass, $x_{it} = \alpha_t + \widehat{\boldsymbol{\beta}}_i' \boldsymbol{f}_t + \varepsilon_{it}$, for $t = 1 \dots T$, and retaining $\widehat{\boldsymbol{f}}_t$. First-stage coefficient estimates map the cross-sectional distribution of predictors to the latent factors. Second-stage cross section regressions use this map to back out estimates of the factors, $\widehat{\boldsymbol{f}}_t$, at each point in time.

Pass 3 The third pass is a single time series forecasting regression of the target variable y_{t+1} on the predictive factors estimated in the second pass, $y_{t+1} = \alpha + \widehat{\boldsymbol{f}}_t' \boldsymbol{\beta} + \varepsilon_{t+1}$. The third-pass fitted value, $\widehat{y}_{t+1}$, is the forecast for period $t + 1$.

The automatic proxy selection algorithm is initialized with the target variable itself $z_1 = \boldsymbol{y}$, and additional proxies based on the prediction error are added iteratively, $z_{k+1} = \boldsymbol{y} - \widehat{\boldsymbol{y}}_k$ for $k = 1 \dots m - 1$, where k is the number of proxies in the model at the given iteration. The authors point out that partial least squares, further analyzed by Groen and Kapetanios (2016), is a special case of the three-pass regression filter.

Bräuning and Koopman (2014) proposed a collapsed dynamic factor model where the factor estimates are established jointly by the predictors $\boldsymbol{x}_t$ and the target variable y_t. The procedure is a two-step process.

Step 1 The first step uses principal component analysis to reduce the dimension of a large panel of macroeconomic predictors as in Stock and Watson (2002a,b).

Step 2 In the second step, the authors use a state space model with a small number of parameters to model the principal components jointly with the target variable y_t. The principal components, $\boldsymbol{f}_{PC,t}$, are treated as dependent variables that are associated exclusively with the factors $\boldsymbol{f}_t$, but the factors $\boldsymbol{f}_t$ enter the equation for the target variable y_t. The unknown parameters are estimated by maximum likelihood, and the Kalman filter is used for signal extraction.

In contrast to the two-step method of Doz et al. (2011), this approach allows for a specific dynamic model for the target variable that may already produce good forecasts for y_t.

Several additional methods originating in the machine learning literature have been used to improve forecasting performance of dynamic factor models, including bagging (Inoue & Kilian, 2008) and boosting (Bai & Ng, 2009), which will be discussed in Chaps. 14 and 16, respectively.

2.7 Hierarchical Dynamic Factor Models

If a panel of data can be organized into blocks using a priori information, then between- and within-block variation in the data can be captured by the hierarchical dynamic factor model framework formalized by Moench, Ng, and Potter (2013). The block structure helps to model covariations that are not sufficiently pervasive to be treated as common factors. For example, in a three-level model, the series i, in a given block b, at each time t can exhibit idiosyncratic, block specific, and common variation

$$\begin{aligned} x_{ibt} &= \boldsymbol{\lambda}_{gib}(L)\boldsymbol{g}_{bt} + e_{xibt} \\ \boldsymbol{g}_{bt} &= \boldsymbol{\Lambda}_{fb}(L)\boldsymbol{f}_t + \boldsymbol{e}_{gbt} \\ \boldsymbol{\Phi}_f(L)\boldsymbol{f}_t &= \boldsymbol{u}_t , \end{aligned}$$

where variables x_{ibt} and x_{jbt} within a block b are correlated because of the common factors $\boldsymbol{f}_t$ or the block-specific shocks e_{gbt}, but correlations between blocks are possible only through $\boldsymbol{f}_t$. Some of the x_{it} may not belong to a block and could be affected by the common factors directly, as in the two-level model (2.5). The idiosyncratic components can be allowed to follow stationary autoregressive processes, $\phi_{xib}(L)e_{xibt} = \boldsymbol{\varepsilon}_{xibt}$ and $\boldsymbol{\Phi}_{gb}(L)\boldsymbol{e}_{gbt} = \boldsymbol{\varepsilon}_{gbt}$.

If we were given data for production, employment, and consumption, then x_{i1t} could be one of the n_1 production series, x_{i2t} could be one of the n_2 employment series, and x_{i3t} could be one of the n_3 consumption series. The correlation between the production, employment, and consumption factors, $\boldsymbol{g}_{1t}, \boldsymbol{g}_{2t}, \boldsymbol{g}_{3t}$, due to economy-wide fluctuations, would be captured by $\boldsymbol{f}_t$. In a multicountry setting, a four-level hierarchical model could account for series-specific, country (subblock), region (block), and global (common) fluctuations, as in Kose, Otrok, and Whiteman (2003). If the country and regional variations were not properly modeled, they would appear as either weak common factors or idiosyncratic errors that would be cross-correlated among series in the same region. Instead of assuming weak cross-sectional correlation as in approximate factor models, the hierarchical model explicitly specifies the block structure, which helps with the interpretation of the factors.

To estimate the model, Moench et al. (2013) extend the Markov chain Monte Carlo method that Otrok and Whiteman (1998) originally applied to a single factor model. Let $\boldsymbol{\Lambda} = (\boldsymbol{\Lambda}_g, \boldsymbol{\Lambda}_f)$, $\boldsymbol{\Phi} = (\boldsymbol{\Phi}_f, \boldsymbol{\Phi}_g, \boldsymbol{\Phi}_x)$, and $\boldsymbol{\Psi} = (\boldsymbol{\Psi}_f, \boldsymbol{\Psi}_g, \boldsymbol{\Psi}_x)$ denote the matrices containing the loadings, coefficients of the lag polynomials, and variances of the innovations, respectively. Organize the data into blocks, $\boldsymbol{x}_{bt}$, and get initial values for $\boldsymbol{g}_t$ and $\boldsymbol{f}_t$ using principal components; use these to produce initial values for $\boldsymbol{\Lambda}, \boldsymbol{\Phi}, \boldsymbol{\Psi}$.

Step 1 Conditional on $\boldsymbol{\Lambda}, \boldsymbol{\Phi}, \boldsymbol{\Psi}, \boldsymbol{f}_t$, and the data $\boldsymbol{x}_{bt}$, draw $\boldsymbol{g}_{bt}$ for all b.
Step 2 Conditional on $\boldsymbol{\Lambda}, \boldsymbol{\Phi}, \boldsymbol{\Psi}$, and $\boldsymbol{g}_{bt}$, draw $\boldsymbol{f}_t$.
Step 3 Conditional on $\boldsymbol{f}_t$ and $\boldsymbol{g}_{bt}$, draw $\boldsymbol{\Lambda}, \boldsymbol{\Phi}, \boldsymbol{\Psi}$, and return to **Step 1**.

The sampling of $\boldsymbol{g}_{bt}$ needs to take into account the correlation across blocks due to $\boldsymbol{f}_t$. As in previously discussed dynamic factor models, the factors and the loadings are not separately identified. To achieve identification, the authors suggest using lower triangular loading matrices, fixed variances of the innovations to the factors, $\boldsymbol{\Psi}_f, \boldsymbol{\Psi}_g$, and imposing additional restrictions on the structure of the lag polynomials. Jackson, Kose, Otrok, and Owyang (2016) survey additional Bayesian estimation methods. Under the assumption that common factors have a direct impact on x, but are uncorrelated with block-specific ones, so that $x_{ibt} = \boldsymbol{\lambda}_{fib}(L)\boldsymbol{f}_t + \boldsymbol{\lambda}_{gib}(L)\boldsymbol{g}_{bt} + e_{xibt}$, Breitung and Eickmeier (2016) propose a sequential least squares estimation approach and compare it to other frequentist estimation methods.

Kose, Otrok, and Whiteman (2008) use a hierarchical model to study international business cycle comovements by decomposing fluctuations in macroeconomic aggregates of G-7 countries into a common factor across all countries, country factors that are common across variables within a country, and idiosyncratic fluctuations. Moench and Ng (2011) apply this model to national and regional housing data to estimate the effects of housing shocks on consumption, while Fu (2007) decomposes house prices in 62 US metropolitan areas into national, regional, and metro-specific idiosyncratic factors. Del Negro and Otrok (2008) and Stock and Watson (2008) use this approach to add stochastic volatility to their models.

2.8 Structural Breaks in Dynamic Factor Models

As evident from the discussion so far, the literature on dynamic factor models has grown tremendously, evolving in many directions. In the remainder of this chapter we will concentrate on two strands of research: dynamic factor models (1) with Markov-switching behavior and (2) with time-varying loadings. In both cases, the aim is to take into account the evolution of macroeconomic conditions over time, either through (1) non-linearities in the dynamics of the factors or (2) the variation of loadings, which measure the intensity of the links between each observable variable and the underlying common factors. This instability seemed indeed particularly important to address after the 2008 global financial crisis and the subsequent slow recovery. These two strands of literature have presented a number of interesting papers in recent years. In what follows, we briefly describe some of them, but we do not provide an exhaustive description of the corresponding literature.

2.8.1 Markov-Switching Dynamic Factor Models

One of the first uses of dynamic factor models was the construction of coincident indexes. The literature soon sought to allow the dynamics of the index to vary according to the phases of the business cycle. Incorporating Markov switching into dynamic factor models (MS-DFM) was first suggested by Kim (1994) and Diebold and Rudebusch (1996).[4] Diebold and Rudebusch (1996) considered a single factor that played the role of a coincident composite index capturing the latent state of the economy. They suggested that the parameters describing the dynamics of the factor may themselves depend on an unobservable two-state Markov-switching

[4]The working paper version appeared in 1994 in NBER Working Papers 4643.

latent variable. More precisely, they modeled the factor the same way as Hamilton (1989) modeled US real GNP to obtain a statistical characterization of business cycle phases. In practice, Diebold and Rudebusch (1996) used a two-step estimation method since they applied Hamilton's model to a previously estimated composite index.

Kim (1994) introduced a very general model in which the dynamics of the factors and the loadings may depend on a state-variable. Using the notation used so far in the current chapter, his model was

$$\begin{aligned} \boldsymbol{y}_t &= \boldsymbol{\Lambda}_{S_t} \boldsymbol{f}_t + \boldsymbol{B}_{S_t} \boldsymbol{x}_t + \boldsymbol{e}_t \\ \boldsymbol{f}_t &= \boldsymbol{\Phi}_{S_t} \boldsymbol{f}_{t-1} + \boldsymbol{\Gamma}_{S_t} \boldsymbol{x}_t + \boldsymbol{H}_{S_t} \boldsymbol{\varepsilon}_t, \end{aligned}$$

where $\boldsymbol{x}_t$ is a vector of exogenous or lagged dependent variables, and where the $\begin{pmatrix} \boldsymbol{e}_t \\ \boldsymbol{\varepsilon}_t \end{pmatrix}$'s are i.i.d. Gaussian with covariance matrix $\begin{pmatrix} \boldsymbol{R} & \mathbf{O} \\ \mathbf{O} & \boldsymbol{Q} \end{pmatrix}$. The underlying state variable S_t can take M possible values, and is Markovian of order one, so that $P(S_t = j|S_{t-1} = i, S_{t-2} = k, \ldots) = P(S_t = j|S_{t-1} = i) = p_{ij}$. He proposed a very powerful approximation in the computation of the Kalman filter and the likelihood and estimated the model in one step by maximum likelihood[5]. Using this approximation, the likelihood can be computed at any point of the parameter space and maximized using a numerical procedure. It must be however emphasized that such a procedure is applicable in practice only when the number of parameters is small, which means that the dimension of $\boldsymbol{x}_t$ must be small. Once the parameters have been estimated, it is possible to obtain the best approximations of $\boldsymbol{f}_t$ and S_t for any t using the Kalman filter and smoother, given the observations $\boldsymbol{x}_1 \ldots \boldsymbol{x}_t$ and $\boldsymbol{x}_1 \ldots \boldsymbol{x}_T$, respectively.

Kim (1994) and Chauvet (1998) estimated a one factor Markov-switching model using this methodology. In both papers, the model is formulated like a classical dynamic factor model, with the factor following an autoregressive process whose constant term depends on a two-state Markov variable S_t:

$$\boldsymbol{x}_t = \boldsymbol{\lambda} f_t + \boldsymbol{e}_t \text{ and } \phi(L) f_t = \beta_{S_t} + \eta_t$$

where $\eta_t \sim$ i.i.d. $N(0, 1)$ and S_t can take two values denoted as 0 and 1, which basically correspond to expansions and recessions: the conditional expectation of the factor is higher during expansions than during recessions. Kim and Yoo (1995) used the same four observable variables as the US Department of Commerce and Stock and Watson (1989, 1991, 1993), whereas Chauvet (1998) considered several sets of observable series over various time periods, taking into account different specifications for the dynamics of the common factor. Both papers obtained

[5]Without this approximation, the Kalman filter would be untractable, since it would be necessary to take the M^T possible trajectories of $S_1, \ldots, S_T$. For further details, see Kim (1994) and the references therein.

posterior recession probabilities and turning points that were very close to official NBER dates. Kim and Nelson (1998) proposed a Gibbs sampling methodology that helped to avoid Kim's (1994) approximation of the likelihood. This Bayesian approach also provided results that were very close to the NBER dates, and allowed tests of business cycle duration dependence. Chauvet and Piger (2008) compared the nonparametric dating algorithm given in Harding and Pagan (2003) with the dating obtained using Markov-switching models similar to those of Chauvet (1998). They showed that both approaches identify the NBER turning point dates in real time with reasonable accuracy, and identify the troughs with more timeliness than the NBER. But they found evidence in favor of MS-DFMs, which identified NBER turning point dates more accurately overall. Chauvet and Senyuz (2016) used a two-factor MS-DFM to represent four series; three related to the yield curve and the fourth, industrial production, representing economic activity. Their model allowed to analyze the lead-lag relationship between the cyclical phases of the two sectors.

Camacho, Perez-Quiros, and Poncela (2014) extended the Kim and Yoo (1995) approach to deal with mixed frequency and/or ragged-edge data. They ran simulations and applied their methodology to real time observations of the four variables used by the US Department of Commerce. They found evidence that taking into account all the available information in this framework yields substantial improvements in the estimated real time probabilities. Camacho, Perez-Quiros, and Poncela (2015) used the same approach and applied the method to a set of thirteen Euro-area series. They obtained a non-linear indicator of the overall economic activity and showed that the associated business cycle dating is very close to the CEPR Committee's dating.

The one-step methods used in all these papers (most of them following Kim's (1994) approximation of the likelihood, others relying on Kim and Nelson's (1998) Gibbs sampling) have been successful in estimating MS-DFMs of very small dimensions. In order to estimate MS-DFM of larger dimensions, it is possible to take advantage of the two-step approach originally applied to a small number of variables by Diebold and Rudebusch (1996). Indeed, it is possible to use a two-step approach similar to the one of Doz et al. (2011) in a standard DFM framework: in the first step, a linear DFM is estimated by principal components, in the second step, a Markov-switching model, as in Hamilton (1989), is specified for the estimated factor(s) and is estimated by maximum likelihood. Camacho et al. (2015) compared this two-step approach to a one-step approach applied to a small dataset of coincident indicators. They concluded that the one-step approach was better at turning point detection when the small dataset contained good quality business cycle indicators, and they also observed a decreasing marginal gain in accuracy when the number of indicators increased. However other authors have obtained satisfying results with the two-step approach. Bessec and Bouabdallah (2015) applied MS-factor MIDAS models[6] to a large dataset of mixed frequency

[6]Their model combines the MS-MIDAS model (Guérin & Marcellino, 2013) and the factor-MIDAS model (Marcellino & Schumacher, 2010).

variables. They ran Monte Carlo simulations and applied their model to US data containing a large number of financial series and real GDP growth: in both cases the model properly detected recessions. Doz and Petronevich (2016) also used the two-step approach: using French data, they compared business cycle dates obtained from a one factor MS-DFM estimated on a small dataset with Kim's (1994) method, to the dates obtained using a one factor MS-DFM estimated in two steps from a large database. The two-step approach successfully predicted the turning point dates released by the OECD. As a complement, Doz and Petronevich (2017) conducted a Monte-Carlo experiment, which provided evidence that the two-step method is asymptotically valid for large N and T and provides good turning points prediction. Thus the relative performances of the one-step and two-step methods under the MS-DFM framework are worth exploring further, both from a turning point detection perspective and a forecasting perspective.

2.8.2 Time Varying Loadings

Another strand of the literature deals with time-varying loadings. The assumption that the links between the economic variables under study and the underlying factors remain stable over long periods of time may be seen as too restrictive. If the common factors are driven by a small number of structural shocks, the observable variables may react to those structural shocks in a time varying fashion. Structural changes in the economy may also lead to changes in the comovements of variables, and in turn require adjustment in the underlying factor model. Such shifts have become particularly relevant after the 2008 financial crisis and the ensuing slow recovery. But the literature had dealt with similar questions even earlier, for instance, during the Great Moderation. The issue is important since assuming constant loadings—when in fact the true relationships experience large structural breaks—can lead to several problems: overestimation of the number of factors, inconsistent estimates of the loadings, and deterioration of the forecasting performance of the factors.

The literature on this topic is rapidly growing, but the empirical results and conclusions vary across authors. We provide an overview of the literature, without covering it exhaustively. This overview can be roughly divided into two parts. In a first group of papers, the loadings are different at each point of time. In a second group of papers, the loadings display one break or a small number of breaks, and tests are proposed for the null hypothesis of no break.

The first paper addressing changes in the loadings is Stock and Watson (2002a). The authors allowed for small-amplitude time variations in the loadings. More precisely, they assumed that $x_{it} = \lambda_{it} f_t + e_{it}$ with $\lambda_{it} = \lambda_{it-1} + g_{it}\zeta_{it}$, where $g_{it} = O\left(T^{-1}\right)$ and ζ_{it} has weak cross-sectional dependence. They proved that PCA estimation of the loadings and factors is consistent, even though the estimation method assumes constant loadings. This result has been confirmed by Stock and Watson (2009), who analyzed a set of 110 US quarterly series spanning 1959 to 2006 and introduce a single break in 1984Q1 (start of the Great Moderation). They found evidence of instability in the loadings for about half of the series, but showed that the factors are well estimated using PCA on the full sample. Bates, Plagborg-Möller, Stock, and Watson (2013) further characterized the type and magnitude of parameter instability that is compatible with the consistency of PCA estimates. They showed that, under an appropriate set of assumptions, the PCA estimated factors are consistent if the loading matrix is decomposed as $\mathbf{\Lambda}_t = \mathbf{\Lambda}_0 + h_{nT}\boldsymbol{\xi}_t$ with $h_{nT} = O\left(T^{-1/2}\right)$, which strengthens the result of Stock and Watson (2002a), obtained for $h_{nT} = O\left(T^{-1}\right)$). They further showed that, for a given number of factors, if $h_{nT} = O\left(1/\min(n^{1/4}T^{1/2}, T^{3/4}\right)$ the estimated factors converge to the true ones (up to an invertible matrix) at the same rate as in Bai and Ng (2002) i.e., $1/\min(n^{1/2}, T^{1/2})$. However, they showed that, if the proportion of series undergoing a break is too high, usual criteria are likely to select too many factors.

In a Monte-Carlo experiment, Banerjee, Marcellino, and Masten (2007) demonstrated that consistent estimation of the factors does not preclude a deterioration of factor based forecasts. Stock and Watson (2009) pointed out that forecast equations may display even more instability than the factor loadings, and they assessed the importance of this instability on the forecast performance. In particular, they showed that the best forecast results are obtained when the factors are estimated from the whole sample, but the forecast equation is only estimated on the sub-sample where its coefficients are stable.

Del Negro and Otrok (2008) proposed a DFM with time-varying factor loadings, but they also included stochastic volatility in both the factors and the idiosyncratic components. Their model aimed at studying the evolution of international business cycles, and their dataset consisted of real GDP growth for nineteen advanced countries. They considered a model with two factors: a world factor and a European factor. The model, with time varying loadings, can be written as

$$y_{it} = a_i + b_{it}^w f_t^w + b_{it}^e f_t^e + \varepsilon_{it} ,$$

where f_t^w and f_t^e are the world and European factors, and where $b_{it}^e = 0$ for non-European countries. The loadings are assumed to follow a random walk without drift: $b_{it} = b_{it-1} + \sigma_{\eta_i}\eta_{it}$, with the underlying idea that the sensitivity of a given country to the factors may evolve over time, and that this evolution is slow but picks up permanent changes in the economy. The factors and the idiosyncratic components have stochastic volatility, which allows variation in the importance of global/regional shocks and country-specific shocks. The model was estimated using Gibbs sampling. The results supported the notion of the Great Moderation in all the countries in the sample, notwithstanding important heterogeneity in the timing and magnitude, and in the sources (domestic or international) of this moderation. This is in line with features later highlighted by Stock and Watson (2012) (see below). Del Negro and Otrok (2008) also showed that the intensity of comovements is time-varying, but that there has been a convergence in the volatility of fluctuations in activity across countries.

Su and Wang (2017) considered a factor model where the number of factors is fixed, and the loadings change smoothly over time, using the following specification: $\lambda_{it} = \lambda_i(t/T)$ where λ_i is an unknown smooth function. They employed a local version of PCA to estimate the factors and the loadings and obtained local versions of the Bai (2003) asymptotic distributions. They used an information criterion similar to the Bai and Ng (2002) information criteria and proved its consistency for large n and T. They also proposed a consistent test of the null hypothesis of constant loadings: the test statistic is a rescaled version of the mean square discrepancy between the common components estimated with time-varying loadings and the common components estimated by PCA with constant loadings, and it is asymptotically Gaussian under the null. Finally, the authors also suggested a bootstrap version of the test in order to improve its size in finite samples. Their simulations showed that the information criteria work well, and that the bootstrap version of the test is more powerful than other existing tests when there is a single break at an unknown date. Finally, using the Stock and Watson (2009) dataset, they clearly rejected the null of constant loadings.

Breitung and Eickmeier (2011) were the first to consider the case where strong breaks may occur in the loadings. They noted that, in this case, the number of common factors has to be increased: for instance, in the case of a single break, two sets of factors are needed to describe the common component before and after the break, which is tantamount to increasing the number of factors in the whole sample. For a known break date, they proposed to test the null hypothesis of constant loadings in individual series, using a Chow test, a Wald test, or an LM test, with the PCA-estimated factors replacing the unknown factors. They also addressed the issue of a structural break at an unknown date: building on Andrews (1993), they proposed a Sup-LM statistic to test the null of constant loadings in an individual series. In both cases, autocorrelation in the factors and idiosyncratic terms are taken into account. Applying an LM-test, and using Stock and Watson 2005 US data, they found evidence of a structural break at the beginning of 1984 (start of the Great Moderation). They also found evidence of structural breaks for the Euro-area, at the beginning of 1992 (Maastricht treaty) and the beginning of 1999 (stage 3 of EMU). Yamamoto and Tanaka (2015) noted that this testing procedure suffers from non-monotonic power, which is widespread in structural change tests. To remedy this issue, they proposed a modified version of the Breitung and Eickmeier (2011) test, taking the maximum of the Sup-Wald test statistics obtained from regressing the variable of interest on each estimated factor. They showed that this new test does not suffer from the non-monotonic power problem.

Stock and Watson (2012) addressed the issue of a potential new factor associated with the 2007Q4 recession and its aftermath. The authors used a large dataset of 132 disaggregated quarterly series, which were transformed to induce stationarity and subsequently "detrended" to eliminate low frequency variations from the data. They estimated six factors and the corresponding loadings by PCA over the period 1959Q1–2007Q3. The factors were extended over 2007Q4–2011Q2 by using the estimated loadings from the pre-recession period to form linear combinations of the observed variables after the onset of the recession. The extended factors available from 1959Q1 to 2011Q2 with constant loadings were denoted the "old factors." The authors showed that these old factors explain most of the variation in the individual time series, which suggests that there was no new factor after the financial crisis. They also tested the null hypothesis of a break in the loadings after 2007Q4 and showed that this hypothesis is rejected only for a small number of series. Further, they investigated the presence of a new factor by testing whether the idiosyncratic residuals display a factor structure, and concluded that there is no evidence of a new factor. Finally, they examined the volatility of the estimated innovations of the factors during different subperiods and found evidence that the recession was associated with exceptionally large unexpected movements in the "old factors." Overall, they concluded that the financial crisis resulted in larger volatility of the factors, but neither did a new factor appear, nor did the response of the series to the factors change, at least for most series. These results are consistent with those obtained by Del Negro and Otrok (2008).

However, many users of DFMs have focused on the breaks-in-the-loadings scheme mentioned above, and several other tests have been proposed to test the null hypothesis of no break. Han and Inoue (2015) focused on the joint null hypothesis that all factor loadings are constant over time against the alternative that a fraction α of the loadings are not. Their test assumes that there is a single break at an unknown date that is identical for all series. They used the fact that if the factors are estimated from the whole sample, their empirical covariance matrix before the break will differ from their empirical covariance matrix after the break. They proposed a Sup-Wald and Sup-LM test, where the supremum is taken over the possible break dates. These tests were shown to be consistent even if the number of factors is overestimated.

Chen, Dolado, and Gonzalo (2014) proposed a test designed to detect big breaks at potentially unknown dates. As previously noticed by Breitung and Eickmeier (2011), in such a situation one can write a model with fixed loadings and a larger number of factors. The test is based on the behavior of the estimated factors before and after the break date if there is one. It relies on a linear regression of one of the estimated factors on the others and tests for a structural break in the coefficients of this regression. If the potential break date is known, the test is a standard Wald test. If it is unknown, the test can be run using the Sup-LM or Sup-Wald tests which have been proposed by Andrews (1993). The authors' Monte-Carlo experiment showed that both tests perform well when $T \geq 100$ and have better power than the tests proposed by Breitung and Eickmeier (2011) or Han and Inoue (2015). The authors also showed that the Sup-Wald test generally behaves better than Sup-LM test in finite samples and confirms that Bai-Ng's criteria overestimate the number of factors when there is a break. Finally, the authors applied the Sup-Wald test to the same dataset as Stock and Watson (2009): the null hypothesis of no break was rejected, and the estimated break date was around 1979–1980, rather than 1984, the break date chosen by Stock and Watson (2009) and usually associated with the start of the Great Moderation.

Cheng, Liao, and Schorfheide (2016) considered the case where the number of factors may change at one, possibly unknown, break date, but adopted a different approach, based on shrinkage estimation. Since it is only the product of factors and loadings, the common component is uniquely identified in a factor model. The authors used a normalization that attributes changes in this product to changes in the loadings. The estimator is based on a penalized least-squares (PLS) criterion function, in which adaptive group-LASSO penalties are attached to pre-break factor loadings and to changes in the factor loadings. This PLS estimator shrinks the small coefficients to zero, but a new factor appears if a column of zero loadings turns into non-zero values after the break. The authors proved the consistency of the estimated number of pre- and post-break factors and the detection of changes in the loadings, under a general set of assumptions. Once the number of pre- and post-break factors has been consistently estimated, the break date can also be consistently estimated. The authors' Monte-Carlo experiment showed that their shrinkage estimator cannot detect small breaks, but is more likely to detect large breaks than Breitung and Eickmeier (2011), Chen et al. (2014), or Han and Inoue (2015). Finally, they applied their procedure to the same dataset as Stock and Watson (2012). The results provided strong evidence for a change in the loadings after 2007, and the emergence of a new factor seems to capture comovements among financial series, but also spills over into real variables.

Corradi and Swanson (2014) looked at the consequences of instability in factor-augmented forecast equations. Forecast failures can result from instability in the loadings, instability in the regression coefficients of forecast equations, or both. They built a test for the joint null hypothesis of structural stability of factor loadings and factor-augmented forecast equation coefficients. The test statistic is based on the difference between the sample covariance of the forecasted variable and the factors estimated on the whole sample, and the sample covariance of the forecasted variable and the factors estimated using a rolling window estimation scheme. The number of factors is fixed according to the Bai and Ng (2002) criterion and is thus overestimated if there is a break in the loadings. Under a general set of assumptions, and if $\sqrt{T}/n \to 0$, the test statistics based on the difference between the two sample covariances has an asymptotic χ^2 distribution under the null. Using this test on an empirical dataset analogous to Stock and Watson (2002a), the authors rejected the null of stability for six forecasted variables (in particular GDP) but did not reject the null for four others.

Baltagi, Kao, and Wang (2017) also addressed the issue of a single break in the number of factors and/or the factor loadings at an unknown date. The number of factors is fixed on the whole sample, without taking the break into account, and the estimation of the break point relies on the discrepancy between the pre- and post-break second moment matrices of the estimated factors. Once the break point is estimated, the authors showed that the number of factors and the factor space are consistently estimated on each sub-sample at the same rate of convergence as in Bai and Ng (2002).

Ma and Su (2018) considered the case where the loadings exhibit an unknown number of breaks. They proposed a three-step procedure to detect the breaks if any and identify the dates when they occur. In the first step, the sample is divided into $J + 1$ intervals, with $T \gg J \gg m$, where m is an upper bound for the number of breaks, and a factor model is estimated by PCA on each interval. In the second step a fused group Lasso is applied to identify intervals containing a break. In the third step, a grid search allows to determine each break inside the corresponding interval. The authors proved that this procedure consistently estimates the number of breaks and their location. Using this method on Stock and Watson (2009) dataset, they identified five breaks in the factor loadings for the 1959–2006 period.

2.9 Conclusion

This chapter reviews the literature on dynamic factor models and several extensions of the basic framework. The modeling and estimation techniques surveyed include static and dynamic representation of small and large scale factor models, non-parametric and maximum likelihood estimation, estimation in the time and frequency domain, accommodating datasets with missing observations, and regime switching and time varying parameter models.

References

Ahn, S. C., & Horenstein, A. R. (2013). Eigenvalue ratio test for the number of factors. *Econometrica, 81*(3), 1203–1227.

Amengual, D., & Watson, M. W. (2007). Consistent estimation of the number of dynamic factors in a large N and T panel. *Journal of Business & Economic Statistics, 25*(1), 91–96.

Anderson, T. (1984). *An Introduction to Multivariate Statistical Analysis* (2nd ed.). New York: Wiley.

Anderson, T., & Rubin, H. (1956). Statistical inference in factor analysis. In *Proceedings of the Third Berkeley Symposium on Mathematical Statistics and Probability*. Contributions to Econometrics, Industrial Research, and Psychometry (Vol. 5).

Andrews, D. W. (1993). Tests for parameter instability and structural change with unknown change point. *Econometrica, 61*(4), 821–856.

Aruoba, S. B., Diebold, F. X., & Scotti, C. (2009). Real-time measurement of business conditions. *Journal of Business & Economic Statistics, 27*(4), 417–427.

Bai, J. (2003). Inferential theory for factor models of large dimensions. *Econometrica, 71*(1), 135–171.

Bai, J. (2004). Estimating cross-section common stochastic trends in nonstationary panel data. *Journal of Econometrics, 122*(1), 137–183.

Bai, J., & Li, K. (2012). Statistical analysis of factor models of high dimension. *The Annals of Statistics, 40*(1), 436–465.

Bai, J., & Li, K. (2016). Maximum likelihood estimation and inference for approximate factor models of high dimension. *Review of Economics and Statistics, 98*(2), 298–309.

Bai, J., & Liao, Y. (2013). Statistical inferences using large estimated covariances for panel data and factor models. *SSRN Electronic Journal*. https://doi.org/10.2139/ssrn.2353396.

Bai, J., & Liao, Y. (2016). Efficient estimation of approximate factor models via penalized maximum likelihood. *Journal of Econometrics, 191*(1), 1–18.

Bai, J., & Ng, S. (2002). Determining the number of factors in approximate factor models. *Econometrica, 70*(1), 191–221.

Bai, J., & Ng, S. (2004). A panic attack on unit roots and cointegration. *Econometrica, 72*(4), 1127–1177.

Bai, J., & Ng, S. (2006). Confidence intervals for diffusion index forecasts and inference for factor-augmented regressions. *Econometrica, 74*(4), 1133–1150.
Bai, J., & Ng, S. (2007). Determining the number of primitive shocks in factor models. *Journal of Business & Economic Statistics, 25*(1), 52–60.
Bai, J., & Ng, S. (2008a). Forecasting economic time series using targeted predictors. *Journal of Econometrics, 146*(2), 304–317.
Bai, J., & Ng, S. (2008b). Large dimensional factor analysis. *Foundations and Trends in Econometrics, 3*(2), 89–163.
Bai, J., & Ng, S. (2009). Boosting diffusion indices. *Journal of Applied Econometrics, 24*(4), 607–629.
Bai, J., & Wang, P. (2016). Econometric analysis of large factor models. *Annual Review of Economics, 8*, 53–80.
Baltagi, B. H., Kao, C., & Wang, F. (2017). Identification and estimation of a large factor model with structural instability. *Journal of Econometrics, 197*(1), 87–100.
Banbura, M., & Modugno, M. (2014). Maximum likelihood estimation of factor models on datasets with arbitrary pattern of missing data. *Journal of Applied Econometrics, 29*(1), 133–160.
Banbura, M., & Rünstler, G. (2011). A look into the factor model black box: Publication lags and the role of hard and soft data in forecasting GDP. *International Journal of Forecasting, 27*(2), 333–346.
Banerjee, A., Marcellino, M., & Masten, I. (2007). Forecasting macroeconomic variables using diffusion indexes in short samples with structural change. In D. Rapach & M. Wohar (Eds.), *Forecasting in the presence of structural breaks and model uncertainty*. Bingley: Emerald Group Publishing.
Barigozzi, M. (2018). *Dynamic factor models*. Lecture notes. London School of Economics.
Bartholomew, D. J. (1987). *Latent Variable Models and Factors Analysis*. Oxford: Oxford University Press.
Barhoumi, K., Darné, O., & Ferrara, L. (2014). Dynamic factor models: A review of the literature. *OECD Journal: Journal of Business Cycle Measurement and Analysis, 2013*(2), 73–107 (OECD Publishing, Centre for International Research on Economic Tendency Surveys).
Bates, B. J., Plagborg-Möller, M., Stock, J. H., & Watson, M.W. (2013). Consistent factor estimation in dynamic factor models with structural instability. *Journal of Econometrics, 177*(2), 289–304.
Bessec, M., & Bouabdallah, O. (2015). Forecasting GDP over the business cycle in a multi-frequency and data-rich environment. *Oxford Bulletin of Economics and Statistics, 77*(3), 360–384.
Boivin, J., & Ng, S. (2006). Are more data always better for factor analysis? *Journal of Econometrics, 132*(1), 169–194.
Bräuning, F., & Koopman, S. J. (2014). Forecasting macroeconomic variables using collapsed dynamic factor analysis. *International Journal of Forecasting, 30*(3), 572–584.
Breitung, J., & Eickmeier, S. (2006). Dynamic factor models. *Allgemeines Statistisches Archiv, 90*(1), 27–42.
Breitung, J., & Eickmeier, S. (2011). Testing for structural breaks in dynamic factor models. *Journal of Econometrics, 163*(1), 71–84.
Breitung, J., & Eickmeier, S. (2016). Analyzing international business and financial cycles using multi-level factor models: A comparison of alternative approaches. In *Dynamic Factor Models* (pp. 177–214). Bingley: Emerald Group Publishing Limited.
Breitung, J., & Tenhofen, J. (2011). GLS estimation of dynamic factor models. *Journal of the American Statistical Association, 106*(495), 1150–1166.
Brillinger, D. R. (1981). *Time series: Data analysis and theory*. Philadelphia: SIAM.
Burns, A. F., & Mitchell, W. C. (1946). *Measuring business cycles*. New York: NBER.
Camacho, M., Perez-Quiros, G., & Poncela, P. (2014). Green shoots and double dips in the euro area: A real time measure. *International Journal of Forecasting, 30*(3), 520–535.
Camacho, M., Perez-Quiros, G., & Poncela, P. (2015). Extracting nonlinear signals from several economic indicators. *Journal of Applied Econometrics, 30*(7), 1073–1089.

Chamberlain, G. (1983). Funds, factors, and diversification in arbitrage pricing models. *Econometrica, 51*(5), 1305–1323.
Chamberlain, G., & Rothschild, M. (1983). Arbitrage, factor structure, and mean-variance analysis on large asset markets. *Econometrica, 51*(5), 1281–1304.
Chauvet, M. (1998). An econometric characterization of business cycle dynamics with factor structure and regime switching. *International Economic Review, 39*(4), 969–996.
Chauvet, M., & Piger, J. (2008). A comparison of the real-time performance of business cycle dating methods. *Journal of Business & Economic Statistics, 26*(1), 42–49.
Chauvet, M., & Senyuz, Z. (2016). A dynamic factor model of the yield curve components as a predictor of the economy. *International Journal of Forecasting, 32*(2), 324–343.
Chen, L., Dolado, J. J., & Gonzalo, J. (2014). Detecting big structural breaks in large factor models. *Journal of Econometrics, 180*(1), 30–48.
Cheng, X., Liao, Z., & Schorfheide, F. (2016). Shrinkage estimation of high dimensional factor models with structural instabilities. *Review of Economic Studies, 83*(4), 1511–1543.
Choi, I. (2012). Efficient estimation of factor models. *Econometric Theory, 28*(2), 274–308.
Connor, G., & Korajczyk, R. A. (1986). Performance measurement with the arbitrage pricing theory: A new framework for analysis. *Journal of Financial Economics, 15*(3), 373–394.
Connor, G., & Korajczyk, R. A. (1988). Risk and return in an equilibrium apt: Application of a new test methodology. *Journal of Financial Economics, 21*(2), 255–289.
Connor, G., & Korajczyk, R. A. (1993). A test for the number of factors in an approximate factor model. *Journal of Finance, 48*(4), 1263–1291.
Corradi, V., & Swanson, N. R. (2014). Testing for structural stability of factor augmented forecasting models. *Journal of Econometrics, 182*(1), 100–118.
Del Negro, M., & Otrok, C. (2008). Dynamic factor models with time-varying parameters: Measuring changes in international business cycles. *Federal Reserve Bank of New York, Staff Report, 326*, 46.
Dempster, A., Laird, N., & Rubin, D. (1977). Maximum likelihood from incomplete data via the EM algorithm. *Journal of the Royal Statistical Society. Series B (Methodological), 39*(1), 1–38.
Diebold, F. X. (2003). Big data dynamic factor models for macroeconomic measurement and forecasting. In L. H. M. Dewatripont & S. Turnovsky (Eds.), *Advances in Economics and Econometrics: Theory and Applications, Eighth World Congress of the Econometric Society* (pp. 115–122).
Diebold, F. X., & Rudebusch, G. D. (1996). Measuring business cycles: A modern perspective. *Review of Economics and Statistics, 78*(1), 67–77.
Doz, C., Giannone, D., & Reichlin, L. (2011). A two-step estimator for large approximate dynamic factor models based on Kalman filtering. *Journal of Econometrics, 164*(1), 188–205.
Doz, C., Giannone, D., & Reichlin, L. (2012). A quasi–maximum likelihood approach for large, approximate dynamic factor models. *Review of Economics and Statistics, 94*(4), 1014–1024.
Doz, C., & Petronevich, A. (2016). Dating business cycle turning points for the french economy: An MS-DFM approach. In *Dynamic Factor Models* (pp. 481–538). Bingley: Emerald Group Publishing Limited.
Doz, C., & Petronevich, A. (2017). *On the consistency of the two-step estimates of the MS-DFM: A Monte Carlo study*. Working paper.
Efron, B., Hastie, T., Johnstone, I., & Tibshirani, R. (2004). Least angle regression. *Annals of Statistics, 32*(2), 407–499.
Engle, R., & Watson, M. (1981). A one-factor multivariate time series model of metropolitan wage rates. *Journal of the American Statistical Association, 76*(376), 774–781.
Favero, C. A., Marcellino, M., & Neglia, F. (2005). Principal components at work: The empirical analysis of monetary policy with large data sets. *Journal of Applied Econometrics, 20*(5), 603–620.
Fiorentini, G., Galesi, A., & Sentana, E. (2018). A spectral EM algorithm for dynamic factor models. *Journal of Econometrics, 205*(1), 249–279.
Forni, M., & Lippi, M. (2011). The general dynamic factor model: One-sided representation results. *Journal of Econometrics, 163*(1), 23–28.

Forni, M., Hallin, M., Lippi, M., & Reichlin, L. (2000). The generalized dynamic factor model: Identification and estimation. *Review of Economics and Statistics, 82*(4), 540–554.

Forni, M., Hallin, M., Lippi, M., & Reichlin, L. (2005). The generalized dynamic factor model: One-sided estimation and forecasting. *Journal of the American Statistical Association, 100*(471), 830–840.

Forni, M., Giannone, D., Lippi, M., & Reichlin, L. (2009). Opening the black box: Structural factor models with large cross sections. *Econometric Theory, 25*(5), 1319–1347.

Forni, M., Hallin, M., Lippi, M., & Zaffaroni, P. (2015). Dynamic factor models with infinite-dimensional factor spaces: One-sided representations. *Journal of Econometrics, 185*(2), 359–371.

Forni, M., Hallin, M., Lippi, M., & Zaffaroni, P. (2017). Dynamic factor models with infinite-dimensional factor space: Asymptotic analysis. *Journal of Econometrics, 199*(1), 74–92.

Forni, M., Giovannelli, A., Lippi, M., & Soccorsi, S. (2018). Dynamic factor model with infinite-dimensional factor space: Forecasting. *Journal of Applied Econometrics, 33*(5), 625–642.

Fu, D. (2007). National, regional and metro-specific factors of the us housing market. In *Federal Reserve Bank of Dallas Working Paper 0707*.

Fuleky, P., & Bonham, C. S. (2015). Forecasting with mixed frequency factor models in the presence of common trends. *Macroeconomic Dynamics, 19*(4), 753–775.

Geweke, J. (1977). The dynamic factor analysis of economic time series. In D. J. Aigner & A. S. Goldberger (Eds.), *Latent variables in socio-economic models*. Amsterdam: North Holland.

Ghysels, E., Santa-Clara, P., & Valkanov, R. (2004). The MIDAS touch: Mixed Data Sampling regression models. CIRANO Working paper.

Giannone, D., Reichlin, L., & Small, D. (2008). Nowcasting: The real-time informational content of macroeconomic data. *Journal of Monetary Economics, 55*(4), 665–676.

Groen, J. J., & Kapetanios, G. (2016). Revisiting useful approaches to data-rich macroeconomic forecasting. *Computational Statistics & Data Analysis, 100*, 221–239.

Guérin, P., & Marcellino, M. (2013). Markov-switching MIDAS models. *Journal of Business & Economic Statistics, 31*(1), 45–56.

Hallin, M., & Lippi, M. (2013). Factor models in high-dimensional time series a time-domain approach. *Stochastic Processes and their Applications, 123*(7), 2678–2695.

Hallin, M., & Liska, R. (2007). The generalized dynamic factor model: Determining the number of factors. *Journal of the American Statistical Association, 102*, 603–617.

Hamilton, J. D. (1989). A new approach to the economic analysis of nonstationary time series and the business cycle. *Econometrica, 57*(2), 357–384.

Han, X., & Inoue, A. (2015). Tests for parameter instability in dynamic factor models. *Econometric Theory, 31*(5), 1117–1152.

Harding, D., & Pagan, A. (2003). A comparison of two business cycle dating methods. *Journal of Economic Dynamics and Control, 27*(9), 1681–1690.

Harvey, A. (1989). *Forecasting, Structural Time Series Models and the Kalman Filter*. Cambridge: Cambridge University Press

Inoue, A., & Kilian, L. (2008). How useful is bagging in forecasting economic time series? A case study of US consumer price inflation. *Journal of the American Statistical Association, 103*(482), 511–522.

Jackson, L. E., Kose, M. A., Otrok, C., & Owyang, M. T. (2016). Specification and estimation of Bayesian dynamic factor models: A Monte Carlo analysis with an application to global house price comovement. In *Dynamic Factor Models* (pp. 361–400). Bingley: Emerald Group Publishing Limited.

Jones, C. S. (2001). Extracting factors from heteroskedastic asset returns. *Journal of Financial Economics, 62*(2), 293–325.

Jöreskog, K. G. (1967). Some contributions to maximum likelihood factor analysis. *Psychometrika, 32*(4), 443–482.

Jungbacker, B., & Koopman, S. J. (2014). Likelihood-based dynamic factor analysis for measurement and forecasting. *Econometrics Journal, 18*(2), C1–C21.

Jungbacker, B., Koopman, S. J., & van der Wel, M. (2011). Maximum likelihood estimation for dynamic factor models with missing data. *Journal of Economic Dynamics and Control, 35*(8), 1358–1368.

Kelly, B., & Pruitt, S. (2015). The three-pass regression filter: A new approach to forecasting using many predictors. *Journal of Econometrics, 186*(2), 294–316.

Kim, C.-J. (1994). Dynamic linear models with Markov-switching. *Journal of Econometrics, 60*(1–2), 1–22.

Kim, C.-J., & Nelson, C. R. (1998). Business cycle turning points, a new coincident index, and tests of duration dependence based on a dynamic factor model with regime switching. *Review of Economics and Statistics, 80*(2), 188–201.

Kim, M.-J., & Yoo, J.-S. (1995). New index of coincident indicators: A multivariate Markov switching factor model approach. *Journal of Monetary Economics, 36*(3), 607–630.

Koopman, S., & Harvey, A. (2003). Computing observation weights for signal extraction and filtering. *Journal of Economic Dynamics and Control, 27*(7), 1317–1333.

Kose, M. A., Otrok, C., & Whiteman, C. H. (2003). International business cycles: World, region, and country-specific factors. *American Economic Review*, 1216–1239.

Kose, M. A., Otrok, C., & Whiteman, C. H. (2008). Understanding the evolution of world business cycles. *Journal of International Economics, 75*(1), 110–130.

Lawley, D. N., & Maxwell, A. E. (1971). *Factor analysis as statistical method* (2nd ed.). London: Butterworths.

Li, H., Li, Q., & Shi, Y. (2017). Determining the number of factors when the number of factors can increase with sample size. *Journal of Econometrics, 197*(1), 76–86.

Lütkepohl, H. (2014). *Structural vector autoregressive analysis in a data rich environment: A survey*. Berlin: Humboldt University.

Ma, S., & Su, L. (2018). Estimation of large dimensional factor models with an unknown number of breaks. *Journal of Econometrics, 207*(1), 1–29.

Magnus, J. R., & Neudecker, H. (2019). *Matrix differential calculus with applications in statistics and econometrics*. New York: Wiley.

Marcellino, M., & Schumacher, C. (2010). Factor MIDAS for nowcasting and forecasting with ragged-edge data: A model comparison for German GDP. *Oxford Bulletin of Economics and Statistics, 72*(4), 518–550.

Mariano, R., & Murasawa, Y. (2003). A new coincident index of business cycles based on monthly and quarterly series. *Journal of Applied Econometrics, 18*(4), 427–443.

Moench, E., & Ng, S. (2011). A hierarchical factor analysis of us housing market dynamics. *Econometrics Journal, 14*(1), C1–C24.

Moench, E., Ng, S., & Potter, S. (2013). Dynamic hierarchical factor models. *Review of Economics and Statistics, 95*(5), 1811–1817.

Onatski, A. (2009). Testing hypotheses about the number of factors in large factor models. *Econometrica, 77*(5), 1447–1479.

Onatski, A. (2010). Determining the number of factors from empirical distribution of eigenvalues. *Review of Economics and Statistics, 92*(4), 1004–1016.

Otrok, C., & Whiteman, C. H. (1998). Bayesian leading indicators: Measuring and predicting economic conditions in Iowa. *International Economic Review, 39*(4), 997–1014.

Peña, D., & Yohai, V. J. (2016). Generalized dynamic principal components. *Journal of the American Statistical Association, 111*(515), 1121–1131.

Pinheiro, M., Rua, A., & Dias, F. (2013). Dynamic factor models with jagged edge panel data: Taking on board the dynamics of the idiosyncratic components. *Oxford Bulletin of Economics and Statistics, 75*(1), 80–102.

Proietti, T., & Moauro, F. (2006). Dynamic factor analysis with non-linear temporal aggregation constraints. *Journal of The Royal Statistical Society Series C, 55*(2), 281–300.

Reis, R., & Watson, M. W. (2010). Relative goods' prices, pure inflation, and the Phillips correlation. *American Economic Journal: Macroeconomics, 2*(3), 128–57.

Sargent, T. J., & Sims, C. A. (1977). Business cycle modeling without pretending to have too much a priori economic theory. *New Methods in Business Cycle Research, 1*, 145–168.

Shumway, R. H., & Stoffer, D. S. (1982). An approach to time series smoothing and forecasting using the EM algorithm. *Journal of Time Series Analysis, 3*(4), 253–264.

Spearman, C. (1904). General intelligence objectively determined and measured. The American *Journal of Psychology, 15*(2), 201–292.

Stock, J. H., & Watson, M. (1989). New indexes of coincident and leading economic indicators. *NBER Macroeconomics Annual, 4*, 351–394.

Stock, J. H., & Watson, M. (1991). A probability model of the coincident economic indicators. *Leading Economic indicators: New approaches and forecasting records, 66*, 63–90.

Stock, J. H., & Watson, M. W. (1993). A procedure for predicting recessions with leading indicators: Econometric issues and recent experience. In *Business cycles, indicators and forecasting* (pp. 95–156). Chicago: University of Chicago Press.

Stock, J. H., & Watson, M. (2002a). Forecasting using principal components from a large number of predictors. *Journal of the American Statistical Association, 97*(460), 1167–1179.

Stock, J. H., & Watson, M. (2002b). Macroeconomic forecasting using diffusion indexes. *Journal of Business and Economic Statistics, 20*(2), 147–162.

Stock, J. H., & Watson, M. W. (2005). *Implications of dynamic factor models for VAR analysis*. Cambridge: National Bureau of Economic Research. Technical report.

Stock, J. H., & Watson, M. (2006). Forecasting with many predictors. *Handbook of Economic Forecasting, 1*, 515.

Stock, J. H., & Watson, M. (2008). The evolution of national and regional factors in US housing construction. In *Volatility and Time Series Econometrics: Essays in Honor of Robert F. Engle*.

Stock, J. H., & Watson, M. (2009). Forecasting in dynamic factor models subject to structural instability. *The Methodology and Practice of Econometrics. A Festschrift in Honour of David F. Hendry, 173*, 205.

Stock, J. H., & Watson, M. W. (2011). Dynamic factor models. In *Handbook on Economic Forecasting*. Oxford: Oxford University Press.

Stock, J. H., & Watson, M. W. (2012). Disentangling the channels of the 2007–09 recession. *Brookings Papers on Economic Activity, 1*, 81–135.

Stock, J. H., & Watson, M.W. (2016). Dynamic factor models, factor-augmented vector autoregressions, and structural vector autoregressions in macroeconomics. *Handbook of Macroeconomics, 2*, 415–525.

Su, L., & Wang, X. (2017). On time-varying factor models: Estimation and testing. *Journal of Econometrics, 198*(1), 84–101.

Watson, M. W., & Engle, R. F. (1983). Alternative algorithms for the estimation of dynamic factor, mimic and varying coefficient regression models. *Journal of Econometrics, 23*(3), 385–400.

Yamamoto, Y., & Tanaka, S. (2015). Testing for factor loading structural change under common breaks. *Journal of Econometrics, 189*(1), 187–206.

Chapter 3
Factor Augmented Vector Autoregressions, Panel VARs, and Global VARs

Martin Feldkircher, Florian Huber, and Michael Pfarrhofer

3.1 Introduction

Statistical institutions and central banks of several major countries now provide large and complete databases on key macroeconomic indicators that calls for appropriate modeling techniques. As examples of such databases, the FRED-MD database (McCracken & Ng, 2016) includes over 150 macroeconomic and financial time series for the USA that are updated on a monthly basis. In Europe, the European Central Bank (ECB) as well as the Bank of England (BoE) maintain similar databases. Efficiently exploiting these large amounts of data is already challenging at the individual country level but even more difficult if the researcher is interested in jointly modeling relations across countries in light of a large number of time series per country. In this chapter, we discuss three popular techniques that allow for including a large number of time series per country and jointly modeling a (potentially) large number of units (i.e., countries). These competing approaches are essentially special cases of large vector autoregressive (VAR) models described in Chap. 4 and dynamic factor models (DFMs), discussed in Chap. 2 of this volume.

The first approach that allows for modeling interactions across units are panel vector autoregressive models (PVARs). Compared to the literature on large VAR models described in Chap. 4 of this book, PVARs feature a panel structure in their

M. Feldkircher (✉)
Oesterreichische Nationalbank (OeNB), Vienna, Austria
e-mail: martin.feldkircher@oenb.at

F. Huber · M. Pfarrhofer
Salzburg Centre of European Union Studies (SCEUS), University of Salzburg, Salzburg, Austria
e-mail: florian.huber@sbg.ac.at; michael.pfarrhofer@sbg.ac.at

P. Fuleky (ed.), *Macroeconomic Forecasting in the Era of Big Data*,
Advanced Studies in Theoretical and Applied Econometrics 52,
https://doi.org/10.1007/978-3-030-31150-6_3

endogenous variables that gives rise to a variety of potential restrictions on the parameters. In this framework, regularization is naturally embedded by pooling coefficients attached to the same variable but stemming from different units of the cross-section. In a seminal paper, Canova and Ciccarelli (2004) develop panel VARs with drifting coefficients in order to exploit a large information set. They find that allowing for time-varying parameters and cross-country linkages substantially improves turning point predictions for output growth for a set of industrialized countries. Koop and Korobilis (2016) propose a panel VAR framework that overcomes the problem of overparameterization by averaging over different restrictions on interdependencies between and heterogeneities across cross-sectional units. More recently, Koop and Korobilis (2019) advocate a particular class of priors that allows for soft clustering of variables or countries arguing that classical shrinkage priors are inappropriate for PVARs. In terms of forecasting, these contributions show that using model selection/averaging techniques help in terms of out-of-sample predictability. Dées and Güntner (2017) forecast inflation for the four largest economies in the euro area, namely France, Germany, Italy, and Spain. They show that the PVAR approach performs well against popular alternatives, especially at a short forecast horizon and relative to standard VAR forecasts based on aggregate inflation data. Bridging single country and panel VARs, Jarociński (2010) estimates country-specific VARs with a prior that pushes coefficients towards a (cross-sectional) mean.

As a special case of a PVAR model, Pesaran, Schuermann, and Weiner (2004) introduce the global vector autoregressive (GVAR) model to deal with a large number of countries in an effective manner. This framework proceeds in two steps. In the first, a set of country-specific VARs is estimated with each model consisting of a system of equations for the domestic macroeconomic variables. Information from the cross-section is included by augmenting the set of domestic variables with weakly exogenous foreign and global control variables. Crucially, the foreign variables are endogenously determined within the full system of country VARs since they are simply constructed as weighted averages of the other countries' domestic variables. The weights should reflect the connectivity of countries and have to be exogenously specified. In a second step, the country models are stacked to yield a system of equations that represents the world economy. By estimating the country VARs separately, the framework ensures that cross-country heterogeneity is fully taken into account. Also there is no need for symmetry in the variable coverage of countries. The GVAR achieves regularization by using weighted averages instead of the full set of foreign variables in each country VAR.

In the context of forecasting with GVAR models, Pesaran, Schuermann, and Smith (2009) show that taking global links across economies into account leads to more accurate out-of-sample predictions than using forecasts based on univariate specifications for output and inflation. Yet for interest rates, the exchange rate and financial variables, the results are less spectacular, and the authors also find strong cross-country heterogeneity in the performance of GVAR forecasts. Employing a GVAR model to forecast macroeconomic variables in five Asian economies, Han and Ng (2011) find that one-step-ahead forecasts from GVAR models outperform those of stand-alone VAR specifications for short-term interest rates and real equity prices. Concentrating on predicted directional changes to evaluate the forecasting performance of GVAR specifications, Greenwood-Nimmo, Nguyen, and Shin (2012) confirm the superiority of GVAR specifications over univariate benchmark models at long-run forecast horizons. Feldkircher and Huber (2016) introduce additional regularization by employing shrinkage priors, borrowed from the large Bayesian VAR literature, in the GVAR country VARs. Building on this work, Crespo Cuaresma, Feldkircher, and Huber (2016) show that GVAR forecasts improve upon forecasts from naive models, a global model without shrinkage on the parameters and country-specific vector autoregressions using shrinkage priors. Dovern, Feldkircher, and Huber (2016) and Huber (2016) introduce different forms of time-variation in the residual variances.[1] Both papers demonstrate that the GVAR equipped with stochastic volatility further improves forecast performance and hence generalize the results of Pesaran et al. (2009). In another recent contribution, Chudik, Grossman, and Pesaran (2016) suggest augmenting the GVAR model with additional proxy equations to control for unobserved global factors and show that including data on Purchasing Managers' Indices improves out-of-sample forecasting accuracy.

Finally, another approach to include information from the cross-section is by using factor augmented vector autoregressions (FAVARs). In a seminal contribution, Bernanke, Boivin, and Eliasz (2005) propose modeling a large number of variables as a function of relatively few observed and unobserved factors and an idiosyncratic noise term. The common factors serve to capture joint movements in the data and, conditional on suitable identification, can be thought of as the underlying driving forces of the variables. Typically, these factors are assumed to follow a VAR process. FAVARs are an attractive means of data reduction and are frequently applied when the information set is numerous. In a recent application, Moench (2008) combines a FAVAR with a term structure model to make predictions of bond yields at different maturities. Technically, the original framework of Bernanke et al. (2005) has been subsequently modified. For example, Banerjee, Marcellino, and Masten (2014) introduce the concept of error correction to the FAVAR framework and Eickmeier, Lemke, and Marcellino (2015) introduce time-variation in the factor loadings and residual variances. Both studies show that these modifications lead to further, improved forecasts relative to standard FAVARs. FAVARs can also be applied to extract and condense information from different countries. Indeed, Eickmeier and Ng (2011) provide evidence of superior forecasts when exploiting international data relative to data-rich approaches that focus on domestic variables only. This is especially true for longer forecast horizons. In general, FAVARs that work with international data do not draw particular attention to the fact that data are of the same type but stem from different countries.

[1] See Crespo Cuaresma, Doppelhofer, Feldkircher, and Huber (2019) for a GVAR specification in the context of monetary policy that allows both parameters and residual variances to change over time.

This chapter is organized as follows. The next section introduces the models more formally while the next section illustrates the merits of the competing approaches by means of a small forecasting exercise for inflation, output, and interest rates for the G7 countries.

3.2 Modeling Relations Across Units

In this section, we introduce the econometric framework for the panel VAR, the global VAR, and the factor augmented VAR. Many of the aspects discussed below, such as cross-sectional restrictions, can easily be combined across all three major model classes. Each description of the respective model is accompanied and enhanced by an overview of applications in the related literature. We focus on mainly discussing potential avenues for modeling panel data using multivariate time series models, while we mostly refrain from presenting details on estimation techniques. Estimation of the models below can be carried out by standard (Bayesian) VAR methods, discussed, for instance, in Chap. 4 of this volume.

3.2.1 Panel VAR Models

The discussion in Sect. 3.1 highlights that capturing linkages between units (i.e., countries, companies, etc.) could be crucial in understanding dynamic processes as well as for macroeconomic forecasting. Considering the literature on VAR models (see Chap. 5 of this book), one potential approach to covering cross-sectional linkages would be to include cross-sectional information in an otherwise standard VAR model.

To set the stage, let $\boldsymbol{y}_{it}$ $(i = 1, \ldots, N)$ be a M-dimensional vector of endogenous variables for unit i and measured in time $t = 1, \ldots, T$.[2] Stacking the $\boldsymbol{y}_{it}$'s yields a K-dimensional vector $\boldsymbol{y}_t = (\boldsymbol{y}'_{1t}, \ldots, \boldsymbol{y}'_{Nt})'$ with $K = MN$. Using a standard VAR approach implies that $\boldsymbol{y}_t$ depends on the P lags of $\boldsymbol{y}_t$[3]

$$\boldsymbol{y}_t = \boldsymbol{A}_1 \boldsymbol{y}_{t-1} + \ldots + \boldsymbol{A}_P \boldsymbol{y}_{t-P} + \boldsymbol{\varepsilon}_t, \tag{3.1}$$

where $\boldsymbol{A}_p$ $(p = 1, \ldots, P)$ denote $(K \times K)$-dimensional matrices of autoregressive coefficients and $\boldsymbol{\varepsilon}_t \sim \mathcal{N}(\boldsymbol{0}, \boldsymbol{\Sigma})$ is a Gaussian shock vector with variance-covariance matrix $\boldsymbol{\Sigma}$ of dimension $K \times K$. Analogous to the endogenous vector, the error term

[2]For simplicity, we assume that each cross-sectional unit features the same set of endogenous variables. This restriction, however, can easily be relaxed.

[3]Note that we exclude intercepts and deterministic terms and the specifications below can straightforwardly be adapted.

is given by $\boldsymbol{\varepsilon}_t = (\boldsymbol{\varepsilon}'_{1t}, \ldots, \boldsymbol{\varepsilon}'_{Nt})'$, which yields a block matrix structure for

$$\boldsymbol{\Sigma} = \begin{bmatrix} \boldsymbol{\Sigma}_{11} & \ldots & \boldsymbol{\Sigma}_{1N} \\ \vdots & \ddots & \vdots \\ \boldsymbol{\Sigma}_{N1} & \ldots & \boldsymbol{\Sigma}_{NN} \end{bmatrix}.$$

The block-diagonal elements $\boldsymbol{\Sigma}_{ii}$ of size $M \times M$ refer to the variance-covariance structure among endogenous variables within unit i. By the given symmetry of $\boldsymbol{\Sigma}$, we have $\boldsymbol{\Sigma}_{ij} = \boldsymbol{\Sigma}_{ji}$ again of size $M \times M$ for $i, j = 1, \ldots, N$. Specifically, $\boldsymbol{\Sigma}_{ij}$ captures covariances between the variables in unit i and j, for $i \neq j$.

Notice that the number of autoregressive coefficients is $k = PK^2$ while the number of free elements in $\boldsymbol{\Sigma}$ is $K(K+1)/2$. This implies that the total number of coefficients often exceeds the number of observations T. This problem usually arises in the high-dimensional VAR case discussed in Chap. 4 and leads to the well-known curse of dimensionality. The key difference between the present framework and a standard VAR is the structure of $\boldsymbol{y}_t$. If $\boldsymbol{y}_t$ is composed of different units (say countries), $\boldsymbol{y}_t$ features a panel structure and elements in $\boldsymbol{A}_p$ might be similar across countries. This leads to several modeling possibilities, potentially alleviating overparameterization issues by reducing the number of free parameters. To illustrate how one could proceed, let us rewrite the equations associated with unit i in Eq. (3.1) as follows:

$$\boldsymbol{y}_{it} = \boldsymbol{\Phi}_i \boldsymbol{x}_{it} + \boldsymbol{\Lambda}_i \boldsymbol{z}_{it} + \boldsymbol{\varepsilon}_{it}, \tag{3.2}$$

with $\boldsymbol{\Phi}_i$ denoting an $(M \times MP)$-matrix of coefficients associated with $\boldsymbol{x}_{it} = (\boldsymbol{y}'_{it-1}, \ldots, \boldsymbol{y}'_{it-P})'$ and $\boldsymbol{\Lambda}_i$ is an $(M \times MP(N-1))$-dimensional matrix related to the lags of $\boldsymbol{y}_{jt}$ $(\forall j \neq i)$, stored in $\boldsymbol{z}_{it} = (\boldsymbol{x}'_{1t}, \ldots, \boldsymbol{x}'_{i-1,t}, \boldsymbol{x}'_{i+1,t}, \ldots, \boldsymbol{x}'_{Nt})'$. Equation (3.2) is the standard panel VAR (PVAR) model commonly found in the literature (for an extensive survey, see also Canova & Ciccarelli, 2013).

To more intuitively motivate the potential parameter restrictions discussed in the following paragraphs for exploiting the panel data structure, we present a small example for $N = 3$ countries, $M = 2$ variables. This recurring simple example will also serve to illuminate commonalities and distinctions across the three types of specifications discussed in this chapter. For the sake of simplicity, we only consider $p = 1$ lag of the endogenous variables. In this special case, the full system of equations may be written as

$$\begin{aligned}
\boldsymbol{y}_{1t} &= \boldsymbol{\Phi}_{11}\boldsymbol{y}_{1t-1} + \boldsymbol{\Lambda}_{11}\boldsymbol{y}_{2t-1} + \boldsymbol{\Lambda}_{12}\boldsymbol{y}_{3t-1} + \boldsymbol{\varepsilon}_{1t}, \quad & \boldsymbol{\varepsilon}_{1t} &\sim \mathcal{N}(\boldsymbol{0}, \boldsymbol{\Sigma}_{11}) \\
\boldsymbol{y}_{2t} &= \boldsymbol{\Phi}_{21}\boldsymbol{y}_{2t-1} + \boldsymbol{\Lambda}_{21}\boldsymbol{y}_{1t-1} + \boldsymbol{\Lambda}_{22}\boldsymbol{y}_{3t-1} + \boldsymbol{\varepsilon}_{2t}, \quad & \boldsymbol{\varepsilon}_{2t} &\sim \mathcal{N}(\boldsymbol{0}, \boldsymbol{\Sigma}_{22}) \\
\boldsymbol{y}_{3t} &= \boldsymbol{\Phi}_{31}\boldsymbol{y}_{3t-1} + \boldsymbol{\Lambda}_{31}\boldsymbol{y}_{1t-1} + \boldsymbol{\Lambda}_{32}\boldsymbol{y}_{2t-1} + \boldsymbol{\varepsilon}_{3t}, \quad & \boldsymbol{\varepsilon}_{3t} &\sim \mathcal{N}(\boldsymbol{0}, \boldsymbol{\Sigma}_{33}).
\end{aligned}$$

Notice that each of the country specific models is a standard VAR model with $\boldsymbol{\Phi}_{i1}$ for $i = 1, \ldots, 3$ capturing the coefficients on the first lag of the domestic endogenous variable, augmented by additional information from the respective lagged foreign endogenous vectors.

Cross-country spillovers or linkages emerge along two dimensions. First, the coefficients in $\boldsymbol{\Lambda}_{i1}$ and $\boldsymbol{\Lambda}_{i2}$ govern *dynamic interdependencies* across countries by capturing effects that propagate through the system based on the lagged foreign variables. Setting $\boldsymbol{\Lambda}_{i1} = \boldsymbol{\Lambda}_{i2} = \mathbf{0}$ results in a set of three unrelated multivariate regression models, if one assumes that $\boldsymbol{\varepsilon}_{it}$ and $\boldsymbol{\varepsilon}_{jt}$ for $i \neq j$ are orthogonal. Second, relaxing orthogonality between country-specific error terms results in *static interdependencies*, capturing the case that shocks across countries might be correlated, potentially providing valuable information on contemporaneous movements across elements in $\boldsymbol{y}_t$ that can be exploited for forecasting or conducting structural analysis. The contemporaneous relationship between shocks to country i and j is given by the corresponding block in the variance-covariance matrix, obtained by computing $\mathrm{E}(\boldsymbol{\varepsilon}_{it}\boldsymbol{\varepsilon}'_{jt}) = \boldsymbol{\Sigma}_{ij}$. Finally, a related yet distinct dimension to be considered based on the specific structures provided by panel data is whether domestic dynamics governed by the $\boldsymbol{\Phi}_{i1}$ are similar across countries, a phenomenon usually termed *cross-sectional homogeneity*. These homogeneity restrictions might be warranted if the countries in the panel are homogeneous or their macroeconomic fundamentals feature similar dynamics. For instance, Wang and Wen (2007) find that the short-run behavior of inflation is highly synchronized and similar across a diverse set of economies. Such information can be incorporated by setting the relevant elements in $\boldsymbol{\Phi}_{i1}$ and $\boldsymbol{\Phi}_{j1}$ equal to each other.

3.2.2 Restrictions for Large-Scale Panel Models

Following this small scale example, we proceed by the general model outlined in Eq. (3.2) that constitutes the standard, unrestricted PVAR model. In the literature, researchers typically introduce several parametric restrictions on $\boldsymbol{\Phi}_i$, $\boldsymbol{\Lambda}_i$, as well as on $\boldsymbol{\Sigma}$. In what follows, we discuss the commonly used restrictions addressing *cross-sectional homogeneity* and *static* and *dynamic interdependencies* in more detail and focus on their merits from a forecasting perspective.

Cross-Sectional Homogeneity

The first restriction, henceforth labeled cross-sectional homogeneity, concerns the relationship between the endogenous quantities in $\boldsymbol{y}_{it}$ and their "own" lags $\boldsymbol{y}_{it-p}$ $(p = 1, \ldots, P)$. Note that this restriction is sometimes alternatively referred to as cross-sectional heterogeneity in its converse formulation. Depending on the specific application at hand, researchers typically assume that each unit features similar domestic dynamics. In a cross-country case, for instance, this implies that several countries share identical reactions to (lagged) movements in domestic quantities, implying that $\boldsymbol{\Phi}_i = \boldsymbol{\Phi}_j = \boldsymbol{\Phi}$. These restrictions might be appropriate if the different units are similar to each other. For instance, in a multi-country context, economies that are similar also tend to feature similar domestic dynamics. However, in case of heterogeneous units, this assumption could be too restrictive. Returning to a multi-country context, it might be difficult to argue that China and the USA display similar dynamics in terms of their domestic fundamentals. Hence, using these restrictions potentially translates into distorted estimates of the VAR coefficients, in turn leading to imprecise forecasts and biased impulse responses. By contrast, for a set of comparatively homogeneous countries, e.g., selected euro area member states, pooling information across units may yield more precise inference and forecasts.

Dynamic Interdependencies

The second set of restrictions relates to lagged relations across units, stored in the matrix $\mathbf{\Lambda}_i$. Here, non-zero elements imply that movements in foreign quantities affect the evolution of domestic variables, that is, effects spill over dynamically. The related literature labels this type of relationship dynamic interdependency. Various types of restrictions may be feasible in this context. The simplest case to impose structure on dynamic interdependencies would be to set $\mathbf{\Lambda}_i = \mathbf{0}$, ruling out dynamic cross-country spillovers. A less restrictive and arguably more realistic assumption would be to asymmetrically rule out such relationships. Monetary policy shocks emitting from the USA, for instance, most likely transmit dynamically to other countries, while it is unlikely that smaller countries impact US developments. Extracting information from such international linkages may improve forecasts by alleviating a potential omitted variable bias.

Static Interdependencies

Finally, the third typical restriction is concerned with how to model contemporaneous relations between the shocks across countries in the system, termed static interdependencies. Such restrictions essentially deal with the question whether shocks are correlated along countries and variable types. This translates into a selection issue related to certain elements of the variance-covariance matrix of the full system. For the small-scale example presented earlier, this implies testing whether $\mathbf{\Sigma}_{ij} = \mathbf{0}$ for i and j or whether selected off-diagonal elements of $\mathbf{\Sigma}_{ii}$ for all i are set equal to zero. Choices in this context can be motivated by resorting to theoretical insights. For instance, there is a broad body of literature that separates fast from slow moving variables in the context of monetary policy shocks. In the absence of additional information, however, model selection along this dimension becomes cumbersome and appropriate techniques are necessary.

Implementing Parametric Restrictions

Analyzing real world phenomena in a multi-country time series context often requires a combination of several types of restrictions stated above. Appropriate modeling choices can be either guided by economic theory, common sense, or by relying on the various approaches set forth in the literature to econometrically estimate or test the specification that provides the best fit for the data at hand.

Estimates implementing the restrictions can be introduced both in the frequentist and the Bayesian estimation context. Depending on the length of the number of available observations and the size of the problem considered, one may employ the generalized method of moments (GMM) approach pioneered in Arellano and Bond (1991), or alternative Bayesian pooled estimators (see, for instance, Canova, 2007). Here, it is worth emphasizing that the bulk of the literature relies on Bayesian techniques to estimate PVAR models. The reason is that all three types of restrictions translate into a huge number of model specification issues that are difficult to tackle using frequentist methods. Bayesian methods, by contrast, provide a natural way of exploring a vast dimensional model space by using Markov chain Monte Carlo (MCMC) algorithms that entail exploring promising regions of the model space.

One strand of the literature introduces parsimony in panel VARs by reducing the dimension of the parameter space via a small number of unobserved latent components. These techniques share several features of dynamic factor models and thus provide a synthesis between PVARs and DFMs, not in terms of observed quantities but in terms of coefficients. In their seminal contribution, Canova and Ciccarelli (2004) and Canova and Ciccarelli (2009) employ a dynamic factor structure on the coefficients of the fully parameterized system, greatly reducing the dimensionality of the regression problem. In particular, they assume that unit-specific parameters are driven by a small set of global, regional, and country-specific factors. One key feature is that this approach controls for movements in the underlying regression coefficients over time in a parsimonious manner. Koop and Korobilis (2019) extend the methods proposed in Canova and Ciccarelli (2009) to also select suitable static interdependencies by modeling the variance-covariance matrix using a similar low-dimensional factor structure. However, evidence for time variation in the VAR coefficients for standard macroeconomic models is limited, and the forecasting literature typically identifies that a successful forecasting model should control for heteroscedasticity in the error variances—especially when interest centers on producing accurate density forecasts. Thus, we leave this strand of the literature aside and, in this chapter, focus exclusively on constant regression coefficients while allowing for time-variation in the error variances.

Taking a fully data-driven perspective to select required restrictions in the PVAR context, Koop and Korobilis (2016) adapt a stochastic search variable selection prior (George & McCulloch, 1993; George, Sun, & Ni, 2008) to account for model uncertainty. This approach centers on introducing latent binary indicators that allow for assessing whether coefficients are similar across countries (i.e., whether cross-sectional homogeneity is present) or whether dynamic and static relations across countries should be pushed to zero (i.e., shutting off any dynamic and static interdependencies). One key advantage of this approach is that a deterministic search of the model space is replaced by a stochastic approach to model uncertainty, effectively constructing a Markov chain that explores relevant regions in the model space.

Finally, another approach closely related to assuming a factor structure on coefficients captures cross-sectional homogeneity by introducing a hierarchical Bayesian model on the coefficients. This approach was first used in a multi-country VAR setting by Jarociński (2010). Here, the key assumption is that country-specific coefficients cluster around a common mean. Any cross-country differences are captured by allowing for deviations from this common mean. This type of prior is closely related to random coefficient and heterogeneity models (see Allenby, Arora, & Ginter, 1998; Frühwirth-Schnatter, Tüchler, & Otter, 2004) pioneered in the marketing literature.

3.2.3 Global Vector Autoregressive Models

The basic structure of the global vector autoregressive model is similar to the PVAR, but differs in its treatment of the foreign country variables (Pesaran et al., 2004; Dées, di Mauro, Pesaran, and Smith, 2007). In contrast to the panel VAR, we include information on the foreign variables by means of the vector $\boldsymbol{y}_{it}^*$. This vector of *weakly exogenous variables* is constructed as a cross-sectional weighted average defined by

$$\boldsymbol{y}_{it}^* = \sum_{j=1}^{N} w_{ij} \boldsymbol{y}_{jt},$$

with w_{ij} denoting pre-specified weights with $w_{ii} = 0$, $w_{ij} \geq 0$ for $i \neq j$ and $\sum_{j=1}^{N} w_{ij} = 1$. To achieve a structure akin to Eq. (3.2), we again use $\boldsymbol{x}_{it} = (\boldsymbol{y}_{it-1}', \ldots, \boldsymbol{y}_{it-P}')'$, and stack the cross-sectional averages $\boldsymbol{y}_{it}^*$ in a vector $\boldsymbol{z}_{it} = (\boldsymbol{y}_{it}'^*, \boldsymbol{y}_{it-1}'^*, \ldots, \boldsymbol{y}_{it-P}'^*)'$. The equation for unit i reads

$$\boldsymbol{y}_{it} = \boldsymbol{\Phi}_i \boldsymbol{x}_{it} + \boldsymbol{\Xi}_i \boldsymbol{z}_{it} + \boldsymbol{\varepsilon}_{it}, \tag{3.3}$$

with $\boldsymbol{\Phi}_i$, again, denoting an $(M \times MP)$ matrix of coefficients associated with unit i's own lags, while $\boldsymbol{\Xi}_i$ is a $(M \times M(P+1))$-dimensional matrix containing the parameters measuring dependencies across units. Due to the presence of the foreign-specific variables, Pesaran et al. (2004) refer to this specification as VARX*. It is worth mentioning that the dimensions of the endogenous variables across units need not necessarily be of the same size, an assumption we use here for simplicity. Notice that while the panel VAR specification resulted in $M^2P(N-1)$ coefficients to be estimated for foreign quantities, imposing structure via the exogenous weights w_{ij} reduces this number to $M^2(P+1)$.

The Full GVAR Model

Constructing the GVAR model from Eq. (3.3) involves stacking the country-specific submodels. For the sake of simplicity, we focus on the case of one lag for both the domestic and foreign variables, and adopt some of the notation used above. The more general specification involving more lags is straightforward to obtain employing the logic presented in the following paragraphs. Based on Eq. (3.3), we consider

$$\boldsymbol{y}_{it} = \boldsymbol{\Phi}_{i1}\boldsymbol{y}_{it-1} + \boldsymbol{\Xi}_{i0}\boldsymbol{y}^*_{it} + \boldsymbol{\Xi}_{i1}\boldsymbol{y}^*_{it-1} + \boldsymbol{\varepsilon}_{it}, \tag{3.4}$$

where $\boldsymbol{\Phi}_{i1}$ is a coefficient matrix associated with the own lag of unit i, $\boldsymbol{\Xi}_{i0}$ captures contemporaneous relations across units, and $\boldsymbol{\Xi}_{i1}$ contains the coefficients corresponding to the lags of the foreign-unit averages. All three matrices are of size $M \times M$. Using a vector $\boldsymbol{z}_{it} = (\boldsymbol{y}_{it}, \boldsymbol{y}^*_{it})'$, Eq. (3.4) is rewritten as

$$\boldsymbol{A}_i\boldsymbol{z}_{it} = \boldsymbol{B}_i\boldsymbol{z}_{it-1} + \boldsymbol{\varepsilon}_{it}, \tag{3.5}$$

with $\boldsymbol{A}_i = (\boldsymbol{I}_M, -\boldsymbol{\Xi}_{i0})$ and $\boldsymbol{B}_i = (\boldsymbol{\Phi}_{i1}, \boldsymbol{\Xi}_{i1})$ and $\boldsymbol{I}_M$ denoting the identity matrix of size M. We proceed by stacking the country-specific variables in the K-dimensional vector $\boldsymbol{x}_t = (\boldsymbol{y}'_{1t}, \ldots, \boldsymbol{y}'_{Nt})'$, henceforth referred to as the global vector.

It is straightforward to construct a weighting matrix $\boldsymbol{W}_i$ such that $\boldsymbol{z}_{it} = \boldsymbol{W}_i\boldsymbol{x}_t$. Exploiting the definition of $\boldsymbol{z}_{it}$ in combination with Eq. (3.5) yields a formulation in terms of the global vector

$$\boldsymbol{A}_i\boldsymbol{W}_i\boldsymbol{x}_t = \boldsymbol{B}_i\boldsymbol{W}_i\boldsymbol{x}_{t-1} + \boldsymbol{\varepsilon}_{it}.$$

The final step involves stacking these equations to yield the full system representation. Here, we first define

$$\boldsymbol{G} = \begin{bmatrix} \boldsymbol{A}_1\boldsymbol{W}_1 \\ \boldsymbol{A}_2\boldsymbol{W}_2 \\ \vdots \\ \boldsymbol{A}_N\boldsymbol{W}_N \end{bmatrix}, \quad \boldsymbol{H} = \begin{bmatrix} \boldsymbol{B}_1\boldsymbol{W}_1 \\ \boldsymbol{B}_2\boldsymbol{W}_2 \\ \vdots \\ \boldsymbol{B}_N\boldsymbol{W}_N \end{bmatrix},$$

and the error vector $\boldsymbol{\epsilon}_t = (\boldsymbol{\epsilon}'_{1t}, \ldots, \boldsymbol{\epsilon}'_{Nt})'$. Using these definitions, we obtain the following global model

$$\boldsymbol{G}\boldsymbol{x}_t = \boldsymbol{H}\boldsymbol{x}_{t-1} + \boldsymbol{\varepsilon}_t.$$

In the general case, it is reasonable to assume a normally distributed error term $\boldsymbol{\epsilon}_t \sim \mathcal{N}(\boldsymbol{0}, \boldsymbol{\Sigma})$ with some variance-covariance matrix $\boldsymbol{\Sigma}$. Under the condition that $\boldsymbol{G}$ is non-singular, we obtain the reduced-form representation of the GVAR model

$$\boldsymbol{x}_t = \boldsymbol{G}^{-1}\boldsymbol{H}\boldsymbol{x}_{t-1} + \boldsymbol{G}^{-1}\boldsymbol{\varepsilon}_t.$$

Let $\boldsymbol{\epsilon}_t = \boldsymbol{G}^{-1}\boldsymbol{\varepsilon}_t$, then it follows that the reduced form variance-covariance matrix is given by $\mathrm{E}(\boldsymbol{\epsilon}_t\boldsymbol{\epsilon}_t') = \boldsymbol{G}^{-1}\boldsymbol{\Sigma}(\boldsymbol{G}^{-1})'$.

The specifics of the linking matrix $\boldsymbol{W}_i$ are yet to be discussed. In the empirical literature using GVARs, different weighting schemes determining the w_{ij} have been employed. The proposed schemes range from using bilateral trade flows or financial linkages for measuring the strength of dependencies between countries, to measures based on geographic distances that are often used in the spatial econometrics literature (see, for instance Eickmeier & Ng, 2015). Bayesian estimation techniques also allow for integrating out uncertainty surrounding the choice of the linkage matrix, for instance, as in Feldkircher and Huber (2016).

It is revealing to consider a simple example for the GVAR for two reasons. First, it directly shows the structure required on the $\boldsymbol{W}_i$. Second, the correspondence between the panel VAR and global VAR restrictions becomes evident. The full system of equations, analogous to the previous example in the context of the panel VAR, for three units can be written as

$$\begin{aligned}
\boldsymbol{y}_{1t} &= \boldsymbol{\Phi}_{11}\boldsymbol{y}_{1t-1} + \boldsymbol{\Xi}_{10}(w_{12}\boldsymbol{y}_{2t} + w_{13}\boldsymbol{y}_{3t}) + \boldsymbol{\Xi}_{11}(w_{12}\boldsymbol{y}_{2t-1} + w_{13}\boldsymbol{y}_{3t-1}) + \boldsymbol{\varepsilon}_{1t},\\
\boldsymbol{y}_{2t} &= \boldsymbol{\Phi}_{21}\boldsymbol{y}_{2t-1} + \boldsymbol{\Xi}_{20}(w_{21}\boldsymbol{y}_{1t} + w_{23}\boldsymbol{y}_{3t}) + \boldsymbol{\Xi}_{21}(w_{21}\boldsymbol{y}_{1t-1} + w_{23}\boldsymbol{y}_{3t-1}) + \boldsymbol{\varepsilon}_{2t},\\
\boldsymbol{y}_{3t} &= \boldsymbol{\Phi}_{31}\boldsymbol{y}_{3t-1} + \boldsymbol{\Xi}_{30}(w_{31}\boldsymbol{y}_{1t} + w_{32}\boldsymbol{y}_{2t}) + \boldsymbol{\Xi}_{31}(w_{31}\boldsymbol{y}_{1t-1} + w_{32}\boldsymbol{y}_{2t-1}) + \boldsymbol{\varepsilon}_{3t}.
\end{aligned}$$

Consequently, to achieve the structure of Eq. (3.5) for the first cross-sectional unit in terms of the global vector $\boldsymbol{y}_t = (\boldsymbol{y}_{1t}', \boldsymbol{y}_{2t}', \boldsymbol{y}_{3t}')'$ specific to this example we require $\boldsymbol{W}_1$ to be given by

$$\boldsymbol{W}_1 = \begin{bmatrix} 1 & 0 & 0 & 0 & 0 & 0\\ 0 & 1 & 0 & 0 & 0 & 0\\ 0 & 0 & w_{12} & 0 & w_{13} & 0\\ 0 & 0 & 0 & w_{12} & 0 & w_{13} \end{bmatrix} = \begin{bmatrix} \boldsymbol{I}_2 & \boldsymbol{0} & \boldsymbol{0}\\ \boldsymbol{0} & w_{12}\boldsymbol{I}_2 & w_{13}\boldsymbol{I}_2 \end{bmatrix}.$$

For the remaining units we have

$$\boldsymbol{W}_2 = \begin{bmatrix} \boldsymbol{0} & \boldsymbol{I}_2 & \boldsymbol{0}\\ w_{21}\boldsymbol{I}_2 & \boldsymbol{0} & w_{23}\boldsymbol{I}_2 \end{bmatrix}, \quad \boldsymbol{W}_3 = \begin{bmatrix} \boldsymbol{0} & \boldsymbol{0} & \boldsymbol{I}_2\\ w_{31}\boldsymbol{I}_2 & w_{32}\boldsymbol{I}_2 & \boldsymbol{0} \end{bmatrix}.$$

To illustrate the correspondence between the panel and global VAR coefficients, consider Eq. (3.2) and observe that $\boldsymbol{\Lambda}_{11} = w_{12}\boldsymbol{\Xi}_{11}$ and $\boldsymbol{\Lambda}_{12} = w_{13}\boldsymbol{\Xi}_{11}$. This implies that by relying on cross-sectional averages constructed using the exogenous weights w_{ij}, one imposes a specific structure and restrictions on the full system coefficients, effectively reducing the number of free parameters to estimate.

Specifically, this specification assumes the coefficients for unit i associated with foreign lags to be proportional across the foreign vectors. Shrinkage towards zero is imposed if w_{ij} is small, that is, in the case of the weights reflecting a situation where the respective units are not closely linked. For instance, assuming $w_{12} = 1$ and by the required restrictions $w_{13} = 0$, one would obtain a bilateral VAR specification, shutting down dynamic interdependencies between unit one and three. This claim is easily verified by observing that in this case we have $\boldsymbol{\Lambda}_{11} = \boldsymbol{\Xi}_{11}$ and $\boldsymbol{\Lambda}_{12} = \mathbf{0}$.

3.2.4 Factor Augmented Vector Autoregressive Models

In contrast to PVAR and GVAR models that are mainly used to explicitly account for interdependencies and spillovers between countries and exploiting the cross-sectional dimension in a time series context, FAVAR models are applicable to an even more diverse set of problems. Here, the main intuition is to reduce dimensionality of the model by extracting a small number of unobserved factors from a large-scale data set that summarize the contained information efficiently. This procedure may be employed for estimating single country models where the information set is large, but can also be used in a multi-country context.

Chapter 2 in this volume and Stock and Watson (2016) provide comprehensive discussions of dynamic factor models. The FAVAR is a synthesis of the dynamic factor approach and the literature on vector autoregressive models. In its general form, as proposed by Bernanke et al. (2005), the FAVAR approach links a large number of time series in $\boldsymbol{x}_t$ of size K to a set of M observed quantities of interest $\boldsymbol{m}_t$ and Q latent factors $\boldsymbol{f}_t$

$$\boldsymbol{x}_t = \boldsymbol{\Psi}_f \boldsymbol{f}_t + \boldsymbol{\Psi}_m \boldsymbol{m}_t + \boldsymbol{\epsilon}_t, \quad \boldsymbol{\epsilon}_t \sim \mathcal{N}(\mathbf{0}, \boldsymbol{\Sigma}_\epsilon). \tag{3.6}$$

Equation 3.6 is the so-called measurement equation, with $\boldsymbol{\Psi}_f$ and $\boldsymbol{\Psi}_m$ denoting $K \times Q$ and $K \times M$ matrices of factor loadings. This equation states that the $\boldsymbol{x}_t$ are driven by a small number ($Q \ll K$) of unobserved and observed factors, with the respective sensitivity of elements in $\boldsymbol{x}_t$ to movements in $\boldsymbol{f}_t$ and $\boldsymbol{m}_t$ governed by the factor loadings. Moreover, such specifications typically include zero mean Gaussian measurement errors, with a diagonal variance-covariance matrix $\boldsymbol{\Sigma}_\epsilon$. This implies that any co-movement between the elements in $\boldsymbol{x}_t$ stem exclusively from the common factors in $\boldsymbol{f}_t$ and $\boldsymbol{m}_t$.

The joint evolution of the latent and manifest driving forces, stacked in a $(M + Q)$-vector $z_t = (\boldsymbol{f}_t', \boldsymbol{m}_t')'$, is determined by the state equation

$$z_t = \boldsymbol{A}_1 z_{t-1} + \ldots + \boldsymbol{A}_P z_{t-P} + \boldsymbol{\eta}_t, \quad \boldsymbol{\eta}_t \sim \mathcal{N}(\mathbf{0}, \boldsymbol{\Sigma}_\eta) \tag{3.7}$$

which is a standard VAR process of order P and coefficient matrices $\boldsymbol{A}_p$ $(p = 1, \ldots, P)$ of dimension $(M + Q) \times (M + Q)$. Again, the state equation errors are assumed to be normally distributed and centered on zero with variance covariance matrix $\boldsymbol{\Sigma}_\eta$. Notice that combining Eqs. (3.6) and (3.7) establishes a simple yet flexible state-space representation with Gaussian error terms.

In the context of modeling large systems of multiple economies, a potential FAVAR specification would be to treat the endogenous variables of country i as the observed factors $\boldsymbol{y}_{it}$, while unobserved factors are extracted from the foreign quantities. In terms of the formulation above, this would imply $\boldsymbol{x}_t = (\boldsymbol{y}'_{1t}, \ldots, \boldsymbol{y}'_{Nt})'$ and $\boldsymbol{m}_t = \boldsymbol{y}_{1t}$.

A simple three country example is provided in the following, focusing on the required setup for the first country (the remaining country models are to be understood analogously). The observation equation is given by

$$\underbrace{\begin{bmatrix} \boldsymbol{y}_{1t} \\ \boldsymbol{y}_{2t} \\ \boldsymbol{y}_{3t} \end{bmatrix}}_{\boldsymbol{x}_t} = \begin{bmatrix} \boldsymbol{I}_M & \boldsymbol{0} \\ \boldsymbol{\Psi}_y & \boldsymbol{\Psi}_f \end{bmatrix} \underbrace{\begin{bmatrix} \boldsymbol{y}_{1t} \\ \boldsymbol{f}_t \end{bmatrix}}_{\boldsymbol{z}_t} + \begin{bmatrix} \boldsymbol{0} \\ \boldsymbol{\epsilon}_{2t} \\ \boldsymbol{\epsilon}_{3t} \end{bmatrix},$$

with the dynamic evolution of $\boldsymbol{z}_t$ governed by the process given in Eq. (3.7). This setup produces multiple sets of interesting results. First, the estimates of the latent factors $\boldsymbol{f}_t$ can be interpreted as international factors, since they carry shared information from quantities foreign to the country of interest. Second, the factor loadings matrices $\boldsymbol{\Psi}_y, \boldsymbol{\Psi}_f$ provide information on how strongly individual countries are linked to the extracted international factors, with the variances of the error terms $\boldsymbol{\epsilon}_{2t}, \boldsymbol{\epsilon}_{3t}$ capturing the magnitude of country-specific idiosyncrasies. Third, the simple VAR structure of the state equation in Eq. (3.7) allows for all common types of inferential analyses and forecasting exercises. Besides forecasts for the observed and unobserved factors, that is, the quantities of the country of interest and the international factors, the observation equation can be used to calculate forecasts for all countries considered.

3.2.5 *Computing Forecasts*

A brief summary of how to obtain forecasts is provided in the following. The interested reader is referred to Hamilton (1994), who provide a much more complete treatment of forecasting based on multivariate time series models. Estimated parameters for the VAR processes underlying the PVAR, GVAR, and FAVAR can be used to generate forecasts in a standard way. All models feature vector autoregressive structures of order p as in Eq. (3.1), which may be written as VAR(1). Let $\boldsymbol{Y}_t = (\boldsymbol{y}'_t, \boldsymbol{y}'_{t-1}, \ldots, \boldsymbol{y}'_{t-p+1})'$, $\boldsymbol{u}_t = (\boldsymbol{\varepsilon}'_t, \boldsymbol{0}, \ldots, \boldsymbol{0})'$, and the companion matrix

$$\boldsymbol{A} = \begin{bmatrix} \boldsymbol{A}_1 & \boldsymbol{A}_2 & \ldots & \boldsymbol{A}_{p-1} & \boldsymbol{A}_p \\ \boldsymbol{I}_K & \boldsymbol{0} & \ldots & \boldsymbol{0} & \boldsymbol{0} \\ \boldsymbol{0} & \boldsymbol{I}_K & \ldots & \boldsymbol{0} & \boldsymbol{0} \\ \vdots & & \ddots & & \vdots \\ \boldsymbol{0} & \ldots & \ldots & \boldsymbol{I}_K & \boldsymbol{0} \end{bmatrix}.$$

The corresponding VAR(1) representation that we use to obtain expressions for the predictive density of h-step ahead forecasts is given by $\boldsymbol{Y}_t = \boldsymbol{A}\boldsymbol{Y}_{t-1} + \boldsymbol{u}_t$ and it follows that

$$\boldsymbol{Y}_{t+h} = \boldsymbol{A}^h\boldsymbol{Y}_t + \sum_{i=0}^{h-1} \boldsymbol{A}^i\boldsymbol{u}_{t+h-i}.$$

Using $\hat{\boldsymbol{Y}}_{t+h}$ to denote the forecast at horizon h, we have

$$\mathrm{E}(\boldsymbol{Y}_{t+h}) = \hat{\boldsymbol{Y}}_{t+h} = \boldsymbol{A}^h\boldsymbol{Y}_t,$$

where the prediction is given by the first K rows of $\hat{\boldsymbol{Y}}_{t+h}$. Let $\boldsymbol{\Omega}$ denote the variance-covariance matrix of $\boldsymbol{u}_t$, then for the variance of the forecasts at h-steps ahead denoted by $\boldsymbol{\Sigma}_{t+h}$ we have

$$\mathrm{E}(\boldsymbol{Y}_{t+h}\boldsymbol{Y}'_{t+h}) = \boldsymbol{\Sigma}_{t+h} = \sum_{i=0}^{h-1} \boldsymbol{A}^i\boldsymbol{\Omega}\boldsymbol{A}'^i,$$

where the relevant submatrix analogous to the point prediction is the upper $K \times K$-block of $\boldsymbol{\Sigma}_{t+h}$. Using these relations, one can easily draw from the predictive density under the assumption that the shocks are normally distributed.

3.3 Empirical Application

In this section we provide a simple forecasting application based on a monthly panel of country-level time series. We evaluate the relative performance of the panel, global, and factor augmented vector autoregressive models against standard VAR specifications and a univariate AR(1) benchmark to underline the value of exploiting information from the cross-sectional dimension for forecasting.

3.3.1 Data and Model Specification

For the empirical application, we use quarterly observations for consumer price inflation, short-term interest rates, and industrial production (as a measure of economic performance) for the G7 countries, that is, data from Canada (CA), Germany (DE), France (FR), the United Kingdom (GB), Italy (IT), Japan (JP), and the United States (USA). The considered period ranges from 1980:Q1 to 2013:Q4. For the GVAR weighting scheme, we rely on bilateral trade weights, constructed by averaging the respective bilateral trade relations over the sample period.

An Illustrative Example

We illustrate the structure of the competing modeling approaches discussed above once again using a simple example. We consider the information set included in the respective equations of the PVAR, GVAR, and FAVAR for the case of the USA. For visualization purposes, we normalize the data beforehand to have zero mean and unit standard deviation. Besides the US measures of inflation, interest rates and output, the PVAR includes all series of the other countries. In the GVAR, all foreign quantities are constructed by relying on cross-sectional averages using trade weights. For the FAVAR, we extract three factors summarizing high-dimensional cross-sectional information on non-US countries. Figure 3.1 shows the resulting (foreign) time series for all three approaches. Inflation is depicted in light gray, the interest rate series are dark gray, while industrial production is black.

In the left panel of Fig. 3.1, the information stemming from foreign countries is shown for the unrestricted PVAR model. A few points are worth noting. Co-movement is observable both across the cross-section, but also across variable types. The impact of the global financial crisis of 2007/2008 is clearly visible, with large drops in output and inflation, while expansionary monetary policies are reflected by a decrease of short-term interest rates in all countries. The middle panel depicts the cross-sectional weighted averages that result in the case of the GVAR. The right panel shows the three extracted factors over time. Factor 1 appears to be closely tracking inflation and interest rates across countries, while for the remaining factors it is less obvious to identify commonalities with observed time series.

This small example provides intuition about how high-dimensional information (as is the case for the PVAR) is compressed in both the FAVAR and the GVAR

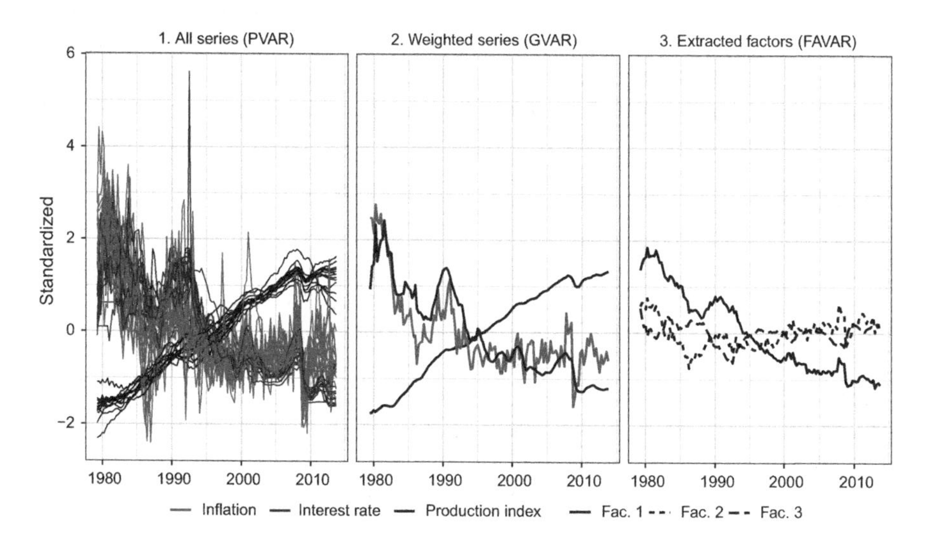

Fig. 3.1 Information set for the United States in the PVAR, GVAR, and FAVAR

case. While the approaches differ in terms of how to achieve a more parsimonious representation of the data, it is apparent that both result in a similar lower dimensional summary of the information set used in the forecasting exercise.

Model Specification

The bulk of the literature on large VARs in general, and multi-country models in particular, uses a Bayesian approach to estimation. This involves choosing prior distributions for all parameters to be estimated in the respective models. In the following, we will sketch the prior setup for all models considered, putting a special emphasis also on discussing possibilities to account for cross-sectional homogeneity, and dynamic and static interdependencies. Moreover, all models are allowed to feature heteroscedastic errors. This feature has proven crucial for forecasting, especially in the context of density forecasts (Carriero, Clark, & Marcellino, 2016). Here, we employ variants of stochastic volatility specifications for all competitors. The models are estimated using Markov chain Monte Carlo (MCMC) simulation methods, with further details on priors and estimation presented in Appendix A.

To assess the merits and comparative strengths and weaknesses of the three approaches discussed here, we provide evidence for two specifications per model class. The set of competing models is complemented by a naive univariate AR(1) process with stochastic volatility and two otherwise standard VAR models with a hierarchical Minnesota shrinkage prior (Litterman, 1986). All competing models we consider are summarized in Table 3.1.

The first model we consider is a simple three-equation VAR specification for each of the seven countries (*single-country Bayesian VAR, BVAR-SC*). The second uses the full information set of 21 variables at once, rendering this specification a large Bayesian VAR (*BVAR*). A similar Minnesota prior is employed for the state equation of the FAVAR models. In this context, we split the information set into domestic and foreign quantities prior to estimation, and extract one (*FAVAR-F1*)

Table 3.1 Competing models, abbreviations, and short description

Model	Description
BVAR-SC	Bayesian VAR, Minnesota prior, country-by-country, stochastic volatility
BVAR	Bayesian VAR, Minnesota prior, all countries jointly, stochastic volatility
PVAR-noDI	Panel VAR, coefficient pooling, no DIs, SIs, stochastic volatility
PVAR-DI	Panel VAR, coefficient pooling, DIs, SIs, stochastic volatility
GVAR-NG	Global VAR, normal-gamma shrinkage prior, stochastic volatility
GVAR-CP	Global VAR, coefficient pooling, stochastic volatility
FAVAR-F1	Factor augmented VAR, one factor, stochastic volatility
FAVAR-F3	Factor augmented VAR, three factors, stochastic volatility
AR(1)	Autoregressive process of order 1, stochastic volatility

Notes: *DIs* dynamic interdependencies, *SIs* static interdependencies and three (*FAVAR-F3*) factors from the foreign variables. The single-country VAR based on the domestic quantities is then augmented by these factors. For the factors, loadings and variance covariance matrices we use standard prior distributions and algorithms for estimation.

For the domestic coefficient matrices of both PVARs (*PVAR without dynamic interdependencies, PVAR-noDI* and *PVAR with dynamic interdependencies, PVAR-DI*) and all coefficients of *GVAR with coefficient pooling, GVAR-CP*, we use a prior designed to exploit the cross-sectional structure by pooling information over the respective countries. A similar prior has, for instance, been used in Jarociński (2010) or Fischer, Huber, and Pfarrhofer (2019). The idea hereby is to assume country-specific coefficients to share a common mean, reflecting cross-sectional homogeneity. The prior used in the forecasting exercise stochastically selects coefficients that are homogeneous, but allows also for heterogeneous coefficients if suggested by the likelihood. The difference between *PVAR-noDI* and *PVAR-DI* is that the first specification rules out dynamic interdependencies a priori, while in the latter case, we impose a normal-gamma shrinkage prior (Griffin & Brown, 2010) to stochastically select non-zero coefficients associated with foreign quantities. The same shrinkage prior is used for all coefficients of the *GVAR normal-gamma, GVAR-NG* specification.

The error terms of the PVARs and GVARs are decomposed using a factor stochastic volatility specification (see, for instance, Aguilar & West, 2000). Here, the stacked error term is decomposed into a lower-dimensional set of dynamic factors that capture co-movements in the cross-section and idiosyncratic stochastic volatility error terms. The factor loadings linking the factors to the high-dimensional errors govern covariances across countries.

3.3.2 Evaluating Forecasting Performance

For the purpose of comparing the three approaches discussed in this chapter and assessing their performance, we use a pseudo out-of-sample forecasting exercise. Here, we split the available data into a training and a holdout sample of length T_{o} that is used to evaluate the predictions in terms of the realized values of the series. The forecasting design is recursive, implying that we start with forecasting the beginning of the holdout sample, indexed by $T_{\mathrm{H}} = T - T_{\mathrm{o}} + 1$. After obtaining the relevant predictive distributions, we expand the initial estimation sample by a single observation and re-estimate all models. This procedure is repeated until no more observations are available in the holdout sample. For the forecasting exercise we use the period from 1999:Q1 to 2013:Q4 as holdout, that is, 60 observations. This holdout serves to calculate root mean squared errors (RMSEs) and log-predictive likelihoods (LPLs) to assess both point and density forecasts.

Precise definitions of the involved performance measures in the following require that we introduce additional notation. Let $\mathcal{I}_{1:t}$ be the information set containing all available time series up to time t. We are interested in computing the h-step ahead point forecast for some variable conditional on the information set, that is the predictive mean $\hat{y}_{t+h} = \mathrm{E}(y_{t+h}|\mathcal{I}_{1:t})$. Moreover, we let $p(y_{t+h}|\mathcal{I}_{1:t})$ denote the predictive density for said variable. Finally, we refer to the realized value of the time series in the holdout sample by y^{o}_{t+h}.

Performance Measures

In terms of point forecasts, we use the well-known measure based on RMSEs. The corresponding definition for the h-step ahead forecast for some model is given by

$$\mathrm{RMSE} = \sqrt{\frac{1}{T_{\mathrm{o}}} \sum_{t=T_{\mathrm{H}}}^{T_{\mathrm{o}}} \left(y^{\mathrm{o}}_{t+h} - \hat{y}_{t+h}\right)^2}.$$

Intuitively, this measure captures the average deviation of the forecast from the realized value over the holdout sample. The smaller the RMSE, the better the out-of-sample predictive performance in terms of point forecasts is the model.

The RMSE measures only how well the model performs in terms of point forecasting, ignoring model performance in terms of higher moments of the predictive distribution. To see how well the predictions of a given model fit the data in the holdout sample, we rely on LPLs (see, for instance, Geweke & Amisano, 2010). The measure at time t is defined as

$$\mathrm{LPL}_{t+h} = \log p\left(y_{t+h} = y^{\mathrm{o}}_{t+h}|\mathcal{I}_{1:t}\right).$$

Here, we evaluate the realized y^{o}_{t+h} under the predictive distribution arising from the respective modeling approach. The magnitude of this value is determined by the mean and variance of the predicted value, accounting both for the point forecast but also how precisely it is estimated.

We consider marginal log-predictive scores (LPSs) used for evaluating the models per variable type, and joint log-predictive scores based on the full predictive vector across variable types and countries. All these measures can be calculated for the h-step ahead predictions across competing econometric models. In particular, based on the available quarterly frequency, we provide measures of forecasting performance for one-quarter ahead, two-quarters ahead, and 1-year ahead in the next section.

3.3.3 Results

In this section, we first consider the performance of the competing models using joint LPSs and average RMSEs. This allows to provide a rough grasp in terms of overall performance. In a second step, we assess the relative merits of the proposed models in terms of marginal LPSs and RMSEs for inflation to investigate model-specific strengths and weaknesses across countries. This second exercise serves to investigate the causes for some of the observed differences across models in overall forecasting performance.

Overall Forecasting Performance

We first present evidence for one-quarter, two-quarters, and 1-year ahead forecasts. The results are summarized in Table 3.2, which presents joint RMSEs calculated across countries and variable types relative to the AR(1) benchmark, implying that values less than 1 indicate a superior forecast performance compared to the benchmark. We also provide LPSs for all competing models summed over the whole out-of-sample period.

One takeaway from Table 3.2 is that performance measures in terms of point and density forecasts do not necessarily agree which model performs best. The best performing model in terms of point forecast at the one-quarter ahead horizon, *PVAR-noDI* improves upon the naive benchmark by roughly 19% in terms of RMSEs. Most of the competing models show improvements over the AR(1) specification, with approximately 10–15% smaller RMSEs, except for *GVAR-CP*. For this model, we obtain higher RMSEs as compared to the univariate model. A similar picture arises for two-quarters ahead forecasts, with slightly worse point forecasts relative to the AR(1) than before and the PVAR without dynamic interdependencies producing the best point forecasts. Interestingly, the *GVAR-CP*, that has been performing poorly at the one-quarter ahead horizon, outperforms the benchmark at this horizon, with the average improvement across multivariate models upon the univariate forecasts being around 10%. For 1-year ahead forecasts, we again observe that the relative performance of multivariate forecasting approaches worsens slightly.

Table 3.2 Overall measures of joint forecasting performance

	One-quarter ahead		Two-quarters ahead		1-year ahead	
Model	RMSE	LPS	RMSE	LPS	RMSE	LPS
BVAR-SC	0.881	4485.71	0.892	3732.35	0.939	2643.46
BVAR	0.852	3989.69	0.868	3514.89	0.907	2513.83
PVAR-noDI	**0.818**	4578.73	**0.831**	4034.75	**0.855**	3304.97
PVAR-DI	0.844	3756.31	0.853	3101.34	0.877	2030.57
GVAR-NG	0.871	4496.99	0.891	3819.25	0.915	2920.52
GVAR-CP	1.017	4353.78	0.925	3948.39	0.950	**3362.14**
FAVAR-F1	0.869	4668.31	0.886	3914.38	0.934	2711.59
FAVAR-F3	0.861	**4893.68**	0.872	4035.68	0.900	2573.57
AR(1)	1.000	4863.63	1.000	**4076.00**	1.000	3311.89

Notes: Model abbreviations as in Table 3.1. LPSs reflect the sum over the whole out-of-sample period. RMSEs are calculated across countries and variable types and relative to the AR(1) process. Respective maximum (LPSs) and minimum values (RMSEs) are in bold

The best performing model *PVAR-noDI* still outperforms the univariate benchmark by roughly 15%. It is worth mentioning that forecasts of the single country BVAR (*BVAR*) outperform forecasts of the AR(1) benchmark. However, exploiting information from the cross-section *and* introducing restrictions of some form seem to improve upon the high-dimensional VAR model.

In terms of density forecasts, the LPSs in Table 3.2 indicate that the AR(1) benchmark is hard to beat across all forecasting horizons considered. Starting with the one-quarter ahead scores, we observe that the FAVAR specifications produce the most accurate density forecasts, displaying the highest value for the three-factor specification. The joint LPS for the univariate competitor is close to the best performing model and outperforms all other multivariate models. It is worth mentioning that multiple approaches extracting information from the cross-section improve upon standard large BVAR methodology. In the two-quarter ahead case we observe an overall lower level of LPSs, and no multivariate model outperforms the benchmark, even though *FAVAR-F3* and *PVAR-noDI* are in close range to the AR(1) model. For the 1-year ahead forecasts, *GVAR-CP* outperforms the AR(1), with *PVAR-noDI* also showing a comparable LPS.

To provide some evidence which periods drive the differences in performance for density forecasts, we present LPSs of the competing models for the one-quarter ahead horizon over time in Fig. 3.2. The left panel shows LPS scores for each point in time, while the right panel depicts the cumulative sum over all periods. A first observation regarding LPS over time is that they appear to be fluctuating more for some models than others. The FAVAR specification using three factors outperforms its competitors in most periods. A second observation refers to the behavior of LPSs

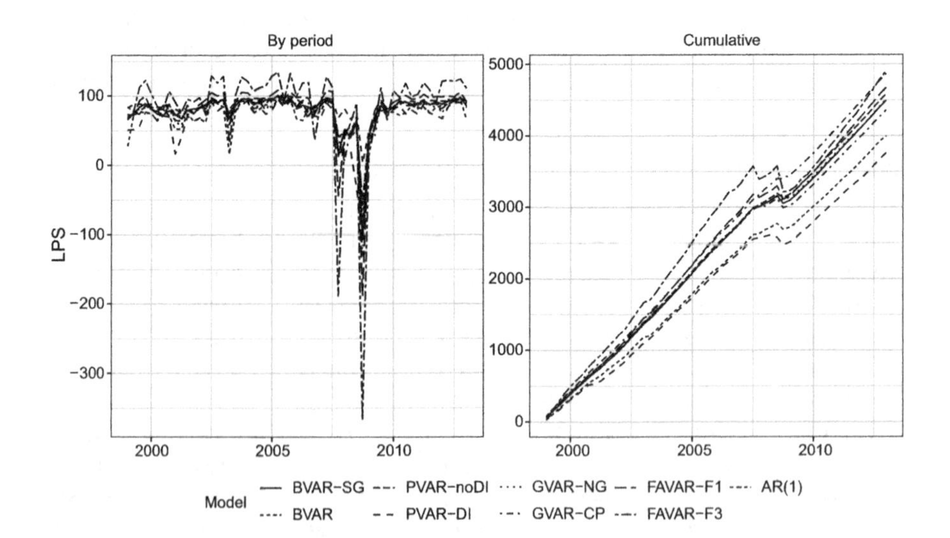

Fig. 3.2 Log-predictive scores of one-quarter ahead forecasts. *Notes*: Model abbreviations as in Table 3.1. The left panel refers to the log-predictive likelihood at the given point in time by period. The right panel depicts the cumulative sum over all periods

during the global financial crisis. All considered models show large decreases in density forecasting performance around this period, with *FAVAR-F3* exhibiting the weakest forecasting performance. Turning to the cumulative scores, two points are worth noting. First, as suggested by examining LPS scores over time, the FAVAR with three factors appears to strongly outperform its competitors up to the outbreak of the global financial crisis. Afterwards, its performance deteriorates considerably relative to the other competing models. The other approaches are similar at a level slightly below the top performers, except for *BVAR* and *PVAR-DI* that display inferior overall log scores.

Forecasts for Individual Countries

Following the discussion of performance measures over time, we now focus on forecasts for inflation per country. To provide an overview of model-specific characteristics of resulting forecasts, we present the forecasts for the USA over the holdout period for the *PVAR-noDI*, *GVAR-NG*, and *FAVAR-F3* alongside the realized time series in Fig. 3.3 for the one-quarter and 1-year ahead horizon. We choose these three models based on their performance relative to the competitors.

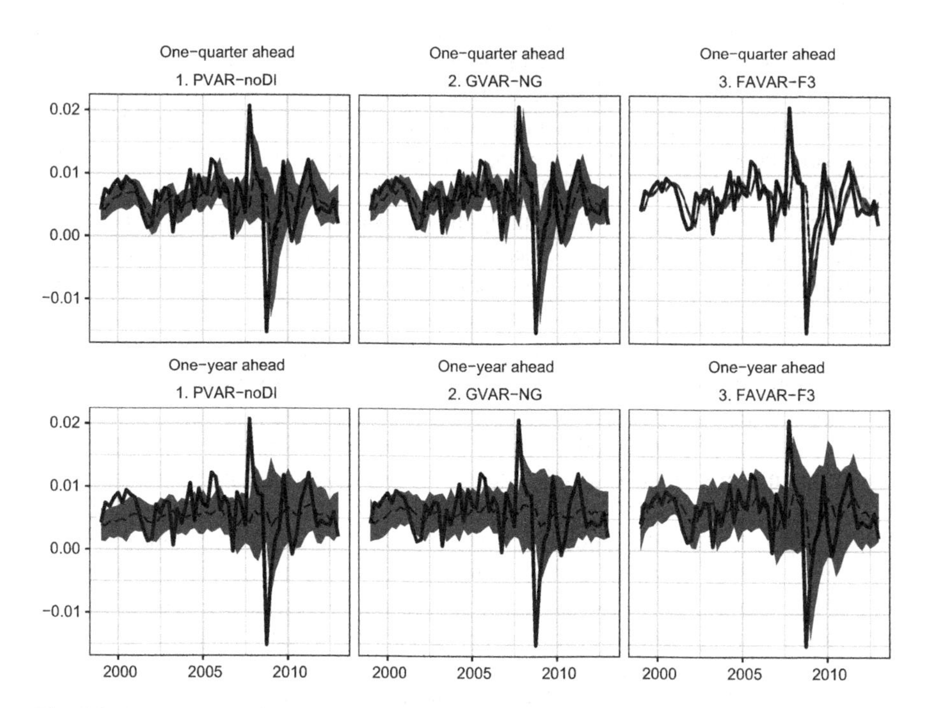

Fig. 3.3 Forecasts for US inflation over the holdout sample. *Notes*: The gray shaded area covers the interval between 16th and 84th percentile of the posterior predictive density, with the dashed line indicating the posterior median. The thicker black line are the realized values for inflation. The upper three panels are based on the one-quarter ahead forecasts, while the lower three refer to the 1-year ahead horizon

Figure 3.3 presents the median of the predictive density alongside the 68% posterior coverage interval in gray. For the one-quarter ahead forecasts, we observe similar predictive densities in terms of the mean and the precision for the PVAR and GVAR. The FAVAR differs drastically in terms of the estimation precision, which can be explained by the fact that for the PVAR and GVAR, we have 21 endogenous variables resulting in a much larger parameter space compared to that of the six endogenous variables of the FAVAR. Time variation in the error variances translates into differences in the forecast precision over time, mirrored by the tighter error bands in the period before the global financial crisis, and wider predictive densities afterwards. For the 1-year ahead horizon, we observe again similar density forecasts for the PVAR and GVAR. Compared to the one-quarter ahead predictions, movements in median predictions are much less pronounced. For the FAVAR specification, we now observe much wider error bands, reflecting higher uncertainty surrounding the forecasts. It is worth mentioning that for the FAVAR, movements in the median predictions appear to more closely track the actual evolution of inflation, especially at the beginning of the holdout sample and during the global financial crisis.

Table 3.3 contains RMSEs for inflation across all countries for the one-quarter and 1-year ahead horizon. One takeaway here is that no specification strictly dominates the others across countries or forecasting horizon. For the one-quarter ahead predictions, we observe improvements upon the univariate benchmark for all countries except the United Kingdom. For Canada, Germany, and the USA, the PVAR without dynamic interdependencies performs best, with improvements of around 10% for the first two countries, and 5% for the latter. For France, Italy, and especially Japan, we observe large improvements against the AR(1) for *PVAR-DI*, with RMSEs for Japan being 21.2% lower than for the benchmark. While the other approaches appear to perform similarly, the *GVAR-CP* performs poorly across all countries, with RMSEs up to twice as large as those of the AR(1) process. Turning to the 1-year ahead horizon, we find that for Germany and France none of the proposed multi-country models is able to outperform the naive benchmark. The GVAR with normal-gamma shrinkage prior performs best relative to the AR(1) specification for Canada and the USA, while we find particularly strong improvements for Japan with *PVAR-noDI* (22.4% lower RMSEs) and France for *PVAR-DI* (15.9% lower).

The final set of results addresses the performance for inflation forecasts across countries in terms of the predictive density. The results are depicted in Table 3.4, again for one-quarter and 1-year ahead. In this context, the multi-country framework apparently does not yield large improvements against the AR(1) benchmark. In fact, for France and the USA, the univariate forecast produces the highest log-predictive score. For Canada, Germany, and Italy, modest improvements are observable. The only somewhat substantive improvement occurs for Japan using the *GVAR-NG* specification. Considering 1-year ahead forecasts, we observe larger improvements for Canada (*GVAR-NG*), the United Kingdom (*PVAR-DI*), and Japan (*PVAR-noDI*). Minor improvements are detectable for France and Italy, while the AR(1) benchmark is superior against all competing models for Germany and the USA at this horizon.

Table 3.3 Root mean squared error for inflation forecasts

	One-quarter ahead						
Model	CA	DE	FR	GB	IT	JP	US
BVAR-SC	0.963	0.995	1.056	1.028	1.019	0.953	0.990
BVAR	0.940	0.989	0.989	1.018	0.964	0.894	0.971
PVAR-noDI	**0.913**	**0.904**	1.004	1.105	0.931	0.862	**0.945**
PVAR-DI	0.934	0.942	**0.933**	1.058	**0.931**	**0.788**	0.965
GVAR-NG	0.947	0.969	0.960	1.058	0.944	0.802	0.949
GVAR-CP	1.384	1.816	1.932	1.921	1.852	1.748	1.308
FAVAR-F1	0.944	1.030	1.056	1.029	1.035	0.965	0.994
FAVAR-F3	0.937	0.973	1.028	1.021	0.996	0.944	1.005
AR(1)	1.000	1.000	1.000	**1.000**	1.000	1.000	1.000
	1-year ahead						
Model	CA	DE	FR	GB	IT	JP	US
BVAR-SC	0.967	1.108	1.092	1.030	1.088	0.938	1.043
BVAR	0.920	1.146	0.964	1.017	0.955	0.834	0.966
PVAR-noDI	0.920	1.077	1.021	1.251	0.926	**0.776**	0.944
PVAR-DI	0.904	1.120	**0.841**	1.027	**0.916**	0.881	0.992
GVAR-NG	**0.901**	1.105	0.932	1.188	0.985	0.788	**0.936**
GVAR-CP	1.133	1.522	1.236	1.475	1.336	1.207	1.090
FAVAR-F1	0.962	1.147	1.085	1.036	1.075	0.914	1.059
FAVAR-F3	0.919	1.074	1.030	1.105	1.024	0.909	1.025
AR(1)	1.000	**1.000**	1.000	**1.000**	1.000	1.000	1.000

Notes: Model abbreviations as in Table 3.1. RMSEs are presented relative to the AR(1) process. Respective minimum values are in bold

Concluding the forecasting exercise, a few points are worth noting. First, forecasting performance measures may yield different rankings of competing models, depending on whether one is interested in point or density forecasts. Second, there is seldom one specification that is strictly superior for all variables and countries over time. Finally, exploiting the cross-sectional structure to obtain more precise estimates in large scale VAR models if applicable usually pays off in terms of forecasting, with gains ranging from substantial to modest depending on the specifics of the forecasting application.

3.4 Summary

In this chapter we have summarized three prominent frameworks to deal with large data repeated from the cross-section, the panel vector autoregressive (PVAR) model, the global vector-autoregressive (GVAR) model, and the factor augmented vector autoregressive (FAVAR) model. We illustrate all three approaches by means of a small forecasting exercise. We find that different forms of multi-country models

Table 3.4 Log-predictive scores for inflation forecasts

Model	CA	DE	FR	GB	IT	JP	US
	One-quarter ahead						
BVAR-SC	222.40	253.98	249.37	250.04	248.9	245.84	219.91
BVAR	222.46	250.12	251.32	**253.75**	254.32	245.11	211.01
PVAR-noDI	**228.91**	**256.69**	250.78	245.72	**256.65**	252.79	212.24
PVAR-DI	218.59	249.12	248.70	246.56	245.00	252.70	201.95
GVAR-NG	225.07	251.01	253.33	250.46	250.47	**252.99**	214.64
GVAR-CP	214.27	238.00	232.81	235.25	226.16	236.06	195.05
FAVAR-F1	223.48	252.98	249.26	249.08	247.15	244.85	216.98
FAVAR-F3	223.47	253.16	249.95	247.21	248.95	246.28	215.83
AR(1)	225.02	255.29	**255.04**	253.07	256.21	244.31	**226.96**
	1-year ahead						
Model	CA	DE	FR	GB	IT	JP	US
BVAR-SC	213.47	239.98	220.66	225.83	222.41	228.33	213.15
BVAR	213.00	234.52	225.41	231.80	227.56	229.74	207.11
PVAR-noDI	218.38	240.74	217.71	217.70	**231.52**	**242.97**	208.58
PVAR-DI	212.34	230.35	228.21	**231.82**	226.27	230.68	190.83
GVAR-NG	**218.93**	238.94	**230.25**	227.35	228.27	240.25	216.53
GVAR-CP	208.51	225.11	219.06	220.23	217.83	220.54	206.63
FAVAR-F1	213.00	239.36	218.45	226.46	223.85	226.37	209.94
FAVAR-F3	215.06	240.75	222.71	223.59	222.44	230.15	212.11
AR(1)	213.91	**241.68**	227.89	225.65	231.10	227.76	**221.30**

Notes: Model abbreviations as in Table 3.1. LPSs reflect the sum over the holdout sample. Respective maximum values are in bold

excel depending on the forecast horizon, the variable set of interest, and the measure of forecast evaluation (point versus density forecasts). More importantly though and irrespective of the preferred multi-country framework, our results demonstrate that taking information from the cross-section into account clearly improves forecasts over models that focus on domestic data only.

Appendix A: Details on Prior Specification

Priors for the BVAR and FAVAR Coefficients

The Minnesota prior pushes the system of equations towards a multivariate random walk, featuring cross-variable and cross-equation shrinkage. We follow a data-driven approach to select the amount of shrinkage applied, by imposing Gamma distributed priors on the hyperparameter governing how tight the prior is on the own lags of a variable, and the hyperparameter related to shrinkage of the lags of other variables in the system. Different ranges of the endogenous variables are reflected in the prior by scaling it based on standard deviations obtained from univariate autoregressive processes of order one for all series.

A similar Minnesota prior is used for estimating the state-equation of the country-specific FAVAR models. We consider extracting one factor (*FAVAR-F1*) and three factors (*FAVAR-F3*) for the forecasting exercise, respectively. For the factors and factor loadings we employ standard Gaussian priors and simulate the related quantities using the forward filtering backward sampling algorithm by Chris and Kohn (1994) and Frühwirth-Schnatter (1994). The errors in the measurement equation are assumed to have zero mean with a diagonal variance-covariance matrix. On the corresponding diagonal elements, we impose independent weakly informative inverse Gamma priors.

Variance Estimation for BVARs and FAVARs

Turning to modeling the variance-covariance matrix of the VAR processes that allows for heteroscedasticity, we employ analogous stochastic volatility specifications for the BVAR, BVAR-SC, FAVAR-F1, and FAVAR-F3. A complete treatment of stochastic volatility models is out of scope of this chapter, the interested reader is referred to Jacquier, Polson, and Rossi (2002). Let $\boldsymbol{\Sigma}_t$ be a generic variance-covariance matrix applicable to all specifications. This matrix may be decomposed into

$$\boldsymbol{\Sigma}_t = \boldsymbol{H}^{-1}\boldsymbol{S}_t(\boldsymbol{H}^{-1})',$$

with $\boldsymbol{H}^{-1}$ denoting a square lower triangular matrix with ones on the main diagonal of appropriate dimension. Time variation stems from the elements of $\boldsymbol{S}_t$, a diagonal matrix with characteristic elements s_{it}. A stochastic volatility specification results assuming the logarithm of s_{it} for all i follows an AR(1) process

$$\ln(s_{it}) = \mu_i + \rho_i\,(\ln(s_{it}) - \mu_i) + v_{it}, \quad v_{it} \sim \mathcal{N}(0, \varsigma_i^2), \tag{3.8}$$

where μ_i, ρ_i, and ς^2 denote the unconditional mean, persistence parameter, and innovation variance of this state equation. For the purposes of this forecasting exercise, we rely on the R-package `stochvol` for estimation and use its default prior settings (Kastner, 2016). For the free elements of $\boldsymbol{H}^{-1}$, weakly informative independent Gaussian priors with zero mean are employed. Combining the likelihood with the respective priors yields conditional posterior distributions to be used in a Gibbs sampler with most of the involved quantities being of standard form (for detailed information, see for instance Koop, 2003).

Priors for the PVAR and GVAR Coefficients

The priors and model specifics for the PVAR and the GVAR specifications are designed to account for dynamic interdependencies using a shrinkage priors and static interdependencies via factor stochastic volatility. Moreover, we introduce a prior to be used for extracting information across countries, reflecting cross-sectional homogeneity. For the domestic coefficients of the PVAR variants (*PVAR* and *PVAR-DI*) and for all coefficients of *GVAR-CP*, we stack the coefficients specific to country i in a column vector $\boldsymbol{a}_i$. We assume the country-specific coefficients to be homogenous across cross-sectional observations with deviations governed by an error $\boldsymbol{w}_i$, cast in regression form as

$$\boldsymbol{a}_i = \boldsymbol{a} + \boldsymbol{w}_i, \quad \boldsymbol{w}_i \sim \mathcal{N}(\boldsymbol{0}, \boldsymbol{V}),$$

with variance-covariance matrix $\boldsymbol{V}$ assumed to be diagonal with characteristic elements v_j. Notice that v_j governs the degree of heterogeneity of coefficients across countries. The specification above may be written in terms of the prior distribution on the country-specific coefficients as $\boldsymbol{a}_i \sim \mathcal{N}(\boldsymbol{a}, \boldsymbol{V})$. As priors on the common mean we use $\boldsymbol{a} \sim \mathcal{N}(\boldsymbol{0}, 10 \times \boldsymbol{I})$, and for $v_j \sim \mathcal{G}(0.01, 0.01)$.

For the non-domestic coefficients of *PVAR-DI* and for all coefficients of *GVAR*, we use a normal-gamma shrinkage (NG, Griffin & Brown, 2010) prior. This prior is among the class of absolutely continuous global-local shrinkage priors and mimics the discrete stochastic search variable selection (SSVS, George & McCulloch, 1993; George et al., 2008) prior discussed in Chap. 4 of this volume. We choose the NG prior rather than the SSVS prior due to its advantageous empirical properties in high-dimensional model spaces (for an application in the VAR context that we base the hyperparameter values on, see Huber & Feldkircher, 2019). The global parameter of the hierarchical prior setup strongly pushes all coefficients towards zero, while local parameters allow for a priori non-zero idiosyncratic coefficients if suggested by the data likelihood. Intuitively, this prior allows for stochastic selection of inclusion and exclusion of VAR coefficients. The specification *PVAR* rules out all dynamic interdependencies a priori.

Factor Stochastic Volatility for Variance Estimation

Rather than decomposing the variance-covariance matrix as in the context of the BVAR and FAVAR specifications, we structure the stacked K-dimensional error vector $\boldsymbol{\varepsilon}_t$ for all countries in the PVAR and GVAR case as follows. We specify

$$\boldsymbol{\varepsilon}_t = \boldsymbol{\Lambda} \boldsymbol{F}_t + \boldsymbol{\eta}_t, \quad \boldsymbol{F}_t \sim \mathcal{N}(\boldsymbol{0}, \boldsymbol{H}_t), \quad \boldsymbol{\eta}_t \sim \mathcal{N}(\boldsymbol{0}, \boldsymbol{\Omega}_t),$$

with $\boldsymbol{F}_t$ denoting a set of Q latent dynamic factors following a Gaussian distribution with zero mean and time-varying diagonal variance-covariance matrix $\boldsymbol{H}_t = \text{diag}(h_{1t}, \dots, h_{Qt})$. $\boldsymbol{\Lambda}$ is a $K \times Q$ matrix of factor loadings, linking the unobserved low-dimensional factors to the high dimensional error term $\boldsymbol{\varepsilon}_t$. The vector $\boldsymbol{\eta}_t$ is a K-vector of idiosyncratic errors that is normally distributed with zero mean and diagonal variance covariance matrix $\boldsymbol{\Omega}_t = \text{diag}(\omega_{1t}, \dots, \omega_{Kt})$. This setup implies that $\text{E}(\boldsymbol{\varepsilon}_t \boldsymbol{\varepsilon}_t') = \boldsymbol{\Lambda} \boldsymbol{H}_t \boldsymbol{\Lambda}' + \boldsymbol{\Omega}_t$. By the assumed diagonal structure of $\boldsymbol{\Omega}_t$, this translates to the covariances being driven by the respective factor loadings in $\boldsymbol{\Lambda}$. Static interdependencies can, for instance, be tested by imposing a suitable shrinkage prior on the elements of this matrix.

It remains to specify the law of motion on the respective logarithm of the diagonal elements of $\boldsymbol{H}_t$ and $\boldsymbol{\Sigma}_t$. Here, we assume

$$\ln(h_{jt}) = \mu_{hj} + \rho_{hj}(\ln(h_{jt}) - \mu_h) + \nu_{hjt}, \quad \nu_{hjt} \sim \mathcal{N}(0, \varsigma_{hj}^2)$$
$$\ln(\omega_{lt}) = \mu_{\omega l} + \rho_{\omega l}(\ln(\omega_{lt}) - \mu_{\omega l}) + \nu_{\omega lt}, \quad \nu_{\omega lt} \sim \mathcal{N}(0, \varsigma_{\omega l}^2),$$

for $j = 1, \ldots, Q$ and $l = 1, \ldots, K$, with the specification and parameters to be understood analogous to Eq. (3.8). Again we employ the R-package `stochvol` for estimation and use its default prior settings (Kastner, 2016) for the AR(1) state equations. For the elements of the factor loadings matrix, we impose independent normally distributed priors with zero mean and unit variance. The latent factors are simulated using a forward filter backward sampling algorithm (Chris & Kohn, 1994; Frühwirth-Schnatter, 1994) similar to the one employed for sampling the factors in the context of the FAVAR specifications.

Acknowledgement The authors gratefully acknowledge financial support by the Austrian Science Fund (FWF): ZK 35-G.

References

Aguilar, O., & West, M. (2000). Bayesian dynamic factor models and portfolio allocation. *Journal of Business & Economic Statistics, 18*(3), 338–357.
Allenby, G. M., Arora, N., & Ginter, J. L. (1998). On the heterogeneity of demand. *Journal of Marketing Research, 35*(3), 384–389.
Arellano, M., & Bond, S. (1991). Some tests of specification for panel data: Monte Carlo evidence and an application to employment equations. *Review of Economic Studies, 58*(2), 277–297.
Banerjee, A., Marcellino, M., & Masten, I. (2014). Forecasting with factor augmented error correction models. *International Journal of Forecasting, 30*(3), 589–612.
Bernanke, B. S., Boivin, J., & Eliasz, P. (2005). Measuring the effects of monetary policy: A factor-augmented vector autoregressive (FAVAR) approach. *Quarterly Journal of Economics, 120*(1), 387–422.
Canova, F. (2007). *Methods for applied macroeconomic research*. Princeton, NJ: Princeton University Press.
Canova, F., & Ciccarelli, M. (2004). Forecasting and turning point predictions in a Bayesian panel VAR model. *Journal of Econometrics, 120*(2), 327–359.
Canova, F., & Ciccarelli, M. (2009). Estimating multicountry VAR models. *International Economic Review, 50*(3), 929–959.
Canova, F.,& Ciccarelli, M. (2013). Panel vector autoregressive models: A survey. In *Var models in macroeconomics–new developments and applications: Essays in honor of Christopher A. Sims* (pp. 205–246). Bingley, UK: Emerald Group Publishing Limited.
Carriero, A., Clark, T. E., & Marcellino, M. (2016). Common drifting volatility in large Bayesian VARs. *Journal of Business & Economic Statistics, 34*(3), 375–390.
Carter, C. K., & Kohn, R. (1994). On Gibbs sampling for state space models. *Biometrika, 81*(3), 541–553.
Chudik, A., Grossman, V., & Pesaran, M. H. (2016). A multi-country approach to forecasting output growth using PMIs. *Journal of Econometrics, 192*(2), 349–365.

Crespo Cuaresma, J., Doppelhofer, G., Feldkircher, M., & Huber, F. (2019). Spillovers from US monetary policy: Evidence from a time varying parameter global vector autoregressive model. *Journal of the Royal Statistical Society: Series A, 182*, 831–861.

Crespo Cuaresma, J., Feldkircher, M., & Huber, F. (2016). Forecasting with global vector autoregressive models: A Bayesian approach. *Journal of Applied Econometrics, 31*(7), 1371–1391.

Dées, S., di Mauro, F., Pesaran, H., & Smith, L. (2007). Exploring the international linkages of the euro area: A global VAR analysis. *Journal of Applied Econometrics, 22*(1), 1–38.

Dées, S., & Güntner, J. (2017). Forecasting inflation across Euro area countries and sectors: A panel VAR approach. *Journal of Forecasting, 36*, 431–453.

Dovern, J., Feldkircher, M., & Huber, F. (2016). Does joint modelling of the world economy pay off? Evaluating global forecasts from a Bayesian GVAR. *Journal of Economic Dynamics and Control, 70*, 86–100.

Eickmeier, S., Lemke, W., & Marcellino, M. (2015). A classical time varying FAVAR model: Estimation, forecasting, and structural analysis. *European Economic Review, 74*, 128–145.

Eickmeier, S., & Ng, T. (2015). How do US credit supply shocks propagate internationally? A GVAR approach. *European Economic Review, 74*, 128–145.

Eickmeier, S., & Ng, T. (2011). Forecasting national activity using lots of international predictors: An application to New Zealand. *International Journal of Forecasting, 27*(2), 496–511.

Feldkircher, M., & Huber, F. (2016). The international transmission of US shocks—evidence from Bayesian global vector autoregressions. *European Economic Review, 81*(100), 167–188.

Fischer, M., Huber, F., & Pfarrhofer, M. (2019). The regional transmission of uncertainty shocks on income inequality in the United States. *Journal of Economic Behavior & Organization*, forthcoming.

Frühwirth-Schnatter, S. (1994). Data augmentation and dynamic linear models. *Journal of Time Series Analysis, 15*(2), 183–202.

Frühwirth-Schnatter, S., Tüchler, R., & Otter, T. (2004). Bayesian analysis of the heterogeneity model. *Journal of Business & Economic Statistics, 22*(1), 2–15.

George, E., & McCulloch, R. (1993). Variable selection via Gibbs sampling. *Journal of the American Statistical Association, 88*, 881–889.

George, E., Sun, D., & Ni, S. (2008). Bayesian stochastic search for VAR model restrictions. *Journal of Econometrics, 142*(1), 553–580.

Geweke, J., & Amisano, G. (2010). Comparing and evaluating Bayesian predictive distributions of asset returns. *International Journal of Forecasting, 26*(2), 216–230.

Greenwood-Nimmo, M., Nguyen, V. H., & Shin, Y. (2012). Probabilistic forecasting of output growth, inflation and the balance of trade in a GVAR framework. *Journal of Applied Econometrics, 27*, 554–573.

Griffin, J. E., & Brown, P. J. (2010). Inference with normal-gamma prior distributions in regression problems. *Bayesian Analysis, 5*(1), 171–188.

Hamilton, J. (1994). *Time series analysis*. Princeton, NJ: Princeton University Press.

Han, F., & Ng, T. H. (2011). ASEAN-5 macroeconomic forecasting using a GVAR model. *Asian Development Bank, Working Series on Regional Economic Integration, 76*.

Huber, F. (2016). Density forecasting using Bayesian global vector autoregressions with stochastic volatility. *International Journal of Forecasting, 32*(3), 818–837.

Huber, F., & Feldkircher, M. (2019). Adaptive Shrinkage in Bayesian Vector Autoregressive Models. *Journal of Business & Economic Statistics, 37*(1), 27–39.

Jacquier, E., Polson, N. G., & Rossi, P. E. (2002). Bayesian analysis of stochastic volatility models. *Journal of Business & Economic Statistics, 20*(1), 69–87.

Jarociński, M. (2010). Responses to monetary policy shocks in the east and the west of Europe: A comparison. *Journal of Applied Econometrics, 25*(5), 833–868.

Kastner, G. (2016). Dealing with stochastic volatility in time series using the R package stochvol. *Journal of Statistical Software, 69*(1), 1–30.

Koop, G. (2003). *Bayesian econometrics*. London: Wiley.

Koop, G., & Korobilis, D. (2016). Model uncertainty in panel vector autoregressive models. *European Economic Review, 81*, 115–131.

Koop, G., & Korobilis, D. (2019). Forecasting with high-dimensional panel VARs. *Oxford Bulletin of Economics and Statistics, 81*(5), 937–959.

Litterman, R. B. (1986). Forecasting with Bayesian vector autoregressions: Five years of experience. *Journal of Business & Economic Statistics, 4*(1), 25–38.

McCracken, M. W., & Ng, S. (2016). FRED-MD: A monthly database for macroeconomic research. *Journal of Business & Economic Statistics, 34*(4), 574–589.

Moench, E. (2008). Forecasting the yield curve in a data-rich environment: A no-arbitrage factor-augmented VAR approach. *Journal of Econometrics, 146*(1), 26–43.

Pesaran, M., Schuermann, T., & Smith, L. (2009). Forecasting economic and financial variables with global VARs. *International Journal of Forecasting, 25*(4), 642–675.

Pesaran, M., Schuermann, T., & Weiner, S. (2004). Modeling regional interdependencies using a global error-correcting macroeconometric model. *Journal of Business & Economic Statistics, 22*(2), 129–162.

Stock, J. H., & Watson, M. W. (2016). Dynamic factor models, factor-augmented vector autoregressions, and structural vector autoregressions in macroeconomics. In J. B. Taylor & H. Uhlig (Eds.), *Handbook of Macroeconomics*. Amsterdam: Elsevier.

Wang, P., & Wen, Y. (2007). Inflation dynamics: A cross-country investigation. *Journal of Monetary Economics, 54*(7), 2004–2031.

Chapter 4
Large Bayesian Vector Autoregressions

Joshua C. C. Chan

4.1 Introduction

Vector autoregressions (VARs) are the workhorse models for empirical macroeconomics. They were introduced to economics by Sims (1980), and have since been widely adopted for macroeconomic forecasting and structural analysis. Despite their simple formulation, VARs often forecast well, and are used as the benchmark for comparing forecast performance of new models and methods. They are also used to better understand the impacts of structural shocks on key macroeconomic variables through the estimation of impulse response functions.

VARs tend to have a lot of parameters. Early works by Doan, Litterman, and Sims (1984) and Litterman (1986) on Bayesian methods that formally incorporate non-data information into informative priors have often been found to greatly improve forecast performance. However, until recently, most empirical work had considered only small systems that rarely include more than a few dependent variables.

This has changed since the seminal work of Banbura, Giannone, and Reichlin (2010), who found that large Bayesian VARs with more than two dozen of dependent variables forecast better than small VARs. This has generated a rapidly expanding literature on using large Bayesian VARs for forecasting and structural analysis; recent papers include Carriero, Kapetanios, and Marcellino (2009), Koop (2013), and Carriero, Clark, and Marcellino (2015a). Large Bayesian VARs thus provide an alternative to factor models that are traditionally used to handle large datasets (e.g., Stock & Watson, 2002; Forni, Hallin, Lippi, and Reichlin, 2003). For more applications and examples that naturally give rise to large VARs, see Chapter 3 on VARs designed for multi-country data.

J. C. C. Chan
Purdue University and UTS, West Lafayette, IN, USA

P. Fuleky (ed.), *Macroeconomic Forecasting in the Era of Big Data*,
Advanced Studies in Theoretical and Applied Econometrics 52,
https://doi.org/10.1007/978-3-030-31150-6_4

There are by now many extensions of small VARs that take into account salient features of macroeconomic data, the most important of which being time-varying volatility (Cogley & Sargent, 2005; Primiceri, 2005). How best to construct large VARs with time-varying volatility is an active research area, and has generated many new approaches, such as Koop and Korobilis (2013), Carriero, Clark, and Marcellino (2015b, 2016), and Chan (2018).

There are two key challenges in estimating large VARs. First, large VARs typically have far more parameters than observations. Without appropriate shrinkage or regularization, parameter uncertainty would make forecasts or any analysis unreliable. Second, estimation of large VARs involves manipulating large matrices and is typically computationally intensive. These two challenges are exacerbated when we extend large VARs to allow for more flexible error covariance structures, such as time-varying volatility.

In what follows, we first study methods to tackle these two challenges in the context of large homoscedastic VARs. We will then discuss a few recent models that incorporate stochastic volatility into large VARs and the associated estimation methods.

4.1.1 Vector Autoregressions

We first consider a standard homoscedastic VAR of order p. Let $\mathbf{y}_t = (y_{1t}, \ldots, y_{nt})'$ denote the $n \times 1$ vector of dependent variables at time t. Then, the basic VAR(p) is given by

$$\mathbf{y}_t = \mathbf{b} + \mathbf{A}_1\mathbf{y}_{t-1} + \cdots + \mathbf{A}_p\mathbf{y}_{t-p} + \boldsymbol{\varepsilon}_t, \tag{4.1}$$

where $\mathbf{b}$ is an $n \times 1$ vector of intercepts, $\mathbf{A}_1, \ldots, \mathbf{A}_p$ are $n \times n$ coefficient matrices, and $\boldsymbol{\varepsilon}_t \sim \mathcal{N}(\mathbf{0}, \boldsymbol{\Sigma})$. In other words, the VAR(p) is simply a multiple-equation regression where the regressors are the lagged dependent variables. Specifically, there are n equations and each equation has $k = np + 1$ regressors—so there are a total of $nk = n^2p + n$ VAR coefficients. With typical quarterly data, the number of VAR coefficients can be more than the number of observations when n is large.

The model in (4.1) runs from $t = 1$ to $t = T$, and it depends on the p initial conditions $\mathbf{y}_{-p+1}, \ldots, \mathbf{y}_0$. In principle these initial conditions can be modeled explicitly. Here all the analysis is done conditioned on these initial conditions. If the series is not too short, both approaches typically give similar results.

There are two common ways to stack the VAR(p) in (4.1) over $t = 1, \ldots, T$. In the first representation, we rewrite the VAR(p) as:

$$\mathbf{y}_t = \mathbf{X}_t\boldsymbol{\beta} + \boldsymbol{\varepsilon}_t,$$

where $\mathbf{X}_t = \mathbf{I}_n \otimes [1, \mathbf{y}'_{t-1}, \ldots, \mathbf{y}'_{t-p}]$ with $\otimes$ denoting the Kronecker product and $\boldsymbol{\beta} = \text{vec}([\mathbf{b}, \mathbf{A}_1, \ldots, \mathbf{A}_p]')$—i.e., the intercepts and VAR coefficient matrices are stacked by rows into a $nk \times 1$ vector. Furthermore, stacking $\mathbf{y} = (\mathbf{y}_1', \ldots, \mathbf{y}_T')'$, we obtain

$$\mathbf{y} = \mathbf{X}\boldsymbol{\beta} + \boldsymbol{\varepsilon}, \tag{4.2}$$

where $\mathbf{X} = (\mathbf{X}_1', \ldots, \mathbf{X}_T')'$ is a $Tn \times nk$ matrix of regressors and $\boldsymbol{\varepsilon} \sim \mathcal{N}(\mathbf{0}, \mathbf{I}_T \otimes \boldsymbol{\Sigma})$.

In the second representation, we first stack the dependent variables into a $T \times n$ matrix $\mathbf{Y}$ so that its t-th row is $\mathbf{y}_t'$. Now, let $\mathbf{Z}$ be a $T \times k$ matrix of regressors, where the t-th row is $\mathbf{x}_t' = (1, \mathbf{y}_{t-1}', \ldots, \mathbf{y}_{t-p}')$. Next, let $\mathbf{A} = (\mathbf{b}, \mathbf{A}_1, \ldots, \mathbf{A}_p)'$ denote the $k \times n$ matrix of VAR coefficients. Then, we can write the VAR(p) as follows:

$$\mathbf{Y} = \mathbf{Z}\mathbf{A} + \mathbf{U}, \tag{4.3}$$

where $\mathbf{U}$ is a $T \times n$ matrix of innovations in which the t-th row is $\boldsymbol{\varepsilon}_t'$. In terms of the first representation in (4.2), $\mathbf{y} = \text{vec}(\mathbf{Y}')$, $\boldsymbol{\beta} = \text{vec}(\mathbf{A})$, and $\boldsymbol{\varepsilon} = \text{vec}(\mathbf{U}')$. It follows that

$$\text{vec}(\mathbf{U}) \sim \mathcal{N}(\mathbf{0}, \boldsymbol{\Sigma} \otimes \mathbf{I}_T). \tag{4.4}$$

4.1.2 Likelihood Functions

Next we derive the likelihood functions implied by the two equivalent representations of the VAR(p), namely (4.2) and (4.3).

Using the first representation of the VAR(p) in (4.2), we have

$$(\mathbf{y} \mid \boldsymbol{\beta}, \boldsymbol{\Sigma}) \sim \mathcal{N}(\mathbf{X}\boldsymbol{\beta}, \mathbf{I}_T \otimes \boldsymbol{\Sigma}).$$

Therefore, the likelihood function is given by

$$\begin{aligned} p(\mathbf{y} \mid \boldsymbol{\beta}, \boldsymbol{\Sigma}) &= (2\pi)^{-\frac{Tn}{2}} |(\mathbf{I}_T \otimes \boldsymbol{\Sigma})|^{-\frac{1}{2}} \mathrm{e}^{-\frac{1}{2}(\mathbf{y}-\mathbf{X}\boldsymbol{\beta})'(\mathbf{I}_T \otimes \boldsymbol{\Sigma})^{-1}(\mathbf{y}-\mathbf{X}\boldsymbol{\beta})} \\ &= (2\pi)^{-\frac{Tn}{2}} |\boldsymbol{\Sigma}|^{-\frac{T}{2}} \mathrm{e}^{-\frac{1}{2}(\mathbf{y}-\mathbf{X}\boldsymbol{\beta})'(\mathbf{I}_T \otimes \boldsymbol{\Sigma}^{-1})(\mathbf{y}-\mathbf{X}\boldsymbol{\beta})}, \end{aligned} \tag{4.5}$$

where the second equality holds because $|\mathbf{I}_T \otimes \boldsymbol{\Sigma}| = |\boldsymbol{\Sigma}|^T$ and $(\mathbf{I}_T \otimes \boldsymbol{\Sigma})^{-1} = \mathbf{I}_T \otimes \boldsymbol{\Sigma}^{-1}$.

Since the two representations of the VAR(p) are equivalent, the likelihood implied by (4.3) should be the same as in (4.5). In what follows we rewrite (4.5) in terms of $\mathbf{Y}$, $\mathbf{Z}$, and $\mathbf{A}$. To do that, we need the following results: for conformable matrices $\mathbf{B}, \mathbf{C}, \mathbf{D}$, we have

$$\text{vec}(\mathbf{BCD}) = (\mathbf{D}' \otimes \mathbf{B})\text{vec}(\mathbf{C}), \tag{4.6}$$

$$\text{tr}(\mathbf{B}'\mathbf{C}) = \text{vec}(\mathbf{B})'\text{vec}(\mathbf{C}), \tag{4.7}$$

$$\text{tr}(\mathbf{BCD}) = \text{tr}(\mathbf{CDB}) = \text{tr}(\mathbf{DBC}), \tag{4.8}$$

where tr(·) is the trace function.

Noting that $\mathbf{y} - \mathbf{X}\boldsymbol{\beta} = \boldsymbol{\varepsilon} = \text{vec}(\mathbf{U}')$, we now rewrite the quadratic form in (4.5) as

$$\begin{aligned}(\mathbf{y} - \mathbf{X}\boldsymbol{\beta})'(\mathbf{I}_T \otimes \boldsymbol{\Sigma}^{-1})(\mathbf{y} - \mathbf{X}\boldsymbol{\beta}) &= \text{vec}(\mathbf{U}')'(\mathbf{I}_T \otimes \boldsymbol{\Sigma}^{-1})\text{vec}(\mathbf{U}') \\ &= \text{vec}(\mathbf{U}')'\text{vec}(\boldsymbol{\Sigma}^{-1}\mathbf{U}') \\ &= \text{tr}(\mathbf{U}\boldsymbol{\Sigma}^{-1}\mathbf{U}') \\ &= \text{tr}(\boldsymbol{\Sigma}^{-1}\mathbf{U}'\mathbf{U}),\end{aligned}$$

where the second equality holds because of (4.6); the third equality holds because of (4.7); and the last equality holds because of (4.8). Using this representation of the quadratic form and $\mathbf{U} = \mathbf{Y} - \mathbf{Z}\mathbf{A}$, the likelihood implied by the second representation in (4.3) is therefore given by

$$p(\mathbf{Y} \mid \mathbf{A}, \boldsymbol{\Sigma}) = (2\pi)^{-\frac{Tn}{2}} |\boldsymbol{\Sigma}|^{-\frac{T}{2}} \text{e}^{-\frac{1}{2}\text{tr}(\boldsymbol{\Sigma}^{-1}(\mathbf{Y}-\mathbf{Z}\mathbf{A})'(\mathbf{Y}-\mathbf{Z}\mathbf{A}))}. \tag{4.9}$$

4.2 Priors for Large Bayesian VARs

What makes Bayesian VARs Bayesian is the use of informative priors that incorporate non-data information. As mentioned in the introduction, VARs tend to have a lot of parameters, and large VARs exacerbate this problem. For example, a VAR(4) with $n = 20$ dependent variables has 1620 VAR coefficients, which is much larger than the number of observations in typical quarterly datasets. Without informative priors or regularization, it is not even possible to estimate the VAR coefficients.

In this section we discuss a range of informative priors that are found useful in the context of large VARs. One common feature of these priors is that they aim to "shrink" an unrestricted VAR to one that is parsimonious and seemingly reasonable. These priors differ in how they achieve this goal, and whether they lead to analytical results or simpler Markov chain Monte Carlo (MCMC) algorithms for estimating the posterior distributions. In addition, they also differ in how easily they can be applied to more flexible VARs, such as VARs with stochastic volatility.

4.2.1 The Minnesota Prior

Shrinkage priors in the context of small VARs were first developed by Doan et al. (1984) and Litterman (1986). Due to their affiliations with the University of Minnesota and the Federal Reserve Bank of Minneapolis at that time, this family of priors is commonly called Minnesota priors. It turns out that Minnesota priors can be directly applied to large VARs. This approach uses an approximation that leads to substantial simplifications in prior elicitation. Below we present a version discussed in Koop and Korobilis (2010).

To introduce the Minnesota priors, we use the first representation of the VAR with likelihood given in (4.5). Here the model parameters consist of two blocks: the VAR coefficients $\boldsymbol{\beta}$ and the error covariance matrix $\boldsymbol{\Sigma}$. Instead of estimating $\boldsymbol{\Sigma}$, the Minnesota prior replaces it with an estimate $\widehat{\boldsymbol{\Sigma}}$ obtained as follows.

We first estimate an AR(p) for each of the n variables separately. Let s_i^2 denote the sample variance of the residuals for the i-th equation. Then, we set $\widehat{\boldsymbol{\Sigma}} = \text{diag}(s_1^2, \ldots, s_n^2)$. As we will see below, the main advantage of this approach is that it simplifies the computations—often MCMC is not needed for posterior analysis or forecasts. One main drawback, however, is that here we replace an unknown quantity $\boldsymbol{\Sigma}$ by a potentially crude estimate $\widehat{\boldsymbol{\Sigma}}$. This approach therefore ignores parameter uncertainty—instead of tackling it by integrating out the unknown parameters with respect to the posterior distribution. As such, this approach often produces inferior density forecasts.

With $\boldsymbol{\Sigma}$ being replaced by an estimate, the only parameters left are the VAR coefficients $\boldsymbol{\beta}$. Now, consider the following normal prior for $\boldsymbol{\beta}$:

$$\boldsymbol{\beta} \sim \mathcal{N}(\boldsymbol{\beta}_{\text{Minn}}, \mathbf{V}_{\text{Minn}}).$$

The Minnesota prior sets sensible values for $\boldsymbol{\beta}_{\text{Minn}}$ and $\mathbf{V}_{\text{Minn}}$ in a systematic manner. To explain the prior elicitation procedure, first note that $\boldsymbol{\beta}$ consists of three groups of parameters: intercepts, coefficients associated with a variable's own lags, and coefficients associated with lags of other variables.

The prior mean $\boldsymbol{\beta}_{\text{Minn}}$ is typically set to zero for growth rates data, such as GDP growth rate or inflation rate. This prior mean provides shrinkage for VAR coefficients, and reflects the prior belief that growth rates data are typically not persistent. For levels data such as money supply or consumption level, $\boldsymbol{\beta}_{\text{Minn}}$ is set to be zero except the coefficients associated with the first own lag, which are set to be one. This prior incorporates the belief that levels data are highly persistent—particularly, it expresses the preference for a random walk specification. Other variants, such as specifying a highly persistent but stationary process, are also commonly used.

The Minnesota prior sets the prior covariance matrix $\mathbf{V}_{\text{Minn}}$ to be diagonal; the exact values of the diagonal elements in turn depend on three key hyperparameters, c_1, c_2, and c_3. Now consider the coefficients in the i-th equation. First, for a coefficient associated with the i-th variable's own lag l, $l = 1, \ldots, p$, its variance is set to be c_1/l^2. That is, the higher the lag length, the higher the degree of shrinkage (either to zero or to unity). Second, for a coefficient associated with the l-th lag of variable j, $j \neq i$, its variance is set to be $c_2 s_i^2/(l^2 s_j^2)$. In other words, in addition to applying higher level of shrinkage to higher lag length, the prior variance also adjusts for the scales of the variables. Lastly, the variance of the intercept is set to be c_3. The Minnesota prior therefore turns a complicated prior elicitation task into setting only three hyperparameters. There are by now many different variants of the

Minnesota prior; see, e.g., Kadiyala and Karlsson (1997) and Karlsson (2013) for additional discussion.

Estimation

Estimation under the Minnesota prior is straightforward; that is, one of the main appeals of the Minnesota prior. Recall that $\boldsymbol{\Sigma}$ is replaced by an estimate $\widehat{\boldsymbol{\Sigma}}$, and we only need to estimate $\boldsymbol{\beta}$. Given the VAR representation in (4.2) and the normal prior $\boldsymbol{\beta} \sim \mathcal{N}(\boldsymbol{\beta}_{\text{Minn}}, \mathbf{V}_{\text{Minn}})$, standard linear regression results give

$$(\boldsymbol{\beta} \,|\, \mathbf{y}) \sim \mathcal{N}(\widehat{\boldsymbol{\beta}}, \mathbf{K}_{\boldsymbol{\beta}}^{-1}),$$

where

$$\mathbf{K}_{\boldsymbol{\beta}} = \mathbf{V}_{\text{Minn}}^{-1} + \mathbf{X}'(\mathbf{I}_T \otimes \widehat{\boldsymbol{\Sigma}}^{-1})\mathbf{X}, \quad \widehat{\boldsymbol{\beta}} = \mathbf{K}_{\boldsymbol{\beta}}^{-1}\left(\mathbf{V}_{\text{Minn}}^{-1}\boldsymbol{\beta}_{\text{Minn}} + \mathbf{X}'(\mathbf{I}_T \otimes \widehat{\boldsymbol{\Sigma}}^{-1})\mathbf{y}\right),$$

and we have replaced $\boldsymbol{\Sigma}$ by the estimate $\widehat{\boldsymbol{\Sigma}}$. In particular, the posterior mean of $\boldsymbol{\beta}$ is $\widehat{\boldsymbol{\beta}}$, and we would only need to compute this once instead of tens of thousands of times within a Gibbs sampler.

When the number of variables n is large, however, computations might still be an issue because $\widehat{\boldsymbol{\beta}}$ is of dimension $nk \times 1$ with $k = np + 1$. In those cases, inverting the $nk \times nk$ precision matrix $\mathbf{K}_{\boldsymbol{\beta}}$ to obtain the covariance matrix $\mathbf{K}_{\boldsymbol{\beta}}^{-1}$ is computationally intensive. It turns out that to obtain $\widehat{\boldsymbol{\beta}}$, one need not compute the inverse $\mathbf{K}_{\boldsymbol{\beta}}^{-1}$ explicitly. To that end, we introduce the following notations: given a non-singular square matrix $\mathbf{B}$ and a conformable vector $\mathbf{c}$, let $\mathbf{B}\backslash\mathbf{c}$ denote the unique solution to the linear system $\mathbf{Bz} = \mathbf{c}$, i.e., $\mathbf{B}\backslash\mathbf{c} = \mathbf{B}^{-1}\mathbf{c}$. When $\mathbf{B}$ is lower triangular, this linear system can be solved quickly by forward substitution. When $\mathbf{B}$ is upper triangular, it can be solved by backward substitution.[1]

Now, we first compute the Cholesky factor $\mathbf{C}_{\mathbf{K}_{\boldsymbol{\beta}}}$ of $\mathbf{K}_{\boldsymbol{\beta}}$ such that $\mathbf{K}_{\boldsymbol{\beta}} = \mathbf{C}_{\mathbf{K}_{\boldsymbol{\beta}}}\mathbf{C}_{\mathbf{K}_{\boldsymbol{\beta}}}'$. Then, compute

$$\mathbf{C}_{\mathbf{K}_{\boldsymbol{\beta}}}'\backslash\left(\mathbf{C}_{\mathbf{K}_{\boldsymbol{\beta}}}\backslash(\mathbf{V}_{\text{Minn}}^{-1}\boldsymbol{\beta}_{\text{Minn}} + \mathbf{X}'(\mathbf{I}_T \otimes \widehat{\boldsymbol{\Sigma}}^{-1})\mathbf{y})\right)$$

by forward then backward substitution.[2] Then, by construction,

$$\begin{aligned}
&(\mathbf{C}_{\mathbf{K}_{\boldsymbol{\beta}}}')^{-1}\mathbf{C}_{\mathbf{K}_{\boldsymbol{\beta}}}^{-1}(\mathbf{V}_{\text{Minn}}^{-1}\boldsymbol{\beta}_{\text{Minn}} + \mathbf{X}'(\mathbf{I}_T \otimes \widehat{\boldsymbol{\Sigma}}^{-1})\mathbf{y}) \\
&\qquad = (\mathbf{C}_{\mathbf{K}_{\boldsymbol{\beta}}}\mathbf{C}_{\mathbf{K}_{\boldsymbol{\beta}}}')^{-1}(\mathbf{V}_{\text{Minn}}^{-1}\boldsymbol{\beta}_{\text{Minn}} + \mathbf{X}'(\mathbf{I}_T \otimes \widehat{\boldsymbol{\Sigma}}^{-1})\mathbf{y}) \\
&\qquad = \widehat{\boldsymbol{\beta}}.
\end{aligned}$$

This alternative way to obtain $\widehat{\boldsymbol{\beta}}$ is substantially faster when n is large.

[1]Forward and backward substitutions are implemented in standard packages such as MATLAB, GAUSS, and R. In MATLAB, for example, it is done by `mldivide(B,c)` or simply $\mathbf{B}\backslash\mathbf{c}$.

[2]Since $\mathbf{V}_{\text{Minn}}$ is diagonal, its inverse is straightforward to compute.

4.2.2 The Natural Conjugate Prior

The original Minnesota prior discussed in Sect. 4.2.1 replaces the error covariance matrix $\mathbf{\Sigma}$ with an estimate—and in doing so ignores parameter uncertainty associated with $\mathbf{\Sigma}$. That approach substantially simplifies the computations at the expense of the quality of density forecasts. In this section we introduce the natural conjugate prior for the VAR coefficients and $\mathbf{\Sigma}$. This prior retains much of the computational tractability of the Minnesota prior, but it explicitly treats $\mathbf{\Sigma}$ to be an unknown quantity to be estimated.

To introduce the natural conjugate prior, we use the second representation of the VAR with likelihood given in (4.9). Now the model parameters consist of two blocks: the error covariance matrix $\mathbf{\Sigma}$ as before and the VAR coefficients organized into the $k \times n$ matrix $\mathbf{A}$. The natural conjugate prior is a joint distribution for $(\text{vec}(\mathbf{A}), \mathbf{\Sigma})$. To describe its specific form, we first need to define the following distributions.

An $n \times n$ random matrix $\mathbf{\Omega}$ is said to have an **inverse-Wishart distribution** with shape parameter $\nu > 0$ and scale matrix $\mathbf{S}$ if its density function is given by

$$f(\mathbf{\Omega}; \nu, \mathbf{S}) = \frac{|\mathbf{S}|^{\nu/2}}{2^{n\nu/2}\Gamma_n(\nu/2)}|\mathbf{\Omega}|^{-\frac{\nu+n+1}{2}}\text{e}^{-\frac{1}{2}\text{tr}(\mathbf{S}\mathbf{\Omega}^{-1})},$$

where Γ_n is the multivariate gamma function. We write $\mathbf{\Omega} \sim \mathcal{IW}(\nu, \mathbf{S})$. For $\nu > m+1$, $\mathbb{E}\mathbf{\Omega} = \mathbf{S}/(\nu - m - 1)$.

Next, an $m \times n$ random matrix $\mathbf{W}$ and an $n \times n$ random matrix $\mathbf{\Omega}$ are said to have a **normal-inverse-Wishart distribution** with parameters $\mathbf{M}, \mathbf{P}, \mathbf{S}$, and ν if $(\text{vec}(\mathbf{W}) \,|\, \mathbf{\Omega}) \sim \mathcal{N}(\text{vec}(\mathbf{M}), \mathbf{\Omega} \otimes \mathbf{P})$ and $\mathbf{\Omega} \sim \mathcal{IW}(\nu, \mathbf{S})$. We write $(\mathbf{W}, \mathbf{\Omega}) \sim \mathcal{NIW}(\mathbf{M}, \mathbf{P}, \nu, \mathbf{S})$. The kernel of the normal-inverse-Wishart density function is given by

$$f(\mathbf{W}, \mathbf{\Omega}; \mathbf{M}, \mathbf{P}, \nu, \mathbf{S}) \propto |\mathbf{\Omega}|^{-\frac{\nu+m+n+1}{2}}\text{e}^{-\frac{1}{2}\text{tr}(\mathbf{\Omega}^{-1}(\mathbf{W}-\mathbf{M})'\mathbf{P}^{-1}(\mathbf{W}-\mathbf{M}))}\text{e}^{-\frac{1}{2}\text{tr}(\mathbf{\Omega}^{-1}\mathbf{S})}. \tag{4.10}$$

To derive this density function from the definition, first note that

$$\begin{aligned}
[\text{vec}(\mathbf{W}-\mathbf{M})]'(\mathbf{\Omega}\otimes\mathbf{P})^{-1}\text{vec}(\mathbf{W}-\mathbf{M}) &= [\text{vec}(\mathbf{W}-\mathbf{M})]'(\mathbf{\Omega}^{-1}\otimes\mathbf{P}^{-1})\text{vec}(\mathbf{W}-\mathbf{M}) \\
&= [\text{vec}(\mathbf{W}-\mathbf{M})]'\text{vec}(\mathbf{P}^{-1}(\mathbf{W}-\mathbf{M})\mathbf{\Omega}^{-1}) \\
&= \text{tr}((\mathbf{W}-\mathbf{M})'\mathbf{P}^{-1}(\mathbf{W}-\mathbf{M})\mathbf{\Omega}^{-1}) \\
&= \text{tr}(\mathbf{\Omega}^{-1}(\mathbf{W}-\mathbf{M})'\mathbf{P}^{-1}(\mathbf{W}-\mathbf{M})).
\end{aligned}$$

In the above derivations, the second equality holds because of (4.6); the third equality holds because of (4.7); and the last equality holds because of (4.8).

Now, from the definition, the joint density function of $(\text{vec}(\mathbf{W}), \boldsymbol{\Omega})$ is given by

$$f(\mathbf{W}, \boldsymbol{\Omega}) \propto |\boldsymbol{\Omega}|^{-\frac{\nu+n+1}{2}} \mathrm{e}^{-\frac{1}{2}\mathrm{tr}(\mathbf{S}\boldsymbol{\Omega}^{-1})} \times |\boldsymbol{\Omega} \otimes \mathbf{P}|^{-\frac{1}{2}} \mathrm{e}^{-\frac{1}{2}[\text{vec}(\mathbf{W}-\mathbf{M})]'(\boldsymbol{\Omega}\otimes\mathbf{P})^{-1}\text{vec}(\mathbf{W}-\mathbf{M})}$$
$$= |\boldsymbol{\Omega}|^{-\frac{\nu+m+n+1}{2}} \mathrm{e}^{-\frac{1}{2}\mathrm{tr}(\boldsymbol{\Omega}^{-1}\mathbf{S})} \mathrm{e}^{-\frac{1}{2}\mathrm{tr}(\boldsymbol{\Omega}^{-1}(\mathbf{W}-\mathbf{M})'\mathbf{P}^{-1}(\mathbf{W}-\mathbf{M}))},$$

where we have used the fact that $|\boldsymbol{\Omega} \otimes \mathbf{P}| = |\boldsymbol{\Omega}|^m |\mathbf{P}|^n$. This proves that the joint density function of $(\text{vec}(\mathbf{W}), \boldsymbol{\Omega})$ has the form given in (4.10).

By construction, the marginal distribution of $\boldsymbol{\Omega}$ is $\mathcal{IW}(\nu, \mathbf{S})$. It turns out that the marginal distribution of $\text{vec}(\mathbf{W})$ unconditional on $\boldsymbol{\Omega}$ is a multivariate t distribution. For more details, see, e.g., Karlsson (2013).

Now we consider the following normal-inverse-Wishart prior on $(\mathbf{A}, \boldsymbol{\Sigma})$:

$$\boldsymbol{\Sigma} \sim \mathcal{IW}(\nu_0, \mathbf{S}_0), \quad (\text{vec}(\mathbf{A}) \mid \boldsymbol{\Sigma}) \sim \mathcal{N}(\text{vec}(\mathbf{A}_0), \boldsymbol{\Sigma} \otimes \mathbf{V}_{\mathbf{A}}).$$

That is, $(\text{vec}(\mathbf{A}), \boldsymbol{\Sigma}) \sim \mathcal{NIW}(\text{vec}(\mathbf{A}_0), \mathbf{V}_{\mathbf{A}}, \nu_0, \mathbf{S}_0)$ with joint density function

$$p(\mathbf{A}, \boldsymbol{\Sigma}) \propto |\boldsymbol{\Sigma}|^{-\frac{\nu_0+n+k+1}{2}} \mathrm{e}^{-\frac{1}{2}\mathrm{tr}(\boldsymbol{\Sigma}^{-1}\mathbf{S}_0)} \mathrm{e}^{-\frac{1}{2}\mathrm{tr}\left(\boldsymbol{\Sigma}^{-1}(\mathbf{A}-\mathbf{A}_0)'\mathbf{V}_{\mathbf{A}}^{-1}(\mathbf{A}-\mathbf{A}_0)\right)}. \tag{4.11}$$

It turns out that the joint posterior distribution of $(\mathbf{A}, \boldsymbol{\Sigma})$ is also a normal-inverse-Wishart distribution, as shown in the next section. Hence, this prior is often called the natural conjugate prior. The hyperparameters of this prior are $\text{vec}(\mathbf{A}_0), \mathbf{V}_{\mathbf{A}}, \nu_0$, and $\mathbf{S}_0$. Below we describe one way to elicit these hyperparameters.

One often sets a small value for ν_0 (say, $n + 2$) so that the prior variance of $\boldsymbol{\Sigma}$ is large—i.e., the prior is relatively uninformative. Given ν_0, one then chooses a value for $\mathbf{S}_0$ to match the desired prior mean of $\boldsymbol{\Sigma}$ via the equality $\mathbb{E}\boldsymbol{\Sigma} = \mathbf{S}_0/(\nu_0 - n - 1)$. As for $\text{vec}(\mathbf{A}_0)$ and $\mathbf{V}_{\mathbf{A}}$, their values are chosen to mimic the Minnesota prior. For example, $\text{vec}(\mathbf{A}_0)$ is typically set to zero for growth rates data. For levels data, $\text{vec}(\mathbf{A}_0)$ is set to be zero except the coefficients associated with the first own lag, which are set to be one.

Finally, to elicit $\mathbf{V}_{\mathbf{A}}$, first note that given $\boldsymbol{\Sigma}$, the prior covariance matrix of $\text{vec}(\mathbf{A})$ is $\boldsymbol{\Sigma} \otimes \mathbf{V}_{\mathbf{A}}$. This Kronecker structure implies cross-equation restrictions on the covariance matrix, which is more restrictive than the covariance matrix $\mathbf{V}_{\text{Minn}}$ under the Minnesota prior. However, the advantage of this Kronecker structure is that it can be exploited to speed up computations, which we will discuss in the next section.

Following the example of the Minnesota prior, we choose $\mathbf{V}_{\mathbf{A}}$ to induce shrinkage. Specifically, $\mathbf{V}_{\mathbf{A}}$ is assumed to be diagonal with diagonal elements $v_{\mathbf{A},ii} = c_1/(l^2 s_r^2)$ for a coefficient associated with the l-th lag of variable r and $v_{\mathbf{A},ii} = c_2$ for an intercept, where s_r^2 is the residual sample variance of an AR(p) model for the variable r. Similar to the Minnesota prior, we apply a higher degree of shrinkage for a coefficient associated with a higher lag length. But contrary to the Minnesota prior, here we cannot have different prior variances for a variable's own lag and the lag of a different variable due to the Kronecker structure.

Estimation

In this section we discuss the estimation of $\mathbf{A}$ and $\boldsymbol{\Sigma}$ under the natural conjugate prior. As mentioned earlier, the posterior distribution of $\mathbf{A}$ and $\boldsymbol{\Sigma}$ turns out to be the normal-inverse-Wishart distribution as well. To see this, we combine the likelihood given in (4.9) and the natural conjugate prior in (4.11) to get

$$\begin{aligned} p(\mathbf{A},\boldsymbol{\Sigma}\,|\,\mathbf{Y}) &\propto p(\mathbf{A},\boldsymbol{\Sigma})p(\mathbf{Y}\,|\,\mathbf{A},\boldsymbol{\Sigma}) \\ &\propto |\boldsymbol{\Sigma}|^{-\frac{\nu_0+n+k+1}{2}}\mathrm{e}^{-\frac{1}{2}\mathrm{tr}(\boldsymbol{\Sigma}^{-1}\mathbf{S}_0)}\mathrm{e}^{-\frac{1}{2}\mathrm{tr}\left(\boldsymbol{\Sigma}^{-1}(\mathbf{A}-\mathbf{A}_0)'\mathbf{V}_{\mathbf{A}}^{-1}(\mathbf{A}-\mathbf{A}_0)\right)} \\ &\quad\times|\boldsymbol{\Sigma}|^{-\frac{T}{2}}\mathrm{e}^{-\frac{1}{2}\mathrm{tr}(\boldsymbol{\Sigma}^{-1}(\mathbf{Y}-\mathbf{Z}\mathbf{A})'(\mathbf{Y}-\mathbf{Z}\mathbf{A}))} \\ &\propto |\boldsymbol{\Sigma}|^{-\frac{\nu_0+n+k+T+1}{2}}\mathrm{e}^{-\frac{1}{2}\mathrm{tr}(\boldsymbol{\Sigma}^{-1}\mathbf{S}_0)}\mathrm{e}^{-\frac{1}{2}\mathrm{tr}\left[\boldsymbol{\Sigma}^{-1}\left((\mathbf{A}-\mathbf{A}_0)'\mathbf{V}_{\mathbf{A}}^{-1}(\mathbf{A}-\mathbf{A}_0)+(\mathbf{Y}-\mathbf{Z}\mathbf{A})'(\mathbf{Y}-\mathbf{Z}\mathbf{A})\right)\right]}. \end{aligned} \tag{4.12}$$

The last line looks almost like the kernel of the normal-inverse-Wishart density function in (4.10)—the only difference is that here we have two quadratic terms involving $\mathbf{A}$ instead of one. If we could somehow write the sum

$$(\mathbf{A}-\mathbf{A}_0)'\mathbf{V}_{\mathbf{A}}^{-1}(\mathbf{A}-\mathbf{A}_0)+(\mathbf{Y}-\mathbf{Z}\mathbf{A})'(\mathbf{Y}-\mathbf{Z}\mathbf{A})$$

as $(\mathbf{A}-\widehat{\mathbf{A}})'\mathbf{K}_{\mathbf{A}}(\mathbf{A}-\widehat{\mathbf{A}})$ for some $k\times n$ matrix $\widehat{\mathbf{A}}$ and $k\times k$ symmetric matrix $\mathbf{K}_{\mathbf{A}}$, then $p(\mathbf{A},\boldsymbol{\Sigma}\,|\,\mathbf{Y})$ is a normal-inverse-Wishart density function.

To that end, below we do a matrix version of "completing the square":

$$\begin{aligned} &(\mathbf{A}-\mathbf{A}_0)'\mathbf{V}_{\mathbf{A}}^{-1}(\mathbf{A}-\mathbf{A}_0)+(\mathbf{Y}-\mathbf{Z}\mathbf{A})'(\mathbf{Y}-\mathbf{Z}\mathbf{A}) \\ &\quad=(\mathbf{A}'\mathbf{V}_{\mathbf{A}}^{-1}\mathbf{A}-2\mathbf{A}'\mathbf{V}_{\mathbf{A}}^{-1}\mathbf{A}_0+\mathbf{A}_0'\mathbf{V}_{\mathbf{A}}^{-1}\mathbf{A}_0)+(\mathbf{A}'\mathbf{Z}'\mathbf{Z}\mathbf{A}-2\mathbf{A}'\mathbf{Z}'\mathbf{Y}+\mathbf{Y}'\mathbf{Y}) \\ &\quad=\mathbf{A}'(\mathbf{V}_{\mathbf{A}}^{-1}+\mathbf{Z}'\mathbf{Z})\mathbf{A}-2\mathbf{A}'(\mathbf{V}_{\mathbf{A}}^{-1}\mathbf{A}_0+\mathbf{Z}'\mathbf{Y})+\widehat{\mathbf{A}}'\mathbf{K}_{\mathbf{A}}\widehat{\mathbf{A}}-\widehat{\mathbf{A}}'\mathbf{K}_{\mathbf{A}}\widehat{\mathbf{A}}+\mathbf{A}_0'\mathbf{V}_{\mathbf{A}}^{-1}\mathbf{A}_0+\mathbf{Y}'\mathbf{Y} \\ &\quad=(\mathbf{A}-\widehat{\mathbf{A}})'\mathbf{K}_{\mathbf{A}}(\mathbf{A}-\widehat{\mathbf{A}})-\widehat{\mathbf{A}}'\mathbf{K}_{\mathbf{A}}\widehat{\mathbf{A}}+\mathbf{A}_0'\mathbf{V}_{\mathbf{A}}^{-1}\mathbf{A}_0+\mathbf{Y}'\mathbf{Y}, \end{aligned} \tag{4.13}$$

where

$$\mathbf{K}_{\mathbf{A}}=\mathbf{V}_{\mathbf{A}}^{-1}+\mathbf{Z}'\mathbf{Z},\quad \widehat{\mathbf{A}}=\mathbf{K}_{\mathbf{A}}^{-1}(\mathbf{V}_{\mathbf{A}}^{-1}\mathbf{A}_0+\mathbf{Z}'\mathbf{Y}).$$

Note that on the right-hand-side of the second equality, we judiciously add and subtract the term $\widehat{\mathbf{A}}'\mathbf{K}_{\mathbf{A}}\widehat{\mathbf{A}}$ so that we obtain one quadratic form in $\mathbf{A}$.

Now, substituting (4.13) into (4.12), we have

$$\begin{aligned} p(\mathbf{A},\boldsymbol{\Sigma}\,|\,\mathbf{Y}) &\propto |\boldsymbol{\Sigma}|^{-\frac{\nu_0+n+k+T+1}{2}}\mathrm{e}^{-\frac{1}{2}\mathrm{tr}(\boldsymbol{\Sigma}^{-1}\mathbf{S}_0)}\mathrm{e}^{-\frac{1}{2}\mathrm{tr}\left[\boldsymbol{\Sigma}^{-1}\left((\mathbf{A}-\widehat{\mathbf{A}})'\mathbf{K}_{\mathbf{A}}(\mathbf{A}-\widehat{\mathbf{A}})-\widehat{\mathbf{A}}'\mathbf{K}_{\mathbf{A}}\widehat{\mathbf{A}}+\mathbf{A}_0'\mathbf{V}_{\mathbf{A}}^{-1}\mathbf{A}_0+\mathbf{Y}'\mathbf{Y}\right)\right]} \\ &= |\boldsymbol{\Sigma}|^{-\frac{\nu_0+n+k+T+1}{2}}\mathrm{e}^{-\frac{1}{2}\mathrm{tr}(\boldsymbol{\Sigma}^{-1}\widehat{\mathbf{S}})}\mathrm{e}^{-\frac{1}{2}\mathrm{tr}\left[\boldsymbol{\Sigma}^{-1}(\mathbf{A}-\widehat{\mathbf{A}})'\mathbf{K}_{\mathbf{A}}(\mathbf{A}-\widehat{\mathbf{A}})\right]}, \end{aligned}$$

where $\widehat{\mathbf{S}} = \mathbf{S}_0 + \mathbf{A}_0'\mathbf{V}_{\mathbf{A}}^{-1}\mathbf{A}_0 + \mathbf{Y}'\mathbf{Y} - \widehat{\mathbf{A}}'\mathbf{K}_{\mathbf{A}}\widehat{\mathbf{A}}$. Comparing this kernel with the normal-inverse-gamma density function in (4.10), we conclude that

$$(\mathbf{A}, \boldsymbol{\Sigma} \,|\, \mathbf{Y}) \sim \mathcal{NIW}(\widehat{\mathbf{A}}, \mathbf{K}_{\mathbf{A}}^{-1}, \nu_0 + T, \widehat{\mathbf{S}}).$$

In particular, the posterior means of $\mathbf{A}$ and $\boldsymbol{\Sigma}$ are respectively $\widehat{\mathbf{A}}$ and $\widehat{\mathbf{S}}/(\nu_0 + T - 1)$. Other posterior moments can often be found by using properties of the normal-inverse-Wishart distribution. When analytical results are not available, we can estimate the quantities of interest by generating draws from the posterior distribution $p(\mathbf{A}, \boldsymbol{\Sigma} \,|\, \mathbf{Y})$. Below we describe a computationally efficient way to obtain posterior draws.

Since $(\mathbf{A}, \boldsymbol{\Sigma} \,|\, \mathbf{Y}) \sim \mathcal{NIW}(\widehat{\mathbf{A}}, \mathbf{K}_{\mathbf{A}}^{-1}, \nu_0 + T, \widehat{\mathbf{S}})$, we can sample $\mathbf{A}$ and $\boldsymbol{\Sigma}$ in two steps. First, we draw $\boldsymbol{\Sigma}$ marginally from $(\boldsymbol{\Sigma} \,|\, \mathbf{Y}) \sim \mathcal{IW}(\nu_0 + T, \widehat{\mathbf{S}})$. Then, given the sampled $\boldsymbol{\Sigma}$, we simulate from the conditional distribution

$$(\text{vec}(\mathbf{A}) \,|\, \mathbf{Y}, \boldsymbol{\Sigma}) \sim \mathcal{N}(\text{vec}(\widehat{\mathbf{A}}), \boldsymbol{\Sigma} \otimes \mathbf{K}_{\mathbf{A}}^{-1}).$$

Here note that the covariance matrix $\boldsymbol{\Sigma} \otimes \mathbf{K}_{\mathbf{A}}^{-1}$ is of dimension $nk = n(np + 1)$, and sampling from this normal distribution using conventional methods—e.g., computing the Cholesky factor of the covariance matrix $\boldsymbol{\Sigma} \otimes \mathbf{K}_{\mathbf{A}}^{-1}$—would involve $\mathcal{O}(n^6)$ operations. This is especially computationally intensive when n is large. Here we consider an alternative method with complexity of the order $\mathcal{O}(n^3)$ only.

This more efficient approach exploits the Kronecker structure $\boldsymbol{\Sigma} \otimes \mathbf{K}_{\mathbf{A}}^{-1}$ to speed up computation. In particular, it is based on an efficient sampling algorithm to draw from the matrix normal distribution.[3] We further improve upon this approach by avoiding the computation of the inverse of the $k \times k$ matrix $\mathbf{K}_{\mathbf{A}}$.

Recall that given a non-singular square matrix $\mathbf{B}$ and a conformable vector $\mathbf{c}$, we use the notation $\mathbf{B}\backslash\mathbf{c}$ to denote the unique solution to the linear system $\mathbf{Bz} = \mathbf{c}$, i.e., $\mathbf{B}\backslash\mathbf{c} = \mathbf{B}^{-1}\mathbf{c}$. Now, we first obtain the Cholesky decomposition $\mathbf{C}_{\mathbf{K}_{\mathbf{A}}}$ of $\mathbf{K}_{\mathbf{A}}$ such that $\mathbf{C}_{\mathbf{K}_{\mathbf{A}}}\mathbf{C}_{\mathbf{K}_{\mathbf{A}}}' = \mathbf{K}_{\mathbf{A}}$. Then compute

$$\mathbf{C}_{\mathbf{K}_{\mathbf{A}}}'\backslash(\mathbf{C}_{\mathbf{K}_{\mathbf{A}}}\backslash(\mathbf{V}_{\mathbf{A}}^{-1}\mathbf{A}_0 + \mathbf{Z}'\mathbf{Y}))$$

by forward followed by backward substitution. By construction,

$$(\mathbf{C}_{\mathbf{K}_{\mathbf{A}}}')^{-1}(\mathbf{C}_{\mathbf{K}_{\mathbf{A}}}^{-1}(\mathbf{V}_{\mathbf{A}}^{-1}\mathbf{A}_0 + \mathbf{Z}'\mathbf{Y})) = (\mathbf{C}_{\mathbf{K}_{\mathbf{A}}}'\mathbf{C}_{\mathbf{K}_{\mathbf{A}}})^{-1}(\mathbf{V}_{\mathbf{A}}^{-1}\mathbf{A}_0 + \mathbf{Z}'\mathbf{Y}) = \widehat{\mathbf{A}}.$$

[3]The algorithm of drawing from the matrix normal distribution is well-known, and is described in the textbook by Bauwens, Lubrano, and Richard (1999, p.320). This algorithm is adapted in Carriero et al. (2016) and Chan (2018) to estimate more flexible large Bayesian VARs.

Next, let $\mathbf{C}_{\mathbf{\Sigma}}$ be the Cholesky decomposition of $\mathbf{\Sigma}$. Then, compute

$$\mathbf{W}_1 = \widehat{\mathbf{A}} + (\mathbf{C}'_{\mathbf{K}_\mathbf{A}} \backslash \mathbf{U})\mathbf{C}'_{\mathbf{\Sigma}},$$

where $\mathbf{U}$ is a $k \times n$ matrix of independent $\mathcal{N}(0, 1)$ random variables. In the Appendix B we show that $\text{vec}(\mathbf{W}_1) \sim \mathcal{N}(\text{vec}(\widehat{\mathbf{A}}), \mathbf{\Sigma} \otimes \mathbf{K}_\mathbf{A}^{-1})$ as desired.

Therefore, we have a computationally efficient way to sample from the posterior distribution $(\mathbf{A}, \mathbf{\Sigma} \mid \mathbf{Y}) \sim \mathcal{NIW}(\widehat{\mathbf{A}}, \mathbf{K}_\mathbf{A}^{-1}, \nu_0 + T, \widehat{\mathbf{S}})$. Note also that the algorithm described above gives us an independent sample—unlike MCMC draws which are correlated by construction.

4.2.3 The Independent Normal and Inverse-Wishart Prior

The main advantage of the natural conjugate prior is that analytical results are available for posterior analysis and simulation is typically not needed. However, it comes at a cost of restricting the form of prior variances on the VAR coefficients. In this section we discuss an alternative joint prior for the VAR coefficients and covariance matrix that is more flexible.

To that end, we use the first representation of the VAR with likelihood given in (4.5). This joint prior on $(\boldsymbol{\beta}, \mathbf{\Sigma})$ is often called the independent normal and inverse-Wishart prior, because it assumes prior independence between $\boldsymbol{\beta}$ and $\mathbf{\Sigma}$, i.e., $p(\boldsymbol{\beta}, \mathbf{\Sigma}) = p(\boldsymbol{\beta})p(\mathbf{\Sigma})$. More specifically, we consider the form

$$\boldsymbol{\beta} \sim \mathcal{N}(\boldsymbol{\beta}_0, \mathbf{V}_{\boldsymbol{\beta}}), \quad \mathbf{\Sigma} \sim \mathcal{IW}(\nu_0, \mathbf{S}_0)$$

with prior densities

$$p(\boldsymbol{\beta}) = (2\pi)^{-\frac{nk}{2}} |\mathbf{V}_{\boldsymbol{\beta}}|^{-\frac{1}{2}} \mathrm{e}^{-\frac{1}{2}(\boldsymbol{\beta}-\boldsymbol{\beta}_0)'\mathbf{V}_{\boldsymbol{\beta}}^{-1}(\boldsymbol{\beta}-\boldsymbol{\beta}_0)}, \tag{4.14}$$

$$p(\mathbf{\Sigma}) = \frac{|\mathbf{S}_0|^{\nu_0/2}}{2^{n\nu_0/2}\Gamma_n(\nu_0/2)} |\mathbf{\Sigma}|^{-\frac{\nu_0+n+1}{2}} \mathrm{e}^{-\frac{1}{2}\text{tr}(\mathbf{S}_0\mathbf{\Sigma}^{-1})}. \tag{4.15}$$

The hyperparameters of this prior are $\boldsymbol{\beta}, \mathbf{V}_{\boldsymbol{\beta}}, \nu_0$, and $\mathbf{S}_0$. The values for ν_0 and $\mathbf{S}_0$ can be chosen the same way as in the case of the natural conjugate prior. For $\boldsymbol{\beta}_0$ and $\mathbf{V}_{\boldsymbol{\beta}}$, we can set them to be the same as the Minnesota prior, i.e., $\boldsymbol{\beta}_0 = \boldsymbol{\beta}_{\text{Minn}}$ and $\mathbf{V}_{\boldsymbol{\beta}} = \mathbf{V}_{\text{Minn}}$. Also note that in contrast to the natural conjugate prior, here $\mathbf{V}_{\boldsymbol{\beta}}$, the prior covariance matrix of the VAR coefficients, is not required to have a Kronecker structure, and is therefore more flexible.

Estimation

As mentioned above, in contrast to the case of the natural conjugate prior, the posterior distribution under the independent normal and inverse-Wishart prior is non-standard, and posterior simulation is needed for estimation and forecasting.

Below we derive a Gibbs sampler to draw from the posterior distribution $p(\boldsymbol{\beta}, \boldsymbol{\Sigma} \,|\, \mathbf{y})$. To that end, we derive the two full conditional distributions $p(\boldsymbol{\beta} \,|\, \mathbf{y}, \boldsymbol{\Sigma})$ and $p(\boldsymbol{\Sigma} \,|\, \mathbf{y}, \boldsymbol{\beta})$.

Using the likelihood given in (4.5) and the prior on $\boldsymbol{\beta}$ given in (4.14), we note that standard linear regression results would apply. In fact, we have

$$(\boldsymbol{\beta} \,|\, \mathbf{y}, \boldsymbol{\Sigma}) \sim \mathcal{N}(\widehat{\boldsymbol{\beta}}, \mathbf{K}_{\boldsymbol{\beta}}^{-1}),$$

where

$$\mathbf{K}_{\boldsymbol{\beta}} = \mathbf{V}_{\boldsymbol{\beta}}^{-1} + \mathbf{X}'(\mathbf{I}_T \otimes \boldsymbol{\Sigma}^{-1})\mathbf{X}, \quad \widehat{\boldsymbol{\beta}} = \mathbf{K}_{\boldsymbol{\beta}}^{-1}\left(\mathbf{V}_{\boldsymbol{\beta}}^{-1}\boldsymbol{\beta}_0 + \mathbf{X}'(\mathbf{I}_T \otimes \boldsymbol{\Sigma}^{-1})\mathbf{y}\right).$$

The main difficulty of obtaining a draw from $\mathcal{N}(\widehat{\boldsymbol{\beta}}, \mathbf{K}_{\boldsymbol{\beta}}^{-1})$ using conventional methods is that computing the $n(np+1) \times n(np+1)$ inverse $\mathbf{K}_{\boldsymbol{\beta}}^{-1}$ is very computationally intensive when n is large. But fortunately we can sample from $\mathcal{N}(\widehat{\boldsymbol{\beta}}, \mathbf{K}_{\boldsymbol{\beta}}^{-1})$ without computing $\mathbf{K}_{\boldsymbol{\beta}}^{-1}$ explicitly. First, $\widehat{\boldsymbol{\beta}}$ can be obtained by forward and backward substitution as before. Second, we can use an alternative algorithm to sample from $\mathcal{N}(\widehat{\boldsymbol{\beta}}, \mathbf{K}_{\boldsymbol{\beta}}^{-1})$ without computing $\mathbf{K}_{\boldsymbol{\beta}}^{-1}$ explicitly.

Algorithm 4.1 **(Sampling from the Normal Distribution Given the Precision Matrix)** To generate R independent draws from $\mathcal{N}(\boldsymbol{\mu}, \mathbf{K}^{-1})$ of dimension m, carry out the following steps:

1. Compute the lower Cholesky factor $\mathbf{B}$ of $\mathbf{K}$ such that $\mathbf{K} = \mathbf{B}\mathbf{B}'$.
2. Generate $\mathbf{U} = (U_1, \ldots, U_m)'$ by drawing $U_1, \ldots, U_m \sim \mathcal{N}(0, 1)$.
3. Return $\mathbf{W} = \boldsymbol{\mu} + (\mathbf{B}')^{-1}\mathbf{U}$.
4. Repeat Steps 2 and 3 independently R times.

To check that $\mathbf{W} \sim \mathcal{N}(\boldsymbol{\mu}, \mathbf{K}^{-1})$, we first note that $\mathbf{W}$ is an affine transformation of the normal random vector $\mathbf{U}$, so it has a normal distribution. It is easy to check that $\mathbb{E}\mathbf{W} = \boldsymbol{\mu}$. The covariance matrix of $\mathbf{W}$ is

$$\mathrm{Cov}(\mathbf{W}) = (\mathbf{B}')^{-1}\mathrm{Cov}(\mathbf{U})((\mathbf{B}')^{-1})' = (\mathbf{B}')^{-1}(\mathbf{B})^{-1} = (\mathbf{B}\mathbf{B}')^{-1} = \mathbf{K}^{-1}.$$

Hence, $\mathbf{W} \sim \mathcal{N}(\boldsymbol{\mu}, \mathbf{K}^{-1})$.

Using this algorithm to sample from $\mathcal{N}(\widehat{\boldsymbol{\beta}}, \mathbf{K}_{\boldsymbol{\beta}}^{-1})$ allows us to avoid the expensive computation of the inverse $\mathbf{K}_{\boldsymbol{\beta}}^{-1}$. However, if $\mathbf{K}_{\boldsymbol{\beta}}$ is a dense matrix, this algorithm still involves $\mathcal{O}(n^6)$ operations. Hence, it is expected to be much slower than simulations under the natural conjugate prior that involves only $\mathcal{O}(n^3)$ operations.

Next, we derive the conditional distribution $p(\boldsymbol{\Sigma} \,|\, \mathbf{y}, \boldsymbol{\beta})$. First note that the likelihood in (4.5) can be equivalently written as

$$p(\mathbf{y} \,|\, \boldsymbol{\beta}, \boldsymbol{\Sigma}) = (2\pi)^{-\frac{Tn}{2}} |\boldsymbol{\Sigma}|^{-\frac{T}{2}} \mathrm{e}^{-\frac{1}{2}\sum_{t=1}^{T}(\mathbf{y}_t - \mathbf{X}_t\boldsymbol{\beta})'\boldsymbol{\Sigma}^{-1}(\mathbf{y}_t - \mathbf{X}_t\boldsymbol{\beta})}. \tag{4.16}$$

Now, combining (4.16) and the prior on $\boldsymbol{\Sigma}$ given in (4.15), we have

$$\begin{aligned}
p(\boldsymbol{\Sigma} \mid \mathbf{y}, \boldsymbol{\beta}) &\propto p(\mathbf{y} \mid \boldsymbol{\beta}, \boldsymbol{\Sigma}) p(\boldsymbol{\Sigma}) \\
&\propto |\boldsymbol{\Sigma}|^{-\frac{T}{2}} \mathrm{e}^{-\frac{1}{2}\sum_{t=1}^{T}(\mathbf{y}_t-\mathbf{X}_t\boldsymbol{\beta})'\boldsymbol{\Sigma}^{-1}(\mathbf{y}_t-\mathbf{X}_t\boldsymbol{\beta})} \times |\boldsymbol{\Sigma}|^{-\frac{\nu_0+n+1}{2}} \mathrm{e}^{-\frac{1}{2}\mathrm{tr}(\mathbf{S}_0\boldsymbol{\Sigma}^{-1})} \\
&= |\boldsymbol{\Sigma}|^{-\frac{\nu_0+n+T+1}{2}} \mathrm{e}^{-\frac{1}{2}\mathrm{tr}(\mathbf{S}_0\boldsymbol{\Sigma}^{-1})} \mathrm{e}^{-\frac{1}{2}\mathrm{tr}\left[\sum_{t=1}^{T}(\mathbf{y}_t-\mathbf{X}_t\boldsymbol{\beta})(\mathbf{y}_t-\mathbf{X}_t\boldsymbol{\beta})'\boldsymbol{\Sigma}^{-1}\right]} \\
&= |\boldsymbol{\Sigma}|^{-\frac{\nu_0+n+T+1}{2}} \mathrm{e}^{-\frac{1}{2}\mathrm{tr}\left[\left(\mathbf{S}_0+\sum_{t=1}^{T}(\mathbf{y}_t-\mathbf{X}_t\boldsymbol{\beta})(\mathbf{y}_t-\mathbf{X}_t\boldsymbol{\beta})'\right)\boldsymbol{\Sigma}^{-1}\right]},
\end{aligned}$$

which is the kernel of an inverse-Wishart density function. In fact, we have

$$(\boldsymbol{\Sigma} \mid \mathbf{y}, \boldsymbol{\beta}) \sim \mathcal{IW}\left(\nu_0 + T, \mathbf{S}_0 + \sum_{t=1}^{T}(\mathbf{y}_t - \mathbf{X}_t\boldsymbol{\beta})(\mathbf{y}_t - \mathbf{X}_t\boldsymbol{\beta})'\right). \tag{4.17}$$

Hence, a Gibbs sampler can be constructed to simulate from the posterior distribution by repeatedly drawing from $p(\boldsymbol{\beta} \mid \mathbf{y}, \boldsymbol{\Sigma})$ and $p(\boldsymbol{\Sigma} \mid \mathbf{y}, \boldsymbol{\beta})$.

4.2.4 The Stochastic Search Variable Selection Prior

Another popular shrinkage prior for the VAR coefficients is the so-called stochastic search variable selection (SSVS) prior considered in George, Sun, and Ni (2008). It is based on the independent normal and inverse-Wishart prior, but it introduces a hierarchical structure for the normal prior on $\boldsymbol{\beta}$. The main idea is to divide, in a data-based manner, the elements in $\boldsymbol{\beta}$ into two groups: in the first group the coefficients are shrunk strongly to zero, whereas they are not shrunk in the second group. In other words, the "variable selection" part is done by setting the coefficients in the first group to be close to zero, and only the variables in the second group are "selected." This partition is done stochastically in each iteration in the MCMC sampler, and hence "stochastic search."

Specifically, the elements of $\boldsymbol{\beta}$ are assumed to be independent, and each element β_j has a two-component mixture distribution with mixture weight $q_j \in (0, 1)$:

$$(\beta_j \mid q_j) \sim (1 - q_j)\phi(\beta_j; 0, \kappa_{0j}) + q_j\phi(\beta_j; 0, \kappa_{1j}),$$

where $\phi(\cdot; \mu, \sigma^2)$ denotes the density function of the $\mathcal{N}(\mu, \sigma^2)$ distribution. The SSVS prior sets the first prior variance κ_{0j} to be "small" and the second prior variance κ_{1j} to be large.

To see the partition more clearly, let us consider an equivalent latent variable representation by introducing the indicator $\gamma_j \in \{0, 1\}$ with success probability q_j, i.e., $\mathbb{P}(\gamma_j = 1 \,|\, q_j) = q_j$. Then, we can rewrite the above prior as

$$(\beta_j \,|\, \gamma_j) \sim (1-\gamma_j)\mathcal{N}(0, \kappa_{0j}) + \gamma_j \mathcal{N}(0, \kappa_{1j}).$$

Hence, when $\gamma_j = 0$, β_j is strongly shrunk to zero; when $\gamma_j = 1$, the prior on β_j is relatively non-informative.

Let $\boldsymbol{\gamma} = (\gamma_1, \ldots, \gamma_{nk})'$. For later reference, we rewrite the joint prior $(\boldsymbol{\beta} \,|\, \boldsymbol{\gamma})$ as:

$$(\boldsymbol{\beta} \,|\, \boldsymbol{\gamma}) \sim \mathcal{N}(\mathbf{0}, \boldsymbol{\Omega}_{\gamma}),$$

where $\boldsymbol{\Omega}_{\gamma}$ is diagonal with diagonal elements $(1-\gamma_j)\kappa_{0j} + \gamma_j\kappa_{1j}$, $j = 1, \ldots, nk$.

It remains to choose values for the prior variances κ_{0j} and κ_{1j}. There are various implementations, here we simply set $\kappa_{1j} = 10$ and κ_{0j} to be the j-th diagonal element of the Minnesota prior covariance matrix $\mathbf{V}_{\text{Minn}}$. As for $\boldsymbol{\Sigma}$, we assume the inverse-Wishart prior:

$$\boldsymbol{\Sigma} \sim \mathcal{IW}(\nu_0, \mathbf{S}_0).$$

It is possible to have a SSVS prior on $\boldsymbol{\Sigma}$ as well. See George et al. (2008) for further details. Finally, we set the mixture weight q_j to be 0.5, so that β_j has equal probabilities in each component. An alternative is to treat q_j as a model parameter to be estimated.

Estimation

Estimation involves only slight modifications of the 2-block Gibbs sampler under the independent normal and inverse-Wishart prior. In particular, here we construct a 3-block sampler to sequentially draw from $p(\boldsymbol{\Sigma} \,|\, \mathbf{y}, \boldsymbol{\gamma}, \boldsymbol{\beta})$, $p(\boldsymbol{\beta} \,|\, \mathbf{y}, \boldsymbol{\gamma}, \boldsymbol{\Sigma})$, and $p(\boldsymbol{\gamma} \,|\, \mathbf{y}, \boldsymbol{\beta}, \boldsymbol{\Sigma})$.

The full conditional distribution of $\boldsymbol{\Sigma}$ is inverse-Wishart, having the exact same form as given in (4.17). Next, the full conditional distribution of $\boldsymbol{\beta}$ is again normal:

$$(\boldsymbol{\beta} \,|\, \mathbf{y}, \boldsymbol{\gamma}, \boldsymbol{\Sigma}) \sim \mathcal{N}(\widehat{\boldsymbol{\beta}}, \mathbf{K}_{\boldsymbol{\beta}}^{-1}),$$

where

$$\mathbf{K}_{\boldsymbol{\beta}} = \boldsymbol{\Omega}_{\gamma}^{-1} + \mathbf{X}'(\mathbf{I}_T \otimes \boldsymbol{\Sigma}^{-1})\mathbf{X}, \quad \widehat{\boldsymbol{\beta}} = \mathbf{K}_{\boldsymbol{\beta}}^{-1}\mathbf{X}'(\mathbf{I}_T \otimes \boldsymbol{\Sigma}^{-1})\mathbf{y}.$$

Sampling from this normal distribution can be done using Algorithm 4.1.

Finally, to draw from $p(\boldsymbol{\gamma} \,|\, \mathbf{y}, \boldsymbol{\beta}, \boldsymbol{\Sigma})$, note that $\gamma_1, \ldots, \gamma_{nk}$ are conditionally independent given the data and other parameters. In fact, we have $p(\boldsymbol{\gamma} \,|\, \mathbf{y}, \boldsymbol{\beta}, \boldsymbol{\Sigma}) = \prod_{j=1}^{nk} p(\gamma_j \,|\, \beta_j)$. Moreover, each γ_j is a Bernoulli random variable and we only need to compute its success probability. To that end, note that

$$\mathbb{P}(\gamma_j = 1 \,|\, \beta_j) \propto q_j \phi(\beta_j; 0, \kappa_{1j})$$

and

$$\mathbb{P}(\gamma_j = 0 \mid \beta_j) \propto (1 - q_j)\phi(\beta_j; 0, \kappa_{0j}).$$

Hence, after normalization, we obtain

$$\mathbb{P}(\gamma_j = 1 \mid \beta_j) = \frac{q_j \phi(\beta_j; 0, \kappa_{1j})}{q_j \phi(\beta_j; 0, \kappa_{1j}) + (1 - q_j)\phi(\beta_j; 0, \kappa_{0j})}.$$

4.3 Large Bayesian VARs with Time-Varying Volatility, Heavy Tails and Serial Dependent Errors

Despite the empirical success of large Bayesian VARs with standard error assumptions (e.g., homoscedastic, Gaussian and serially independent), there is a lot of recent work in developing flexible VARs with more general error distributions. These more flexible VARs are motivated by the empirical observations that features like time-varying volatility and non-Gaussian errors are useful for modeling a variety of macroeconomic time series.

In this section we study a few of these more flexible VARs, including VARs with heteroscedastic, non-Gaussian, and serially correlated errors. To that end, we focus on the second representation of the VAR(p), which we reproduce below for convenience:

$$\mathbf{Y} = \mathbf{Z}\mathbf{A} + \mathbf{U}, \qquad \text{vec}(\mathbf{U}) \sim \mathcal{N}(\mathbf{0}, \mathbf{\Sigma} \otimes \mathbf{I}_T).$$

Note we can equivalently write the error specification as $\mathbf{u}_t \sim \mathcal{N}(\mathbf{0}, \mathbf{\Sigma})$, $t = 1, \ldots, T$. That is, the errors here are assumed to be independent, homoscedastic, and Gaussian. Below we consider a variety of extensions of this basic VAR.

To motivate the framework, recall that the main difficulty in doing posterior simulation for large Bayesian VARs is the large number of VAR coefficients in $\mathbf{A}$. One key advantage of the natural conjugate prior on $(\mathbf{A}, \mathbf{\Sigma})$ is that the conditional distribution of $\mathbf{A}$ given $\mathbf{\Sigma}$ is Gaussian and its covariance matrix has a Kronecker product structure. This special feature can be exploited to dramatically speed up computation from $O(n^6)$ to $O(n^3)$, as described in section "Estimation."

It turns out that this Kronecker product structure in the conditional covariance matrix of $\mathbf{A}$ can be preserved for a wide class of flexible models. Specifically, Chan (2018) proposes the following VAR with a more general covariance structure:

$$\mathbf{Y} = \mathbf{Z}\mathbf{A} + \mathbf{U}, \qquad \text{vec}(\mathbf{U}) \sim \mathcal{N}(\mathbf{0}, \mathbf{\Sigma} \otimes \mathbf{\Omega}), \tag{4.18}$$

where $\mathbf{\Omega}$ is a $T \times T$ covariance matrix. Obviously, if $\mathbf{\Omega} = \mathbf{I}_T$, (4.18) reduces to the standard VAR. Here the covariance matrix of vec($\mathbf{U}$) is assumed to have the Kronecker product structure $\boldsymbol{\Sigma} \otimes \boldsymbol{\Omega}$. Intuitively, it separately models the cross-sectional and serial covariance structures of $\mathbf{U}$, which are governed by $\boldsymbol{\Sigma}$ and $\boldsymbol{\Omega}$ respectively.

In the next few subsections, we first show that by choosing a suitable serial covariance structure $\boldsymbol{\Omega}$, the model in (4.18) includes a wide variety of flexible specifications. Section 4.3.4 then shows that the form of the error covariance matrix, namely $\boldsymbol{\Sigma} \otimes \boldsymbol{\Omega}$, leads to a Kronecker product structure in the conditional covariance matrix of $\mathbf{A}$. Again this special feature is used to dramatically speed up computation. The presentation below follows Chan (2018).

4.3.1 Common Stochastic Volatility

One of the most useful features for modeling macroeconomic time series is time-varying volatility. For example, the volatilities of a wide range of macroeconomic variables were substantially reduced at the start of the Great Moderation in the early 1980s. Models with homoscedastic errors would not be able to capture this feature of the data.

To allow for heteroscedastic errors, Carriero et al. (2016) introduce a large Bayesian VAR with a common stochastic volatility. In their setup, the error covariance matrix is scaled by a common, time-varying factor that can be interpreted as the overall macroeconomic volatility. More specifically, consider $\mathbf{u}_t \sim \mathcal{N}(\mathbf{0}, \mathrm{e}^{h_t}\boldsymbol{\Sigma})$ with the common volatility e^{h_t}. The log volatility h_t in turn follows a stationary AR(1) process:

$$h_t = \rho h_{t-1} + \varepsilon_t^h, \tag{4.19}$$

where $\varepsilon_t^h \sim \mathcal{N}(0, \sigma_h^2)$ and $|\rho| < 1$. Note that for identification purposes, this AR(1) process is assumed to have a zero unconditional mean.

One drawback of this setup is that the volatility specification is somewhat restrictive—all variances are scaled by a single factor and, consequently, they are always proportional to each other. On the other hand, there is empirical evidence, as shown in Carriero et al. (2016), that the volatilities of macroeconomic variables tend to move together. And specifying a common stochastic volatility is a parsimonious way to model that feature.

This common stochastic volatility model falls within the framework in (4.18) with $\boldsymbol{\Omega} = \mathrm{diag}(\mathrm{e}^{h_1}, \ldots, \mathrm{e}^{h_T})$. Empirical applications that use this common stochastic volatility include Mumtaz (2016), Mumtaz and Theodoridis (2018), and Poon (2018).

4.3.2 Non-Gaussian Errors

Gaussian errors are often assumed for convenience rather than for deep theoretical reasons. In fact, some recent work has found that VARs with heavy-tailed error distributions, such as the t distribution, often forecast better than their counterparts with Gaussian errors (see, e.g., Cross & Poon, 2016; Chiu, Mumtaz, & Pinter, 2017).

Since many distributions can be written as a scale mixture of Gaussian distributions, the framework in (4.18) can accommodate various commonly-used non-Gaussian distributions. To see this, let $\mathbf{\Omega} = \text{diag}(\lambda_1, \ldots, \lambda_T)$. If each λ_t follows independently an inverse-gamma distribution $(\lambda_t \mid \nu) \sim \mathcal{IG}(\nu/2, \nu/2)$, then marginally $\mathbf{u}_t$ has a multivariate t distribution with mean vector $\mathbf{0}$, scale matrix $\mathbf{\Sigma}$, and degree of freedom parameter ν (see, e.g., Geweke, 1993).

If each λ_t has an independent exponential distribution with mean α, then marginally $\mathbf{u}_t$ has a multivariate Laplace distribution with mean vector $\mathbf{0}$ and covariance matrix $\alpha\mathbf{\Sigma}$ (Eltoft, Kim, and Lee, 2006). Other scale mixtures of Gaussian distributions can be defined similarly. For additional examples, see, e.g., Eltoft, Kim, and Lee (2006a).

4.3.3 Serially Dependent Errors

Instead of the conventional assumption of serially independent errors, the framework in (4.18) can also handle serially correlated errors, such as errors that follow an ARMA(p, q) process.

For a concrete example, suppose $\mathbf{u}_t$ follows the following MA(2) process:

$$\mathbf{u}_t = \boldsymbol{\varepsilon}_t + \psi_1 \boldsymbol{\varepsilon}_{t-1} + \psi_2 \boldsymbol{\varepsilon}_{t-2},$$

where $\boldsymbol{\varepsilon}_t \sim \mathcal{N}(\mathbf{0}, \mathbf{\Sigma})$, ψ_1 and ψ_2 satisfy the invertibility conditions. This is nested within the general framework with

$$\mathbf{\Omega} = \begin{pmatrix} \omega_0 & \omega_1 & \omega_2 & 0 & \cdots & 0 \\ \omega_1 & \omega_0 & \omega_1 & \ddots & \ddots & \vdots \\ \omega_2 & \omega_1 & \omega_0 & \ddots & \ddots & 0 \\ 0 & \ddots & \ddots & \ddots & \ddots & \omega_2 \\ \vdots & \ddots & \ddots & \ddots & \ddots & \omega_1 \\ 0 & \cdots & 0 & \omega_2 & \omega_1 & \omega_0 \end{pmatrix},$$

where $\omega_0 = 1 + \psi_1^2 + \psi_2^2$, $\omega_1 = \psi_1(1 + \psi_2)$, and $\omega_2 = \psi_2$.

One drawback of the above MA(2) specification is that each element of $\mathbf{u}_t$ must have the same MA coefficients (although their variances can be different). Put it differently, the framework in (4.18) cannot accommodate, for example, a general MA(2) process of the form

$$\mathbf{u}_t = \boldsymbol{\varepsilon}_t + \boldsymbol{\Psi}_1\boldsymbol{\varepsilon}_{t-1} + \boldsymbol{\Psi}_2\boldsymbol{\varepsilon}_{t-2},$$

where $\boldsymbol{\Psi}_1$ and $\boldsymbol{\Psi}_2$ are $n \times n$ matrices of coefficients. This is because in this case the covariance matrix of vec($\mathbf{U}$) does not have a Kronecker structure—i.e., it cannot be written as $\boldsymbol{\Sigma} \otimes \boldsymbol{\Omega}$. Nevertheless, this restricted form of serial correlation might still be useful to capture persistence in the data.

Other more elaborate covariance structures can be constructed by combining different examples in previous sections. For example, suppose $\mathbf{u}_t$ follows an MA(1) stochastic volatility process of the form:

$$\mathbf{u}_t = \boldsymbol{\varepsilon}_t + \psi_1\boldsymbol{\varepsilon}_{t-1},$$

where $\boldsymbol{\varepsilon}_t \sim \mathcal{N}(\mathbf{0}, \mathrm{e}^{h_t}\boldsymbol{\Sigma})$ and h_t has an AR(1) process as in (4.19). This is a multivariate generalization of the univariate moving average stochastic volatility models considered in Chan (2013). This model is a special case of the flexible Bayesian VAR in (4.18) with

$$\boldsymbol{\Omega} = \begin{pmatrix} (1+\psi_1^2)\mathrm{e}^{h_1} & \psi_1\mathrm{e}^{h_1} & 0 & \cdots & 0 \\ \psi_1\mathrm{e}^{h_1} & \psi_1^2\mathrm{e}^{h_1}+\mathrm{e}^{h_2} & \ddots & \ddots & \vdots \\ 0 & \ddots & \ddots & \ddots & 0 \\ \vdots & \ddots & \ddots & \psi_1^2\mathrm{e}^{h_{T-2}}+\mathrm{e}^{h_{T-1}} & \psi_1\mathrm{e}^{h_{T-1}} \\ 0 & \cdots & 0 & \psi_1\mathrm{e}^{h_{T-1}} & \psi_1^2\mathrm{e}^{h_{T-1}}+\mathrm{e}^{h_T} \end{pmatrix}.$$

4.3.4 Estimation

Next we discuss the estimation of the Bayesian VAR in (4.18) using MCMC methods. To keep the discussion general, we leave $\boldsymbol{\Omega}$ unspecified and focus on the key step of jointly sampling both the VAR coefficients $\mathbf{A}$ and the cross-sectional covariance matrix $\boldsymbol{\Sigma}$. Then, we take up various examples of $\boldsymbol{\Omega}$ and provide estimation details for tackling each case.

Using a similar derivation as in Sect. 4.1.2, one can show that the likelihood of the VAR in (4.18) is given by

$$p(\mathbf{Y} \mid \mathbf{A}, \boldsymbol{\Sigma}, \boldsymbol{\Omega}) = (2\pi)^{-\frac{Tn}{2}}|\boldsymbol{\Sigma}|^{-\frac{T}{2}}|\boldsymbol{\Omega}|^{-\frac{n}{2}}\mathrm{e}^{-\frac{1}{2}\mathrm{tr}(\boldsymbol{\Sigma}^{-1}(\mathbf{Y}-\mathbf{XA})'\boldsymbol{\Omega}^{-1}(\mathbf{Y}-\mathbf{XA}))}. \quad (4.20)$$

Next, we assume a prior of the form $p(\mathbf{A}, \boldsymbol{\Sigma}, \boldsymbol{\Omega}) = p(\mathbf{A}, \boldsymbol{\Sigma})p(\boldsymbol{\Omega})$, i.e., the parameter blocks $(\mathbf{A}, \boldsymbol{\Sigma})$ and $\boldsymbol{\Omega}$ are a priori independent. For $(\mathbf{A}, \boldsymbol{\Sigma})$, we adopt the natural conjugate prior:

$$\boldsymbol{\Sigma} \sim \mathcal{IW}(\nu_0, \mathbf{S}_0), \quad (\text{vec}(\mathbf{A}) \mid \boldsymbol{\Sigma}) \sim \mathcal{N}(\text{vec}(\mathbf{A}_0), \boldsymbol{\Sigma} \otimes \mathbf{V}_{\mathbf{A}})$$

with joint density function given in (4.11).

Given the prior $p(\mathbf{A}, \boldsymbol{\Sigma}, \boldsymbol{\Omega}) = p(\mathbf{A}, \boldsymbol{\Sigma})p(\boldsymbol{\Omega})$, posterior draws can be obtained by sequentially sampling from: (1) $p(\mathbf{A}, \boldsymbol{\Sigma} \mid \mathbf{Y}, \boldsymbol{\Omega})$ and (2) $p(\boldsymbol{\Omega} \mid \mathbf{Y}, \mathbf{A}, \boldsymbol{\Sigma})$. Here we first describe how one can implement Step 1 efficiently. Depending on the covariance structure $\boldsymbol{\Omega}$, additional blocks might be needed to sample some extra hierarchical parameters. These steps are typically easy to implement as they amount to fitting a univariate time series model. We will discuss various examples below.

When $\boldsymbol{\Omega} = \mathbf{I}_T$, the Bayesian VAR in (4.18) reduces to the conventional VAR with the natural conjugate prior. And in Sect. 4.2.2 we showed that $(\mathbf{A}, \boldsymbol{\Sigma} \mid \mathbf{Y})$ has a normal-inverse-Wishart distribution. There we also discussed how we can draw from the normal-inverse-Wishart distribution efficiently. It turns out that similar derivations go through even with an arbitrary covariance matrix $\boldsymbol{\Omega}$. More specifically, it follows from (4.20) and (4.11) that

$$\begin{aligned}
p(\mathbf{A}, \boldsymbol{\Sigma} \mid \mathbf{Y}, \boldsymbol{\Omega}) \propto & |\boldsymbol{\Sigma}|^{-\frac{\nu_0+n+k+T}{2}} \mathrm{e}^{-\frac{1}{2}\mathrm{tr}(\boldsymbol{\Sigma}^{-1}\mathbf{S}_0)} \\
& \times \mathrm{e}^{-\frac{1}{2}\mathrm{tr}\left(\boldsymbol{\Sigma}^{-1}((\mathbf{A}-\mathbf{A}_0)'\mathbf{V}_{\mathbf{A}}^{-1}(\mathbf{A}-\mathbf{A}_0)+(\mathbf{Y}-\mathbf{Z}\mathbf{A})'\boldsymbol{\Omega}^{-1}(\mathbf{Y}-\mathbf{Z}\mathbf{A}))\right)} \\
= & |\boldsymbol{\Sigma}|^{-\frac{\nu_0+n+k+T}{2}} \mathrm{e}^{-\frac{1}{2}\mathrm{tr}(\boldsymbol{\Sigma}^{-1}\mathbf{S}_0)} \mathrm{e}^{-\frac{1}{2}\mathrm{tr}\left(\boldsymbol{\Sigma}^{-1}(\mathbf{A}_0'\mathbf{V}_{\mathbf{A}}^{-1}\mathbf{A}_0+\mathbf{Y}'\boldsymbol{\Omega}^{-1}\mathbf{Y}-\widehat{\mathbf{A}}'\mathbf{K}_{\mathbf{A}}\widehat{\mathbf{A}})\right)} \\
& \times \mathrm{e}^{-\frac{1}{2}\mathrm{tr}(\boldsymbol{\Sigma}^{-1}(\mathbf{A}-\widehat{\mathbf{A}})'\mathbf{K}_{\mathbf{A}}(\mathbf{A}-\widehat{\mathbf{A}}))},
\end{aligned}$$

where $\mathbf{K}_{\mathbf{A}} = \mathbf{V}_{\mathbf{A}}^{-1} + \mathbf{Z}'\boldsymbol{\Omega}^{-1}\mathbf{Z}$ and $\widehat{\mathbf{A}} = \mathbf{K}_{\mathbf{A}}^{-1}(\mathbf{V}_{\mathbf{A}}^{-1}\mathbf{A}_0 + \mathbf{Z}'\boldsymbol{\Omega}^{-1}\mathbf{Y})$. In the above derivations, we have "completed the square" and obtained:

$$\begin{aligned}
(\mathbf{A} - \mathbf{A}_0)'\mathbf{V}_{\mathbf{A}}^{-1}(\mathbf{A} - \mathbf{A}_0) + (\mathbf{Y} - \mathbf{Z}\mathbf{A})'\boldsymbol{\Omega}^{-1}(\mathbf{Y} - \mathbf{Z}\mathbf{A}) \\
= (\mathbf{A} - \widehat{\mathbf{A}})'\mathbf{K}_{\mathbf{A}}(\mathbf{A} - \widehat{\mathbf{A}}) + \mathbf{A}_0'\mathbf{V}_{\mathbf{A}}^{-1}\mathbf{A}_0 + \mathbf{Y}'\boldsymbol{\Omega}^{-1}\mathbf{Y} - \widehat{\mathbf{A}}'\mathbf{K}_{\mathbf{A}}\widehat{\mathbf{A}}.
\end{aligned}$$

If we let

$$\widehat{\mathbf{S}} = \mathbf{S}_0 + \mathbf{A}_0'\mathbf{V}_{\mathbf{A}}^{-1}\mathbf{A}_0 + \mathbf{Y}'\boldsymbol{\Omega}^{-1}\mathbf{Y} - \widehat{\mathbf{A}}'\mathbf{K}_{\mathbf{A}}\widehat{\mathbf{A}},$$

then $(\mathbf{A}, \boldsymbol{\Sigma} \mid \mathbf{Y}, \boldsymbol{\Omega})$ has a normal-inverse-Wishart distribution with parameters $\nu_0 + T$, $\widehat{\mathbf{S}}$, $\widehat{\mathbf{A}}$, and $\mathbf{K}_{\mathbf{A}}^{-1}$. We can then sample $(\mathbf{A}, \boldsymbol{\Sigma} \mid \mathbf{Y}, \boldsymbol{\Omega})$ in two steps. First, sample $\boldsymbol{\Sigma}$ marginally from $(\boldsymbol{\Sigma} \mid \mathbf{Y}, \boldsymbol{\Omega}) \sim \mathcal{IW}(\nu_0 + T, \widehat{\mathbf{S}})$. Second, given the $\boldsymbol{\Sigma}$ sampled, simulate

$$(\text{vec}(\mathbf{A}) \mid \mathbf{Y}, \boldsymbol{\Sigma}, \boldsymbol{\Omega}) \sim \mathcal{N}(\text{vec}(\widehat{\mathbf{A}}), \boldsymbol{\Sigma} \otimes \mathbf{K}_{\mathbf{A}}^{-1}).$$

As discussed in Sect. 4.2.2, we can sample from this normal distribution efficiently without explicitly computing the inverse $\mathbf{K}_{\mathbf{A}}^{-1}$.

Here we comment on a few computational details. Again, we need not compute the $T \times T$ inverse $\boldsymbol{\Omega}^{-1}$ to obtain $\mathbf{K}_{\mathbf{A}}$, $\widehat{\mathbf{A}}$, or $\widehat{\mathbf{S}}$. As an example, consider computing the quadratic form $\mathbf{Z}'\boldsymbol{\Omega}^{-1}\mathbf{Z}$. Let $\mathbf{C}_{\boldsymbol{\Omega}}$ be the Cholesky factor of $\boldsymbol{\Omega}$ such that $\mathbf{C}_{\boldsymbol{\Omega}}\mathbf{C}_{\boldsymbol{\Omega}}' = \boldsymbol{\Omega}$. Then, $\mathbf{Z}'\boldsymbol{\Omega}^{-1}\mathbf{Z}$ can be obtained via $\widetilde{\mathbf{Z}}'\widetilde{\mathbf{Z}}$, where $\widetilde{\mathbf{Z}} = \mathbf{C}_{\boldsymbol{\Omega}}\backslash\mathbf{Z}$.

This approach would work fine for an arbitrary $\boldsymbol{\Omega}$ with dimension, say, less than 1000. For larger T, computing the Cholesky factor of $\boldsymbol{\Omega}$ and performing the forward and backward substitution is likely to be time-consuming. Fortunately, for most models, $\boldsymbol{\Omega}$ or $\boldsymbol{\Omega}^{-1}$ are band matrices—i.e., sparse matrices whose nonzero elements are confined to a diagonal band. For example, $\boldsymbol{\Omega}$ is diagonal—hence banded—for both the common stochastic volatility model and the t errors model. Moreover, $\boldsymbol{\Omega}$ is banded for VARs with MA errors and $\boldsymbol{\Omega}^{-1}$ is banded for AR errors.

This special structure of $\boldsymbol{\Omega}$ or $\boldsymbol{\Omega}^{-1}$ can be exploited to speed up computation. For instance, obtaining the Cholesky factor of a band $T \times T$ matrix with fixed bandwidth involves only $O(T)$ operations (e.g., Golub & van Loan, 1983, p.156) as opposed to $O(T^3)$ for a dense matrix of the same size. Similar computational savings can be obtained for operations such as multiplication, forward, and backward substitution by using band matrix routines. We refer the readers to Chan (2013) for a more detailed discussion on computation involving band matrices.

Next, we take up various examples of $\boldsymbol{\Omega}$ and provide the corresponding estimation details.

***t* Errors**

As discussed in Sect. 4.3.2, a VAR with iid t errors falls within the framework in (4.18) with $\boldsymbol{\Omega} = \text{diag}(\lambda_1, \ldots, \lambda_T)$, where each λ_t follows an inverse-gamma distribution $(\lambda_t \mid \nu) \sim \mathcal{IG}(\nu/2, \nu/2)$. Unconditional on λ_t, $\mathbf{u}_t$ has a t distribution with degree of freedom parameter ν. Note that in this case $\boldsymbol{\Omega}$ is diagonal and $\boldsymbol{\Omega}^{-1} = \text{diag}(\lambda_1^{-1}, \ldots, \lambda_T^{-1})$.

Let $p(\nu)$ denote the prior density function of ν. Then, posterior draws can be obtained by sequentially sampling from: (1) $p(\mathbf{A}, \boldsymbol{\Sigma} \mid \mathbf{Y}, \boldsymbol{\Omega}, \nu)$; (2) $p(\boldsymbol{\Omega} \mid \mathbf{Y}, \mathbf{A}, \boldsymbol{\Sigma}, \nu)$; and (3) $p(\nu \mid \mathbf{Y}, \mathbf{A}, \boldsymbol{\Sigma}, \boldsymbol{\Omega})$. Step 1 can be implemented exactly as before. For Step 2, note that

$$p(\boldsymbol{\Omega} \mid \mathbf{Y}, \mathbf{A}, \boldsymbol{\Sigma}, \nu) = \prod_{t=1}^{T} p(\lambda_t \mid \mathbf{Y}, \mathbf{A}, \boldsymbol{\Sigma}, \nu) \propto \prod_{t=1}^{T} \lambda_t^{-\frac{n}{2}} \mathrm{e}^{-\frac{1}{2\lambda_t}\mathbf{u}_t'\boldsymbol{\Sigma}^{-1}\mathbf{u}_t} \times \lambda_t^{-(\frac{\nu}{2}+1)} \mathrm{e}^{-\frac{\nu}{2\lambda_t}}$$

In other words, each λ_t is conditionally independent given other parameters and has an inverse-gamma distribution: $(\lambda_t \mid \mathbf{Y}, \mathbf{A}, \boldsymbol{\Sigma}, \nu) \sim \mathcal{IG}((n+\nu)/2, (\mathbf{u}_t'\boldsymbol{\Sigma}^{-1}\mathbf{u}_t + \nu)/2)$.

Lastly, ν can be sampled by an independence-chain Metropolis–Hastings step with the proposal distribution $\mathcal{N}(\widehat{\nu}, K_\nu^{-1})$, where $\widehat{\nu}$ is the mode of $\log p(\nu \mid \mathbf{Y}, \mathbf{A}, \boldsymbol{\Sigma}, \boldsymbol{\Omega})$ and K_ν is the negative Hessian evaluated at the mode. For implementation details of this step, see Chan and Hsiao (2014).

Common Stochastic Volatility

Now, consider the common stochastic volatility model proposed in Carriero et al. (2016): $\mathbf{u}_t \sim \mathcal{N}(\mathbf{0}, \mathrm{e}^{h_t}\boldsymbol{\Sigma})$, where h_t follows an AR(1) process in (4.19). In this case $\boldsymbol{\Omega} = \mathrm{diag}(\mathrm{e}^{h_1}, \ldots, \mathrm{e}^{h_T})$, which is also diagonal.

We assume independent truncated normal and inverse-gamma priors for ρ and σ_h^2: $\rho \sim \mathcal{N}(\rho_0, V_\rho)1(|\rho| < 1)$ and $\sigma_h^2 \sim \mathcal{IG}(\nu_{h0}, S_{h0})$. Then, posterior draws can be obtained by sampling from: (1) $p(\mathbf{A}, \boldsymbol{\Sigma} \,|\, \mathbf{Y}, \boldsymbol{\Omega}, \rho, \sigma_h^2)$; (2) $p(\boldsymbol{\Omega} \,|\, \mathbf{Y}, \mathbf{A}, \boldsymbol{\Sigma}, \rho, \sigma_h^2)$; (3) $p(\rho \,|\, \mathbf{Y}, \mathbf{A}, \boldsymbol{\Sigma}, \boldsymbol{\Omega}, \sigma_h^2)$; and (4) $p(\sigma_h^2 \,|\, \mathbf{Y}, \mathbf{A}, \boldsymbol{\Sigma}, \boldsymbol{\Omega}, \rho)$.

Step 1 again can be implemented exactly as before. For Step 2, note that

$$p(\boldsymbol{\Omega} \,|\, \mathbf{Y}, \mathbf{A}, \boldsymbol{\Sigma}, \rho, \sigma_h^2) = p(\mathbf{h} \,|\, \mathbf{Y}, \mathbf{A}, \boldsymbol{\Sigma}, \rho, \sigma_h^2) \propto p(\mathbf{h} \,|\, \rho, \sigma_h^2) \prod_{t=1}^{T} p(\mathbf{y}_t \,|\, \mathbf{A}, \boldsymbol{\Sigma}, h_t),$$

where $p(\mathbf{h} \,|\, \rho, \sigma_h^2)$ is a Gaussian density implied by the state equation,

$$\log p(\mathbf{y}_t \,|\, \mathbf{A}, \boldsymbol{\Sigma}, h_t) = c_t - \frac{n}{2}h_t - \frac{1}{2}\mathrm{e}^{-h_t}\mathbf{u}_t'\boldsymbol{\Sigma}^{-1}\mathbf{u}_t$$

and c_t is a constant not dependent on h_t. It is easy to check that

$$\frac{\partial}{\partial h_t}\log p(\mathbf{y}_t \,|\, \mathbf{A}, \boldsymbol{\Sigma}, h_t) = -\frac{n}{2} + \frac{1}{2}\mathrm{e}^{-h_t}\mathbf{u}_t'\boldsymbol{\Sigma}^{-1}\mathbf{u}_t,$$

$$\frac{\partial^2}{\partial h_t^2}\log p(\mathbf{y}_t \,|\, \mathbf{A}, \boldsymbol{\Sigma}, h_t) = -\frac{1}{2}\mathrm{e}^{-h_t}\mathbf{u}_t'\boldsymbol{\Sigma}^{-1}\mathbf{u}_t.$$

Then, one can implement a Newton-Raphson algorithm to obtain the mode of $\log p(\mathbf{h} \,|\, \mathbf{Y}, \mathbf{A}, \boldsymbol{\Sigma}, \rho, \sigma_h^2)$ and compute the negative Hessian evaluated at the mode, which are denoted as $\widehat{\mathbf{h}}$ and $\mathbf{K}_{\mathbf{h}}$, respectively. Using $\mathcal{N}(\widehat{\mathbf{h}}, \mathbf{K}_{\mathbf{h}}^{-1})$ as a proposal distribution, one can sample $\mathbf{h}$ directly using an acceptance-rejection Metropolis–Hastings step. We refer the readers to Chan (2017) and Chan and Jeliazkov (2009) for details. Finally, Steps 3 and 4 are standard and can be easily implemented (see, e.g., Chan & Hsiao, 2014).

MA(1) Errors

We now consider an example where $\boldsymbol{\Omega}$ is not diagonal and we construct $\boldsymbol{\Omega}$ using band matrices. More specifically, suppose each element of $\mathbf{u}_t$ follows the same MA(1) process:

$$u_{it} = \eta_{it} + \psi\eta_{i,t-1},$$

where $|\psi| < 1$, $\eta_{it} \sim \mathcal{N}(0, 1)$, and the process is initialized with $u_{i1} \sim \mathcal{N}(0, 1 + \psi^2)$. Stacking $\mathbf{u}_i = (u_{i1}, \ldots, u_{iT})'$ and $\boldsymbol{\eta}_i = (\eta_{i1}, \ldots, \eta_{iT})'$, we can rewrite the

MA(1) process as

$$\mathbf{u}_i = \mathbf{H}_\psi \boldsymbol{\eta}_i,$$

where $\boldsymbol{\eta}_i \sim \mathcal{N}(\mathbf{0}, \mathbf{O}_\psi)$ with $\mathbf{O}_\psi = \text{diag}(1+\psi^2, 1, \ldots, 1)$, and

$$\mathbf{H}_\psi = \begin{pmatrix} 1 & 0 & \cdots & 0 \\ \psi & 1 & \cdots & 0 \\ \vdots & \ddots & \ddots & \vdots \\ 0 & \cdots & \psi & 1 \end{pmatrix}.$$

It follows that the covariance matrix of $\mathbf{u}_i$ is $\mathbf{H}_\psi \mathbf{O}_\psi \mathbf{H}'_\psi$. That is, $\boldsymbol{\Omega} = \mathbf{H}_\psi \mathbf{O}_\psi \mathbf{H}'_\psi$ is a function of ψ only. Moreover, both $\mathbf{O}_\psi$ and $\mathbf{H}_\psi$ are band matrices. Notice also that for a general MA(q) process, one only needs to redefine $\mathbf{H}_\psi$ and $\mathbf{O}_\psi$ appropriately and the same procedure would apply.

Let $p(\psi)$ be the prior for ψ. Then, posterior draws can be obtained by sequentially sampling from: (1) $p(\mathbf{A}, \boldsymbol{\Sigma} \,|\, \mathbf{Y}, \psi)$ and (2) $p(\psi \,|\, \mathbf{Y}, \mathbf{A}, \boldsymbol{\Sigma})$. Again, Step 1 can be carried out exactly the same as before. In implementing Step 1, we emphasize that products of the form $\mathbf{Z}'\boldsymbol{\Omega}^{-1}\mathbf{Z}$ or $\mathbf{Z}'\boldsymbol{\Omega}^{-1}\mathbf{Y}$ can be obtained without explicitly computing the inverse $\boldsymbol{\Omega}^{-1}$. Instead, since in this case $\boldsymbol{\Omega}$ is a band matrix, its Cholesky factor $\mathbf{C}_{\boldsymbol{\Omega}}$ can be obtained in $O(T)$ operations. Then, to compute $\mathbf{Z}'\boldsymbol{\Omega}^{-1}\mathbf{Z}$, one simply returns $\widetilde{\mathbf{Z}}'\widetilde{\mathbf{Z}}$, where $\widetilde{\mathbf{Z}} = \mathbf{C}_{\boldsymbol{\Omega}} \backslash \mathbf{Z}$.

For Step 2, $p(\psi \,|\, \mathbf{Y}, \mathbf{A}, \boldsymbol{\Sigma})$ is non-standard, but it can be evaluated quickly using the direct method in Chan (2013), which is more efficient than using the Kalman filter. Specifically, since the determinant $|\mathbf{H}_\psi| = 1$, it follows from (4) that the likelihood is given by

$$p(\mathbf{Y} \,|\, \mathbf{A}, \boldsymbol{\Sigma}, \psi) = (2\pi)^{-\frac{Tn}{2}} |\boldsymbol{\Sigma}|^{-\frac{T}{2}} (1+\psi^2)^{-\frac{n}{2}} e^{-\frac{1}{2}\text{tr}\left(\boldsymbol{\Sigma}^{-1}\widetilde{\mathbf{U}}'\mathbf{O}_\psi^{-1}\widetilde{\mathbf{U}}\right)},$$

where $\widetilde{\mathbf{U}} = \mathbf{H}_\psi^{-1}(\mathbf{Y} - \mathbf{Z}\mathbf{A})$, which can be obtained in $O(T)$ operations since $\mathbf{H}_\psi$ is a band matrix. Therefore, $p(\psi \,|\, \mathbf{Y}, \mathbf{A}, \boldsymbol{\Sigma}) \propto p(\mathbf{Y} \,|\, \mathbf{A}, \boldsymbol{\Sigma}, \psi) p(\psi)$ can be evaluated quickly. Then, ψ is sampled using an independence-chain Metropolis–Hastings step as in Chan (2013).

AR(1) Errors

Here we consider an example where $\boldsymbol{\Omega}$ is a full matrix, but $\boldsymbol{\Omega}^{-1}$ is banded. Specifically, suppose each element of $\mathbf{u}_t$ follows the same AR(1) process:

$$u_{it} = \phi u_{i,t-1} + \eta_{it},$$

where $|\phi| < 1$, $\eta_{it} \sim \mathcal{N}(0, 1)$, and the process is initialized with $u_{i1} \sim \mathcal{N}(0, 1/(1-\phi^2))$. Stacking $\mathbf{u}_i = (u_{i1}, \ldots, u_{iT})'$ and $\boldsymbol{\eta}_i = (\eta_{i1}, \ldots, \eta_{iT})'$, we can rewrite the

AR(1) process as

$$\mathbf{H}_\phi \mathbf{u}_i = \boldsymbol{\eta}_i,$$

where $\boldsymbol{\eta}_i \sim \mathcal{N}(\mathbf{0}, \mathbf{O}_\phi)$ with $\mathbf{O}_\phi = \text{diag}(1/(1-\phi^2), 1, \ldots, 1)$, and

$$\mathbf{H}_\phi = \begin{pmatrix} 1 & 0 & \cdots & 0 \\ -\phi & 1 & \cdots & 0 \\ \vdots & \ddots & \ddots & \vdots \\ 0 & \cdots & -\phi & 1 \end{pmatrix}.$$

Since the determinant $|\mathbf{H}_\phi| = 1 \neq 0$, $\mathbf{H}_\phi$ is invertible. It follows that the covariance matrix of $\mathbf{u}_i$ is $\mathbf{H}_\phi^{-1}\mathbf{O}_\phi(\mathbf{H}_\phi')^{-1}$, or $\boldsymbol{\Omega}^{-1} = \mathbf{H}_\phi'\mathbf{O}_\phi^{-1}\mathbf{H}_\phi$, where both $\mathbf{O}_\phi$ and $\mathbf{H}_\phi$ are band matrices.

Suppose we assume the truncated normal prior ϕ: $\phi \sim \mathcal{N}(\phi_0, V_\phi)1(|\phi| < 1)$. Then, posterior draws can be obtained by sampling from: (1) $p(\mathbf{A}, \boldsymbol{\Sigma} \mid \mathbf{Y}, \phi)$; and (2) $p(\phi \mid \mathbf{Y}, \mathbf{A}, \boldsymbol{\Sigma})$. In implementing Step 1, products of the form $\mathbf{Z}'\boldsymbol{\Omega}^{-1}\mathbf{Z}$ can be computed easily as the inverse $\boldsymbol{\Omega}^{-1}$ is a band matrix.

For Step 2, $p(\phi \mid \mathbf{Y}, \mathbf{A}, \boldsymbol{\Sigma})$ is non-standard, but a good approximation can be obtained easily without numerical optimization. To that end, recall that

$$\mathbf{u}_t = \phi \mathbf{u}_{t-1} + \boldsymbol{\varepsilon}_t,$$

where $\boldsymbol{\varepsilon}_t \sim \mathcal{N}(\mathbf{0}, \boldsymbol{\Sigma})$, and the process is initialized by $\mathbf{u}_1 \sim \mathcal{N}(\mathbf{0}, \boldsymbol{\Sigma}/(1-\phi^2))$. Then, consider the Gaussian proposal $\mathcal{N}(\widehat{\phi}, K_\phi^{-1})$, where $K_\phi = 1/V_\phi + \sum_{t=2}^T \mathbf{u}_{t-1}'\boldsymbol{\Sigma}^{-1}\mathbf{u}_{t-1}$ and $\widehat{\phi} = K_\phi^{-1}(\phi_0/V_\phi + \sum_{t=2}^T \mathbf{u}_{t-1}'\boldsymbol{\Sigma}^{-1}\mathbf{u}_t)$. With this proposal distribution, we can then implement an independence-chain Metropolis–Hastings step to sample ϕ.

4.4 Empirical Application: Forecasting with Large Bayesian VARs

In this section we consider a real-time macroeconomic forecasting exercise to illustrate the large Bayesian VARs and the associated estimation methods discussed in Sect. 4.3.

4.4.1 Data, Models, and Priors

In our empirical application we use a real-time dataset considered in Chan (2018) that consists of 20 variables at quarterly frequency. The dataset includes a variety of standard macroeconomic and financial variables such as GDP, industrial production, inflation, interest rates, and unemployment. The data are sourced from the Federal Reserve Bank of Philadelphia and the sample period is from 1964Q1 to 2015Q4. These variables are commonly used in applied work and are similar to the variables included in the large VARs in Banbura et al. (2010) and Koop (2013). A detailed description of the variables and their transformations are provided in Appendix A.

We include a range of large Bayesian VARs combined with different prior specifications. For comparison, we also include a small Bayesian VAR using only four core variables: real GDP growth, industrial production, unemployment rate, and PCE inflation. The full description of other models is given in Table 4.1.

Whenever possible, we choose the same priors for common parameters across models. For the Minnesota prior, we set $\boldsymbol{\beta}_{\text{Minn}} = \mathbf{0}$ and the three hyperparameters for $\mathbf{V}_{\text{Minn}}$ are set to be $c_1 = 0.2^2, c_2 = 0.1^2$, and $c_3 = 10^2$. For the natural conjugate prior, we set $\mathbf{A}_0 = \mathbf{0}$ and the two hyperparameters for the covariance matrix $\mathbf{V}_{\mathbf{A}}$ are assumed to be $c_1 = 0.2^2$ and $c_2 = 10^2$. Moreover we set $\nu_0 = n + 3$, $\mathbf{S}_0 = \text{diag}(s_1^2, \ldots, s_n^2)$, where s_i^2 denotes the standard OLS estimate of the error variance for the i-th equation.

For the common stochastic volatility model, we assume independent priors for σ_h^2 and ρ: $\sigma_h^2 \sim \mathcal{IG}(\nu_{h0}, S_{h0})$ and $\rho \sim \mathcal{N}(\rho_0, V_\rho)\mathbb{1}(|\rho| < 1)$, where we set $\nu_{h0} = 5$, $S_{h0} = 0.04$, $\rho_0 = 0.9$, and $V_\rho = 0.2^2$. These values imply that the prior mean of σ_h^2 is 0.1^2 and ρ is centered at 0.9. For the degree of freedom parameter ν under the t model, we consider a uniform prior on $(2, 50)$, i.e., $\nu \sim \mathcal{U}(2, 50)$. For the MA coefficient ψ under the MA model, we assume the truncated normal prior $\psi \sim \mathcal{N}(\psi_0, V_\psi)\mathbb{1}(|\psi| < 1)$ so that the MA process is invertible. We set $\psi_0 = 0$ and $V_\psi = 1$. The prior thus centers around 0 and has support within the interval $(-1, 1)$. Given the large prior variance, it is also relatively noninformative.

Table 4.1 A list of competing models

Model	Description
BVAR-small	4-variable VAR with the Minnesota prior
BVAR-Minn	20-variable VAR with the Minnesota prior
BVAR-NCP	20-variable VAR with the natural conjugate prior
BVAR-IP	20-variable VAR with the independent prior
BVAR-SSVS	20-variable VAR with the SSVS prior
BVAR-CSV	20-variable VAR with a common stochastic volatility
BVAR-CSV-t	20-variable VAR with a common SV and t errors
BVAR-CSV-t-MA	20-variable VAR with a common SV and MA(1) t errors

4.4.2 *Forecast Evaluation Metrics*

We perform a recursive out-of-sample forecasting exercise to evaluate the performance of the Bayesian VARs with different priors in terms of both point and density forecasts. We focus on four main variables: real GDP growth, industrial production, unemployment rate, and PCE inflation.

We use each of the Bayesian VARs listed in Table 4.1 to produce both point and density m-step-ahead iterated forecasts with $m = 1$ and $m = 2$. Due to reporting lags, the real-time data vintage available at time t contains observations only up to quarter $t - 1$. Hence, the forecasts are current quarter nowcasts and one-quarter-ahead forecasts. The evaluation period is from 1975Q1 to 2015Q4, and we use the 2017Q3 vintage to compute the actual outcomes.

Given the data up to time t, denoted as $\mathbf{Y}_{1:t}$, we obtain posterior draws given $\mathbf{Y}_{1:t}$. We then compute the predictive mean $\mathbb{E}(y_{i,t+m} \,|\, \mathbf{Y}_{1:t})$ as the point forecast for variable i, and the predictive density $p(y_{i,t+m} \,|\, \mathbf{Y}_{1:t})$ as the density forecast for the same variable. For many Bayesian VARs considered, neither the predictive mean nor the predictive density of $y_{i,t+m}$ can be computed analytically. If that is the case, we obtain them using predictive simulation. Next, we move one period forward and repeat the whole exercise with data $\mathbf{Y}_{1:t+1}$, and so on. These forecasts are then evaluated for $t = t_0, \ldots, T - m$.

For forecast evaluation metrics, let $y^{\text{o}}_{i,t+m}$ denote the actual value of the variable $y_{i,t+m}$. The metric used to evaluate the point forecasts is the root mean squared forecast error (RMSFE) defined as

$$\text{RMSFE} = \sqrt{\frac{\sum_{t=t_0}^{T-m}(y^{\text{o}}_{i,t+m} - \mathbb{E}(y_{i,t+m} \,|\, \mathbf{Y}_{1:t}))^2}{T - m - t_0 + 1}}.$$

To evaluate the density forecast $p(y_{i,t+m} \,|\, \mathbf{Y}_{1:t})$, one natural measure is the predictive likelihood $p(y_{i,t+m} = y^{\text{o}}_{i,t+m} \,|\, \mathbf{Y}_{1:t})$, i.e., the predictive density of $y_{i,t+m}$ evaluated at the actual value $y^{\text{o}}_{i,t+m}$. If the actual outcome $y^{\text{o}}_{i,t+m}$ is likely under the density forecast, the value of the predictive likelihood will be large, and vice versa. See, e.g., Geweke and Amisano (2011) for a more detailed discussion of the predictive likelihood and its connection to the marginal likelihood. We evaluate the density forecasts using the average of log predictive likelihoods (ALPL):

$$\text{ALPL} = \frac{1}{T - m - t_0 + 1} \sum_{t=t_0}^{T-m} \log p(y_{i,t+m} = y^{\text{o}}_{i,t+m} \,|\, \mathbf{Y}_{1:t}).$$

For this metric, a larger value indicates better forecast performance.

4.4.3 Forecasting Results

For easy comparison, we report below the ratios of RMSFEs of a given model to those of the 4-variable Bayesian VAR using the core variables: real GDP growth, industrial production, unemployment rate, and PCE inflation. Hence, values smaller than unity indicate better forecast performance than the benchmark. For the average of log predictive likelihoods, we report differences from that of the 4-variable Bayesian VAR. In this case, positive values indicate better forecast performance than the benchmark.

Tables 4.2, 4.3, 4.4, and 4.5 report the point and density forecast results for the four core variables. No single models or priors can outperform others for all variables in all horizons. However, there are a few consistent patterns in the

Table 4.2 Forecast performance relative to a 4-variable Bayesian VAR; GDP growth

	Relative RMSFE		Relative ALPL	
	$m = 1$	$m = 2$	$m = 1$	$m = 2$
BVAR-Minn	0.95	0.98	−0.04	−0.10
BVAR-NCP	0.92	0.98	0.04	0.03
BVAR-IP	1.01	0.96	0.03	0.05
BVAR-SSVS	0.92	1.02	0.01	0.00
BVAR-CSV	0.95	0.94	0.13	0.09
BVAR-CSV-t	0.93	0.95	0.13	0.09
BVAR-CSV-t-MA	0.93	0.93	0.13	0.10

Table 4.3 Forecast performance relative to a 4-variable Bayesian VAR; industrial production

	Relative RMSFE		Relative ALPL	
	$m = 1$	$m = 2$	$m = 1$	$m = 2$
BVAR-Minn	0.97	0.94	0.02	−0.02
BVAR-NCP	0.96	0.95	0.15	0.09
BVAR-IP	0.94	0.90	0.10	0.10
BVAR-SSVS	0.99	0.96	0.08	0.07
BVAR-CSV	0.89	0.90	0.27	0.17
BVAR-CSV-t	0.88	0.89	0.26	0.17
BVAR-CSV-t-MA	0.87	0.89	0.27	0.17

Table 4.4 Forecast performance relative to a 4-variable Bayesian VAR; unemployment rate

	Relative RMSFE		Relative ALPL	
	$m = 1$	$m = 2$	$m = 1$	$m = 2$
BVAR-Minn	0.99	0.96	0.08	0.33
BVAR-NCP	0.99	0.96	0.11	0.30
BVAR-IP	1.02	0.99	−0.01	−0.03
BVAR-SSVS	1.02	1.01	−0.03	−0.07
BVAR-CSV	1.00	0.95	0.18	0.43
BVAR-CSV-t	0.99	0.95	0.16	0.40
BVAR-CSV-t-MA	0.98	0.96	0.16	0.37

Table 4.5 Forecast performance relative to a 4-variable Bayesian VAR; PCE inflation

	Relative RMSFE		Relative ALPL	
	$m=1$	$m=2$	$m=1$	$m=2$
BVAR-Minn	1.04	1.06	−0.01	0.02
BVAR-NCP	1.04	1.06	−0.02	−0.01
BVAR-IP	1.02	1.00	−0.02	0.00
BVAR-SSVS	1.00	1.02	0.00	0.02
BVAR-CSV	1.04	1.04	0.09	0.11
BVAR-CSV-t	1.03	1.03	0.10	0.10
BVAR-CSV-t-MA	1.03	1.04	0.09	0.08

forecasting results. First, consistent with the results in Banbura et al. (2010) and Koop (2013), large VARs tend to forecast real variables better than the small VAR, whereas the small VAR does better than large models for PCE inflation in terms of point forecasts (see also Stock & Watson, 2007).

Second, among the four priors for large Bayesian VARs, the natural conjugate prior seems to perform well—even when it is not the best among the four, its performance is close to the best. See also a similar comparison in Koop (2013). Given that the natural conjugate prior can substantially speed up computations in posterior simulation, it might be justified to be used as the default in large systems.

Third, the results also show that large Bayesian VARs with more flexible error covariance structures tend to outperform the standard VARs. This is especially so for density forecasts. Our results are consistent with those in numerous studies, such as Clark (2011), D'Agostino, Gambetti, and Giannone (2013) and Clark and Ravazzolo (2015), which find that small Bayesian VARs with stochastic volatility outperform their counterparts with only constant variance. Fourth, even though BVAR-CSV tends to forecast very well, in many instances its forecast performance can be further improved by using the t error distribution or adding an MA component.

Overall, these forecasting results show that large Bayesian VARs tend to forecast well relative to small systems. Moreover, their forecast performance can be further enhanced by allowing for stochastic volatility, heavy-tailed, and serially correlated errors.

4.5 Further Reading

Koop and Korobilis (2010) and Karlsson (2013) are two excellent review papers that cover many of the topics discussed in Sect. 4.2. The presentation of the large Bayesian VARs with time-varying volatility, heavy-tailed distributions, and serial dependent errors in Sect. 4.3 closely follows Chan (2018).

Developing large, flexible Bayesian VARs is an active research area and there are many different approaches. For instance, Koop and Korobilis (2013) consider an approximate method for forecasting using large time-varying parameter Bayesian VARs. Chan, Eisenstat, and Koop (2016) estimate a Bayesian VARMA containing

12 variables. Carriero et al. (2015b) propose an efficient method to estimate a 125-variable VAR with a standard stochastic volatility specification. Koop, Korobilis, and Pettenuzzo (2019) consider compressed VARs based on the random projection method. Ahelegbey, Billio, and Casarin (2016a,b) develop Bayesian graphical models for large VARs. Gefang, Koop, and Poon (2019) use variational approximation for estimating large Bayesian VARs with stochastic volatility.

Appendix A: Data

The real-time dataset for our forecasting application includes 13 macroeconomic variables that are frequently revised and 7 financial or survey variables that are not revised. The list of variables is given in Table 4.6. They are sourced from the Federal Reserve Bank of Philadelphia and cover the quarters from 1964Q1 to 2015Q4. All monthly variables are converted to quarterly frequency by averaging the three monthly values within the quarter.

Table 4.6 Description of variables used in the recursive forecasting exercise

Variable	Transformation
Real GNP/GDP	400Δ log
Real personal consumption expenditures: total	400Δ log
Real gross private domestic investment: nonresidential	400Δ log
Real gross private domestic investment: residential	400Δ log
Real net exports of goods and services	No transformation
Nominal personal income	400Δ log
Industrial production index: total	400Δ log
Unemployment rate	No transformation
Nonfarm payroll employment	400Δ log
Indexes of aggregate weekly hours: total	400Δ log
Housing starts	400Δ log
Price index for personal consumption expenditures, constructed	400Δ log
Price index for imports of goods and services	400Δ log
Effective federal funds rate	No transformation
1-year treasury constant maturity rate	No transformation
10-year treasury constant maturity rate	No transformation
Moody's seasoned baa corporate bond minus federal funds rate	No transformation
ISM manufacturing: PMI composite index	No transformation
ISM manufacturing: new orders index	No transformation
S&P 500	400Δ log

Appendix B: Sampling from the Matrix Normal Distribution

Suppose we wish to sample from $\mathcal{N}(\text{vec}(\widehat{\mathbf{A}}), \mathbf{\Sigma} \otimes \mathbf{K}_{\mathbf{A}}^{-1})$. Let $\mathbf{C}_{\mathbf{K}_{\mathbf{A}}}$ and $\mathbf{C}_{\mathbf{\Sigma}}$ be the Cholesky decompositions of $\mathbf{K}_{\mathbf{A}}$ and $\mathbf{\Sigma}$ respectively. We wish to show that if we construct

$$\mathbf{W}_1 = \widehat{\mathbf{A}} + (\mathbf{C}'_{\mathbf{K}_{\mathbf{A}}} \backslash \mathbf{U})\mathbf{C}'_{\mathbf{\Sigma}},$$

where $\mathbf{U}$ is a $k \times n$ matrix of independent $\mathcal{N}(0, 1)$ random variables, then $\text{vec}(\mathbf{W}_1)$ has the desired distribution. To that end, we make use of some standard results on the matrix normal distribution (see, e.g., Bauwens et al., 1999, pp. 301–302).

A $p \times q$ random matrix $\mathbf{W}$ is said to have a **matrix normal distribution** $\mathcal{MN}(\mathbf{M}, \mathbf{Q} \otimes \mathbf{P})$ for covariance matrices $\mathbf{P}$ and $\mathbf{Q}$ of dimensions $p \times p$ and $q \times q$, respectively, if $\text{vec}(\mathbf{W}) \sim \mathcal{N}(\text{vec}(\mathbf{M}), \mathbf{Q} \otimes \mathbf{P})$. Now suppose $\mathbf{W} \sim \mathcal{MN}(\mathbf{M}, \mathbf{Q} \otimes \mathbf{P})$ and define $\mathbf{V} = \mathbf{CWD} + \mathbf{E}$. Then, $\mathbf{V} \sim \mathcal{MN}(\mathbf{CMD} + \mathbf{E}, (\mathbf{D}'\mathbf{QD}) \otimes (\mathbf{CPC}'))$.

Recall that $\mathbf{U}$ is a $k \times n$ matrix of independent $\mathcal{N}(0, 1)$ random variables. Hence, $\mathbf{U} \sim \mathcal{MN}(\mathbf{0}, \mathbf{I}_n \otimes \mathbf{I}_k)$. Using the previous result with $\mathbf{C} = (\mathbf{C}'_{\mathbf{K}_{\mathbf{A}}})^{-1}$, $\mathbf{D} = \mathbf{C}'_{\mathbf{\Sigma}}$, and $\mathbf{E} = \widehat{\mathbf{A}}$, it is easy to see that $\mathbf{W}_1 \sim \mathcal{MN}(\widehat{\mathbf{A}}, \mathbf{\Sigma} \otimes \mathbf{K}_{\mathbf{A}}^{-1})$. Finally, by definition we have $\text{vec}(\mathbf{W}_1) \sim \mathcal{N}(\text{vec}(\widehat{\mathbf{A}}), \mathbf{\Sigma} \otimes \mathbf{K}_{\mathbf{A}}^{-1})$.

References

Ahelegbey, D. F., Billio, M., & Casarin, R. (2016a). Bayesian graphical models for structural vector autoregressive processes. *Journal of Applied Econometrics, 31*(2), 357–386.

Ahelegbey, D. F., Billio, M., & Casarin, R. (2016b). Sparse graphical multivariate autoregression: A Bayesian approach. *Annals of Economics and Statistics, 123/124*, 1–30.

Banbura, M., Giannone, D., & Reichlin, L. (2010). Large Bayesian vector auto regressions. *Journal of Applied Econometrics, 25*(1), 71–92.

Bauwens, L., Lubrano, M., & Richard, J. (1999). *Bayesian inference in dynamic econometric models*. New York, NY: Oxford University Press.

Carriero, A., Clark, T. E., & Marcellino, M. (2015a). Bayesian VARs: Specification choices and forecast accuracy. *Journal of Applied Econometrics, 30*(1), 46–73.

Carriero, A., Clark, T. E., & Marcellino, M. (2015b). Large vector autoregressions with asymmetric priors and time-varying volatilities. *Working Paper, School of Economics and Finance, Queen Mary University of London.*

Carriero, A., Clark, T. E., & Marcellino, M. (2016). Common drifting volatility in large Bayesian VARs. *Journal of Business and Economic Statistics, 34*(3), 375–390.

Carriero, A., Kapetanios, G., & Marcellino, M. (2009). Forecasting exchange rates with a large Bayesian VAR. *International Journal of Forecasting, 25*(2), 400–417.

Chan, J. C. C. (2013). Moving average stochastic volatility models with application to inflation forecast. *Journal of Econometrics, 176*(2), 162–172.

Chan, J. C. C. (2017). The stochastic volatility in mean model with time-varying parameters: An application to inflation modeling. *Journal of Business and Economic Statistics, 35*(1), 17–28.

Chan, J. C. C. (2018). Large Bayesian VARs: A flexible Kronecker error covariance structure. *Journal of Business and Economic Statistics*. https://doi.org/10.1080/07350015.2018.1451336

Chan, J. C. C., Eisenstat, E., & Koop, G. (2016). Large Bayesian VARMAs. *Journal of Econometrics, 192*(2), 374–390.
Chan, J. C. C., & Hsiao, C. Y. L. (2014). Estimation of stochastic volatility models with heavy tails and serial dependence. In I. Jeliazkov & X.-S. Yang (Eds.), *Bayesian inference in the social sciences*. Hoboken, NJ: Wiley.
Chan, J. C. C., & Jeliazkov, I. (2009). Efficient simulation and integrated likelihood estimation in state space models. *International Journal of Mathematical Modelling and Numerical Optimisation, 1*, 101–120.
Chiu, C. J., Mumtaz, H., & Pinter, G. (2017). Forecasting with VAR models: Fat tails and stochastic volatility. *International Journal of Forecasting, 33*(4), 1124–1143.
Clark, T. E. (2011). Real-time density forecasts from Bayesian vector autoregressions with stochastic volatility. *Journal of Business & Economic Statistics, 29*(3), 327–341.
Clark, T. E., & Ravazzolo, F. (2015). Macroeconomic forecasting performance under alternative specifications of time-varying volatility. *Journal of Applied Econometrics, 30*(4), 551–575.
Cogley, T., & Sargent, T. J. (2005). Drifts and volatilities: Monetary policies and outcomes in the post WWII US. *Review of Economic Dynamics, 8*(2), 262–302.
Cross, J., & Poon, A. (2016). Forecasting structural change and fat-tailed events in Australian macroeconomic variables. *Economic Modelling, 58*, 34–51.
D'Agostino, A., Gambetti, L., & Giannone, D. (2013). Macroeconomic forecasting and structural change. *Journal of Applied Econometrics, 28*, 82–101.
Doan, T., Litterman, R., & Sims, C. (1984). Forecasting and conditional projection using realistic prior distributions. *Econometric Reviews, 3*(1), 1–100.
Eltoft, T., Kim, T., & Lee, T. (2006a). Multivariate scale mixture of Gaussians modeling. In J. Rosca, D. Erdogmus, J. Principe, & S. Haykin (Eds.), *Independent component analysis and blind signal separation* (Vol. 3889, pp. 799–806). Lecture Notes in Computer Science. Berlin: Springer.
Eltoft, T., Kim, T., & Lee, T. (2006b). On the multivariate Laplace distribution. *IEEE Signal Processing Letters, 13*(5), 300–303.
Forni, M., Hallin, M., Lippi, M., & Reichlin, L. (2003). Do financial variables help forecasting inflation and real activity in the euro area? *Journal of Monetary Economics, 50*(6), 1243–1255.
Gefang, D., Koop, G., & Poon, A. (2019). Variational Bayesian inference in large vector autoregressions with hierarchical shrinkage. *CAMA Working Paper*.
George, E. I., Sun, D., & Ni, S. (2008). Bayesian stochastic search for var model restrictions. *Journal of Econometrics, 142*(1), 553–580.
Geweke, J. (1993). Bayesian treatment of the independent Student-t linear model. *Journal of Applied Econometrics, 8*(S1), S19–S40.
Geweke, J., & Amisano, G. (2011). Hierarchical Markov normal mixture models with applications to financial asset returns. *Journal of Applied Econometrics, 26*, 1–29.
Golub, G. H., & van Loan, C. F. (1983). *Matrix computations*. Baltimore, MD: Johns Hopkins University Press.
Kadiyala, K., & Karlsson, S. (1997). Numerical methods for estimation and inference in Bayesian VAR-models. *Journal of Applied Econometrics, 12*(2), 99–132.
Karlsson, S. (2013). Forecasting with Bayesian vector autoregressions. In G. Elliott & A. Timmermann (Eds.), *Handbook of economic forecasting* (Vol. 2, pp. 791–897). Handbook of Economic Forecasting. Amsterdam: Elsevier.
Koop, G. (2013). Forecasting with medium and large Bayesian VARs. *Journal of Applied Econometrics, 28*(2), 177–203.
Koop, G., & Korobilis, D. (2010). Bayesian multivariate time series methods for empirical macroeconomics. *Foundations and Trends in Econometrics, 3*(4), 267–358.
Koop, G., & Korobilis, D. (2013). Large time-varying parameter VARs. *Journal of Econometrics, 177*(2), 185–198.
Koop, G., Korobilis, D., & Pettenuzzo, D. (2019). Bayesian compressed vector autoregression. *Journal of Econometrics, 210*(1), 135–154.

Litterman, R. (1986). Forecasting with Bayesian vector autoregressions—five years of experience. *Journal of Business and Economic Statistics, 4*, 25–38.
Mumtaz, H. (2016). The evolving transmission of uncertainty shocks in the United Kingdom. *Econometrics, 4*(1), 16.
Mumtaz, H., & Theodoridis, K. (2018). The changing transmission of uncertainty shocks in the U.S. *Journal of Business and Economic Statistics, 36*(2), 239–252.
Poon, A. (2018). Assessing the synchronicity and nature of Australian state business cycles. *Economic Record, 94*(307), 372–390.
Primiceri, G. E. (2005). Time varying structural vector autoregressions and monetary policy. *Review of Economic Studies, 72*(3), 821–852.
Sims, C. A. (1980). Macroeconomics and reality. *Econometrica, 48*, 1–48.
Stock, J. H., & Watson, M. W. (2002). Macroeconomic forecasting using diffusion indexes. *Journal of Business and Economic Statistics, 20*, 147–162.
Stock, J. H., & Watson, M. W. (2007). Why has U.S. inflation become harder to forecast? *Journal of Money Credit and Banking, 39*, 3–33.

Chapter 5
Volatility Forecasting in a Data Rich Environment

Mauro Bernardi, Giovanni Bonaccolto, Massimiliano Caporin, and Michele Costola

5.1 Introduction

Volatility forecasting has a central role in many economic and financial frameworks, ranging from asset allocation, risk management and trading to macroeconomic forecasting. Dozens of authors introduced and discussed econometric models to better capture the stylized features of observed data to provide optimal predictions of the variance for either a single variable or a vector of variables of interest. The early works on models for volatility forecasting appeared after 1980, starting from the seminal contribution of Engle (1982). Nowadays, the various models might be clustered into several families according to either the data used to estimate and then forecast the volatility or to the model structure.

When focusing on variance (or volatility), covariance and correlation forecasting, the first model that is usually mentioned is the Generalized Auto Regressive Conditional Heteroskedasticity (GARCH) model. The modelling of variance dynamics

M. Bernardi
Department of Statistical Sciences, University of Padova and Istituto per le Applicazioni del Calcolo "Mauro Picone" - CNR, Roma, Italy
e-mail: mauro.bernardi@unipd.it

G. Bonaccolto
School of Economics and Law, University of Enna Kore, Enna, Italy
e-mail: giovanni.bonaccolto@unikore.it

M. Caporin (✉)
Department of Statistical Sciences, University of Padova, Padova, Italy
e-mail: massimiliano.caporin@unipd.it

M. Costola
SAFE, Goethe University Frankfurt, Frankfurt, Germany
e-mail: costola@safe.uni-frankfurt.de

P. Fuleky (ed.), *Macroeconomic Forecasting in the Era of Big Data*,
Advanced Studies in Theoretical and Applied Econometrics 52,
https://doi.org/10.1007/978-3-030-31150-6_5

evolved from the seminal works of Engle (1982) and Bollerslev (1986) both in terms of GARCH-type specifications, either univariate or multivariate, and through the introduction of additional models. The family of volatility models now includes models for both latent and observable (co-)variance sequences (Caporin & McAleer, 2010). Among the former, beside the GARCH models, we might include stochastic volatility models, as pioneered by Taylor (1986), and multivariate models based on latent factors. The latter includes models based on realized volatility and realized covariance sequences. These model classes have been reviewed by several authors, including Bollerslev, Chou, and Kroner (1992), Asai, McAleer, and Yu (2006), Bauwens, Laurent, and Rombouts (2006) and McAleer and Medeiros (2006). In the last decade the literature moved from the development of novel and more flexible specifications to a few aspects of volatility modelling. These include the derivation of the asymptotic behaviour of parameter estimators, the introduction of multivariate model specification feasible with large cross-sectional dimensions (either by proper model structures or by tailored estimation approaches), i.e. in the presence of the so-called curse of dimensionality (see, for instance, Caporin and Paruolo, 2015; Dhaene, Sercu, and Wu, 2017; Noureldin, Shephard, and Sheppard, 2014) and the development and introduction of models for realized covariances (Golosnoy, Gribisch, & Liesenfeld, 2012). The development and use of volatility models with large cross-sectional dimensions remains a challenging topic, particularly when we adopt those models to produce forecasts in portfolio allocation, risk management, hedging and in many other areas not limited to a financial framework. One key issue is that models need to remain feasible when the number of assets increases, and have a specification flexible enough to allow for an economic interpretation of the model parameters; see Silvennoinen and Teräsvirta (2009). The increased availability of high-frequency data points at the possibility of using high-frequency data to estimate realized covariance sequences. These, in turn, might be later directly modelled or integrated within GARCH-type models; see Shepard and Sheppard (2010). Finally, the availability of huge data sources on financial companies, including not just market-related and balance-sheet-related variables but also web-based searches and textual data on news and tweets, as well as the possibility of recovering interdependence structures (i.e. networks) among target variables, call for additional efforts in model development; see, among others, Caporin and Poli (2017) and Billio, Caporin, Frattarolo, and Pelizzon (2018).

The various models that allow forecasting volatility, covariance and correlations share a common feature: recovering forecasts from these models is rather simple. The complexity is in the model estimation, particularly when the cross-sectional dimension is large. The curse of dimensionality, a typical problem in conditional covariance modelling, occurs due to the increase in the number of parameters. Coping with both issues remains a challenging topic. In this chapter, we review the volatility modelling literature starting from the univariate models and moving toward the multivariate specifications with a focus on how the current models allow for the management of large cross-sectional dimensions or the integration of different data sources. We will address these issues by focusing on three main model families, namely, the classical GARCH-type specifications, the stochastic volatility model class and the models based on high-frequency data, i.e. the Realized Volatility model class. We will not discuss in depth the tools for forecast evaluation; instead, we refer the readers to Chapter 24 or to the survey by Violante and Laurent (2012) and the references cited therein.

5.2 Classical Tools for Volatility Forecasting: ARCH Models

Volatility clustering, i.e. the presence of serial dependence in the second order moment, is one of the most known stylized facts that characterizes financial returns time series. The GARCH model class, introduced by Engle (1982) and Bollerslev (1986), allows capturing this feature by modelling the conditional variance with linear or non-linear specifications and has become a reference tool widely used in risk management and option pricing. In this section, we first briefly review the GARCH class both at the univariate and multivariate levels, pointing at the most relevant specifications. Second, we discuss the specifications and approaches addressing the issues that might be encountered within a data rich environment when the final purpose is volatility (or covariance) forecasting. For a detailed review and discussion on univariate GARCH models, see Bollerslev, Engle, and Nelson (1994) and Teräsvirta (2009), while for multivariate GARCH models we refer the reader to Bauwens et al. (2006), Silvennoinen and Teräsvirta (2009) and Caporin and McAleer (2012).

5.2.1 *Univariate GARCH Models*

Among the several univariate GARCH models (see the survey by Bollerslev, 2010), we choose only a few cases, the most relevant ones, and we limit ourselves to the simplest specifications. Let x_t define the sequence of the logarithmic difference for the price P_t of a given asset; then, the conditional mean is characterized as follows:

$$x_t = E(x_t|\Omega_{t-1}) + \varepsilon_t,\ \varepsilon_t \equiv h_t^{\frac{1}{2}} z_t,\ z_t \sim iid\,(0,1),$$

where Ω_{t-1} is the information set at time $t-1$, h_t is the conditional variance and z_t is an innovation term with zero mean and unit variance following an i.i.d distribution. We might then consider alternative specifications for h_t.

The GARCH model of Bollerslev (1986) provides the following dynamic for the conditional variances:

$$h_t = \omega + \beta h_{t-1} + \alpha \varepsilon_{t-1}^2.$$

The parameters must satisfy the constraints $\omega > 0, \alpha \geq 0, \beta \geq 0$ and $\alpha + \beta < 1$ to ensure the positivity and the stationarity of the conditional variance h_t.

The Exponential GARCH (EGARCH) model by Nelson (1991) defines the dynamic on the logarithm of the conditional variance. Therefore, it does not require restrictions on the parameters for the positivity of h_t. The EGARCH model is defined as follows:

$$\log h_t = \omega + \beta \log h_{t-1} + \alpha z_{t-1} + \kappa \left(|z_{t-1}| - E|z_{t-1}| \right).$$

A sufficient condition for covariance stationarity is that $|\beta| < 1$. The EGARCH model allows for different effects of negative and positive shocks on the conditional variance, which depend on α and κ.

The GJR-GARCH model of Glosten, Jagannathan, and Runkle (1993) adds a dummy for negative shocks with respect to the GARCH model:

$$h_t = \omega + \beta h_{t-1} + \alpha \varepsilon_{t-1}^2 + \gamma \varepsilon_{t-1}^2 \mathbb{1}_{[\varepsilon_{t-1} < 0]},$$

where $\mathbb{1}_{(.)}$ is the indicator function. The coefficient γ measures the degree of asymmetry of negative shocks. Sufficient conditions for the positivity of conditional variances (h_t) require that $\omega \geq 0$, $\alpha \geq 0$, $\beta \geq 0$, $\gamma \geq 0$, while for covariance stationarity, under the additional assumption of symmetry for the density of $z_t = h_t^{-1/2} \varepsilon_t$, we need $\alpha + \beta + \frac{\gamma}{2} < 1$.

Finally, the Asymmetric Power ARCH (APARCH) model of Ding, Granger, and Engle (1993) can be viewed as a non-linear generalization of the GARCH and GJR-GARCH models. The parameter $\delta > 0$ drives the non-linearity and represents the power of the conditional volatility over which we define the dynamic. The model reads as follows:

$$h_t^\delta = \omega + \beta h_{t-1}^\delta + \alpha \left(|\varepsilon_{t-1}| - \varphi \varepsilon_{t-1} \right)^\delta,$$

where $|\varphi| \leq 1$ and $\delta > 0$. An excellent discussion for necessary and sufficient moment conditions is provided by Ling and McAleer (2002). The APARCH model nests several univariate GARCH specifications.

Forecasting volatility from GARCH-type models requires limited efforts as, by default, one-step-ahead forecasts are a bi-product of model estimation. For instance, for the GARCH model, we have:

$$h_{T+1} = \hat{\omega} + \hat{\beta} h_T + \hat{\alpha} \varepsilon_T^2,$$

where we assume we estimate parameters using a sample from 1 to T. Multi-step-ahead forecasts might be easily obtained by recursive substitutions, replacing unknown conditional variances with their predicted values.

These models, as well as many other univariate GARCH specifications, can be easily extended with the introduction of exogenous information, leading to GARCHX models (see, among many others, Fleming, Kirby, and Ostdiek, 2008).

Referring to the simple GARCH model, we could have

$$h_t = \omega + \beta h_{t-1} + \alpha \varepsilon_{t-1}^2 + \delta' \mathbf{X}_{t-1},$$

where $\mathbf{X}_t$ is the vector of conditioning information, and the parameter vector δ might be constrained to ensure the positivity of the conditional variance. Issues associated with the data dimension emerge in GARCH models when they include conditioning variables. In fact, the number of potentially relevant drivers we could include in $\mathbf{X}_t$ might be huge. For instance, we might introduce a large number of macroeconomic variables that could impact on the uncertainty of a given variable of interest, or we might collect several financial indicators that could impact on the volatility of a specific financial instrument. This clearly challenges the estimation of even univariate models and, consequently, their use from a forecasting perspective. However, there are two approaches we might consider to deal with the dimensionality (of covariates) issue. The first makes use of dimension reduction techniques, where we summarize the informative content of covariates into a few indicators. The classical example is principal component analysis, but other dimension reduction approaches might be used. The second approach points at regularization methods, for instance, the Least Absolute Shrinkge and Selection Operator (LASSO) (Tibshirani, 1996), that is introducing in the model estimation step a penalty component that helps in selecting the most relevant covariates.

5.2.2 *Multivariate GARCH Models*

Multivariate GARCH (MGARCH) models are a natural extension of the univariate specifications for the modelling of covariance and/or correlation matrices. The mean equation becomes

$$\mathbf{x}_t = E(\mathbf{x}_t|\Omega_{t-1}) + \boldsymbol{\varepsilon}_t, \; \boldsymbol{\varepsilon}_t \equiv \mathbf{H}_t^{\frac{1}{2}} \mathbf{z}_t, \; \mathbf{z}_t \sim iid\,(0, \mathbf{I}),$$

where now $\mathbf{x}_t$ is a $(N \times 1)$ vector and $\mathbf{H}_t$ is a $N \times N$ positive definite matrix, the conditional covariance matrix. Following a standard practice, we cluster MGARCH models into two families: the conditional covariance models and the conditional correlations models. The two groups differ, as the first explicitly models conditional covariances, while the second explicitly models conditional correlations. We briefly review here the most common models, and we refer the reader to Bauwens et al. (2006) and Silvennoinen and Teräsvirta (2009) for additional details on the following models as well as on additional specifications. For all the models we consider, recovering the forecasts of conditional covariances and correlations is relatively simple because, by construction, the models implicitly provide the one-step-ahead predictions as a bi-product of model estimation. Multi-step-ahead forecasts can be easily obtained by recursive substitutions.

Conditional Covariance Models

The prototype model in this class is the VECH model proposed by Bollerslev, Engle, and Wooldridge (1988), where each element of the matrix $\mathbf{H}_t$ is a linear function of past values of $\mathbf{H}_t$ and the lagged and square products of mean innovations errors:

$$vech(\mathbf{H}_t) = \omega + \mathbf{A}vech(\varepsilon_{t-1}\varepsilon_{t-1}') + \mathbf{B}vech(\mathbf{H}_{t-1}),$$

where $vech(\cdot)$ denotes the vectorization operator, which stacks the lower portion of a $N \times N$ symmetric matrix in a $m = N \times (N+1)/2 \times 1$ vector, and ω is the vector of the same dimension for the constant terms. Finally, $\mathbf{A}$ and $\mathbf{B}$ are square matrices of parameters with dimensions $m \times m$. This model is the most flexible in this class. Nevertheless, it is highly exposed to the curse of dimensionality, as the number of parameters to be estimated is of order $O\left(N^4\right)$ and parameters must be non-linearly constrained to ensure a positive definite conditional covariance. This makes the model infeasible even in small cross-sectional dimensions.

A relevant alternative is the BEKK model of Engle and Kroner (1995), which reads as follows:

$$\mathbf{H}_t = \mathbf{C}\mathbf{C}' + \mathbf{A}\varepsilon_{t-1}\varepsilon_{t-1}'\mathbf{C}' + \mathbf{B}\mathbf{H}_{t-1}\mathbf{B}',$$

where $\mathbf{A}$ and $\mathbf{B}$ are now $N \times N$ matrices (symmetry is not required) and $\mathbf{C}$ is lower triangular. This model is less exposed to the curse of dimensionality, which is, however, still present as the parameter number is of order $O\left(N^2\right)$. Nevertheless, the absence of relevant parameter constraints (apart from a simple positivity constraint on parameters of position $(1, 1)$ in $\mathbf{A}$ and $\mathbf{B}$) makes the model feasible in small or moderate cross-sectional dimensions.

A third relevant model is the Orthogonal GARCH (OGARCH) of Alexander (2001), in which we first rotate the mean innovations moving to the principal components and then we fit univariate GARCH:

$$\begin{aligned}
&E\left(\varepsilon_t\varepsilon_t'\right) = \Sigma = \mathbf{L}\mathbf{D}\mathbf{L}',\\
&\mathbf{u}_t = \mathbf{L}'\varepsilon_t, \quad E\left(\mathbf{u}_t\mathbf{u}_t'|I_{t-1}\right) = diag\left(\mathbf{h}_t\right),\\
&h_{i,t} = \left(1-\alpha_i-\beta_i\right)d_{i,i} + \alpha_i u_{i,t-1}^2 + \beta_i h_{i,t-1},\\
&E\left(\varepsilon_t\varepsilon_t'|I_{t-1}\right) = \mathbf{L}diag\left(\mathbf{h}_t\right)\mathbf{L}',
\end{aligned}$$

where $\mathbf{L}$ is the matrix of eigenvectors, $\mathbf{D}$ is the diagonal matrix of eigenvalues (with $d_{i,i}$ being a single element of $\mathbf{D}$) and $\mathbf{u}_t$ is the vector of principal components, with a conditional covariance that is equal to a diagonal matrix, in which the elements in the main diagonal are equal to $\mathbf{h}_t$. Furthermore, the conditional variances of the principal components follow a GARCH(1,1) with the intercept targeted to the specific eigenvalue and the dynamic driven by the parameters α_i and β_i. The OGARCH model requires the use of a sample estimator to recover the principal components. It then moves to the univariate GARCH estimation, where parameter constraints are required only in the last step to ensure positivity of the conditional variances. This procedure allows fitting the model even in large cross-sectional dimensions. However, from a forecasting perspective, the models make a relevant implicit assumption, that is, that the loadings matrix (the eigenvector matrix) is time-invariant.

Conditional Correlation Models

Multivariate GARCH models based on conditional variances and correlations require fewer parameters compared to models for conditional covariances and are thus feasible in moderate or even large cross-sectional dimensions. The first model in this family is the Constant Conditional Correlation (CCC) proposed by Bollerslev (1990), where univariate conditional variance specifications are accompanied by an unconditional correlation matrix:

$$\mathbf{H}_t = \mathbf{D}_t \mathbf{R} \mathbf{D}_t,$$

where $\mathbf{D}_t = \left(h_{11t}^{1/2}, \ldots, h_{NNt}^{1/2}\right)$, h_{ii} is a conditional variance obtained from a univariate GARCH process on ε_t and R is unconditional correlation matrix of $\mathbf{v}_t = \mathbb{D}_t^{-1}\varepsilon_t$.

The Dynamic Conditional Correlation (DCC) model introduced by Engle (2002) generalizes the CCC model, allowing for dynamic correlations and leading to

$$\mathbf{H}_t = \mathbf{D}_t \mathbf{R}_t \mathbf{D}_t,$$

where

$$R_t = \bar{\mathbf{Q}}_t^{-1/2} \mathbf{Q}_t \bar{\mathbf{Q}}_t^{-1/2}, \quad \bar{\mathbf{Q}}_t = \operatorname{diag}\left(q_{11,t}, \ldots, q_{NN,t}\right) \tag{5.1}$$

and $\mathbf{Q}_t$ is equal to

$$\mathbf{Q}_t = (1 - \alpha - \beta)\mathbf{S} + \alpha \mathbf{u}_{t-1}\mathbf{u}_{t-1}' + \beta \mathbf{Q}_{t-1}, \tag{5.2}$$

where $u_{it} = \varepsilon_{i,t} / \sqrt{h_{ii,t}}$, $\mathbf{S}$ is a symmetric positive definite matrix close to the unconditional correlation matrix (Aielli, 2013) while α and β are scalar parameters satisfying $\alpha + \beta < 1$ and $\alpha, \beta > 0$ to guarantee covariance stationarity.

In both the CCC and DCC models, the curse of dimensionality has a limited impact since the model estimation requires a collection of steps pointing either at univariate GARCH estimation or at the adoption of sample moments. However, this flexibility comes with a relevant drawback, as the models completely neglect the presence of spillovers or interdependence among conditional variances, covariances and correlations.

5.2.3 Dealing with Large Dimension in Multivariate Models

When dealing with MGARCH models, the curse of dimensionality represents a crucial issue. While on the one side conditional correlation models might represent a solution, on the other side we do have interest in recovering model structures that, besides being feasible, also lead to economically relevant relationships, and thus there is a need to include spillovers. In turn, these elements would allow for better covariance forecasts, where the latter can be easily obtained from the model's dynamic structures. In MGARCH specifications, similarly to univariate GARCH, one-step-ahead forecasts are a by-product of model estimation while multi-step forecasts derive from simple recursions. Consequently, obtaining forecasts is not complex once we have an estimated model; the complicated step is the estimation of a useful, from an economic point of view, model specification.

Several proposals have been introduced in the financial econometrics literature since the introduction of the DCC model in 2002. These proposals follow two main research lines. The first highlights the introduction of model parametrizations that limit the curse of dimensionality or cope with the curse of dimensionality while trying to recover some elements from the interpretation point of view or in terms of variance spillovers. The second strand focuses on the introduction of estimation methods that try to circumvent the dimensionality issue.

With respect to the introduction of restricted parametrizations, several authors focused on the VECH and BEKK models, introducing scalar and diagonal specifications, where the parameter matrices capturing the interdependence between variances and covariances include a much smaller number of parameters. While this choice makes the models feasible in moderate cross-sectional dimensions, the curse of dimensionality is still present, as it depends on the covariance intercepts that still include $O\left(N^2\right)$ parameters. In addition, by moving to scalar or diagonal specifications, the models completely lose the presence of spillovers, making them less relevant from an empirical point of view. Examples are given by Attanasio (1991), Marcus and Minc (1992) and Ding and Engle (2001), where, for instance, we might choose one of the following structures for the $\mathbf{A}$ matrix of a BEKK model (and similarly for the B matrix or for the VECH model):

$$\mathbf{A} = \alpha I, \quad \mathbf{A} = diag\left(\mathbf{a}\right), \quad \mathbf{A} = \mathbf{a}\mathbf{a}', \tag{5.3}$$

where the N-dimensional vector $\mathbf{a}$ might include N different parameters or a smaller number leading to common parameters for groups of variables.

Opposite to the choice of a restricted parameterization, in the DCC model case that, in its most common specification, includes scalar parameters, the literature attempts to re-introduce flexibility by proposing diagonal or similar structures; see, among many others, Billio, Caporin, and Gobbo (2006), Cappiello, Engle, and Sheppard (2006) and Caporin and McAleer (2012). This leads to parameter matrices structured as in Eq. (5.3).

A different approach, potentially crossing all models, makes use of the known interdependence relationships among the modelled variables to provide restricted and interpretable model specifications (see Caporin and Paruolo, 2015 and Billio et al., 2018). In these approaches, the parameter matrices have a structure driven, for instance, by proximity relationships or observed networks linking the variables of interest, leading, in the BEKK case, to the following form for the parameter matrix **A** (and similarly for **B**):

$$\mathbf{A} = diag\left(\mathbf{a}_0\right) + diag\left(\mathbf{a}_0\right) W,$$

where **W** is the adjacency matrix of an observed network, and the matrix **A** depends on two vector or parameters. The network proxies for the interdependence allow for the presence of spillovers with an additional, but limited, set of parameters compared to diagonal specifications.

Finally, we mention models that cope with the curse of dimensionality by introducing a rotation of the observed variables before estimating a multivariate dynamic model. These models are closely related to the OGARCH but try to overcome its limitations. We mention here the approach of Van der Weide (2002) and the more recent RARCH model of Noureldin et al. (2014) in which a rotation of the variables is followed by a BEKK-type model with a restricted intercept, thus potentially solving the curse of dimensionality if the parameter matrices are set to be diagonal.

A second set of papers focuses on estimation methods. On the one side we have approaches dealing with the dimensionality issue by resorting to composite likelihood approaches (Bauwens, Grigoryeva, & Ortega, 2016; Pakel, Shephard, Sheppard, & Engle, 2017). Despite the appeal of these suggestions, the estimated models might be affected by the absence of spillovers due to the adoption of a diagonal specification for the parameter matrices. A different approach is taken by Dhaene et al. (2017), who introduced a LASSO penalization in the estimation of the BEKK model parameters. This allows for the presence of spillovers, but the estimation in large cross-sectional dimensions would be complex due to the presence of the model intercept.

5.3 Stochastic Volatility Models

In recent decades, stochastic volatility models have gained increasing interest in mathematical finance, financial econometrics and asset and risk management (see, among others, Hull & White, 1987; Heston, 1993; Ghysels, Harvey, & Renault, 1996). Particularly, they have proven to be useful tools for pricing financial assets and derivatives and estimating efficient portfolio allocations. Like the GARCH models, they go beyond Black and Scholes (1973)'s restrictive assumption that volatility is constant, trying to capture the main stylized facts characterizing financial time series, such as volatility clustering and fat-tailed distributions (see, e.g., Cont, 2001), or the leverage effect highlighted by Black (1976). Stochastic volatility models represent a valid alternative to GARCH models, which are relatively easy to estimate but have relevant drawbacks. For instance, they impose parameter restrictions that are often violated by the estimated coefficients (Nelson, 1991). While returns volatility evolves according to a specific function in GARCH models, it follows an unobserved random process in stochastic volatility models (see, e.g., Melino and Turnbull, 1990; Harvey, Ruiz, and Shephard, 1994; Jacquier, Polson, and Rossi, 1994; Taylor, 1994). Stochastic volatility models can be defined in either a discrete or a continuous time setting. The models belonging to the first category (discrete) are mainly used by econometricians for risk and portfolio management, whereas the ones included in the second category (continuous) are mainly used by mathematicians for pricing derivatives (Meucci, 2010). Here, we focus on discrete-time stochastic volatility models, whereas we use the continuous time setting to derive realized volatility measures in Sect. 5.4.1.

5.3.1 Univariate Stochastic Volatility Models

Starting from the univariate setting, a standard specification (see, e.g., Taylor, 1986; Taylor, 1994; Jacquier et al., 1994; Kim, Shephard, & Chib, 1998b; Liesenfeld & Richard, 2003; Hautsch & Ou, 2008 reads as follows:

$$r_t = \exp\left(\frac{h_t}{2}\right)\epsilon_t \tag{5.4a}$$

$$h_t = \alpha + \beta h_{t-1} + \eta_t, \tag{5.4b}$$

where r_t and h_t are, respectively, the log-return and the log-volatility of a financial asset at time t, for $t = 1, 2, \ldots, T$, whereas the error terms $\epsilon_t \sim \mathcal{N}(0, 1)$ and $\eta_t \sim \mathcal{N}(0, \sigma_\eta^2)$ are mutually independent Gaussian white noise processes. Therefore, the model in Eqs. (5.4a)–(5.4b) treats the log-volatility as a latent, unobserved variable following an AR(1) process. Under the condition $|\beta| < 1$, the process is stationary.

Unfortunately, the likelihood function of r_t does not have an analytical expression and, therefore, is difficult to evaluate. As a result, the estimation of the parameters in Eqs. (5.4a)–(5.4b) requires nontrivial econometric techniques. Among them, we mention the generalized method of moments (see, e.g., Andersen & Sorensen, 1996; Jacquier et al., 1994; Melino & Turnbull, 1990), quasi-maximum likelihood estimation (see, e.g., Harvey et al., 1994; Melino & Turnbull, 1990) and Markov Chain Monte Carlo methods (see, e.g., Kim et al., 1998b; Melino & Turnbull, 1990).

Harvey et al. (1994) and Chib, Nardari, and Shephard (2002), among others, specified a Student-t distribution for the error term ϵ_t. In addition, it is also possible to include a jump component in either Eq. (5.4a) or (5.4b) or both (see, e.g., Bates, 2000; Duffie, Pan, and Singleton, 2000). To capture the leverage effect, Shephard (1996) and Jacquier, Polson, and Rossi (2004) introduced a negative correlation between the contemporaneous error terms ϵ_t and η_t in Eqs. (5.4a) and (5.4b).

5.3.2 Multivariate Stochastic Volatility Models

Univariate stochastic volatility models have the drawback of not capturing the co-movements of different assets, which become particularly relevant during periods of financial distress. As a result, they are of little use in applications where the correlation structure of financial assets plays a relevant role, such as in asset allocation and risk management. This drawback has stimulated the development of multivariate stochastic volatility models.

The first multivariate stochastic volatility model introduced in the literature was proposed by Harvey et al. (1994) and is described below. Let $\mathbf{r}_t = (r_{1,t}, r_{2,t}, \ldots, r_{N,t})'$ be the $N \times 1$ vector including the log-returns of N stocks at time t, where $r_{i,t} = \exp(h_{i,t}/2)\epsilon_{i,t}$ and $h_{i,t} = \alpha_i + \beta_i h_{i,t-1} + \eta_{i,t}$, for $i = 1, 2, \ldots, N$ and $t = 1, 2, \ldots, T$. Here, we assume that the vectors of error terms $\boldsymbol{\epsilon}_t = (\epsilon_{1,t}, \epsilon_{2,t}, \ldots, \epsilon_{N,t})'$ and $\boldsymbol{\eta}_t = (\eta_{1,t}, \eta_{2,t}, \ldots, \eta_{N,t})'$ are serially and mutually independent, being distributed as follows:

$$\begin{bmatrix} \boldsymbol{\epsilon}_t \\ \boldsymbol{\eta}_t \end{bmatrix} \sim \mathcal{N}\left(\begin{bmatrix} \mathbf{0} \\ \mathbf{0} \end{bmatrix}, \begin{bmatrix} \boldsymbol{\Sigma}_\epsilon & \mathbf{0} \\ \mathbf{0} & \boldsymbol{\Sigma}_\eta \end{bmatrix} \right),$$

where $\boldsymbol{\Sigma}_\eta$ is a positive definite covariance matrix, whereas $\boldsymbol{\Sigma}_\epsilon$ defines the correlation matrix of assets returns.

Then, the model proposed by Harvey et al. (1994) reads as follows:

$$\mathbf{r}_t = \mathbf{H}_t^{1/2} \boldsymbol{\epsilon}_t \tag{5.5}$$

for $t = 1, 2, \ldots, T$, where $\mathbf{H}_t = \text{diag}\{\exp(h_{1,t}), \exp(h_{2,t}), \ldots, \exp(h_{N,t})\}$ is the diagonal matrix of time-varying volatilities, whose elements follow a restricted VAR(1) process, that is:

$$\mathbf{h}_t = \boldsymbol{\alpha} + \boldsymbol{\beta} \mathbf{h}_{t-1} + \boldsymbol{\eta}_t,$$

where $\mathbf{h}_t = (h_{1,t}, h_{2,t}, \ldots, h_{N,t})'$ is an $N \times 1$ vector, $\boldsymbol{\alpha} = (\alpha_1, \alpha_2, \ldots, \alpha_N)'$ is an $N \times 1$ vector of intercepts, whereas $\boldsymbol{\beta} = \text{diag}\{\beta_1, \beta_2, \ldots, \beta_N\}$ is an $N \times N$ diagonal matrix.

As a result, the assets returns have the following conditional distribution:

$$\mathbf{r}_t | \mathbf{h}_t \sim \mathcal{N}\left(\mathbf{0}, \mathbf{H}_t^{1/2} \boldsymbol{\Sigma}_\epsilon \mathbf{H}_t^{1/2}\right).$$

As highlighted by Harvey et al. (1994), the correlations are constant similarly to the model of Bollerslev (1990). The assumption of constant correlations is reasonable in empirical applications according to Bollerslev (1990) and Harvey et al. (1994). Nevertheless, other contributions question these assumptions (see, e.g., Diebold and Nerlove, 1989).

In the estimation process, Harvey et al. (1994) focused on the particular case in which $\mathbf{h}_t$ follows a multivariate random walk, such that it is possible to emphasize the persistence of volatility over time. Particularly, after linearizing (5.5) and setting $w_{i,t} = \log r_{i,t}^2$, Harvey et al. (1994) considered the following specification:

$$\mathbf{w}_t = (-1.27)\boldsymbol{\iota} + \mathbf{h}_t + \boldsymbol{\xi}_t \tag{5.6a}$$

$$\mathbf{h}_t = \mathbf{h}_{t-1} + \boldsymbol{\eta}_t, \tag{5.6b}$$

where $\boldsymbol{\iota}$ is an $N \times 1$ vector, the elements of which are equal to 1, $\mathbf{w}_t = (w_{1,t}, w_{2,t}, \ldots, w_{N,t})'$, whereas $\boldsymbol{\xi}_t = (\xi_{1,t}, \xi_{2,t}, \ldots, \xi_{N,t})'$, with $\xi_{i,t} = \log \epsilon_{i,t}^2 + 1.27$ being i.i.d. and having a covariance matrix that is a known function of $\boldsymbol{\Sigma}_\epsilon$ (Harvey et al., 1994).

By treating the system of Eq. (5.6) as a Gaussian state-space model and using the Kalman filter, Harvey et al. (1994) retrieved the corresponding quasi-maximum likelihood estimators. Later, So, Li, and Lam (1997) and Danielsson (1998) focused on specifications that resemble the basic model introduced by Harvey et al. (1994). Particularly, So et al. (1997) used a quasi-maximum likelihood method in a context in which the off-diagonal elements of $\boldsymbol{\beta}$ are not necessarily equal to zero. In contrast, Danielsson (1998) used a simulated maximum likelihood method. Another tool that has attracted interest for estimating multivariate stochastic volatility models is represented by Markov chain Monte Carlo (MCMC) methods (see, e.g., Jacquier et al. (1994), Chib and Greenberg (1996) and Smith and Pitts (2006); see the next section). Several extensions of the basic multivariate stochastic volatility model of Harvey et al. (1994), described above, have been proposed. For instance, as highlighted by Harvey et al. (1994), a possible generalization allows for a more complex specification for $\mathbf{h}_t$ in place of a simple VAR(1) process, such as a multivariate autoregressive moving average (ARMA) or a stationary vector autoregressive fractionally integrated moving average (ARFIMA) process, as suggested by So and Kwok (2006). Another well-known extension includes latent common factors (see, e.g., Harvey et al., 1994; Jacquier, Polson, and Rossi, 1999b; Aguilar and West, 2000). Similarly to the univariate setting described in Sect. 5.3.1, it is possible to define multivariate stochastic volatility models that capture stylized facts of financial time series. For instance, among others, we mention Asai and McAleer (2006a), who proposed a model that reproduces the leverage effect, and Yu and Meyer (2006a), who developed a model in which $\boldsymbol{\epsilon}_t$ follows a multivariate Student-t distribution to capture the excess kurtosis of financial returns.

5.3.3 Improvements on Classical Models

Since the seminal paper of Harvey et al. (1994), the literature on multivariate stochastic volatility models has been enriched by many interesting contributions from either the perspective of parsimonious modelling high dimensional vectors or focusing on the parameters estimation issue. In this section, we consider the major contributions dealing with multivariate extensions of the basic stochastic volatility and ARCH-type models that are able to capture additional features of observed time series. We further provide an additional investigation of recent contributions for large dimensional models in the next section.

Given the large number of existing contributions on multivariate SV models, a complete overview is beyond the scope of the present contribution. Instead, we summarize the most relevant approaches from several categories while referring to previous up-to-date reviews for details when available.

The most important extension of basic SV models is the Multivariate Factor Stochastic Volatility (MFSV) model first introduced by Pitt and Shephard (1999); see also Kim, Shephard, and Chib (1998a) and Chib, Nardari, and Shephard (2006). The model generates time-varying correlations by factorizing the variance-covariance matrix in terms of a small number of latent processes following SV dynamics:

$$\mathbf{y}_t = \mathbf{B}_t\mathbf{f}_t + \mathbf{K}_t\mathbf{q}_t + \mathbf{V}_t^{1/2}\boldsymbol{\Lambda}_t^{-1}\boldsymbol{\varepsilon}_t, \qquad \boldsymbol{\varepsilon}_t \sim \mathcal{N}_p\left(0, \mathbf{I}_p\right) \tag{5.7a}$$

$$\mathbf{f}_t = \mathbf{c} + \mathbf{A}\mathbf{f}_{t-1} + \mathbf{D}_t^{1/2}\boldsymbol{\gamma}_t, \qquad \boldsymbol{\gamma}_t \sim \mathcal{N}_q\left(0, \mathbf{I}_p\right) \tag{5.7b}$$

$$\mathbf{h}_t = \boldsymbol{\mu} + \boldsymbol{\Phi}\left(\mathbf{h}_t - \boldsymbol{\mu}\right) + \boldsymbol{\eta}_t, \qquad \boldsymbol{\eta}_t \sim \mathcal{N}_{p+q}\left(0, \boldsymbol{\Sigma}_\eta\right), \tag{5.7c}$$

where $\mathbf{f}_t = \left(f_{1,t}, f_{2,t}, \ldots, f_{q,t}\right)^{\mathsf{T}}$ is a vector of latent factors,

$$\boldsymbol{\Lambda}_t = \operatorname{diag}\left\{\lambda_{1,t}, \ldots, \lambda_{p,t}\right\} \tag{5.8a}$$

$$\mathbf{K} = \operatorname{diag}\left\{k_{1,t}, \ldots, k_{p,t}\right\} \tag{5.8b}$$

$$\mathbf{q}_t = \left(q_{1,t}, \ldots, q_{p,t}\right)' \tag{5.8c}$$

$$\mathbf{h}_t = \left(h_{1,t}, \ldots, h_{p,t}, h_{p+1,t}, \ldots, h_{p+q,t}\right) \tag{5.8d}$$

$$\mathbf{V}_t = \operatorname{diag}\left\{\exp\left(h_{1,t}\right), \ldots, \exp\left(h_{p,t}\right)\right\} \tag{5.8e}$$

$$\mathbf{D}_t = \operatorname{diag}\left\{\exp\left(h_{p+1,t}\right), \ldots, \exp\left(h_{p+q,t}\right)\right\} \tag{5.8f}$$

$$\boldsymbol{\Phi} = \operatorname{diag}\left\{\phi_1, \ldots, \phi_{p+q}\right\} \tag{5.8g}$$

$$\boldsymbol{\Sigma}_\eta = \operatorname{diag}\left\{\sigma_{\eta,1}, \ldots, \sigma_{\eta,p+q}\right\}, \tag{5.8h}$$

$\lambda_{j,t} \sim \mathcal{G}\left(\frac{\nu_j}{2}, \frac{\nu_j}{2}\right)$, $k_{j,t}$ are jump sizes for $j = 1, 2, \ldots, p$ and $q_{j,t} \sim \mathcal{B}er\left(\kappa_j\right)$. Hereafter, $x \sim \mathcal{B}er\left(\pi\right)$ denotes a Bernoulli random variable with a success probability equal to π,i.e. $\mathbb{P}\left(x = 1\right) = \pi$. For identification purposes, the $p \times q$ loading matrices $\mathbf{B}$ are assumed to be such that $b_{ij,t} = 0$, for $(i < j, i \leq q)$ and $b_{ii,t} = 1$ for $(i \leq q)$ with all other elements unrestricted. Thus, in this model, each of the factors and each of the errors evolve according to univariate SV models.

The MFSV model defined in Eqs. (5.7)–(5.8) accounts for heavy-tails, jumps-in mean as well as specific dynamic evolution for the conditional volatility and the latent factor, providing a general framework in which both the conditional variances and correlations evolve over time following a stochastic process. Given its general formulation, the MFSV model nests several alternative specifications previously proposed in literature. The jump components $\mathbf{q}_t$ were introduced by Chib et al. (2006), who also considered a simpler evolution for the factors that correspond to the following parameter restrictions: $\mathbf{c} = \mathbf{0}_q$, $\mathbf{A} = \mathbf{0}_{q\times q}$ and time-invariant loading matrix $\mathbf{B}_t = \mathbf{B}$. Chib et al. (2006) also introduced the diagonal scaling matrix $\mathbf{\Lambda}_t$ in order to account for the non-Gaussian nature of financial returns that is heavily documented in literature. Indeed, by marginalizing out the latent factors $\mathbf{\Lambda}_{j,t}$, we retrieve a standard Student-t distribution for the error terms $\lambda_{j,t}^{-1}\varepsilon_{j,t} \sim \mathcal{S}t_{\nu_j}(0, 1)$. If we further impose the constraints that $\mathbf{K}_t = \mathbf{0}_{p\times p}$ and $\mathbf{\Lambda}_t = \mathbf{0}_{p\times p}$, we get the specification of Kim et al. (1998a) and Pitt and Shephard (1999). Jacquier, Polson, and Rossi (1999a) and Liesenfeld and Richard (2003) considered a similar FSV model with the additional restriction that $\mathbf{V}_t$ is not time-varying and estimate the parameter using single-move Gibbs sampling and efficient importance sampling schemes to approximate the likelihood, respectively. Pitt and Shephard (1999) also employed a MCMC-based approach for the general model formulation discussed above, sampling the log-volatilities along the lines of Shephard and Pitt (2004). An extensive discussion of Bayesian inference in the context of FSV is presented by Chib et al. (2006) in the context of a fat-tailed FSV with jumps.

Yu and Meyer (2006b) proposed a bivariate alternative to FSV models with a dynamic equation for correlation evolution. However, the model generalization to larger dimensions is complex.

Tsay (2010) proposed a Cholesky decomposition SV model that routinely applies a Cholesky decomposition of the time-varying covariance matrix to preserve the positive-definiteness. However, the model implies a sequential ordering for the outcome variables. To prevent this issue, Asai and McAleer (2006b) exploited the matrix exponential operator to get a compact SV model representation:

$$\mathbf{y}_t|\mathbf{A}_t \sim \mathcal{N}_p\left(0, \mathbf{\Sigma}_t\right) \tag{5.9a}$$

$$\mathbf{\Sigma}_t = \exp\left(\mathbf{A}_t\right) = \mathbf{E}_t \exp\left(\mathbf{V}_t\right) \mathbf{E}_t^{\mathsf{T}} \tag{5.9b}$$

$$\boldsymbol{\alpha}_t = \operatorname{vech}\left(\mathbf{A}_t\right) \tag{5.9c}$$

$$\boldsymbol{\alpha}_{t+1} = \boldsymbol{\mu} + \mathbf{\Phi}\left(\boldsymbol{\alpha}_t - \boldsymbol{\mu}\right) + \boldsymbol{\eta}_t, \qquad \boldsymbol{\eta}_t \sim \mathcal{N}_{p(p+1)/2}\left(0, \mathbf{\Sigma}_\eta\right), \tag{5.9d}$$

where $\boldsymbol{\mu} \in \mathbb{R}^p$, $\mathbf{\Phi}$ is the autoregressive matrix of dimension $p \times p$ and $\mathbf{E}_t$ and $\mathbf{V}_t$ denote the spectral decomposition of the matrix $\mathbf{A}_t$. Despite the elegance of the solution, the matrix exponentiation is computationally intensive in large dimensions since it involves the spectral decomposition of the variance-covariance matrix in Eq. (5.9b) at each time t. Philipov and Glickman (2006b) allowed for dynamic correlations by assuming that the conditional covariance matrix $\mathbf{\Sigma}_t$ follows an

Inverted Wishart distribution with parameters that depend on the past covariance matrix $\mathbf{\Sigma}_{t-1}$. Particularly,

$$\mathbf{y}_t|\mathbf{\Sigma}_t \sim \mathcal{N}_p(0, \mathbf{\Sigma}_t) \tag{5.10a}$$

$$\mathbf{\Sigma}_t|\nu, \mathbf{S}_{t-1}, \mathbf{\Sigma}_{t-1} \sim \mathcal{IW}_q(\nu, \mathbf{S}_{t-1}) \tag{5.10b}$$

$$\mathbf{S}_{t-1} = \frac{1}{\nu}\left(\mathbf{\Sigma}_{t-1}^{-1}\right)^{d/2} \mathbf{A}\left(\mathbf{\Sigma}_{t-1}^{-d/2}\right)', \tag{5.10c}$$

where $\mathbf{\Sigma}_{t-1}^{-d/2} = \left(\mathbf{\Sigma}_{t-1}^{-1}\right)^{d/2}$ and $\mathcal{IW}_q(\nu_0, \mathbf{Q}_0)$ denote an Inverted Wishart distribution with parameters $(\nu_0, \mathbf{Q}_0)$, and for convenience in Eq. (5.10c), we used the parametrization proposed by Asai and McAleer (2009a). Philipov and Glickman (2006b) estimated this model in a Bayesian setting with an MCMC algorithm. Gourieroux, Jasiak, and Sufana (2009a) used an alternative approach and derived a Wishart autoregressive process.

The approach of Gribisch (2016) generalizes the basic Wishart MSV model of Philipov and Glickman (2006b) and Asai and Asai and McAleer (2009a) to encompass a regime-switching behaviour. The latent state variable is driven by a first-order Markov process. The model allows for state-dependent covariance and correlation levels and state-dependent volatility spillover effects. Parameter estimates are obtained using Bayesian Markov Chain Monte Carlo procedures and filtered estimates of the latent variances, and covariances are generated by particle filter techniques. The model is defined as in Eq. (5.10) with the following minor modifications:

$$\mathbf{S}_{t-1} = \frac{1}{\nu}\left(\mathbf{\Sigma}_{t-1}^{-1}\right)^{d/2} \mathbf{A}_{S_t}\left(\mathbf{\Sigma}_{t-1}^{-d/2}\right)',$$

while $\{S_t, t = 1, 2, \ldots, T\}$ is a Markov chain.

5.3.4 Dealing with Large Dimensional Models

One of the main challenges in modern financial econometrics involves building reliable multivariate volatility models in large dimensions. When the dimension p is large, improvements over traditional models to efficiently deal with the curse of dimensionality problem can come from two different directions. The first is through the building and validation of more parsimonious models that are able to account for most of the stylized facts and dependences among assets, while the second relevant way improves upon the existing estimation methodologies. In this latter direction, one promising estimation method has been recently proposed by Kastner (2018) for the multivariate Student-t FSV model of Chib et al. (2006) with constant factor loadings and without the jump component. Therefore their model is similar to that specified in (5.7), with the additional restrictions $\mathbf{K}_t = \mathbf{0}$

and $\mathbf{B}_t = \mathbf{0}$, for all t, $\mathbf{c} = \mathbf{0}$ and $\mathbf{A} = \mathbf{0}$. Kastner, Frühwirth-Schnatter, and Lopes (2017) discussed efficient Bayesian estimation of a dynamic factor stochastic volatility model. They proposed two interweaving MCMC strategies (see Yu & Meng, 2011) to substantially accelerate the convergence and mixing of standard approaches by exploiting the non-identifiability issues that arise in factor models. Their simulation experiments show that the proposed interweaving methods boost estimation efficiency by several orders of magnitude.

On the same stream of literature, Bitto and Frühwirth-Schnatter (2019) and Kastner (2018) accounted for the curse of dimensionality on time-varying parameters (TVP) models with stochastic volatility and FSV, respectively. Specifically, Bitto and Frühwirth-Schnatter (2019) proposed a Normal-Gamma prior (see, e.g., Griffin & Brown, 2010) for variable selection in time-varying parameters regression (TVP) models by exploiting the non-centred parameterization of Frühwirth-Schnatter and Wagner (2010). Although originally proposed for dealing with variable selection in a dynamic regression context, the general framework can be easily extended to include time-varying conditional covariance in multivariate models, by exploiting the Cholesky decomposition of Lopes, McCulloch, and Tsay (2014). More specifically, let $\mathbf{y}_t \sim \mathcal{N}(0, \boldsymbol{\Sigma}_t)$, with time-varying variance covariance $\boldsymbol{\Sigma}_t = \mathbf{A}_t \mathbf{D}_t \mathbf{A}_t$, where $\mathbf{A}_t \mathbf{D}_t^{1/2}$ is the lower triangular Cholesky decomposition of $\boldsymbol{\Sigma}_t$; then, $\mathbf{A}_t^{-1} \mathbf{y}_t \sim \mathcal{N}(0, \mathbf{D}_t)$ and $\mathbf{y}_t$ can be expressed as

$$\mathbf{y}_t = \mathcal{N}(\mathbf{B}\mathbf{x}_t, \mathbf{D}_t),$$

where

$$\mathbf{B}_t = \begin{pmatrix} -\Phi_{2,1,t} & 0 & 0 & \cdots & 0 & 0 \\ -\Phi_{3,1,t} & -\Phi_{3,2,t} & 0 & \cdots & 0 & 0 \\ \cdots & \cdots & \cdots & \cdots & \cdots & \cdots \\ -\Phi_{p,1,t} & -\Phi_{p,2,t} & -\Phi_{p,3,t} & \cdots & -\Phi_{p,p-1,t} & 0 \end{pmatrix}, \tag{5.11}$$

with $\Phi_{ij,t}$ for $j < i$ being the elements of the matrix $\mathbf{A}_t^{-1}$, and $\mathbf{x}_t = (y_{1,t}, y_{2,t}, \ldots, y_{p-1,t})^{\mathsf{T}}$ being a regressor derived from $\mathbf{y}_t$. Moreover, to capture the conditional heteroskedasticity the diagonal matrix $\mathbf{D}_t$ can be parameterized in terms of a vector of conditional volatilities $\mathbf{D}_t = \operatorname{diag}\{e^{h_{1,t}}, \ldots, e^{h_{p,t}}\}$, having autoregressive dynamics as in Eq. (5.7c). Therefore, placing the Normal-Gamma prior distribution on the unrestricted elements of the matrix of loading factors $\mathbf{B}_t$ in Eq. (5.11) induces sparsity into the correlation matrix of $\mathbf{y}_t$. Another recent interesting contribution on estimating large dimensional MFSV models is that of Kastner (2018). Quoting Kastner (2018), the main aim of his paper is '*to strike the indispensable balance between the necessary flexibility and parameter parsimony by using a factor stochastic volatility (SV) model in combination with a global-local shrinkage prior*'. Indeed, Kastner (2018) proposed an MFSV model that alleviates the curse of dimensionality by combining the parsimonious factor representation with sparsity. As for Bitto and Frühwirth-Schnatter (2019), sparsity is obtained by imposing computationally efficient absolutely continuous shrinkage priors (see Griffin & Brown, 2010) on the factor loadings. The FSV model specified by Kastner (2018) can be cast within the general FSV formulation in Eqs. (5.7)–(5.8). Again, the factor loadings $\mathbf{B}_t = \mathbf{B}$ are assumed to be constant over time, the jump matrix of jump intensities is null, $\mathbf{K}_t = \mathbf{0}_{p\times p}$ and the matrix $\mathbf{\Lambda}_t$ is also fixed to one, although a more complex fat-tailed distribution can be considered. Moreover, the common factor dynamics specified in Eq. (5.7b) impose constraints on both the vector $\mathbf{c}$ and the loading matrix $\mathbf{A}$, where $\mathbf{c} = \mathbf{0}$ and $\mathbf{A} = \mathbf{0}$. Moreover, the autoregressive dynamics for the factor stochastic volatilities are assumed to have zero means. Therefore, an additional restriction is that $\left(\mu_1, \mu_2, \ldots, \mu_p\right)^\mathsf{T} = \mathbf{0}$ in Eq. (5.7c). The distinguishing feature of the FSV approach proposed by Kastner (2018) is that it introduces sparse estimation on each unrestricted element of the matrix of factor loadings $\mathbf{B}$ by assuming the following Normal-Gamma hierarchical prior, which further shrinks unimportant elements of the factor loadings matrix to zero in an automatic way:

$$b_{ij}|\tau_{ij}^2 \sim \mathcal{N}\left(0, \tau_{ij}^2\right) \tag{5.12a}$$

$$\tau_{ij}|\lambda_i^2 \sim \mathcal{G}\left(a_i, a_i\lambda_i^2/2\right) \tag{5.12b}$$

$$\lambda_i^2|\lambda_i^2 \sim \mathcal{G}\left(c_i, d_i\right), \tag{5.12c}$$

where $\mathcal{G}(\cdot,\cdot)$ denotes a Gamma distribution. The hierarchical prior distribution specified in Eq. (5.12) extends the usual prior for factor loadings, in which $\tau_{ij}^2 = \tau^2$ for all the elements of the matrix (see, e.g., Chib et al., 2006; Kastner et al., 2017; Pitt & Shephard, 1999), to achieve more shrinkage. This approach is related to that of Bhattacharya and Dunson (2011) for the class of non-parametric static factor models.

With regard to the WSV approach, Asai and McAleer (2009b) proposed further extensions of the basic multivariate SV model that can be useful in high dimensions, which are strongly related to the DCC model of Engle (2002) (see section "Conditional Correlation Models"). Indeed, Asai and McAleer (2009b) modelled the vector of log-returns as in Eq. (5.1), where the dynamic evolution of the correlation in Eq. (5.2) is replaced by:

$$\mathbf{Q}_{t+1} = (1-\psi)\,\mathbf{A}r\mathbf{Q} + \psi\mathbf{Q}_t + \Xi_t, \qquad \Xi_t \sim \mathcal{W}_p(\nu, \mathbf{\Lambda}),$$

where $\mathbf{A}r\mathbf{Q}$ is a $p \times p$ matrix assumed to be positive definite and $\psi \in (-1, 1)$. An alternative specification provided by Asai and McAleer (2009b) is closely related to the approach of Philipov and Glickman (2006a,b) discussed in the previous section. Despite the various advancements in recent years, this model class still suffers from the limited presence of spillovers across variances, covariances and correlations because, in many cases, the (time-varying) parameter matrices are diagonal.

5.4 Volatility Forecasting with High Frequency Data

A huge stimulus to the advancements in volatility forecasting came from the increased availability of high-frequency data in the late 1990s. Several works in the area emerged, starting from the studies focused on the analysis of high-frequency data features or on the modelling of high-frequency data with GARCh-type specifications (see, among many others, Andersen and Bollerslev, 1997, 1998; Engle and Sokalska, 2012; Engle and Russell, 1998; Ghysels and Jasiak, 1998). Later, we observed the surge of a more general and well-defined class of models and tools. In this group, we include both the approaches estimating daily variances, covariances and correlations starting from high-frequency data (let us call these, *measurement* papers) as well as the works proposing models for realized variances, covariances and correlation sequences (*modelling* papers). Both measuring and modelling research lines clash with the dimensionality issues. In this section, after briefly reviewing the univariate and multivariate measuring and modelling standard approaches, we will show how they cope with a data rich environment.

5.4.1 Measuring Realized Variances

Since the seminal contribution of Andersen, Bollerslev, and Diebold (2003), the realized volatility literature focused on the measurement issue has evolved, both in terms of the availability of asymptotic results regarding volatility estimator properties (allowing, for instance, the design of several tests for the occurrence of jumps in equity prices) as well as for the possibility of handling a number of empirical data features, with a particular reference to the microstructural noise (i.e. the noise due to, for instance, infrequent trading and bid-ask bounce).

To introduce realized volatility measuring, we must make a hypothesis about the process generating the equity prices. Let p_t be the logarithmic price of a financial asset at time t and assume it follows a Brownian semimartingale process, which reads as follows:

$$p_t = p_0 + \int_0^t \mu_u du + \int_0^t \sigma_u dW_u, \quad t \geq 0, \tag{5.13}$$

where the drift $\mu = (\mu_t)_{t\geq 0}$ is locally bounded and predictable, whereas $\sigma = (\sigma_t)_{t\geq 0}$ is a strictly positive process, independent of the standard Brownian Motion $W = (W_t)_{t\geq 0}$ and càdlag.

We now focus on the quadratic variation. If each trading day equals the interval $[0, 1]$ and is divided into M subintervals having the same width, the quadratic variation at the t-th day is defined as follows:

$$QV_t = \operatorname*{plim}_{M\to\infty} \sum_{i=2}^{M} (p_{t,i} - p_{t,(i-1)})^2$$

where $p_{t,i}$ is the price of a given asset we record at the i-th intraday interval of the t-th day.

Hull and White (1987) showed that under our hypothesis for the price process, and under additional regularity conditions for μ_u and σ_u, the quadratic variation equals the integrated volatility (IV), i.e. the integral $\int_0^t \sigma_u^2 du$. Building on this result, we can estimate the volatility for day t by means of the realized variance:

$$RV_t = \sum_{i=1}^{M-1} r_{t,i}^2,$$

where $r_{t,i} = p_{t,(i+1)} - p_{t,i}$.

Barndorff-Nielsen and Shephard (2002) highlighted that the realized variance is an unbiased estimator of the quadratic variation and converges to it as the sampling frequency tends to infinity. Furthermore, Barndorff-Nielsen and Shephard (2002) showed that:

$$M^{1/2}\left(RV_t - \int_0^1 \sigma_u^2 du\right) \xrightarrow{d} \mathcal{MN}\left(0, 2\int_0^1 \sigma_u^4 du\right)$$

where $\mathcal{MN}$ denotes the mixed normal distribution, whereas $\int_0^1 \sigma_u^4 du$ is the integrated quarticity.

Several studies have addressed the impact of microstructure noise on the properties of RV_t (see, e.g., Hasbrouck, 2006). Bandi and Russel (2006) suggested limiting the sampling frequency and proposed the rule of thumb of using 5-min returns, optimizing the trade-off between bias and variance. In contrast, Zhang, Mykland, and Ait-Sahalia (2005) introduced the two time-scales realized variance, whereas Hansen and Lunde (2006) used a kernel approach for realized volatility estimation.

In addition to the microstructure noise, the volatility of financial returns is affected by rapid and large increments, as highlighted by several contributions in the literature (see, e.g., Ball and Torous, 1983; Jarrow and Rosenfeld, 1984; Jorion, 1988; Duffie et al., 2000; Eraker, Johannes, and Polson, 2003). Several testing methods have been developed to detect the presence and deal with the occurrence of jumps in stock prices. Among others, we mention here Barndorff-Nielsen and Shephard (2006), Andersen, Bollerslev, and Dobrev (2007), Lee and Mykland (2008), Andersen, Bollerslev, Frederiksen, and Ørregaard Nielsen (2010) and Corsi, Pirino, and Renò (2010).

5.4.2 *Realized Variance Modelling and Forecasting*

The availability of realized variance sequences opens the door to a vast set of modelling strategies, starting from ARMA-type specifications, whose final purpose is variance forecasting. However, the empirical evidence suggests that RV shows long-memory features, thus requiring some care in model design. Corsi (2009) introduced a successful model, that is, the Heterogeneous Auto Regressive (HAR) model, with the purpose of approximating the long-memory behaviour of RV with a restricted auto-regressive structure:

$$RV_t = \beta_0 + \beta_1 RV_{t-1} + \beta_2 RV_{t-1:t-5} + \beta_3 RV_{t-1:t-20} + \epsilon_t \tag{5.14}$$

where $RV_{t-1:t-s} = \frac{1}{s}\sum_{j=1}^{k} RV_{t-j}$ is the mean of the realized volatility in the last s days, whereas ϵ_t is an error term with an expected value equal to zero.

The HAR model includes three volatility determinants that reflect the behaviour of different financial operators according to the horizon characterizing their activity (Corsi, 2009). Particularly, RV_{t-1} refers to short-term operators, the activity of which has a daily or intra-daily frequency, such as dealers, market makers and intraday speculators. $RV_{t-1:t-5}$ captures the behaviour of medium-term operators, such as portfolio managers who typically rebalance their portfolios weekly. $RV_{t-1:t-20}$ reflects long-term operators, such as insurance companies and pension funds who trade much less frequently and possibly for larger amounts. Therefore, the HAR model postulates that operators having different time horizons perceive, react to and cause different types of volatility components (Corsi, 2009), consistent with the heterogeneous market hypothesis discussed in Muller et al. (1993). The HAR model has different advantages. For instance, it is easy to implement and to estimate with the standard ordinary least squares (OLS), as all the variables in (5.14) are observed. Moreover, by using simulated data, the HAR-RV model is able to reproduce the typical features of financial time series, such as long memory and fat tails, and empirical results highlight its excellent forecasting performance (Corsi, 2009). As the HAR model might be red as a restricted AR specification, variance forecasting from the HAR model inherits all the features and properties of AR model forecasting.

The HAR model of Corsi (2009) becomes the prototype for several more flexible specifications. For instance, Bollerslev, Patton, and Quaedvlieg (2016) highlighted the fact that RV_t includes a measurement error in addition to the true latent integrated volatility, and therefore the forecasts arising from OLS estimates are affected by the so-called errors-in-variables problem. Notably, this often leads to an attenuation bias, with the directly observable RV process being less persistent than the latent IV process (Bollerslev et al., 2016). Bollerslev et al. (2016) addressed this issue by developing the so-called HARQ-F model, which builds on time-varying coefficients. Empirical analyses of the S&P 500 equity index and the constituents of the Dow Jones Industrial Average show that the HARQ-F model provides more responsive forecasts, with significant benefits in terms of predictive accuracy. An alternative specification is proposed in Bekierman and Manner (2018), where the autoregressive parameters of the HAR-RV model are allowed to be driven by a latent Gaussian autoregressive process and are estimated by means of maximum likelihood using the Kalman filter.

Another interesting approach that we can use to forecast realized variance measures involves the mixed data sampling model (MIDAS-RV) of Ghysels, Santa-Clara, and Valkanov (2004), Ghysels and Valkanov (2006). The main advantage provided by the MIDAS is that it makes it possible to incorporate different levels of sample frequency. In doing so, we use an approach that preserves the information included in high-frequency data, without computing daily aggregates, such as the realized variance, to use as regressors. In fact, the standard procedure of pre-filtering data implies the risk of losing important information. Like the HAR-RV model, the MIDAS is able to reproduce the persistence of volatility over time, with a reduced number of parameters to be estimated when compared to standard econometric models. Particularly, the general specification of the MIDAS variance model reads as follows:

$$RV_{t,v} = \mu + \phi \sum_{j=0}^{J} \mathbf{b}(j, \boldsymbol{\theta})\mathbf{Z}_{t-j} + \epsilon_t,$$

where $RV_{t,v} = RV_t + \ldots + RV_{t+v}$ quantifies the integrated variance over a v-interval, whereas $\mathbf{Z}_{t-j}$ includes a set of covariates that can be sampled at any frequency; for instance, Forsberg and Ghysels (2007) analysed different choices for $\mathbf{Z}_{t-j}$. In Ghysels and Valkanov (2006), the weighting function $\mathbf{b}(j, \boldsymbol{\theta})$ builds on the Beta function with parameter $\boldsymbol{\theta}$ to reproduce a gradually declining and hump-shaped behaviour.

Note that the HAR-RV and the MIDAS-RV models are two of the econometric tools that we can use to forecast the realized measures of volatility. Nevertheless, additional different methods are available in the literature. For instance, Andersen et al. (2003) used long-memory Gaussian vector autoregressive (VAR) models to forecast realized volatilities. Shephard and Sheppard (2010) introduced the high-frequency-based volatility (HEAVY) models, in which one equation models the volatility of returns, building on a realized measure that is specified in the second equation. Christiansen, Schmeling, and Schrimpf (2012) predicted the asset return volatility in a Bayesian model averaging framework, while Hansen, Huang, and Shek (2012) added a realized volatility component to the GARCH structure leading to the Realized GARCH model. Caporin and Velo (2015) used an HAR model with asymmetric effects with respect to the volatility and the return and GARCH and GJR GARCH specifications for the variance equation. Bonaccolto and Caporin (2016) predicted realized measures of volatility in regression quantiles.

Engle and Gallo (2006) introduced an approach alternative to the HAR for modelling and forecasting realized sequences, the multiplicative error model (MEM). The model read as follows:

$$RV_t = h_t \eta_t$$

and

$$h_t = \omega + \alpha RV_{t-1} + \beta h_{t-1}.$$

The model estimation requires a proper distributional assumption for the innovations η_t, usually a Gamma. The MEM is somewhat less used than the HAR due to the need of resorting to likelihood methods for model estimation.

For both approaches based on realized variances, the construction of forecasts, similarly to univariate GARCH models, is a bi-product of model estimation for the one-step-ahead case, while for forecasts horizons larger than one period, we might adopt recursive substitution.

5.4.3 Measuring and Modelling Realized Covariances

In the framework of high-frequency data, multivariate analyses requires the estimation of additional realized measures, that is, realized covariances. The pioneering works in this area are those of Andersen et al. (2003) and Barndorff-Nielsen and Shephard (2004). Let $\mathbf{r}_{t,i} = \mathbf{p}_{t,i} - \mathbf{p}_{t,(i-1)}$ be the $N \times 1$ vector of intraday returns of N stocks, recorded at the i-th intraday interval (for $i = 2, \dots M$) of the t-th trading day, with $\mathbf{p}_{t,(i)}$ and $\mathbf{p}_{t,(i-1)}$ being the corresponding log-prices observed, respectively, at the end and at the beginning of that intraday interval. The realized covariance matrix is then computed on the t-th trading day, as follows:

$$\mathbf{RCOV}_t = \sum_{i=2}^{M} \mathbf{r}_{t,i}\mathbf{r}'_{t,i} \tag{5.15}$$

After specifying a continuous time diffusion process for the logarithmic prices of the N stocks—the multivariate extension of (5.13)—Barndorff-Nielsen and Shephard (2004) showed that **RCOV** is a consistent estimator of the daily covariance matrix $\boldsymbol{\Sigma}_t$:

$$\sqrt{M}\left[\text{vech}\left(\mathbf{RCOV}\right) - \text{vech}\left(\left(\int_0^1 \boldsymbol{\Sigma}_t(u)du\right)\right)\right] \xrightarrow{d} \mathcal{N}(\mathbf{0}, \boldsymbol{\Pi}_t),$$

where vech is the half-vectorization operator, whereas $\boldsymbol{\Pi}_t$ is a positive definite matrix (see Barndorff-Nielsen and Shephard, 2004).

The advantage of using the realized covariance matrix in (5.15) consists of exploiting the additional information about the high-frequency co-movements of N stocks. Nevertheless, as highlighted by Sheppard (2006), frequent sampling is recommended if prices are error free. Unfortunately, high-frequency data are contaminated by market microstructure noise, and their presence affects the performance of realized covariance estimators. Particularly, Bandi and Russell (2005) showed that (5.15) is not consistent in the presence of microstructure noise. As highlighted by Bandi and Russell (2005), higher sampling frequencies produce a bias due to the accumulation of microstructure noise. On the other hand, lower sampling frequencies reduce the accuracy of the estimates. In order to optimize the trade-off between bias and efficiency, Bandi and Russell (2005) introduced a method to determine the optimal sampling frequency.

When moving from the estimation of realized variances to the estimation of realized covariances, we need to take into account an additional issue, which is particularly critical when using ultra-high-frequency data: the so-called Epps effect. Epps (1979) documented the fact that stock correlations tend to decrease as the sampling frequency increases. Indeed, information arrives at different frequencies for the different N stocks (non-synchronous trading), and this is an additional source of bias when estimating realized covariances. Several approaches have been proposed in the literature to address the nonsynchronicity issue. Among them, we mention Barndorff-Nielsen, Hansen, Lunde, and Shephard (2011), who proposed a multivariate realized kernel to estimate ex-post covariations. Notably, the estimator introduced by Barndorff-Nielsen et al. (2011) is consistent in the presence of both microstructure effects and asynchronous trading.

Similarly to the univariate case in which we focus on individual realized variances, a typical approach to model and then forecast the realized covariance matrix consists in building multivariate econometric models. For this purpose, for instance, Bauer and Vorkink (2011) used a multivariate HAR model, inspired by Corsi (2009), whereas Chiriac and Voev (2011) used a multivariate ARFIMA model. Here, we suppose that $\mathbf{RCOV}_t$ is the realized covariance matrix on the t-th day, which we estimate either from (5.15) or by using other estimators proposed in the literature. Then, after defining the variable $\mathbf{z}_t = \text{vech}(\mathbf{RCOV}_t)$, we can estimate a simple vector autoregression of order p, defined as follows:

$$\mathbf{z}_t = \boldsymbol{\alpha} + \sum_{i=1}^{p} \boldsymbol{\beta}_i \mathbf{z}_{t-i} + \boldsymbol{\epsilon}_t, \tag{5.16}$$

where $\boldsymbol{\epsilon}_t$ is a zero-mean random noise, $\boldsymbol{\beta}_i$, for $i = 1, \ldots, p$, are the $k \times k$ parameter matrices, with $k = N(N+1)/2$, whereas $\boldsymbol{\alpha}$ is a $k \times 1$ vector of intercepts.

A slightly different approach for joint modelling the dynamic evolution of the matrices of realized covariances has been recently proposed by Gourieroux (2006) and Gourieroux, Jasiak, and Sufana (2009b) to derive a Wishart autoregressive process. Specifically, they define a Wishart autoregressive process of order 1 (WAR(1)) as a matrix process having the following conditional Laplace transform:

$$\Psi_t(\boldsymbol{\Gamma}) = \mathbb{E}_{\mathbf{y}_{t+1}}\left(\exp\left(\text{tr}\left(\boldsymbol{\Gamma}\mathbf{y}_{t+1}\right)\right)\right) = \frac{\exp\left[\text{tr}\left(\mathbf{M}^{\intercal}\boldsymbol{\Gamma}\left(\mathbf{I}_p - 2\boldsymbol{\Sigma}\boldsymbol{\Gamma}\right)^{-1}\mathbf{M}\mathbf{y}_t\right)\right]}{|\mathbf{I}_p - 2\boldsymbol{\Sigma}\boldsymbol{\Gamma}|^{k/2}},$$

where $\mathbf{y}_t$ is a stochastic matrix of order $p \times p$, $\mathbb{E}_{\mathbf{y}_{t+1}}(\cdot)$ denotes the conditional expectation of $\mathbf{y}_{t-1}$ given the information set up to time t,i.e. $\{\mathbf{y}_t, \mathbf{y}_{t-1}, \ldots\}$, k is the scalar degree of freedom, with $k > p-1$, $\mathbf{M}$ is the $p \times p$ matrix of autoregressive parameters and $\boldsymbol{\Sigma}$ is a $p \times p$ symmetric, positive definite matrix. The matrix $\boldsymbol{\Gamma}$ has to be such that $\|\boldsymbol{\Sigma}^{1/2}\boldsymbol{\Gamma}\boldsymbol{\Sigma}^{1/2}\| < 1$, to ensure that the Laplace transform is well defined. Gourieroux et al. (2009b) provided a closed-form expression for the first two conditional moments of the Wishart autoregressive process, which are useful to express the dynamic evolution:

$$\mathbb{E}_{\mathbf{y}_{t+1}}(\mathbf{y}_{t+1}) = \mathbf{M}\mathbf{y}_t\mathbf{M}^\mathsf{T} + k\boldsymbol{\Sigma}.$$

Therefore, the dynamics for $\mathbf{y}_{t+1}$ can be defined as follows:

$$\mathbf{y}_{t+1} = \mathbf{M}\mathbf{y}_t\mathbf{M}^\mathsf{T} + k\boldsymbol{\Sigma} + \boldsymbol{\eta}_t,$$

where $\boldsymbol{\eta}_t$ is an innovation such that its expectation is zero, i.e. $\mathbb{E}\left(\boldsymbol{\eta}_t\right) = \underline{0}$. The conditional probability density function of $\mathbf{y}_{t+1}$ is given by

$$f\left(\mathbf{y}_{t+1}|\mathbf{y}_t\right) = \frac{|\mathbf{y}_{t+1}|^{(k-p-1)/2}}{2^{kp/2}\Gamma_p(k/2)|\boldsymbol{\Sigma}|^{k/2}} \exp\left[-\frac{1}{2}\operatorname{tr}\left(\boldsymbol{\Sigma}^{-1}\left(\mathbf{y}_{t+1} + \mathbf{M}\mathbf{y}_t\mathbf{M}^\mathsf{T}\right)\right)\right] \times \quad (5.17)$$
$$ {}_0F_1\left(k/2, 1/4\mathbf{M}\mathbf{y}_t\mathbf{M}^\mathsf{T}\mathbf{y}_{t+1}\right),$$

where $\Gamma_p\left(k/2\right) = \int_{\mathbf{A}\gg 0} \exp\left(\operatorname{tr}\left(-\mathbf{A}\right)\right)|\mathbf{A}|^{(k-n-1)/2}d\mathbf{A}$ is the multidimensional gamma function, ${}_0F_1$ is the hypergeometric function of matrix argument and the density is defined on positive definite matrices. Applications of the WAR(1) process can be found in Gourieroux and Sufana (2010) and Gourieroux and Sufana (2011); see also Bonato, Caporin, and Ranaldo (2012, 2013).

5.4.4 Realized (Co)variance Tools for Large Dimensional Settings

The increased availability of high-frequency data, both with respect to the cross-sectional dimension as well as in terms of the data frequency (higher and higher, going to the millisecond level and pointing at nanoseconds), challenges the measurement as well as the modelling of realized covariance sequences.

When focusing on the measurement issue, the recent literature has focused on the need to estimate a large realized covariance matrix starting from a framework where the number of intra-daily returns is smaller than the cross-sectional dimension. Obviously, to cope with this issue, the sampling frequency might be increased to enlarge the degrees of freedom, moving for instance to tick data. However, when increasing the data frequency, the impact of microstructure noise is larger, and we lose synchronicity among observed prices in the cross-section. Among the possible solutions proposed in recent years, we mention Tao, Wang, Yao, and Zou (2011), Tao, Wang, and Chen (2013) and Tao, Wang, and Zhou (2013), who introduced estimators based either on thresholding techniques or regularization approaches that are robust to microstructure noise, asynchronicity and that are valid when the cross-sectional dimension is larger than the temporal dimension. Similarly, Lam,

Feng, and Hu (2017) proposed a non-linear shrinkage approach for cross-sectional dimensions of the same order of the temporal dimensions, while Morimoto and Nagata (2017) proposed a robust covariance estimator building on the features of the realized covariance eigenvalues. A different research line follows from Ait-Sahalia and Xiu (2017) and Ait-Sahalia and Xiu (2019), who used factor structures to introduce realized covariance estimators.

On the modelling side, estimating a linear model such as that in (5.16) remains feasible when the cross-sectional dimension is large, under the assumption that the time series of realized covariances are sufficiently long. Nevertheless, financial operators typically deal with large portfolios in their real-world activity and need to carefully address the dimensionality issue, which becomes relevant in two different cases. First, we might have time series of realized covariances of limited length, relative to the cross-sectional dimension. Second, the number of parameters in models like (5.16) increases exponentially as N grows, with a possible impact on the estimators' efficiency as well as on inferential aspects.

The latter might be due to the empirical evidence showing that assets returns are typically highly correlated, and, when $N > T$, estimates derived from standard econometric techniques, such as OLS, are poorly determined and exhibit high variance. As a result, as N increases, the consequent accumulation of estimation errors becomes a problem that must be addressed. By considering the dimensionality issue, Callot, Kock, and Medeiros (2017) proposed a penalized model that builds on the least absolute shrinkage and selection operator (LASSO), introduced by Tibshirani (1996). As highlighted by Callot et al. (2017), an advantage of this approach is that it does not reduce dimensionality by transforming variables. In doing so, it is possible to keep the interpretability of the individual original variables. Particularly, in a first step, Callot et al. (2017) rewrote (5.16) in a stacked form, such that it is possible to specify and estimate a regression model for each of the k variables included in $\mathbf{z}_t$. For instance, the regression model specified for the j-th variable in $\mathbf{z}_t$, for $j = 1, \ldots, k$, reads as follows:

$$\mathbf{z}_j = \mathbf{Z}_j \boldsymbol{\beta}_j + \boldsymbol{\epsilon}_j, \tag{5.18}$$

where $\mathbf{z}_j = [z_{1,j}, \ldots, z_{T,j}]'$ is the $T \times 1$ time series vector for the j-th variable included in $\mathbf{z}_t$, whereas $\mathbf{Z}_j$ is the $T \times (kp + 1)$ matrix of covariates for the j-th equation (Callot et al., 2017).

Then, in a second step, Callot et al. (2017) included an ℓ_1-norm penalty for each regression (5.18), focusing on the following objective function:

$$L(\boldsymbol{\beta}_j) = \frac{1}{T} ||\mathbf{z}_j - \mathbf{Z}_j \boldsymbol{\beta}_j||^2 + \lambda ||\boldsymbol{\beta}_j||_1, \tag{5.19}$$

where λ is a tuning parameter that reflects the magnitude of the penalization. Indeed, the greater λ, the sparser the solutions yielded by (5.19), with an increasing number of coefficients that approach zero. Along the solutions path, the optimal value of λ could be determined by means of well-known statistical techniques, such as cross-validation.

A different research line builds on factor structures to model and forecast realized covariances. Among the papers belonging to this group, we cite Fan, Furger, and Xiu (2016), Shen, Yao, and Li (2018), Jin, Maheu, and Yang (2019) and Sheppard and Xu (2019).

5.4.5 Bayesian Tools

Conditional autoregressive Wishart processes (see Gourieroux et al., 2009b; Philipov & Glickman, 2006a,b) have also been considered for directly modelling realized volatility (RV) and realized correlations (RCOV). Golosnoy et al. (2012) proposed a Conditional Autoregressive Wishart (CAW) model for the analysis of realized covariance matrices of asset returns, which was further generalized by Yu, Li, and Ng (2017). The Generalized Conditional Autoregressive Wishart (GCAW) is a generalization of the existing models and accounts for the symmetry and positive definiteness of RCOV matrices without imposing any parametric restriction. In what follows, we briefly describe these two approaches.

Golosnoy et al. (2012) assumed an autoregressive moving average structure for the scale matrix of the Wishart distribution. Specifically, let $\mathbf{R}_t$ be the $p \times p$ matrix of realized covariances at time t, then the CAW(r, s) model can be expressed as

$$\mathbf{R}_t | \mathcal{F}_{t-1} \sim \mathcal{W}_p\left(\nu, \frac{\mathbf{S}_t}{\nu}\right) \tag{5.20a}$$

$$\mathbf{S}_t = \mathbf{c}\mathbf{c}^\mathsf{T} + \sum_{i=1}^{r} \mathbf{B}_i \mathbf{S}_{t-i} \mathbf{B}_i^\mathsf{T} + \sum_{i=1}^{s} \mathbf{A}_j \mathbf{R}_{t-j} \mathbf{A}_j^\mathsf{T}, \tag{5.20b}$$

where $\mathcal{F}_{t-1}$ denotes past information up to time $t-1$, $\nu > p$ is the degree of freedom scalar parameter, $\mathbf{S}_t$ is a symmetric and definite positive scale matrix and $\mathcal{W}_p(\nu_0, \mathbf{Q}_0)$ denotes a Wishart distribution with parameters $(\nu_0, \mathbf{Q}_0)$. Equation (5.20b) introduces a matrix-variate autoregressive dynamics for the RCOV, which is parameterized in terms of a lower triangular $p \times p$ matrix $\mathbf{c}$ and the unconstrained matrices of dimension $p \times p$, $\mathbf{B}_i$ and $\mathbf{A}_j$. The CAW(r, s) in Eq. (5.20) is not identified. Sufficient parameter restrictions that ensure identification are provided by Golosnoy et al. (2012). Specifically, the main diagonal elements of $\mathbf{c}$ and the first diagonal element for each of the matrices $\mathbf{A}_j$ and $\mathbf{B}_i$ are restricted to be positive. The CAW(r, s) model involves $p(p+1)/2 + (r+s)p^2 + 1$ parameters. However, the number of CAW parameters can be reduced by imposing restrictions on the matrices $(\mathbf{B}_i, \mathbf{A}_j)$. A natural restriction is to impose a diagonal structure on the dynamics of $\mathbf{S}_t$ by assuming that $\mathbf{B}_i$ and $\mathbf{A}_j$ are diagonal matrices that considerably reduce the number of parameters to $p(p+1)/2 + (r+s)p + 1$.

The recursion in Eq. (5.20b) resembles the BEKK–GARCH(p, q) specification of Engle and Kroner (1995) for the conditional covariance in models for multivariate returns, and it has the appealing property of guaranteeing the symmetry and positive definiteness of the conditional mean $\mathbf{S}_t$ essentially without imposing parametric restrictions on $\left(\mathbf{c}, \mathbf{A}_j, \mathbf{B}_i\right)$ as long as the initial matrices are symmetric and positive definite. The CAW(r, s) is also related to the Wishart Autoregressive (WAR) model of Gourieroux et al. (2009b), which is based on a conditional non-central Wishart distribution for $\mathbf{R}_t$. As recognized by Golosnoy et al. (2012), the WAR provides a dynamic model for the matrix of non-centrality parameters of the Wishart distribution, which is assumed to depend on lagged values of $\mathbf{R}_t$ rather than on the scale matrix (as under the CAW model). Therefore, the CAW(r, s) model can be interpreted as a generalization of the $WAR(r)$ model. Golosnoy et al. (2012) also proposed extensions of the CAW model obtained by including a Mixed Data Sampling (MIDAS) component and Heterogeneous Autoregressive (HAR) dynamics for long-run fluctuations. For those extensions, interested readers can refer to the original paper.

The CAW model in Eq. (5.20) was further generalized to the GCAW specification by Yu et al. (2017), which replaces Eq. (5.20a) by

$$\mathbf{R}_t|\mathcal{F}_{t-1} \sim \mathcal{W}_p\left(\nu, \mathbf{\Lambda}_t, \mathbf{S}_t\right),$$

where $\mathbf{\Lambda}_t$ is the $p \times p$ symmetric and positive semidefinite noncentrality matrix, while ν and $\mathbf{S}_t$ are defined as in Golosnoy et al. (2012). Yu et al. (2017) proposed the same dynamics of $\mathbf{S}_t$ proposed by Golosnoy et al. (2012) in Eq. (5.20b) and introduced an additional evolution for $\mathbf{\Lambda}_t$:

$$\mathbf{\Lambda}_t = \sum_{k=1}^{K} \mathbf{M}_k \mathbf{R}_{t-k} \mathbf{M}_k,$$

where, again, $\mathbf{M}_k$ is a $p \times p$ matrix of parameters and accounts for the autoregressive property of high-frequency returns. The same identifiability restrictions of Golosnoy et al. (2012) can be applied to the GCAW model.

An interesting approach that jointly models the conditional volatility, the realized volatilities and the pairwise realized correlations have been recently proposed by Yamauchi and Omori (2018). The basic MSV model is given by Eqs. (5.7)–(5.8), with $\mathbf{B}_t = \mathbf{0}$, $\mathbf{K}_t = \mathbf{0}$ and $\mathbf{\Lambda}_t = \text{diag}\{\iota\}$, which is enriched by including the vectors of RV and RCOV having their own dynamics. Thus, they introduced the additional measurement equations based on realized measures $R_{i,t} = \log(RV_{i,t})$ and $C_t = \log\left(\frac{1+RCOV_{ij,t}}{1-RCOV_{ij,t}}\right)$

$$R_{i,t} = \xi_i + h_{i,t} + \varepsilon^R_{i,t}, \qquad \varepsilon^R_{i,t} \overset{i.i.d.}{\sim} \mathcal{N}\left(0, \sigma^2_{i,t}\right)$$

$$C_{ij,t} = \delta_{ij} + g_{ij,t} + \varepsilon^C_{ij,t}, \qquad \varepsilon^C_{ij,t} \overset{i.i.d.}{\sim} \mathcal{N}\left(0, \sigma^2_{ij,t}\right),$$

where the terms ξ_j and δ_{ij}s are included to adjust for the biases resulting from microstructure noise, non-trading hours, non-synchronous trading and so forth (see Yamauchi and Omori, 2018).

5.5 Conclusion

In this chapter, we reviewed the main classes of models that allow the construction of volatility, covariances and correlations, with a focus on the most recent advancements in the financial econometrics literature and on the challenges posed by the increased availability of data. All models share some limits when the cross-sectional dimension starts to diverge, unless strong restrictions are imposed on the model dynamic. In the latter case, the models might become feasible, but the economic intuition we could recover from the model fit is reduced. In turn, this could have a negative impact on the forecast analyses if our interest is not only limited to the derivation of a forecast but also includes the identification of the drivers of the forecast as well as the elements that impact the forecast performances.

References

Aguilar, O., & West, M. (2000). Bayesian dynamic factor models and portfolio allocation. *Journal of Business & Economic Statistics, 18*(3), 338–357.

Aielli, G. P. (2013). Dynamic conditional correlation: On properties and estimation. *Journal of Business & Economic Statistics, 31*(3), 282–299.

Ait-Sahalia, Y., & Xiu, D. (2017). Using principal component analysis to estimate a high dimensional factor model with high-frequency data. *Journal of Econometrics, 201*(2), 384–399.

Ait-Sahalia, Y., & Xiu, D. (2019). Principal component analysis of high-frequency data. *Journal of the American Statistical Association, 114*, 287–303.

Alexander, C. (2001). *Orthogonal GARCH*. Upper Saddle River, NJ: Financial Times-Prentice Hall.

Andersen, T. G., & Bollerslev, T. (1997). Intraday periodicity and volatility persistence in financial markets. *Journal of Empirical Finance, 4*(2–3), 115–158.

Andersen, T. G., & Bollerslev, T. (1998). Answering the skeptics: Yes, standard volatility models do provide accurate forecasts. *International Economic Review, 39*(4), 885–905.

Andersen, T. G., Bollerslev, T., Diebold, F. X., & Labys, P. (2003). Modeling and forecasting realized volatility. *Econometrica, 71*(2), 579–625.

Andersen, T. G., Bollerslev, T., & Dobrev, D. (2007). No-arbitrage semi-martingale restrictions for continuous-time volatility models subject to leverage effects, jumps and i.i.d. noise: Theory and testable distributional implications. *Journal of Econometrics, 138*(1), 125–180.

Andersen, T. G., Bollerslev, T., Frederiksen, P., & Ørregaard Nielsen, M. (2010). Continuous-time models, realizedvolatilities, and testable distributional implications for daily stock returns. *Journal of Applied Econometrics, 25*(2), 233–261.

Andersen, T. G., & Sorensen, B. E. (1996). Gmm estimation of a stochastic volatility model: A monte carlo study. *Journal of Business & Economic Statistics, 14*(3), 328–352.

Asai, M., & McAleer, M. (2006a). Asymmetric multivariate stochastic volatility. *Econometric Reviews, 25*(2–3), 453–473.
Asai, M., & McAleer, M. (2006b). Asymmetric multivariate stochastic volatility. *Econometric Reviews, 25*(2–3), 453–473.
Asai, M., & McAleer, M. (2009a). The structure of dynamic correlations in multivariate stochastic volatility models. *Journal Econometrics, 150*(2), 182–192.
Asai, M., & McAleer, M. (2009b). The structure of dynamic correlations in multivariate stochastic volatility models. *Journal Econometrics, 150*(2), 182–192.
Asai, M., McAleer, M., & Yu, J. (2006). Multivariate stochastic volatility: A review. *Econometric Reviews, 25*(2-3), 145–175.
Attanasio, O. P. (1991). Risk, time-varying second moments and market efficiency. *The Review of Economic Studies, 58*(3), 479–494.
Ball, C. A., & Torous, W. N. (1983). A simplified jump process for common stock returns. *Journal of Financial and Quantitative Analysis, 18*(1), 53–65.
Bandi, F. M., & Russel, J. R. (2006). Separating microstructure noise from volatility. *Journal of Financial Economics, 79*(3), 655–692.
Bandi, F. M., & Russell, J. R. (2005). *Realized covariation, realized beta and microstructure noise*. Unpublished paper.
Barndorff-Nielsen, O. E., Hansen, P. R., Lunde, A., & Shephard, N. (2011). Multivariate realised kernels: Consistent positive semi-definite estimators of the covariation of equity prices with noise and non-synchronous trading. *Journal of Econometrics, 162*(2), 149–169.
Barndorff-Nielsen, O. E., & Shephard, N. (2002). Econometric analysis of realized volatility and its use in estimating stochastic volatility models. *Journal of the Royal Statistical Society: Series B, 64*(2), 253–280.
Barndorff-Nielsen, O. E., & Shephard, N. (2004). Econometric analysis of realized covariation: High frequency based covariance, regression, and correlation in financial economics. *Econometrica, 72*(3), 885–925.
Barndorff-Nielsen, O. E., & Shephard, N. (2006). Econometrics of testing for jumps in financial economics using bipower variation. *Journal of Financial Econometrics, 4*(1), 1–30.
Bates, D. S. (2000). Post- '87 crash fears in the s & p 500 futures option market. *Journal of Econometrics, 94*(1–2), 181–238.
Bauer, G., & Vorkink, K. (2011). Forecasting multivariate realized stock market volatility. *Journal of Econometrics, 2*(1), 93–101.
Bauwens, L., Grigoryeva, L., & Ortega, J.-P. (2016). Estimation and empirical performance of non-scalar dynamic conditional correlation models. *Computational Statistics and Data Analysis, 100*, 17–36.
Bauwens, L., Laurent, S., & Rombouts, J. V. K. (2006). Multivariate GARCH models: A survey. *Journal of Applied Econometrics, 21*(1), 79–109.
Bekierman, J., & Manner, H. (2018). Forecasting realized variance measures using time-varying coefficient models. *International Journal of Forecasting, 34*(2), 276–287.
Bhattacharya, A., & Dunson, D. B. (2011). Sparse Bayesian infinite factor models. *Biometrika, 98*(2), 291–306.
Billio, M., Caporin, M., Frattarolo, L., & Pelizzon, L. (2018). *Networks in risk spillovers: A multivariate GARCH perspective*. SAFE.
Billio, M., Caporin, M., & Gobbo, M. (2006). Flexible dynamic conditional correlation multivariate GARCH models for asset allocation. *Applied Financial Economics Letters, 2*(2), 123–130.
Bitto, A., & Frühwirth-Schnatter, S. (2019). Achieving shrinkage in a time-varying parameter model framework. *Journal of Econometrics, 210*(1), 75–97.
Black, F. (1976). Studies of stock price volatility changes. In *Proceedings of the 1976 Meetings of the American Statistical Association* (pp. 171–181).
Black, F., & Scholes, M. (1973). The pricing of options and corporate liabilities. *The Journal of Political Economy, 81*(3), 637–654.
Bollerslev, T. (1986). Generalized autoregressive conditional heteroskedasticity. *Journal of Econometrics, 31*(3), 307–327.

Bollerslev, T. (1990). Modelling the coherence in short-run nominal exchange rates: A multivariate generalized arch model. *The Review of Economics and Statistics, 72*(3), 498–505.

Bollerslev, T. (2010). Glossary to ARCH GARCH. In R. Engle, M. Watson, T. Bollerslev, & J. Russell (Eds.), *Volatility and time series econometrics: Essays in honour of Robert F. Engle*. Advanced Texts in Econometrics. Oxford: Oxford University Press.

Bollerslev, T., Chou, R. Y., & Kroner, K. F. (1992). ARCH modeling in finance: A review of the theory and empirical evidence. *Journal of Econometrics, 52*(1–2), 5–59.

Bollerslev, T., Engle, R. F., & Nelson, D. B. (1994). Arch models. *Handbook of Econometrics, 4*, 2959–3038.

Bollerslev, T., Engle, R. F., & Wooldridge, J. M. (1988). A capital asset pricing model with time-varying covariances. *Journal of Political Economy, 96*(1), 116–131.

Bollerslev, T., Patton, A. J., & Quaedvlieg, R. (2016). Exploiting the errors: A simple approach for improved volatility forecasting. *Journal of Econometrics, 192*(1), 1–18.

Bonaccolto, G., & Caporin, M. (2016). The determinants of equity risk and their forecasting implications: A quantile regression perspective. *Journal of Risk and Financial Management, 9*(3), 8.

Bonato, M., Caporin, M., & Ranaldo, A. (2012). A forecast-based comparison of restricted wishart autoregressive models for realized covariance matrices. *The European Journal of Finance, 18*(9), 761–774.

Bonato, M., Caporin, M., & Ranaldo, A. (2013). Risk spillovers in international equity portfolios. *Journal of Empirical Finance, 24*, 121–137.

Callot, L. A. F., Kock, A. B., & Medeiros, M. C. (2017). Modeling and forecasting large realized covariance matrices and portfolio choice. *Journal of Applied Econometrics, 32*(1), 140–158.

Caporin, M., & McAleer, M. (2010). A scientific classification of volatility models. *Journal of Economic Surveys, 24*(1), 191–195.

Caporin, M., & McAleer, M. (2012). Do we really need both BEKK and DCC? A tale of two multivariate GARCH models. *Journal of Economic Surveys, 26*(4), 736–751.

Caporin, M., & Paruolo, P. (2015). Proximity-structured multivariate volatility models. *Econometric Reviews, 34*(5), 559–593.

Caporin, M., & Poli, F. (2017). Building news measures from textual data and an application to volatility forecasting. *Econometrics, 5*(35), 1–46.

Caporin, M., & Velo, G. G. (2015). Realized range volatility forecasting: Dynamic features and predictive variables. *International Review of Economics & Finance, 40*, 98–112.

Cappiello, L., Engle, R. F., & Sheppard, K. (2006). Asymmetric dynamics in the correlations of global equity and bond returns. *Journal of Financial Econometrics, 4*(4), 537–572.

Chib, S., & Greenberg, E. (1996). Markov chain monte carlo simulation methods in econometrics. *Econometric Theory, 12*(3), 409–431.

Chib, S., Nardari, F., & Shephard, N. (2002). Markov chain monte carlo methods for stochastic volatility models. *Journal of Econometrics, 108*(2), 281–316.

Chib, S., Nardari, F., & Shephard, N. (2006). Analysis of high dimensional multivariate stochastic volatility models. *Journal of Econometrics, 134*(2), 341–371.

Chiriac, R., & Voev, V. (2011). Modelling and forecasting multivariate realized volatility. *Journal of Applied Econometrics, 26*(6), 893–1057.

Christiansen, C., Schmeling, M., & Schrimpf, A. (2012). A comprehensive look at financial volatility prediction by economic variables. *Journal of Applied Econometrics, 27*(6), 956–977.

Cont, R. (2001). Empirical properties of asset returns: Stylized facts and statistical issues. *Quantitative Finance, 1*, 223–236.

Corsi, F. (2009). A simple approximate long-memory model of realized volatility. *Journal of Financial Econometrics, 7*(2), 174–196.

Corsi, F., Pirino, D., & Renò, R. (2010). Threshold bipower variation and the impact of jumps on volatility forecasting. *Journal of Econometrics, 159*(2), 276–288.

Danielsson, J. (1998). Multivariate stochastic volatility models: Estimation and a comparison with VGARCH models. *Journal of Empirical Finance, 5*(2), 155–173.

Dhaene, G., Sercu, P., & Wu, J. (2017). *Multi-market volatility spillovers in equities: A sparse multivariate GARCH approach*. Working paper.
Diebold, F., & Nerlove, M. (1989). The dynamics of exchange rate volatility: A multivariate latent factor arch model. *Journal of Applied Econometrics, 4*(1), 1–21.
Ding, Z., & Engle, R. F. (2001). Large scale conditional covariance matrix modeling, estimation and testing. *Academia Economic Papers, 29*, 157–184.
Ding, Z., Granger, C. W., & Engle, R. F. (1993). A long memory property of stock market returns and a new model. *Journal of Empirical Finance, 1*(1), 83–106.
Duffie, D., Pan, J., & Singleton, K. (2000). Transform analysis and asset pricing for affine jump-diffusions. *Econometrica, 68*(6), 1343–1376.
Engle, R. (2002). Dynamic conditional correlation: A simple class of multivariate generalized autoregressive conditional heteroskedasticity models. *Journal of Business & Economic Statistics, 20*(3), 339–350.
Engle, R. F. (1982). Autoregressive conditional heteroscedasticity with estimates of the variance of United Kingdom inflation. *Econometrica, 50*(4), 987–1007.
Engle, R. F., & Gallo, G. M. (2006). A multiple indicators model for volatility using intra-daily data. *Journal of Econometrics, 131*(1–2), 3–27.
Engle, R. F., & Kroner, K. F. (1995). Multivariate simultaneous generalized arch. *Econometric Theory, 11*(1), 122–150.
Engle, R. F., & Russell, J. R. (1998). Autoregressive conditional duration: A new model for irregularly spaced transaction data. *Econometrica, 66(5)*, 1127–1162.
Engle, R. F., & Sokalska, M. E. (2012). Forecasting intraday volatility in the US equity market. Multiplicative component GARCH. *Journal of Financial Econometrics, 10*(1), 54–83.
Epps, T. W. (1979). Comovements in stock prices in the very short run. *Journal of the American Statistical Association, 74*(366), 291–298.
Eraker, B., Johannes, M. S., & Polson, N. G. (2003). The impact of jumps in returns and volatility. *Journal of Finance, 53*(3), 1269–1300.
Fan, J., Furger, A., & Xiu, D. (2016). Incorporating gloabl industrial classification standard into portfolio allocation: A simple factor-based large covariance matrix estimator with high-frequency data. *Journal of Business and Economic Statistics, 34*, 489–503.
Fleming, J., Kirby, C., & Ostdiek, B. (2008). The specification of GARCH models with stochastic covariates. *Journal of Future Markets, 28*(10), 922–934.
Forsberg, L., & Ghysels, E. (2007). Why do absolute returns predict volatility so well? *Journal of Financial Econometrics, 5*(1), 31–67.
Frühwirth-Schnatter, S., & Wagner, H. (2010). Stochastic model specification search for Gaussian and partial non-Gaussian state space models. *Journal of Econometrics, 154*(1), 85–100.
Ghysels, E., Harvey, A., & Renault, E. (1996). Stochastic volatility. In G. S. Maddala & C. R. Rao (Eds.), *Handbook of statistics statistical methods in finance* (Vol. 14). Amsterdam: North-Holland.
Ghysels, E., & Jasiak, J. (1998). GARCH for irregularly spaced financial data: The ACD-GARCH model. *Studies in Nonlinear Dynamics & Econometrics, 2*(4).
Ghysels, E., P., S.-C., & Valkanov, R. (2006). Predicting volatility: Getting the most out of return data sampled at different frequencies. *Journal of Econometrics, 131*(1–2), 59–95.
Ghysels, E., Santa-Clara, P., & Valkanov, R. (2004). *The MIDAS touch: Mixed data sampling regression models*. Montreal, QC: CIRANO.
Glosten, L. R., Jagannathan, R., & Runkle, D. E. (1993). On the relation between the expected value and the volatility of the nominal excess return on stocks. *The Journal of Finance, 48*(5), 1779–1801.
Golosnoy, V., Gribisch, B., & Liesenfeld, R. (2012). The conditional autoregressive Wishart model for multivariate stock market volatility. *Journal of Econometrics, 167*(1), 211–223.
Gourieroux, C. (2006). Continuous time Wishart process for stochastic risk. *Economic Review, 25*(2–3), 177–217.
Gourieroux, C., Jasiak, J., & Sufana, R. (2009a). The Wishart autoregressive process of multivariate stochastic volatility. *Journal of Econometrics, 150*(2), 167–181.

Gourieroux, C., Jasiak, J., & Sufana, R. (2009b). The Wishart autoregressive process of multivariate stochastic volatility. *Journal of Econometrics, 150*(2), 167–181.
Gourieroux, C., & Sufana, R. (2010). Derivative pricing with Wishart multivariate stochastic volatility. *Journal of Business & Economic Statistics, 28*(3), 438–451.
Gourieroux, C., & Sufana, R. (2011). Discrete time Wishart term structure models. *Journal of Economic Dynamics and Control, 35*(6), 815–824.
Gribisch, B. (2016). Multivariate Wishart stochastic volatility and changes in regime. *AStA Advances in Statistical Analysis, 100*(4), 443–473.
Griffin, J. E., & Brown, P. J. (2010). Inference with normal-gamma prior distributions in regression problems. *Bayesian Analysis, 5*(1), 171–188.
Hansen, P. R., Huang, Z., & Shek, H. H. (2012). Realized GARCH: A joint model for returns and realized measures of volatility. *Journal of Applied Econometrics, 27*(6), 877–906.
Hansen, P. R., & Lunde, A. (2006). Realized variance and market microstructure noise. *Journal of Business and Economic Statistics, 24*(2), 127–161.
Harvey, A., Ruiz, E., & Shephard, N. (1994). Multivariate stochastic variance models. *The Review of Economic Studies, 61*(2), 247–264.
Hasbrouck, J. (2006). *Empirical market microstructure. The institutions, economics, and econometrics of securities trading*. New York, NY: Oxford University Press.
Hautsch, N., & Ou, Y. (2008). *Discrete-time stochastic volatility models and MCMC-based statistical inference*. SFB 649 Discussion Papers, Humboldt University, Berlin, Germany.
Heston, S. L. (1993). A closed-form solution for options with stochastic volatility with applications to bond and currency options. *The Review of Financial Studies, 6*(2), 327–343.
Hull, J., & White, A. (1987). The pricing of options on assets with stochastic volatilities. *The Journal of Finance, 42*(2), 281–300.
Jacquier, E., Polson, N. G., & Rossi, P. (1999a). Stochastic volatility: Univariate and multivariate extensions. *Computing in Economics and Finance, 1999*, 112.
Jacquier, E., Polson, N. G., & Rossi, P. (2004). Bayesian analysis of stochastic volatility models with fat-tails and correlated errors. *Journal of Econometrics, 122*(1), 185–212.
Jacquier, E., Polson, N. G., & Rossi, P. E. (1994). Bayesian analysis of stochastic volatility models. *Journal of Business & Economic Statistics, 12*(4), 371–389.
Jacquier, E., Polson, N., & Rossi, P. (1999b). *Stochastic volatility: Univariate and multivariate extensions* (Working paper No. 99). Montreal, QC: CIRANO.
Jarrow, R. A., & Rosenfeld, E. R. (1984). Jump risks and the intertemporal capital asset pricing model. *Journal of Business, 57*, 337–351.
Jin, X., Maheu, J. M., & Yang, Q. (2019). Bayesian parametric and semiparametric factor models for large realized covariance matrices. *Journal of Applied Econometrics, 34*(5), 641–660.
Jorion, P. (1988). On jump processes in the foreign exchange and stock markets. *Review of Financial Studies, 1*(4), 427–445.
Kastner, G. (2018). *Sparse Bayesian time-varying covariance estimation in many dimensions*. Journal of Econometrics.
Kastner, G., Frühwirth-Schnatter, S., & Lopes, H. F. (2017). Efficient Bayesian inference for multivariate factor stochastic volatility models. *Journal of Computational and Graphical Statistics, 26*(4), 905–917.
Kim, S., Shephard, N., & Chib, S. (1998a). Stochastic volatility: Likelihood inference and comparison with ARCH models. *The Review of Economic Studies, 65*(3), 361–393.
Kim, S., Shephard, N., & Chib, S. (1998b). Stochastic volatility: Likelihood inference and comparison with ARCH models. *Review of Economic Studies, 65*(3), 361–393.
Lam, C., Feng, P., & Hu, C. (2017). Nonlinear shrinkage estimation of large integrated covariance matrices. *Biometrika, 104*, 481–488.
Lee, S. S., & Mykland, P. A. (2008). Jumps in financial markets: A new nonparametric test and jump dynamics. *Review of Financial Studies, 21*(6), 2535–2563.
Liesenfeld, R., & Richard, J.-f. (2003). Univariate and multivariate stochastic volatility models: Estimation and diagnostics. *Journal of Empirical Finance, 10*(4), 505–531.

Ling, S., & McAleer, M. (2002). Necessary and sufficient moment conditions for the GARCH(r, s) and asymmetric power GARCH(r, s) models. *Econometric Theory, 18*(3), 722–729.

Lopes, H. F., McCulloch, R. E., & Tsay, R. S. (2014). *Parsimony inducing priors for large scale state-space models*. Technical report.

Marcus, M., & Minc, H. (1992). A survey of matrix theory and matrix inequalities (Vol. 14). New York, NY: Dover.

McAleer, M., & Medeiros, M. (2006). Realized volatility: A review. *Econometric Reviews, 27*(1–3), 10–45.

Melino, A., & Turnbull, S. M. (1990). Pricing foreign currency options with stochastic volatility. *Journal of Econometrics, 45*(1–2), 239–265.

Meucci, A. (2010). *Review of discrete and continuous processes in finance: Theory and applications*. Working paper.

Morimoto, T., & Nagata, S. (2017). Robust estimation of a high-dimensional integrated covariance matrix. *Communications in Statistics - Simulation and Computation, 46*(2), 1102–1112.

Muller, U., Dacorogna, M., Dav, R., Pictet, O., Olsen, R., & Ward, J. (1993). Fractals and intrinsic time - A challenge to econometricians. In *XXXIXth International AEA Conference on Real Time Econometrics*, Luxembourg.

Nelson, D. B. (1991). Conditional heteroskedasticity in asset returns: A new approach. *Econometrica, 59*(2), 347–370.

Noureldin, D., Shephard, N., & Sheppard, K. (2014). Multivariate rotated arch models. *Journal of Econometrics, 179*(1), 16–30.

Pakel, C., Shephard, N., Sheppard, K., & Engle, R. F. (2017). *Fitting vast dimensional time-varying covariance models*. NYU Working paper.

Philipov, A., & Glickman, M. E. (2006a). Factor multivariate stochastic volatility via Wishart processes. *Econometric Review, 25*(2–3), 311–334.

Philipov, A., & Glickman, M. E. (2006b). Multivariate stochastic volatility via Wishart processes. *Journal of Business & Economic Statistics, 24*(3), 313–328.

Pitt, M. K., & Shephard, N. (1999). Time-varying covariances: A factor stochastic volatility approach. In *Bayesian statistics* (pp. 547–570). Oxford: Oxford University Press.

Shen, K., Yao, J., & Li, W. K. (2018). Forecasting high-dimensional realized volatility matrices using a factor model. *Quantitative Finance*, forthcoming.

Shepard, N., & Sheppard, K. (2010). Realising the future: Forecasting with high-frequency-based volatility (heavy) models. *Journal of Applied Econometrics, 25*, 197–231.

Shephard, N. (1996). Statistical aspects of arch and stochastic volatility. In *Time series models in econometrics* (pp. 1–67). London: Chapman & Hall.

Shephard, N., & Pitt, M. K. (2004). Erratum: Likelihood analysis of non-Gaussian measurement time series'. *Biometrika, 91*(1), 249–250.

Shephard, N., & Sheppard, K. (2010). Realising the future: Forecasting with high-frequency-based volatility (heavy) models. *Journal of Applied Econometrics, 25*(2), 197–231.

Sheppard, K. (2006). *Realized covariance and scrambling*. Unpublished manuscript.

Sheppard, K., & Xu, W. (2019). Factor high-frequency-based volatility (heavy) models. *Journal of Financial Econometrics, 17*, 33–65.

Silvennoinen, A., & Teräsvirta, T. (2009). Multivariate GARCH models. In *Handbook of financial time series* (pp. 201–229). Berlin: Springer.

Smith, M., & Pitts, A. (2006). Foreign exchange intervention by the bank of Japan: Bayesian analysis using a bivariate stochastic volatility model. *Econometric Reviews, 25*(2–3), 425–451.

So, M. K. P., & Kwok, S. W. Y. (2006). A multivariate long memory stochastic volatility model. *Physica A: Statistical Mechanics and its Applications, 362*(2), 450–464.

So, M. K. P., Li, W. K., & Lam, K. (1997). Multivariate modelling of the autoregressive random variance process. *Journal of Time Series Analysis, 18*(4), 429–446.

Tao, M., Wang, Y., & Chen, X. (2013). Fast convergence rates in estimating large volatility matrices using high-frequency financial data. *Econometric Theory, 29*, 838–856.

Tao, M., Wang, Y., Yao, Q., & Zou, J. (2011). Large volatility matrix inference via combining low-frequency and high-frequency approaches. *Journal of the American Statistical Association, 106*, 1025–1040.

Tao, M., Wang, Y., & Zhou, H. H. (2013). Optimal sparse volatility matrix estimation for high-dimensional Ito processes with measurement error. *Annals of Statistics, 41*, 1816–1864.

Taylor, S. J. (1986). *Modelling financial time series*. Chichester: Wiley.

Taylor, S. J. (1994). Modelling stochastic volatility: A review and comparative study. *Mathematical Finance, 4*(2), 183–204.

Teräsvirta, T. (2009). An introduction to univariate GARCH models. In *Handbook of financial time series* (pp. 17–42).

Tibshirani, R. (1996). Regression shrinkage and selection via the lasso. *Journal of the Royal Statistical Society Series B, 58*(1), 267–288.

Tsay, R. S. (2010). *Analysis of financial time series* (3rd ed.). Wiley Series in Probability and Statistics. Hoboken, NJ: Wiley.

Van der Weide, R. (2002). GO-GARCH: A multivariate generalized orthogonal GARCH model. *Journal of Applied Econometrics, 17*(5), 549–564.

Violante, F., & Laurent, S. (2012). Volatility forecasts evaluation and comparison. In L. Bauwens, C. Hafner, & S. Laurent (Eds.), *Handbook of volatility models and their applications*. London: Wiley.

Yamauchi, Y., & Omori, Y. (2018). Multivariate stochastic volatility model with realized volatilities and pairwise realized correlations. *Journal of Business and Economic Statistics*, forthcoming.

Yu, J., & Meyer, R. (2006a). Multivariate stochastic volatility models: Bayesian estimation and model comparison. *Econometric Reviews, 25*(2–3), 361–384.

Yu, J., & Meyer, R. (2006b). Multivariate stochastic volatility models: Bayesian estimation and model comparison. *Econometric Review, 25*(2–3), 361–384.

Yu, P. L. H., Li, W. K., & Ng, F. C. (2017). The generalized conditional autoregressive Wishart model for multivariate realized volatility. *Journal of Business & Economic Statistics, 35*(4), 513–527.

Yu, Y., & Meng, X.-l. (2011). To center or not to center: That is not the question—an ancillarity-sufficiency interweaving strategy (ASIS) for boosting MCMC efficiency. *Journal of Computational and Graphical Statistics, 20*(3), 531–570.

Zhang, L., Mykland, P. A., & Ait-Sahalia, Y. (2005). A tale of two time scales. *The Journal of the American Statistical Association, 100*(472), 1394–1411.

Chapter 6
Neural Networks

Thomas R. Cook

6.1 Introduction

Neural networks have emerged in the last 10 years as a powerful and versatile set of machine learning tools. They have been deployed to produce art, write novels, read handwriting, translate languages, caption images, interpret MRIs, and many other tasks. In this chapter, we will introduce neural networks and their application to forecasting.

Though neural networks have recently become very popular, they are an old technology. Early work on neurons as computing units dates as far back as 1943 (McCulloch and Pitts) with early commercial applications arising in the late 1950s and early 1960s. The development of neural networks since then, however, has been rocky. In the late 1960s, work by Minsky and Papert showed that perceptrons (an elemental form of neural network) were incapable of emulating the exclusive-or (XOR) function. This led to a sharp decline in neural network research that lasted through the mid-1980s. From the mid-1980s through the end of the century, neural networks were a productive but niche area of computer science research. Starting in the mid-2000s however, neural networks have seen widespread adoption as a powerful machine learning method. This surge in popularity has been attributable

The views expressed are those of the author and do not necessarily reflect the views of the Federal Reserve Bank of Kansas City or the Federal Reserve System.

T. R. Cook
Federal Reserve Bank of Kansas City, Kansas City, MO, USA
e-mail: thomas.cook@kc.frb.org

P. Fuleky (ed.), *Macroeconomic Forecasting in the Era of Big Data*,
Advanced Studies in Theoretical and Applied Econometrics 52,
https://doi.org/10.1007/978-3-030-31150-6_6

largely to a confluence of factors: algorithmic developments that made neural networks useful for practical applications; advances in processing power that made model training feasible; the rise of "big data"; and a few high-profile successes in areas such as computer vision. At this point in time, neural networks have gained mainstream appeal in areas far beyond computer science such as bioinformatics, geology, medicine, chemistry, and others.

In economics and finance, neural networks have been used since the early 1990s, mostly in the context of microeconomics and finance. Much of the early work focused on bankruptcy prediction (see Altman, Marco, & Varetto, 1994; Odom & Sharda, 1990; Tam, 1991). Additional research using neural networks to predict creditworthiness was performed around this time and there is a growing appetite among banks to use artificial intelligence for credit underwriting. More recently, in the area of finance, neural networks have been successfully used for market forecasting (see Dixon, Klabjan, & Bang, 2017; Heaton, Polson, & Witte, 2016; Kristjanpoller & Minutolo, 2015; McNelis, 2005 for examples). There has been some limited use of neural networks in macroeconomic research (see Dijk, Teräsvirta, & Franses, 2002; Terasvirta & Anderson, 1992, as examples), but much of this research seems to have occurred prior to the major resurgence of neural networks in the 2010s.

Although there are already many capable tools in the econometric toolkit, neural networks are a worthy addition because of their versatility of use and because they are universal function approximators. This is established by the theorem of universal approximation, first put forth by Cybenko (1989) with similar findings offered by Hornik, Stinchcombe, and White (1989) and further generalized by Hornik (1991). In summary, the theorem states, that for any continuous function $f : \mathbb{R}^m \mapsto \mathbb{R}^n$, there exists a neural network with one hidden layer, G, that can approximate f to an arbitrary level of accuracy (i.e., $|f(\boldsymbol{x}) - G(\boldsymbol{x})| < \epsilon$ for any $\epsilon > 0$). While there are other algorithms that can be used as universal function approximators, neural networks require few assumptions (inductive biases), have a tendency to generalize well, and scale well to the size of the input space in ways that other methods do not.

Neural networks are often associated with "big data." The reason for this is that people often associate neural networks with complex modeling tasks that are difficult/impractical with other types of models. For example we often hear of neural networks in reference to computer vision, speech translation and drug discovery. Each of these types of task produces high-dimensional, complex outputs (and likely takes equivalently complex inputs). And, like any type of model, the amount of data needed to train a neural network typically scales to the dimensionality of its inputs/outputs. Take, for example, the Inception neural network model (Szegedy et al., 2015). This is an image classification model that learns a distribution over about 1000 categories that are then used to classify an image. The capabilities of this model are impressive, but to get the network to learn such a large distribution of possible image categories, researchers made use of a dataset containing over one million labeled images.[1]

The remainder of this chapter will proceed as follows. In the remainder of this section the technical aspects of neural networks will be presented, focusing on the fully connected network as a point of reference. Section 6.2 will discuss neural network model design considerations. Sections 6.3 and 6.4 will introduce recurrent networks and encoder-decoder networks. Section 6.5 will provide an applied example in the form of unemployment forecasting.

6.1.1 Fully Connected Networks

A fully connected neural network, sometimes called a multi-layer perceptron, is among the most straightforward types of neural network models. It consists of several interconnected layers of neurons that translate inputs into a target output.

The fully connected neural network, and neural networks generally, are fundamentally comprised of neurons. A neuron is simply a linear combination of inputs, plus a constant term (called a *bias*), and transformed through a function (called an *activation function*),

$$f(\boldsymbol{x}\boldsymbol{\beta} + \alpha),$$

where $\boldsymbol{x}$ is an n-length vector of inputs, $\boldsymbol{\beta}$ is a corresponding vector of weights, and α is a scalar bias term.

Neurons are typically stacked into layers. Layers can have various forms, but the most simple is called a dense, or fully connected layer. For a layer with p neurons, let $\boldsymbol{B} = (\boldsymbol{\beta}_1 \dots \boldsymbol{\beta}_p)$ so that $\boldsymbol{B}$ has the dimensions $(n \times p)$, and let $\boldsymbol{\alpha} = (\alpha_1 \dots \alpha_p)$. The matrix $\boldsymbol{B}$ supplies weights for each term in the input vector to each of the p neurons while $\boldsymbol{\alpha}$ supplies the bias for each neuron. Given an n-length input vector $\boldsymbol{x}$, we can write a dense layer with p neurons as,

$$\begin{aligned} g(\boldsymbol{x}) &= f(\boldsymbol{x}\boldsymbol{B} + \boldsymbol{\alpha}) \\ &= \begin{bmatrix} f(\boldsymbol{x}\boldsymbol{\beta}_1 + \boldsymbol{\alpha}_1) \\ f(\boldsymbol{x}\boldsymbol{\beta}_2 + \boldsymbol{\alpha}_2) \\ \vdots \\ f(\boldsymbol{x}\boldsymbol{\beta}_p + \boldsymbol{\alpha}_p) \end{bmatrix}^T . \end{aligned} \tag{6.1}$$

[1] Specifically, a subset of the imagenet dataset. See Russakovsky et al. (2015).

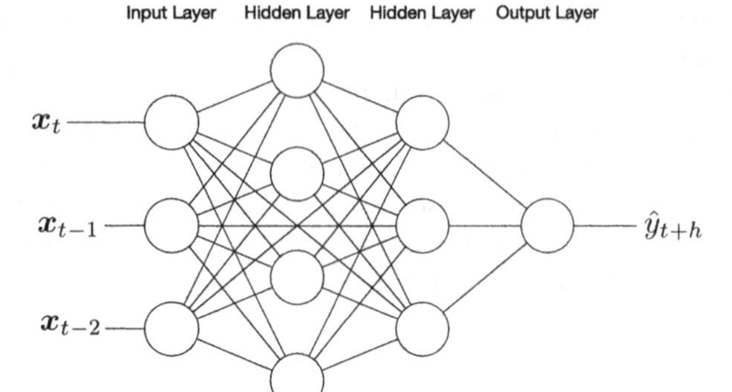

Fig. 6.1 Diagram of a fully connected network as might be constructed for a forecasting task. The network takes in several lags of the input vector $\boldsymbol{x}$ and returns an estimate of the target at the desired forecast horizon $t + h$. This network has two hidden layers with three and four neurons respectively. The output layer is a single neuron, returning a one-dimensional output

As should be clear from this expression, each element in the input vector bears some influence on (or connection to) each of the p neurons, which is why we call this a fully connected layer.

The layer described in Eq. (6.1) can also accept higher-order input such as an $m \times n$ matrix of several observations, $\boldsymbol{X} = (\boldsymbol{x}_1, \boldsymbol{x}_2, \boldsymbol{x}_m)'$, in which case,

$$\begin{aligned} g(\boldsymbol{X}) &= f(\boldsymbol{X}\boldsymbol{B} + \boldsymbol{\alpha}) \\ &= \begin{bmatrix} f(\boldsymbol{x}_1\boldsymbol{\beta}_1 + \boldsymbol{\alpha}_1) & \dots & f(\boldsymbol{x}_1\boldsymbol{\beta}_p + \boldsymbol{\alpha}_p) \\ \vdots & \ddots & \vdots \\ f(\boldsymbol{x}_m\boldsymbol{\beta}_1 + \boldsymbol{\alpha}_1) & \dots & f(\boldsymbol{x}_m\boldsymbol{\beta}_p + \boldsymbol{\alpha}_p) \end{bmatrix}. \end{aligned}$$

A fully connected, feed forward network (Fig. 6.1), with K layers is formed by connecting dense layers together so that the output of the preceding layer serves as the input for the current layer. Let k index a given layer, then the output of the k-th layer is

$$\begin{aligned} g_k(\boldsymbol{X}) &= f_k(g_{k-1}(\boldsymbol{X})\boldsymbol{B}_k + \boldsymbol{\alpha}_k) \\ g_0(\boldsymbol{X}) &= \boldsymbol{X}, \end{aligned}$$

The parameters of the network are all elements $\boldsymbol{B}_k, \boldsymbol{\alpha}_k$ for $k \in (1 \dots .K)$. For simplicity, denote these parameters by $\boldsymbol{\theta}$, where $\boldsymbol{\theta}_k = (\boldsymbol{B}_k, \boldsymbol{\alpha}_k)$. Further, for simplicity, denote the final output of the network $G(\boldsymbol{X}; \boldsymbol{\theta}) = g_K(\boldsymbol{X})$.

In the context of a supervised learning problem (such as a forecasting problem), we have a known target, $\boldsymbol{y}$, estimated as $\hat{\boldsymbol{y}} = G(\boldsymbol{X}; \boldsymbol{\theta})$, and we can define a loss function, $L(\boldsymbol{y}; \boldsymbol{\theta})$ to summarize the discrepancy between our estimate and target. To estimate the model, we simply find $\boldsymbol{\theta}$ that minimizes $L(\boldsymbol{y}; \boldsymbol{\theta})$. The estimation procedure will be discussed in greater detail below.

6.1.2 Estimation

To fit a neural network, we follow a modified variation of gradient descent. Gradient descent is an iterative procedure. For each of $\boldsymbol{\theta}_k \in \boldsymbol{\theta}$, we calculate the gradient of the loss, $\nabla_{\boldsymbol{\theta}_k} L(\boldsymbol{y}; \boldsymbol{\theta})$. The negative gradient tells us the direction of steepest descent and the direction in which to adjust $\boldsymbol{\theta}_k$ to reduce $L(\boldsymbol{y}; \boldsymbol{\theta})$. After calculating $\nabla_{\boldsymbol{\theta}_k} L(\boldsymbol{y}; \boldsymbol{\theta})$, we perform an update,

$$\boldsymbol{\theta}_k \leftarrow \boldsymbol{\theta}_k - \gamma \nabla_{\boldsymbol{\theta}_k} L(\boldsymbol{y}; \boldsymbol{\theta}).$$

where γ controls the size of the update and is sometimes referred to as the learning rate. After updates are computed for all $\boldsymbol{\theta}_k \in \boldsymbol{\theta}$, $L(\boldsymbol{y}; \boldsymbol{\theta})$ is recomputed. These steps are repeated until a stopping rule has been reached (e.g., the $L(\boldsymbol{y}; \boldsymbol{\theta})$ falls below a preset threshold).

The computation of $\nabla_{\boldsymbol{\theta}} L(\boldsymbol{y}; \boldsymbol{\theta})$ is costly and increases with the size of $\boldsymbol{X}$ and $\boldsymbol{y}$. To reduce this cost, and the overall computation time needed, we turn to stochastic gradient descent (SGD). This is a modification of gradient descent in which updates to $\boldsymbol{\theta}$ are calculated using only one observation at a time. For each iteration of the procedure, one observation, $\{\boldsymbol{x}, y\}$, is chosen, then updates to $\boldsymbol{\theta}$ are calculated and applied as described for gradient descent, and a new observation is chosen for use in the next iteration of the procedure. By using SGD, we reduce the time needed for each iteration of the optimization procedure, but increase the expected number of iterations needed to reach performance equivalent to gradient descent.

In many cases the speed of optimization can be further boosted through Mini-batch SGD. This is a modification of SGD in which updates are calculated using several observations at a time. Mini-batch SGD should generally require fewer iterations than SGD, but computing the updates for each iteration will be more computationally costly. The per-iteration cost of calculating updates to $\boldsymbol{\theta}$, however, should be lower than for gradient descent. Mini-batch SGD is by far the most popular procedure for fitting a neural network.

Fitting a neural network is a non-convex optimization problem. It is possible and quite easy for a mini-batch SGD procedure to get stuck at local minima or saddle points (Dauphin et al., 2014). To overcome this, a number of modified optimization algorithms have been proposed. These include RMSprop and adaGrad (Duchi, Hazan, & Singer, 2011). Generally, these modifications employ adaptive learning rates and/or notions of momentum to encourage the optimization algorithm to choose appropriate learning rates and avoid suboptimal local minima (see Ruder, 2016 for a review).

More recently, Adam (Kingma & Ba, 2014) has emerged as a popular variation of gradient descent and as argued in Ruder (2016), "Adam may be the best overall choice [of optimizer]." Adam modifies vanilla gradient descent by scaling the learning rates of individual parameters using the estimated first and second moments of the gradient. Let $\boldsymbol{\theta}_k$ be the estimated value of $\boldsymbol{\theta}_k$ at the current step in the Adam optimization procedure, then we can estimate the first and second moments of the gradient of $\boldsymbol{\theta}_k$ via exponential moving average,

$$\begin{aligned}\boldsymbol{\mu}_t &= \boldsymbol{\mu}_{t-1}\gamma_\mu + (1-\gamma_\mu)(\nabla_{\boldsymbol{\theta}_k} L)\\ \boldsymbol{\nu}_t &= \boldsymbol{\nu}_{t-1}\gamma_\nu + (1-\gamma_\nu)(\nabla_{\boldsymbol{\theta}_k} L)^2\\ \boldsymbol{\mu}_0 &= \mathbf{0}\\ \boldsymbol{\nu}_0 &= \mathbf{0},\end{aligned}$$

where arguments to the loss function are suppressed for readability. Both γ_μ and γ_ν are hyper parameters that control the pace at which $\boldsymbol{\mu}$ and $\boldsymbol{\nu}$ change. With $\boldsymbol{\mu}$ and $\boldsymbol{\nu}$, we can assemble an approximate signal to noise ratio of the gradient and use that ratio as the basis for the update step:

$$\begin{aligned}\hat{\boldsymbol{\mu}}_t &= \frac{\boldsymbol{\mu}_t}{1-(\gamma_\mu)^t}\\ \hat{\boldsymbol{\nu}}_t &= \frac{\boldsymbol{\nu}_t}{1-(\gamma_\nu)^t}\\ \boldsymbol{\theta}_k &\leftarrow \boldsymbol{\theta}_k + \gamma\frac{\hat{\boldsymbol{\mu}}_t}{\sqrt{\hat{\boldsymbol{\nu}}_t+\epsilon}},\end{aligned}$$

where $\hat{\boldsymbol{\mu}}_t$ and $\hat{\boldsymbol{\nu}}_t$ correct for the bias induced by the initialization of $\boldsymbol{\mu}$ and $\boldsymbol{\nu}$ to zero, γ is the maximum step-size for any iteration of the procedure (the learning rate), and where division should be understood in this context as element-wise division. By constructing the update from a signal to noise ratio the path of gradient descent becomes smoother. That is, the algorithm is encouraged to take large step sizes along dimensions of the gradient that are steep and relatively stable; it is cautioned to take small step sizes along dimensions of the gradient that are shallow or relatively volatile. As a result, parameter updates are less volatile. The authors of the algorithm suggest values of $\gamma_\mu = 0.9999$, $\gamma_\nu = 0.9$ and $\epsilon = 1e-8$.

Gradient Estimation

Each of the optimization routines described in the previous section rely upon the computation of $\nabla_{\boldsymbol{\theta}_k} L(\boldsymbol{y}; \boldsymbol{\theta})$ for all $\boldsymbol{\theta}_k \in \boldsymbol{\theta}$. This is achieved through the backpropagation algorithm (Rumelhart, Hinton, & Williams, 1986), which is a generalization of the chain rule from calculus. Consider, for example, a network $G(\boldsymbol{X}; \boldsymbol{\theta})$ with an accompanying loss $L(\boldsymbol{y}; \boldsymbol{\theta}) = \frac{1}{2m}\|G(\boldsymbol{X}; \boldsymbol{\theta}) - \boldsymbol{y}\|_2^2 = \frac{1}{2m}\sum_i^m (G(\boldsymbol{X}; \boldsymbol{\theta})_i - y_i)^2$. Then

$$\nabla_{G(\boldsymbol{X};\boldsymbol{\theta})} L(\boldsymbol{y}; \boldsymbol{\theta}) = \frac{1}{m}(G(\boldsymbol{X}; \boldsymbol{\theta}) - \boldsymbol{y}).$$

To derive $\nabla_{\boldsymbol{\theta}_K} L(\boldsymbol{y}; \boldsymbol{\theta})$, we simply apply chain rule to the above equation, suppressing arguments to G and g for notational simplicity:

$$\nabla_{\boldsymbol{\theta}_K} L(\boldsymbol{y}; \boldsymbol{\theta}) = \frac{\partial G}{\partial \boldsymbol{\theta}_K}^T \nabla_G L(\boldsymbol{y})$$

$$= \begin{bmatrix} (f'(g_{k-1}\boldsymbol{B}_k + \boldsymbol{\alpha}_k)g_{k-1})^T \frac{1}{m}(G-\boldsymbol{y}) \\ (f'(g_{k-1}\boldsymbol{B}_k + \boldsymbol{\alpha}_k))^T \frac{1}{m}(G-\boldsymbol{y}), \end{bmatrix}^T$$

where $\frac{\partial G}{\boldsymbol{\theta}_K}$ is a generalized form of a Jacobian matrix, capable of representing higher-order tensors, and f' indicates the first derivative of f with respect to its argument.

Collecting right-hand-side gradients into Jacobian matrices, we can extend the application of backpropagation to calculate the gradient of the loss with respect to any of the set of parameters $\boldsymbol{\theta}_k$:

$$\nabla_{\boldsymbol{\theta}_k} L(\boldsymbol{y}; \boldsymbol{\theta}) = \frac{\partial L(\boldsymbol{y}; \boldsymbol{\theta})}{\partial G_k} \frac{\partial g_K}{\partial g_{K-1}} \frac{\partial g_{K-1}}{\partial g_{K-2}} \cdots \frac{\partial g_{k+1}}{g_k} \frac{\partial g_k}{\partial \boldsymbol{\theta}_k}.$$

6.1.3 Example: XOR Network

To illustrate the concepts discussed thus far, we will review a simple, well known network that illustrates the construction of a neural network from end to end. This network is known as the XOR network (Minsky & Papert, 1969; Rumelhart, Hinton, & Williams, 1985). It was an important hurdle in the development of neural networks.

Consider a dataset with labels $\boldsymbol{y}$ whose values depend on features, $\boldsymbol{X}$:

$$\boldsymbol{X} = \begin{bmatrix} 1 & 0 \\ 0 & 0 \\ 1 & 1 \\ 0 & 1 \end{bmatrix} \quad \boldsymbol{y} = \begin{bmatrix} 1 \\ 0 \\ 0 \\ 1 \end{bmatrix}.$$

The label of any given observation follows the logic of the exclusive-or operation – $y_i = 1$ only if $\boldsymbol{x}_i$ contains *exactly* one non-zero element.

We can build a fully connected network, $G(\boldsymbol{X}; \boldsymbol{\theta})$ that perfectly represents this relationship using only two layers and three neurons (two in the first layer and one in the last layer):

$$g_1(\boldsymbol{X}) = f(\boldsymbol{X}\boldsymbol{B}_1 + \boldsymbol{\alpha}_1) \tag{6.2}$$

$$g_2(\boldsymbol{X}) = f(g_1(\boldsymbol{X})\boldsymbol{B}_2 + \boldsymbol{\alpha}_2) \tag{6.3}$$

$$\boldsymbol{B}_1 = \begin{bmatrix} \beta_{11} & \beta_{12} \\ \beta_{13} & \beta_{14} \end{bmatrix} \quad \boldsymbol{B}_2 = \begin{bmatrix} \beta_{21} \\ \beta_{22} \end{bmatrix}$$

$$f(a) = \frac{1}{1 + e^{-a}}.$$

The structure of this network is illustrated in Fig. 6.2.

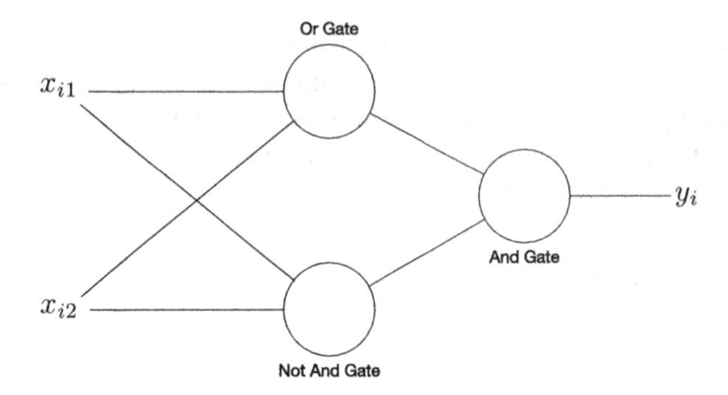

Fig. 6.2 The XOR network. This network is sufficiently simple that each neuron can be labeled according to the logical function it performs

Because this is a classification problem, we will measure loss by log-loss (i.e., negative log-likelihood)[2]:

$$\begin{aligned} L(\boldsymbol{y};\boldsymbol{\theta}) &= -\sum_i^m y_i\, log(G(\boldsymbol{X};\boldsymbol{\theta})_i) + (1-y_i)\, log(1-G(\boldsymbol{X};\boldsymbol{\theta})_i) \\ &= -\,(\boldsymbol{y}\, log(G(\boldsymbol{X};\boldsymbol{\theta})) + (1-\boldsymbol{y}) log(1-G(\boldsymbol{X};\boldsymbol{\theta}))). \end{aligned} \tag{6.4}$$

Calculation of gradients for the final layer yields

$$\frac{\partial L(\boldsymbol{y};\boldsymbol{\theta})}{\partial G(\boldsymbol{X};\boldsymbol{\theta})} = \frac{\boldsymbol{y} - G(\boldsymbol{X};\boldsymbol{\theta})}{(G(\boldsymbol{X};\boldsymbol{\theta})-1)G(\boldsymbol{X};\boldsymbol{\theta})}$$

and application of chain rule provides gradients for $\boldsymbol{\theta}_1$ and $\boldsymbol{\theta}_2$,

$$\begin{aligned} \nabla_{\boldsymbol{\theta}_2} L(\boldsymbol{y};\boldsymbol{\theta}) &= \begin{bmatrix} \frac{\partial L(\boldsymbol{y};\boldsymbol{\theta})}{\partial G(\boldsymbol{X};\boldsymbol{\theta})} \frac{\partial G(\boldsymbol{X};\boldsymbol{\theta})}{\partial \boldsymbol{B}_2} \\ \frac{\partial L(\boldsymbol{y};\boldsymbol{\theta})}{\partial G(\boldsymbol{X};\boldsymbol{\theta})} \frac{\partial G(\boldsymbol{X};\boldsymbol{\theta})}{\partial \boldsymbol{\alpha}_2} \end{bmatrix} \\ &= \begin{bmatrix} g_1(X)^T(\boldsymbol{y} - G(\boldsymbol{X};\boldsymbol{\theta})) \\ (\boldsymbol{y} - G(\boldsymbol{X};\boldsymbol{\theta})) \end{bmatrix} \end{aligned} \tag{6.5}$$

$$\begin{aligned} \nabla_{\boldsymbol{\theta}_1} L(\boldsymbol{y};\boldsymbol{\theta}) &= \begin{bmatrix} \frac{\partial L(\boldsymbol{y};\boldsymbol{\theta})}{\partial G(\boldsymbol{X};\boldsymbol{\theta})} \frac{\partial G(\boldsymbol{X};\boldsymbol{\theta})}{\partial g_1(X)} \frac{\partial g_1(X)}{\partial \boldsymbol{B}_1} \\ \frac{\partial L(\boldsymbol{y};\boldsymbol{\theta})}{\partial G(\boldsymbol{X};\boldsymbol{\theta})} \frac{\partial G(\boldsymbol{X};\boldsymbol{\theta})}{\partial g_1(X)} \frac{\partial g_1(X)}{\partial \alpha_1} \end{bmatrix} \\ &= \begin{bmatrix} X^T(\boldsymbol{y} - G(\boldsymbol{X};\boldsymbol{\theta}))\boldsymbol{B}_2^T \odot f'(\boldsymbol{X}\boldsymbol{B}_1 + \boldsymbol{\alpha}_1) \\ (\boldsymbol{y} - G(\boldsymbol{X};\boldsymbol{\theta}))\boldsymbol{B}_2^T \odot f'(\boldsymbol{X}\boldsymbol{B}_1 + \boldsymbol{\alpha}_1), \end{bmatrix} \end{aligned} \tag{6.6}$$

[2]We use log-loss because it is the convention (in both the machine learning and statistical literature) for this type of categorization problem. Other loss functions, including mean squared error would likely work as well.

Algorithm 1 Gradient descent to fit the XOR network

Data: $\boldsymbol{X}, \boldsymbol{y}$
Input: γ, stop_rule
Initialize $\boldsymbol{\theta} = B_1, B_2, \alpha_1, \alpha_2$ to random values
while stop_rule not met **do**
  Forward pass:
  calculate $\hat{\boldsymbol{y}} = G(\boldsymbol{X}; \boldsymbol{\theta})$ as in (6.2)–(6.3)
  calculate $L(\boldsymbol{y}; \boldsymbol{\theta})$ by log-loss($\hat{\boldsymbol{y}}, \boldsymbol{y}$) as in (6.4)
  Backward pass:
  calculate $\nabla_{\boldsymbol{\theta}_2} = \nabla_{\boldsymbol{\theta}_2} L(\boldsymbol{y}; \boldsymbol{\theta})$ as in (6.5)
  calculate $\nabla_{\boldsymbol{\theta}_1} = \nabla_{\boldsymbol{\theta}_1} L(\boldsymbol{y}; \boldsymbol{\theta})$ as in (6.6)
  Update:
  $\boldsymbol{\theta}_1 \leftarrow \boldsymbol{\theta}_1 - \gamma \nabla_{\boldsymbol{\theta}_1}$
  $\boldsymbol{\theta}_2 \leftarrow \boldsymbol{\theta}_2 - \gamma \nabla_{\boldsymbol{\theta}_2}$
end while

where f' indicates the first derivative of the activation function (i.e., $f'(a) = f(a) \odot (1 - f(a))$). To fit (or *train*) this model, we minimize $L(\boldsymbol{y}; \boldsymbol{\theta})$ via vanilla gradient descent as described in Algorithm 1.

Figure 6.3 illustrates the results of this training process. It shows that, as the number of training iterations increases, the model output predicts the correct classification of each element in $\boldsymbol{y}$.

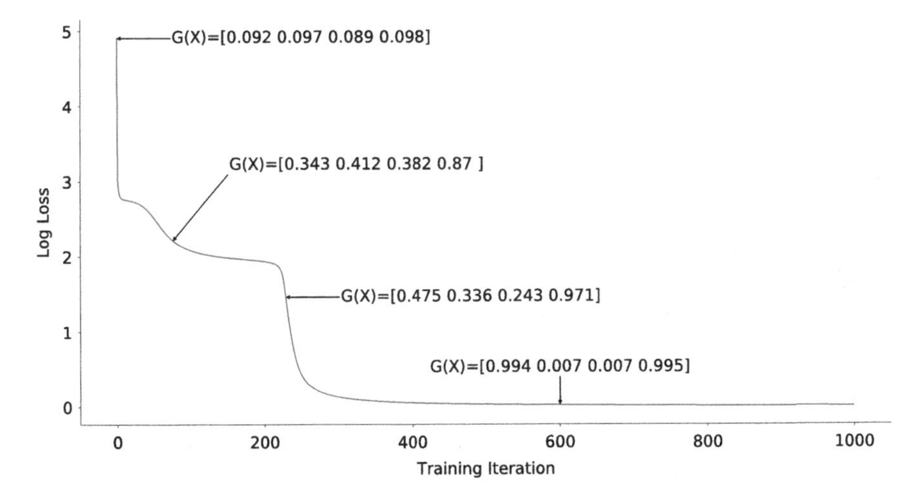

Fig. 6.3 The path of the loss function over the training process. Annotations indicate the model prediction at various points during the training process. The model target is $\boldsymbol{y} = [1, 0, 0, 1]$

6.2 Design Considerations

The XOR neural network is an example of a network that is very deliberately designed in the way that one might design a circuit or economic model. Each neuron in the network carries out a specific and identifiable task. The neurons in g_1 learn to emulate OR and NOT AND gates on the input, while the g_2 neuron learns to emulate an AND gate on the output of g_1.[3] It is somewhat unusual to design neural network models this explicitly. Moreover explicitly designing a neural network this way obviates one of the central advantages of neural network models: a network model with sufficient number of neurons and an appropriate amount of training data can learn to approximate any function without being *ex ante* and explicitly designed to approximate that function.

The typical process for designing a neural network occurs without guidance from an explicit, substantive theory. Instead, the process of designing a neural network is usually functional in nature. As such, when designing a neural network, we are usually left with many design decisions or, alternatively stated, a large space of hyper parameters to explore. Finding the optimal set of hyper parameters needed to make a neural network work effectively for a given problem is one of the biggest challenges to building a successful model. Efficient, automatic processes to optimize model hyper parameters is an active area of research. In this section, we will discuss some of the common design decisions that we must make when designing a neural network model.

6.2.1 Activation Functions

Activation functions are what enable neural networks to approximate non-linear functions. Any differentiable function can be used as an activation function. Moreover, some non-differentiable functions can also be used, as long as there are relatively few points of non-differentiability. Activation functions also tend to be monotonic, though this is not required. The influence of an activation function on model performance is inherently related to the structure of the network model, the method of weight initialization, and idiosyncrasies in the data.

For model training to be successful, the codomain of the final layer activation function must admit the range of possible target values, $\boldsymbol{y}$. For many forecasting tasks, then, the most appropriate final layer activation function is the identify function, $f(a) = a$.

Generally for hidden layers, we want to choose activation functions that return the value of the input (i.e., approximate identity) when the value of the input is near

[3]See Bland (1998), Rumelhart et al. (1985) for further discussion.

Table 6.1 Common activation functions

Sigmoid	$f(a) = \frac{1}{1+e^a}$
ReLU (Nair & Hinton, 2010)	$f(a) = \begin{cases} a & a > 0 \\ 0 & a \leq 0 \end{cases}$
Leaky ReLU (Maas, Hannun, & Ng, 2013)	$f(a) = \begin{cases} a & a > 0 \\ a\,\alpha & a \leq 0 \end{cases} \;\&\; 0 < \alpha \ll 1$
Hyperbolic tangent (Karlik & Olgac, 2011)	$f(a) = \frac{e^a - 1}{e^a + 1}$
Swish	$f(x\beta) = x\,\frac{1}{1+e^{x\beta}}$

zero. This is a desirable property because it removes complications with weight initialization (Sussillo & Abbott, 2014).

Additionally, we want to choose activations that are unbounded (in at least one direction). This helps to prevent neuron saturation (which occurs when gradients approach zero). In turn, this helps prevent the problem of vanishing gradients (Bengio, Simard, & Frasconi, 1994; Glorot & Bengio, 2010) in which early network layers update very slowly. The severity of this problem scales to the depth of the network (assuming the same, bounded, activation function is used for every layer in the network). In the extreme, this can cause adjustments to model weights to effectively stop very early in the training process. It is largely because of the vanishing gradient problem that sigmoid and hyperbolic tangent (see Table 6.1 below) have fallen out of favor for general use.

Table 6.1 provides a list of some common activation functions. Sigmoid and Hyperbolic Tangent activation functions were commonly used in the early development of neural networks, but in recent years the Rectified Linear Unit (ReLU) has become the most popular choice for activation function. Other activation functions such as Swish have emerged more recently and, while they have not found the same widespread adoption, recent research suggests that they may perform better than ReLU in general settings (Ramachandran, Zoph, & Le, 2017).

6.2.2 *Model Shape*

Cybenko (1989) provides the universal approximation theorem, which establishes that a feed forward network with a single hidden layer can approximate any continuous function. Hornik et al. (1989) provides a related and contemporaneous result. As a matter of theory then, no network should need to be larger than two layers (an output layer and a hidden layer) to predict a target from a given input. This, however, requires that each layer (especially the hidden layer) contain sufficient neurons to approximate the desired function. Indeed, the hidden layer in a two layer network may require as many neurons as the number of training samples,

N, to effectively approximate a desired function (Huang, 2003; Huang & Babri, 1997).

Additional hidden layers can drastically reduce the parameter space without impinging the expressiveness of the model (Hastad, 1986; Telgarsky, 2016). For example, results from Huang (2003) show that a three layer network with m output neurons can exactly fit the target data when the first layer contains $\sqrt{(m+2)N} + 2\sqrt{N/(m+2)}$ neurons and the second layer contains $m\sqrt{N/(m+2)}$ neurons. Combined, this three layer network has $2\sqrt{(m+2)N} \ll N$ neurons. This result establishes the size of a three layer network that is needed to considerably *overfit* the training data. As such, it establishes an upper bound to the parameterization of a three layer network. Note that increasing the number of layers does not serve to improve the performance of the model *per se*, but rather lowers the number of neurons required to fit the model. Further, difficulties with weight initialization and vanishing/exploding gradients increase with depth.

There are few well-established rules for determining *ex ante* how many layers a network should have or how many neurons should go in each layer. Broadly speaking, over-parameterization of a network will not impact the model's accuracy as long as an appropriate training methodology is used (Zou, Cao, Zhou, & Gu, 2018). But over-parameterization will increase computational costs and it may increase the likelihood that training becomes prematurely stuck in a suboptimal minima. Under-parameterization, on the other hand will limit the expressiveness of a network and yield an under-performing model. The most obvious, heuristic strategy to determining the appropriate size and shape of a network is to begin with a small network and successively adjust its depth (the number of layers) or width (the number of neurons in each layer) in small increments to improve training accuracy.

6.2.3 Weight Initialization

While gradient descent and backpropagation provide a method to optimize parameters in a neural network, we must set the initial values for the parameters. Caution must be taken when initializing weights as bad initializations can cause gradients to saturate (i.e., reduce to small values near zero) prematurely. When this happens, the associated neuron will produce the same output regardless of variation in its input. These neurons are called "dead neurons." In practice, a few dead neurons will not influence the accuracy of a model if the network layer is large. If however, most or all of the neurons in a layer die, then gradient descent will lose the ability to update earlier layers and the network will become effectively unresponsive to its input. Poor weight initialization can also cause volatility in the training process, and may prevent gradient descent from finding an ideal set of parameters.

One might suspect that i.i.d. random draws from a distribution would be sufficient to initialize all weights in a network. For example, we might initialize all weights in a network with a random draw from a standard normal distribution. Indeed this was a common approach with early neural networks. For small networks, this will work.

However for deep neural networks, this is inadequate and will tend to encourage the problems described in the preceding paragraph. Indeed, it is the inadequacy of random initialization that led researchers to conclude that deep neural networks performed worse than simple ones (Bengio, Lamblin, Popovici, & Larochelle, 2007).

Early breakthroughs in weight initialization came in 2006 and 2007 (Bengio et al., 2007; Hinton, Osindero, & Teh, 2006) in the form of network pre-training. This is a method where the network is built iteratively, one layer at a time. We begin with a single-layer network with weights initialized to random values. Then train that single-layer network. When training is complete, recover the weights for the layer as the initialization weights for that layer. Then add an additional layer and repeat the process until the network is complete. This process is still occasionally employed, but it is time-consuming for large networks.

Instead, consider the method put forth in Glorot and Bengio (2010). This paper observes that the tendency for gradients to vanish (or explode) is somewhat controlled by keeping variances consistent across layers. To avoid vanishing gradients, we want to initialize weights so that the variance of the output of each layer is roughly consistent with the variance of the output of the preceding layer (and ultimately the variance of the input). To achieve this, the authors suggest initializing all weights $\beta_i \in \boldsymbol{B}_k$ as,

$$\beta_i \sim N\left(0, \frac{2}{p_k + p_{k-1}}\right),$$

where p_k is the number of output neurons for layer k, and p_{k-1} is the number of output neurons from the preceding layer (i.e., the number of input neurons to the current layer). This approach has been widely adopted in the neural network community as it tends to produce good results.

6.2.4 Regularization

To build models that generalize well, it is necessary prevent overfitting. This can partially be accomplished by adopting a training regime that uses out of sample data to determine when gradient descent should stop. We can further prevent overfitting by limiting the complexity of a neural network. To do this, we engage in the process of regularization. There are a number of approaches to regularization; we will discuss two of the more commonly used forms: weight decay and dropout.

Weight decay, or alternatively L2 regularization, applies a loss penalty to each weight in a layer according to its L2 norm: $\frac{\lambda_k}{2} \|\boldsymbol{B}_k\|_2^2$. The hyper parameter λ_k controls the magnitude of the penalty. When weight decay is employed, it is typically applied identically to each layer. Consider a network G containing no bias terms, so that all of the network weights can be represented in a single vector $\boldsymbol{\theta} = (\text{vec}(\boldsymbol{B}_1) \ldots \text{vec}(\boldsymbol{B}_K))$, and where, for each layer $\lambda_k = \lambda$. Then we can rewrite the model's objective function[4] J to incorporate the loss function, L along with the penalty as

$$J(\boldsymbol{\theta}) = L(\boldsymbol{y}; \boldsymbol{\theta}) + \frac{\lambda_k}{2} \sum_{k}^{K} \|\boldsymbol{B}_k\|_2^2$$

$$= L(\boldsymbol{y}; \boldsymbol{\theta}) + \frac{\lambda}{2} \|\boldsymbol{\theta}\|_2^2$$

with a gradient

$$\nabla_{\boldsymbol{\theta}} J(\boldsymbol{\theta}) = \nabla_{\boldsymbol{\theta}} L(\boldsymbol{y}; \boldsymbol{\theta}) + \lambda \boldsymbol{\theta}.$$

Through some rearrangement of terms in the (vanilla gradient descent) update step, it becomes clear why this type of regularization is called weight decay:

$$\boldsymbol{\theta} \leftarrow (1 - \gamma\lambda)\boldsymbol{\theta} - \gamma \nabla_{\boldsymbol{\theta}} J(\boldsymbol{\theta}).$$

That is, by applying an L2 regularization penalty, we are imposing a reduction in $\boldsymbol{\theta}$ by a factor of $(1 - \gamma\lambda)$ at each iteration of the training process. For a given non-zero $\beta_i \in \boldsymbol{\theta}$, if during the training process $\nabla_{\boldsymbol{\theta}} L(\boldsymbol{y}; \boldsymbol{\theta})$ does not encourage movement in the direction of β_i, then it will decay towards zero. In the aggregate, then, the application of weight decay will produce parameter estimates that emphasize the parameters that represent significant contribution to the reduction of the objective function (Goodfellow, Bengio, & Courville, 2016). At the same time, the application of weight decay discourages the model fitting procedure from overreacting to non-systematic variation in the model target (Krogh & Hertz, 1992). Note that for weight decay to work properly, λ must be set so that $\gamma\lambda < 1$

Outside of weight decay, a common approach to regularization is a process called dropout (Srivastava, Hinton, Krizhevsky, Sutskever, & Salakhutdinov, 2014). Consider a network $G(\boldsymbol{X})$ with a layer k and its preceding layer $k-1$ with p neurons. The application of dropout to layer k draws a p-length vector $\boldsymbol{r} \sim Bernoulli(\pi)$ at each step in the training process. It then modifies the input to layer k,

$$g_k(\boldsymbol{X}) = f(r \odot g_{k-1}(\boldsymbol{x})\boldsymbol{B}_k + \alpha_k).$$

This modification is only applied during the training process. After the model has been trained,

$$g_k(\boldsymbol{X}) = f(g_{k-1}(\boldsymbol{x})\boldsymbol{B}_k + \alpha_k).$$

[4] In this setting, the goal of the model fitting process would be to minimize this objective function.

The application of r to the output of layer $k-1$ effectively turns some of the neurons in the network off. The dropout procedure accomplishes two things.

First, dropout limits overfitting by breaking heavily correlated updates of connected neurons (co-adaption). Updates become heavily correlated when one neuron updates to compensate for the output of a connected neuron. This is undesirable as it tends to correspond to fitting idiosyncrasies in the data and thus overfitting (see discussion in Srivastava et al., 2014). Dropout introduces instability in the inputs to a layer, thus breaking the ability of a neuron in that layer to become overly dependent on the output from any given neuron in the preceding layer. This breaks co-adaptation and thus reduces the propensity for overfitting.

Second, dropout allows us to approximate many models at once. Since dropout will set the output of a random number of neurons to zero, it achieves the effect of removing those neurons (briefly) from the network. With the neurons removed, we can consider the network to be an example of a sparse network sampled from G. Srivastava et al. (2014) argue that this interpretation suggests that training a network with dropout provides estimates that approximate a model averaging over many sparse networks. Gal and Ghahramani (2016) extend this view to argue that models with dropout can be interpreted as Bayesian models. Specifically, they argue that dropout in a deep neural network is equivalent to variational inference with a Gaussian process. By applying dropout during inference as well as estimation, we can generate uncertainty estimates via bootstrap simulation.

6.2.5 Data Preprocessing

Neural network models do not require strong assumptions about the data generating process. As a matter of practice however, neural network models are quite sensitive to several properties of the data.

When feeding a model with more than one feature, it is important that the features are at roughly similar scales (to within about an order of magnitude). In theory, a neural network should be able to adjust to inputs of differing scales. But in the initial iterations of training, larger-scaled inputs will dominate gradients and thus parameter adjustments. This can lead to premature saturation of the neurons or very slow model convergence. Pre-scaling the model inputs to have similar scales will alleviate this problem. Typical approaches include scaling inputs to standard normal distribution (normalization), and scaling inputs to the interval $(0, 1]$ through the following affine transformation:

$$x^* = \frac{x - min(\boldsymbol{x})}{max(\boldsymbol{x}) - min(\boldsymbol{x})}.$$

Beyond scaling the data, it is important to consider its bounds. Neural networks excel at generating predictions that generally lie within the boundaries of the training data. Out-of-bounds predictions are subject to more error. In some cases,

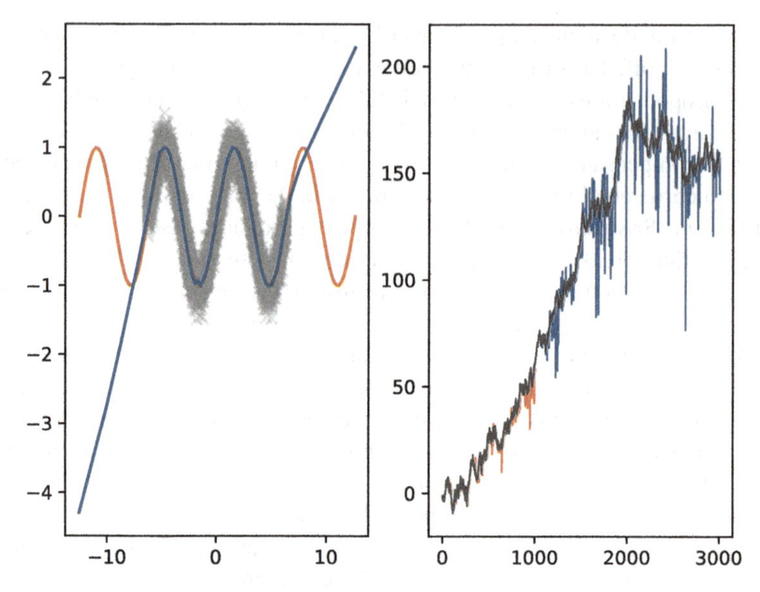

Fig. 6.4 Left: neural network predictions of a sine wave. The training data (in gray) is randomly distributed about the sin curve on the interval $[-2\pi, 2\pi]$. Trained model estimates (blue) are shown for the interval $[-4\pi, 4\pi]$. The sin function (orange) is provided for reference. Right: neural network predictions for a random walk with drift. Training data consists of the first 1000 observations of the walk. In-bounds model predictions (orange) are shown for the first 1000 observations. Out-of-bounds model predictions (blue) are shown for observations beyond observation 1040

this error will be severe. See, for example the left panel in Fig. 6.4. A neural network was given scaler training values $x \in [-2\pi, 2\pi]$, and trained to predict corresponding values of y distributed about $sin(x)$. After training, the model faithfully reproduces $sin(x)$ within the interval represented by the training data. Predictions outside of this interval (i.e., out-of-bounds) do not conform to a sine wave and resemble a linear extrapolation from the model predictions of the nearest training data.

In other cases, out-of-bounds predictions may present errors that are less severe. The right panel of Fig. 6.4 shows neural network predictions of a random walk with drift. A fully-connected network was trained on the first 1000 observations. For each observation y_t, the network was provided with prior observations, $(y_{t-1}, y_{t-2} \dots y_{t-30})$, as input. The figure shows predictions on test-data (i.e., data not used for model training, but generated from the same random walk process). The network can fit in-bounds observations (the first 1000 observations) quite closely. Predictions for out-bounds predictions follow the general trend of the random walk, but are subject to considerably more error. The size of the error tends to grow with distance from the training data.

For economists, the issues posed by out-of-bounds predictions will most likely create complications in dealing with non-stationary data. To reduce the potential for errors in model prediction, researchers can transform data into a mean-reverting (or as nearly mean-reverting as possible) form using standard econometric tools. An alternative technique that has seen success in recent years is to employ wavelet networks for forecasting with non-stationary data. Wavelet networks refer to networks that operate on data that has been preprocessed through a wavelet decomposition (see Jothimani, Yadav, & Shankar, 2015; Lineesh, Minu, & John, 2010; Minu, Lineesh, & John, 2010, as examples).[5]

6.3 RNNs and LSTM

For purposes of forecasting we are almost always making use of time-series data or data that is in some other way sequential. We can incorporate the temporal dependencies of our data into fully connected networks by structuring model inputs as in a distributed lag model. This approach, however, increases the model input space and requires corresponding increases to the size of model's parameter space. It also requires that all inputs to the model be of the same size and will require us to drop one observation per lag in our data.

Recurrent neural networks (RNNs) are a type of neural network that is designed for sequence data; in the context of forecasting, these type of networks can be a good alternative to a fully connected model. Unlike a fully connected network, a recurrent neural network layer imposes an ordering on its inputs and considers them as a sequence. Consider a sequence[6] $\boldsymbol{x} = (x_1, x_2 \ldots x_T)$. We can write a basic RNN (Fig. 6.5) model as $G(\boldsymbol{x}; \boldsymbol{\theta})$, with the output of any given layer written as:

$$\begin{aligned} g_t(\boldsymbol{x}) &= f(x_t \boldsymbol{B}_x + g_{t-1}(\boldsymbol{x})\boldsymbol{B}_g) \\ g_0(\boldsymbol{x}) &= \boldsymbol{0}, \end{aligned}$$

where $\boldsymbol{B}_x$ is a $1 \times p$ matrix of weights, $\boldsymbol{B}_g$ is a $p \times p$ matrix of weights, and the resulting $g_t(\boldsymbol{x})$ is a p-length vector.

This model diverges substantially from the fully connected architecture discussed in Sect. 6.1.1. All layers share a single set of weights, $(\boldsymbol{B}_x, \boldsymbol{B}_g)$. Further, while each layer $g_t(\boldsymbol{x})$ receives input from the preceding layer $g_{t-1}(\boldsymbol{x})$, each layer also receives external input from the t-th element in $\boldsymbol{x}$. Because each layer includes a new input and because each layer's output is taken as input to the subsequent layer, we can think of $g_t(\boldsymbol{x})$ as representing the *state* of the model at a specific point in the sequence. The state of the model at t is an accumulation of the model response

[5] An alternative form of the wavelet neural network uses wavelet functions as activation functions for hidden nodes in the network. This form of wavelet network, however, is designed to improve optimization speeds, create self-assembling networks, or achieve ends other than accommodating non-stationary data.

[6] We focus here on a sequence of scalar values. All discussion in this section extends to sequences of multi-dimensional input (e.g., a sequence of vectors).

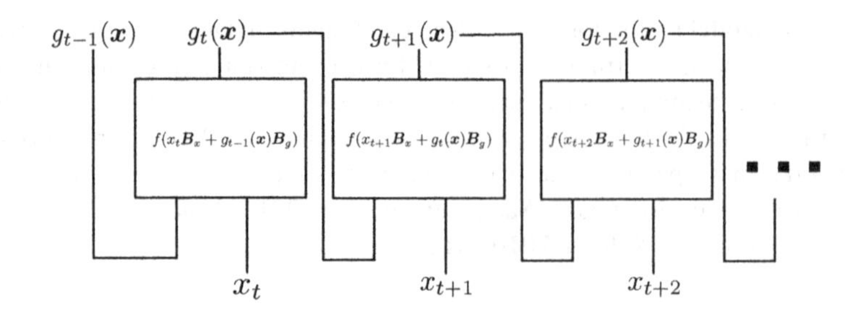

Fig. 6.5 Comparison of cell-based and unrolled implementations of an RNN. The top network represents the unrolled conceptualization of the RNN. The bottom network illustrates a network containing an RNN cell

to all items in $\boldsymbol{x}$ prior to t and as such, can be thought of as a representation of the network's *memory*.

In a forecasting framework, we might only be primarily interested in the final layer output $g_T(\boldsymbol{x})$, which we could treat as an estimate of a target variable at a specified forecast horizon, y_{T+h}. However, because of the structure of this network and the fact that its parameters are shared, we can collect the output of each layer as $\boldsymbol{g} = (g_1(\boldsymbol{x}), g_2(\boldsymbol{x}) \dots g_T(\boldsymbol{x}))$ in which case the network becomes a mapping $G : \boldsymbol{x} \to \boldsymbol{g}$.

To this point, we have been discussing the model $G(\boldsymbol{x}; \boldsymbol{\theta})$ as a set of T network layers. This conception of an RNN is called the "unrolled" form of an RNN. It is useful and more intuitive to illustrate the concept of an RNN in the context of its unrolled form. In practice, however, it is often more efficient to program the RNN as a special network object called a cell. A cell implements a for-loop in the computational graph of the model. Implemented as a cell, the RNN produces the entire sequence $\boldsymbol{g}$ from the inputs $\boldsymbol{x}$ and occupies the same space in a network as a single layer. This makes it easier to embed the RNN into larger networks and brings computational benefits in terms of memory efficiency.

A simple RNN, such as the one described above, illustrates the concept of an RNN, but it will not perform well will lengthy input sequences. As discussed by Bengio et al. (1994), they will have difficulty learning long-term time-dependencies. For example, a simple RNN may have difficulty learning that the impact of a shock in an input time series leads the response in the output series by several periods. When simple RNN models do learn long-term dependencies, they usually suffer from vanishing gradients. When this occurs, the model parameters become established based upon only early portions of the input sequence and later portions of the input sequence have no effect on parameter updates.

The Long Short Term Memory (LSTM) network (Fig. 6.6) (Hochreiter & Schmidhuber, 1997) has emerged as a variant of the RNN that does not suffer from vanishing gradients and is capable of learning long-term dependencies. This type of network incorporates both long-run state information (long-run memory) as well as short-term state information (short-term memory). This type of network also includes mechanisms for resetting the long-run memory and thereby helping to avoid the vanishing gradients problem (Gers, Schmidhuber, & Cummins, 2000).

An LSTM network is a collection of several equations that take the current input, x_t, the network output generated at the previous timestep, $\boldsymbol{h}_{t-1}$, and the network state $\boldsymbol{s}_{t-1}$, which is responsible for its long-run memory, and produce a new output, $\boldsymbol{h}_t$, and an updated version of the network state, $\boldsymbol{s}_t$. Collecting $\boldsymbol{Z}_t = [x_t, \boldsymbol{h}_{t-1}]$, and writing the inverse logit function as σ, we can represent an LSTM cell with two equations,

$$\boldsymbol{s}_t = \overbrace{\underbrace{\boldsymbol{s}_{t-1}}_{\text{old state}} \odot \underbrace{\sigma(\boldsymbol{Z}_t \boldsymbol{B}_d)}_{\text{delete selection}}}^{\text{forget}} + \overbrace{\underbrace{\sigma(\boldsymbol{Z}_t \boldsymbol{B}_i)}_{\text{modification selection}} \odot \underbrace{\tanh(\boldsymbol{Z}_t \boldsymbol{B}_c)}_{\text{modification magnitude}}}^{\text{modify}} \quad (6.7)$$

$$\boldsymbol{h}_t = \sigma(\boldsymbol{Z}_t \boldsymbol{B}_o) \odot \tanh(s_t). \quad (6.8)$$

Equation (6.7) updates the LSTM cell's memory. It is comprised of two components. The first component is a forget step, which selects which components of the cell's memory to delete. The second component is a modification step which identifies which portions of the state should be modified and the extent of modification. The raw cell output, $\boldsymbol{h}_t$, is a representation of the cell state (memory) filtered through an output gate based on the current input and previous cell output. The use of hyperbolic tangent (tanh) activation functions serves to maintain the scale of values in the state and cell outputs. This helps to prevent gradients from vanishing or exploding.

With the raw output, $\boldsymbol{h}_t$ from an LSTM cell, we typically add an additional layer to transform it into a direct prediction that is compatible with our target variable, $\hat{y}_t = f(\boldsymbol{h}_t \boldsymbol{\beta}_y)$.

Note, the parameters $\boldsymbol{\theta} = [\boldsymbol{B}_d, \boldsymbol{B}_i, \boldsymbol{B}_c, \boldsymbol{B}_o, \boldsymbol{\beta}_y]$ are shared across timesteps. At the same time, note that the LSTM cell passes the raw output $\boldsymbol{h}_t$ and long-run state, $\boldsymbol{s}_t$, from one timestep to the next. Thus, even though the LSTM cell parameters are shared across timesteps, the computation of the gradients for the parameters requires iterating backward through the timesteps (see Werbos, 1990) to compute

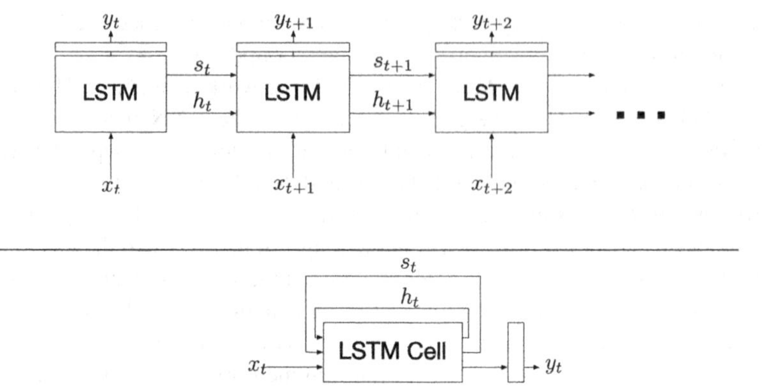

Fig. 6.6 Illustration of rolled and unrolled versions of an LSTM cell. This figure is similar to Fig. 6.4, the top network represents the unrolled conceptualization of the LSTM. The bottom network illustrates a network containing an LSTM cell. The operations within the layers labeled "LSTM" and "LSTM Cell" are provided in Eqs. (6.7)–(6.8). The networks are shown with a final layer that transforms output h into its final form, y

the intermediate gradients for the state and raw output:

$$\frac{\partial L(y_t;\boldsymbol{\theta})}{\partial \boldsymbol{h}_t} = \frac{\partial L(y_t;\boldsymbol{\theta})}{\partial y_t}\boldsymbol{\beta}_y + \frac{\partial L(y_t;\boldsymbol{\theta})}{\partial \boldsymbol{h}_{t+1}}\frac{\partial \boldsymbol{h}_{t+1}}{\partial \boldsymbol{h}_t}$$
$$\frac{\partial L(y_t;\boldsymbol{\theta})}{\partial \boldsymbol{s}_t} = \frac{\partial L(y_t;\boldsymbol{\theta})}{\partial \boldsymbol{h}_t} \odot \tanh'(\boldsymbol{s}_t) + \frac{\partial L(y_t;\boldsymbol{\theta})}{\partial \boldsymbol{s}_{t+1}}\frac{\partial \boldsymbol{s}_{t+1}}{\partial \boldsymbol{s}_t}.$$

We can recover these gradients by observing that $\frac{\partial L(y_t;\boldsymbol{\theta})}{\partial \boldsymbol{s}_{T+1}} = \frac{\partial L(y_t;\boldsymbol{\theta})}{\partial \boldsymbol{h}_{T+1}} = \boldsymbol{0}$ and that

$$\begin{aligned}\frac{\partial L(y_t;\boldsymbol{\theta})}{\partial \boldsymbol{h}_{t-1}} =& \frac{\partial L(y_t;\boldsymbol{\theta})}{\partial \sigma(\boldsymbol{Z}_t\boldsymbol{B}_d)}\sigma'(\boldsymbol{Z}_t\boldsymbol{B}_d)\dot{\boldsymbol{B}}_d + \frac{\partial L(y_t;\boldsymbol{\theta})}{\partial \sigma(\boldsymbol{Z}_t\boldsymbol{B}_i)}\sigma'(\boldsymbol{Z}_t\boldsymbol{B}_i)\dot{\boldsymbol{B}}_i + \\ & \frac{\partial L(y_t;\boldsymbol{\theta})}{\partial \tanh(\boldsymbol{Z}_t\boldsymbol{B}_c)}\tanh'(\boldsymbol{Z}_t\boldsymbol{B}_c)\dot{\boldsymbol{B}}_c + \frac{\partial L(y_t;\boldsymbol{\theta})}{\partial \sigma(\boldsymbol{Z}_t\boldsymbol{B}_o)}\sigma'(\boldsymbol{Z}_t\boldsymbol{B}_o)\dot{\boldsymbol{B}}_o \\ \frac{\partial L(y_t;\boldsymbol{\theta})}{\partial \boldsymbol{s}_{t-1}} =& \frac{\partial L(y_t;\boldsymbol{\theta})}{\partial \boldsymbol{s}_t} \odot \sigma(\boldsymbol{Z}_t\boldsymbol{B}_d),\end{aligned}$$

where $\dot{\boldsymbol{B}}$ indicates the portion of the parameter that is multiplied by $\boldsymbol{h}_{t-1}$ in $Z_t\boldsymbol{B} = (x_t, h_{t-1})\boldsymbol{B}$.

With these intermediate gradients, we can calculate gradients for each item in $\boldsymbol{\theta}$,

$$\frac{\partial L(y_t;\boldsymbol{\theta})}{\partial \boldsymbol{\beta}_y} = \boldsymbol{h}\frac{\partial L(y_t;\boldsymbol{\theta})}{\partial \boldsymbol{y}}$$

$$\frac{\partial L(y_t;\boldsymbol{\theta})}{\partial \boldsymbol{B}_d} = Z_t\left(\frac{\partial L(y_t;\boldsymbol{\theta})}{\partial s_t} \odot \boldsymbol{s}_{t-1} \odot \sigma'(Z_t\boldsymbol{B}_d)\right)$$

$$\frac{\partial L(y_t;\boldsymbol{\theta})}{\partial \boldsymbol{B}_i} = Z_t\left(\frac{\partial L(y_t;\boldsymbol{\theta})}{\partial s_t} \odot \tanh(Z_t,\boldsymbol{B}_c) \odot \sigma'(Z_t\boldsymbol{B}_i)\right)$$

$$\frac{\partial L(y_t;\boldsymbol{\theta})}{\partial \boldsymbol{B}_c} = Z_t\left(\frac{\partial L(y_t;\boldsymbol{\theta})}{\partial s_t} \odot \sigma(Z_t,\boldsymbol{B}_i) \odot \tanh'(Z_t\boldsymbol{B}_c)\right)$$

$$\frac{\partial L(y_t;\boldsymbol{\theta})}{\partial \boldsymbol{B}_o} = Z_t\left(\frac{\partial L(y_t;\boldsymbol{\theta})}{\partial h_t} \odot \tanh(\boldsymbol{s}_t) \odot \sigma'(Z_t\boldsymbol{B}_o)\right).$$

With the gradients calculated, we can fit the model via gradient descent.

6.4 Encoder-Decoder

The LSTM model can be used in the context of forecasting as follows. Consider a time series $\boldsymbol{X} = (x_1 \ldots x_T)$, a target series corresponding to an h-step ahead forecast horizon $\boldsymbol{y} = (y_{1+h}, y_{2+h} \ldots y_{T+h})$, and an LSTM model $G(\boldsymbol{x};\boldsymbol{\theta}) = \hat{y}_{T+h}$. The estimate produced by the LSTM model would be analogous to a direct forecast (see Marcellino, Stock, & Watson, 2006). An iterative forecast could be generated, but the fundamental LSTM model would remain unchanged.

Instead, we can make use of an *encoder-decoder* network (Cho et al., 2014; Sutskever, Vinyals, & Le, 2014). This type of network is a member of a broader class of networks called sequence-to-sequence networks. The encoder-decoder architecture was initially developed to facilitate language modeling tasks (e.g., translation). Specifically, it was developed to allow a model to predict words in the output while considering the context of individual words in the input along with the context of the words that have already been predicted in the output.

The model is comprised of two components, aptly named the encoder and the decoder. The encoder consists of the RNN model from the previous section, $G(\boldsymbol{x};\boldsymbol{\theta})$. For the purposes of our discussion here, consider the encoder to be an LSTM cell with an accompanying fully connected final layer. The encoder takes the sequence $\boldsymbol{x}$ and returns a fixed-length representation. Conventionally, we specify this fixed-length representation as the final output from the models, $g_T(\boldsymbol{x})$. We also recover from $G(\boldsymbol{x};\boldsymbol{\theta})$ the RNN cell's final state, $\boldsymbol{s}_T$.

The second component of the model is called the decoder. It consists of an RNN network and a final, fully connected layer. Whereas the encoder began with a variable length sequence and produced a fixed-length output $g_T(\boldsymbol{x})$, the decoder

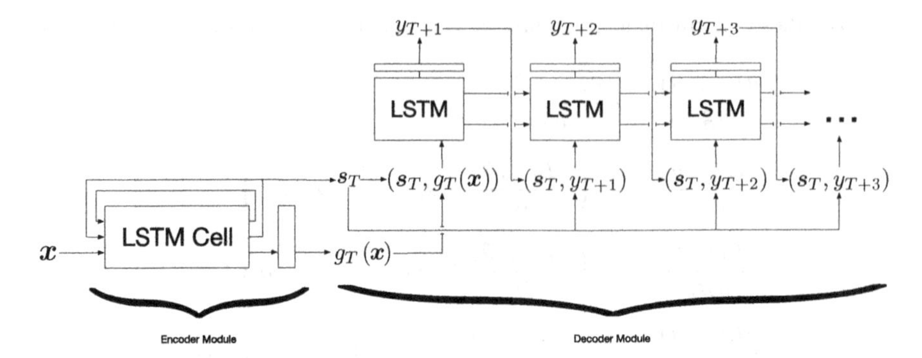

Fig. 6.7 Illustration of an encoder-decoder network. The figure shows an LSTM cell encoding inputs $\boldsymbol{x}$ into a fixed-length representation via an LSTM cell. The fixed-length representation is then processed through an unrolled LSTM network (the decoder module) to produce a variable length sequence $\boldsymbol{y}$. Network weights for the encoder module are shared across timesteps; weights for the decoder module are also shared across timesteps

begins with a fixed-length input $g_T(\boldsymbol{x})$ and produces a variable length output. It does this by taking output of the previous timestep as input to produce output for the current timestep. In practice, for use in forecasting, we would fix the length of the decoder output to correspond to the desired forecast horizon.

As a specific implementation, consider the decoder, $D(\boldsymbol{g}, \boldsymbol{s}_T; \boldsymbol{\theta})$, as an LSTM network with a fully connected final layer. Following Cho et al. (2014), the decoder takes in the final encoder state, s_T, as part of its input at every timestep. Denote by $\boldsymbol{h}_t$ the raw output[7] of the LSTM at timestep t and as produced by d_t. We can write the decoder output of each timestep along the forecast horizon, $h \in (1, 2, \ldots, H)$, as

$$
\begin{aligned}
y_{T+h} &= f(d_{T+h}([y_{T+h-1}, s_T], \boldsymbol{h}_{T+h-1})\boldsymbol{\beta}_y) \\
y_{T+0} &= f(d_{T+0}([g_T(\boldsymbol{x}), s_T], 0)\boldsymbol{\beta}_y),
\end{aligned}
$$

where f is the decoder's final layer activation function and where $\boldsymbol{\beta}_y$ is the vector of weights for corresponding to the decoder's final layer. Note that just as with the RNN cell, the parameters $\boldsymbol{\beta}_y$ are shared across timesteps. The reason for this is to ensure that the raw output from the RNN cell is converted into a target output in a consistent fashion for each timestep. Figure 6.7 provides an illustration of this entire encoder-decoder network.

Gradients for this model are derived in the same fashion as they are for RNNs, via backpropagation through time. As with the other neural network models discussed in this chapter, we train this model using gradient descent.

[7] In other words, the output of the decoder LSTM prior to the final, fully connected layer.

6.5 Empirical Application: Unemployment Forecasting

In this section we will examine the performance of the three neural network architectures (Fully Connected, LSTM, Encoder-Decoder) as applied to the task of unemployment forecasting. This analysis will closely follow Cook and Hall (2017).

6.5.1 Data

To test the performance of the neural network approach, we trained each of the models presented above to predict the civilian unemployment rate. This measure is collected monthly by the US Bureau of Labor and Statistics. It measures the percentage of the labor force that is currently unemployed. The unemployment rate only measures unemployment in the US. At the time of this writing, data for the unemployment rate is available as far back as 1948, and as recently as last month.

Unemployment is a useful indicator to target for this exercise for a few reasons. First, unemployment is a substantively meaningful indicator to forecast; the Federal Reserve works to manage the unemployment rate as part of its dual mandate, and it is closely monitored by economic actors and scholars across a variety of sectors.

Second, in contrast to GDP, unemployment usually undergoes limited revision after its initial release. This is an important consideration since it allows us to generally sidestep the problems of collecting and assembling appropriate "vintages" of the data. We use the last release of the unemployment rate for all training and testing. To be clear, the largest discrepancy between the original vintage of the data and final release of the data is about 23 basis points, with the average discrepancy being nine basis points. We will assume the impact of these discrepancies on the predictive accuracy of our forecasts to be negligible.

For this exercise, we will target 1, 3, 6, 9, and 12 month forecast horizons for the target. For each forecast horizon, we train each of the three models presented above, yielding 15 total model variants for training.

The target will be the sole series used as input for each of the models. For each observation, the model inputs are the previous 36 monthly values of the target, along with first and second order differences in the target. In theory, the model could identify and extract the first and second order differences of the input data, but we supply them directly because (1) we can be reasonably certain that they will supply the model with useful information and (2) because it allows us to reduce the training time and simplify the model structure.

It is possible and relatively easy to add additional series to these models and there should be performance gains from doing so. We will refrain from adding additional series here, however, as this will simplify our discussion of the model.

6.5.2 Model Specification

The fully connected model is comprised of one hidden layer, with a 32 neurons, and a final output layer consisting of a single neuron. The ReLU activation function is applied to each neuron in the hidden layer. The output layer neuron uses a linear (i.e., the identity) activation function. Dropout is applied to all layers with a probability of dropout set to 10%. Weight decay is also applied to all layer with a value of 0.0009. Each of the hyperparameters was chosen via hand tuning.

The LSTM model is comprised of a single LSTM cell with state and output sizes set to twelve. Due to complexities with the LSTM cell, it does not employ dropout or weight decay. The output layer of the LSTM model, is a single neuron with a linear activation function. We could apply the output layer to all outputs from the LSTM cell yielding $G(\boldsymbol{x}|\boldsymbol{\theta}) = (f(\boldsymbol{h}_1\boldsymbol{\beta}_y), f(\boldsymbol{h}_2\boldsymbol{\beta}_y), \ldots f(\boldsymbol{h}_T\boldsymbol{\beta}_y))$. However, since we only care about the final output from the sequence, we discard the output of all earlier timesteps and apply the output layer to only the output from the final timestep, yielding our model output $G(\boldsymbol{x}|\boldsymbol{\theta}) = f(\boldsymbol{h}_T\boldsymbol{\beta}_y)$. This reduces the computational cost of model training by reducing the complexity of calculating $\frac{\partial L(y_t;\boldsymbol{\theta})}{\partial \boldsymbol{\beta}_y}$.

The Encoder-Decoder model uses two LSTM cells and a final, fully connected output layer. The encoder module is identical to the LSTM model just described. The decoder module consists of an LSTM module with a state size of twelve. A final output layer consisting of a single neuron with a linear activation function is applied to the output of each timestep. The parameters of this output layer are shared across all timesteps. As described Eq. (6.4), the initial input to the decoder is the output from the encoder module. At every subsequent timestep, the input to the decoder is the decoder output from the previous timestep.

6.5.3 Model Training

We construct a training data set from the unemployment rate data from 1963 to 1996. Every tenth observation in this period is sequestered into a validation dataset. We use the validation dataset to evaluate the performance of the model and implement early stopping in the training process. The remainder of the data, from 1997 to 2015, is sequestered into a testing dataset. We use this dataset to assess the performance of the trained model.

The training process is subject to stochasticity. The initial weights for each model network are randomly distributed using Xavier initialization. Random weight initialization drives stochasticity in the training process. Beyond this, there are a few other sources of stochasticity in the training process, including dropout and the optimization routine itself (mini-batch Adam).

As a consequence of the stochasticity inherent to the model training process, repeated runs of the same model will yield trained networks that vary in their weights and, consequently, in forecasts. To accommodate this variance, we train 30

Table 6.2 Performance metrics for DARM and neural network models at 0–4 quarter prediction horizons

Horizon		Fully connected	LSTM	Encoder decoder	DARM
1 Months	Mean MAE	20.61	4.10	4.02	11.7
	St. Dev.	4.22	0.12	0.07	
3 Months	Mean MAE	25.38	15.43	15.53	32.8
	St. Dev.	5.50	0.08	0.21	
6 Months	Mean MAE	34.64	28.76	29.00	49.3
	St. Dev.	4.45	0.61	0.25	
9 Months	Mean MAE	47.69	44.99	44.79	65.8
	St. Dev.	3.26	1.93	0.88	
12 Months	Mean MAE	63.45	63.06	61.01	90.7
	St. Dev.	3.04	3.28	1.70	

All metrics presented as hundredths of one percent

instances of each model. This allows us to assess expected model performance as well as assess the variance in performance across repeated runs of the same model.

All model variants trained in less than 5 min.

6.5.4 Results

Model performance is provided in Table 6.2. Each of the first three columns describe the performance of a model in terms of test mean absolute error (MAE), aggregated across repeated iterations. The mean MAE indicates the average model performance. The standard deviation of the MAE gives some sense of the distribution in model performance across repeated trainings of a model. The final column provides performance metrics against a benchmark model.

As a benchmark, we consider a direct[8] autoregressive model (DARM) that uses monthly data. The model is specified as follows:

$$\hat{y}_{t+h} = \sum_{i=1}^{k} \beta_i y_{t-i}, \tag{6.9}$$

where t indexes the time of forecast, k is the number of lags, and n indicates the forecast horizon. In this paper, we use the DARM model estimates published by the SPF (Stark, 2017).

[8]This is to be contrasted with an iterative model, in which the next-step-ahead is forecast and then iterative extrapolation is used to generate a prediction for the desired forecast horizon.

Broadly speaking, each of the neural network models outperform the benchmark model, with the exception of the fully connected model at the 1 month horizon. The encoder-decoder and LSTM models outperform the fully connected models quite strongly at the early horizons. At the 9 and 12 month horizons, the models converge in performance. It is notable, however, that the standard deviation of the mean absolute forecasting error is considerably lower for the LSTM and encoder-decoder models, with the encoder-decoder model having the lowest variance in performance at most horizons.

6.6 Conclusion

This chapter has discussed the fundamentals of neural network models with a primary focus on their application to supervised, predictive tasks. Through this discussion, it showed the flexibility of neural networks and their potential for application to econometric tasks such as forecasting. Yet this chapter is by no means a complete description of the potential of neural networks in econometric settings. Macroeconomists might find additional uses for neural networks in unsupervised econometric applications (e.g., interpreting textual data or generating low dimensional representations of large datasets), or agent-based applications (where neural networks might be used in the context of reinforcement learning). Moreover, as new sources of "Big Data" emerge, economists will be able to train networks to produce increasingly sophisticated outputs or to operate on increasingly complex inputs. Lastly, it is important to note that neural networks represent an area of rapid methodological research and innovation. For example, strong efforts are afoot to adapt neural networks for use within the framework of causal inference. As these efforts develop, so will the utility of neural networks in macroeconomic analysis.

References

Altman, E. I., Marco, G., & Varetto, F. (1994). Corporate distress diagnosis: Comparisons using linear discriminant analysis and neural networks (the Italian experience). *Journal of Banking & Finance, 18*(3), 505–529.

Bengio, Y., Lamblin, P., Popovici, D., & Larochelle, H. (2007). Greedy layer-wise training of deep networks. In *Advances in Neural Information Processing Systems* (pp. 153–160).

Bengio, Y., Simard, P., & Frasconi, P. (1994). Learning long-term dependencies with gradient descent is difficult. *IEEE Transactions on Neural Networks, 5*(2), 157–166. Retrieved from http://www.comp.hkbu.edu.hk/~markus/teaching/comp7650/tnn-%2094-gradient.pdf

Bland, R. (1998). *Learning xor: Exploring the space of a classic problem*. Stirling: Department of Computing Science and Mathematics, University of Stirling.

Cho, K., Van Merriënboer, B., Gulcehre, C., Bahdanau, D., Bougares, F., Schwenk, H., & Bengio, Y. (2014). Learning phrase representations using RNN encoder-decoder for statistical machine translation. arXiv preprint, 1406.1078.

Cook, T. R., & Hall, A. S. (2017). Macroeconomic indicator forecasting with deep neural networks. *Federal Reserve Bank of Kansas City Research Working Paper* (pp. 17-11).

Cybenko, G. (1989). Approximations by superpositions of a sigmoidal function. *Mathematics of Control, Signals and Systems, 2*, 183–192.

Dauphin, Y. N., Pascanu, R., Gulcehre, C., Cho, K., Ganguli, S., & Bengio, Y. (2014). Identifying and attacking the saddle point problem in high-dimensional non-convex optimization. In *Advances in Neural Information Processing Systems* (pp. 2933–2941).

Dijk, D. v., Teräsvirta, T., & Franses, P. H. (2002). Smooth transition autoregressive models—A survey of recent developments. *Econometric Reviews, 21*(1), 1–47.

Dixon, M., Klabjan, D., & Bang, J. H. (2017). Classification-based financial markets prediction using deep neural networks. *Algorithmic Finance, 6*(3–4), 67–77.

Duchi, J., Hazan, E., & Singer, Y. (2011). Adaptive subgradient methods for online learning and stochastic optimization. *Journal of Machine Learning Research, 12*(7), 2121–2159.

Gal, Y., & Ghahramani, Z. (2016). Dropout as a Bayesian approximation. In *Proceedings of the 33rd International Conference on Machine Learning* (Vol. 3, pp. 1661–1680).

Gers, F. A., Schmidhuber, J., & Cummins, F. (2000). Learning to forget: Continual prediction with LSTM. *Neural Computation, 12*(10), 2451–2471.

Glorot, X., & Bengio, Y. (2010). Understanding the difficulty of training deep feed-forward neural networks. In *Proceedings of the Thirteenth International Conference on Artificial Intelligence and Statistics* (pp. 249–256). Retrieved from http://proceedings.mlr.press/v9/glorot10a/glorot10a.pdf?%20hc_location=ufi

Goodfellow, I., Bengio, Y., & Courville, A. (2016). *Deep learning*. Cambridge: MIT Press. http://www.deeplearningbook.org

Hastad, J. (1986). Almost optimal lower bounds for small depth circuits. In *Proceedings of the Eighteenth Annual ACM Symposium on Theory of Computing* (pp. 6–20).

Heaton, J., Polson, N. G., & Witte, J. H. (2016). Deep learning in finance. arXiv preprint, 1602.06561.

Hinton, G. E., Osindero, S., & Teh, Y.-W. (2006). A fast learning algorithm for deep belief nets. *Neural Computation, 18*(7), 1527–1554.

Hochreiter, S., & Schmidhuber, J. (1997). Long short-term memory. *Neural Computation, 9*(8), 1735–1780.

Hornik, K. (1991). Approximation capabilities of multilayer feedforward networks. *Neural Networks, 4*(2), 251–257.

Hornik, K., Stinchcombe, M., & White, H. (1989). Multilayer feedforward networks are universal approximators. *Neural Networks, 2*(5), 359–366.

Huang, G.-B. (2003). Learning capability and storage capacity of two-hidden-layer feedforward networks. *IEEE Transactions on Neural Networks, 14*(2), 274–281.

Huang, G.-B., & Babri, H. A. (1997). General approximation theorem on feedforward networks. In *Proceedings of the 1997 International Conference on Information, Communications and Signal Processing* (Vol. 2, pp. 698–702). Piscataway: IEEE.

Jothimani, D., Yadav, S. S., & Shankar, R. (2015). Discrete wavelet transform-based prediction of stock index: A study on national stock exchange fifty index. *Journal of Financial Management and Analysis, 28*(2), 35–42.

Karlik, B., & Olgac, A. V. (2011). Performance analysis of various activation functions in generalized MLP architectures of neural networks. *International Journal of Artificial Intelligence and Expert Systems, 1*(4), 111–122. Retrieved from https://www.researchgate.net/publication/%20228813985_Performance_Analysis_of_Various_Activation_Functions_in_Generalized_MLP_Architectures_of_Neural_Networks

Kingma, D. P., & Ba, J. (2014). Adam: A method for stochastic optimization. arXiv preprint, 1412.6980.

Kristjanpoller, W., & Minutolo, M. C. (2015). Gold price volatility: A forecasting approach using the artificial neural network–GARCH model. *Expert Systems with Applications, 42*(20), 7245–7251.

Krogh, A., & Hertz, J. A. (1992). A simple weight decay can improve generalization. In *Advances in Neural Information Processing Systems* (pp. 950–957).

Lineesh, M., Minu, K., & John, C. J. (2010). Analysis of nonstationary nonlinear economic time series of gold price: A comparative study. In *International Mathematical Forum* (Vol. 5, 34, pp. 1673–1683). Citeseer.

Maas, A. L., Hannun, A. Y., & Ng, A. Y. (2013). Rectifier nonlinearities improve neural network acoustic models. In *Proceedings of the 30th International Conference on Machine Learning* (Vol. 30, *1*, p. 3). Retrieved from http://robotics.stanford.edu/~amaas/papers/%20relu_hybrid_icml2013_final.pdf

Marcellino, M., Stock, J. H., & Watson, M. W. (2006). A comparison of direct and iterated multistep AR methods for forecasting macroeconomic time series. *Journal of Econometrics, 135*, 499–526.

McCulloch, W. S., & Pitts, W. (1943). A logical calculus of the ideas immanent in nervous activity. *The Bulletin of Mathematical Biophysics, 5*(4), 115–133.

McNelis, P. (2005). *Neural networks in finance: Gaining predictive edge in the market*. Amsterdam: Elsevier.

Minsky, M., & Papert, S. (1969). *Perceptrons: An introduction to computation geometry* (Vol. 200, pp. 355–368). Cambridge: MIT Press.

Minu, K., Lineesh, M., & John, C. J. (2010). Wavelet neural networks for nonlinear time series analysis. *Applied Mathematical Sciences, 4*(50), 2485–2495.

Nair, V., & Hinton, G. E. (2010). Rectified linear units improve restricted Boltzmann machines. In *Proceedings of the 27th International Conference on Machine Learning* (pp. 807–814). Retrieved from http://www.cs.toronto.edu/~fritz/absps/reluICML.pdf

Odom, M. D., & Sharda, R. (1990). A neural network model for bankruptcy prediction. In *Proceedings of the 1990 International Joint Conference on Neural Networks* (pp. 163–168). Piscataway: IEEE.

Ramachandran, P., Zoph, B., & Le, Q. V. (2017). Searching for activation functions. arXiv preprint, 1710.05941. Retrieved from https://arxiv.org/pdf/1710.05941

Ruder, S. (2016). An overview of gradient descent optimization algorithms. arXiv preprint, 1609.04747.

Rumelhart, D. E., Hinton, G. E., & Williams, R. J. (1985). *Learning internal representations by error propagation*. San Diego: California University, La Jolla Institute for Cognitive Science.

Rumelhart, D. E., Hinton, G. E., & Williams, R. J. (1986). Learning representations by back-propagating errors. *Nature, 323*(6088), 533–536. Retrieved from http://www.cs.toronto.edu/~hinton/absps/naturebp.pdf

Russakovsky, O., Deng, J., Su, H., Krause, J., Satheesh, S., Ma, S.,... Bernstein, M., et al. (2015). Imagenet large scale visual recognition challenge. *International Journal of Computer Vision, 115*(3), 211–252.

Srivastava, N., Hinton, G., Krizhevsky, A., Sutskever, I., & Salakhutdinov, R. (2014). Dropout: A simple way to prevent neural networks from overfitting. *The Journal of Machine Learning Research, 15*(1), 1929–1958.

Stark, T. (2017). *Error statistics for the survey of professional forecasters for unemployment rate*. Philadelphia: Federal Reserve Bank of Philadelphia. Retrieved from https://www.philadelphiafed.org/-/media/research-and-data/%20real-time-center/survey-of-professional-forecasters/data-%20files/unemp/spf_error_statistics_unemp_1_aic.pdf?la=en

Sussillo, D., & Abbott, L. (2014). Random walk initialization for training very deep feedforward networks. arXiv preprint, 1412.6558.

Sutskever, I., Vinyals, O., & Le, Q. V. (2014). Sequence to sequence learning with neural networks. In *Advances in Neural Information Processing Systems* (pp. 3104–3112).

Szegedy, C., Liu, W., Jia, Y., Sermanet, P., Reed, S., Anguelov, D.,... Rabinovich, A. (2015). Going deeper with convolutions. In *Proceedings of the IEEE Conference on Computer Vision and Pattern Recognition* (pp. 1–9).

Tam, K. Y. (1991). Neural network models and the prediction of bank bankruptcy. *Omega, 19*(5), 429–445.

Telgarsky, M. (2016). Benefits of depth in neural networks. arXiv preprint, 1602.04485.
Terasvirta, T., & Anderson, H. M. (1992). Characterizing nonlinearities in business cycles using smooth transition autoregressive models. *Journal of Applied Econometrics, 7*(S1), S119–S136.
Werbos, P. J. (1990). Backpropagation through time: What it does and how to do it. *Proceedings of the IEEE, 78*(10), 1550–1560.
Zou, D., Cao, Y., Zhou, D., & Gu, Q. (2018). Stochastic gradient descent optimizes over-parameterized deep ReLU networks. arXiv preprint, 1811.08888.

Part III
Seeking Parsimony

Chapter 7
Penalized Time Series Regression

Anders Bredahl Kock, Marcelo Medeiros, and Gabriel Vasconcelos

7.1 Introduction

Penalized regression methods have become an important estimation and model selection tool for applied researchers. With the availability of vast datasets in the era of *Big Data*, selecting the relevant variables in a regression model is of great importance. In this chapter we review the most commonly used penalized regression methods in the framework of time series models. We pay special attention to Ridge Regression (Hoerl & Kennard, 1970), the Least Absolute Shrinkage and Selection Operator (Lasso) (Tibshirani, 1996), the Elastic Net (Zou & Hastie, 2005), the adaptive Lasso (Zou, 2006), the adaptive Elastic Net (Zou & Zhang, 2009), and the group Lasso (Yuan & Lin, 2006).

The main contents of this chapter are as follows: We first consider a linear regression model which nests three commonly used linear time series models that we are going to focus on. The three nested models are the autoregressive (AR), the autoregressive distributed lag (ADL) with strongly exogenous regressors, and the vector autoregressive (VAR) model. For this general model we review the penalties mentioned in the previous paragraph. We do not claim to make an exhaustive review and mainly focus on those penalty functions that have been used in the context of time series models and for which theoretical performance guarantees have been established in that context. Furthermore, we focus solely on covariance stationary

A. B. Kock (✉)
University of Oxford, Oxford, UK

Aarhus University and CREATES, Aarhus, Denmark
e-mail: anders.kock@economics.ox.ac.uk

M. Medeiros · G. Vasconcelos
Pontifical Catholic University of Rio de Janeiro, Rio de Janeiro, Brazil
e-mail: mcm@econ.puc-rio.br

P. Fuleky (ed.), *Macroeconomic Forecasting in the Era of Big Data*,
Advanced Studies in Theoretical and Applied Econometrics 52,
https://doi.org/10.1007/978-3-030-31150-6_7

models and instead refer to Chap. 17 of this book for a treatment of unit roots and cointegration (in the context of high-dimensional data).

The remainder of this chapter is organized as follows: We define notation used throughout the chapter in Sect. 7.2. In Sect. 7.3 we define the linear model used throughout the chapter and in Sect. 7.4 we review commonly used penalty functions. The main theoretical developments for the penalized methods considered are discussed in Sect. 7.5 and we give practical recommendations concerning the selection of the penalty parameters and software implementations in Sect. 7.6. In Sect. 7.7 we include a Monte Carlo simulation study comparing the presented methods in a controlled environment. We compare the penalties in terms of parameter estimation, model selection, and forecasting ability. Finally, we provide a small empirical application to inflation forecasting in Sect. 7.8.

7.2 Notation

For any $\boldsymbol{x} \in \mathbb{R}^n$, let $||\boldsymbol{x}|| = \sqrt{\sum_{i=1}^n x_i^2}$ be the ℓ_2-norm of $\boldsymbol{x}$ and let $||\boldsymbol{x}||_{\ell_1} = \sum_{i=1}^n |x_i|$ be the ℓ_1-norm. $||\boldsymbol{x}||_{\ell_0} = \sum_{i=1}^n 1_{\{x_i \neq 0\}}$ denotes the ℓ_0-"norm" of $\boldsymbol{x}$, i.e., the number of non-zero elements of $\boldsymbol{x}$. For any $A \subseteq \mathbb{R}^n$, let $|A|$ denote its cardinality. Furthermore, $\boldsymbol{x}_A$ denotes the vector of length $|A|$ consisting of those entries of $\boldsymbol{x}$ whose indices belong to A. For any $n \times n$ matrix $\boldsymbol{B}$, $\boldsymbol{B}_A$ denotes the $|A| \times |A|$ matrix consisting only of those rows and columns of $\boldsymbol{B}$ that have indices in A.

Throughout this chapter, all random variables are defined on a probability space $(\Omega, \mathcal{F}, \mathbb{P})$ where we denote the expectation with respect to $\mathbb{P}$ by $\mathbb{E}$. Convergence in distribution will be denoted by $\xrightarrow{d}$.

7.3 Linear Models

In this section we shall consider time series models that are special cases of the classic linear regression model

$$y_t = \boldsymbol{\beta}'\boldsymbol{x}_t + \epsilon_t, \qquad t = 1, \ldots, T \tag{7.1}$$

for some $T \in \mathbb{N}$ and where $\boldsymbol{\beta}$ is a $k \times 1$ vector of unknown parameters. The subscript t for the observations is used as we are concerned with time series models in the sequel. Furthermore, we shall refer to $\boldsymbol{\beta}^0$ as the *true* parameter vector which we assume to be unique throughout this chapter. Let $\mathcal{A} = \{i : \boldsymbol{\beta}_i^0 \neq 0\}$ be the indices of the non-zero coefficients with cardinality $s_0 = |\mathcal{A}|$. If s_0 is much smaller than k, then $\boldsymbol{\beta}^0$ is said to be *sparse*.

Many commonly used linear time series models are of the form (7.1). To illustrate this point, consider the following examples.

7.3.1 Autoregressive Models

In the autoregressive model of order p, AR(p), the data is generated by the recursion

$$y_t = \phi_0 + \phi_1 y_{t-1} + \cdots + \phi_p y_{t-p} + \epsilon_t, \qquad t = 1, \ldots, T \tag{7.2}$$

with p initial values, which, for concreteness, one can set $y_0 = \ldots = y_{-p+1} = 0$.[1] Thus, $k = p + 1$ and $\boldsymbol{\beta} = (\phi_0, \ldots, \phi_p)$' in (7.1). Other deterministic terms than an intercept are handled in the usual manner.

7.3.2 Autoregressive Distributed Lag Models

An autoregressive distributed lag (ADL) model is written as

$$y_t = \phi_0 + \phi_1 y_{t-1} + \cdots + \phi_p y_{t-p} + \boldsymbol{\theta}_1' \boldsymbol{x}_{t-1} + \cdots + \boldsymbol{\theta}_q' \boldsymbol{x}_{t-q} + \epsilon_t, \qquad t = 1, \ldots, T, \tag{7.3}$$

where $\boldsymbol{x}_t$ is an $m \times 1$ vector of weakly exogenous covariates, that is, $\mathbb{E}(\epsilon_t | \boldsymbol{x}_{t-1}, \ldots, \boldsymbol{x}_1) = 0$. In this case, we have $k = (p + 1) + q \times m$ and $\boldsymbol{\beta} = (\phi_0, \phi_1, \ldots, \phi_p, \boldsymbol{\theta}_1', \ldots, \boldsymbol{\theta}_q')$' in (7.1).

7.3.3 Vector Autoregressive Models

For $m \times 1$ vectors $\boldsymbol{y}_t = (y_{t,1}, \ldots, y_{t,m})$ and $\boldsymbol{\epsilon}_t = (\epsilon_{t,1}, \ldots, \epsilon_{t,m})$, a vector autoregressive model of order p, VAR(p), is defined as

$$\boldsymbol{y}_t = \boldsymbol{\phi}_0 + \boldsymbol{\Phi}_1 \boldsymbol{y}_{t-1} + \ldots + \boldsymbol{\Phi}_p \boldsymbol{y}_{t-p} + \boldsymbol{\varepsilon}_t, \qquad t = 1, \ldots, T. \tag{7.4}$$

For any $1 \leq i \leq m$ note that the ith equation in (7.4) is of the form (7.1). In particular, letting $\boldsymbol{\Phi}_l^{(i)}$ denote the ith row of $\boldsymbol{\Phi}_l$, one has $k = 1 + pm$ and $\boldsymbol{\beta} = \boldsymbol{\beta}_i = (\boldsymbol{\phi}_{0,i}, \boldsymbol{\Phi}_1^{(i)}, \ldots, \boldsymbol{\Phi}_p^{(i)})'$ for the ith equation in (7.4).

7.3.4 Further Models

The above three examples reveal that all models that are linear in $\boldsymbol{\beta}$ fall within the framework of (7.1). In particular, we have made no/few assumptions on probabilistic

[1] Unless mentioned otherwise, we shall throughout assume that all necessary initial values equal 0.

properties of the error terms (and exogenous regressors).[2] Thus, these can in principle be non-independent and (conditionally) heteroskedastic. Such assumptions come into play once one wishes to establish the theoretical properties of the estimator under study. Before turning to theoretical properties of various penalized estimators in Sect. 7.5, we introduce some commonly used penalty functions.

7.4 Penalized Regression and Penalties

Let $\boldsymbol{y} = \boldsymbol{X}\boldsymbol{\beta} + \boldsymbol{\epsilon}$ be the usual matrix form of (7.1) and $\boldsymbol{Z} = (\boldsymbol{y}, \boldsymbol{X})$. A *penalized* regression estimator $\hat{\boldsymbol{\beta}}$ is obtained as

$$\widehat{\boldsymbol{\beta}} \in \underset{\boldsymbol{\beta} \in \mathbb{R}^k}{\operatorname{argmin}} \left[\|\boldsymbol{y} - \boldsymbol{X}\boldsymbol{\beta}\|^2 + \lambda \mathsf{p}(\boldsymbol{\beta}, \boldsymbol{\alpha}, \boldsymbol{Z}) \right], \tag{7.5}$$

where $\lambda \geq 0$ is a penalty/tuning parameter and $\mathsf{p} : \mathbb{R}^k \times \mathbb{R}^d \times \mathbb{R}^{T \times (1+k)} \to [0, \infty)$ is called the *penalty function* as it penalizes the entries of $\boldsymbol{\beta}$ for being different from zero. $\boldsymbol{\alpha}$ is a d-dimensional tuning parameter to be chosen by the user. Often p does not depend on any tuning parameter. Note that p can also depend on the observed data $\boldsymbol{Z}$ which is relevant below for the adaptive versions of the Lasso and the Elastic Net.

In all examples of p below, for all $(\boldsymbol{\alpha}, \boldsymbol{B}) \in \mathbb{R}^d \times \mathbb{R}^{T \times (1+k)}$, one has $\mathsf{p}(\boldsymbol{\beta}, \boldsymbol{\alpha}, \boldsymbol{B}) = 0$ if and only if $\boldsymbol{\beta} = 0$. Thus, p penalizes $\boldsymbol{\beta}$ for being non-zero and an estimator $\hat{\boldsymbol{\beta}}$ obtained from (7.5) is called *penalized* since it minimizes a combination of the usual least squares objective function and the penalty function p. The larger the $\lambda \geq 0$ is, the more weight is put on the penalty function and a solution $\hat{\boldsymbol{\beta}}$ to (7.5) will tend to have smaller entries (in absolute value) than the least squares estimator. Often one considers the sample average of the least squares part in (7.5), $\frac{1}{T}\|\boldsymbol{y} - \boldsymbol{X}\boldsymbol{\beta}\|^2$. This is of no importance and merely results in a rescaling of λ by a factor of $\frac{1}{T}$ as well.

Note also that in (7.5), $\boldsymbol{\beta}$ is allowed to vary over all of $\mathbb{R}^k$ which can, of course, be restricted to any subset of $\mathbb{R}^k$. However, in order to establish desirable theoretical properties of a (penalized) estimator, it is usually assumed that $\boldsymbol{\beta}^0$ belongs to the subset of $\mathbb{R}^k$ minimized over. Finally, the existence of a unique minimizer in (7.5) depends on the properties of $\boldsymbol{X}$ and p. If these are such that the objective function is strictly convex, the minimizer is guaranteed to be unique. We now turn to exhibiting concrete examples of the penalty function p. It is important to note that all penalties mentioned result in estimators which are applicable even when $k > T$.[3] Thus, they work in the context of high-dimensional data.

[2] Hence, strictly speaking, the model is not yet fully specified.

[3] For the adaptive Lasso and the adaptive Elastic Net this is of course conditional on choosing an initial estimator that is applicable when $k > T$.

7.4.1 Ridge Regression

Ridge Regression, as introduced in Hoerl and Kennard (1970), results from using (note that p only depends on $\boldsymbol{\beta}$)

$$\mathsf{p}(\boldsymbol{\beta}, \alpha, \boldsymbol{Z}) = \sum_{i=1}^{k} \beta_i^2 \tag{7.6}$$

in (7.5). In this case one has the closed form solution $\hat{\boldsymbol{\beta}}_{ridge}(\lambda) = \left(\boldsymbol{X}'\boldsymbol{X} + \lambda \boldsymbol{I}_k\right)^{-1} \boldsymbol{X}'\boldsymbol{y}$. Note that $\hat{\boldsymbol{\beta}}_{ridge}(0) = \hat{\boldsymbol{\beta}}_{OLS}$ whenever these are well-defined. Ridge Regression is useful when $\boldsymbol{X}'\boldsymbol{X}$ is (nearly) singular since the least squares estimator is (1) not unique when $\boldsymbol{X}'\boldsymbol{X}$ is singular, and (2) has high variance when $\boldsymbol{X}'\boldsymbol{X}$ is nearly singular. Despite being unbiased (in cross sectional data) this high variance results in a high mean square error. Thus, even though $\hat{\boldsymbol{\beta}}_{ridge}$ is biased for all $\lambda > 0$, there always exists a $\lambda > 0$ such that the mean square error of $\hat{\boldsymbol{\beta}}_{ridge}(\lambda)$ is strictly lower than the one of the least square estimator, cf. Hoerl and Kennard (1970) Theorem 4.3 for a precise statement, due to the lower variance of the ridge estimator. While these results are proven in the context of cross sectional data, they motivate the use of Ridge Regression also for time series data. We note that despite the fact that Ridge Regression shrinks the parameter estimates to zero compared to the least squares estimator in the sense that $||\hat{\boldsymbol{\beta}}_{ridge}(\lambda)||^2 < ||\hat{\boldsymbol{\beta}}_{OLS}||^2$ for $\lambda > 0$, $\hat{\boldsymbol{\beta}}_{ridge}(\lambda)$ will not have entries that are *exactly* equal to zero and it is thus not directly useful for variable selection.

7.4.2 Least Absolute Shrinkage and Selection Operator (Lasso)

The Lasso was proposed by Tibshirani (1996) and uses (note that p only depends on $\boldsymbol{\beta}$)

$$\mathsf{p}(\boldsymbol{\beta}, \alpha, \boldsymbol{Z}) = \sum_{i=1}^{k} |\beta_i| \tag{7.7}$$

in (7.5). Thus, while Ridge Regression uses the squared ℓ_2-norm as penalty function for $\boldsymbol{\beta}$, the Lasso uses the ℓ_1-norm. The appealing feature of the Lasso is that $\hat{\boldsymbol{\beta}}_{Lasso}(\lambda)$ can contain *exact* zeros for λ sufficiently large in (7.5). We refer to Lemma 2.1 in Bühlmann and van de Geer (2011) for necessary and sufficient conditions for an entry of $\hat{\boldsymbol{\beta}}_{Lasso}$ to equal zero. The important thing to note is that the Lasso performs estimation and variable selection in one step. This is in stark contrast to traditional procedures which would estimate $\boldsymbol{\beta}^0$ by, say, least squares (when applicable) and then decide which entries of $\boldsymbol{\beta}^0$ are (non)-zero by means of hypothesis tests. However, the final model one arrives at by such a testing procedure depends heavily on the order in which such tests are carried out. For example, one could gradually test down the model by testing the significance of individual coefficients by means of t-tests stopping at the first rejection of the null of no significance. Alternatively, one could use joint tests or combinations of joint and individual tests—the point being that the final model depends heavily on the type and order of tests being used. Another alternative to the Lasso is to use information criteria such as AIC, BIC, or HQ for model selection.[4] However, just like the above test procedures, these are only applicable to the least squares estimator when $k < T$ unless one resorts to ad hoc ways of splitting the regressors into subgroups of size less than T. Furthermore, the computational burden of these information criteria increases exponentially in k. That being said, the Lasso is only variable selection consistent under rather stringent assumptions, which are rarely satisfied for time series, cf. Zou (2006) (Theorem 1 and Corollary 1) and Zhao and Yu (2006) (Theorems 1 and 2). As we shall see, the Lasso can often still estimate $\boldsymbol{\beta}^0$ precisely though. The Lasso has by now become a very popular estimator in high-dimensional models where k can potentially be much larger than T.

7.4.3 Adaptive Lasso

The adaptive Lasso was introduced by Zou (2006) and uses (note that p depends only on $\boldsymbol{\beta}$ and $\mathbf{Z}$)

$$\mathsf{p}(\boldsymbol{\beta}, \alpha, \mathbf{Z}) = \sum_{i=1}^{k} \frac{1}{|\hat{\beta}_{OLS,i}|} |\beta_i|, \tag{7.8}$$

where $\hat{\beta}_{OLS,i}$ is the least squares estimator of β_i^0. Note that in contrast to the Ridge and Lasso estimator, the adaptive Lasso penalty depends on the data $\mathbf{Z} = (\boldsymbol{y}, \boldsymbol{X})$ through the initial estimator $\hat{\boldsymbol{\beta}}_{OLS}(\mathbf{Z}) = (\boldsymbol{X}'\boldsymbol{X})^{-1}\boldsymbol{X}'\boldsymbol{y}$. The underlying idea of the adaptive Lasso penalty is to use more "tailored" weights than the Lasso, which penalizes all parameters by the same factor λ. As long as the least squares estimator is consistent (which is the case under mild regularity conditions) this will lead to penalizing truly zero coefficients more than the non-zero coefficients (for T sufficiently large). To see this, assume first that $\beta_i^0 = 0$. By the consistency of $\hat{\beta}_{OLS,i}$ it will be close to zero in large samples. Thus, $\frac{1}{|\hat{\beta}_{OLS,i}|}$ will tend to infinity and it will be costly to choose a non-zero value of β_i in the minimization problem (7.5). If,

[4] Section 7.6.1 contains a discussion of these, where, however, they are used for selection of the tuning parameter λ rather than *directly* the model.

on the other hand, $\beta_i^0 \neq 0$, then by the same reasoning as above $\frac{1}{|\hat{\beta}_{OLS,i}|}$ remains bounded in probability. Thus, the zero coefficients are penalized much more than the non-zero ones and we shall later see (cf. Table 7.1) that the adaptive Lasso has seemingly good properties under mild regularity conditions. Note also that Zou (2006) originally proposed

$$\mathsf{p}(\boldsymbol{\beta}, \alpha, \mathbf{Z}) = \sum_{i=1}^{k} \frac{1}{|\hat{\beta}_{OLS,i}|^{\alpha}} |\beta_i|, \tag{7.9}$$

for an $\alpha > 0$. However, our feeling is that the choice of $\alpha = 1$ is most common in practice and it also avoids an additional tuning parameter. Thus, we will focus on $\alpha = 1$ in the remainder of this paper.

Finally, it should be mentioned that other estimators than the least squares estimator can be used as initial estimator for the adaptive Lasso: in fact the least squares estimator is not even unique when $k > T$ as $\boldsymbol{X}'\boldsymbol{X}$ is singular in that case. In this case one could use, e.g., the Lasso as initial estimator instead such that[5]

$$\mathsf{p}(\boldsymbol{\beta}, \alpha, \mathbf{Z}) = \sum_{i:\hat{\beta}_{Lasso,i} \neq 0} \frac{1}{|\hat{\beta}_{Lasso,i}|} |\beta_i|,$$

excluding all variables for which $\hat{\beta}_{Lasso,i} = 0$. In general, one only needs to know the rate of convergence for the initial estimator in order to establish "good" theoretical properties of the adaptive Lasso. Thus, the adaptive Lasso is also applicable in high-dimensional models with more explanatory variables than observations.

7.4.4 *Elastic Net*

The elastic net penalty was proposed by Zou and Hastie (2005) and is (note that p depends only on $\boldsymbol{\beta}$ and $\alpha \in [0, 1]$)

$$\mathsf{p}(\boldsymbol{\beta}, \alpha, \mathbf{Z}) = \alpha \sum_{i=1}^{k} \beta_i^2 + (1 - \alpha) \sum_{i=1}^{k} |\beta_i|, \tag{7.10}$$

for $\alpha \in [0, 1]$. The elastic net penalty is a convex combination of the ridge penalty ($\alpha = 1$) and the Lasso penalty ($\alpha = 0$) and the idea of the elastic net is to combine the strengths of these two estimators. In particular, as argued in Zou and Hastie

[5] In case the initial estimator, $\hat{\boldsymbol{\beta}}_{initial}$ say, has entries equal to zero, which can be the case for the Lasso, it is a common practice to simply leave the corresponding variables out of the second step adaptive Lasso estimation.

(2005), in case a group of explanatory variables is highly correlated (which is often the case in time series applications), the Lasso selects only one of the variables in the group. Whether this is a desirable property or not may depend on the application at hand. In case one thinks of the highly correlated variable as forming a group of (relevant) variables, then one may want to select all the relevant variables instead of only one of the group members. Hence, one may prefer the elastic net to the Lasso in such a situation.

7.4.5 Adaptive Elastic Net

The adaptive elastic net as introduced in Zou and Zhang (2009) uses

$$\mathsf{p}(\boldsymbol{\beta}, \boldsymbol{\alpha}, \mathbf{Z}) = \left(\alpha \sum_{i=1}^{k} \frac{1}{|\hat{\beta}_{EN,i}|^{\gamma}} \beta_i^2 + (1-\alpha) \sum_{i=1}^{k} \frac{1}{|\hat{\beta}_{EN,i}|^{\gamma}} |\beta_i| \right), \tag{7.11}$$

where $\boldsymbol{\alpha} = (\alpha, \gamma) \in [0, 1] \times (0, \infty)$ and $\hat{\boldsymbol{\beta}}_{EN}$ is the elastic net estimator (as remarked for the adaptive Lasso above, other estimators can in principle be used as initial estimators). The motivation for Zou and Zhang (2009) to introduce the adaptive elastic net was that it can combine the tailored weights of the adaptive Lasso with the potentially good performance in highly correlated designs of the elastic net.

7.4.6 Group Lasso

The group Lasso was introduced by Yuan and Lin (2006). It divides the full parameter vector $\boldsymbol{\beta}$ into $\boldsymbol{\beta}' = (\boldsymbol{\beta}'_1, \ldots, \boldsymbol{\beta}'_J)'$, where the length of $\boldsymbol{\beta}_j$ is k_j for $j = 1, \ldots, J$ $(J \leq k)$, and uses (note that p only depends on $\boldsymbol{\beta}$)

$$\mathsf{p}(\boldsymbol{\beta}, \boldsymbol{\alpha}, \mathbf{Z}) = \sum_{j=1}^{J} \sqrt{k_j} ||\boldsymbol{\beta}_j||.$$

The name *group* Lasso stems from the fact that the parameter vector has been divided into J disjoint groups by the user. This is particularly relevant in macroeconometrics where the explanatory variables often belong to natural groups such as the group of house price variables, the group of interest rate variables, or the group of exchange rate variables. The group Lasso contains the plain Lasso as a special case upon choosing $J = k$ (such that $k_j = 1$ for $j = 1, \ldots, k$). In the general case, the group Lasso penalizes the parameter vector of each group by its ℓ_2-norm. The group Lasso may be of interest since it has been shown, cf. Proposition 1 of Yuan and Lin (2006), that its penalty can result in the whole estimator for a group being zero. This is useful as an interpretation of this event is that a group of variables is irrelevant. Finally, we note that an adaptive version of the group Lasso was proposed in Wang and Leng (2008). The (adaptive) group Lasso is popular in additive models and has, e.g., been used by Huang, Horowitz, and Wei (2010) to select the components in an expansion of such models.

7.4.7 Other penalties and methods

We stress that the penalty functions p discussed above are just a subset of those introduced in the literature. We have focused on these as they are commonly used. Let us, however, mention that also penalties such as the SCAD (Fan and Li, 2001) and Bridge (Frank and Friedman, 1993) are common. The Dantzig selector of Candes and Tao (2007) is also rather popular, while the non-negative garrote is an early contribution (Breiman, 1995).

A procedure developed explicitly with time series in mind is FARM of Fan, Ke, and Wang (2018) who decorrelate the covariates by a factor model before using a penalized estimator in the second step. See also Kneip and Sarda (2011) for a precursor to this work.

7.5 Theoretical Properties

In this section we discuss some desirable properties that one may want a (penalized) estimator $\hat{\boldsymbol{\beta}}$ of $\boldsymbol{\beta}^0$ to possess. Recall that $\mathcal{A} = \{i : \beta_i^0 \neq 0\}$ and let $\hat{\mathcal{A}} = \{i : \hat{\beta}_i \neq 0\}$. We begin by discussing three *asymptotic* properties of a sequence of estimators.

1. *Consistency*: $\hat{\boldsymbol{\beta}}$ is consistent for $\boldsymbol{\beta}^0$ if $\mathbb{P}(||\hat{\boldsymbol{\beta}} - \boldsymbol{\beta}^0|| > \delta) \to 0$ as $T \to \infty$ for all $\delta > 0$. Other norms than the ℓ_2-norm may be relevant in the case where $k \to \infty$ as $T \to \infty$.
2. *Variable selection consistency*: $\hat{\boldsymbol{\beta}}$ is said to be variable selection consistent if $\mathbb{P}(\hat{\mathcal{A}} = \mathcal{A}) \to 1$ as $T \to \infty$. This property is also sometimes called *sparsistency* and entails that all relevant variables (those with non-zero coefficients) are asymptotically retained in the model and all irrelevant variables (those with a coefficient of zero) are discarded from the model.
3. *Oracle property*: In the context of linear models as in (7.1) and in which the least squares estimator converges at rate $\sqrt{T}$, $\hat{\boldsymbol{\beta}}$ is said to possess the *oracle property* or, alternatively, to be *oracle efficient* if

   (a) it is variable selection consistent, cf. 2 above.
   (b) $\sqrt{T}(\hat{\boldsymbol{\beta}}_{\mathcal{A}} - \boldsymbol{\beta}^0_{\mathcal{A}})$ has the same limit in distribution as $\sqrt{T}(\hat{\boldsymbol{\beta}}_{OLS,\mathcal{A}} - \boldsymbol{\beta}^0_{\mathcal{A}})$ (assumed to exist). In words, this entails that the limiting distribution of

the properly centered and scaled estimator of the non-zero coefficients is the same as if only the relevant variables had been included in the model from the outset and their coefficients had been estimated by least squares. Put differently, the limiting distribution of the estimator of the non-zero coefficients is the same as if an *oracle* had revealed the relevant variables prior to estimation and one had only included these.

Of course the exact form of the limiting distribution of $\sqrt{T}(\hat{\boldsymbol{\beta}}_{OLS,\mathcal{A}} - \boldsymbol{\beta}^0_{\mathcal{A}})$ depends on the model under consideration. Under the mild regularity conditions that (i) $\frac{1}{T}\sum_{t=1}^T \boldsymbol{x}_t\boldsymbol{x}_t' \to \boldsymbol{\Sigma}_X$ in probability for a positive definite matrix $\boldsymbol{\Sigma}_X$, which typically is the limit of $\frac{1}{T}\sum_{t=1}^T \mathbb{E}(\boldsymbol{x}_t\boldsymbol{x}_t')$ (assumed to exist), and (ii) $\frac{1}{\sqrt{T}}\sum_{t=1}^T \boldsymbol{x}_t\epsilon_t \xrightarrow{d} N(0, \boldsymbol{\Sigma})$ for some positive definite matrix $\boldsymbol{\Sigma}$, which typically is the limit of $\frac{1}{T}\sum_{t=1}^T \mathbb{E}(\epsilon_t^2\boldsymbol{x}_t\boldsymbol{x}_t')$ (assumed to exist), one has that[6]

$$\sqrt{T}(\hat{\boldsymbol{\beta}}_{OLS,\mathcal{A}} - \boldsymbol{\beta}^0_{\mathcal{A}}) \xrightarrow{d} N_{|\mathcal{A}|}(\mathbf{0}, [\boldsymbol{\Sigma}_{X,\mathcal{A}}]^{-1}\boldsymbol{\Sigma}_{\mathcal{A}}[\boldsymbol{\Sigma}_{X,\mathcal{A}}]^{-1}). \tag{7.12}$$

In the classic setting of $(\boldsymbol{x}_t, \epsilon_t)$ being i.i.d. with $\boldsymbol{x}_1$ and ϵ_1 independent and $\mathbb{E}(\epsilon_1) = 0$, this reduces to $\boldsymbol{\Sigma}_X = \mathbb{E}(\boldsymbol{x}_1\boldsymbol{x}_1')$ and $\boldsymbol{\Sigma} = \sigma^2\boldsymbol{\Sigma}_X$, where $\sigma^2 = \mathbb{E}(\epsilon_1^2)$ such that

$$\sqrt{T}(\hat{\boldsymbol{\beta}}_{OLS,\mathcal{A}} - \boldsymbol{\beta}^0_{\mathcal{A}}) \xrightarrow{d} N_{|\mathcal{A}|}(\mathbf{0}, \sigma^2[\boldsymbol{\Sigma}_{X,\mathcal{A}}]^{-1}).$$

Finally, we note that in order to establish that $\sqrt{T}(\hat{\boldsymbol{\beta}}_{\mathcal{A}} - \boldsymbol{\beta}^0_{\mathcal{A}})$ and $\sqrt{T}(\hat{\boldsymbol{\beta}}_{OLS,\mathcal{A}} - \boldsymbol{\beta}^0_{\mathcal{A}})$ have the same limiting distribution, one often shows the slightly stronger property $|\sqrt{T}(\hat{\boldsymbol{\beta}}_{\mathcal{A}} - \hat{\boldsymbol{\beta}}_{OLS,\mathcal{A}})| \to 0$ in probability.

Before introducing a desirable *non-asymptotic* property of $\hat{\boldsymbol{\beta}}$, it is worth making a few remarks to provide some perspective on the oracle property.

Remark 7.1 The oracle property, while looking appealing, is nothing new in the sense that it is satisfied by well-known procedures in simple models. For example, one could use BIC to consistently select a model estimated by least squares under standard assumptions. Therefore, with probability tending to one, the resulting estimator of the non-zero coefficients coincides with the least squares estimator only including the relevant variables, thus proving the oracle property (of course using BIC has a higher computational cost). Similarly, letting $\tilde{\mathcal{A}} = \{i : |\hat{\beta}_{OLS,i}| > T^{-1/4}\}$ in a setting where the least squares estimator converges at rate $\sqrt{T}$, implementing least squares only including the variables in $\tilde{\mathcal{A}}$ results in an oracle efficient estimator.

---

[6]Of course one only needs to impose (i) and (ii) on the relevant variables (those indexed by $\mathcal{A}$) in order to ensure that the least squares estimator has the limit distribution in (7.12).

Remark 7.2 The oracle property sounds almost too good to be true—and in some sense it is. As pointed out in Leeb and Pötscher (2005) and Leeb and Pötscher (2008), *any* estimator that is variable selection consistent (and thus classifies truly zero coefficients as zero with probability tending to one) must also classify coefficients as zero that induce a sequence of models (probability measures) contiguous to the model with a zero coefficient. In many commonly used (linear time series) models, including AR, MA, (fractional) ARIMA, VAR, and VARMA models, one has such contiguity, cf. Chapter 2 of Taniguchi and Kakizawa (2000) for details.[7] This implies that local to zero (yet non-zero!) parameters of the form $\frac{c}{\sqrt{T}}$ for $c \in \mathbb{R}$ will be classified as zero with probability tending to one. Thus, many sequences of non-zero coefficients must be classified as zero by any procedure that is variable selection consistent. This can be shown to imply that for *any* variable selection consistent estimator $\hat{\boldsymbol{\beta}}$ and any $M > 0$

$$\sup_{\boldsymbol{\beta}^0 \in \mathbb{R}^k} \mathbb{P}_{\boldsymbol{\beta}^0}\left(T \left\| \hat{\boldsymbol{\beta}} - \boldsymbol{\beta}^0 \right\|^2 \geq M\right) \to 1, \tag{7.13}$$

where $\mathbb{P}_{\boldsymbol{\beta}^0}$ is the sequence of measures induced by the model (7.1).[8] This lack of uniform boundedness of the scaled ℓ_2-estimation error of a variable selection consistent estimator is in stark contrast to the behavior of the least squares estimator. More precisely, since $\hat{\boldsymbol{\beta}}_{OLS} - \boldsymbol{\beta}^0 = (\boldsymbol{X}'\boldsymbol{X})^{-1}\boldsymbol{X}'\boldsymbol{\epsilon}$ does not depend on $\boldsymbol{\beta}^0$, one has that

$$T \left\| \hat{\boldsymbol{\beta}}_{OLS} - \boldsymbol{\beta}^0 \right\|^2 = \text{trace}\left[\left(\frac{\boldsymbol{X}'\boldsymbol{X}}{T}\right)^{-1} \frac{\boldsymbol{X}'\boldsymbol{\epsilon}\boldsymbol{\epsilon}'\boldsymbol{X}}{T} \left(\frac{\boldsymbol{X}'\boldsymbol{X}}{T}\right)^{-1}\right]$$

is uniformly bounded in probability (over $\boldsymbol{\beta}^0 \in \mathbb{R}^k$) if, for example, $\frac{\boldsymbol{X}'\boldsymbol{X}}{T}$ converges in probability to some positive definite $\boldsymbol{\Sigma}_X$ and $\frac{\boldsymbol{X}'\boldsymbol{\epsilon}\boldsymbol{\epsilon}'\boldsymbol{X}}{T}$ converges in probability to some positive definite $\boldsymbol{\Sigma}$. This, in turn, only requires a law of large numbers to apply to $\frac{\boldsymbol{X}'\boldsymbol{X}}{T}$ and $\frac{\boldsymbol{X}'\boldsymbol{\epsilon}\boldsymbol{\epsilon}'\boldsymbol{X}}{T}$ which is the case for many dependence structures of $\boldsymbol{X}$ and $\boldsymbol{\epsilon}$. Therefore, in contrast to (7.13), one has for $M := \text{trace}(\boldsymbol{\Sigma}_X^{-1}\boldsymbol{\Sigma}\boldsymbol{\Sigma}_X^{-1}) + 1$ that

$$\sup_{\beta^0 \in \mathbb{R}^k} \mathbb{P}_{\boldsymbol{\beta}^0}\left(T \left\| \hat{\boldsymbol{\beta}}_{OLS} - \boldsymbol{\beta}^0 \right\|^2 \geq M\right) \to 0.$$

[7] Taniguchi and Kakizawa (2000) actually prove local asymptotic normality in the mentioned models. This implies contiguity by Le Cam's 1st Lemma.

[8] See Theorem 2.1 in Leeb and Pötscher (2008) for a precise statement of an "expectation" version of this statement.

Thus, in terms of asymptotic maximal estimation error, the least squares estimator performs better than any variable selection consistent estimator (and thus better than any oracle efficient estimator).

Recall that we are considering models of the form (7.1) and let $s_0 = |\mathcal{A}|$ be the number of relevant variables. A *non-asymptotic* property that is often discussed is the so-called *oracle inequality*. An oracle inequality essentially guarantees an upper bound on the estimation error of an estimator (in some norm) to hold with at least a certain probability. To set the stage, assume that we knew which variables are relevant in (7.1) (but not the value of their non-zero coefficients). An obvious way to estimate $\boldsymbol{\beta}^0_{\mathcal{A}}$ would be to use least squares in

$$y_t = \boldsymbol{\beta}'_{\mathcal{A}} \boldsymbol{x}_{t,\mathcal{A}} + \epsilon_t.$$

$||\hat{\boldsymbol{\beta}}_{OLS} - \boldsymbol{\beta}^0||$ including only the s_0 relevant variables (and setting $\hat{\boldsymbol{\beta}}_{OLS,\mathcal{A}^c} = \mathbf{0}$) is typically of order $\sqrt{\frac{s_0}{T}}$ (in probability). An oracle inequality is then a statement of the form:

Without using any knowledge of $\boldsymbol{\beta}^0$, the estimator $\hat{\boldsymbol{\beta}}$ of $\boldsymbol{\beta}^0$ satisfies

$$||\hat{\boldsymbol{\beta}} - \boldsymbol{\beta}^0|| \leq C\sqrt{\frac{s_0 \log(k)}{T}} \tag{7.14}$$

with probability at least $1 - \delta$ *for some* $\delta, C > 0$.[9]

Remark 7.3 The term *oracle inequality* comes from the fact that the order of the ℓ_2-estimation error of an estimator satisfying an oracle inequality is only larger by a factor $\sqrt{\log(k)}$ than the one of the least squares estimator that *knows* the relevant variables. Thus, even when the total number of variables, k, is much larger than the number of relevant variables, s_0, an estimator that obeys an oracle inequality only pays a price of $\sqrt{\log(k)}$ for not knowing from the outset which variables are relevant. Put differently, the estimator is almost as precise as if an oracle had revealed the relevant variables prior to estimation and one had acted on that knowledge by only including these. We shall see that, e.g., the Lasso often satisfies an oracle inequality with a rate as in (7.14).

While the oracle inequality in (7.14) is stated for the ℓ_2-norm one can also consider other norms. As a rule of thumb, only the dependence on s_0 changes and is generally $s_0^{1/q}$ for the ℓ_q-norm.

[9]The proviso "Without using any knowledge of $\boldsymbol{\beta}^0$" should go without saying as $\hat{\boldsymbol{\beta}}$ is assumed to be an estimator and can hence not depend on $\boldsymbol{\beta}^0$. However, we stress it here to underscore the point that without knowing which variables are relevant, one can obtain almost the same estimation error (in terms of rates) as if one did. For an oracle inequality to be of interest, one typically thinks of δ of being close to zero.

Remark 7.4 The results in Raskutti, Wainwright, and Yu (2011) show the rate in (7.14) is (near) minimax optimal over $\mathcal{B}_{\ell_0}(s_0) := \{\boldsymbol{\beta} \in \mathbb{R}^k : ||\boldsymbol{\beta}||_{\ell_0} \leq s_0\}$. Thus, there exists no estimator which generally has a lower rate of maximal estimation error over $\mathcal{B}_{\ell_0}(s_0)$, i.e., sparse parameter vectors, than the Lasso.

Remark 7.5 A one-step-ahead forecast of y_{T+1} from the model

$$y_t = \boldsymbol{x}'_{t-1}\boldsymbol{\beta} + \epsilon_t$$

is typically created as $\hat{y}_{T+1|T} = \hat{\boldsymbol{\beta}}'_T \boldsymbol{x}_T$, where $\hat{\boldsymbol{\beta}}_T$ is an estimator of $\boldsymbol{\beta}^0$ based on data up to time T, i.e., on $\boldsymbol{Z}_T := \boldsymbol{Z}$. The forecast error of $\hat{y}_{T+1|T}$ is

$$|\hat{y}_{T+1|T} - y_{T+1}| = |(\hat{\boldsymbol{\beta}}_T - \boldsymbol{\beta}^0)'\boldsymbol{x}_T - \epsilon_{T+1}| \leq ||\hat{\boldsymbol{\beta}}_T - \boldsymbol{\beta}^0||||\boldsymbol{x}_T|| + |\epsilon_{T+1}|,$$

where the estimate follows by the Cauchy–Schwarz inequality. The above display implies that an oracle inequality for $\hat{\boldsymbol{\beta}}_T$ implies a guarantee on the forecast precision of $\hat{y}_{T+1|T}$ as long as $\boldsymbol{x}_T$ does not have a "too large" ℓ_2-norm and $|\epsilon_{T+1}|$ not being "too large." When k is fixed, $||\boldsymbol{x}_T||$ is typically bounded in probability while it may be increasing if $k \to \infty$ as $T \to \infty$. However, as long as $||\boldsymbol{x}_T||||\hat{\boldsymbol{\beta}}_T - \boldsymbol{\beta}^0||$ tends to zero, the forecast error can still be guaranteed to be asymptotically no larger than the one of the infeasible $\tilde{y}_{T+1|T} = \boldsymbol{\beta}^{0'}\boldsymbol{x}_T$ requiring knowledge of $\boldsymbol{\beta}^0$.

We now return to the linear models discussed in Sect. 7.3 (which all are special cases of (7.1)) and indicate some known results on the properties of the penalized estimators discussed in Sect. 7.4. The results known to us on these estimators in AR, ADL, and VAR models can be found in Table 7.1. Each entry indicates which estimator(s) has the given property in a given model along with a reference to exact assumptions on the model.

7.6 Practical Recommendations

7.6.1 *Selection of the Penalty Parameters*

The estimation of linear models by the penalized regression methods described in Sect. 7.4 involves the choice of the penalty (tuning) parameter λ. In some cases, the econometrician also needs to specify other tuning parameters, such as α for the Elastic Net. In general our feeling is that specifying a value of the penalty parameter(s) with theoretical performance guarantees (e.g., the oracle property or oracle inequalities) is still an open problem for most time series models. The penalty parameter λ is directly related to the number of variables in the model and thus is a key quantity for correct model selection. We next review the most used methods to select the penalty parameter(s).

Table 7.1 Literature comparison

	Model		
	AR	ADL	VAR
Consistent	–Lasso Nardi and Rinaldo (2011)		–Lasso Kock and Callot (2015) –Lasso Basu and Michailidis (2015)
Sparsistency	Lasso Nardi and Rinaldo (2011)		
Oracle property	–Adaptive Lasso Kock (2016)	–Weighted Lasso Medeiros and Mendes (2016) –Adaptive Lasso Audrino and Camponovo (2018)	–Adaptive Lasso Ren and Zhang (2010) –Adaptive Group Lasso Callot and Kock (2014)
Oracle inequality			–Lasso Kock and Callot (2015) –Lasso Basu and Michailidis (2015)

For the model mentioned in a column, the table shows which estimator has been shown to possess the property in the corresponding row along with a reference to a paper providing detailed assumptions. A reference is only included in "its greatest generality." Thus if a paper has, e.g., established the oracle property in a VAR model, the reference will not be mentioned in the context of AR models. Neither will it be mentioned in the context of consistent estimation in VAR models. Audrino and Camponovo (2018) also consider a class of non-linear time series models

Cross-Validation

One of the methods that is most used for model (variable) selection is cross-validation (CV). In the context of penalized regressions, CV methods have been used to select the penalty parameters. The idea of CV methods is to split the sample into two disjoint subsets: the training set ("in-sample") and the validation set ("out-of-sample"). The parameters of the model are estimated using solely the training set and the performance of the model using the estimated parameters is tested on the validation set. Let Λ and A be the sets potential values of λ and $\boldsymbol{\alpha}$. Furthermore, let $\mathsf{V} \subseteq \{1, \ldots, T\}$ be the indices of the observations in the validation set and $\mathsf{T} \subseteq \{1, \ldots, T\}$ be the indices of the observations in the training set. Often, but not always, $\mathsf{T} := \mathsf{V}^c$. Let $\widehat{\boldsymbol{\beta}}_{\mathsf{T}}(\lambda, \boldsymbol{\alpha})$ be the parameter estimate based on the training data T using the tuning parameters $(\lambda, \boldsymbol{\alpha}) \in \Lambda \times A$. For each $(\lambda, \boldsymbol{\alpha})$

$$CV(\lambda, \boldsymbol{\alpha}, \mathsf{V}) = \sum_{t \in \mathsf{V}} (y_t - \boldsymbol{x}_t' \widehat{\boldsymbol{\beta}}_{\mathsf{T}}(\lambda, \boldsymbol{\alpha}))^2$$

denotes the corresponding prediction error over the validation set V. Let $\mathcal{V} = \{\mathsf{V}_1, \ldots, \mathsf{V}_B\}$ be a user specified collection of validation sets (with corresponding training sets $\{\mathsf{T}_1, \ldots, \mathsf{T}_B\}$), which we will say more about shortly. The cross-validation error for the parameter combination $(\lambda, \boldsymbol{\alpha})$ is then calculated as

$$CV(\lambda, \boldsymbol{\alpha}) = \sum_{i=1}^{B} CV(\lambda, \boldsymbol{\alpha}, \mathsf{V}_i)$$

and one chooses a set of tuning parameters

$$(\hat{\lambda}, \hat{\boldsymbol{\alpha}}) \in \underset{(\lambda, \boldsymbol{\alpha}) \in \Lambda \times A}{\operatorname{argmin}} \; CV(\lambda, \boldsymbol{\alpha}).$$

The final parameter estimate is then found as $\hat{\boldsymbol{\beta}}(\hat{\lambda}, \hat{\boldsymbol{\alpha}})$ based on all observations $1, \ldots, T$.

We now turn to the choice of B and corresponding $\{\mathsf{V}_1, \ldots, \mathsf{V}_B\}$. There are two classic ways to split the sample: *exhaustive* and *non-exhaustive* CV; see Arlot and Celisse (2010) for a recent survey on CV methods for variable selection. For the exhaustive class of CV methods, leave-v-out CV is the most common one. The idea is to use v observations as the validation set and the remaining observations to estimate the parameters of the model. This is done for all possible ways of choosing v observations out of T. Thus, $B = \binom{T}{v}$ with each V_i having cardinality v in the above general setting. $\mathsf{T}_i = \mathsf{V}_i^c$, $i = 1, \ldots, B$. A popular choice is to set $v = 1$ such that $B = T$ and $\mathsf{V}_i = \{i\}$. This is known as the *leave-one-out CV*.

Exhaustive CV is computationally very intensive as it performs cross-validation over many sample splits. A remedy to the computational intensity is to use non-exhaustive CV methods. Among these, *B-fold CV* is the most popular. In B-fold CV, the sample is (randomly) partitioned into B subsamples with "approximately" the same number of observations. Thus, $\mathsf{V}_1, \ldots, \mathsf{V}_B$ have roughly the same cardinality.[10] In many implementations the B validation groups are disjoint such that each observation belongs to one group only. Again, $\mathsf{T}_i = \mathsf{V}_i^c$, $i = 1, \ldots, B$. B equal to 5 or 10 are typical choices.

As mentioned before, CV methods have been used extensively in econometrics and statistics for variable selection in linear models. In the framework of penalized regressions the variables selected depend on the choice of the penalty parameters. However, for time series data, the CV procedures described above are not guaranteed to be variable selection consistent as this property has mainly been established under the assumption of independent observations; see Shao (1993) and Racine (2000) for a discussion. A potential solution is h-block CV. Here the idea is to remove h observations before and after the ith observation and only use the remaining observations for estimation. Thus, $\mathsf{T}_i = \{1, \ldots, i-h-1\} \cup \{i+h+1, \ldots, T\}$, where $\{a, \ldots, b\} = \emptyset$ if $a > b$, $i = 1, \ldots, T$. One uses $\mathsf{V}_i = \{i\}$.

However, Racine (2000) showed that the h-block CV is also model selection inconsistent for time-dependent data and instead proposed the hv-block CV. Here

[10] In fact, one typically chooses each V_i to have T/B observations (up to rounding) such that $\max_{1 \le i \le j \le B} \big| |\mathsf{V}_i| - |\mathsf{V}_j| \big| \le 1$.

the idea is to first remove v observations around $i = 1, \ldots, T - v$ and use $\mathsf{V}_i = \{i - v, i + v\}$. Removing further h observations on each side of i one sets $\mathsf{T}_i = \{1, \ldots, i - v - h - 1\} \cup \{i + v + h + 1, \ldots, T\}$, where $\{a, \ldots, b\} = \emptyset$ if $a > b$. We refer to Racine (2000) for choices of v and h and note that his results focus on the least squares estimator.

Information Criteria

The penalty parameter is directly related to the degrees of freedom of the models. Classic information criteria such as the ones of Akaike (AIC, Akaike, 1974), Schwarz (BIC, Schwarz et al., 1978) as well as Hannan and Quinn (HQ, Hannan and Quinn, 1979) can be used to select λ.

Letting $\widehat{\sigma}^2(\lambda, \boldsymbol{\alpha}) = \frac{1}{T}\sum_{t=1}^{T}(y_t - \boldsymbol{x}_t'\hat{\boldsymbol{\beta}}(\lambda, \boldsymbol{\alpha}))^2$ for an estimator $\hat{\boldsymbol{\beta}}(\lambda, \boldsymbol{\alpha})$ pertaining to a specific value of tuning parameters $(\lambda, \boldsymbol{\alpha})$, the *AIC*, *BIC*, and *HQ* can be written, respectively, as:

$$\begin{aligned} AIC(\lambda, \boldsymbol{\alpha}) &= \log(\widehat{\sigma}^2(\lambda, \boldsymbol{\alpha})) + \mathsf{df}(\lambda, \boldsymbol{\alpha})\frac{2}{T} \\ BIC(\lambda, \boldsymbol{\alpha}) &= \log(\widehat{\sigma}^2(\lambda, \boldsymbol{\alpha})) + \mathsf{df}(\lambda, \boldsymbol{\alpha})\frac{\log(T)}{T} \\ HQ(\lambda, \boldsymbol{\alpha}) &= \log(\widehat{\sigma}^2(\lambda, \boldsymbol{\alpha})) + \mathsf{df}(\lambda, \boldsymbol{\alpha})\frac{\log\log(T)}{T}, \end{aligned} \tag{7.15}$$

where $\mathsf{df}(\lambda, \boldsymbol{\alpha})$ is the number of degrees of freedom of the used estimation method. For the sparsity inducing methods such as the Lasso, adaptive Lasso, Elastic Net, and adaptive Elastic Net, one typically has that $\mathsf{df}(\lambda, \boldsymbol{\alpha})$ is the number of estimated non-zero parameters for a given choice of $(\lambda, \boldsymbol{\alpha})$. For Ridge Regression, $\mathsf{df}(\lambda, \boldsymbol{\alpha}) = \mathsf{tr}[\boldsymbol{X}(\boldsymbol{X}'\boldsymbol{X} + \lambda \boldsymbol{I})^{-1}\boldsymbol{X}']$; see Hastie, Tibshirami, and Friedman (2001, page 68). For each information criterion one chooses the combination $(\hat{\lambda}, \hat{\boldsymbol{\alpha}})$ that minimizes it. The resulting estimator is $\hat{\boldsymbol{\beta}}(\hat{\lambda}, \hat{\boldsymbol{\alpha}})$.

7.6.2 Computer Implementations

All the models presented in this chapter are implemented in libraries in R. The `glmnet` package (Friedman, Hastie & Tibshirani, 2010) can implement the Lasso, Ridge, Elastic Net, and the respective adaptive versions. As for the selection of λ, the glmnet package is implemented only by cross-validation. Information criteria must be programmed manually. When estimating a model by a penalized estimator with `glmnet`, the output is the entire regularization path from the largest possible model to the intercept-only model. `glmnet` is well documented and the help files are downloaded automatically during installation. The documentation contains examples with codes that can be reproduced and adjusted. Finally, generalized linear models are also implemented in the package and we refer to the reference manual and vignette for full details.[11]

Although the `glmnet` is the most popular package for implementing shrinkage estimators, there are several alternatives in other R packages. In January 2019 alone, 6 packages related to the Lasso and shrinkage estimators were published in the Comprehensive R Archive Network (CRAN). A good example of an alternative package is `lars`. For restrictions on coefficients there is the `nnlasso` package. If one is dealing with large multi-gigabyte datasets one can use the `biglasso` package.

7.7 Simulations

Consider the data generating process

$$\begin{aligned} y_t &= 0.6y_{t-1} + \boldsymbol{x}'_{1,t-1}\boldsymbol{\beta}^0 + \varepsilon_t, \\ \varepsilon_t &= h_t^{\frac{1}{2}} u_t, \; u_t \overset{\text{iid}}{\sim} \text{t}^*(5), \\ h_t &= 5 \times 10^{-4} + 0.9h_{t-1} + 0.05\varepsilon_{t-1}^2 \\ \boldsymbol{x}_t &= \begin{bmatrix} \boldsymbol{x}_{1,t} \\ \boldsymbol{x}_{2,t} \end{bmatrix} = \boldsymbol{A}_1 \begin{bmatrix} \boldsymbol{x}_{1,t-1} \\ \boldsymbol{x}_{2,t-1} \end{bmatrix} + \boldsymbol{A}_4 \begin{bmatrix} \boldsymbol{x}_{1,t-4} \\ \boldsymbol{x}_{2,t-4} \end{bmatrix} + \boldsymbol{v}_t, \; \boldsymbol{v}_t \overset{\text{iid}}{\sim} \text{t}^*(5), \end{aligned} \tag{7.16}$$

Medeiros and Mendes (2016), where each element of $\boldsymbol{\beta}^0$ is given by $\beta_i^0 = \frac{1}{\sqrt{s_0}}(-1)^i$, $i = 1, \ldots, s_0 - 1$. $\boldsymbol{x}_{1,t}$ is a $(s_0 - 1) \times 1$ vector of relevant variables, and $\{u_t\}$ and $\{\boldsymbol{v}_t\}$ are independent iid sequences of t-distributed random variables with 5 degrees of freedom that have been normalized to have zero mean and unit variance. Furthermore, all entries of $\boldsymbol{v}_t$ are independent. The vector $\boldsymbol{x}_t = (\boldsymbol{x}'_{1,t}, \boldsymbol{x}'_{2,t})' \in \mathbb{R}^{(k-1)}$ has $k - s$ irrelevant variables, $\boldsymbol{x}_{2,t}$, and follows a fourth-order VAR model. The matrices $\boldsymbol{A}_1$ and $\boldsymbol{A}_2$ are block diagonal with each block of dimension 5×5 and consisting solely of elements equal to 0.15 and -0.1, respectively. The described setting is a rather adverse one as the errors are not normal, fat-tailed, and conditionally heteroskedastic. The unconditional variance of ε_t is given by $5 \times 10^{-4}/(1 - 0.90 - 0.05) = 0.01$.

In order to evaluate the effects of non-Gaussianity and heteroskedasticity, we also simulate the case where all the error distributions are Gaussian and homoskedastic. The variance of $\boldsymbol{v}_t$ is set to identity and the variance of ε_t is set to 0.01.

We simulate from (7.16) with $T = \{100, 500\}$, $k = \{100, 500, 1000\}$, and $s_0 = \{5, 10, 20\}$. The models are estimated by the Lasso, adaptive Lasso, Elastic Net,

[11]The reference manual as well as the vignette files can be found at https://cran.r-project.org/web/packages/glmnet/index.html.

and adaptive Elastic Net. The weights of the adaptive versions of the Lasso and the Elastic Net are the inverse of the absolute value of the estimated coefficients of their non-adaptive counterparts plus $1/\sqrt{T}$. Adding $1/\sqrt{T}$ to the weights described in Sect. 7.4 gives variables excluded by the initial estimator a second chance. An alternative often used is to exclude from the second step those variables excluded in the first step, cf. footnote 5. λ is always chosen by the BIC and the α parameter in the Elastic Net and adaptive Elastic Net is set to 0.5. The α parameter in the adaptive Lasso, Eq. (7.8), and the γ parameter in the adaptive Elastic Net, equation (7.11), are both set to 1. The number of Monte Carlo replications is 1000.

Table 7.2 shows the average bias and the average mean squared error (MSE) for the estimators over the Monte Carlo replications and the candidate variables, i.e.,

$$\text{Bias} = \frac{1}{1000k}\sum_{j=1}^{1000}\left[\widehat{\phi}^{(j)} - 0.6 + \sum_{i=1}^{n-1}\left(\widehat{\beta}_i^{(j)} - \beta_i^0\right)\right],$$

and

$$\text{MSE} = \frac{1}{1000k}\sum_{j=1}^{1000}\left[\left(\widehat{\phi}^{(j)} - 0.6\right)^2 + \sum_{i=1}^{n-1}\left(\widehat{\beta}_i^{(j)} - \beta_i^0\right)^2\right],$$

where $\widehat{\phi}^{(j)}$ and $\widehat{\boldsymbol{\beta}}^{(j)}$ denote the estimators in the jth Monte Carlo replication. Several facts emerge from Table 7.2. First, as expected, the bias and the MSE of all different methods decrease as the sample size grows. The MSE grows also when the number of relevant regressors, s_0, increases, while the bias shows no pattern. Neither bias nor MSE shows a pattern in k. The Ridge Regression is the method with the highest bias and MSE. The Elastic Net is not superior to the Lasso. On the other hand, the adaptive versions of the Lasso and the Elastic Net are clearly superior to their non-adaptive versions. This is not surprising since the penalties in the adaptive versions are more "intelligent" than in the non-adaptive versions.

Table 7.3 presents model selection results. Panel (a) presents the fraction of replications in which the correct model has been selected, i.e., $\hat{\mathcal{A}} = \mathcal{A}$ in the terminology of Sect. 7.5. First, it is clear that the performance of the Ridge Regression is not satisfactory in terms of correct model selection. This is by construction since Ridge Regression does not produce *any* zero estimates (in general). The Lasso and the Elastic Net also deliver a poor performance in terms of detection of the correct sparsity pattern. This can be explained by the fact that the regressors are highly correlated, cf. Zhao and Yu (2006). On the other hand, the adaptive versions provide good results for larger samples in particular. The adaptive Lasso seems marginally superior to the adaptive Elastic Net in terms of model selection capability. Panel (b) shows the fraction of replications in which the relevant variables are all included. Again, by construction, Ridge always includes all relevant variables since it includes *all* variables. One interesting result is that adaptive or non-adaptive methods achieve almost the same results. The results deteriorate as the number of variables in the

Table 7.2 Parameter estimates: descriptive statistics

	BIAS						MSE					
	$T = 100$			$T = 500$			$T = 100$			$T = 500$		
$s_0 \backslash k$	100	500	1000	100	500	1000	100	500	1000	100	500	1000
	Ridge											
5	−1.494	−1.086	−0.563	−0.448	−0.243	−0.523	2.281	2.096	1.082	0.276	0.427	1.012
10	−1.332	−0.580	−0.295	−0.513	−0.235	−0.288	2.801	2.307	1.186	0.358	0.520	1.118
20	−1.474	−0.734	−0.373	−0.580	−0.292	−0.367	3.010	2.404	1.236	0.395	0.562	1.167
	Lasso											
5	−0.178	−0.052	−0.035	−0.073	−0.020	−0.011	0.043	0.016	0.013	0.008	0.003	0.002
10	−0.133	−0.065	−0.058	−0.053	−0.014	−0.008	0.068	0.062	0.103	0.012	0.005	0.003
20	−0.104	−0.154	−0.145	−0.037	−0.013	−0.009	0.139	0.820	0.787	0.017	0.008	0.007
	Elastic net											
5	−0.240	−0.098	−0.083	−0.092	−0.028	−0.016	0.076	0.067	0.073	0.013	0.005	0.003
10	−0.187	−0.133	−0.115	−0.071	−0.022	−0.013	0.100	0.217	0.272	0.015	0.007	0.005
20	−0.127	−0.239	−0.178	−0.055	−0.020	−0.014	0.181	1.079	0.848	0.020	0.010	0.011
	Adaptive Lasso											
5	−0.043	−0.014	−0.008	−0.022	−0.004	−0.007	0.007	0.002	0.001	0.001	0.000	0.001
10	−0.027	−0.018	−0.022	−0.013	−0.002	−0.006	0.016	0.025	0.047	0.003	0.001	0.006
20	−0.039	−0.067	−0.095	−0.006	−0.001	−0.003	0.044	0.710	0.734	0.005	0.001	0.037
	Adaptive elastic net											
5	−0.046	−0.014	−0.009	−0.022	−0.004	−0.007	0.007	0.002	0.003	0.001	0.000	0.001
10	−0.030	−0.024	−0.033	−0.013	−0.002	−0.006	0.016	0.046	0.077	0.003	0.001	0.006
20	−0.043	−0.082	−0.106	−0.005	−0.001	−0.003	0.046	0.860	0.772	0.005	0.001	0.038

The table reports for different sample sizes, the average bias and the average mean squared error (MSE) over all parameter estimates and Monte Carlo simulations. k is the number of candidate variables, whereas s_0 is the number of relevant regressors

Table 7.3 Model selection: descriptive statistics

	Panel (a)						Panel (b)						Panel (c)						Panel (d)					
	Correct sparsity pattern						True model included						Fraction of relevant variables included						Fraction of irrelevant variables excluded					
	T=100			T=500			T=100			T=500			T=100			T=500			T=100			T=500		
s_0/k	100	500	1000	100	500	1000	100	500	1000	100	500	1000	100	500	1000	100	500	1000	100	500	1000	100	500	1000
	Ridge																							
5	0.000	0.000	0.000	0.000	0.000	0.000	1.000	1.000	1.000	1.000	1.000	1.000	1.000	1.000	1.000	1.000	1.000	1.000	0.000	0.000	0.000	0.000	0.000	0.000
10	0.000	0.000	0.000	0.000	0.000	0.000	1.000	1.000	1.000	1.000	1.000	1.000	1.000	1.000	1.000	1.000	1.000	1.000	0.000	0.000	0.000	0.000	0.000	0.000
20	0.000	0.000	0.000	0.000	0.000	0.000	1.000	1.000	1.000	1.000	1.000	1.000	1.000	1.000	1.000	1.000	1.000	1.000	0.000	0.000	0.000	0.000	0.000	0.000
	Lasso																							
5	0.009	0.002	0.002	0.055	0.033	0.019	1.000	1.000	1.000	1.000	1.000	1.000	1.000	1.000	1.000	1.000	1.000	1.000	0.916	0.969	0.978	0.955	0.988	0.993
10	0.000	0.000	0.000	0.002	0.000	0.001	1.000	0.994	0.937	1.000	1.000	1.000	1.000	0.999	0.985	1.000	1.000	1.000	0.821	0.932	0.949	0.909	0.974	0.987
20	0.000	0.000	0.000	0.000	0.000	0.000	1.000	0.433	0.052	1.000	1.000	1.000	1.000	0.902	0.703	1.000	1.000	1.000	0.601	0.867	0.917	0.833	0.952	0.980
	Elastic net																							
5	0.000	0.000	0.000	0.002	0.000	0.001	1.000	1.000	0.999	1.000	1.000	1.000	1.000	1.000	0.999	1.000	1.000	1.000	0.802	0.916	0.941	0.896	0.969	0.982
10	0.000	0.000	0.000	0.000	0.000	0.000	1.000	0.981	0.869	1.000	1.000	1.000	1.000	0.995	0.968	1.000	1.000	1.000	0.699	0.887	0.923	0.852	0.955	0.977
20	0.000	0.000	0.000	0.000	0.000	0.000	1.000	0.306	0.023	1.000	1.000	1.000	1.000	0.850	0.666	1.000	1.000	1.000	0.485	0.854	0.914	0.778	0.929	0.970
	Adaptive Lasso																							
5	0.766	0.961	0.952	1.000	1.000	1.000	1.000	1.000	1.000	1.000	1.000	1.000	1.000	1.000	1.000	1.000	1.000	1.000	0.996	1.000	1.000	1.000	1.000	1.000
10	0.536	0.949	0.830	0.998	0.992	1.000	1.000	0.993	0.944	1.000	1.000	1.000	1.000	0.998	0.982	1.000	1.000	1.000	0.989	1.000	0.998	1.000	1.000	1.000
20	0.109	0.163	0.011	0.771	0.717	1.000	1.000	0.392	0.047	1.000	1.000	1.000	1.000	0.830	0.589	1.000	1.000	1.000	0.948	0.972	0.969	0.996	0.999	1.000
	Adaptive elastic net																							
5	0.745	0.906	0.839	1.000	1.000	1.000	1.000	1.000	1.000	1.000	1.000	1.000	1.000	1.000	1.000	1.000	1.000	1.000	0.995	1.000	1.000	1.000	1.000	1.000
10	0.522	0.791	0.495	0.994	0.983	1.000	1.000	0.984	0.907	1.000	1.000	1.000	1.000	0.996	0.970	1.000	1.000	1.000	0.989	0.998	0.994	1.000	1.000	1.000
20	0.081	0.044	0.002	0.764	0.728	1.000	1.000	0.260	0.025	1.000	1.000	1.000	1.000	0.773	0.555	1.000	1.000	1.000	0.943	0.965	0.968	0.996	0.999	1.000

The table reports for different sample sizes, several statistics concerning model selection. Panel (a) presents the fraction of replications where the correct model has been selected. Panel (b) shows the fraction of replications where the relevant variables are all included. Panel (c) presents the fraction of relevant variables included. Panel (d) shows the fraction of irrelevant variables excluded model increases. Note also that in some cases the adaptive version includes a higher fraction of relevant variables than the non-adaptive counterparts. This is possible as we are adding $\frac{1}{\sqrt{T}}$ to the estimated coefficients in the first step. Therefore, a variable that was removed in the first step by the Lasso or the Elastic Net will have a penalty equal to $\sqrt{T}$ and will have a chance of being included in the final model. Panel (c) presents the fraction of relevant variables included. All methods perform equally well. The difference between adaptive and non-adaptive version becomes evident in Panel (d), which shows the fraction of irrelevant variables excluded from the model. Both the adaptive Lasso and the adaptive Elastic Net discard many more irrelevant variables than their non-adaptive versions.

Table 7.4 shows the MSE for one-step-ahead out-of-sample forecasts. We consider a total of 100 out-of-sample observations. We compare the different procedures to the oracle procedure. By oracle procedure we mean the model that only includes $\boldsymbol{x}_{1,t}$ and estimates its parameters by least squares. As expected, the results for $T = 500$ are superior to the ones for $T = 100$. Furthermore, all the methods deteriorate as the number of either relevant or irrelevant variables increases, with the worst performance being observed for large s_0 and large k. Finally, the adaptive versions are clearly superior to their non-adaptive counterparts. We also evaluate the case where the hypothesis of sparsity is violated by making $s_0 = k$. In this case the performance of all methods, including the Ridge, deteriorates substantially.

The results with respect to the Gaussian and homoskedastic case are depicted in Tables 7.5, 7.6, and 7.7. The overall conclusions do not change much.

7.8 Empirical Example: Inflation Forecasting

7.8.1 Overview

Inflation forecasting is of great importance in rational economic decision-making. For example, central banks rely on inflation forecasts not only to inform monetary policy but also to anchor inflation expectations and thus enhance policy efficacy. Indeed, as part of an effort to improve economic decision-making, many central banks release their inflation forecasts on a regular basis.

Despite the benefits of forecasting inflation accurately, improving upon simple benchmark models has proven to be a challenge for both academics and practitioners; see Faust and Wright (2013) for a survey. In this section we investigate how penalized estimators can improve upon two traditional benchmarks in the literature, namely the random walk (RW) and the autoregressive (AR) models.

Table 7.4 Forecasting: descriptive statistics

	$MSE \times 10$											
	T = 100			T = 500			T = 100			T = 500		
s_0/k	100	500	1000	100	500	1000	100	500	1000	100	500	1000
	Ridge						Lasso					
5	3.781	14.875	13.084	0.411	2.855	11.407	0.151	0.159	0.265	0.109	0.122	0.116
10	4.999	17.541	18.361	0.530	7.699	16.031	0.195	0.441	1.584	0.112	0.161	0.132
20	5.599	30.848	20.921	0.589	4.227	18.270	0.336	2.427	12.767	0.118	0.137	0.174
k	6.041	22.359	23.080	0.633	5.755	20.900	6.789	21.573	23.465	0.129	6.402	13.567
	Elastic net						Adaptive Lasso					
5	0.203	0.307	1.136	0.114	0.144	0.134	0.106	0.111	0.115	0.102	0.103	0.104
10	0.253	1.945	4.152	0.116	0.181	0.156	0.119	0.195	0.706	0.101	0.122	0.160
20	0.420	5.300	13.720	0.122	0.164	0.214	0.163	1.540	11.116	0.104	0.102	0.469
k	6.286	20.171	22.373	0.129	6.169	14.212	8.591	21.035	23.097	0.130	10.804	13.133
	Adaptive elastic net						Oracle					
5	0.107	0.111	0.135	0.102	0.103	0.104	0.103	0.111	0.107	0.101	0.100	0.099
10	0.120	0.223	1.069	0.101	0.122	0.161	0.111	0.075	0.112	0.100	0.116	0.102
20	0.166	2.147	11.640	0.104	0.102	0.477	0.129	0.131	0.129	0.103	0.100	0.103
k	8.350	19.124	21.850	0.130	10.887	13.131	–	–	–	–	–	–

The table reports for each different sample sizes, the one-step-ahead mean squared error (MSE) for the adaLasso and the Oracle estimators. k is the number of candidate variables, whereas s_0 is the number of relevant regressors

Table 7.5 Parameter estimates: descriptive statistics—Gaussian model

	BIAS						MSE					
	$T = 100$			$T = 500$			$T = 100$			$T = 500$		
$s_0 \backslash k$	100	500	1000	100	500	1000	100	500	1000	100	500	1000
	Ridge											
5	−1.399	−1.103	−0.570	−0.449	−0.267	−0.527	2.258	2.097	1.083	0.270	0.426	1.011
10	−1.380	−0.584	−0.276	−0.515	−0.283	−0.291	2.755	2.304	1.186	0.352	0.525	1.118
20	−1.315	−0.731	−0.371	−0.564	−0.291	−0.369	2.995	2.405	1.235	0.388	0.559	1.166
	Lasso											
5	−0.193	−0.055	−0.038	−0.072	−0.021	−0.012	0.041	0.017	0.012	0.008	0.003	0.002
10	−0.130	−0.065	−0.051	−0.050	−0.015	−0.009	0.070	0.059	0.104	0.012	0.005	0.003
20	−0.073	−0.198	−0.148	−0.031	−0.010	−0.008	0.133	0.846	0.797	0.017	0.008	0.007
	Elastic net											
5	−0.248	−0.118	−0.090	−0.092	−0.029	−0.017	0.076	0.071	0.077	0.013	0.005	0.003
10	−0.177	−0.126	−0.097	−0.068	−0.023	−0.014	0.099	0.207	0.282	0.015	0.007	0.005
20	−0.110	−0.265	−0.186	−0.048	−0.018	−0.014	0.171	1.119	0.855	0.020	0.010	0.010
	Adaptive Lasso											
5	−0.042	−0.013	−0.007	−0.020	−0.005	−0.006	0.006	0.002	0.001	0.001	0.000	0.001
10	−0.033	−0.020	−0.021	−0.008	−0.002	−0.006	0.016	0.022	0.050	0.003	0.001	0.006
20	−0.011	−0.110	−0.099	−0.000	−0.001	−0.003	0.044	0.720	0.747	0.005	0.001	0.035
	Adaptive elastic net											
5	−0.045	−0.013	−0.008	−0.020	−0.005	−0.006	0.007	0.002	0.002	0.001	0.000	0.001
10	−0.035	−0.022	−0.027	−0.008	−0.002	−0.006	0.016	0.037	0.082	0.003	0.001	0.006
20	−0.012	−0.113	−0.105	−0.000	−0.001	−0.003	0.045	0.881	0.780	0.005	0.001	0.036

The table reports for different sample sizes, the average bias and the average mean squared error (MSE) over all parameter estimates and Monte Carlo simulations. k is the number of candidate variables, whereas s_0 is the number of relevant regressors

Table 7.6 Model selection: descriptive statistics—Gaussian model

	Panel (a)						Panel (b)						Panel (c)						Panel (d)					
	Correct sparsity pattern						True model included						Fraction of relevant variables included						Fraction of irrelevant variables excluded					
	T=100			T=500			T=100			T=500			T=100			T=500			T=100			T=500		
s0/k	100	500	1000	100	500	1000	100	500	1000	100	500	1000	100	500	1000	100	500	1000	100	500	1000	100	500	1000
	Ridge																							
5	0.000	0.000	0.000	0.000	0.000	0.000	1.000	1.000	1.000	1.000	1.000	1.000	1.000	1.000	1.000	1.000	1.000	1.000	0.000	0.000	0.000	0.000	0.000	0.000
10	0.000	0.000	0.000	0.000	0.000	0.000	1.000	1.000	1.000	1.000	1.000	1.000	1.000	1.000	1.000	1.000	1.000	1.000	0.000	0.000	0.000	0.000	0.000	0.000
20	0.000	0.000	0.000	0.000	0.000	0.000	1.000	1.000	1.000	1.000	1.000	1.000	1.000	1.000	1.000	1.000	1.000	1.000	0.000	0.000	0.000	0.000	0.000	0.000
	Lasso																							
5	0.009	0.002	0.002	0.055	0.033	0.019	1.000	1.000	1.000	1.000	1.000	1.000	1.000	1.000	1.000	1.000	1.000	1.000	0.916	0.969	0.978	0.955	0.988	0.993
10	0.000	0.000	0.000	0.002	0.000	0.001	1.000	0.994	0.937	1.000	1.000	1.000	1.000	0.999	0.985	1.000	1.000	1.000	0.821	0.932	0.949	0.909	0.974	0.987
20	0.000	0.000	0.000	0.000	0.000	0.000	1.000	0.433	0.052	1.000	1.000	1.000	1.000	0.902	0.703	1.000	1.000	1.000	0.601	0.867	0.917	0.833	0.952	0.980
	Elastic net																							
5	0.000	0.000	0.000	0.002	0.000	0.001	1.000	1.000	0.999	1.000	1.000	1.000	1.000	1.000	0.999	1.000	1.000	1.000	0.802	0.916	0.941	0.896	0.969	0.982
10	0.000	0.000	0.000	0.000	0.000	0.000	1.000	0.981	0.869	1.000	1.000	1.000	1.000	0.995	0.968	1.000	1.000	1.000	0.699	0.887	0.923	0.852	0.955	0.977
20	0.000	0.000	0.000	0.000	0.000	0.000	1.000	0.306	0.023	1.000	1.000	1.000	1.000	0.850	0.666	1.000	1.000	1.000	0.485	0.854	0.914	0.778	0.929	0.970
	Adaptive Lasso																							
5	0.766	0.961	0.952	1.000	1.000	1.000	1.000	1.000	1.000	1.000	1.000	1.000	1.000	1.000	1.000	1.000	1.000	1.000	0.996	1.000	1.000	1.000	1.000	1.000
10	0.536	0.949	0.830	0.998	0.992	1.000	1.000	0.993	0.944	1.000	1.000	1.000	1.000	0.998	0.982	1.000	1.000	1.000	0.989	1.000	0.998	1.000	1.000	1.000
20	0.109	0.163	0.011	0.771	0.717	1.000	1.000	0.392	0.047	1.000	1.000	1.000	1.000	0.830	0.589	1.000	1.000	1.000	0.948	0.972	0.969	0.996	0.999	1.000
	Adaptive elastic net																							
5	0.745	0.906	0.839	1.000	1.000	1.000	1.000	1.000	1.000	1.000	1.000	1.000	1.000	1.000	1.000	1.000	1.000	1.000	0.995	1.000	1.000	1.000	1.000	1.000
10	0.522	0.791	0.495	0.994	0.983	1.000	1.000	0.984	0.907	1.000	1.000	1.000	1.000	0.996	0.970	1.000	1.000	1.000	0.989	0.998	0.994	1.000	1.000	1.000
20	0.081	0.044	0.002	0.764	0.728	1.000	1.000	0.260	0.025	1.000	1.000	1.000	1.000	0.773	0.555	1.000	1.000	1.000	0.943	0.965	0.968	0.996	0.999	1.000

The table reports for different sample sizes, several statistics concerning model selection. Panel (a) presents the fraction of replications where the correct model has been selected. Panel (b) shows the fraction of replications where the relevant variables are all included. Panel (c) presents the fraction of relevant variables included. Panel (d) shows the fraction of irrelevant variables excluded

Table 7.7 Forecasting: descriptive statistics—Gaussian model

	$MSE \times 10$											
	T=100			T=500			T=100			T=500		
s_0/k	100	500	1000	100	500	1000	100	500	1000	100	500	1000
	Ridge						Lasso					
5	3.791	12.489	13.227	0.406	3.525	11.450	0.152	0.207	0.261	0.109	0.117	0.118
10	4.925	17.469	18.061	0.526	4.655	16.140	0.198	0.515	1.682	0.113	0.126	0.134
20	5.604	19.877	20.638	0.584	5.086	18.925	0.325	6.860	12.744	0.121	0.144	0.176
k	5.943	22.641	23.363	0.627	5.674	20.873	6.838	21.187	22.856	0.132	6.483	13.595
	Elastic net						Adaptive Lasso					
5	0.208	0.606	1.193	0.114	0.130	0.136	0.110	0.114	0.114	0.101	0.103	0.107
10	0.252	1.651	4.295	0.117	0.140	0.157	0.122	0.219	0.813	0.103	0.104	0.160
20	0.402	9.039	13.645	0.124	0.159	0.216	0.164	5.120	11.305	0.106	0.107	0.462
k	6.315	19.963	21.910	0.132	6.199	14.240	8.601	20.625	22.724	0.133	10.802	13.204
	Adaptive elastic net						Oracle					
5	0.110	0.115	0.122	0.101	0.103	0.107	0.107	0.107	0.107	0.101	0.103	0.101
10	0.122	0.312	1.203	0.103	0.104	0.161	0.113	0.113	0.113	0.102	0.103	0.103
20	0.166	6.266	11.656	0.106	0.107	0.469	0.132	0.132	0.132	0.104	0.105	0.105
k	8.365	18.889	21.460	0.133	10.823	13.198	–	–	–	–	–	–

The table reports for each different sample sizes, the one-step-ahead mean squared error (MSE) for the adaLasso and the Oracle estimators. k is the number of candidate variables, whereas s_0 is the number of relevant regressors

7.8.2 Data

Our data consist of variables from the FRED-MD database, which is a large monthly macroeconomic dataset designed for empirical analysis in data-rich environments. The dataset is updated in real-time through the FRED database and is available from Michael McCracken's webpage.[12] For further details, we refer to McCracken and Ng (2016).

In this chapter, we use the vintage of the data as of January 2016. Our sample starts in January 1960 and ends in December 2015 (672 observations). Only variables without missing observations in the entire sample period are used (122 variables). In addition, we include as potential predictors the four principal component factors computed from this set of variables. We include four lags of all explanatory variables, including the four factors, and four lags of inflation. Therefore, the estimated models have $508(= (122 + 4) \cdot 4 + 4)$ potential predictors plus an intercept. The out-of-sample window is from January 1990 to December 2015. All variables are transformed as described in McCracken and Ng (2016). The only exceptions are the price indices which are log-differenced only once. π_t denotes the inflation in month t computed as $\pi_t = \log(\mathsf{P}_t) - \log(\mathsf{P}_{t-1})$, where P_t is the price index in month t. The baseline price index is the personal consumption expenditures (PCEs).

We compare performances not only across models in the out-of-sample window but also in two different subsample periods, namely January 1990 to December 2000 (132 out-of-sample observations) and January 2001 to December 2015 (180 out-of-sample observations). The first sample corresponds to a period of low inflation volatility ($\hat{\sigma} = 0.17\%$), while in the second sample, inflation is more volatile ($\hat{\sigma} = 0.32\%$). However, on average, inflation is higher during 1990–2000 than 2001–2015. Relative to the 1990–2000 period, inflation was more volatile near the recession in the early 1990s. Figure 7.1 shows the time series evolution of the PCE inflation during the entire out-of-sample period.

7.8.3 Methodology

We shall create direct forecasts of the h-period ahead inflation from the linear model

$$\pi_{t+h} = \boldsymbol{\beta}_h' \boldsymbol{x}_t + \varepsilon_{t+h}, \quad h = 1, \ldots, 12, \quad t = 1, \ldots, T, \tag{7.17}$$

where π_{t+h} is the inflation in month $t + h$; $\boldsymbol{x}_t$ contains the 508 explanatory variables; $\boldsymbol{\beta}_h$ is a horizon-specific vector of parameters; and ε_{t+h} is a zero-mean random error.

[12] https://research.stlouisfed.org/econ/mccracken/fred-databases/.

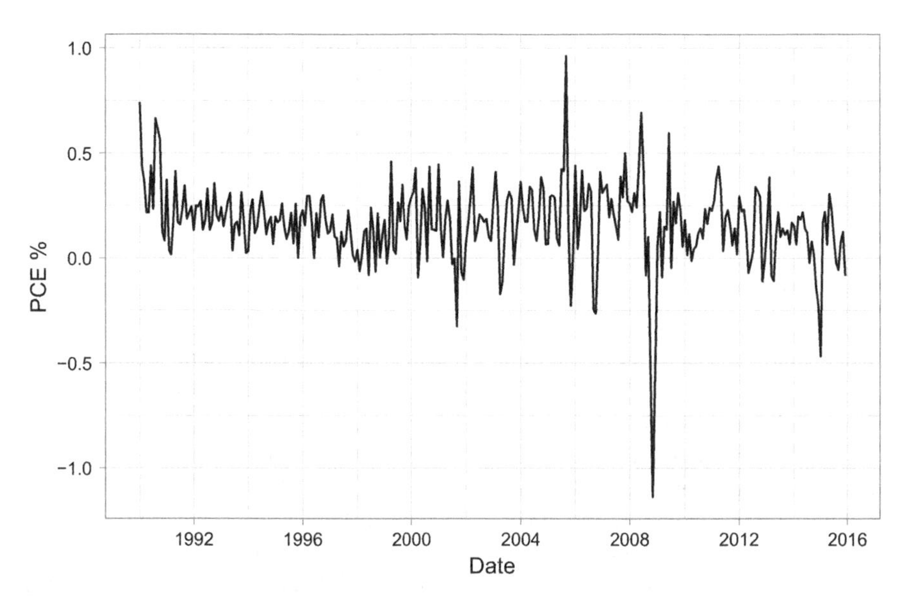

Fig. 7.1 Personal consumption expenditure inflation rate from 1990 to 2015. The figure shows the time evolution of the personal consumption expenditures (PCEs) inflation rate from January 1990 to December 2015 (312 observations). Inflation is computed as $\pi_t = \log(\mathsf{P}_t) - \log(\mathsf{P}_{t-1})$, where P_t represents the PCE price index

The direct forecasts at horizon $h = 1, \ldots, 12$ are created as

$$\widehat{\pi}_{t+h|t} = \widehat{\boldsymbol{\beta}}'_{h,t-R_h+1:t} \boldsymbol{x}_t, \tag{7.18}$$

where $\widehat{\boldsymbol{\beta}}_{h,t-R_h+1:t}$ is the estimated vector of parameters based on data from time $t - R_h + 1$ up to t and R_h is the window size used for estimation of $\boldsymbol{\beta}_h$. R_h varies according to the forecasting horizon: for the 1990–2000 period, the number of observations is $R_h = 360 - h - p - 1$ (where $p = 4$ is the number of lags in the model), while for 2001–2015, $R_h = 492 - h - 4 - 1$. Thus, the forecasts are based on a rolling window framework of fixed length. However, the actual in-sample number of observations depends on the forecasting horizon. We consider direct forecasts as opposed to recursive/iterative forecasts in order to avoid having to forecast the covariates.

In addition to three benchmark specifications (No-change (Random Walk) and AR(4) models), we consider the following shrinkage estimators: the Lasso, the adaptive Lasso, the Elastic Net, the adaptive Elastic Net, and Ridge Regression.

The penalty term for all the penalized estimators is chosen by the Bayesian Information Criterion (BIC) as described in Sect. 7.6.1. The α parameter for the Elastic Net and the adaptive Elastic Net is set to 0.5. Finally, the adaptive versions of the Lasso and the Elastic Net use the inverse of the absolute non-adaptive estimates plus $1/\sqrt{T}$ as weights.

7.8.4 Results

The main results are shown in Table 7.8. Panels (a) and (b) present the results for the 1990–2000 and 2001–2015 periods, respectively, while Panel (c) shows the results for the full out-of-sample period. The table reports the RMSEs and, in parentheses, the MAEs for all models relative to the RW specification. The error measures were calculated from 132 rolling windows covering the 1990–2000 period and 180 rolling windows covering the 2001–2015 period. Values in bold denote the most accurate model at each horizon. Cells in gray (blue) show the models included in the 50% model confidence sets (MCSs) of Hansen, Lunde, and Nason (2011) using the squared error (absolute error) as the loss function. The MCSs were constructed based on the maximum t statistic. A MCS is a set of models that is constructed such that it will contain the model with the best forecasting performance with a given level of confidence (50% confidence in our case) and is based on a sequence of tests of equal predictive ability.

Several conclusions emerge from the tables and we start by considering the out-of-sample period from January 1990 until December 2000. It is clear from the results that the RW model can be easily beaten by the AR(4) specification as well as by the penalized regression alternatives. The best performing shrinkage method is the Ridge Regression with gains of more than 20% over the RW benchmark. The other penalized regressions have gains up to 20%. On the other hand, apart from the first two horizons, it is evident from the table that is quite difficult to beat the AR model. Finally, in terms of point forecast performance, the Ridge Regression attains the lowest average squared errors for eight out of twelve horizons and the lowest mean absolute errors for six horizons. In terms of MCSs, the RW is rarely included in the set and Ridge is the penalized estimator that is included in the MCS most frequently. Among the benchmarks, the AR(4) model is the most competitive one and, apart from the two first horizons, is always included in the MCS.

We now turn to the out-of-sample from January 2001 to December 2015. This is the period of higher inflation volatility. In this case, both the RW and AR(4) benchmarks are beaten by the penalized estimators. Now the Lasso, the Elastic Net, and Ridge Regression are all very competitive. The performance gains over the RW benchmark can be higher than 25%. The RW model is never included in the MCS. The AR(4) alternative is included only for three different horizons.

Table 7.8 Forecasting errors for the PCE

Forecasting horizon												
RMSE/(MAE)	1	2	3	4	5	6	7	8	9	10	11	12
Panel (a): personal consumer expenditure 1990–2000												
RW	1.00	1.00	1.00	1.00	1.00	1.00	1.00	1.00	1.00	1.00	1.00	1.00
	(1.00)	(1.00)	(1.00)	(1.00)	(1.00)	(1.00)	(1.00)	(1.00)	(1.00)	(1.00)	(1.00)	(1.00)
AR	0.84	0.80	0.86	**0.82**	0.79	**0.80**	**0.83**	**0.84**	**0.90**	0.86	0.95	0.92
	(0.86)	(0.79)	(0.88)	**(0.85)**	**(0.79)**	**(0.84)**	**(0.89)**	**(0.80)**	**(0.89)**	(0.90)	(1.00)	(0.98)
Lasso	0.83	0.83	0.86	0.89	0.85	0.88	0.89	0.89	0.96	0.88	0.95	0.88
	(0.83)	(0.82)	(0.87)	(0.94)	(0.89)	(0.93)	(1.00)	(0.86)	(0.96)	(0.91)	(1.02)	(0.95)
adaLasso	0.84	0.84	0.86	0.85	0.81	0.83	0.86	0.88	0.93	**0.83**	**0.90**	0.87
	(0.84)	(0.83)	(0.87)	(0.87)	(0.82)	(0.85)	(0.92)	(0.83)	(0.91)	**(0.85)**	**(0.94)**	(0.92)
ElNet	**0.80**	0.83	0.87	0.90	0.86	0.89	0.92	0.91	0.95	0.88	1.00	0.91
	(0.81)	(0.83)	(0.89)	(0.97)	(0.92)	(0.96)	(1.02)	(0.88)	(0.96)	(0.92)	(1.08)	(0.98)
adaElnet	0.85	0.84	0.86	0.86	0.80	0.84	0.86	0.88	0.95	0.84	0.92	0.88
	(0.86)	(0.84)	(0.87)	(0.90)	(0.82)	(0.87)	(0.93)	(0.84)	(0.93)	(0.87)	(0.97)	(0.94)
Ridge	0.82	**0.77**	**0.85**	0.83	**0.78**	0.82	**0.83**	**0.84**	**0.90**	**0.83**	0.91	**0.84**
	(0.82)	**(0.75)**	**(0.86)**	**(0.85)**	(0.82)	(0.87)	**(0.90)**	**(0.80)**	(0.90)	(0.87)	(0.96)	**(0.87)**

(continued)

Table 7.8 (continued)

Forecasting horizon												
RMSE/(MAE)	1	2	3	4	5	6	7	8	9	10	11	12
Panel (b): personal consumer expenditure 2001–2015												
RW	1.00	1.00	1.00	1.00	1.00	1.00	1.00	1.00	1.00	1.00	1.00	1.00
	(1.00)	(1.00)	(1.00)	(1.00)	(1.00)	(1.00)	(1.00)	(1.00)	(1.00)	(1.00)	(1.00)	(1.00)
AR	0.91	0.84	0.81	0.84	0.82	0.80	0.79	0.77	0.80	0.83	0.85	0.77
	(0.88)	(0.82)	(0.75)	(0.82)	(0.83)	(0.79)	(0.74)	(0.72)	(0.78)	(0.81)	(0.83)	(0.74)
Lasso	**0.83**	0.77	0.73	0.77	**0.76**	0.76	0.75	0.75	0.78	0.80	0.81	0.73
	(0.79)	**(0.74)**	**(0.67)**	(0.74)	(0.77)	**(0.72)**	(0.69)	**(0.68)**	**(0.73)**	(0.76)	(0.78)	**(0.68)**
adaLasso	0.84	0.77	0.74	0.78	0.78	0.77	0.76	0.77	0.79	0.81	0.83	0.73
	(0.80)	(0.76)	(0.69)	(0.77)	(0.78)	(0.74)	(0.71)	(0.70)	(0.74)	(0.78)	(0.81)	(0.70)
ElNet	0.84	**0.76**	**0.72**	**0.76**	**0.76**	**0.75**	**0.74**	0.74	0.78	0.80	0.81	0.72
	(0.80)	**(0.74)**	**(0.67)**	**(0.73)**	**(0.76)**	**(0.72)**	**(0.68)**	**(0.67)**	(0.74)	(0.76)	(0.78)	**(0.68)**
adaElnet	0.84	0.78	0.74	0.78	0.79	0.77	0.76	0.76	0.79	0.81	0.82	0.73
	(0.81)	(0.76)	(0.68)	(0.76)	(0.78)	(0.74)	(0.71)	(0.69)	(0.75)	(0.78)	(0.81)	(0.70)
Ridge	0.87	**0.76**	0.73	0.77	**0.76**	**0.75**	**0.74**	**0.73**	**0.76**	**0.77**	**0.78**	**0.71**
	(0.84)	**(0.74)**	**(0.67)**	(0.75)	(0.78)	(0.75)	(0.69)	**(0.68)**	(0.74)	**(0.74)**	**(0.76)**	(0.70)

Panel (c): personal consumer expenditure 1990–2015												
RW	1.00	1.00	1.00	1.00	1.00	1.00	1.00	1.00	1.00	1.00	1.00	1.00
	(1.00)	(1.00)	(1.00)	(1.00)	(1.00)	(1.00)	(1.00)	(1.00)	(1.00)	(1.00)	(1.00)	(1.00)
AR	0.89	0.83	0.82	0.84	0.82	0.80	0.79	0.78	0.82	0.84	0.87	0.80
	(0.87)	(0.81)	(0.79)	(0.83)	(0.82)	(0.81)	(0.79)	(0.75)	(0.82)	(0.84)	(0.89)	(0.82)
Lasso	**0.83**	0.78	**0.75**	**0.79**	0.78	0.78	0.78	0.77	0.81	0.82	0.84	0.76
	(0.80)	(0.77)	**(0.73)**	(0.80)	(0.81)	(0.79)	(0.78)	(0.73)	(0.80)	(0.81)	(0.86)	(0.77)
adaLasso	0.84	0.79	0.77	0.80	0.79	0.78	0.78	0.79	0.82	0.82	0.84	0.76
	(0.82)	(0.78)	(0.74)	(0.80)	(0.79)	**(0.77)**	(0.77)	(0.74)	(0.80)	(0.80)	(0.85)	(0.77)
ElNet	**0.83**	0.78	**0.75**	**0.79**	0.78	0.78	0.77	0.77	0.81	0.82	0.85	0.76
	(0.80)	(0.77)	**(0.73)**	(0.81)	(0.81)	(0.80)	(0.78)	(0.74)	(0.81)	(0.81)	(0.87)	(0.78)
adaElnet	0.84	0.79	0.76	0.80	0.79	0.78	0.78	0.78	0.82	0.82	0.84	0.76
	(0.83)	(0.79)	(0.74)	(0.80)	(0.80)	(0.78)	(0.77)	(0.74)	(0.80)	(0.81)	(0.86)	(0.77)
Ridge	0.85	**0.76**	0.76	**0.79**	0.76	**0.77**	**0.76**	0.75	**0.79**	**0.78**	**0.81**	**0.74**
	(0.83)	**(0.75)**	**(0.73)**	**(0.78)**	(0.79)	(0.78)	**(0.75)**	**(0.72)**	**(0.79)**	**(0.78)**	**(0.83)**	**(0.75)**

The table shows the root mean squared error (RMSE) and, between parentheses, the mean absolute errors (MAEs) for all models relative to the Random Walk (RW). The error measures were calculated from 132 rolling windows covering the 2001–2015 period and 180 rolling windows covering the 2001–2015 period. Values in bold show the most accurate model in each horizon. Cells in gray (blue) show the models included in the 50% model confidence set (MCS) using the squared error (absolute error) as loss function. The MCSs were constructed based on the maximum t-statistic

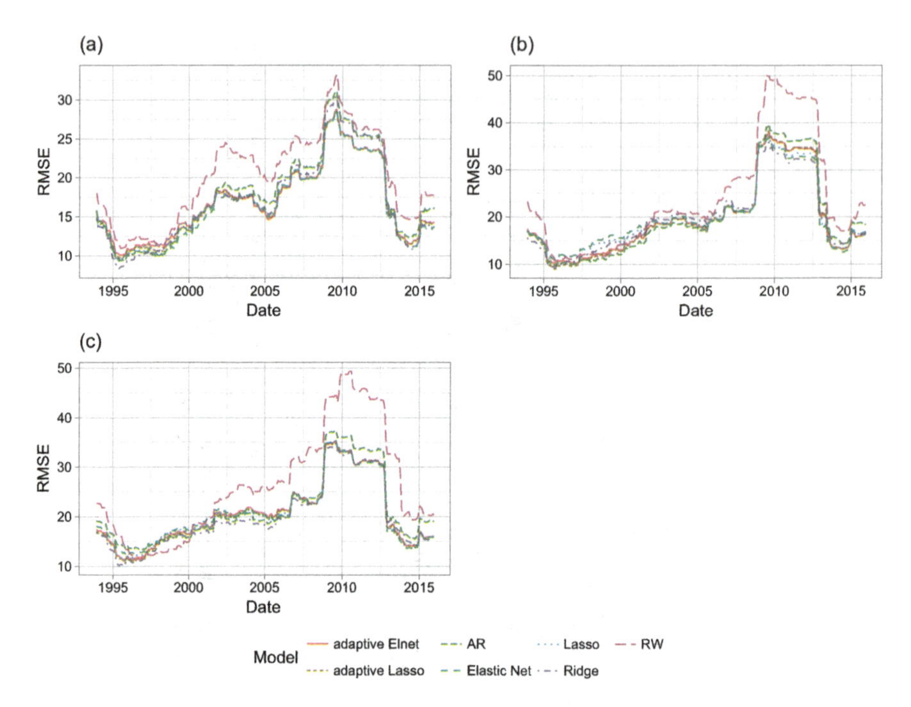

Fig. 7.2 Rolling root mean squared error (RMSE). The figure displays the root mean squared errors (RMSEs) computed over rolling windows of 48 observations. Panel (a)–(c) display, respectively, the results for 1-, 6-, and 12-month-ahead forecasts. (**a**) Rolling RMSE h = 1. (**b**) Rolling RMSE h = 6. (**c**) Rolling RMSE h = 12

Finally, analyzing the results for the full out-of-sample period, it is evident that the penalized estimators outperform the benchmarks. Ridge Regression seems to be the best estimator. However, in terms of MCSs it is difficult to discriminate among different shrinkage methods.

In order to check the robustness of the results over the out-of-sample period, we compute rolling RMSEs and MAEs over windows of 48 observations. The results are shown in Figs. 7.2 and 7.3. As can be seen from the figures, the RW is systematically dominated by the other models. The only exception is a short period in the beginning of the sample when the RW specification is comparable to the other procedures.

To shed some light on the estimated models we report a measure of variable importance in Fig. 7.4. As there are a large number of predictors we group them into ten different classes: (1) AR terms (lags of inflation); (2) output and income; (3) labor market; (4) housing; (5) consumption, orders, and inventories; (6) money and credit; (7) interest and exchange rates; (8) prices; (9) stock market; and (10) factors. For each forecasting horizon the figure shows the average estimated coefficient across the rolling windows for each group. Prior to averaging, the coefficients are multiplied by the variables standard deviation in order to be comparable. The final

Fig. 7.3 Rolling mean absolute error (MAE). The figure displays the mean absolute errors (MAEs) computed over rolling windows of 48 observations. Panel (a)–(c) display, respectively, the results for 1-, 6-, and 12-month-ahead forecasts. (**a**) Rolling MAE h = 1. (**b**) Rolling MAE h = 6. (**c**) Rolling MAE h = 12

measures of importance are rescaled to sum one. As the number of variables in each group is very different, we also divide the importance measures by the number of variables in the group. The most important variables in the Ridge are different from the other penalized regression methods. First, for the Ridge, the pattern across forecasting horizons is quite stable. On the other hand, the other methods display a more erratic behavior. For the Ridge, AR terms and other price measures are the most important variables, followed by money, employment, output-income, and stocks. Factors, housing, interest-exchange have almost no relevance in the model. For the other penalized regressions, apart from AR terms, output-income is the most important class of variables. Interestingly, for 4- and 5-months-ahead forecasts, the AR terms lose their importance.

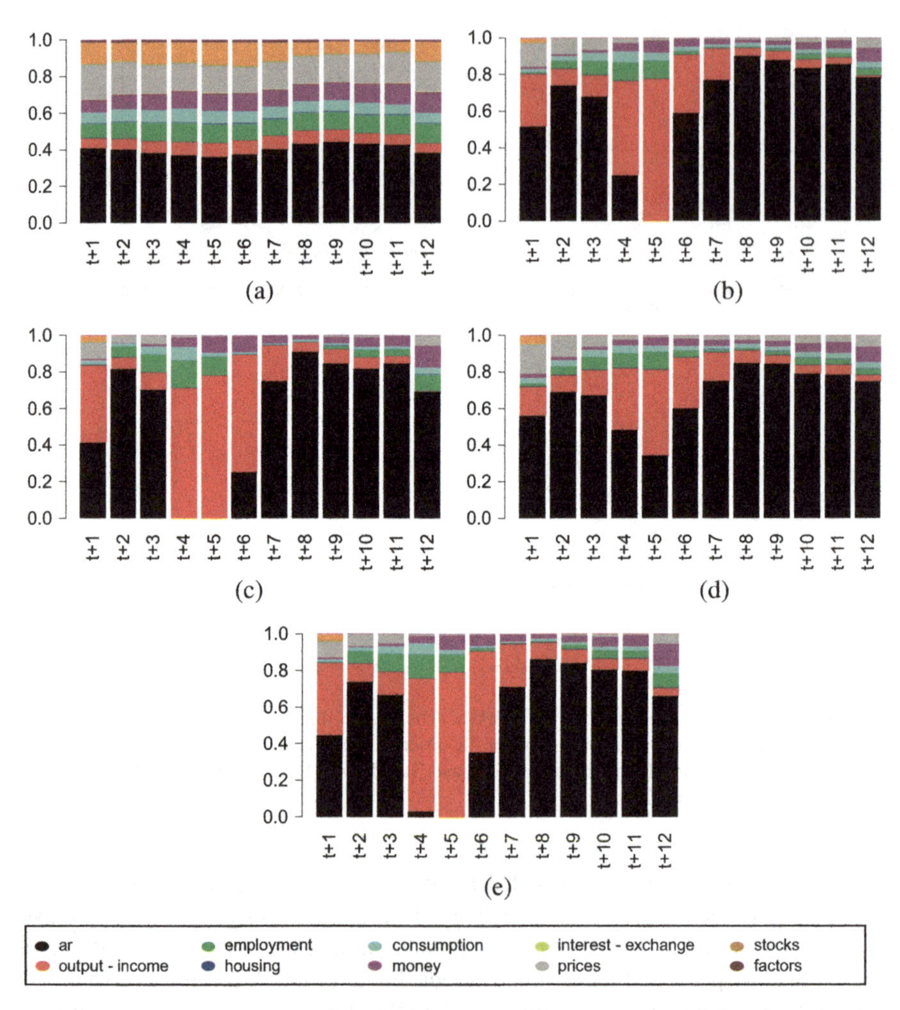

Fig. 7.4 Variable selection. The figure displays the relative importance of each class of variables as measured by the standardized estimated coefficients for horizon 1–12. The importance of each variable is given by the sum of the absolute value of its coefficients in the standardized scale across all rolling windows. The importance of a class of variable is the sum of the importance of all variables in the class divided by the number of variables in the class. The results were rescaled to sum one. (**a**) Ridge. (**b**) Lasso. (**c**) Adaptive Lasso. (**d**) Elastic net. (**e**) Adaptive elastic net

7.9 Conclusions

In this chapter we reviewed some of the most used penalized estimators in regression models for time series forecasting. We paid special attention to the Lasso, the adaptive Lasso, the Elastic Net, and the adaptive Elastic Net as well as the Ridge Regression. We highlighted the main theoretical results for the methods. We also illustrated the different approaches in a simulated study and in an empirical exercise forecasting monthly US inflation.

References

Akaike, H. (1974). A new look at the statistical model identification. *IEEE Transactions on Automatic Control, 19*(6), 716–723.

Arlot, S., & Celisse, A. (2010). A survey of cross-validation procedures for model selection. *Statistics Surveys, 4*, 40–79.

Audrino, F., & Camponovo, L. (2018). Oracle properties, bias correction, and bootstrap inference for adaptive lasso for time series m-estimators. *Journal of Time Series Analysis, 39*(2), 111–128.

Basu, S., & Michailidis, G. (2015). Regularized estimation in sparse high-dimensional time series models. *The Annals of Statistics, 43*(4), 1535–1567.

Breiman, L. (1995). Better subset regression using the nonnegative garrote. *Technometrics, 37*(4), 373–384.

Bühlmann, P., & van de Geer, S. (2011). *Statistics for high-dimensional data: Methods, theory and applications*. Berlin: Springer.

Callot, L. A., & Kock, A. B. (2014). Oracle efficient estimation and forecasting with the adaptive lasso and the adaptive group lasso in vector autoregressions. Essays in Nonlinear Time Series Econometrics, pp. 238–268.

Candes, E., & Tao, T. (2007). The Dantzig selector: Statistical estimation when p is much larger than n. *The Annals of Statistics, 35*(6), 2313–2351.

Fan, J., Ke, Y., & Wang, K. (2018). Factor-adjusted regularized model selection. Available at SSRN 3248047.

Fan, J., & Li, R. (2001). Variable selection via nonconcave penalized likelihood and its oracle properties. *Journal of the American Statistical Association, 96*(456), 1348–1360.

Faust, J., & Wright, J. (2013). Forecasting inflation. In G. Elliott & A. Timmermann (Eds.), *Handbook of economic forecasting* (vol. 2A). Amsterdam: Elsevier.

Frank, L. E., & Friedman, J. H. (1993). A statistical view of some chemometrics regression tools. *Technometrics, 35*(2), 109–135.

Friedman, J., Hastie, T., & Tibshirani, R. (2010). Regularization paths for generalized linear models via coordinate descent. *Journal of Statistical Software, 33*(1), 1.

Hannan, E. J., & Quinn, B. G. (1979). The determination of the order of an autoregression. *Journal of the Royal Statistical Society. Series B (Methodological)*, 190–195.

Hansen, P. R., Lunde, A., & Nason, J. M. (2011). The model confidence set. *Econometrica, 79*(2), 453–497.

Hastie, T., Tibshirami, R., & Friedman, J. (2001). *The elements of statistical learning; data mining, inference and prediction*. Berlin: Springer.

Hoerl, A. E., & Kennard, R. W. (1970). Ridge regression: Biased estimation for nonorthogonal problems. *Technometrics, 12*(1), 55–67.

Huang, J., Horowitz, J. L., & Wei, F. (2010). Variable selection in nonparametric additive models. *Annals of Statistics, 38*(4), 2282–2313.

Kneip, A., & Sarda, P. (2011). Factor models and variable selection in high-dimensional regression analysis. *The Annals of Statistics, 39*(5), 2410–2447.

Kock, A. B. (2016). Consistent and conservative model selection with the adaptive lasso in stationary and nonstationary autoregressions. *Econometric Theory, 32*(1), 243–259.

Kock, A. B., & Callot, L. (2015). Oracle inequalities for high dimensional vector autoregressions. *Journal of Econometrics, 186*(2), 325–344.

Leeb, H., & Pötscher, B. M. (2005). Model selection and inference: Facts and fiction. *Econometric Theory, 21*(1), 21–59.
Leeb, H., & Pötscher B. M. (2008). Sparse estimators and the oracle property, or the return of Hodges' estimator. *Journal of Econometrics, 142*(1), 201–211.
McCracken, M., & Ng, S. (2016). FRED-MD: A monthly database for macroeconomic research. *Journal of Business and Economic Statistics, 34*, 574–589.
Medeiros, M. C., & Mendes, E. F. (2016). ℓ_1-regularization of high-dimensional time-series models with non-Gaussian and heteroskedastic errors. *Journal of Econometrics, 191*(1), 255–271.
Nardi, Y., & Rinaldo, A. (2011). Autoregressive process modeling via the lasso procedure. *Journal of Multivariate Analysis, 102*(3), 528–549.
Racine, J. (2000). Consistent cross-validatory model-selection for dependent data: hv-block cross-validation. *Journal of Econometrics, 99*, 39–61.
Raskutti, G., Wainwright, M. J., & Yu, B. (2011). Minimax rates of estimation for high-dimensional linear regression over ℓ_q-balls. *IEEE Transactions on Information Theory, 57*(10), 6976–6994.
Ren, Y., & Zhang, X. (2010). Subset selection for vector autoregressive processes via adaptive lasso. *Statistics & Probability Letters, 80*(23–24), 1705–1712.
Schwarz, G. (1978). Estimating the dimension of a model. *The Annals of Statistics, 6*(2), 461–464.
Shao, J. (1993). Linear model selection by cross-validation. *Journal of the American Statistical Association, 88*, 486–495.
Taniguchi, M., & Kakizawa, Y. (2000). *Asymptotic theory of statistical inference for time series*. Berlin: Springer.
Tibshirani, R. (1996). Regression shrinkage and selection via the lasso. *Journal of the Royal Statistical Society. Series B (Methodological), 58*, 267–288.
Wang, H., & Leng, C. (2008). A note on adaptive group lasso. *Computational Statistics & Data Analysis, 52*(12), 5277–5286.
Yuan, M., & Lin, Y. (2006). Model selection and estimation in regression with grouped variables. *Journal of the Royal Statistical Society: Series B (Statistical Methodology), 68*(1), 49–67.
Zhao, P., & Yu, B. (2006). On model selection consistency of lasso. *Journal of Machine Learning Research, 7*, 2541–2563.
Zou, H. (2006). The adaptive lasso and its oracle properties. *Journal of the American Statistical Association, 101*(476), 1418–1429.
Zou, H., & Hastie, T. (2005). Regularization and variable selection via the elastic net. *Journal of the Royal Statistical Society: Series B (Statistical Methodology), 67*(2), 301–320.
Zou, H., & Zhang, H. H. (2009). On the adaptive elastic-net with a diverging number of parameters. *Annals of Statistics, 37*(4), 1733.

Chapter 8
Principal Component and Static Factor Analysis

Jianfei Cao, Chris Gu, and Yike Wang

8.1 Principal Component Analysis

Principal component analysis (PCA) and factor analysis are closely related techniques. Given a T-dimensional dataset, both techniques aim to capture the variation in the data using low dimensional representations. In particular, principal component analysis captures the major variations in the data covariance matrix using an "important" subset of its eigenvectors. Factor analysis, on the other hand, assumes an explicit model with a factor structure, which leads to the estimation of a set of common factors that capture the data variation. We will see in Sect. 8.2.1 that principal components can be used to form estimators of factor models. In this section, we focus on principal component analysis, which sheds light upon the rationale of factor analysis.

J. Cao
University of Chicago Booth School of Business, Chicago, IL, USA
e-mail: jcao0@chicagobooth.edu

C. Gu
Scheller College of Business, Georgia Institute of Technology, Atlanta, GA, USA
e-mail: chris.gu@scheller.gatech.edu

Y. Wang (✉)
Department of Economics, London School of Economics, London, UK
e-mail: y.wang379@lse.ac.uk

P. Fuleky (ed.), *Macroeconomic Forecasting in the Era of Big Data*,
Advanced Studies in Theoretical and Applied Econometrics 52,
https://doi.org/10.1007/978-3-030-31150-6_8

8.1.1 Introduction

The goal of principal component analysis is to find a low dimensional approximation by which most of the variation in the data is retained.

Principal component analysis was first proposed by Pearson (1901) and Hotelling (1933). They adopted different approaches: Pearson (1901) considers the low dimensional linear subspace that captures the data variation, while Hotelling (1933) uses orthogonal transformations to approximate the data. Since then there have been a large number of theoretical developments and successful applications in various disciplines. See Jolliffe (2002) and Jolliffe and Cadima (2016) for a detailed and lucid account of the history of principal component analysis as well as the recent developments.

We will first present the approach of Hotelling (1933), and then that of Pearson (1901), using Muirhead (2009) and Hastie, Tibshirani, and Wainwright (2015) as respective major sources of reference.

8.1.2 Variance Maximization

Suppose we have a random vector $\mathbf{x} \in \mathbb{R}^T$ with mean $\boldsymbol{\mu}$ and a positive definite covariance matrix $\boldsymbol{\Sigma}$. Let $\mathbf{UDU}'$ be the eigenvalue decomposition of $\boldsymbol{\Sigma}$, where $\mathbf{D} = \operatorname{diag}(d_1, \ldots, d_T)$, $d_1 \geq \cdots \geq d_T > 0$, and $\mathbf{U}$ contains orthonormal columns $\mathbf{u}_t$, $t = 1, \ldots, T$. We want to look for various linear functions $\mathbf{h}_t'\mathbf{x}$ of $\mathbf{x}$, such that they satisfy certain conditions:

- $\mathbf{h}_1'\mathbf{x}$ has the largest variance,
- $\mathbf{h}_2'\mathbf{x}$ has the second largest variance and is uncorrelated with $\mathbf{h}_1'\mathbf{x}$,
- $\mathbf{h}_3'\mathbf{x}$ has the third largest variance and is uncorrelated with $\mathbf{h}_1'\mathbf{x}$ and $\mathbf{h}_2'\mathbf{x}$,
- and so on.

Note that the variance of an arbitrary linear function $\boldsymbol{\alpha}'\mathbf{x}$ of $\mathbf{x}$ is $\operatorname{var}\left(\boldsymbol{\alpha}'\mathbf{x}\right) = \boldsymbol{\alpha}'\boldsymbol{\Sigma}\boldsymbol{\alpha}$.

To solve the first problem, we want to maximize $\operatorname{var}\left(\mathbf{h}_1'\mathbf{x}\right) = \mathbf{h}_1'\boldsymbol{\Sigma}\mathbf{h}_1$ over $\mathbf{h}_1$ such that $\mathbf{h}_1'\mathbf{h}_1 = 1$. The normalization is necessary because otherwise we do not have a finite solution. Let $\boldsymbol{\beta} = \mathbf{U}'\mathbf{h}_1 = (\beta_1, \ldots, \beta_T)'$. Since $\mathbf{h}_1'\mathbf{h}_1 = 1$ and $\mathbf{UU}' = \mathbf{I}_T$, we have $\boldsymbol{\beta}'\boldsymbol{\beta} = 1$. Then we have

$$\mathbf{h}_1'\boldsymbol{\Sigma}\mathbf{h}_1 = \boldsymbol{\beta}'\mathbf{D}\boldsymbol{\beta} = \sum_{t=1}^{T} d_t\beta_t^2 \leq d_1 \sum_{r=1}^{T} \beta_t^2 = d_1,$$

with equality when $\boldsymbol{\beta} = [1, 0, \ldots, 0]'$. This means that $\mathbf{h}_1 = \mathbf{u}_1$, again because $\mathbf{U}'\mathbf{U} = \mathbf{I}_T$. Therefore, $\mathbf{h}_1'\boldsymbol{\Sigma}\mathbf{h}_1 = d_1$.

For the second problem, the condition of uncorrelatedness of $\mathbf{h}_2'\mathbf{x}$ and $\mathbf{h}_1'\mathbf{x}$ implies that

$$\begin{aligned} 0 &= \text{cov}\left(\mathbf{h}_2'\mathbf{x}, \mathbf{u}_1'\mathbf{x}\right) \\ &= \mathbf{h}_2'\boldsymbol{\Sigma}\mathbf{u}_1 \\ &= d_1\mathbf{h}_2'\mathbf{u}_1, \end{aligned}$$

where the last line is due to the eigenvalue decomposition. In other words, we have $\mathbf{h}_2'\mathbf{u}_1 = 0$. In general, we will show that $\mathbf{h}_t$ is the eigenvector corresponding to the t-th largest eigenvalue d_t, and the condition of the uncorrelatedness of $\mathbf{h}_t'\mathbf{x}$ and $\mathbf{h}_s'\mathbf{x}$ for $s = 1, \ldots, t-1$ can all be restated as $\mathbf{h}_t'\mathbf{u}_s = 0$.

The solution for the t-th linear function is established by the following theorem.

Theorem 8.1 *Let* $\mathbf{UDU}'$ *be the eigenvalue decomposition of the covariance matrix* $\boldsymbol{\Sigma}$*. Then, for* $t = 2, \ldots, T$*, the* t*-th linear transformation* $\mathbf{h}_t$ *is the solution to the following problem:*

$$\begin{aligned} d_t = &\max_{\alpha \in \mathbb{R}^T} \boldsymbol{\alpha}'\boldsymbol{\Sigma}\boldsymbol{\alpha} \\ &\text{s.t. } \boldsymbol{\alpha}'\boldsymbol{\alpha} = 1 \\ &\qquad \boldsymbol{\alpha}'\mathbf{u}_s = 0, \textit{ for } s = 1, \ldots, t-1, \end{aligned}$$

and the maximizer is the eigenvector $\mathbf{u}_t$ *corresponding to the* t*-th eigenvalue* d_t*.*

Proof Let $\boldsymbol{\beta} = \mathbf{U}'\boldsymbol{\alpha} = (\beta_1, \ldots, \beta_T)'$. For $t = 2$, we have $\boldsymbol{\alpha}'\mathbf{u}_1 = 0$, so $\boldsymbol{\beta}'\mathbf{U}'\mathbf{u}_1 = \boldsymbol{\alpha}'\mathbf{UU}'\mathbf{u}_1 = \boldsymbol{\alpha}'\mathbf{u}_1 = 0$. Note that $\mathbf{U}$ is orthonormal, so $\mathbf{U}'\mathbf{u}_1 = (1, 0, \ldots, 0)'$. This implies that $\beta_1 = 0$.

Since $\boldsymbol{\alpha}'\boldsymbol{\alpha} = 1$, we have $\boldsymbol{\beta}'\boldsymbol{\beta} = \boldsymbol{\alpha}'\mathbf{UU}'\boldsymbol{\alpha} = \boldsymbol{\alpha}'\boldsymbol{\alpha} = 1$. This yields

$$\boldsymbol{\alpha}'\boldsymbol{\Sigma}\boldsymbol{\alpha} = \boldsymbol{\alpha}'\mathbf{UDU}'\boldsymbol{\alpha} = \boldsymbol{\beta}'\mathbf{D}\boldsymbol{\beta} = \sum_{t=1}^{T} d_t\beta_t^2 = \sum_{t=2}^{T} d_t\beta_d^2 \leq d_2\sum_{t=2}^{T} \beta_t^2,$$

with equality when $\boldsymbol{\beta} = (0, 1, 0, \ldots, 0)'$. The third equality is because $\mathbf{D}$ is diagonal. The fourth equality is because $\beta_1 = 0$. The inequality is because $d_2 \geq d_3 \geq \cdots \geq d_T > 0$. Therefore, maximum variance is achieved when $\boldsymbol{\alpha} = \mathbf{UU}'\boldsymbol{\alpha} = \mathbf{U}(0, 1, 0, \ldots, 0)' = \mathbf{u}_2$.

The rest of the proof proceeds in exactly the same way. □

We call $\mathbf{u}_t'(\mathbf{x} - \boldsymbol{\mu})$ the t-th principal component of $\mathbf{x}$. We can find up to T principal components, but dimension reduction can be conducted when most of the variation in the random vector could be captured by R principal components, where $R < T$.

What we have presented above is the population principal components. To obtain sample principal components, suppose we are given a collection of data

$\mathbf{x}_1, \mathbf{x}_2, \ldots, \mathbf{x}_N$, we can define $\mathbf{S} = \sum_{i=1}^{N} (\mathbf{x}_i - \bar{\mathbf{x}}) (\mathbf{x}_i - \bar{\mathbf{x}})' / N$ to be the sample covariance matrix. Then we can form the sample version of the principal components by using the sample version of the corresponding quantities.

8.1.3 Reconstruction Error Minimization

Next, we present the approach of Pearson (1901), which aims to find the best low dimensional linear fit to the data. Namely, we minimize the reconstruction error of the data using rank R subspace by the minimization problem

$$\min_{\boldsymbol{\mu}, \{\boldsymbol{\lambda}_i\}_{i=1}^N, \mathbf{H}_R} \frac{1}{N} \sum_{i=1}^{N} \|\mathbf{x}_i - \mathbf{H}_R \boldsymbol{\lambda}_i - \boldsymbol{\mu}\|^2, \qquad (8.1)$$

$$s.t. \ \mathbf{H}_R' \mathbf{H}_R = \mathbf{I}_R,$$

where $\boldsymbol{\mu} \in \mathbb{R}^T$ is the location parameter, $\mathbf{H}_R \in \mathbb{R}^{T \times R}$ is a matrix with R orthonormal columns, $\mathbf{I}_R$ is the identity matrix of size R, and $\boldsymbol{\lambda}_i \in \mathbb{R}^R$ is a coefficient vector that linearly combines the orthonormal columns of $\mathbf{H}_R$ for every $i \in \{1, \ldots, N\}$. For the sample covariance matrix $\mathbf{S} = \sum_{i=1}^{N} (\mathbf{x}_i - \bar{\mathbf{x}})(\mathbf{x}_i - \bar{\mathbf{x}})'/N$, let $\mathbf{S} = \mathbf{UDU}'$ be the eigenvalue decomposition of $\mathbf{S}$, where $\mathbf{U} = [\mathbf{u}_1, \ldots, \mathbf{u}_T]$ and $\mathbf{D} = \text{diag}(d_1, \ldots, d_T)$ with $d_1 \geq \cdots \geq d_T \geq 0$.

Theorem 8.2 *The solution to the reconstruction error minimization problem* (8.1) *is*

$$\widehat{\boldsymbol{\mu}} = \bar{\mathbf{x}},$$

$$\widehat{\mathbf{H}}_R = [\mathbf{u}_1 \cdots \mathbf{u}_R],$$

$$\widehat{\boldsymbol{\lambda}}_i = \widehat{\mathbf{H}}_R' (\mathbf{x}_i - \bar{\mathbf{x}}),$$

for $i = 1, \ldots, N$.

Proof Holding $\mathbf{H}_R$ fixed, we can minimize over $\boldsymbol{\mu}$ and $\boldsymbol{\lambda}_i$ to obtain

$$\widehat{\boldsymbol{\mu}} = \bar{\mathbf{x}},$$

$$\widehat{\boldsymbol{\lambda}}_i = \mathbf{H}_R' (\mathbf{x}_i - \bar{\mathbf{x}}).$$

After plugging in $\boldsymbol{\mu}$ and $\boldsymbol{\lambda}_i$ as functions of $\mathbf{H}_R$, we can minimize the following over $\mathbf{H}_R$:

$$\min_{\mathbf{H}_R} \frac{1}{N} \sum_{i=1}^{N} \left\| (\mathbf{x}_i - \bar{\mathbf{x}}) - \mathbf{H}_R \mathbf{H}_R' (\mathbf{x}_i - \bar{\mathbf{x}}) \right\|^2.$$

Letting $\tilde{\mathbf{X}} \in \mathbb{R}^{N \times T}$ be the demeaned data matrix, we can rewrite the above problem as

$$\begin{aligned}&\min_{\mathbf{H}_R} \frac{1}{N}\mathrm{tr}\left(\left(\tilde{\mathbf{X}} - \tilde{\mathbf{X}}\mathbf{H}_R\mathbf{H}_R'\right)\left(\tilde{\mathbf{X}}' - \mathbf{H}_R\mathbf{H}_R'\tilde{\mathbf{X}}'\right)\right)\\ =&\min_{\mathbf{H}_R} \mathrm{tr}\left(\mathbf{S}\left(\mathbf{I}_T - \mathbf{H}_R\mathbf{H}_R'\right)\right).\end{aligned}$$

Using the eigenvalue decomposition of $\mathbf{S}$, we can further simplify the problem as

$$\begin{aligned}&\min_{\mathbf{H}_R} \mathrm{tr}\left(\mathbf{S}\left(\mathbf{I}_T - \mathbf{H}_R\mathbf{H}_R'\right)\right)\\ =&\min_{\mathbf{H}_R} \mathrm{tr}\left(\mathbf{D}\mathbf{U}'\left(\mathbf{I}_T - \mathbf{H}_R\mathbf{H}_R'\right)\mathbf{U}\right)\\ =&\min_{\mathbf{H}_R} (d_1 + \cdots + d_T) - \mathrm{tr}\left(\mathbf{D}\mathbf{U}'\mathbf{H}_R\mathbf{H}_R'\mathbf{U}\right).\end{aligned}$$

Therefore, to compute the optimal solution $\widehat{\mathbf{H}}_R$, it is the same as solving the maximization problem below:

$$\max_{\mathbf{H}_R} \mathrm{tr}\left(\mathbf{D}\mathbf{U}'\mathbf{H}_R\mathbf{H}_R'\mathbf{U}\right) \tag{8.2}$$

$$s.t.\ \mathbf{H}_R'\mathbf{H}_R = \mathbf{I}_R. \tag{8.3}$$

Problem 8.2 has the same optimal objective function value as the following problem:

$$\max_{\mathbf{H}_R} \mathrm{tr}\left(\mathbf{D}\mathbf{H}_R\mathbf{H}_R'\right) \tag{8.4}$$

$$s.t.\ \mathbf{H}_R'\mathbf{H}_R = \mathbf{I}_R. \tag{8.5}$$

This can be easily seen by rewriting problem 8.2 as an optimization problem over $\mathbf{U}'\mathbf{H}_R$, and we could simply replace $\mathbf{U}'\mathbf{H}_R$ in the problem by $\mathbf{H}_R$ since $\mathbf{U}$ is an orthogonal matrix.

Given any matrix $\mathbf{H}_R$ satisfying $\mathbf{H}_R'\mathbf{H}_R = \mathbf{I}_R$, we define $\mathbf{A} = \mathbf{H}_R\mathbf{H}_R'$ with diagonal elements $\mathbf{A}_{tt}$ for $t = 1, \ldots, T$. Then, the objective function becomes $\mathrm{tr}\left(\mathbf{D}\mathbf{H}_R\mathbf{H}_R'\right) = \sum_{t=1}^{T} \mathbf{A}_{tt} d_t$. Also, we have that $\sum_{t=1}^{T} \mathbf{A}_{tt} = \mathrm{tr}\left(\mathbf{H}_R\mathbf{H}_R'\right) = R$. Since $\mathbf{A}_{tt}$ is the ℓ_2-norm of the t-th row of $\mathbf{H}_R$, we have that $\mathbf{A}_{tt} \geq 0$. To find an upper bound for $\mathbf{A}_{tt}$, let $\mathbf{H}_T$ be a $T \times T$ square matrix, in which the first R columns are $\mathbf{H}_R$, and the remaining columns are chosen so that all the columns of $\mathbf{H}_T$ are orthonormal. Since $\mathbf{A}_{tt}$ is bounded above by the ℓ_2-norm of the t-th row of $\mathbf{H}_T$, and $\mathbf{H}_T$ is an orthonormal matrix, so $\mathbf{H}_T\mathbf{H}_T' = \mathbf{I}_T$, and $\mathbf{A}_{tt} \leq 1$. Therefore, the optimal objective function value of problem (8.4) is dominated by (i.e., no larger than) the optimal value of the following problem:

$$\max_{\mathbf{A}_{tt}\in[0,1]} \sum_{t=1}^{T} \mathbf{A}_{tt} d_t$$

$$s.t. \sum_{t=1}^{T} \mathbf{A}_{tt} = R.$$

Provided that $d_1 \geq \cdots \geq d_T \geq 0$, an optimal solution of the above problem is that $\mathbf{A}_{tt} = 1$ for $t \leq R$ and $\mathbf{A}_{tt} = 0$ for $R < t \leq T$. This is the unique solution if d_t is strictly ordered. So, the optimal objective function value of the above problem is $\sum_{t=1}^{R} d_t$. It is easy to see that this objective function value can be achieved if $\mathbf{H}_R$ in problem (8.2) is set to be the first R columns of $\mathbf{U}$: $[\mathbf{u}_1 \cdots \mathbf{u}_R]$, which thus is the optimal solution of the original problem, $\widehat{\mathbf{H}}_R$. □

Note that this method aligns with the first approach as in Sect. 8.1.2 in the sense that they generate the same optimal solutions.

8.1.4 Related Methods

In this subsection we introduced two methods that are related to PCA: the independent component analysis and the sparse principal component analysis.

Independent Component Analysis

As discussed in Sect. 8.1.2, PCA generates a linear transformation of the random variable $\mathbf{x} \in \mathbb{R}^T$: $\mathbf{s}_1 = \mathbf{h}_1'\mathbf{x}$, $\mathbf{s}_2 = \mathbf{h}_2'\mathbf{x}$, ..., $\mathbf{s}_R = \mathbf{h}_R'\mathbf{x}$, where $\mathbf{s}_1$, ..., $\mathbf{s}_R$ are orthogonal of each other. In this subsection, we introduce independent component analysis (ICA), which aims at generating a linear transformation, $\mathbf{s}_1 = \mathbf{w}_1'\mathbf{x}$, $\mathbf{s}_2 = \mathbf{w}_2'\mathbf{x}, \ldots, \mathbf{s}_R = \mathbf{w}_R'\mathbf{x}$, so that $\mathbf{s}_1, \ldots, \mathbf{s}_R$ are not only orthogonal but also independent from each other. Independence is a stronger condition than orthogonality, but it is plausible in various applications.

For instance, consider the setting when we have several microphones located in different places, and there are different people talking at the same time. The mixed speech signals recorded by the microphones (denoted by $\mathbf{x} \in \mathbb{R}^T$) can be considered as a linear transformation of the original speech signals (denoted by $\mathbf{s} \in \mathbb{R}^R$). The goal here is to find a linear transformation matrix $\mathbf{W} \in \mathbb{R}^{T\times R}$ to recover the original signals through $\mathbf{s} = \mathbf{W}'\mathbf{x}$. Usually, it is plausible to assume that the original speech signals are *independently* generated. Next, we present the ICA method following the study by Hyvärinen and Oja (2000).

First, to quantify the dependency level of the components in $\mathbf{s}$, the following measure of *mutual information* is introduced:

$$I(\mathbf{s}) = \sum_{r=1}^{R} H(\mathbf{s}_r) - H(\mathbf{s}),$$

where $H(\mathbf{y}) \equiv -E[\log(f(\mathbf{y}))]$ is the Shannon entropy of the random variable $\mathbf{y}$, which depends on the probability distribution $f(\mathbf{y})$.

Shannon entropy is a concept in information theory which measures the expected amount of information in an event drawn from a distribution, and it gives a lower bound on the number of bits needed on average to encode symbols drawn from the distribution. The more deterministic a distribution is, the lower its entropy. When the components of $\mathbf{s}$ are independent, coding $\mathbf{s}$ as a random vector requires the same amount of code length than coding $\mathbf{s}_i$ separately, and thus $\sum_{r=1}^{R} H(\mathbf{s}_r) = H(\mathbf{s})$ and $I(\mathbf{s}) = 0$. While when $\mathbf{s}_i$ are dependent, coding $\mathbf{s}$ as a whole requires less code length than coding $\mathbf{s}_i$ separately, which implies that $I(\mathbf{s}) > 0$. Therefore, $I(\mathbf{s})$ is a non-negative quantity which measures the amount of dependence among $\mathbf{s}_i$. Thus, the goal of ICA is to find a matrix $\mathbf{W}$ which minimizes $I(\mathbf{s})$.

In Hyvärinen and Oja (2000), it is derived that the mutual information measure can be rewritten as:

$$I(\mathbf{s}) = c - \sum_{r=1}^{R} J(\mathbf{s}_r),$$

where c is a constant, and $J(\mathbf{y}) \equiv H(\mathbf{y}_{\text{gauss}}) - H(\mathbf{y})$, in which $\mathbf{y}_{\text{gauss}}$ is a Gaussian random variable of the same covariance matrix as $\mathbf{y}$. $J(\mathbf{y})$ is called negentropy, and it measures the level of non-Gaussianity of the random variable $\mathbf{y}$. Notice that Gaussian random variable has the highest entropy, and the more $\mathbf{y}$ deviates from the Gaussian variable, the less its entropy is and thus the higher negentropy is. Therefore, ICA can also be interpreted as an approach to maximize the level of non-Gaussianity of each of the source signals.

In practice, computing $J(\mathbf{s}_r)$ is challenging since it depends on an estimate of the probability distribution of $\mathbf{x}$. To see this, recall that $H(\mathbf{s}_r)$ by definition involves the probability distribution $f_{\mathbf{s}_r}(\mathbf{s}_r)$, which in turn depends on the probability distribution of $\mathbf{x}$ since $\mathbf{s}_r = \mathbf{w}_r'\mathbf{x}$. Therefore, to avoid computing $J(\mathbf{s}_r)$ directly, Hyvärinen and Oja (2000) show that the negentropy measure can be approximated as follows:

$$J(\mathbf{s}_r) \propto (E[G(\mathbf{s}_r)] - E[G(v)])^2, \tag{8.6}$$

where v follows the standard normal distribution and G is a non-quadratic function. In particular, a good choice of $G(u)$ is $a_1^{-1} \log\cosh a_1 u$, where $1 \leq a_1 \leq 2$, and its derivative is $g(u) = \tanh(a_1 u)$.

In Hyvärinen (1999) and Hyvärinen and Oja (2000), an efficient algorithm called FastICA is introduced. This algorithm maximizes the objective function (8.6) using approximate Newton iterations, while it decorrelates the output in each iteration step. The FastICA algorithm can be implemented as follows:

Step 1 Prewhiten the data matrix $\mathbf{X} = (\mathbf{x}_1, \ldots, \mathbf{x}_N)' \in \mathbb{R}^{N \times T}$ such that columns of $\mathbf{X}$ have mean zero, variance one, and are uncorrelated. Uncorrelated columns can be obtained through PCA.

Step 2 For $r = 1, \ldots, R$: initialize $\mathbf{w}_r$ by some random vector of length T. Given the set of weights $\{\mathbf{w}_1, \ldots, \mathbf{w}_{r-1}\}$, find $\mathbf{w}_r$ by iterating the following steps until convergence:

(i) Let

$$\mathbf{w}_r \leftarrow \frac{1}{N}\mathbf{X}' g(\mathbf{X}\mathbf{w}_r) - \frac{1}{N}[h(\mathbf{X}\mathbf{w}_r)'\iota_N]\mathbf{w}_r$$

with $g(u) = \tanh(u)$, $h(u) = 1 - \tanh^2(u)$, and ι_N being a column vector of 1's of dimension N.

(ii) Achieve decorrelation by

$$\mathbf{w}_r \leftarrow \mathbf{w}_r - \sum_{t=1}^{r-1}(\mathbf{w}_r'\mathbf{w}_t)\mathbf{w}_t.$$

(iii) Normalize $\mathbf{w}_r$ by

$$\mathbf{w}_r \leftarrow \frac{\mathbf{w}_r}{\|\mathbf{w}_r\|}.$$

Step 3 Let $\mathbf{W} = [\mathbf{w}_1, \ldots, \mathbf{w}_R]$. The extracted independent signals are given by $\mathbf{XW}$.

ICA is widely used in signal recognition and biomedical research to discover patterns from noisy data. See Bartlett, Movellan, & Sejnowski (2002) and Brown, Yamada, and Sejnowski (2001) for examples. It is first used by Back and Weigend (1997) to extract structures from stock returns. In addition, Kim and Swanson (2018) investigate the performance of ICA in predicting macroeconomic data, which we will discuss in Sect. 8.5.4. For formal treatment of ICA, Tong, Liu, and Soon (1991) and Stone (2004) provide detailed discussions on the theory and algorithms.

Sparse Principal Component Analysis

One limitation of PCA is that its results lack interpretability, which can be improved by using the sparse principal component analysis (SPCA). To see this, let $\mathbf{X}$ be an $N \times T$ matrix, and $\mathbf{X} = \mathbf{VDU}'$ be the singular value decomposition of $\mathbf{X}$, where $\mathbf{V}$

is $N \times T$, $\mathbf{D}$ is $T \times T$, and $\mathbf{U}$ is $T \times T$. Then $\mathbf{F} = \mathbf{VD}$ are the principal components and $\mathbf{U}$ are the corresponding loadings. That is, each dimension of the data matrix can be written as a linear combination of the learned principal components $\mathbf{F}$ with weights given in each row of $\mathbf{U}$. Standard PCA typically gives nonzero weights to all principal components, which makes it hard to interpret the relationship between principal components and the observed data. SPCA, as proposed by Zou, Hastie, and Tibshirani (2006), solves this problem by introducing sparsity on the loadings.

SPCA can be performed using the following procedure as proposed in Zou et al. (2006):

Step 1 Setting $\mathbf{A} = [\mathbf{u}_1, \ldots, \mathbf{u}_R]$, we compute $\mathbf{w}_r$ using the following elastic net problem for $r = 1, \ldots, R$:

$$\mathbf{w}_r = \underset{\mathbf{w} \in \mathbb{R}^T}{\arg\min}\ (\mathbf{u}_r - \mathbf{w})'\mathbf{X}'\mathbf{X}(\mathbf{u}_r - \mathbf{w}) + \lambda \|\mathbf{w}\|^2 + \lambda_{1,r} \|\mathbf{w}\|_1.$$

Step 2 Given $\mathbf{W} = [\mathbf{w}_1, \ldots, \mathbf{w}_R]$, we compute the singular value decomposition, $\mathbf{X}'\mathbf{X}\mathbf{W} = \mathbf{U}\mathbf{D}\mathbf{V}'$, and let $\mathbf{A} = \mathbf{U}\mathbf{V}'$.

Step 3 Repeat Steps 1–2 until convergence.

Step 4 After normalization, $\mathbf{w}_r / \|\mathbf{w}_r\|$ is the estimate of the rth sparse principal component loading vector, for $r = 1, \ldots, R$.

SPCA has been extensively studied in the literature. For example, Cai, Ma, and Wu (2013) establish optimal rates of convergence and propose an adaptive estimation procedure. Detailed discussions can be found in Hastie, Tibshirani, and Friedman (2009). In Sect. 8.5.4, we will present the performance of SPCA in forecasting macroeconomic aggregates, following the study by Kim and Swanson (2018).

8.2 Factor Analysis with Large Datasets

Factor models have been the workhorse models for macroeconomic forecasting with prominent performances demonstrated in, e.g., Stock and Watson (1999) and Stock and Watson (2002b). In this section, we discuss the estimation and inference problems of factor models, which rely on large-scale datasets containing both a large number of units and a large number of time periods.

Throughout this section, we consider a model with a factor structure such that for each $i = 1, \ldots, N$ and each $t = 1, \ldots, T$,

$$x_{it} = \boldsymbol{\lambda}_i' \mathbf{f}_t + u_{it}, \tag{8.7}$$

where $\boldsymbol{\lambda}_i$ is an $r \times 1$ non-random vector and $\mathbf{f}_t$ is an $r \times 1$ random vector. Here $\boldsymbol{\lambda}_i$ is the individual factor loading and $\mathbf{f}_t$ is the underlying common factor. The last term u_{it} is the idiosyncratic error term. In this factor model, only x_{it} is observed.

In a vector notation where we stack all t's for each i, Eq. (8.7) can be written as

$$\mathbf{x}_i = \mathbf{F}\boldsymbol{\lambda}_i + \mathbf{u}_i,$$

where $\mathbf{x}_i = (x_{i1}, \ldots, x_{iT})'$ is a $T \times 1$ vector, $\mathbf{F} = (\mathbf{f}_1, \ldots, \mathbf{f}_T)'$ is a $T \times r$ matrix, and $\mathbf{u}_i = (u_{i1}, \ldots, u_{iT})'$ is a $T \times 1$ vector. By further stacking, we can write the factor model in a matrix notation such that

$$\mathbf{X} = \mathbf{F}\boldsymbol{\Lambda}' + \mathbf{U}, \tag{8.8}$$

where $\mathbf{X} = (\mathbf{x}_1, \ldots, \mathbf{x}_N)$ is a $T \times N$ matrix, $\boldsymbol{\Lambda} = (\boldsymbol{\lambda}_1, \ldots, \boldsymbol{\lambda}_N)'$ is an $N \times r$ matrix, and $\mathbf{U} = (\mathbf{u}_1, \ldots, \mathbf{u}_N)$ is a $T \times N$ matrix.

In this section, we consider the asymptotic framework where both N and T go to infinity, and we focus on the case where the number of common factors r is fixed. We will discuss how both the factor loadings and common factors can be consistently estimated, and the asymptotic behavior of the estimator. We will also discuss how the relative rate of growth of N and T affects our results.

Section 8.2.1 introduces the principal component method, which is a predominant technique for the factor model estimation. We discuss the properties of the estimator and explain how this knowledge can be used to analyze forecasting. In Sect. 8.3, we introduce some machine learning methods and discuss how the usage of these methods can facilitate macroeconomic forecasting with large factor models. In Sect. 8.4, we discuss factor analysis in the context of policy evaluation.

8.2.1 Factor Model Estimation by the Principal Component Method

Despite the existence of other methods such as the maximum likelihood estimation, the principal component method has been the predominant way of performing factor model estimations because of its transparent estimation procedures and fast implementations. In this subsection, we discuss the asymptotic properties of this method and explain how to perform macroeconomic forecasting using the estimates from the factor model estimations. Rigorous illustrations can be found in Bai and Ng (2002), Bai (2003), and Bai and Ng (2006).

Estimation

For clarification, we use superscript 0 to denote the true parameters. Namely, the matrix form of the data generating process of the factor model is now

$$\mathbf{X} = \mathbf{F}^0\boldsymbol{\Lambda}^{0'} + \mathbf{U}.$$

As in Bai and Ng (2002), Bai (2003), and Bai and Ng (2006), we consider the optimization problem

$$\min_{\mathbf{\Lambda}\in\mathbb{R}^{N\times r},\mathbf{F}\in\mathbb{R}^{T\times r}} \frac{1}{NT}\sum_{i=1}^{N}\sum_{t=1}^{T}(x_{it}-\boldsymbol{\lambda}_i{}'\mathbf{f}_t)^2, \tag{8.9}$$

where we minimize the mean squared loss over both $\mathbf{\Lambda}$ and $\mathbf{F}$. Under the normalization that $\mathbf{F}'\mathbf{F}/T$ is an identity matrix, the solution $\widehat{\mathbf{F}}$ is $\sqrt{T}$ times the eigenvectors corresponding to the r leading eigenvalues of the matrix $\mathbf{X}\mathbf{X}'$. Therefore, this estimation procedure lines up with the PCA method as introduced in Sect. 8.1.

Estimate the Number of Factors

We have assumed so far that the number of factors r is known. In practice, this is usually infeasible, since we observe neither the true common factors $\mathbf{F}^0$ nor the true factor loadings $\mathbf{\Lambda}^0$. Bai and Ng (2002) introduce a procedure that provides a consistent estimate of the true number of factors r.

For some $\bar{k}$ that is an upper bound of the possible number of factors, let k be such that $0 \le k \le \bar{k}$. Then, for each k, the corresponding value of the objective function is

$$V(k) = \min_{\mathbf{\Lambda}\in\mathbb{R}^{N\times k},\mathbf{F}\in\mathbb{R}^{T\times k}} \frac{1}{NT}\sum_{i=1}^{N}\sum_{t=1}^{T}(x_{it}-\boldsymbol{\lambda}_i'\mathbf{f}_t)^2.$$

Following Bai and Ng (2002), define the panel C_p criteria by

$$PC(k) = V(k) + kg(N,T),$$

and the information criteria by

$$IC(k) = \log(V(k)) + kg(N,T),$$

where $g(N,T)$ is a function of both N and T that scales the penalty term. It is shown in Bai and Ng (2002) that under a general set of assumptions of the underlying factor model, the minimizer of either the penal C_p criterion or the information criterion consistently estimates the true number of factors, if (i) $g(N,T) \to 0$ and (ii) $\min\{N,T\}\cdot g(N,T) \to \infty$ as $N,T \to \infty$. That is, for $\widehat{k} = \arg\min_{0\le k\le \bar{k}} PC(k)$ or $\widehat{k} = \arg\min_{0\le k\le \bar{k}} IC(k)$,

$$\Pr(\widehat{k} = r) \to 1$$

as $N, T \to \infty$, given that the function $g(N, T)$ is chosen appropriately. A set of viable criteria includes

$$PC_{p1}(k) = V(k) + k\widehat{\sigma}^2 \left(\frac{N+T}{NT}\right) \log\left(\frac{NT}{N+T}\right),$$

$$PC_{p2}(k) = V(k) + k\widehat{\sigma}^2 \left(\frac{N+T}{NT}\right) \log\left(\min\{N, T\}\right),$$

$$PC_{p3}(k) = V(k) + k\widehat{\sigma}^2 \left(\frac{\log\left(\min\{N, T\}\right)}{\min\{N, T\}}\right),$$

where $\widehat{\sigma}$ is a consistent estimate of $(NT)^{-1} \sum_{i=1}^{N} \sum_{t=1}^{T} E[u_{it}]^2$, and

$$IC_{p1}(k) = \log(V(k)) + k \left(\frac{N+T}{NT}\right) \log\left(\frac{NT}{N+T}\right), \qquad (8.10)$$

$$IC_{p2}(k) = \log(V(k)) + k \left(\frac{N+T}{NT}\right) \log\left(\min\{N, T\}\right),$$

$$IC_{p3}(k) = \log(V(k)) + k \left(\frac{\log\left(\min\{N, T\}\right)}{\min\{N, T\}}\right).$$

One valid choice of $\widehat{\boldsymbol{\sigma}}^2$ is $V(\bar{k})$.

All the six above methods are valid and have good finite sample performance. In practice, the information criteria are often preferred because they do not depend on the unknown scaling term, $\widehat{\sigma}^2$. The usage of IC_{p1} in (8.10) is suggested by Bai (2003).

Note that we assume the true number of factors r is fixed throughout our discussion. Li, Li, and Shi (2017) propose a procedure to consistently estimate the number of factors when r grows to infinity. This is of particular use in applications where new factors might emerge.

Rate of Convergence and Asymptotic Distribution

Knowing the rate of convergence, or what is better, the asymptotic distribution can help us understand the uncertainty of the factor model estimation, which is of essential importance when we use the estimates from factor models to form forecasting. In this subsection, we discuss the asymptotic properties of the factor model estimation.

Following the literature of factor models as in, e.g., Bai and Ng (2002), Bai (2003), and Bai and Ng (2006), we assume that there exist positive definite matrices, $\boldsymbol{\Sigma}_{\mathbf{F}}$ and $\boldsymbol{\Sigma}_{\boldsymbol{\Lambda}}$ such that

$$\boldsymbol{\Sigma}_{\mathbf{F}} = \operatorname*{plim}_{T\to\infty} \mathbf{F}^{0\prime}\mathbf{F}^{0}/T$$

and

$$\mathbf{\Sigma}_{\mathbf{\Lambda}} = \lim_{N\to\infty} \mathbf{\Lambda}^{0\prime}\mathbf{\Lambda}^{0}/N.$$

As in Bai (2003), let $\mathbf{\Upsilon V \Upsilon}'$ be the eigenvalue decomposition of $\mathbf{\Sigma}_{\mathbf{\Lambda}}^{1/2}\mathbf{\Sigma}_{\mathbf{F}}\mathbf{\Sigma}_{\mathbf{\Lambda}}^{1/2}$ such that $\mathbf{V} = \text{diag}(v_1, v_2, \ldots, v_r)$, and assume that $v_1 > v_2 > \cdots > v_r > 0$. Further define $\mathbf{Q} = \mathbf{V}^{1/2}\mathbf{\Upsilon}'\mathbf{\Sigma}_{\mathbf{\Lambda}}^{-1/2}$ and an $r \times r$ invertible matrix, $\mathbf{H}$, as follows:

$$\mathbf{H} = (\mathbf{\Lambda}^{0\prime}\mathbf{\Lambda}^{0}/N)(\mathbf{F}^{0\prime}\widehat{\mathbf{F}}/T)\mathbf{V}_{NT}^{-1},$$

where $\mathbf{V}_{NT}$ is a diagonal matrix whose diagonal entries are the first r eigenvalues of $\mathbf{XX}'/(NT)$ in decreasing order.

Then, the following results are established in Bai (2003):

1. if $\sqrt{N}/T \to 0$, then for each t,

$$\sqrt{N}(\widehat{\mathbf{f}}_t - \mathbf{H}'\mathbf{f}_t^0) \xrightarrow{d} N(0, \mathbf{V}^{-1}\mathbf{Q}\mathbf{\Gamma}_t\mathbf{Q}'\mathbf{V}^{-1}), \tag{8.11}$$

   where $\mathbf{\Gamma}_t = \lim_{N\to\infty} N^{-1}\sum_{i=1}^{N}\sum_{j=1}^{N} \mathbf{\lambda}_i^0\mathbf{\lambda}_j^{0\prime} E[u_{it}u_{jt}]$;
2. if $\liminf \sqrt{N}/T \geq \tau$ for some $\tau > 0$, then for each t

$$T(\widehat{\mathbf{f}}_t - \mathbf{H}'\mathbf{f}_t^0) = O_p(1); \tag{8.12}$$

3. if $\sqrt{T}/N \to 0$, then for each i,

$$\sqrt{T}(\widehat{\mathbf{\lambda}}_i - \mathbf{H}^{-1}\mathbf{\lambda}_i^0) \xrightarrow{d} N(0, (\mathbf{Q}')^{-1}\mathbf{\Phi}_i\mathbf{Q}^{-1}), \tag{8.13}$$

   where $\mathbf{\Phi}_i = \text{plim}_{T\to\infty} T^{-1}\sum_{s=1}^{T}\sum_{t=1}^{T} E[\mathbf{f}_t^0\mathbf{f}_s^{0\prime}u_{is}u_{it}]$;
4. if $\liminf \sqrt{T}/N \geq \tau$ for some $\tau > 0$, then

$$N(\widehat{\mathbf{\lambda}}_i - \mathbf{H}^{-1}\mathbf{\lambda}_i^0) = O_p(1); \tag{8.14}$$

5. for each (i, t), the common components $\widehat{c}_{it} = \widehat{\mathbf{\lambda}}_i{}'\widehat{\mathbf{f}}_t$ and $c_{it}^0 = \mathbf{\lambda}_i^{0\prime}\mathbf{f}_t^0$ satisfy

$$\frac{\widehat{c}_{it} - c_{it}^0}{\sqrt{v_{it}/N + w_{it}/T}} \xrightarrow{d} N(0, 1), \tag{8.15}$$

   where $v_{it} = \mathbf{\lambda}_i^{0\prime}\mathbf{\Sigma}_{\mathbf{\Lambda}}^{-1}\mathbf{\Gamma}_t\mathbf{\Sigma}_{\mathbf{\Lambda}}^{-1}\mathbf{\lambda}_i^0$ and $w_{it} = \mathbf{f}_t^{0\prime}\mathbf{\Sigma}_{\mathbf{F}}^{-1}\mathbf{\Phi}_i\mathbf{\Sigma}_{\mathbf{F}}^{-1}\mathbf{f}_t^0$.

Note that $\widehat{\mathbf{f}}_t$ can only consistently estimate $\mathbf{H}'\mathbf{f}_t^0$, a transformation of the original factor vector. Thus, $\mathbf{F}^0$ is only identified up to this transformation. However, learning $\mathbf{F}^0\mathbf{H}$ is good enough in many empirical settings. For example, the space spanned by the column vectors of $\mathbf{F}^0\mathbf{H}$ is the same with that of $\mathbf{F}^0$, so they yield the same linear predictor.

The above results shed light on the rate of convergence. Results (8.11) and (8.12) show that the rate of convergence for the common factor estimator, $\widehat{\mathbf{f}}_t$, is $\min\{\sqrt{N}, T\}$, (8.13) and (8.14) show that the rate of convergence for the factor loading estimator, $\widehat{\boldsymbol{\lambda}}_i$, is $\min\{\sqrt{T}, N\}$, and (8.15) shows that the rate of convergence for the common component estimator, $\widehat{c}_{it}$, is $\min\{\sqrt{N}, \sqrt{T}\}$.

Finally, the asymptotic covariance matrix of $\widehat{\mathbf{f}}_t$ can be estimated by

$$\widehat{\boldsymbol{\Pi}}_t = \mathbf{V}_{NT}^{-1}\left(\frac{1}{N}\sum_{i=1}^{N}\widehat{u}_{it}^2\widehat{\boldsymbol{\lambda}}_i\widehat{\boldsymbol{\lambda}}_i'\right)\mathbf{V}_{NT}^{-1},$$

where $\mathbf{V}_{NT}$ is defined as in $\mathbf{H}$.

Factor-Augmented Regression

The estimated common factors can be used to assist forecasting. To formalize this idea, we present the framework in Bai and Ng (2006) to illustrate the "factor-augmented" regression, and we discuss the theoretical results that lead to statistical inference.

Consider a forecasting problem where we observe $\{y_t, \mathbf{w}_t, \mathbf{x}_t\}_{t=1}^T$ with $\mathbf{w}_t = (w_{1t}, \ldots, w_{Kt})'$ and $\mathbf{x}_t = (x_{1t}, \ldots, x_{Nt})'$. The goal is to forecast an outcome variable y_{T+h} with h being the horizon of the forecast. We assume the following linear regression function for y_{T+h}:

$$y_{t+h} = \boldsymbol{\alpha}'\mathbf{f}_t + \boldsymbol{\beta}'\mathbf{w}_t + \varepsilon_{t+h}, \tag{8.16}$$

and $\mathbf{f}_t$ is assumed to appear in the following factor model:

$$x_{it} = \boldsymbol{\lambda}_i'\mathbf{f}_t + u_{it}. \tag{8.17}$$

We cannot estimate Eq. (8.16) directly since we do not observe $\mathbf{f}_t$. Instead, we could estimate the factor model (8.17) first and use the estimated factors to assist the regression estimation. Namely, we could obtain the common factor estimates $\{\widehat{\mathbf{f}}_t\}_{t=1}^T$ by estimating Model (8.17) using the principal component method, and then we could regress y_{t+h} on both $\widehat{\mathbf{f}}_t$ and $\mathbf{w}_t$ to obtain the least squares estimates $\widehat{\alpha}$ and $\widehat{\beta}$. The forecast of y_{t+h} is then given by

$$\widehat{y}_{T+h|T} = \widehat{\boldsymbol{\alpha}}'\widehat{\mathbf{f}}_t + \widehat{\boldsymbol{\beta}}'\mathbf{w}_t.$$

As discussed in the previous discussion, $\widehat{\mathbf{f}}_t$ can only identify an invertible transformation of $\mathbf{f}_t$, but this is sufficient for the purpose of forecasting or recovering β alone.

Let $\mathbf{z}_t = (\mathbf{f}_t', \mathbf{w}_t')'$, $\widehat{\mathbf{z}}_t = (\widehat{\mathbf{f}}_t', \mathbf{w}_t')'$, $\boldsymbol{\delta} = ((\mathbf{H}^{-1}\boldsymbol{\alpha})', \boldsymbol{\beta}')'$, $\widehat{\boldsymbol{\delta}} = (\widehat{\boldsymbol{\alpha}}', \widehat{\boldsymbol{\beta}}')'$, and $\widehat{\varepsilon}_{t+h} = \widehat{y}_{t+h|T} - y_{t+h}$. Then, it is shown in Bai and Ng (2006) that if $\sqrt{T}/N \to 0$, we have that

$$\sqrt{T}(\widehat{\boldsymbol{\delta}} - \boldsymbol{\delta}) \xrightarrow{d} N(0, \boldsymbol{\Sigma}_{\delta}),$$

where $\boldsymbol{\Sigma}_{\delta}$ can be estimated by

$$\widehat{\boldsymbol{\Sigma}}_{\delta} = \left(\frac{1}{T}\sum_{t=1}^{T-h}\widehat{\mathbf{z}}_t\widehat{\mathbf{z}}_t'\right)^{-1}\left(\frac{1}{T}\sum_{t=1}^{T-h}\widehat{\mathbf{z}}_t\widehat{\mathbf{z}}_t'\widehat{\varepsilon}_{t+h}^2\right)\left(\frac{1}{T}\sum_{t=1}^{T-h}\widehat{\mathbf{z}}_t\widehat{\mathbf{z}}_t'\right)^{-1}.$$

Assume that given the information at $t = T$, ε_{T+h} has mean zero. Then, the conditional expectation of y_{T+h} is simply

$$y_{T+h|T} = \boldsymbol{\alpha}'\mathbf{f}_T + \boldsymbol{\beta}'\mathbf{w}_T.$$

Under both $\sqrt{T}/N \to 0$ and $\sqrt{N}/T \to 0$, Bai and Ng (2006) show that the forecast $\widehat{y}_{T+h|T}$ satisfies

$$\frac{\widehat{y}_{T+h|T} - y_{T+h|T}}{\widehat{\sigma}_y} \xrightarrow{d} N(0, 1),$$

where

$$\widehat{\sigma}_y^2 = \frac{1}{T}\widehat{\mathbf{z}}_T'\widehat{\boldsymbol{\Sigma}}_{\delta}\widehat{\mathbf{z}}_T + \frac{1}{N}\widehat{\boldsymbol{\alpha}}'\widehat{\boldsymbol{\Pi}}_T\widehat{\boldsymbol{\alpha}}.$$

The reason for having two terms in $\widehat{\sigma}_y^2$ is that we need to account for the uncertainty from both the common factors estimation and the least squares estimation. A confidence interval for $y_{T+h|T}$ with $1-\alpha$ confidence level is then

$$[\widehat{y}_{T+h|T} - z_{1-\alpha/2}\widehat{\sigma}_y, \widehat{y}_{T+h|T} + z_{1-\alpha/2}\widehat{\sigma}_y], \tag{8.18}$$

where $z_{1-\alpha/2}$ is the $1-\alpha/2$ quantile of the standard normal distribution.

Notice that (8.18) is the confidence interval for the conditional expectation of y_{T+h} at time T. If one is willing to assume normality of the error term ε_t, one can form a confidence interval for the forecasting variable y_{T+h} as well. Following Bai and Ng (2006), let $\varepsilon_t \sim N(0, \sigma_{\varepsilon}^2)$. Then a confidence interval for y_{T+h} with $1-\alpha$ confidence level is

$$[\widehat{y}_{T+h|T} - z_{1-\alpha/2}(\widehat{\sigma}_{\varepsilon}^2 + \widehat{\sigma}_y^2), \widehat{y}_{T+h|T} + z_{1-\alpha/2}(\widehat{\sigma}_{\varepsilon}^2 + \widehat{\sigma}_y^2)],$$

where $\widehat{\sigma}_{\varepsilon}^2 = T^{-1}\sum_{t=1}^{T}\widehat{\varepsilon}_t^2$, and $\widehat{\sigma}_y^2$ and $z_{1-\alpha/2}$ are as defined above.

8.3 Regularization and Machine Learning in Factor Models

Machine learning methods have been widely used in macroeconomic forecasting with large datasets, where there are both a large number of cross-sectional units, and a large number of repeated observations. The benefit of using large-scale datasets also comes with the danger of overfitting. Therefore, some level of regularization is needed to prevent the model from being overly complex. Many machine learning methods are designed such that the degree of model complexity is optimally tuned towards better forecasting.

This is also the case in the context of factor-augmented regression, which uses the estimated factors and potentially their lags as the regressors. Provided that the estimation of common factors is not targeted at forecasting, not all the estimated factors are necessarily useful in forecasting. Nevertheless, machine learning methods can be applied to mitigate the influence of those unimportant regressors.

8.3.1 Machine Learning Methods

In this section, we consider a general form of the factor-augmented forecasting model

$$y_{t+h} = g(\mathbf{z}_t) + \varepsilon_{t+h},$$

where $\mathbf{z}_t$ is a vector of regressors including the estimated common factors $\widehat{\mathbf{f}}_t$, current outcome variable y_t, extra control variables $\mathbf{w}_t$, and any possible lags of them. A linear form of this model is

$$y_{t+h} = \boldsymbol{\alpha}(L)'\widehat{\mathbf{f}}_t + \boldsymbol{\beta}(L)'\mathbf{w}_t + \boldsymbol{\gamma}(L)'\mathbf{y}_t + \varepsilon_{t+h} = \boldsymbol{\pi}'\mathbf{z}_t + \varepsilon_{t+h}, \tag{8.19}$$

where $(\boldsymbol{\alpha}(L), \boldsymbol{\beta}(L), \boldsymbol{\gamma}(L))$ are conformable polynomials of lags. We introduce some machine learning methods that estimate $g(\cdot)$ by $\widehat{g}(\cdot)$, and thus the forecast for y_{T+h} is given by $\widehat{g}(\mathbf{z}_T)$. Note that although we introduce the general prediction formulation which includes both static and dynamic regressors, we focus on the static setting. See Chap. 2 for more detailed discussion on dynamic factor analysis. Also, see Chap. 7 for discussion on the statistical properties of the ridge regression, lasso, and elastic net, and Chap. 14 for boosting.

Ridge Regression Ridge regression is a shrinkage method that imposes the $\ell 2$-regularization. Instead of minimizing the mean squared prediction error as in the least squares estimation, the ridge regression solves the following optimization problem:

$$\min_{\pi}\left(\frac{1}{T}\sum_{t=1}^{T-h}(y_{t+h}-\boldsymbol{\pi}'\mathbf{z}_t)^2\right)+\left(\lambda_R\sum_{j=1}^{d_\pi}\pi_j^2\right),$$

where λ_R is the penalty parameter and usually is chosen via cross-validation, and d_π is the dimensionality of π. Intuitively, the ridge regression simultaneously "shrinks" each element of the least squares estimate towards zero, and the amount of shrinkage is determined by the penalty parameter λ_R. In practice, the ridge regression estimator is easy to calculate and has a closed-form solution.

Lasso As introduced in Tibshirani (1996), the least absolute shrinkage and selection operator (Lasso) is a linear regression method that is similar to the ridge regression, except that instead of penalizing the $\ell 2$-norm of the coefficients, it penalizes the $\ell 1$-norm. In particular, the Lasso estimator solves the optimization problem

$$\min_{\pi}\left(\frac{1}{T}\sum_{t=1}^{T-h}(y_{t+h}-\boldsymbol{\pi}'\mathbf{z}_t)^2\right)+\left(\lambda_L\sum_{j=1}^{d_\pi}|\pi_j|\right).$$

In principle, the Lasso estimator effectively "shrinks" each element of the least squares estimate towards zero by a certain amount, which is determined by the penalty level λ_L. If the absolute values of some entries of the least squares estimate are sufficiently small, then the Lasso estimator sets them to zero. This is a feature of the $\ell 1$-penalty function, and thus the Lasso estimator can perform model selection by assigning zero coefficients to some unimportant regressors. Computationally, the Lasso problem is convex and can be solved efficiently.

Elastic Net The ridge regression and Lasso have their own strengths and weaknesses. The idea of the elastic net as introduced in Zou and Hastie (2005) is to combine the two methods such that the underlying model can be more flexible. The elastic net estimator solves the optimization problem

$$\min_{\pi}\left(\frac{1}{T}\sum_{t=1}^{T-h}(y_{t+h}-\boldsymbol{\pi}'\mathbf{z}_t)^2\right)+\left(\lambda_R\sum_{j=1}^{d_\pi}\pi_j^2\right)+\left(\lambda_L\sum_{j=1}^{d_\pi}|\pi_j|\right).$$

One way to determine the penalty parameters is to let $\lambda_R=\lambda_E(1-\alpha)/2$ and $\lambda_L=\lambda_E\alpha$, where α is chosen by the researcher and λ_E is chosen via cross-validation. By doing this, the ridge regression and Lasso become special cases of the elastic net.

Boosting The boosting algorithm uses the idea of combining results of different methods in order to improve forecasting. The goal here is to estimate a model $\widehat{G}(\cdot)$ between the outcome variable y_{t+h} and observables $\mathbf{z}_t$. We present the boosting algorithm below following Bai and Ng (2009). First, we initialize the estimator by letting $\widehat{G}^{(0)}(\cdot) = \bar{y}$. Then, we update $\widehat{G}^{(m)}(\cdot)$ iteratively for $m = 1, \ldots, M$, using the following steps:

Step 1 Form the stepwise regression residual $\widehat{u}_t^{(m-1)}$ by

$$\widehat{u}_t^{(m-1)} = y_{t+h} - \widehat{G}^{(m-1)}(\mathbf{z}_t).$$

Step 2 Fit a model $\widehat{g}^{(m)}(\cdot)$ between $\widehat{u}_t^{(m-1)}$ and $\mathbf{z}_t$. For example, run Lasso of $\widehat{u}_t^{(m-1)}$ on $\mathbf{z}_t$, where the resulting predictor $\widehat{g}^{(m)}(\mathbf{z}_t)$ is a linear predictor for $\widehat{u}_t^{(m-1)}$.

Step 3 Update the forecasting model by letting $\widehat{G}^{(m)}(\cdot) = \widehat{G}^{(m-1)}(\cdot) + \nu\widehat{g}^{(m)}(\cdot)$, where $0 < \nu < 1$ is the step length.

The forecasting for y_{T+h} is then $\widehat{G}^M(\mathbf{z}_t)$. Note that in Step 2, we can use different methods for different iterations. The idea is that some source of variation that is not learned by one method might be captured by another method, and combining them may result in better forecasting. It is shown in Bai and Ng (2009) that boosting can sometimes outperform the standard factor-augmented forecasts and is far superior to the autoregressive forecast. See Chap. 14 of this book for more details.

Other machine learning methods have also been used in macroeconomic forecasting. For example, the least angle regression (LARS) is used in Bai and Ng (2008). An extensive review of the machine learning algorithms can be found in Hastie et al. (2009), and a horse race among various forecasting methods using machine learning techniques has been conducted by Kim and Swanson (2018), which we will present in details in Sect. 8.5.

8.3.2 Model Selection Targeted at Prediction

One issue about using estimated factors in forecasting problems is that not all the factors are necessarily useful for forecasting. One plausible reason is that the estimation of the common factors is not targeted at predicting the outcome variable but at explaining the auxiliary variables. In the standard factor-augmented regression as described in Eqs. (8.16) and (8.17), we choose the factors that can capture x_{it} well, rather than predicting y_{t+h} well. In this subsection, we present two sets of methods that consider prediction in forming factors.

Targeted Predictor

One way to overcome this issue is to perform model selection before factor model estimation. In the following, we present the "targeted predictor" method as introduced in Bai and Ng (2008). In particular, let $\mathbf{x}_t = (x_{1t}, \ldots, x_{Nt})'$ and consider the linear model below

$$y_{t+h} = \tilde{\boldsymbol{\alpha}}'\mathbf{x}_t + \boldsymbol{\beta}'\mathbf{w}_t + \tilde{\varepsilon}_{t+h}.$$

One can perform variable selection with respect to $\mathbf{x}_t$ using this regression function. For example, one can impose the L_1 penalty in the spirit of Lasso and solve the following problem:

$$\min_{\tilde{\boldsymbol{\alpha}},\boldsymbol{\beta}} \left(\frac{1}{T} \sum_{t=1}^{T-h} (y_{t+h} - \tilde{\boldsymbol{\alpha}}'\mathbf{x}_t - \boldsymbol{\beta}'\mathbf{w}_t)^2 \right) + \left(\lambda_L \sum_{i=1}^{N} |\tilde{\alpha}_i| \right).$$

Let $\tilde{\mathbf{x}}_t$ be the collection of x_{it} that are assigned nonzero coefficients through this procedure. Then, the factor model estimation can be performed based on Model (8.17) using $\tilde{\mathbf{x}}_t$ instead of $\mathbf{x}_t$, which is followed by the factor-augmented regression estimation using Model (8.16). Bai and Ng (2008) show that the "targeted predictor" method often generates better forecasting outcomes than the standard factor-augmented regression, which we will discuss further in Sect. 8.5.2.

Partial Least Squares and Sparse Partial Least Squares

Another approach that considers prediction at the stage of forming factors is the partial linear squares regression (PLS) method. The idea behind the PLS estimator is very similar to the idea behind the factor-augmented regression (the PCA-type regression). Both methods assume that the target variable y_{t+h} and the predictors $\mathbf{x}_t$ are driven by some unobserved common factors. The difference is that the PLS approach extracts factors using the information of both the target variable and the predictors, while the PCA method constructs factors based on the information of the predictors only.

Let the matrix of the predictor variables be $\mathbf{X} \in \mathbb{R}^{T\times N}$ and the vector of the target variable be $\mathbf{y} \in \mathbb{R}^{T\times 1}$. Let R be the desired number of factors. Following Chun and Keleş (2010), the PLS-type regression can be conducted in the following steps:

Step 1 For $r = 1, \ldots, R$: given $\{\mathbf{w}_1, \ldots, \mathbf{w}_{r-1}\}$, compute $\mathbf{w}_r \in \mathbb{R}^{N \times 1}$ using the optimization problem

$$\begin{aligned} \mathbf{w}_r = \arg\max_{\mathbf{w} \in \mathbb{R}^N} \quad & \mathbf{w}'\mathbf{X}'\mathbf{y}\mathbf{y}'\mathbf{X}\mathbf{w} \\ \text{s.t.} \quad & \mathbf{w}'\mathbf{w} = 1 \\ & \mathbf{w}'\mathbf{X}'\mathbf{X}\mathbf{w}_j = 0, \forall j = 1, \ldots, r-1. \end{aligned}$$

Let $\mathbf{W} = [\mathbf{w}_1, \ldots, \mathbf{w}_R]$.

Step 2 Compute $\widehat{\mathbf{q}}$ using the OLS estimation:

$$\widehat{\mathbf{q}} = \arg\min_{\mathbf{q} \in \mathbb{R}^{R \times 1}} \|\mathbf{y} - \mathbf{X}\mathbf{W}\mathbf{q}\|_2.$$

The resulting PLS estimate is $\widehat{\boldsymbol{\beta}}^{PLS} = \mathbf{W}\widehat{\mathbf{q}}$.

The PCA and PLS approaches use different objective functions to extract the common components. The former uses $\mathbf{w}'\mathbf{X}'\mathbf{X}\mathbf{w}$ to capture the variations in the predictors $\mathbf{X}$, whereas the latter uses $\mathbf{w}'\mathbf{X}'\mathbf{y}\mathbf{y}'\mathbf{X}\mathbf{w}$ to account for the correlation between the predictors $\mathbf{X}$ and the target variable $\mathbf{y}$.

By a similar motivation of using SPCA instead of PCA, Chun and Keleş (2010) also propose a penalized version of PLS called sparse partial least squares regression (SPLS). For some fixed parameter κ, the SPLS estimator can be implemented using the following procedure:

Step 1 Initialize $\widehat{\boldsymbol{\beta}}^{SPLS} = \mathbf{0}$, the active set $\mathcal{A} = \{\}$, $r = 1$, and residuals $\widehat{\mathbf{u}} = \mathbf{y}$.

Step 2 While $r \leq R$,

(i) compute $\widehat{\mathbf{w}}$ using the optimization problem

$$\min_{\mathbf{w} \in \mathbb{R}^N, \mathbf{c} \in \mathbb{R}^N} \quad -\kappa \mathbf{w}'\mathbf{X}'\widehat{\mathbf{u}}\widehat{\mathbf{u}}'\mathbf{X}\mathbf{w} + (1-\kappa)(\mathbf{c}-\mathbf{w})'\mathbf{X}'\widehat{\mathbf{u}}\widehat{\mathbf{u}}'\mathbf{X}(\mathbf{c}-\mathbf{w}) + \lambda_1\|\mathbf{c}\|_1 + \lambda_2\|\mathbf{c}\|_2^2;$$

(ii) update $\mathcal{A}$ as $\{i : \widehat{\mathbf{w}}_i \neq 0\} \cup \{i : \widehat{\boldsymbol{\beta}}_i^{SPLS} \neq 0\}$;

(iii) run Step 2 in the PLS procedure with $\mathbf{y}$, $\mathbf{X}_{\mathcal{A}}$, and $\mathbf{W} = [\mathbf{w}_1, \ldots, \mathbf{w}_r]$ and obtain the new estimate $\widehat{\boldsymbol{\beta}}^{SPLS}$;

(iv) update r with $r + 1$ and $\widehat{\mathbf{u}}$ with $\mathbf{y} - \mathbf{X}\widehat{\boldsymbol{\beta}}^{SPLS}$.

In particular, the performance of the sparse PLS estimator in macroeconomic forecasting has been investigated by Fuentes, Poncela, and Rodríguez (2015) and will be presented in details in Sect. 8.5.3.

8.4 Policy Evaluation with Factor Model

Policy evaluation can often be transformed into a problem of forecasting the counter-factual outcome. In this section, we discuss the prominent policy evaluation problems where the data generating process of the counter-factual outcome follows a factor structure. Empirical applications of the methods introduced in this section will be presented in Sect. 8.6.

8.4.1 Rubin's Model and ATT

Suppose we want to evaluate the effect of some policy d on some outcome y. We can observe a panel with N individuals and T time periods. Individuals with indexes from 1 to N_0 are the control group and those with indexes from $N_0 + 1$ to N are the treatment group. Let $N_1 = N - N_0$. The policy starts to take effect at some point between time T_0 and $T_0 + 1$, where $T_0 < T$. That is, any period after T_0 is considered a post-treatment period. Let $T_1 = T - T_0$.

Let d_{it} denote the treatment status, where $d_{it} = 1$ only if $N_0 + 1 \leq i \leq N$ and $T_0 + 1 \leq t \leq T$, and $d_{it} = 0$ otherwise. Rubin's model defines

$$y_{it} = \begin{cases} y_{it}(1), & \text{if } d_{it} = 1 \\ y_{it}(0), & \text{otherwise,} \end{cases}$$

and the treatment effect on this individual i at time t is defined by

$$\alpha_{it} = y_{it}(1) - y_{it}(0).$$

The parameter of interest is the average treatment effect on the treated units (ATT):

$$\text{ATT} = \frac{1}{N_1 T_1} \sum_{i=N_0+1}^{N} \sum_{t=T_0+1}^{T} \alpha_{it} = \frac{1}{N_1 T_1} \sum_{i=N_0+1}^{N} \sum_{t=T_0+1}^{T} (y_{it}(1) - y_{it}(0)).$$

Note that we do not observe $y_{it}(0)$ for the treated units in the post-treatment time periods. Therefore, the problem becomes a forecasting problem where we use the observed data of the untreated units to predict the counter-factual outcome of the treated units in the post-treatment time periods.

8.4.2 Interactive Fixed-Effects Model

The interactive fixed-effects model is formally studied by Bai (2009), which can be applied in policy evaluation. Consider the following model:

$$y_{it} = \alpha d_{it} + \boldsymbol{\beta}'\mathbf{x}_{it} + \boldsymbol{\lambda}_i'\mathbf{f}_t + u_{it}, \tag{8.20}$$

where, for simplicity, we assume homogeneous treatment effect which is captured by α. The researchers only observe $(y_{it}, d_{it}, \mathbf{x}_{it})$ for $i = 1, \ldots, N$ and $t = 1, \ldots, T$. Throughout this section, assume both N and T are large. An estimator of Model (8.20) is called the interactive fixed-effects estimator (IFE), which treats the unobserved variables $\boldsymbol{\lambda}_i$ and $\mathbf{f}_t$ as unknown parameters to recover. Model (8.20) is a generalization of the additive fixed-effects model, for which we simply set $\boldsymbol{\lambda}_i = (1, \tilde{\lambda}_i)'$ and $\mathbf{f}_t = (\tilde{f}_t, 1)'$. It is also a generalization of the factor model by allowing for the possible causal effects of the observed characteristics $\mathbf{x}_{it}$. The underlying counter-factual process is

$$y_{it}(0) = \boldsymbol{\beta}'\mathbf{x}_{it} + \boldsymbol{\lambda}_i'\mathbf{f}_t + u_{it}.$$

To estimate Model (8.20), Bai (2009) performs principal component analysis and least squares estimation in iterations. First, initialize the estimate for $(\alpha, \boldsymbol{\beta})$ and denote it by $(\widehat{\alpha}^{(0)}, \widehat{\boldsymbol{\beta}}^{(0)})$. For example, one can use the least squares estimate

$$(\widehat{\alpha}^{(0)}, \widehat{\boldsymbol{\beta}}^{(0)}) = \arg\min_{\alpha, \boldsymbol{\beta}} \sum_{i=1}^{N} \sum_{t=1}^{T} (y_{it} - \alpha d_{it} - \boldsymbol{\beta}'\mathbf{x}_{it})^2.$$

Then, construct $(\widehat{\alpha}^{(m)}, \widehat{\boldsymbol{\beta}}^{(m)}, \widehat{\boldsymbol{\lambda}}_i^{(m)}, \widehat{\mathbf{f}}_t^{(m)})$ for $m = 1, \ldots, M$ iteratively using Steps 1 and 2 below until numerical convergence:

Step 1 Given $(\widehat{\alpha}^{(m-1)}, \widehat{\boldsymbol{\beta}}^{(m-1)})$, perform principal component analysis as introduced in Sect. 8.2.1 with respect to the model

$$y_{it} - \alpha d_{it} - \boldsymbol{\beta}'\mathbf{x}_{it} = \boldsymbol{\lambda}_i'\mathbf{f}_t + u_{it}.$$

That is, compute $(\widehat{\boldsymbol{\lambda}}_i^{(m)}, \widehat{\mathbf{f}}_t^{(m)})$ by treating $y_{it} - \widehat{\alpha}^{(m-1)} d_{it} - (\widehat{\boldsymbol{\beta}}^{(m-1)})'\mathbf{x}_{it}$ as the observable variable in the factor model estimation.

Step 2 Given $(\widehat{\boldsymbol{\lambda}}_i^{(m)}, \widehat{\mathbf{f}}_t^{(m)})$, perform least squares estimation to the model

$$y_{it} - \boldsymbol{\lambda}_i'\mathbf{f}_t = \alpha d_{it} + \boldsymbol{\beta}'\mathbf{x}_{it} + u_{it}$$

by computing

$$(\widehat{\alpha}^{(m)}, \widehat{\boldsymbol{\beta}}^{(m)}) = \arg\min_{\alpha,\boldsymbol{\beta}} \sum_{i=1}^{N}\sum_{t=1}^{T}(y_{it} - (\widehat{\boldsymbol{\lambda}}_i^{(m)})'\widehat{\mathbf{f}}_t^{(m)} - \alpha d_{it} - \boldsymbol{\beta}'\mathbf{x}_{it})^2.$$

Then, the estimate of ATT is given by

$$\widehat{\mathrm{ATT}}_{IFE} = \widehat{\alpha}^{(M)}.$$

It is shown in Bai (2009) that the IFE estimator is consistent and asymptotically normal under a set of moderate conditions.

In Sect. 8.6, we will discuss in details the empirical applications of the IFE estimator in policy evaluation, which include the studies by Gobillon and Magnac (2016) and Kim and Oka (2014).

8.4.3 Synthetic Control Method

The synthetic control method is commonly used to perform program evaluation in event studies. In this section, we present the synthetic control method in estimating the average treatment effect on the treated. References of discussing the synthetic control method with multiple treated units include Firpo and Possebom (2018), Robbins, Saunders, and Kilmer (2017), and Xu (2017).

Consider the following model:

$$y_{it} = \delta_t + \alpha d_{it} + \boldsymbol{\lambda}_i'\mathbf{f}_t + u_{it},$$

where we only observe y_{it} and d_{it}. One motivation of applying the synthetic control method (SCM) to estimate this model compared with the IFE method is that this approach allows for non-stationary time fixed effects, δ_t, while the IFE approach does not. On the other hand, the IFE estimator can account for the effects of observable characteristics, $\mathbf{x}_{it}$. The underlying counter-factual process is

$$y_{it}(0) = \delta_t + \boldsymbol{\lambda}_i'\mathbf{f}_t + u_{it}.$$

The idea of the synthetic control method is to use a convex combination of the untreated units (plus a constant) to estimate the counter-factual outcome of the treated units as if they were not treated. The weights are assigned such that the pre-treatment data fit well. Namely, for each $i = N_0 + 1, \ldots, N$, we compute

$$\widehat{\mathbf{w}}_i = \arg\min_{\mathbf{w}\in W} \sum_{t=1}^{T_0}\left(y_{it} - w_0 - \sum_{j=1}^{N_0} w_j y_{jt}\right)^2,$$

where $W = \{\mathbf{w} = (w_0, w_1, \ldots, w_{N_0}) \in \mathbb{R}^{N_0+1} | \sum_{j=1}^{N_0} w_j = 1 \text{ and } w_j \geq 0, \forall j = 1, \ldots, N_0\}$. That is, W is the set of weights that are positive and sum up to one, with no restrictions on the intercept w_0. For each $i = N_0 + 1, \ldots, N$ and $t = T_0 + 1, \ldots, T$, the estimate of the counter-factual outcome is

$$\widehat{y}_{it}(0) = \widehat{w}_0 + \sum_{j=1}^{N_0} \widehat{w}_j y_{jt},$$

and the corresponding treatment effect estimate is

$$\widehat{\alpha}_{it} = y_{it} - \widehat{y}_{it}(0).$$

Finally, the estimate of ATT is given by

$$\widehat{\text{ATT}}_{SC} = \frac{1}{N_1 T_1} \sum_{i=N_0+1}^{N} \sum_{t=T_0+1}^{T} \widehat{\alpha}_{it}.$$

Empirical applications of applying the synthetic control method in policy evaluation can be found in, e.g., Gobillon and Magnac (2016) and Hsiao, Ching, and Wan (2012), which we will present in details in Sect. 8.6.

8.5 Empirical Application: Forecasting in Macroeconomics

In this section, we discuss the empirical applications of factor models in forecasting macroeconomic variables. Regarding the application of forecasting macroeconomic variables, we first summarize the performance of the diffusion index approach. Then, we present the improvements of integrating the diffusion index method with machine learning techniques.

8.5.1 Forecasting with Diffusion Index Method

In a series of early pioneering studies by Stock and Watson, e.g., Stock and Watson (1999, 2002a,b), the researchers have demonstrated that applying the factor analysis can substantially improve the forecasting capabilities of the models on a variety of key macroeconomic aggregates. In this section, following the classical paper of Stock and Watson (2002b), we present in details the empirical performance of the diffusion index method, which is a method based on principal component analysis and has been introduced in Sect. 8.2.1.

Forecasting Procedures

The literature of macroeconomic forecasting focuses on examining model predictability for important macroeconomic variables that reflect real economic activities and financial conditions. For instance, the goal of Stock and Watson (2002b) is to predict 8 monthly US macroeconomic time series. Among the eight time series, four variables are related to real economic activities, which are the measures of total industrial production (ip), real personal income less transfers (gmyxpq), real manufacturing and trade sales (msmtq), and the number of employees on nonagricultural payrolls (lpnag). The other four variables are the indexes of prices, which include the consumer price index (punew), the personal consumption expenditure implicit price deflator (gmdc), the consumer price index less food and energy (puxx), and the producer price index for finished goods (pwfsa). As discussed in Kim and Swanson (2018), the ability to precisely forecast these measures is of important economic relevance, because the Federal Reserve relies on these measures to formulate the national monetary policy.

The study in Stock and Watson (2002b) contains 215 predictors which represent 14 main categories of macroeconomic time series. Using these predictors, which are monthly time series for the US from 1959:1 (i.e., January 1959) to 1998:12 (i.e., December 1998), the empirical goal of the study is to construct 6-, 12-, and 24-month-ahead forecasts for the eight macroeconomic aggregates from 1970:1 to 1998:12. The forecasting procedure applies the following "recursive" scheme. First, the researchers use the data from 1959:1 to 1970:1 to form an out-of-sample forecast for 1970:1+h, where h is 6, 12, and 24, respectively, depending on the forecasting horizons. Then, the data from 1959:1 to 1970:2 are used to forecast for 1970:2+h and so on. Finally, the last forecast is based on the data from 1959:1 to 1998:12-h for the period of 1998:12.[1]

Benchmark Models

In the study of Stock and Watson (2002b), four benchmark models are applied as comparisons with the diffusion index method. First, the univariate autoregression (AR) model might be one of the simplest time-series tools for forecasting, which is shown in the following:

$$\widehat{y}_{T+h|T} = \widehat{\alpha}_{h0} + \sum_{j=1}^{p} \widehat{\gamma}_{hj} y_{T-j+1},$$

[1]Another popular forecasting strategy is the so-called "rolling" scheme, which, in each step, drops the earliest observation in the current forecast window while adding a new one. The relative performance between the recursive and rolling schemes can be found in, for example, Kim and Swanson (2018).

where we forecast the future observations for a given variable using its historical realizations. The number of lags, p, is set using BIC with $0 \leq p \leq 6$. The second benchmark model in the study is the vector autoregression (VAR) model, where the following three variables are used in the multivariate regressions: a measure of the real-activity-monthly growth, a measure of the variation in monthly inflation, and the change in the 90-day US treasury bill rate. The number of lags is fixed at 4, which provides better forecasting performance than the one selected through BIC. Third, this study considers the model using leading indicators (LI). In particular, 11 leading indicators are used for forecasting the real-activity measures, and 8 leading indicators are applied for predicting the price indexes. These leading indicators performed well individually in former forecasting studies. Finally, the Phillips curve (PC) is considered in this study, due to its reliability for forecasting inflation. The predictors of the PC method include the unemployment rate, the relative price of goods and energy, and a measure controlling for the imposition and removal of the Nixon wage and price controls. In both the LI and PC methods, lags of the control variables and lags of y_t are included, and the number of lags is selected with BIC.

Diffusion Index Models

Stock and Watson (2002b) present the forecasting performance of the diffusion index method. In this method, the factors are first extracted from the 215 predictors using the principal component analysis, and then the factors are used for forecasting the eight macroeconomic time series. We present the diffusion index model in the following:

$$\widehat{y}_{T+h|T} = \widehat{\alpha}_h + \sum_{j=1}^{m} \widehat{\boldsymbol{\beta}}_{hj}' \widehat{\mathbf{f}}_{T-j+1} + \sum_{j=1}^{p} \widehat{\gamma}_{hj} y_{T-j+1}, \tag{8.21}$$

where $\widehat{\mathbf{f}}_t$ is the r-dimensional estimated factors. Also, Stock and Watson (2002b) consider three versions of this model. First, the DI-AR-Lag model includes lags of the factors and lags of y_t, where the number of factors, r, the number of lags of $\widehat{\mathbf{f}}_t$, m, and the number of lags of y_t, p, are selected by BIC, with $1 \leq r \leq 4$, $1 \leq m \leq 3$, and $0 \leq p \leq 6$. Then, the DI-AR model includes only the contemporaneous factors and lags of y_t. So, $m = 1$, and r and p are chosen via BIC with $1 \leq r \leq 12$ and $0 \leq p \leq 6$. And finally, the DI model contains the contemporaneous factors only. So $m = 1$ and $p = 0$, and r is selected by BIC, where $1 \leq r \leq 12$.

Forecasting Performance

In Table 8.1, we summarize the performance of the four benchmark models and the three diffusion index models for forecasting the eight macroeconomic time series 12 months ahead. In particular, each cell of the table reports the relative ratio of the

Table 8.1 Relative mean squared forecasting errors: benchmark vs diffusion index methods

	Real variables				Price inflation			
	Industrial production	Personal income	Mfg and trade sales	Nonag. employment	CPI	Consumption deflator	CPI exc. food and Energy	Producer price index
Benchmarks								
AR	1.00	1.00	1.00	1.00	1.00	1.00	1.00	1.00
VAR	0.97 (0.07)	0.98 (0.05)	0.98 (0.04)	1.05 (0.09)	0.91 (0.09)	1.02 (0.06)	0.99 (0.05)	1.29 (0.14)
LI	0.86 (0.27)	0.97 (0.21)	0.82 (0.25)	**0.89 (0.23)**	0.79 (0.15)	0.95 (0.12)	1.00 (0.16)	**0.82 (0.15)**
PC					0.82 (0.13)	0.92 (0.10)	**0.79 (0.18)**	0.87 (0.14)
DI models								
DI-AR-Lag	0.57 (0.27)	**0.77 (0.14)**	**0.48 (0.25)**	0.91 (0.13)	0.72 (0.14)	**0.90 (0.09)**	0.84 (0.15)	0.83 (0.13)
DI-AR	0.63 (0.25)	0.86 (0.16)	0.57 (0.24)	0.99 (0.31)	**0.71 (0.16)**	**0.90 (0.10)**	0.85 (0.15)	**0.82 (0.14)**
DI	**0.52 (0.26)**	0.86 (0.16)	0.56 (0.23)	0.92 (0.26)	1.30 (0.16)	1.40 (0.16)	1.55 (0.31)	2.40 (0.88)

Note: This table summarizes the results in Tables 1 and 3 of Stock and Watson (2002b)

out-of-sample mean squared errors of each method with respect to the ones of the benchmark AR method. For every targeted macroeconomic variable, we highlight the result generating the smallest forecasting errors in bold. The standard errors are reported in parentheses.

We first discuss the forecasting performance for the four real-activity measures in Table 8.1. As can be seen from the table, in most cases the diffusion index methods outperform the benchmark methods substantially, with the forecasts of employment being the only exception. The enhancements over the benchmark methods are substantial for the measure of industrial production, real personal income less transfers, and real manufacturing and trade sales. For instance, considering the measure of real manufacturing and trade sales, the forecasting MSE of the DI-AR-Lag model is 48% of that of the AR model and 58% of that of the LI model. The forecasting capabilities of the DI method and the DI-AR-Lag method are similar, which suggests that mostly the estimated factors, rather than y_t and the lags of $\widehat{\mathbf{f}}_t$, account for the forecasts. The 6-month- and 24-month-ahead forecasts are also presented in Stock and Watson (2002b), which shows that the longer the forecasting horizon is, the larger the relative improvements of the diffusion index methods are.

Moreover, as can be seen from Table 8.1, in terms of forecasting the inflation measures, the DI-AR-Lag and DI-AR methods outperform the benchmark methods less often and the relative improvements are smaller than the case of forecasting the real-activity measures. This result suggests that there is more room for the inflation forecasts to improve, and accordingly in the following we present the study in Bai and Ng (2008), which introduces refinements to the current diffusion index methods using machine learning techniques and demonstrates the improvements on the inflation forecasts.

8.5.2 Forecasting Augmented with Machine Learning Methods

In this section, we summarize the study in Bai and Ng (2008) to show that machine learning methods can augment the diffusion index methods and enhance forecasting performance. The intuition of the research is that instead of extracting factors from all the available predictors, it could be useful to only extract factors from the predictors that are informative for the outcome variables to predict. As a result, this study applies machine learning methods, including "hard" and "soft" thresholdings, to select the informative predictors. This approach is called the "targeted diffusion index" method.

The dataset used in Bai and Ng (2008) contains 132 monthly time series from 1960:1 to 2003:12, which are the potential predictors. The focus of this study is to predict CPI, which has been considered as a challenging task in the previous studies as is shown in, e.g., Stock and Watson (2002b). This study applies the recursive scheme to make forecasts 1, 6, 12, and 24 months ahead. Provided that the forecasting performance may vary over sample period due to variations in the underlying economic situations, this paper considers the following seven forecast subsamples separately: 70:3-80:12, 80:3-90:12, 90:3-00:12, 70:3-90:12, 70:3-00:12, 80:3-00:12, and 70:3-03:12.

Diffusion Index Models

We describe the forecasting models used in this study. First, this paper considers the original principal component method, denoted as PC, as the benchmark model, which extracts factors from all the 132 predictors. Denote each of the 132 predictors using x_{it}. Second, the squared principal component (SPC) method exacts factors from 264 predictors, which contain both x_{it} and the squared ones, x_{it}^2. Third, the squared factors method (PC^2) extracts factors from the 132 predictors and uses both the estimated factors and the squares of the estimated factors to make forecasts.

Hard Thresholding Models

In addition, this paper introduces three methods that apply hard thresholding to select targeted predictors, and then extract targeted diffusion indexes from the targeted predictors to make forecasts. Using hard thresholding, each potential predictor is evaluated individually, and if a given predictor's marginal predictive power for the inflation variable is above a certain threshold, it is selected as a targeted predictor, and not otherwise. The marginal predictive power is measured as the absolute t-statistic of a predictor in the regression controlling for some other variables, e.g., lags of the inflation measure. And this paper considers three thresholding levels: 1.28, 1.75, and 2.58, which are the critical values at the level of 10%, 5%, and 1%, respectively. The first hard thresholding method, TPC, selects targeted predictors from the 132 predictors using the hard thresholding rule, and then extracts factors from the targeted predictors. The second method, TSPC, selects targeted predictors from the 264 predictors which include x_{it} and x_{it}^2. And the third method, TPC^2, uses the same targeted diffusion indexes as the TSPC method does, and uses also the squares of the targeted diffusion indexes for forecasting.

Soft Thresholding Models

A concern of the hard thresholding method is that this approach only considers each predictor individually without taking into account the influences of other predictors. As an alternative, Bai and Ng (2008) also consider the soft thresholding methods with particular attention paid to the "least angle regression" (LARS) approach. As is introduced in Efron, Hastie, Johnstone and Tibshirani (2004), LARS is a variable selection algorithm that approximates the optimal solutions of the Lasso problem. Specifically, LARS starts from a zero coefficient vector. Then, for the predictor that is most correlated with the dependent variable, say x_j, the algorithm moves its coefficient towards the direction of the OLS coefficient and computes the residuals along the way. Until some other predictor, say x_k, is found to have as much correlation with the residuals as x_j has, the algorithm adds x_k to the active set of predictors. Then, the algorithm moves the coefficients of x_j and x_k jointly towards the least squares direction until another predictor, say x_m, is found to have much correlation with the residuals. This process proceeds in this way until all the predictors are added sequentially to the active set.

Bai and Ng (2008) consider four methods using LARS. The first method is LA(PC), which selects the 30 best predictors from the 132 predictors using LARS, and then extracts factors from the 30 selected predictors. The second method, LA(SPC), instead selects the 30 best predictors using LARS from the 264 predictors, which include both x_{it} and x_{it}^2, and then extracts factors. Moreover, this paper also considers LA(5), which uses the 5 best predictors selected by LARS to make forecasts, and LA(k^*), which uses the k^* best predictors selected by LARS, where k^* is chosen through BIC. Note that both LA(5) and LA(k^*) do not extract factors and thus they are not diffusion index methods.

Empirical Findings

In the following, we present the average number of selected predictors and the forecasting performances of each method. To save space, we only report the results for the 12-month-ahead forecasts with the hard-thresholding cutoff value being 1.65 and the LARS tuning parameter value being 0.5. Details for other forecasting horizons, tuning parameter values, and outcome variables (e.g., personal income, retail sales, industrial production, and total employment) can be found in Bai and Ng (2008), which in general are coherent with the main findings we present here.

In Table 8.2, we report the average number of predictors used across different forecasting methods. By design, the number of predictors is fixed ex-ante for PC, SPC, PC^2, LA(PC), LA(5), and LA(SPC). We see that the number of targeted predictors selected by hard thresholding is much larger than the number of predictors selected via BIC under LARS.

Table 8.2 Average number of selected predictors across different forecasting methods

	PC	SPC	TPC	TSPC	PC^2	TPC^2	LA(PC)	LA(5)	LA(k*)	LA(SPC)
70.3–80.12	132	264	64.2	122.8	132	64.2	30	5	15.0	30
80.3–90.12	132	264	87.3	169.7	132	87.3	30	5	10.7	30
90.3–00.12	132	264	89.0	171.4	132	89.0	30	5	11.8	30
70.3–90.12	132	264	75.7	146.2	132	75.7	30	5	12.9	30
70.3–00.12	132	264	80.1	154.5	132	80.1	30	5	12.5	30
80.3–00.12	132	264	88.1	170.6	132	88.1	30	5	11.2	30
70.3–03.12	132	264	80.6	155.6	132	80.6	30	5	12.8	30

Note: This table summarizes the results in Table 3 of Bai and Ng (2008)

Table 8.3 Relative mean squared forecasting errors across different methods

	PC	SPC	TPC	TSPC	PC^2	TPC^2	LA(PC)	LA(5)	LA(k*)	LA(SPC)
70.3–80.12	0.631	0.595	0.659	0.654	0.644	0.612	0.599	0.623	0.691	**0.562**
80.3–90.12	0.575	0.582	0.689	0.573	0.633	0.661	0.569	0.566	0.702	**0.477**
90.3–00.12	0.723	0.699	0.616	0.698	0.703	**0.613**	0.681	0.665	1.088	0.675
70.3–90.12	0.603	0.589	0.675	0.613	0.639	0.638	0.584	0.594	0.698	**0.519**
70.3–00.12	0.611	0.597	0.665	0.618	0.642	0.631	0.590	0.597	0.733	**0.531**
80.3–00.12	0.594	0.597	0.669	0.589	0.639	0.644	0.583	0.576	0.764	**0.506**
70.3–03.12	0.609	0.597	0.665	0.615	0.639	0.632	0.587	0.597	0.751	**0.531**

Note: This table summarizes the results in Table 6 of Bai and Ng (2008). Bold fonts indicate the method which generates the smallest relative mean squared errors for each forecast subsample.

In addition, in Table 8.3, we present the ratio of the forecasting mean squared errors of each method with the ones using the univariate autoregression model with 4 lags, i.e., AR(4). For each forecast subsample, we highlight the method that generates the smallest relative mean squared errors. As can be seen from this table, most of the entries are below 1, which confirms the findings in the previous literature that the diffusion index methods in general outperform the benchmark AR models. More importantly, the results suggest that the targeted diffusion index methods have substantially enhanced the original diffusion index methods. The soft thresholding method, LA(SPC), is the best approach in most cases, with only one exception in which the hard thresholding method, TPC^2, wins. Overall, the soft thresholding procedures outperform the hard thresholding ones. Also, it is noticeable that adding the squares of the predictors from which to select the targeted predictors further reduces the forecast errors, which suggests that introducing non-linearity between the predictors and principal components is beneficial.

In the current literature, more and more research has been conducted to combine the diffusion index methods with the state-of-the-art machine learning techniques. For instance, Bai and Ng (2009) introduce some boosting methods to select the predictors in the factor-augmented autoregressive estimations, and Kim and Swanson (2018) combine the diffusion index methods with the machine learning techniques, which include bagging, boosting, ridge regression, least angle regression, elastic net, and non-negative garotte. Finally, it is also noteworthy that machine learning techniques may not always improve the diffusion index methods. As demonstrated in Stock and Watson (2012), the forecasts based on some shrinkage representations using, e.g., pretest methods, Bayesian model averaging, empirical Bayes, and bagging, tend to fall behind the traditional dynamic factor model forecasts.

8.5.3 Forecasting with PLS and Sparse PLS

One approach that is closely related to the targeted predictor model is the partial least squares (PLS) method. As introduced in Fuentes et al. (2015), the PLS approach also takes into account the target variable (i.e., outcome variable to

Table 8.4 Relative mean squared forecasting errors using PLS and SPLS

	Bai and Ng (2008)	PLS			SPLS		
Period	LA(SPC)	a (k=2)	b (k=2)	c (k=2)	a (k=2)	b (k=2)	c (k=2)
70.3–80.12	0.562	0.462	0.890	0.472	0.472	0.765	**0.458**
80.3–90.12	0.477	0.486	0.995	0.476	**0.440**	0.784	0.441
90.3–00.12	0.675	0.532	1.200	0.621	**0.521**	0.820	0.623
70.3–90.12	0.519	0.465	0.947	0.465	0.469	0.780	**0.465**
70.3–00.12	0.531	**0.466**	0.957	0.473	0.470	0.823	0.473
80.3–00.12	0.506	0.484	1.005	0.487	**0.461**	0.858	0.473
70.3–03.12	0.531	**0.469**	0.977	0.479	0.472	0.845	0.478

Note: This table summarizes the results in Table 6 of Bai and Ng (2008) and Table 2 of Fuentes et al. (2015). Bold fonts indicate the method which generates the smallest relative mean squared errors for each forecast subsample.

forecast) when it forms the principal components. Moreover, the corresponding shrinkage version of the model, the sparse partial least squares (SPLS) method, imposes the $\ell 1$-penalty on the direction vectors so that it allows the latent factors to depend on a small subset of the predictors. In Table 8.4, we report the forecasting performances based on PLS and SPLS, and compare them with the ones based on targeted predictors in Bai and Ng (2008).

Following the study by Fuentes et al. (2015), Table 8.4 presents the relative mean squared forecasting errors of the PLS and SPLS methods, using AR(4) as the benchmark model. In the same table, the relative performances of the targeted predictor method, LA(SPC), as discussed in Bai and Ng (2008), are presented as well to serve as a comparison. To save space, we only show the results for the 12-month forecasting horizon, while similar conclusions can be drawn for other forecasting horizons. Both PLS and SPLS consider the following three setups: the baseline version (i.e., setup (a)) extracts orthogonal latent components based on the covariance between the predictors, $\mathbf{X} \in \mathbb{R}^{N \times T}$, and the target variable $\mathbf{Y} \in \mathbb{R}^{T}$: $\mathbf{X}'\mathbf{Y}$; setup (b) relies on the covariance between an enlarged set of the predictors, including both $\mathbf{X}$ and the lags of the target variable, and the target variable $\mathbf{Y}$; and setup (c) uses the covariance between the predictors, $\mathbf{X}$, and the residuals from regressing the target variable, $\mathbf{Y}$, on its own lags using AR(p).

Table 8.4 suggests that the most preferred versions of PLS or SPLS, which are indicated in bold in the table, provide less forecasting errors in general than the LA(SPC) method does. Overall, setups (a) and (c) forecast better than setup (b), and the best SPLS setup performs slightly better than the best PLS setup for most of the forecasting subsamples.

8.5.4 Forecasting with ICA and Sparse PCA

The forecasting performances using ICA and sparse PCA (SPCA) are studied in Kim and Swanson (2018). The authors provide some evidence that, comparing

Table 8.5 Relative forecasting mean squared errors using PCA, ICA, and SPCA

	UR	PI	TB10Y	CPI	PPI	NPE	HS	IPX	M2	SNP	GDP
PCA	**0.780**	0.870	0.940	0.875	0.943	0.811	0.900	0.800	**0.939**	0.976	**0.916**
ICA	0.897	0.920	0.931	**0.840**	**0.843**	0.802	0.901	0.574	0.965	0.920	**0.916**
SPCA	0.827	**0.789**	**0.409**	0.870	0.858	**0.706**	**0.542**	**0.268**	0.969	**0.897**	**0.916**

Note: This table summarizes the results in Table 3 of Kim and Swanson (2018) for specification type 1 (SP1) and 1-month ahead forecasting horizon (h=1). Bold fonts highlight the method which generates the smallest relative mean squared errors for each outcome measure.

with the standard PCA method, ICA or SPCA tends to generate less forecasting errors for short-term predictions, e.g., 1-month ahead forecasts, whereas PCA has better performances than ICA and SPCA for longer-period predictions, e.g., 6 month and 12 month forecasts. Theoretical explanations for the different forecasting performances over the various methods and forecasting horizons await future research. In Table 8.5, we present the 1-month ahead forecasting results using PCA, ICA, and SPCA based on the study of Kim and Swanson (2018), which shows that among 9 out of 11 target variables, ICA or SPCA outperforms PCA.[2]

8.6 Empirical Application: Policy Evaluation with Interactive Effects

In this section, we consider the setting of policy evaluation. In practice, it is important for policy makers to understand the consequences of implementing a policy in order to make informed decisions. It has been emphasized in some recent literature, e.g., Gobillon and Magnac (2016), Hsiao et al. (2012), and Kim and Oka (2014), that controlling for the latent factor structure can facilitate the empirical study of policy evaluation. In the following, we consider the same data generating process that has been introduced in Sect. 8.4, which assumes that the values of the outcome variable with and without treatment are the following:

$$y_{it}(0) = \boldsymbol{\beta}'\mathbf{x}_{it} + \boldsymbol{\lambda}_i'\mathbf{f}_t + \varepsilon_{it} \tag{8.22}$$

$$y_{it}(1) = y_{it}(0) + \alpha_{it}. \tag{8.23}$$

[2]Readers need to be cautious to understand and interpret the results of comparing different forecasting methods. For instance, each entry in Table 8.5 corresponds to the best performance of a given method, say PCA, ICA, or SPCA, across a variety of machine learning models, which are used to forecast the target variables using the extracted factors. This implies that the reported forecasting errors in the table already take into account the data to forecast, due to the selection over the machine learning models. However, the relative forecasting performances across PCA, ICA, and SPCA may be different when we "truly" forecast out of sample.

In this framework, the factor structure allows for a number of unobserved economic shocks, which can be time varying and are represented by the factor vector, $\mathbf{f}_t$. Also, different cross-sectional observations can react to the latent shocks differently, which is captured by the i-specific factor loading vector, $\boldsymbol{\lambda}_i$, for every observation i. More importantly, this framework is flexible to allow the treatment variable to be freely correlated with the latent factors and factor loadings. For example, in Gobillon and Magnac (2016), the researchers are interested in studying the effect of implementing the enterprise zone program in France in 1997 on the regional employment levels. In this context, the factor structure can incorporate not only the additive region specific effects and time effects, but also unobserved economic shocks and region specific reactions to these shocks. The unobserved economic shocks may include, but are not limited to, time varying business cycles, technological shocks, and sector specific economic shocks. The enterprise zone program provides financial incentives for the municipalities to hire the local labor force, and thus the decision of implementing the policy may also depend on the unobserved local economic conditions.

8.6.1 Findings Based on Monte Carlo Experiments

In the following, we present the Monte Carlo simulation experiments in Gobillon and Magnac (2016) to illustrate that when the multidimensional factor structure exists, the popular difference-in-differences method is generically biased and the synthetic control method is also possible to fail, whereas the interactive fixed-effects estimator tends to be well-performed. In particular, the simulation exercise uses (8.22) and (8.23) as the data generating process, where the treatment effect coefficient is set to be $\alpha_{it} = 0.3$ and no explanatory variable is included. We report the simulation results using three factors. The elements in the first factor are restricted to be ones, and thus the model contains the additive individual effects. The entry in the second factor is set to be $a\sin(\pi t/T)$ with $a > 0$, and thus it is a function of time. Also, each element in the third factor is an i.i.d draw from the uniform distribution on [0,1].

Moreover, the simulation exercise contains three possibilities for generating the factor loadings. In the baseline model, the factor loadings are random draws from the uniform distribution on the support of [0,1]. Then, this study considers overlapping supports for the treated and untreated units, where the factor loadings of the treated units are shifted by 0.5 from the baseline counterparts. Finally, this paper considers the case with disjoint supports, in which the factor loadings of the untreated units are draws from the uniform distribution on [0,1], while those of the treated units from the uniform distribution on [1,2]. Therefore, the treatment variable is independent of the factor loadings in the baseline scenario, while the correlation between the factor loadings and the treatment dummy is 0.446 and 0.706 in the second and third cases, respectively.

Table 8.6 Biases of estimated treatment effects using different methods in Monte Carlo simulations

Support difference	0	0.5	1
IFE	0.002	−0.009	−0.002
	(0.143)	(0.154)	(0.209)
SC	0.010	0.633	1.420
	(0.102)	(0.120)	(0.206)
DID	−0.087	0.209	0.518
	(0.134)	(0.134)	(0.137)

Note: This table summarizes Table 2 of Gobillon and Magnac (2016)

Then, we report the estimated treatment effects using the interactive fixed effects, synthetic controls, and difference-in-differences methods in Table 8.6. As can be seen, there is little estimation bias in the baseline setting for all the methods, which is not surprising since the endogeneity issue does not occur in this setting provided that the treatment dummy is independent of the factors and factor loadings. Also, the synthetic control method is more efficient than the interactive fixed-effects estimator, since the synthetic control method imposes additional constraints that are aligned with the true data generating process. In addition, considering the settings with overlapping and disjoint supports, the interactive fixed-effects estimator is unbiased and is able to account for the endogenous factor structure. However, the synthetic control method imposes the constraints that the factor loadings of each treated unit are linear combinations of the factor loadings of the untreated units, which do not hold in general in these settings and thus the synthetic control estimates are fairly biased. Also, it is expected to see that the difference-in-differences method is rather biased in these settings since it can only account for additive fixed effects. As a summary, the interactive fixed-effects estimator outperforms the synthetic controls and the difference-in-differences methods when an endogenous factor structure is present, which is of significant relevance in various empirical settings of policy evaluation.

8.6.2 *Empirical Findings*

The empirical findings in Gobillon and Magnac (2016) also suggest that the interactive fixed-effects approach is relatively more plausible for the empirical setting. In their study, the researchers evaluate the enterprise zone program, which exempts the firms' contributions to the national health insurance and pension system if the employers hire at least 20% of their labor force locally. The tax reduction is as high as around 30% of the total labor expenses. Therefore, it is important to understand if the policy would facilitate the local residents to find a job. The researchers use a flow sample of unemployment spells in France from July 1989 to June 2003 to examine the effect of the enterprise zone program introduced on January 1, 1997. The researchers consider the impact on the number of exits from unemployment to a job, and the number of exits from unemployment for unknown reasons. For both measures, the findings based on the interactive fixed effects and the difference-in-differences methods are similar: the program has a positive and significant effect on the number of exits from unemployment to employment, and an insignificant effect on the amount of exits for unknown reasons. Nevertheless, the synthetic control estimates find a negative and insignificant effect on the exits to employment, and a positive and significant effect on leaving unemployment for unknown reasons. The synthetic control estimates seem to be counterintuitive, since the financial incentives of the program are likely to induce a higher effect on the exits to a job than the exits for unknown reasons. As a result, the researchers conclude that the interactive fixed-effects estimator behaves well relative to its competitors in both the Monte Carlo simulations and the empirical application.

In the recent empirical literature, a larger number of studies have started to apply the interactive fixed-effects approach for policy evaluations. For instance, Hsiao et al. (2012) analyze the influences of political and economic integration of Hong Kong with Mainland China on the growth of the Hong Kong economy. Using a panel dataset of 24 countries, the latent factor structure is flexible to control for the potential endogenous economic and political shocks and country specific reactions to these shocks. Also, Kim and Oka (2014) evaluate the effects of implementing the unilateral divorce law on divorce rates in the USA using a panel of state-level divorce rates, where the researchers are motivated to control for the multidimensional unobserved social and economic shocks and the state specific reactions to the shocks. It is a recent trend that the potential existence of the endogenous factor structure becomes more emphasized by the researchers and thus the interactive fixed-effects approach becomes more popular in policy evaluation to control for that influence.

References

Back, A. D., & Weigend, A. S. (1997). A first application of independent component analysis to extracting structure from stock returns. *International Journal of Neural Systems, 8*(4), 473–484.

Bai, J. (2003). Inferential theory for factor models of large dimensions. *Econometrica, 71*(1), 135–171.

Bai, J. (2009). Panel data models with interactive fixed effects. *Econometrica, 77*(4), 1229–1279.

Bai, J., & Ng, S. (2002). Determining the number of factors in approximate factor models. *Econometrica, 70*(1), 191–221.

Bai, J., & Ng, S. (2006). Confidence intervals for diffusion index forecasts and inference for factor-augmented regressions. *Econometrica, 74*(4), 1133–1150.

Bai, J., & Ng, S. (2008). Forecasting economic time series using targeted predictors. *Journal of Econometrics, 146*(2), 304–317.

Bai, J., & Ng, S. (2009). Boosting diffusion indices. *Journal of Applied Econometrics, 24*(4), 607–629.

Bartlett, M. S., Movellan, J. R., & Sejnowski, T. J. (2002). Face recognition by independent component analysis. *IEEE Transactions on Neural Networks, 13*(6), 1450–1464.

Brown, G. D., Yamada, S., & Sejnowski, T. J. (2001). Independent component analysis at the neural cocktail party. *Trends in Neurosciences, 24*(1), 54–63.

Cai, T. T., Ma, Z., & Wu, Y. (2013). Sparse PCA: Optimal rates and adaptive estimation. *The Annals of Statistics, 41*(6), 3074–3110.
Chun, H., & Keleş, S. (2010). Sparse partial least squares regression for simultaneous dimension reduction and variable selection. *Journal of the Royal Statistical Society: Series B (Statistical Methodology), 72*(1), 3–25.
Efron, B., Hastie, T., Johnstone, I., & Tibshirani, R. (2004). Least angle regression. *The Annals of Statistics, 32*(2), 407–499.
Firpo, S., & Possebom, V. (2018). Synthetic control method: Inference, sensitivity analysis and confidence sets. *Journal of Causal Inference, 6*(2), 1–26.
Fuentes, J., Poncela, P., & Rodríguez, J. (2015). Sparse partial least squares in time series for macroeconomic forecasting. *Journal of Applied Econometrics, 30*(4), 576–595.
Gobillon, L., & Magnac, T. (2016). Regional policy evaluation: Interactive fixed effects and synthetic controls. *Review of Economics and Statistics, 98*(3), 535–551.
Hastie, T., Tibshirani, R., & Friedman, J. (2009). *The elements of statistical learning: Data mining, inference, and prediction*. Berlin: Springer.
Hastie, T., Tibshirani, R., & Wainwright, M. (2015). *Statistical learning with sparsity: The lasso and generalizations*. London: Chapman and Hall/CRC.
Hotelling, H. (1933). Analysis of a complex of statistical variables into principal components. *Journal of Educational Psychology, 24*(6), 417–441.
Hsiao, C., Ching, H. S., & Wan, S. K. (2012). A panel data approach for program evaluation: Measuring the benefits of political and economic integration of Hong Kong with mainland China. *Journal of Applied Econometrics, 27*(5), 705–740.
Hyvärinen, A. (1999). Fast and robust fixed-point algorithms for independent component analysis. *IEEE Transactions on Neural Networks, 10*(3), 626–634.
Hyvärinen, A., & Oja, E. (2000). Independent component analysis: Algorithms and applications. *Neural Networks, 13*(4–5), 411–430.
Jolliffe, I. T. (2002). *Principal component analysis*. Berlin: Springer.
Jolliffe, I. T., & Cadima, J. (2016). Principal component analysis: A review and recent developments. *Philosophical Transactions of the Royal Society A: Mathematical, Physical and Engineering Sciences, 374*(2065), 20150202.
Kim, D., & Oka, T. (2014). Divorce law reforms and divorce rates in the USA: An interactive fixed-effects approach. *Journal of Applied Econometrics, 29*(2), 231–245.
Kim, H. H., & Swanson, N. R. (2018). Mining big data using parsimonious factor, machine learning, variable selection and shrinkage methods. *International Journal of Forecasting, 34*(2), 339–354.
Li, H., Li, Q., & Shi, Y. (2017). Determining the number of factors when the number of factors can increase with sample size. *Journal of Econometrics, 197*(1), 76–86.
Muirhead, R. J. (2009). *Aspects of multivariate statistical theory*. Hoboken: Wiley.
Pearson, K. (1901). On lines and planes of closest fit to systems of points in space. *The London, Edinburgh, and Dublin Philosophical Magazine and Journal of Science, 2*(11), 559–572.
Robbins, M. W., Saunders, J., & Kilmer B. (2017). A framework for synthetic control methods with high-dimensional, micro-level data: Evaluating a neighborhood-specific crime intervention. *Journal of the American Statistical Association, 112*(517), 109–126.
Stock, J. H., & Watson, M. W. (1999). Forecasting inflation. *Journal of Monetary Economics, 44*(2), 293–335.
Stock, J. H., & Watson, M. W. (2002a). Forecasting using principal components from a large number of predictors. *Journal of the American Statistical Association, 97*(460), 1167–1179.
Stock, J. H., & Watson, M. W. (2002b). Macroeconomic forecasting using diffusion indexes. *Journal of Business & Economic Statistics, 20*(2), 147–162.
Stock, J. H., & Watson, M. W. (2012). Generalized shrinkage methods for forecasting using many predictors. *Journal of Business & Economic Statistics, 30*(4), 481–493.
Stone, J. V. (2004). *Independent component analysis: A tutorial introduction*. Cambridge: MIT Press.

Tibshirani, R. (1996). Regression shrinkage and selection via the lasso. *Journal of the Royal Statistical Society Series B, 58*(1), 267–288.

Tong, L., Liu, R. W., Soon, V. C., & Huang, Y. F. (1991). Indeterminacy and identifiability of blind identification. *IEEE Transactions on Circuits and Systems, 38*(5), 499–509.

Xu, Y. (2017). Generalized synthetic control method: Causal inference with interactive fixed effects models. *Political Analysis, 25*(1), 57–76.

Zou, H., & Hastie, T. (2005). Regularization and variable selection via the elastic net. *Journal of the Royal Statistical Society: Series B (Statistical Methodology), 67*, 301–320.

Zou, H., Hastie, T., & Tibshirani, R. (2006). Sparse principal component analysis. *Journal of Computational and Graphical Statistics, 15*(2), 265–286.

Chapter 9
Subspace Methods

Tom Boot and Didier Nibbering

9.1 Introduction

With a limited number of observations and a large number of variables, dimension reduction is necessary to obtain accurate macroeconomic forecasts. This can be done by identifying a small set of variables that appear most relevant to the variable of interest by variable selection (Ng, 2013) or shrinkage methods as the lasso (Tibshirani, 1996). Alternatively, one can transform the high dimensional data set into a small number of factors that capture most of the available information, as in factor models by Stock and Watson (2002, 2006) and Bai and Ng (2006, 2008), among others. In both cases, the low measurement frequency of macroeconomic data and the weak forecast relations make it a challenging task to find an accurate low-dimensional representation of the high-dimensional data. In fact, the selection methods themselves can add considerable noise to the forecast, potentially offsetting some of their gains.

Instead of trying to find the most informative part of the data, subspace methods reduce the dimension of the data in a data-oblivious fashion. For example, they draw a small set of variables at random, based on which a forecast can be constructed. Repeating this many times, and combining the forecasts using a simple average, has been found to be highly effective when constructing macroeconomic forecasts. The exact way in which these methods achieve such high forecast accuracy is not

T. Boot (✉)
Department of Economics, Econometrics and Finance, University of Groningen, Groningen, The Netherlands
e-mail: t.boot@rug.nl

D. Nibbering
Department of Econometrics and Business Statistics, Monash University, Clayton, VIC, Australia
e-mail: didier.nibbering@monash.edu

P. Fuleky (ed.), *Macroeconomic Forecasting in the Era of Big Data*,
Advanced Studies in Theoretical and Applied Econometrics 52,
https://doi.org/10.1007/978-3-030-31150-6_9

yet fully understood. Upper bounds on the mean squared forecast error indicate that high correlations in the data limit the omitted variables bias, while the small subspace dimension greatly reduces the forecast variance.

In this chapter, we provide an introduction to four different subspace methods: complete subset regression (CSR), random subset regression (RS), random projection regression (RP), and compressed regression (CR). The first is introduced by Elliott, Gargano, and Timmermann (2013) who propose to average over individual forecasts constructed from all possible linear regression models with a fixed number of predictors. Since using all available subsets is infeasible in high-dimensional settings, Elliott, Gargano, and Timmermann (2015) take arbitrary subsets as an approximation. The latter approach is further analyzed by Boot and Nibbering (2019), who discuss two different approaches to construct a random subspace in a macroeconomic forecasting setting. Random subset regression uses randomly selected subsets of predictors to estimate many low-dimensional approximations to the original model. Random projection regression forms a low-dimensional subspace by averaging over predictors using random weights drawn from a standard normal distribution. Koop, Korobilis, and Pettenuzzo (2019) exploit the computational gains from using sparse random projection matrices in a Bayesian compressed regression.

While applications of random subspace methods to economic forecasting are very recent, the ideas underlying random subspace methods have been used in a number of applications in the statistics and machine learning literature. Bootstrap aggregation methods aggregate the forecasts of models fitted on subsets of the data to a final prediction. Breiman (1996) and Breiman (1999) use bootstrap aggregation to draw subsets of observations instead of predictors, a method known as bagging. Feature bagging is a form of bootstrap aggregation where one draws subsets of predictors instead of observations. Within computer science, the performance of decision trees is often improved by training trees using random subsets of the available observations or variables at each split point, and then averaging the predictions from each of these trees. Random subset regression is a simple form of feature bagging aiming to reduce the dimension of the set of predictors in a linear regression model. Random projection regression and compressed regression are part of a large machine learning literature on constructing a new set of predictors (Frieze, Kannan, and Vempala, 2004; Mahoney and Drineas, 2009). The justification for the use of random projections is often derived from the Johnson–Lindenstrauss lemma (Johnson and Lindenstrauss, 1984). In a linear regression model, Kabán (2014) shows that the results can be sharpened without invoking the Johnson–Lindenstrauss lemma.

This chapter is organized as follows: Sect. 9.2 establishes notation and sets the forecasting framework. Section 9.3 briefly discusses two fundamental ideas behind subspace methods: forecast combinations and factor models. Section 9.4 introduces the random subspace methods. Section 9.5 surveys the performance of the random subspace methods in empirical applications reported in the literature. Theoretical guarantees on the forecast performance of random subspace methods are reviewed in Sect. 9.6. Section 9.7 illustrates the use of these methods in two empirical applications to macroeconomic forecasting with many predictors, and Sect. 9.8 concludes with recommendations for future research.

9.2 Notation

We denote by y_{t+1} the macroeconomic variable of interest. To forecast this variable, we make use of a set of p_w must-have predictors $\boldsymbol{w}_t$, such as a constant and lags of y_{t+1}, and a set of p_x possibly relevant predictors $\boldsymbol{x}_t$. The case of interest is where p_x is large, so that including all predictors in a single model leads to inefficient forecasts. When it is not clear which variables to include in $\boldsymbol{w}_t$, this set can be left empty. Both $\boldsymbol{x}_t$ and $\boldsymbol{w}_t$ can include lags of y_{t+1}.

We focus on the linear model

$$y_{t+1} = \boldsymbol{w}_t'\boldsymbol{\beta}_w + \boldsymbol{x}_t'\boldsymbol{\beta}_x + \varepsilon_{t+1}, \tag{9.1}$$

where the forecast error is denoted by ε_{t+1} and the time index t runs from $t = 0, \ldots, T$. The predictors $z_t = (\boldsymbol{w}_t', \boldsymbol{x}_t')'$, with $t = 0, \ldots, T-1$, are used in the estimation of the $p \times 1$ parameter vector $\boldsymbol{\beta} = (\boldsymbol{\beta}_w', \boldsymbol{\beta}_x')'$, and $z_T = (\boldsymbol{w}_T', \boldsymbol{x}_T')'$ is only used for the construction of point forecasts $\hat{y}_{T+1}$ for y_{T+1}. Theoretically, the linearity assumption is not too restrictive, as in many cases we can approximate a nonlinear function $f(\boldsymbol{x}_t, \boldsymbol{\beta})$ using a first order Taylor expansion.

9.3 Two Different Approaches to Macroeconomic Forecasting

9.3.1 *Forecast Combinations*

A consistent finding in macroeconomic forecasting is that averaging forecasts from multiple forecasting models increases prediction accuracy (Timmermann, 2006). When the forecasts from the individual models are unbiased, combining multiple forecasts lowers the variance, and hence the mean squared forecast error. When individual models are misspecified, averaging might lower the bias if the misspecification error cancels.

A linear forecast combination pools over N individual forecasts $\hat{y}_{T+1,i}$ to obtain a forecast $\hat{y}_{T+1}$ for y_{T+1} by

$$\hat{y}_{T+1} = \sum_{i=1}^{N} \omega_i \hat{y}_{T+1,i},$$

where ω_i is the weight on $\hat{y}_{T+1,i}$.

Ideally, different models are combined using a weighted average, with weights that reflect the reliability of each forecast. In practice, estimation of combination weights introduces additional variance in the forecast. This can offset a large part of the variance reduction by combination. It turns out that taking a simple average over different models is generally hard to beat (Claeskens, Magnus, Vasnev, and Wang, 2016; Elliott and Timmermann, 2013). This is especially true if all models perform roughly the same, in which case equal weighting is close to being optimal.

The literature on forecast combinations generally starts by assuming that a number of different forecasts are available, but is silent on where these forecasts actually come from. The subspace methods discussed in this paper generate a sequence of forecasts that can then be combined to yield an accurate forecast.

9.3.2 Principal Component Analysis, Diffusion Indices, Factor Models

Factor models reflect the idea that a small number of latent components drive most variation in macroeconomic series. We observe a vector of regressors x_t that satisfies

$$\boldsymbol{x}_t = \boldsymbol{\Lambda}' \boldsymbol{f}_t + \boldsymbol{e}_t,$$

where $\boldsymbol{f}_t$ is a $r \times 1$ vector, $\boldsymbol{\Lambda}$ is a $r \times p$ matrix of loadings, and e_t is an idiosyncratic component. The factors are driving the dependent variable according to

$$y_{t+1} = \boldsymbol{w}_t' \boldsymbol{\beta}_w + \boldsymbol{f}_t' \boldsymbol{\beta}_f + \varepsilon_{t+1}.$$

It is generally assumed that the factors are orthonormal, such that $\frac{1}{T} \sum_{t=1}^{T} \boldsymbol{f}_t \boldsymbol{f}_t' = \boldsymbol{I}_r$. Taking a singular value decomposition of $\boldsymbol{x}_t$, we have

$$\boldsymbol{x}_t = \boldsymbol{U} \boldsymbol{S} \boldsymbol{v}_t = \boldsymbol{\Lambda}' \boldsymbol{f}_t + \boldsymbol{e}_t.$$

If the idiosyncratic component $e_t = 0$, the factors equal the left singular vectors (up to a rotation), and the model essentially applies a linear dimension reduction by replacing x_t as

$$\boldsymbol{x}_t' \rightarrow \boldsymbol{x}_t' \boldsymbol{U} \boldsymbol{S}^{-1} = \boldsymbol{f}_t'. \tag{9.2}$$

It turns out that this is still a good strategy if $\boldsymbol{e}_t \neq 0$, as the left singular vectors are consistent for the space spanned by the unknown factors Stock and Watson (2002). Forecasts based on (estimated) factors are generally precise, and often an interpretation of the factors can be derived from the loadings.

Random subspace methods apply a linear dimension reduction step as in (9.2), but instead of multiplication with a data dependent matrix that is explicitly designed to find factors, it replaces $\boldsymbol{U}\boldsymbol{S}^{-1}$ by a random matrix $\boldsymbol{R}$. Intuitively, suppose that all predictors strongly correlate with a single factor that drives the dependent variable. Then including a randomly selected subset of predictors, where $\boldsymbol{R}$ is a random permutation matrix, or a set of randomly weighted averages of the predictors, where $\boldsymbol{R}$ contains independent standard normally distributed elements, will not lead to a large omitted variable bias. When the dimension of these sets is small, there can however be a substantial reduction in variance compared to the case where one includes all available predictors. The theoretical results later in this chapter also point in this direction. To avoid that the random selection/weighting induces additional variance itself, we can average over forecasts obtained under many different random selections/weights.

9.4 Subspace Methods

Subspace methods can be seen as a combination of the ideas outlined in the previous section. First, a linear dimension reduction step as in a factor model is performed to obtain a low-dimensional approximation to the data. That is, we post-multiply the regressors x_t with a $p_x \times k$ matrix $\boldsymbol{R}_i$. The subspace dimension $k < p_x$ equals the number of predictors in the subspace. A commonly used deterministic matrix $\boldsymbol{R}_i$ contains the principal component loadings corresponding to the k largest eigenvalues of the sample covariance matrix of the predictors. If the elements of $\boldsymbol{R}_i$ are drawn from a known probability distribution, then this opens up the possibility to generate many low-dimensional approximations to the data. Inspired by the forecast combination literature, the forecasts based on the individual approximations are then combined to improve the forecast accuracy.

Replacing the full regressor matrix with the regressors projected onto the subspace leads to the approximating model

$$y_{t+1} = \boldsymbol{w}_t'\boldsymbol{\beta}_{w,i} + \boldsymbol{x}_t'\boldsymbol{R}_i\boldsymbol{\beta}_{x,i} + u_{t+1}, \tag{9.3}$$

which can be rewritten to

$$y_{t+1} = \boldsymbol{z}_t'\boldsymbol{S}_i\boldsymbol{\beta}_i + u_{t+1}, \text{ with } \boldsymbol{S}_i = \begin{pmatrix} \boldsymbol{I}_{p_w} & \boldsymbol{O} \\ \boldsymbol{O} & \boldsymbol{R}_i \end{pmatrix},$$

with z_t as defined below (9.1). Instead of estimating $\boldsymbol{\beta}$ in the high-dimensional model, where all $p_w + p_x$ predictors are included, we estimate the $p_w + k$ parameters in $\boldsymbol{\beta}_i$ in this low-dimensional model.

Under the assumption that $T > (p_w + k)$, the least squares estimator of $\boldsymbol{\beta}_i$ is given by

$$\hat{\boldsymbol{\beta}}_i = (\boldsymbol{S}_i'\boldsymbol{Z}'\boldsymbol{Z}\boldsymbol{S}_i)^{-1}\boldsymbol{S}_i'\boldsymbol{Z}'\boldsymbol{y}.$$

Using this estimate, we construct an individual forecast for y_{T+1} as

$$\hat{y}_{T+1,i} = z_T' S_i \hat{\boldsymbol{\beta}}_i.$$

Based on N different realizations of the dimension reduction matrix $\boldsymbol{R}_i$, N different forecasts are obtained. These forecasts are subsequently averaged to obtain the prediction

$$\hat{y}_{T+1} = \frac{1}{N} \sum_{i=1}^{N} \hat{y}_{T+1,i}.$$

As we see here, each forecast receives an equal weight.

We now discuss three subspace methods that all have recently been applied to economic data.

9.4.1 Complete Subset Regression

Complete subset regression (Elliott et al., 2013) is a subspace method in which the individual forecasts are constructed from all possible linear regression models with a fixed number of k predictors. The matrix $\boldsymbol{R}_i$ in (9.3) selects one of the N unique combinations of k predictors out of the p_x available predictors. The selection matrix $\boldsymbol{R}_i$ is data-independent and has $\binom{p_x}{k}$ possible realizations. The set of models following from all possible realizations of $\boldsymbol{R}_i$ is called a complete subset.

To construct a more formal definition of complete subset regression, define an index $l = 1, \ldots k$ with k the dimension of the subspace, and a scalar $c(l)$ such that $1 \leq c(l) \leq p_x$. Denote by $\boldsymbol{e}_{c(l)}$ a p_x-dimensional vector with its $c(l)$-th entry equal to one, then complete subset regressions are based on the matrices $\boldsymbol{R}_i$ of the form

$$\left[\boldsymbol{e}_{c(1)}, \ldots, \boldsymbol{e}_{c(k)}\right], \quad \boldsymbol{e}_{c(m)} \neq \boldsymbol{e}_{c(n)} \text{ if } m \neq n,$$

where $c(1), \ldots, c(k)$ is one of the $N = \binom{p_x}{k}$ combinations of k predictors.

A special case of complete subset regressions are forecast combinations of univariate models for each available predictor. It is generally found however that using larger values of k leads to lower omitted variable bias at the expense of a relatively small increase in the variance.

Subspace Dimension

Optimal selection of the subspace dimension is a largely unexplored area. The value of k is the number regressors in each regression, but the variance of the averaged forecast generally does not scale linearly with k. For example, when all regressors are orthonormal, the variance scales as $k \cdot \frac{k}{p}$ (see Sect. 9.6). An appropriate information criterion to select k is not yet available.

Elliott et al. (2013) show that under the assumption that y_{t+1} and z_t are independent and identically distributed, the value of k can be set to minimize the mean squared forecast error for given values of $\boldsymbol{\beta}$. However, both the i.i.d. assumption and available information on the values in $\boldsymbol{\beta}$ do generally not hold in macroeconomic forecasting.

In practice, the subspace dimension can be recursively selected by evaluating the forecast accuracy of each dimension in the current information set and selecting the best performing k. Empirical results in Boot and Nibbering (2019) and Elliott et al. (2015) indicate that the performance is not heavily dependent on the precise value of k.

Weighting Schemes

Since there is little empirical evidence that alternative weighting schemes outperform equal weighted forecast combinations, random subspace methods use an equal weighted average. Since macroeconomic data typically has a small number of observations, estimating sophisticated weighting schemes is particularly difficult in this setting. Elliott et al. (2013) experimented with a weighting scheme that is based on the Bayesian Information Criterion (BIC). The weight of each model is proportional to the exponential of its BIC, which results in larger weights for models with a high likelihood value. They do not find structural improvements over using equal weights.

Limitations

When the number of available predictors is large, considering all combinations of predictors for a given k becomes infeasible. As a solution, we can average over forecasts from a smaller set of models. This smaller set of models can be selected by a stochastic search. Elliott et al. (2013, 2015) implement a Markov Chain and a Shotgun approach but find similar performance for randomly drawing a set of models. They draw with uniform probability a feasible number of models without replacement from the complete subspace regressions. Randomly drawing models is simple, fast, and assigns, similar to equal weighting, the same probability to each submodel. This brings us to a feasible alternative to complete subset regressions in macroeconomic forecasting settings.

9.4.2 Random Subset Regression

Random subset regression uses randomly selected subsets of predictors to estimate many low-dimensional approximations to the original model. The forecasts from these submodels are then combined to reduce the forecast variance while maintaining most of the signal. Note that this method is slightly different from randomly drawing models from the set of complete subset regressions. Instead of drawing a set of models without replacement, random subset regression draws a set of predictors without replacement for each submodel.

In random subset regression, the matrix $\boldsymbol{R}_i$ is a random selection matrix that selects a random set of k predictors out of the original p_x available predictors. For example, if $p_x = 5$ and $k = 3$, a possible realization of $\boldsymbol{R}_i$ is

$$\begin{pmatrix} 0 & 1 & 0 \\ 0 & 0 & 0 \\ 1 & 0 & 0 \\ 0 & 0 & 1 \\ 0 & 0 & 0 \end{pmatrix}.$$

The same formulation as for complete subset regression can be used to represent a general matrix $\boldsymbol{R}_i$. We have a p_x-dimensional vector $\boldsymbol{e}_{c(l)}$ with its $c(l)$-th entry equal to one, where $l = 1, \ldots k$ is an index and $c(l)$ a scalar such that $1 \leq c(l) \leq p_x$. Then random subset regression is based on random matrices

$$\left[\boldsymbol{e}_{c(1)}, \ldots, \boldsymbol{e}_{c(k)}\right], \quad \boldsymbol{e}_{c(m)} \neq \boldsymbol{e}_{c(n)} \text{ if } m \neq n.$$

The intuition why random subset regression is effective in macroeconomic forecasting is the following. One might generally worry that selecting predictors at random results in irrelevant predictors being present in each submodel. However, in macroeconomic data, the predictors are often highly correlated. This means that even if a predictor is irrelevant, it is most likely correlated with a predictor that is relevant. As such, even predictors that are irrelevant when all variables are included are relevant when only small sets of variables are included. With strong correlation present, the use of random subsets reduces the variance of the forecast by at least a factor $\frac{k}{p_x}$, while the increase in bias is only small.

9.4.3 Random Projection Regression

An alternative for using a random selection matrix is to use a random weighting matrix. Instead of selecting individual regressors, this gives a new set of predictors which are weighted averages of the original predictors. This can be easily implemented by drawing a matrix $\boldsymbol{R}_i$ of which each element is distributed independent and identically following some probability distribution. The most common choice is

$$[\boldsymbol{R}_i]_{hj} \sim N(0, 1/\sqrt{k}), \tag{9.4}$$

with $h = 1, \ldots, p_x$ and $j = 1, \ldots, k$. Repeating the same procedure as for random subset regression, only now with the matrix $\boldsymbol{R}_i$ defined in (9.4) yields random projection regression.

Random projections are originally motivated by data sets that contain too many observations to allow a calculation of the covariance matrix $\frac{1}{T}\boldsymbol{X}'\boldsymbol{X}$ in a reasonable time. To reduce the dimension T, we can note that $\mathrm{E}_R[\frac{1}{T}\boldsymbol{X}'\boldsymbol{R}\boldsymbol{R}'\boldsymbol{X}] = \frac{1}{T}\boldsymbol{X}'\boldsymbol{X}$. If $\frac{1}{T}\boldsymbol{X}'\boldsymbol{R}\boldsymbol{R}'\boldsymbol{X}$ is sufficiently close to its expectation, then we can calculate the covariance matrix accurately using the smaller matrix $\boldsymbol{X}'\boldsymbol{R}$. Such a result is generally shown through the Johnson–Lindenstrauss lemma (Achlioptas, 2003; Dasgupta and Gupta, 2003). If we use (9.4), this still requires the multiplication of $\boldsymbol{X}'\boldsymbol{R}$. Hence, subsequent work focused on showing the Johnson–Lindenstrauss lemma for matrices $\boldsymbol{R}$ that contain many zeros (Achlioptas, 2003; Li, Hastie, and Church, 2006).

Since we apply dimension reduction to the large number of predictors, instead of reducing a large number of observations, the above motivation for random projection does not align with our application here. The problem we aim to solve is more statistical in nature, with a large variance caused by the inclusion of many predictors, rather than computational. This direction has received substantially less attention. We discuss the relevant empirical literature in Sect. 9.5, and theoretical results in Sect. 9.6.

9.4.4 Compressed Regression

To speed up calculations, sparse random projection matrices have been proposed. The prime example is to choose

$$R_{ij} = \sqrt{s} \begin{cases} -1 \text{ with probability } \frac{1}{2s}, \\ 0 \text{ with probability } 1 - \frac{1}{s}, \\ +1 \text{ with probability } \frac{1}{2s}. \end{cases}$$

This specification was proposed by Achlioptas (2003), who shows that the Johnson–Lindenstrauss lemma applies to this matrix for $s = 3$. This was extended to $s = O(\sqrt{n})$ by Li et al. (2006). As s governs the sparsity of $\boldsymbol{R}$, setting s large can lead to significant computational gains, without loss in accuracy.

Guhaniyogi and Dunson (2015) use an alternative specification for the random matrix,

$$R_{ij} = \sqrt{s}\begin{cases} -1 \text{ with probability } \frac{1}{s^2}, \\ \ \ 0 \text{ with probability } 2(1-\frac{1}{s})\frac{1}{s}, \\ +1 \text{ with probability } (1-\frac{1}{s})^2, \end{cases}$$

where $1/s \sim U[0.1, 0.9]$, and subsequently the columns of R are orthonormalized using the Gram–Schmidt algorithm. The subspace dimension k is drawn uniformly over $[2\log p_x, \min(n, p_x)]$. Guhaniyogi and Dunson (2015) apply the dimension reduction in a Bayesian framework, and use Bayesian model averaging to weight the forecasts. Under normal-inverse gamma priors, and a diffuse prior on the models, the weights can be analytically calculated. Furthermore, they provide a set of sufficient conditions for the predictive density under dimension reduction to converge to the true predictive density.

9.5 Empirical Applications of Subspace Methods

Subspace methods are relatively new to the economics literature. This section lists early applications in the field of macroeconomic forecasting, but also in microeconomics and finance. Since subspace methods have their roots in the machine learning, we also refer to some applications in this literature.

9.5.1 *Macroeconomics*

The first applications of subspace methods in the macroeconomic literature use random subset regression to forecast economic indicators with a large number of possible predictors. Elliott et al. (2015) forecast quarterly US unemployment, GDP growth, and inflation and find more accurate point forecasts by random subset regression than dynamic factor models or univariate regressions over multiple forecast horizons. Although the paper presents the subspace method as complete subset regressions, the methods randomly sample low-dimensional models to form a final forecast in practice.

Leroux, Kotchoni, and Stevanovic (2017) draw similar conclusions in an extensive forecasting exercise on monthly US data. The authors compare the forecast accuracy of a large number of different forecasting models on industrial production, employment, inflation, the stock market index, and exchange rates. They find that random subset regression shows often better performance in terms of mean squared forecast error than univariate models, factor-augmented regressions, dynamic factor models, and standard forecast combinations for industrial production and employment growth. The forecast accuracy for other series does not seem to benefit from any data-rich model.

Garcia, Medeiros, and Vasconcelos (2017) also confirm the strong performance of random subset regression in a real-time forecasting exercise for Brazilian inflation. They use several lasso-type models, target factors, random forests, Bayesian vector autoregressions, standard time series models, random subset regression, and expert forecasts to produce forecasts for twelve different forecast horizons. Apart from the first two horizons, complete subset regression dominates all other forecasting methods on mean squared forecasting error. The same holds for density forecasts based on bootstrap resampling and for forecasts of accumulated inflation.

Koop et al. (2019) use compressed regression in a Bayesian vector autoregressive model to forecast 7 monthly macroeconomic variables with up to 129 dependent variables and 13 lags. They compare the forecast performance to univariate autoregressive models, dynamic factor models, factor augmented vector autoregressions, and the Minnesota prior vector autoregressions. The forecast accuracy of the Bayesian compressed vector autoregressions is similar or better than either factor methods or large vector autoregression methods involving prior shrinkage, for 1–12 months ahead forecast horizons.

Pick and Carpay (2018) analyze in detail multi-step ahead forecasting performance on 14 variables using a large vector autoregression to which the random subspace methods are applied, as well as a number of popular dimension reduction methods. They find that the random subspace methods and the lasso algorithm provide the most accurate forecasts. This is in line with the findings for one-step-ahead forecasts in Boot and Nibbering (2019).

9.5.2 Microeconomics

Schneider and Gupta (2016) use random projection regression to forecast 1-week-ahead sales for existing and newly introduced tablet computers. They produce the forecast with a bag-of-words model based on consumer reviews. This model counts the number of occurrences of words in a particular piece of text. Since many words can occur, the number of parameters is equally large. Dimension reduction by random projection results in improved forecast accuracy compared to a baseline model that ignores the consumer reviews and a support vector machine based on the reviews.

Chiong and Shum (2018) use random projections to estimate aggregate discrete-choice models with large choice sets. However, they project the large number of different choice categories on a low-dimensional subspace instead of reducing the variable space. The paper focuses on parameter estimation in several simulations and in an application to supermarket scanner data, but does not perform a forecasting exercise.

9.5.3 Finance

Elliott et al. (2013) propose complete subset regressions and illustrate the methods in a forecasting exercise on the monthly US stock returns. They find an increase in forecast accuracy for the one-step-ahead forecasts for the excess return of the S&P 500 by complete subset regressions relative to standard forecast combination methods, univariate forecasting models, lasso, ridge regression, bagging, and Bayesian model averaging.

Meligkotsidou, Panopoulou, Vrontos and Vrontos (2019) extend the complete subset regressions framework to a quantile regression setting. After constructing the complete subset combination of quantile forecasts, they employ a recursive algorithm that selects the best complete subset for each predictive regression quantile. The method is applied to forecasting the quarterly S&P 500 equity premium. Complete quantile subset regression outperforms both the historical average benchmark and the complete subset regressions in forecast accuracy and economic value.

Gillen (2016) uses a strategy very similar to complete subset regressions to obtain efficient portfolios. He selects a random subset of securities and determines the optimal portfolio weights for these securities. By generating many different random subsets of securities, one obtains multiple weights for each security. These weights are averaged to obtain the final portfolio weight for each security. The expected out-of-sample performance of this strategy dominates both the $1/N$ strategy rule and sample-based optimization of the portfolio. Shen and Wang (2017) propose a similar approach and also conclude that the out-of-sample performance is superior to various competing strategies on diversified data sets.

9.5.4 Machine Learning

Random subspace methods have their roots in the machine learning literature. Although this literature commonly does not consider macroeconomic applications, it does focus on forecasting and big data. We briefly discuss some of the contributions in the machine learning literature to forecasting with random subspace methods.

Ho (1998) introduces random subspace methods in decision trees. Multiple trees are constructed by training on random subsets of the available explanatory variables. Decisions of the trees are combined in a decision forest by averaging the estimates of the class probabilities in each tree. Compared to decision trees that are trained on the full set of explanatory variables, the subspace method shows significant improvements in forecast accuracy in a collection of different data sets.

Bay (1998) applies the same idea to nearest neighbor classifiers. He combines multiple nearest neighbor classifiers that only use a random subset of the explanatory variables. The method outperforms standard nearest neighbor classifiers, k nearest neighbor classifiers, and neighbor classifiers with forward and backward selection of explanatory variables.

Bryll, Gutierrez-Osuna, and Quek (2003) develop a wrapper method that can be used with combinations of any type of classifier trained on random subsets of explanatory variables. The method selects the size of the random subset by comparing classification accuracy of different sizes of subsets. Moreover, the method only uses the best performing subsets to form a final prediction. The proposed method improves performance for bagging and several single-classifier algorithms on an image recognition data set.

Bingham and Mannila (2001) apply random projection as a dimensionality reduction tool to image and text data. They find that randomly projecting the data on a low-dimensional subspace performs similarly to principal component regression. However, they only assess the distortion created by the dimension reduction tool, that is, the Euclidean distance between two dimensionality reduced data vectors relative to their Euclidean distance in the original high-dimensional space, and the computational costs.

Fradkin and Madigan (2003) evaluate the performance of random projections in classification. They apply classifiers as decision trees, nearest neighbor algorithms, and support vector machines to several data sets after dimension reduction by principal component analysis and random projections. They find in all cases that principal component analysis results in better predictive performance than random projections. However, the results are based on a single draw of the random projection matrix instead of averaging over multiple predictions based on different draws of the random matrix. Cannings and Samworth (2017) also evaluate the performance of random projections in high dimensional classification. They provide a framework in which any type of classifier is applied to a low-dimensional subspace constructed by random projections of the original explanatory variables. The final prediction is based on an average over random projections, but only over random projections that provide the lowest error. The random projections approach outperforms a collection of benchmarks classifiers on a number of empirical data sets in terms of forecast accuracy.

Guhaniyogi and Dunson (2015) introduce Bayesian compressed regression with an application to data from a molecular epidemiology study. They find better performance in terms of mean square prediction error relative to benchmark models in predicting individual's sensitivity to DNA damage and individual's repair rate after DNA damage.

9.6 Theoretical Results: Forecast Accuracy

This section provides theoretical results on the performance of the random subspace methods for point forecasts $\hat{y}_{T+1}$ for y_{T+1}. We consider the forecasts by random subset regression and random projection regression, two subspace methods that are feasible in high-dimensional macroeconomic forecasting settings.

A large part of this section is based on Boot and Nibbering (2019) and Elliott et al. (2013). The results are based on the assumptions that the number of regressors p_x is large, but fixed. Then, asymptotic results are established in the limit where $T \to \infty$. Estimation uncertainty is introduced by scaling the coefficients with a factor $T^{-1/2}$, which excludes the possibility to estimate these coefficients consistently, even in the asymptotic limit.

9.6.1 Mean Squared Forecast Error

In this section, we analyze the asymptotic mean squared forecast error. The results are valid for a range of weakly dependent time series models. For any estimator $\tilde{\beta}$ for the true parameter β, the asymptotic MSFE is defined as

$$\rho = \lim_{T\to\infty} T\mathrm{E}\left[\left(y_{T+1} - z_T'\tilde{\boldsymbol{\beta}}\right)^2 - \sigma^2\right] = \lim_{T\to\infty} T\mathrm{E}\left[(z_T'\boldsymbol{\beta} - z_T'\tilde{\boldsymbol{\beta}})^2\right].$$

The error variance σ^2 is subtracted as it arises from the error ε_{T+1}, which is unpredictable by any method. This expression is very similar to the one in Hansen (2010), who divides the expression by the error variance σ^2.

Denote the ordinary least squares (OLS) estimator without dimension reduction as $\hat{\beta}$. This estimator is based on all predictors in $z_t = (\boldsymbol{w}_t', \boldsymbol{x}_t')'$, and yields the following forecast:

$$\hat{y}_{T+1}^{\mathrm{OLS}} = z_T'\hat{\boldsymbol{\beta}} = z_T'(\boldsymbol{Z}'\boldsymbol{Z})^{-1}\boldsymbol{Z}'\boldsymbol{y},$$

where $\boldsymbol{y} = (y_1, \ldots, y_T)'$, $\boldsymbol{Z} = (z_0, \ldots, z_{T-1})'$. Under standard regularity assumptions that ensure the convergence of $\sqrt{T}(\hat{\boldsymbol{\beta}} - \boldsymbol{\beta})$ to a mean zero, normally distributed random vector, the asymptotic mean squared forecast error is

$$\begin{aligned} \rho(p_w, p_x) &= \lim_{T\to\infty} T\mathrm{E}\left[(\hat{\boldsymbol{\beta}} - \boldsymbol{\beta})' z_T z_T' (\hat{\boldsymbol{\beta}} - \boldsymbol{\beta})\right] \\ &= \sigma^2(p_w + p_x), \end{aligned} \tag{9.5}$$

using that $\mathrm{E}[z_t z_t'] = \boldsymbol{\Sigma}_z$ for $t = 0, \ldots, T$. The fact that the MSFE increases with p_x is the motivation to consider k-dimensional subspaces with $k < p_x$. This will reduce the variance from $\sigma^2 p_x$ to (at most) $\sigma^2 k$, but it will also induce a squared bias term into the MSFE. The trade-off between reducing the variance and increasing the bias is determined by the subspace dimension k.

When the coefficients $\boldsymbol{\beta}_x$ are fixed, the bias in the asymptotic mean squared forecast error is infinite because of the multiplication with T. In practice, one would expect the bias and the variance to be roughly of the same order. To make sure that the asymptotic theory reflects this finite sample situation, a convenient assumption is that the coefficients $\boldsymbol{\beta}_x$ are local-to-zero, that is, $\boldsymbol{\beta}_x = \boldsymbol{\beta}_{x,0}/\sqrt{T}$. This renders a finite bias in the asymptotic limit. The forecast under the random subspace methods is given by

$$\hat{y}_{T+1} = \sum_{i=1}^{N} [\hat{y}_{T+1,S_i}],$$

with $\hat{y}_{T+1,S_i} = \boldsymbol{z}_T' \boldsymbol{S}_i \hat{\boldsymbol{\beta}}_{S_i}$, $\hat{\boldsymbol{\beta}}_{S_i} = (\boldsymbol{S}_i' \boldsymbol{Z}' \boldsymbol{Z} \boldsymbol{S}_i)^{-1} \boldsymbol{S}_i' \boldsymbol{Z}' \boldsymbol{y}$, and $\boldsymbol{S}_i = \begin{pmatrix} \boldsymbol{I}_{p_w} & \boldsymbol{O} \\ \boldsymbol{O} & \boldsymbol{R}_i \end{pmatrix}$.

Boot and Nibbering (2019) show that the sum over i can be replaced by an expectation with a negligible effect on the MSFE when the number of draws of the random matrix $\boldsymbol{R}$ is as large as $N = O(p_x \log p_x)$.

Identity Covariance Matrix

We can get a sense of the bias variance trade-off by making the assumption that $\boldsymbol{\Sigma}_x = \boldsymbol{I}_{p_x}$. In this case, the exact mean squared error for random subset regression and random projection regression is

$$\rho(p_w, k) = \sigma^2 \left[p_w + k \frac{k}{p_x} \right] + \left[1 - \frac{k}{p_x} \right]^2 \boldsymbol{\beta}_{x,0}' \boldsymbol{\beta}_{x,0}.$$

This shows that although the subspace dimension is k, the variance is in fact even lower, namely of $k \frac{k}{p_x}$. The optimal value of k is given by

$$k^* = \frac{\eta}{1+\eta} p_x, \qquad \eta = \frac{\boldsymbol{\beta}_{x,0}' \boldsymbol{\beta}_{x,0}}{\sigma^2 \cdot p_x},$$

for which we have

$$\rho^*(p_w, k^*) = \sigma^2 \left(p_w + \frac{\boldsymbol{\beta}_{x,0}' \boldsymbol{\beta}_{x,0}}{\sigma^2 p_x + \boldsymbol{\beta}_{x,0}' \boldsymbol{\beta}_{x,0}} p_x \right).$$

Comparing this to (9.5), we see that the factor $\sigma^2 p_x$ is reduced by a factor $\eta/(1+\eta)$ with η the signal-to-noise ratio per parameter. If the informational content in each predictor is therefore small, gains can be expected from the dimension reduction procedure.

Intuitively, one might expect that when strong correlations between variables are present, the random subspace methods are increasingly effective. In this case, omitting one variable, but including a strongly correlated one would prevent a large omitted variable bias. However, when the covariance matrix $\boldsymbol{\Sigma}_x$ is of general form, there is no analytic expression for the MSFE. To circumvent this problem, there have been several studies that derive bounds on the in-sample mean squared error under i.i.d. observations, starting with Kabán (2014) and refined by Thanei, Heinze, and Meinshausen (2017). Boot and Nibbering (2019) provide expressions for the asymptotic mean squared forecast error under general assumptions that allow for a broad range of time series models. We discuss these results in some detail to explain part of the success of random subspace methods in macroeconomic forecasting settings.

9.6.2 Mean Squared Forecast Error Bounds

For random subset regression, Boot and Nibbering (2019) show that the asymptotic MSFE can be upper bounded by

$$\rho(p_w, k) \leq \sigma^2(p_w + k) + \boldsymbol{\beta}'_{x,0}\boldsymbol{\Sigma}\boldsymbol{\beta}_{x,0} - \frac{k}{p_x}\boldsymbol{\beta}'_{x,0}\boldsymbol{\Sigma}\left[w_s\boldsymbol{\Sigma} + (1 - w_s)\boldsymbol{D}_{\Sigma}\right]^{-1}\boldsymbol{\Sigma}\boldsymbol{\beta}_{x,0}.$$

For random projection, the upper bound is

$$\rho(p_w, k) \leq \sigma^2(p_w + k) + \boldsymbol{\beta}'_{x,0}\boldsymbol{\Sigma}\boldsymbol{\beta}_{x,0} - \frac{k}{p_x}\boldsymbol{\beta}'_{x,0}\boldsymbol{\Sigma}\left[w_p\boldsymbol{\Sigma} + (1 - w_p)\frac{tr(\boldsymbol{\Sigma})}{p_x}\boldsymbol{I}_{p_x}\right]^{-1}\boldsymbol{\Sigma}\boldsymbol{\beta}_{x,0},$$

where $w_s = \frac{k-1}{p_x-1}$, $\boldsymbol{D}_{\Sigma}$ is a diagonal matrix with $[\boldsymbol{D}_{\Sigma}]_{ii} = [\boldsymbol{\Sigma}]_{ii}$, $w_p = \frac{p_x(k+1)-2}{(p_x+2)(p_x-1)}$, and $\text{tr}(\boldsymbol{\Sigma})$ denotes the trace of $\boldsymbol{\Sigma} = \text{plim}_{T\to\infty}\boldsymbol{X}'\boldsymbol{M}_W\boldsymbol{X}$, where $\boldsymbol{M}_W = \boldsymbol{I}_T - \boldsymbol{W}(\boldsymbol{W}'\boldsymbol{W})^{-1}\boldsymbol{W}'$.

The first term reflects the variance, while the second and third term reflect the bias induced by selecting/projection to a low-dimensional subspace. As expected, the variance is reduced from $\sigma^2 p_x$ to $\sigma^2 k$. This variance reduction comes at the expense of a bias term. For random subset regression the bias is a weighted average of the covariance matrix $\boldsymbol{\Sigma}$ and its diagonal elements $\boldsymbol{D}_{\Sigma}$. When $k = 1$, all weight is put on the diagonal matrix and any information on cross-correlations is lost in the low-dimensional subspace. When $k = p_x$, we obtain the OLS mean squared forecast error (9.5). The bound for random projection regression depends on a weighted average of $\boldsymbol{\Sigma}$ and the constant diagonal matrix $\frac{\text{tr}(\boldsymbol{\Sigma})}{p_x}\boldsymbol{I}_{p_x}$. When $k = 1$, nearly all weight is put on the diagonal matrix. When $k = p_x$, we again obtain the OLS mean squared forecast error (9.5).

To get further insight in the bounds, decompose the covariance matrix $\boldsymbol{\Sigma}_x = \boldsymbol{V}\boldsymbol{\Lambda}\boldsymbol{V}'$, where the orthogonal matrix $\boldsymbol{V}$ contains the eigenvectors of $\boldsymbol{\Sigma}_x$ and $\boldsymbol{\Lambda}$ is a diagonal matrix with eigenvalues sorted in decreasing order. We assume that the predictors are standardized, such that $[\boldsymbol{\Sigma}_x]_{ii} = 1$ for $i = 1, \ldots, p_x$. The data generating process (9.1) can be rewritten in terms of the principal components $\boldsymbol{f}_t' = \boldsymbol{x}_t'\boldsymbol{V}$ as

$$y_{t+1} = \boldsymbol{f}_t'\boldsymbol{\gamma} + \varepsilon_{t+1}.$$

The bounds for both random subspace methods reduce to

$$\rho(0,k) \leq \sigma^2 k + \sum_{i=1}^{p_f} \gamma_{0,i}^2 \lambda_i \left(1 - \frac{k}{p_x} \frac{\lambda_i}{w_r(\lambda_i - 1) + 1}\right), \tag{9.6}$$

with $w_r = w_s$ for random subset regression and $w_r = w_p$ for random projection regression.

In macroeconomic forecasting, it is generally assumed that a small number of principal components, associated with the largest eigenvalues of $\boldsymbol{\Sigma}_x$, drive the dependent variable. In other words, the nonzero $\gamma_{0,i}$ corresponds to $\lambda_i \gg 1$. In settings with p_x and $k < p_x$ sufficiently large, the weight w_r is close to k/p_x for both random subspace methods, and $w_r\lambda_i \gg 1$ for each $\lambda_i \gg 1$. The bound in (9.6) shows that in this setting a subspace dimension $k < p_x$ reduces the variance without inducing a large bias in the forecast. However, when $\gamma_{0,i}$ is nonzero for $\lambda_i < 1$, both methods should set k close to p_x to avoid bias.

9.6.3 Theoretical Results in the Literature

Random projections are historically motivated by the Johnson–Lindenstrauss lemma (Johnson and Lindenstrauss, 1984), for which Dasgupta and Gupta (2003) provide a simple proof. The lemma shows that random projections reduce the dimensionality of a set of points in Euclidean space while approximately preserving pairwise distances. Achlioptas (2003) and Li et al. (2006) show that this lemma also holds for random projections with very low computational costs. Given that they preserve pairwise distances, random projections can then, for example, be used to perform k-means clustering in a low-dimensional space with provably accurate assignment of observations to clusters (Boutsidis, Zouzias, Mahoney, and Drineas, 2015). Vempala (2005) gives an overview of the theoretical results for random projections.

Upper bounds on the in-sample mean squared error under random projections have been established by Maillard and Munos (2009) based on the Johnson–Lindenstrauss lemma, and improved upon by Kabán (2014). Thanei et al. (2017) showed that further improvements can be made by including a minimization step in the derivation, which uniformly lowers the upper bound. Boot and Nibbering (2019) improve these bounds further by recognizing that the random projection matrix, which is not exactly orthogonal, can be replaced by an exactly orthogonal matrix. They also provide the corresponding bound for random subset regression. The bounds show that the bias resulting from components associated with large eigenvectors is expected to be small. This is also the conclusion of Slawski et al. (2018), who use a different route to arrive at an upper bound, where the bias is shown to depend on the variation in the data associated with small eigenvalues. Elliott et al. (2013) provide an exact expression for the MSFE which contains the sum over all subsets and therefore does not lend itself for interpretation. Exact results for the mean squared forecast error for both random projections and random subset regression are not yet available.

In a Bayesian framework, Guhaniyogi and Dunson (2015) show that the predictive density under random projection converges to the true predictive density when the model is near sparse, i.e., the sum of the absolute coefficients remains finite when the sample size and the number of predictors increase.

9.7 Empirical Illustrations

This section illustrates the performance of the random subspace methods in forecasting macroeconomic indicators of the US economy. The results in the first application are taken from Boot and Nibbering (2019), where here we add the results for random compression. The second application provides new results for multi-step-ahead forecasts using the data from Stock and Watson (2002).

9.7.1 Empirical Application: FRED-MD

A detailed description of the data and methods can be found in Boot and Nibbering (2019). We forecast industrial production and consumer price index (as transformed by McCracken and Ng (2016)) from January 1980 to December 2014. The model is a linear AR(4) model, treating the remaining 129 series in the FRED-MD data set, as well as lags five and six, as potentially relevant predictors x_t. We compare the forecast performance as measured by mean squared forecast error of random subset regression (RS), random projection regression (RP), compressed regression (CR), principal component regression (PC), partial least squares (PL), ridge regression (RI), and lasso (LA). Compressed regression is implemented as described above, with the only exception that, like the other methods, the subspace dimension k is determined through historical forecast performance.

Results

Table 9.1 shows the MSFE of the selected forecasts relative to the AR(4) model, for forecasting the industrial production index (INDP) and inflation (CPI). The subspace dimension, number of factors, and regularization constants are determined based on past performance. Random subset regression is most accurate for forecast-

Table 9.1 FRED-MD: relative MSFE on industrial production and inflation

	RS	RP	CR	PC	PL	RI	LA
INDP	0.814	0.840	0.836	0.890	0.898	0.844	0.826
CPI	0.887	0.868	0.938	0.962	0.872	0.901	0.897

Note: This table shows the MSFE relative to the AR(4) model for Industrial Production (INDP) and the Consumer Price Index (CPI) for different methods. The tuning parameters are selected based on past predictive performance

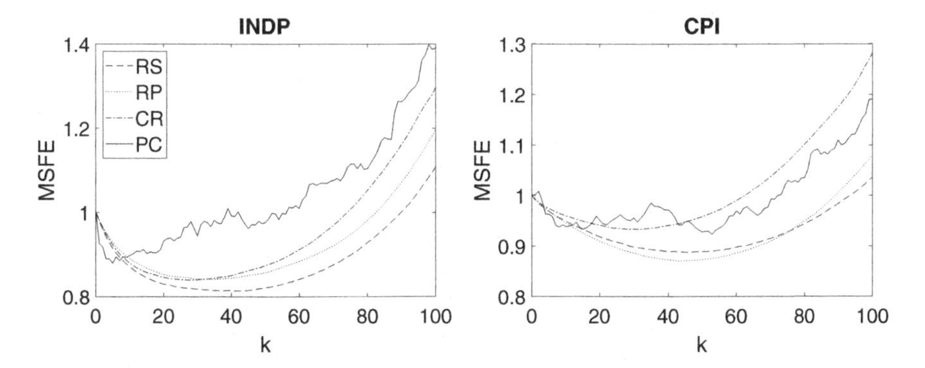

Fig. 9.1 Forecast accuracy for different subspace dimensions

ing the industrial production index and random projection regression for forecasting inflation.

Figure 9.1 shows how the MSFE depends on the subspace dimension k, and on the number of principal components, if the same dimension is selected throughout the forecast period. For industrial production a much smaller number of principal components is selected compared to CPI. The bounds on the MSFE of the random subspace methods discussed above suggest that random subset regression is more accurate if only principal components associated with the largest eigenvalues drive the dependent variable. The empirical result here, although based on only two series, agrees with this observation. We further note that the dependence on the subspace dimension is relatively mild.

9.7.2 Empirical Application: Stock and Watson (2002)

Stock and Watson (2002) study the performance of factor models on macroeconomic data for multi-step-ahead forecasts. Here, we repeat their analysis and compare the forecast error of the diffusion index models based on principal component analysis to random subset regression, random projection regression, and compressed regression.

Methods

Details of the data and appropriate transformations can be found in Stock and Watson (2002). The goal is to forecast a set of four real and four price variables. The real variables are: industrial production (IP), personal income (PI), manufacturing (MFG), and nonagricultural employment (NE). The price variables are consumer price index (CPI), consumption deflator (CD), CPI excluding food and energy (CFE), and the producer price index (PPI). For each of these variables, we consider $h = \{6, 12, 24\}$-step ahead forecasts for the period 1970:1+h, until 1998:12.

Stock and Watson (2002) consider three versions of a diffusion index model: a model with only diffusion indices as explanatory variables (DI), a model that adds a number of lags of the dependent variable to DI (DI-AR), and a model that also includes lags of the diffusion indices themselves (DI-AR, Lag). The first two models lower the dimension of the space spanned by the predictors by considering $\boldsymbol{f}_t' = \boldsymbol{x}_t'\boldsymbol{E}$ with $\boldsymbol{E}$ the matrix of eigenvector loadings of the correlation matrix of the predictors. The random subspace methods instead take $\boldsymbol{f}_t^i = \boldsymbol{x}_t'\boldsymbol{R}_i$, with $\boldsymbol{R}_i$ a random matrix, and average over a large number of forecasts obtained from using different realizations $\boldsymbol{R}_i$. The relative forecast performance then allows us to compare whether information is extracted more efficiently from $\boldsymbol{x}_t$ by using diffusion indices or random subspaces. For the DI-AR, Lag model, the random subspace analogue is less clear. For example, for random subset regression, we could select a number of variables and also include lags of these variables. Alternatively, we could expand $\boldsymbol{x}_t$ by lags of the data, and select directly from this enlarged matrix. We do not pursue these options here.

The forecasts of the diffusion index models are based on

$$\hat{y}_{T+h|T}^h = \hat{\alpha}_h + \sum_{j=1}^{m} \hat{\boldsymbol{\beta}}_{hj}' \hat{\boldsymbol{f}}_{T-j+1} + \sum_{j=1}^{p} \hat{\gamma}_{hj} y_{T-j+1},$$

where the factors $\hat{\boldsymbol{f}}_t$ are the first k principal components of the standardized matrix of predictors $\boldsymbol{X}$. The DI model takes $p = 0$, $m = 1$, and $0 \leq k \leq 12$. The DI-AR model takes $m = 1$, $0 \leq p \leq 6$, and $0 \leq k \leq 12$. The DI-AR, Lag model takes $0 \leq k \leq 4$, $1 \leq m \leq 3$, and $0 \leq p \leq 6$. The number of factors and lags are selected by BIC.

The number of lags and the subspace dimension for the subspace methods are based on past predictive performance, with a burn-in of 36 months. The maximum subspace dimension is set as $k_{\max} = 70$, which is the largest number feasible for the first forecasts.

Results

For the real variables, Table 9.2 reports the root mean squared forecast error for the AR model, and the diffusion index models DI, DI-AR, and DI-AR, Lag. For the

Table 9.2 Stock and Watson (2002): forecasting results real variables

	IP			PI			MS			NE		
Horizon	6	12	24	6	12	24	6	12	24	6	12	24
DI-AR, Lag	0.73	0.66	0.57	0.79	0.82	0.76	0.66	0.56	0.64	0.93	0.87	0.74
DI-AR	0.78	0.67	0.58	0.81	0.92	0.83	0.76	0.61	0.66	0.97	0.88	0.76
DI	0.74	0.59	0.58	0.81	0.92	0.83	0.68	0.57	0.66	0.95	0.84	0.75
RS-AR	0.59	0.55	0.72	0.72	0.76	0.90	0.61	0.53	0.82	0.76	0.73	0.86
RP-AR	0.61	0.57	0.64	0.72	0.77	0.90	0.62	0.56	0.79	0.77	0.72	0.83
CR-AR	0.65	0.59	0.67	0.73	0.77	0.92	0.66	0.56	0.80	0.78	0.75	0.82
RS	0.60	0.53	0.64	0.71	0.77	0.87	0.63	0.54	0.79	0.76	0.72	0.87
RP	0.60	0.56	0.64	0.71	0.77	0.89	0.64	0.56	0.78	0.76	0.73	0.84
CR	0.61	0.57	0.64	0.72	0.80	0.87	0.65	0.57	0.79	0.80	0.81	0.88
RMSE, AR	0.030	0.049	0.075	0.016	0.027	0.046	0.028	0.045	0.070	0.008	0.017	0.031

Note: This table shows the mean squared forecast error relative to the AR model for the real variables industrial production (IP), personal income (PI), manufacturing and sales (MS), nonagricultural employment (NE), for forecast horizons $h = \{6, 12, 24\}$, for a model with lags of the dependent variable and lags of the diffusion indices (DI-AR, Lag), with diffusion indices and lags of the dependent variable (DI-AR), and with diffusion indices only (DI). We also report the relative mean squared forecast error of the random subset regression (RS), random projection regression (RP), and compression regression (CR) with and without lags of the dependent variable

DI-AR and DI model, the table also reports the relative mean squared error of the three random subspace methods.

We see a clear pattern where the subspace methods outperform the diffusion indices for $h = \{6, 12\}$, but are outperformed for $h = 24$. We also see that the differences between the three random subspace methods are generally small. The ranking of the three methods on 6 months and 1 year horizons is relatively stable as: random subset regression, random projection regression, and random compression. For 2-year ahead forecasts random projection seems to be the most stable.

In Table 9.3, we report the results for the price variables. Here, the results are more diffuse. For CPI and PPI, the best diffusion index models outperform the random subspace methods at $h = \{6, 12\}$. For CD and CPI-FE there is no clear winner. Surprisingly, on the price variables, random subset regression is the weakest random subspace methods. The difference between random projection regression and random compression is generally small.

9.8 Discussion

This chapter examines random subspace methods in a macroeconomic forecasting framework. Since the number of variables in this setting approaches the number of observations, parameter uncertainty is an important source of forecast inaccuracy. Random subspace methods form a new class of dimension reduction methods that can potentially address these concerns. This chapter summarizes empirical and theoretical results on the use of random subspace methods in forecasting. There is much scope for expanding these results, and below we briefly discuss four possible directions for future research.

There are no formal results on the effect of instabilities on the forecast performance under random subspace methods. There is a wealth of empirical evidence for unstable forecasting relations in economic forecasting models. This leaves open the possibility that the success of random subspace methods in macroeconomic forecasting is due to the fact that they somehow reduce adversary effects of structural instabilities.

This chapter only focuses on point forecasts under mean squared error loss. The performance of random subspace methods under alternative loss functions for point forecasts is not well studied. Moving away from point forecasts, it would be interesting to see whether the combination of big data and subspace methods can contribute to improvements in density forecasting.

Theoretical results on random subspace forecasts are limited in two important ways. We lack easy to interpret exact expressions of the mean squared forecast error, even when the number of variables grows slowly compared to the number of observations. This leaves open the question whether these methods are in some sense optimal, or at least optimal for a particular class of data generating processes.

Another main case of interest is when the number of variables is much larger than the number of observations. Given that subspace methods can easily be implemented

Table 9.3 Stock and Watson (2002): forecasting results price inflation

	CPI			CD			CPI-FE			PPI		
Horizon	6	12	24	6	12	24	6	12	24	6	12	24
DI-AR, Lag	0.79	0.70	0.59	0.97	0.90	0.67	0.85	0.84	0.84	0.91	0.86	0.76
DI-AR	0.78	0.69	0.70	0.95	0.87	0.70	0.85	0.85	0.87	0.91	0.85	0.86
DI	1.59	1.30	1.07	1.64	1.34	1.07	1.74	1.57	1.46	2.41	2.43	2.11
RS-AR	0.80	0.77	0.67	0.93	0.89	0.65	0.86	0.79	0.91	0.95	0.90	0.79
RP-AR	0.79	0.75	0.64	0.93	0.86	0.65	0.85	0.82	0.78	0.94	0.87	0.76
CR-AR	0.80	0.73	0.67	0.91	0.87	0.67	0.85	0.78	0.79	0.94	0.88	0.80
RS	1.61	1.31	1.10	1.51	1.27	1.03	1.76	1.55	1.37	2.42	2.45	2.05
RP	1.44	1.17	0.95	1.39	1.16	0.92	1.44	1.31	1.18	2.05	1.92	1.58
CR	1.46	1.17	0.95	1.39	1.16	0.92	1.45	1.31	1.19	2.04	1.91	1.58
RMSE, AR	0.010	0.021	0.052	0.007	0.016	0.038	0.009	0.019	0.046	0.017	0.033	0.078

Note: This table shows the mean squared forecast error relative to the AR model for the price inflation variables consumer price index (CPI), consumption deflator (CD), consumer price index excluding food and energy (CPI-FE), producer price index (PPI). For further information, see the note following Table 9.2

in scenarios where $p_x \gg T$, theoretical results under the assumption that $p_x/T \rightarrow c$ with $c \in (0, \infty)$ or even $p_x/T \rightarrow \infty$ would be highly desirable. However, we are not aware of any results in this regime.

References

Achlioptas, D. (2003). Database-friendly random projections: Johnson-Lindenstrauss with binary coins. *Journal of Computer and System Sciences, 66*(4), 671–687.

Bai, J., & Ng, S. (2006). Confidence intervals for diffusion index forecasts and inference for factor-augmented regressions. *Econometrica, 74*(4), 1133–1150.

Bai, J., & Ng, S. (2008). Forecasting economic time series using targeted predictors. *Journal of Econometrics, 146*(2), 304–317.

Bay, S. D. (1998). Combining nearest neighbor classifiers through multiple feature subsets. In *Proceedings of the Fifteenth International Conference on Machine Learning ICML* (vol. 98, pp. 37–45). San Francisco: Morgan Kaufmann.

Bingham, E., & Mannila, H. (2001). Random projection in dimensionality reduction: Applications to image and text data. In *Proceedings of the Seventh ACM SIGKDD International Conference on Knowledge Discovery and Data Mining* (pp. 245–250). New York: ACM.

Boot, T., & Nibbering, D. (2019). Forecasting using random subspace methods. *Journal of Econometrics, 209*(2), 391–406. https://doi.org/10.1016/j.jeconom.2019.01.009.

Boutsidis, C., Zouzias, A., Mahoney, M. W., & Drineas, P. (2015). Randomized dimensionality reduction for k-means clustering. *IEEE Transactions on Information Theory, 61*(2), 1045–1062.

Breiman, L. (1996). Bagging predictors. *Machine Learning, 24*(2), 123–140.

Breiman, L. (1999). Pasting small votes for classification in large databases and on-line. *Machine Learning, 36*(1–2), 85–103.

Bryll, R., Gutierrez-Osuna, R., & Quek, F. (2003). Attribute bagging: Improving accuracy of classifier ensembles by using random feature subsets. *Pattern Recognition, 36*(6), 1291–1302.

Cannings, T. I., & Samworth, R. J. (2017). Random-projection ensemble classification. *Journal of the Royal Statistical Society: Series B (Statistical Methodology), 79*(4), 959–1035.

Chiong, K. X., & Shum, M. (2018). Random projection estimation of discrete-choice models with large choice sets. *Management Science, 65*, 1–457.

Claeskens, G., Magnus, J. R., Vasnev, A. L., & Wang, W. (2016). The forecast combination puzzle: a simple theoretical explanation. *International Journal of Forecasting, 32*(3), 754–762.

Dasgupta, S., & Gupta, A. (2003). An elementary proof of a theorem of Johnson and Lindenstrauss. *Random Structures & Algorithms, 22*(1), 60–65.

Elliott, G., Gargano, A., & Timmermann, A. (2013). Complete subset regressions. *Journal of Econometrics, 177*(2), 357–373.

Elliott, G., Gargano, A., & Timmermann, A. (2015). Complete subset regressions with large-dimensional sets of predictors. *Journal of Economic Dynamics and Control, 54*, 86–110.

Elliott, G., & Timmermann, A. (2013). *Handbook of economic forecasting.* Amsterdam: Elsevier.

Fradkin, D., & Madigan, D. (2003). Experiments with random projections for machine learning. In *Proceedings of the Ninth ACM SIGKDD International Conference on Knowledge Discovery and Data Mining* (pp. 517–522). New York: ACM.

Frieze, A., Kannan, R., & Vempala, S. (2004). Fast Monte-Carlo algorithms for finding low-rank approximations. *Journal of the Association for Computing Machinery, 51*(6), 1025–1041.

Garcia, M. G., Medeiros, M. C., & Vasconcelos, G. F. (2017). Real-time inflation forecasting with high-dimensional models: The case of Brazil. *International Journal of Forecasting, 33*(3), 679–693.

Gillen, B. J. (2016). Subset optimization for asset allocation. Social Science Working Paper, 1421, California Institute of Technology, Pasadena

Guhaniyogi, R., & Dunson, D. B. (2015). Bayesian compressed regression. *Journal of the American Statistical Association, 110*(512), 1500–1514.

Hansen, B. E. (2010). Averaging estimators for autoregressions with a near unit root. *Journal of Econometrics, 158*(1), 142–155.

Ho, T. K. (1998). The random subspace method for constructing decision forests. *IEEE Transactions on Pattern Analysis and Machine Intelligence, 20*(8), 832–844.

Johnson, W. B., & Lindenstrauss, J. (1984). Extensions of Lipschitz mappings into a Hilbert space. *Contemporary Mathematics, 26*(189–206), 1.

Kabán, A. (2014). New bounds on compressive linear least squares regression. In *Artificial intelligence and statistics* (pp. 448–456). Boston: Addison-Wesley.

Koop, G., Korobilis, D., & Pettenuzzo, D. (2019). Bayesian compressed vector autoregressions. *Journal of Econometrics, 210*(1), 135–154.

Leroux, M., Kotchoni, R., & Stevanovic, D. (2017). Forecasting economic activity in data-rich environment. University of Paris Nanterre, EconomiX.

Li, P., Hastie, T. J., & Church, K. W. (2006). Very sparse random projections. In *Proceedings of the 12th ACM SIGKDD International Conference on Knowledge Discovery and Data Mining* (pp. 287–296). New York: ACM.

Mahoney, M. W., & Drineas, P. (2009). CUR matrix decompositions for improved data analysis. *Proceedings of the National Academy of Sciences, 106*(3), 697–702.

Maillard, O., & Munos, R. (2009). Compressed least-squares regression. In *Advances in neural information processing systems* (pp. 1213–1221). Cambridge: MIT Press.

McCracken, M. W., & Ng, S. (2016). FRED-MD: A monthly database for macroeconomic research. *Journal of Business & Economic Statistics, 34*(4), 574–589.

Meligkotsidou, L., Panopoulou, E., Vrontos, I. D., & Vrontos, S. D. (2019). Out-of-sample equity premium prediction: A complete subset quantile regression approach. *European Journal of Finance*, 1–26.

Ng, S. (2013). Variable selection in predictive regressions. *Handbook of Economic Forecasting, 2*(Part B), 752–789.

Pick, A., & Carpay, M. (2018). Multi-step forecasting with large vector autoregressions. Working Paper.

Schneider, M. J., & Gupta, S. (2016). Forecasting sales of new and existing products using consumer reviews: A random projections approach. *International Journal of Forecasting, 32*(2), 243–256.

Shen, W., & Wang, J. (2017). Portfolio selection via subset resampling. In *Thirty-First AAAI Conference on Artificial Intelligence* (pp. 1517–1523).

Slawski, M. et al. (2018). On principal components regression, random projections, and column subsampling. *Electronic Journal of Statistics, 12*(2), 3673–3712.

Stock, J. H., & Watson, M. W. (2002). Forecasting using principal components from a large number of predictors. *Journal of the American Statistical Association, 97*(460), 1167–1179.

Stock, J. H., & Watson, M. W. (2006). Forecasting with many predictors. *Handbook of economic forecasting* (vol. 1, pp. 515–554). Amsterdam: Elsevier.

Thanei, G.-A., Heinze, C., & Meinshausen, N. (2017). Random projections for large-scale regression. In *Big and complex data analysis* (pp. 51–68). Berlin: Springer.

Tibshirani, R. (1996). Regression shrinkage and selection via the lasso. *Journal of the Royal Statistical Society. Series B (Methodological) 58,*(1), 267–288.

Timmermann, A. (2006). Forecast combinations. In *Handbook of economic forecasting* (vol. 1, pp. 135–196). Amsterdam: Elsevier.

Vempala, S. S. (2005). *The random projection method. Series in discrete mathematics and theoretical computer science*. Providence: American Mathematical Society.

Chapter 10
Variable Selection and Feature Screening

Wanjun Liu and Runze Li

10.1 Introduction

With the advent of modern technology for data collection, ultra-high dimensional datasets are widely encountered in machine learning, statistics, genomics, medicine, finance, marketing, etc. For example, in biomedical studies, huge numbers of magnetic resonance images (MRI) and functional MRI data are collected for each subject. Financial data is also of a high dimensional nature. Hundreds of thousands of financial instruments can be measured and tracked over time at very fine time intervals for use in high frequency trading. This ultra-high dimensionality causes challenges in both computation and methodology. Scalability is the major challenge to ultra-high dimensional data analysis. Many traditional methods that perform well for low-dimensional data do not scale to ultra-high dimensional data. Other issues such as high collinearity, spurious correlation, and noise accumulation (Fan & Lv, 2008, 2010) bring in additional challenges. Therefore, variable selection and feature screening have been a fundamental problem in the analysis of ultra-high dimensional data. For example, the issue of spurious correlation is illustrated by a simple example in Fan and Lv (2008). Suppose we have a $n \times p$ dataset with sample size n and the p predictors independently follow the standard normal distribution. When $p \gg n$, the maximum absolute value of sample correlation

---

W. Liu
Department of Statistics, The Pennsylvania State University, State College, PA, USA
e-mail: wxl204@psu.edu

R. Li (✉)
Department of Statistics and The Methodology Center, The Pennsylvania State University, State College, PA, USA
e-mail: rzli@psu.edu

P. Fuleky (ed.), *Macroeconomic Forecasting in the Era of Big Data*,
Advanced Studies in Theoretical and Applied Econometrics 52,
https://doi.org/10.1007/978-3-030-31150-6_10

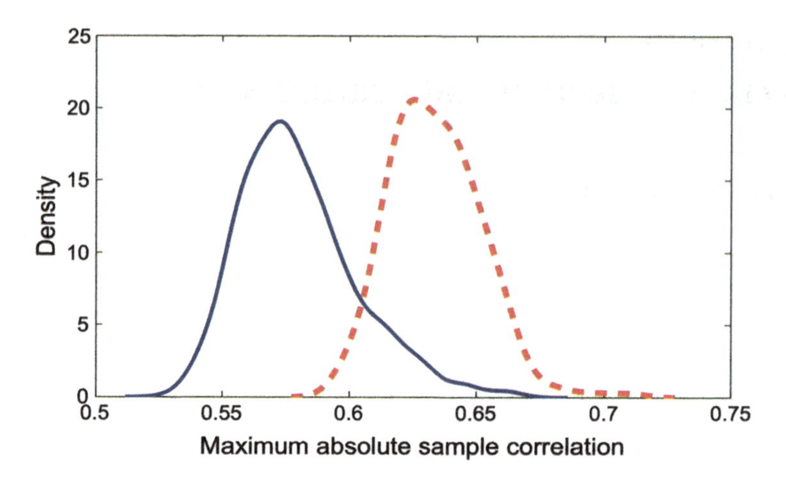

Fig. 10.1 Distributions of the maximum absolute sample correlation coefficient when $n = 60, p = 1000$ (solid curve) and $n = 60, p = 5000$ (dashed curve)

coefficient among predictors can be very large. Figure 10.1 shows the distributions of the maximum absolute sample correlation with $n = 60$ and $p = 1000, 5000$. Though the predictors are generated independently, some of them can be highly correlated due to high-dimensionality.

Over the past two decades, a large amount of variable selection approaches based on regularized M-estimation have been developed. These approaches include the Lasso (Tibshirani, 1996), the SCAD (Fan & Li, 2001), the Dantzig selector (Candes & Tao, 2007), and the MCP (Zhang, 2010), among others. However, these regularization methods may not perform well for ultra-high dimensional data due to the simultaneous challenges of computational expediency, statistical accuracy, and algorithmic stability (Fan, Samworth, & Wu, 2009). To improve the statistical performance of regularization methods and reduce computational cost, a class of two stage approaches is proposed. In the first stage, we reduce the number of features from a very large scale to a moderate size in a computationally fast way. Then in the second stage, we further implement refined variable selection algorithms such as regularization methods to the features selected from the first stage. Ideally, we select all the important features and may allow a few unimportant features entering our model in the first stage. The first stage is referred to as the feature screening stage. We will only focus on the feature screening stage in this chapter.

Suppose we have p features $X_1, \ldots, X_p$ in the feature space and denote the true index set of important variables by $\mathcal{M}_\star$. The definition of $\mathcal{M}_\star$ may vary across different models. For example, in a parametric model associated with true parameters $\boldsymbol{\beta}^\star = (\beta_1^\star, \ldots, \beta_p^\star)^\top$, $\mathcal{M}_\star$ is typically defined to be

$$\mathcal{M}_\star = \{1 \le j \le p : \beta_j^\star \neq 0\}.$$

Our goal in the feature screening stage is to select a submodel $\widehat{\mathcal{M}} \subset \{1, \ldots, p\}$ with little computational cost such that $\mathcal{M}_\star \subset \widehat{\mathcal{M}}$ with high probability. This is referred to as the sure screening property.

Definition 10.1 (Sure Screening) Let $\mathcal{M}_\star$ be the true index set of important features and $\widehat{\mathcal{M}}$ be the index set of selected important variables by some feature screening procedure based on a sample of size n, then this feature screening procedure has the sure screening property if

$$\Pr(\mathcal{M}_\star \subset \widehat{\mathcal{M}}) \to 1 \text{ as } n \to \infty.$$

The sure screening property ensures that all the important features will be included in the selected submodel with probability approaching to 1 as the sample size goes to infinity. A trivial but less interesting choice of $\widehat{\mathcal{M}}$ is $\widehat{\mathcal{M}} = \{1, \ldots, p\}$, which always satisfies the definition of sure screening. Here we assume the number of true important features is much smaller than p. This kind of assumption is also known as sparsity assumption in the sense that most of the entries in the true parameter $\boldsymbol{\beta}^\star$ are zero. Of interest is to find a $\widehat{\mathcal{M}}$ whose cardinality is much smaller than p (i.e., $|\widehat{\mathcal{M}}| \ll p$) and meanwhile the sure screening holds.

10.2 Marginal, Iterative, and Joint Feature Screening

10.2.1 Marginal Feature Screening

The most popular feature screening method is the marginal feature screening, which ranks the importance of features based on marginal utility and thus is computationally attractive. More specifically, the marginal feature screening procedure assigns an index, say $\widehat{\omega}_j$, to the feature X_j for $j = 1, \ldots, p$. This index $\widehat{\omega}_j$ measures the dependence between the jth feature and the response variable. Then we can rank the importance of all features according to $\widehat{\omega}_j$ and include the features ranked on the top in the submodel. For example, in the setting of linear regression, the index $\widehat{\omega}_j$ is chosen to be the absolute value of marginal Pearson correlation between the jth feature and the response (Fan & Lv, 2008). Features with larger absolute values of $\widehat{\omega}_j$ are more relevant to the response and thus are ranked on the top. As a result, we include the top d_n features in the submodel,

$$\widehat{\mathcal{M}}_{d_n} = \{1 \leq j \leq p : \widehat{\omega}_j \text{ is among the top } d_n \text{ ones}\},$$

where d_n is some pre-specified threshold. Note that the marginal feature screening procedure only uses the information of jth feature and the response without looking at all other features and thus it can be carried out in a very efficient way. A large amount of literature have studied the sure screening property of various marginal feature screening methods, see Fan and Lv (2008); Fan et al. (2009); Fan, Feng, and Song (2011); Fan, Ma, and Dai (2014); Li, Zhong, and Zhu (2012).

10.2.2 *Iterative Feature Screening*

As pointed out in Fan and Lv (2008), the marginal feature screening procedure may suffer from the following two issues:

1. Some unimportant features that are highly correlated with important features can have higher rankings than other important features that are relatively weakly related to the response.
2. An important feature that is marginally independent but jointly dependent on the response tends to have lower ranking.

The first issue says that the marginal feature screening has chance to include some unimportant features in the submodel. This is not a big issue for the purpose of feature screening. The second one is a bigger issue, which indicates that the marginal feature screening may fail to include all the important feature if it is marginally independent of the response. Absence of any important feature may lead to a biased estimation. To overcome the two aforementioned issues, one can apply an iterative feature screening procedure by iteratively carrying out the marginal screening procedure. This iterative procedure was first introduced by Fan and Lv (2008) and can be viewed as a natural extension of the marginal feature screening. At the kth iteration, we apply marginal feature screening to the features survived from the previous step and is typically followed by a regularization methods if a regression model is specified. Let $\widehat{\mathcal{M}}_k$ be the selected index set of important variables at the kth iteration and the final selected index set of important variables is given by $\widehat{\mathcal{M}} = \widehat{\mathcal{M}}_1 \cup \widehat{\mathcal{M}}_2 \cup \ldots$, the union of all selected index sets. For example, Fan and Lv (2008) uses the residuals computed from linear regression as the new response and iteratively applies marginal feature screening based on Pearson correlation. The iterative feature screening can significantly improve the simple marginal screening, but it can also be much more computationally expensive.

10.2.3 *Joint Feature Screening*

Another approach to improve the marginal screening is known as the joint screening (Xu & Chen, 2014; Yang, Yu, Li, & Buu, 2016). Many regularization methods involve solving an optimization problem of the following form:

$$\min_{\boldsymbol{\beta}} \frac{1}{n} \sum_{i=1}^{n} \ell(\mathbf{x}_i, \boldsymbol{\beta}) \quad \text{subject to } \|\boldsymbol{\beta}\|_0 \leq k, \tag{10.1}$$

where $\ell(\cdot, \cdot)$ is some loss function of negative log-likelihood function. It is quite challenging to solve the minimization problem in (10.1) especially in the ultra-high dimensional setting. The joint screening approach approximates the objective function by its Taylor's expansion and replaces the possibly singular Hessian matrix with some invertible matrix. After the approximation, one can solve such optimization problem iteratively in a fast manner. In many applications, one can obtain a closed form at each iteration for the joint screening approach.

10.2.4 Notations and Organization

We introduce some notations used in this chapter. Let $Y \in \mathbb{R}$ be the univariate response variable and $\mathbf{x} = (X_1, \ldots, X_p)^\top \in \mathbb{R}^p$ be the p-dimensional features. We observe a sample $\{(\mathbf{x}_i, Y_i)\}, i = 1, \ldots, n$ from the population $(\mathbf{x}, Y)$ with $\mathbf{x}_i = (X_{i1}, \ldots, X_{ip})^\top$. Let $\mathbf{y} = (Y_1, \ldots, Y_n)^\top$ be the response vector and $\mathbf{X} = (\mathbf{x}_1, \ldots, \mathbf{x}_n)^\top$ be the design matrix. We use $\mathbf{x}_{(j)}$ to denote the jth column of $\mathbf{X}$ and use $\mathbf{1}(\cdot)$ to denote the indicator function. For a vector $\boldsymbol{\beta} = (\beta_1, \ldots, \beta_p)^\top \in \mathbb{R}^p$, $\|\boldsymbol{\beta}\|_q = (\sum_{j=1}^p |\beta_j|^q)^{1/q}$ denotes its ℓ_q norm for $0 \leq q \leq \infty$. In particular, $\|\boldsymbol{\beta}\|_0 = \sum_{j=1}^p \mathbf{1}(|\beta_j| \neq 0)$ is the number of non-zero elements in $\boldsymbol{\beta}$ and $\|\boldsymbol{\beta}\|_\infty = \max_{1 \leq j \leq p} |\beta_j|$. For a symmetric matrix $\mathbf{M} \in \mathbb{R}^{p \times p}$, we use $\|\mathbf{M}\|_F$ and $\|\mathbf{M}\|_\infty$ to denote the Frobenius norm and supremum norm respectively. Let $\lambda_{\min}(\mathbf{M})$ and $\lambda_{\max}(\mathbf{M})$ be the smallest and largest eigenvalue of $\mathbf{M}$. Let $\mathcal{M}$ be a subset of $\{1, \ldots, p\}$ and $\boldsymbol{\beta}_{\mathcal{M}}$, a sub-vector of $\boldsymbol{\beta}$, consists of β_j for all $j \in \mathcal{M}$. We use $\mathcal{M}_\star$ to denote the true index set of important features and $\boldsymbol{\beta}^\star = (\beta_1^\star, \ldots, \beta_p^\star)^\top$ denote the true parameter. We assume $|\mathcal{M}_\star| = s$ throughout this chapter, where $|\mathcal{M}_\star|$ denotes the cardinality of the set $\mathcal{M}_\star$.

In the rest of this chapter, we spend most of the efforts reviewing the marginal feature screening methods as the marginal feature screening is the most popular screening method. The iterative feature screening can be viewed as a natural extension of marginal feature screening. We will discuss the details on the iterative and joint screening methods in one or two particular examples.

The rest of this chapter is organized as follows: In Sect. 10.3, we introduce the feature screening methods for independent and identically distributed data, which is the most common assumption in statistical modeling. Many different models have been developed for such data, including linear model, generalized linear model, additive model, varying-coefficient model, etc. However, this assumption is usually violated in areas such as finance and economics. In Sect. 10.4, we review the feature screening methods that are developed for longitudinal data, that is, data is collected over a period of time for each subject. In Sect. 10.5, we review the feature screening methods for survival data, which is widely seen in reliability analysis in engineering, duration analysis in economics, and event history analysis in sociology, etc.

10.3 Independent and Identically Distributed Data

Independent and identically distributed (IID) data is the most common assumption in statistical literature and a large amount of feature screening methods have been developed for IID data. In this section, we review some of the widely used feature screening methods for such data. Throughout this section, we assume that $\{(\mathbf{x}_i, Y_i)\}, i = 1, \dots, n$ is a random sample from the population $(\mathbf{x}, Y)$.

10.3.1 Linear Model

Let us consider the linear regression model

$$Y = \beta_0 + \mathbf{x}^\top \boldsymbol{\beta} + \epsilon, \tag{10.2}$$

where β_0 is the intercept, $\boldsymbol{\beta} = (\beta_1, \dots, \beta_p)^\top$ is a p-dimensional regression coefficient vector, and ϵ is the error term. In the ultra-high dimensional setting, the true regression coefficient vector $\boldsymbol{\beta}^\star = (\beta_1^\star, \dots, \beta_p^\star)^\top$ is assumed to be sparse, meaning that most of the coefficients $\beta_j^\star$ are 0. The true index set of the model is defined as

$$\mathcal{M}_\star = \{1 \le j \le p : \beta_j^\star \ne 0\}.$$

We call the features with indices in the set $\mathcal{M}_\star$ important features. Fan and Lv (2008) suggests ranking all features according to the marginal Pearson correlation coefficient between individual feature and the response and select the top features which have strong correlation with the response as important features. For a pre-specified value $\nu_n (0 < \nu_n < 1)$, the index set of selected features is given by

$$\widehat{\mathcal{M}}_{\nu_n} = \{1 \le j \le p : |\widehat{\text{corr}}(\mathbf{x}_{(j)}, \mathbf{y})| \text{ is among the top } \lfloor \nu_n n \rfloor \text{ largest ones}\},$$

where $\mathbf{x}_{(j)}$ is the jth column of $\mathbf{X}$, $\widehat{\text{corr}}$ denotes the sample Pearson correlation, and $\lfloor \nu_n n \rfloor$ is the integer part of $\nu_n n$. This procedure achieves the goal of feature screening since it reduces the ultra-high dimensionality down to a relatively moderate scale $\lfloor \nu_n n \rfloor$. This procedure is referred to as the sure independence screening (SIS). Then appropriate regularization methods such as Lasso, SCAD, and Dantzig selector can be further applied to the selected important features. The corresponding methods are referred to as SIS-LASSO, SIS-SCAD, and SIS-DS. This feature screening procedure is based on Pearson correlation and can be carried out in an extremely simple way at very low computational cost. In addition to the computational advantage, this SIS enjoys the sure screening property. Assume that the error is normally distributed and the following conditions hold:

(A1) $\min_{j\in\mathcal{M}_\star} \beta_j^\star \geq c_1 n^{-\kappa}$ and $\min_{j\in\mathcal{M}_\star} |\text{cov}(\beta_j^{\star-1} Y, X_j)| \geq c_2$, for some $\kappa > 0$ and $c_1, c_2 > 0$.
(A2) There exists $\tau \geq 0$ and $c_3 > 0$ such that $\lambda_{\max}(\boldsymbol{\Sigma}) \leq c_3 n^\tau$, where $\boldsymbol{\Sigma} = \text{cov}(\mathbf{x})$ is the covariance matrix of $\mathbf{x}$ and $\lambda_{\max}(\boldsymbol{\Sigma})$ is the largest eigenvalue of $\boldsymbol{\Sigma}$.
(A3) $p > n$ and $\log p = O(n^\xi)$ for some $\xi \in (0, 1-2\kappa)$.

Fan and Lv (2008) showed that if $2\kappa + \tau < 1$, then with the choice of $\nu_n = cn^{-\theta}$ for some $0 < \theta < 1 - 2\kappa - \tau$ and $c > 0$, we have for some $C > 0$

$$\Pr(\mathcal{M}_\star \subset \widehat{\mathcal{M}}_{\nu_n}) \geq 1 - O(\exp\{-Cn^{1-2\kappa} / \log n\}). \tag{10.3}$$

Conditions (A1) requires certain order of minimal signal among the important features, condition (A2) rules out the case of strong collinearity, and condition (A3) allows p grows exponentially with sample size n. Equation (10.3) shows that the SIS can reduce the exponentially growing dimension p down to a relatively small scale $d_n = \lfloor \nu_n n \rfloor = O(n^{1-\theta}) < n$, while include all important features in the submodel with high probability. The optimal choice of d_n relies on unknown parameters. It is common to assume $s/n \to 0$ where s is the number of important features. In practice, one can conservatively set $d_n = n - 1$ or require $d_n/n \to 0$ with $d_n = n/\log n$. See more details in Fan and Lv (2008).

Marginal Pearson correlation is employed to rank the importance of features; SIS may suffer from the potential issues with marginal screening. On the one hand, SIS may fail to select the important feature when it is jointly correlated but marginally uncorrelated with the response. On the other hand, the SIS tends to select unimportant features which are jointly uncorrelated but highly marginally correlated with the response. To address these issues, Fan and Lv (2008) also introduced an iterative SIS procedure (ISIS) by iteratively replacing the response with the residuals obtained from the linear regression using the selected features from the previous step. The ISIS works as follows: In the first iteration, we select a subset of k_1 features $\mathcal{A}_1 = \{X_{i_1}, \ldots, X_{i_{k_1}}\}$ using an SIS based model selection method such as SIS-LASSO or SIS-SCAD. Then we regress the response Y over the selected features $\mathcal{A}_1$ and obtain the residuals. We treat the residuals as the new responses and apply the same method to the remaining $k_2 = p - k_1$ features $\mathcal{A}_2 = \{X_{j_1}, \ldots, X_{j_{k_2}}\}$. We keep doing this until we get l disjoint subsets $\mathcal{A}_1, \ldots, \mathcal{A}_l$ such that $d = \sum_{i=1}^{l} |\mathcal{A}_i| < n$. We use the union $\mathcal{A} = \cup_{i=1}^{\ell} \mathcal{A}_i$ as the set of selected features. In practical implementation, we can choose, for example, the largest l such that $|\mathcal{A}| < n$. This iterative procedure makes those important features that are missed in the previous step possible to re-enter the selected model. In fact, after features in $\mathcal{A}_1$ entering into the model, those that are marginally weakly correlated with Y purely due to the presence of variables in $\mathcal{A}_1$ should now be correlated with the residuals.

10.3.2 Generalized Linear Model and Beyond

A natural extension of SIS is applying the feature screening procedure to generalized linear models. Assume that the response Y is from an exponential family with the following canonical form:

$$f_Y(y,\theta) = \exp\{y\theta - b(\theta) + c(y)\},$$

for some known functions $b(\cdot)$, $c(\cdot)$, and unknown parameter θ. Consider the following generalized linear model:

$$E(Y|\mathbf{x}) = g^{-1}(\beta_0 + \mathbf{x}^\top \boldsymbol{\beta}), \tag{10.4}$$

where $g(\cdot)$ is the link function, β_0 is an unknown scalar, and $\boldsymbol{\beta} = (\beta_1, \ldots, \beta_p)^\top$ is a p-dimensional unknown vector. The linear regression model in (10.2) is just a special case of (10.4) by taking $g(\mu) = \mu$. Without loss of generality, we assume that all the features are standardized to have mean zero and standard deviation one. Fan and Song (2010) proposes a feature screening procedure for (10.4) by ranking the maximum marginal likelihood estimator (MMLE). For each $1 \le j \le p$, the MMLE $\widehat{\boldsymbol{\beta}}_j^M$ is a 2-dimensional vector and defined as

$$\widehat{\boldsymbol{\beta}}_j^M = (\widehat{\beta}_{j0}^M, \widehat{\beta}_{j1}^M)^\top = \arg\min_{\beta_{j0},\beta_{j1}} \frac{1}{n}\sum_{i=1}^n \ell(Y_i, \beta_{j0} + \beta_{j1}X_{ij}), \tag{10.5}$$

where $\ell(y,\theta) = -y\theta + b(\theta) - c(y)$ is the negative log-likelihood function. The minimization problem in (10.5) can be rapidly computed and its implementation is robust since it only involves two parameters. Such a feature screening procedure ranks the importance of features according to their magnitude of marginal regression coefficients. The set of important features is defined as

$$\widehat{\mathcal{M}}_{\nu_n} = \{1 \le j \le p : |\widehat{\beta}_{j1}^M| > \nu_n\},$$

where ν_n is some pre-specified threshold. As a result, we dramatically decrease the dimension from p to a moderate size by choosing a large ν_n and hence the computation is much more feasible after screening. Although the interpretations and implications of the marginal models are biased from the full model, it is suitable for the purpose of variable screening. In the linear regression setting, the MMLE ranking is equivalent to the marginal correlation ranking. However, the MMLE screening does not rely on the normality assumption and can be more easily applied to other models. Under proper regularity conditions, Fan and Song (2010) established the sure screening property of the MMLE ranking. By taking $\nu_n = cn^{1-2\kappa}$ for some $0 < \kappa < 1/2$ and $c > 0$, we have

$$\Pr(\mathcal{M}_\star \subset \widehat{\mathcal{M}}_{\nu_n}) \to 1 \quad \text{as } n \to \infty.$$

See more about the details about the conditions in Fan and Song (2010). This MMLE procedure can handle the NP-dimensionality of order

$$\log p = o(n^{(1-2\kappa)\alpha/(\alpha+2)}),$$

where α is some positive parameter that characterizes how fast the tail of distribution of features decay. For instance, $\alpha = 2$ corresponds to normal features and $\alpha = \infty$ corresponds to features that are bounded. When features are normal ($\alpha = 2$), the MMLE gives a weaker result than that of the SIS which permits $\log p = o(n^{1-2\kappa})$. However, MMLE allows non-normal features and other error distributions.

Fan et al. (2009) studied a very general pseudo-likelihood framework in which the aim is to find the parameter vector $\boldsymbol{\beta} = (\beta_1, \ldots, \beta_p)^\top$ that is sparse and minimizes an objective function of the form

$$Q(\beta_0, \boldsymbol{\beta}) = \frac{1}{n} \sum_{i=1}^{n} \ell(Y_i, \beta_0 + \boldsymbol{\beta}^\top \mathbf{x}_i), \tag{10.6}$$

where the function $\ell(\cdot, \cdot)$ can be some loss function or negative log-likelihood function. This formulation in (10.6) includes a lot of important statistical models including

1. **Generalized linear models**: All generalized linear models, including logistic regression and Poisson log-linear models, fit very naturally into the framework.
2. **Classification**: Some common approaches to classification assume the response takes values in $\{-1, 1\}$ also fit the framework. For instance, support vector machine (Vapnik, 2013) uses the hinge loss function $\ell(Y_i, \beta_0 + \mathbf{x}_i^\top \boldsymbol{\beta}) = (1 - Y_i(\beta_0 + \mathbf{x}_i^\top \boldsymbol{\beta}))_+$, while the boosting algorithm AdaBoost (Freund & Schapire, 1997) uses $\ell(Y_i, \beta_0 + \mathbf{x}_i^\top \boldsymbol{\beta}) = \exp\{-Y_i(\beta_0 + \mathbf{x}_i^\top \boldsymbol{\beta})\}$.
3. **Robust fitting**: Instead of the conventional least squares loss function, one may prefer a robust loss function such as the ℓ_1 loss $\ell(Y_i, \beta_0 + \mathbf{x}_i^\top \boldsymbol{\beta}) = |Y_i - \beta_0 - \mathbf{x}_i^\top \boldsymbol{\beta}|$ or the Huber loss (Huber, 1964), which also fits into the framework.

Fan et al. (2009) suggests to rank the importance of features according to their marginal contributions to the magnitude of the likelihood function. This method can be viewed as a marginal likelihood ratio screening, as it builds on the increments of the log-likelihood. The marginal utility of the jth feature X_j is quantified by

$$L_j = \min_{\beta_0, \beta_j} n^{-1} \sum_{i=1}^{n} \ell(Y_i, \beta_0 + X_{ij}\beta_j).$$

The idea is to compute the vector of marginal utilities $\mathbf{L} = (L_1, \ldots, L_p)^\top$ and rank the features according to the marginal utilities: the smaller L_j is, the more important X_j is. Note that in order to compute L_j, we only need to fit a model with two parameters, β_0 and β_j, so computing the vector $\mathbf{L}$ can be done very quickly and stably, even for an ultra-high dimensional problem. The feature X_j is selected if the corresponding utility L_j is among the d_n smallest components of $\mathbf{L}$. Typically, we may take $d_n = \lfloor n/\log n \rfloor$. When d_n is large enough, it has high probability of selecting all of the important features. The marginal likelihood screening and the MMLE screening share a common computation procedure as both procedures solve p optimization problems over a two-dimensional parameter space. Fan and Song (2010) showed that these two procedures are actually equivalent in the sense that they both possess the sure screening property and that the number of selected variables of the two methods is of the same order of magnitude.

Fan et al. (2009) also proposes an iterative feature screening procedure, which consists of the following steps:

Step 1. Compute the vector of marginal utilities $\mathbf{L} = (L_1, \ldots, L_p)^\top$ and select the set $\widehat{\mathcal{A}}_1 = \{1 \le j \le p : L_j$ is among the first k_1 smallest ones$\}$. Then apply a penalized (pseudo)-likelihood, such as Lasso and SCAD, to select a subset $\widehat{\mathcal{M}}$.

Step 2. For each $j \in \{1, \ldots, p\}/\widehat{\mathcal{M}}$, compute

$$L_j^{(2)} = \min_{\beta_0, \beta_j, \boldsymbol{\beta}_{\widehat{\mathcal{M}}}} \frac{1}{n} \sum_{i=1}^{n} L(Y_i, \beta_0 + \mathbf{x}_{i,\widehat{\mathcal{M}}}^\top \boldsymbol{\beta}_{\widehat{\mathcal{M}}} + X_{ij}\beta_j), \tag{10.7}$$

where $\mathbf{x}_{i,\widehat{\mathcal{M}}}$ denotes the sub-vector of $\mathbf{x}_i$ consisting of those elements in $\widehat{\mathcal{M}}$. Then select the set

$$\widehat{\mathcal{A}}_2 = \{j \in \{1, \ldots, p\}/\widehat{\mathcal{M}} : L_j^{(2)} \text{ is among the first } k_2 \text{ smallest ones}\}.$$

Step 3. Use penalized likelihood to the features in set $\widehat{\mathcal{M}} \cup \widehat{\mathcal{A}}_2$,

$$\widehat{\boldsymbol{\beta}}_2 = \mathop{\arg\min}_{\beta_0, \boldsymbol{\beta}_{\widehat{\mathcal{A}}_2}, \boldsymbol{\beta}_{\widehat{\mathcal{M}}}} \frac{1}{n} \sum_{i=1}^{n} \ell(Y_i, \beta_0 + \mathbf{x}_{i,\widehat{\mathcal{M}}}^\top \boldsymbol{\beta}_{\widehat{\mathcal{M}}} + \mathbf{x}_{i,\widehat{\mathcal{A}}_2}^\top \boldsymbol{\beta}_{\widehat{\mathcal{A}}_2}) + \sum_{j \in \widehat{\mathcal{M}} \cup \widehat{\mathcal{A}}_2} p_\lambda(|\beta_j|),$$

where $p_\lambda(\cdot)$ is some penalty function such as Lasso or SCAD. The indices of $\widehat{\boldsymbol{\beta}}_2$ that are non-zero yield a new estimated set $\widehat{\mathcal{M}}$.

Step 4. Repeat Step 2 and Step 3 and stop once $|\widehat{\mathcal{M}}| \ge d_n$.

Note that $L_j^{(2)}$ can be interpreted as the additional contribution of feature X_j given the presence of features in $\widehat{\mathcal{M}}$. The optimization problem in Step 2 is a low-dimensional problem which can be solved efficiently. An alternative approach in Step 2 is to substitute the fitted value $\widehat{\boldsymbol{\beta}}_{\widehat{\mathcal{M}}_1}$ from the Step 1 into (10.7). Then the optimization in (10.7) only involves two parameters and is exactly an extension of Fan and Lv (2008). To see this, let $r_i = Y_i - \mathbf{x}_{i,\widehat{\mathcal{M}}}^\top \boldsymbol{\beta}_{\widehat{\mathcal{M}}}$ denote the residual from the previous step and we choose the square loss function, then

$$\ell(Y_i, \beta_0 + \mathbf{x}_{i,\widehat{\mathcal{M}}}^\top \boldsymbol{\beta}_{\widehat{\mathcal{M}}} + X_{ij}\beta_j) = (r_i - \beta_0 - \beta_j X_{ij})^2.$$

Without explicit definition of residuals, the idea of considering additional contribution to the response can be applied to a much more general framework.

10.3.3 Nonparametric Regression Models

Fan et al. (2011) proposes a nonparametric independence screening (NIS) for ultra-high dimensional additive model of the following form:

$$Y = \sum_{j=1}^{p} m_j(X_j) + \epsilon, \tag{10.8}$$

where $m_j(X_j)$ is assumed to have mean zero for identifiability. The true index set of important features is defined as

$$\mathcal{M}_\star = \{1 \leq j \leq p : Em_j^2(X_j) > 0\}.$$

To identify the important features in (10.8), Fan et al. (2011) considers the following p marginal nonparametric regression problems

$$\min_{f_j \in L_2(P)} E(Y - f_j(X_j))^2, \tag{10.9}$$

where P denotes the joint distribution of $(\mathbf{x}, Y)$ and $L_2(P)$ is the family of square integrable functions under the measure P. The minimizer of (10.9) is $f_j = E(Y|X_j)$ and hence $Ef_j^2(X_j)$ can be used as marginal utility to measure the importance of feature X_j at population level. Given a random sample $\{(\mathbf{x}_i, Y_i)\}, i = 1, \ldots, n$, $f_j(x)$ can be estimated by a set of B-spline basis. Let $\mathbf{B}(x) = (B_1(x), \ldots, B_L(x))^\top$ be a B-spline basis and $\boldsymbol{\beta}_j = (\beta_{j1}, \ldots, \beta_{jL})^\top$ be the corresponding coefficients for the B-spline basis associated with feature X_j. Consider the following least squares,

$$\widehat{\boldsymbol{\beta}}_j = \arg\min_{\boldsymbol{\beta}_j} \frac{1}{n} \sum_{i=1}^{n} (Y_i - \boldsymbol{\beta}_j^\top \mathbf{B}(X_{ij}))^2.$$

Thus $f_j(x)$ can be estimated by $\widehat{f}_j(x) = \widehat{\boldsymbol{\beta}}_j^\top \mathbf{B}(x)$. The index set of selected submodel is given by

$$\widehat{\mathcal{M}}_{\nu_n} = \{1 \leq j \leq p : \|\widehat{f}_j\|_n^2 \geq \nu_n\},$$

where $\|\widehat{f}_j\|_n^2 = n^{-1} \sum_{i=1}^n \widehat{f}_j(X_{ij})^2$ and ν_n is some pre-specified threshold. The NIS ranks the importance according to the marginal strength of the marginal nonparametric regression. Under the regularity conditions, Fan et al. (2011) shows that by taking $\nu_n = c_1 L n^{-2\kappa}$, we have

$$\Pr(\mathcal{M}_\star \subset \widehat{\mathcal{M}}_{\nu_n}) \geq 1 - sL[(8+2L)\exp(-c_2 n^{1-4\kappa} L^{-3}) + 6L\exp(-c_3 n L^{-3})],$$

where L is the number of B-spline basis, $s = |\mathcal{M}_\star|$ and c_2, c_3 are some positive constants. It follows that if

$$\log p = o(n^{1-4\kappa} L^{-3} + nL^{-3}), \tag{10.10}$$

then $\Pr(\mathcal{M}_\star \subset \widehat{\mathcal{M}}_{\nu_n}) \to 1$. It is worthwhile to point out that the number of spline bases L affects the order of dimensionality. Equation (10.10) shows that the smaller the number of basis functions, the higher the dimensionality that the NIS can handle. However, the number of basis functions cannot be too small since the approximation error would be too large if we only use a small number of basis functions. After the feature screening, a natural next step is to use penalized method for additive model such as penGAM proposed in Meier, Van de Geer and Bühlmann (2009) to further select important features. Similar to the iterative procedure in Fan et al. (2009), Fan et al. (2011) also introduces an iterative version of NIS, namely INIS-penGAM, by carrying out the NIS procedure and penGAM alternatively. We omit the details here.

Varying-coefficient model is another important nonparametric statistical model that allows us to examine how the effects of features vary with some exposure variable. It is a natural extension of classical linear models with good interpretability and flexibility. Varying-coefficient model arises frequently in economics, finance, epidemiology, medical science, ecology, among others. For an overview, see Fan and Zhang (2008). An example of varying-coefficient model is the analysis of cross-country growth. Linear model is often used in the standard growth analysis. However, a particular country's growth rate will depend on its state of development and it would make much more sense if we treat the coefficients as functions of the state of development, which leads to a standard varying-coefficient model (Fan & Zhang, 2008). In this example, state of development is the exposure variable.

Consider the following varying-coefficient model,

$$Y = \sum_{j=1}^{p} \beta_j(U) X_j + \epsilon, \tag{10.11}$$

where U is some observable univariate exposure variable and the coefficient $\beta_j(\cdot)$ is a smooth function of variable U. In the form of (10.11), the features X_j enter the model linearly. Such nonparametric formulation allows nonlinear interactions between the exposure variable and the features. The true index set of important features is defined as

$$\mathcal{M}_\star = \{1 \leq j \leq p : E(\beta_j^2(U)) > 0\},$$

with model size $s = |\mathcal{M}_\star|$. Fan et al. (2014) considered a nonparametric screening procedure by ranking a measure of the marginal nonparametric contribution of each feature given the exposure variable. For each feature X_j, $j = 1, \ldots, p$, consider the following marginal regression:

$$\min_{a_j, b_j} E[(Y - a_j - b_j X_j)^2 | U]. \quad (10.12)$$

Let $a_j(U)$ and $b_j(U)$ be the solution to (10.12) and we have

$$b_j(U) = \frac{\text{Cov}[X_j, Y|U]}{\text{Var}[X_j|U]} \text{ and } a_j(U) = E(Y|U) - b_j(U)E(X_j|U).$$

The marginal contribution of X_j for the response can be characterized by

$$\omega_j = \|a_j(U) + b_j(U)X_j)\|^2 - \|a_0(U)\|^2, \quad (10.13)$$

where $a_0(U) = E[Y|U]$ and $\|f\|^2 = Ef^2$. By some algebra, it can be seen that

$$\omega_j = E\left[\frac{(\text{Cov}[X_j, Y|U])^2}{\text{Var}[X_j|U]}\right].$$

This marginal utility ω_j is closely related to the conditional correlation between X_j and Y since $\omega_j = 0$ if and only if $\text{Cov}[X_j, Y|U] = 0$. On the other hand, if we assume $\text{Var}[X_j|U] = 1$, then the marginal utility ω_j is the same as the measure of marginal functional coefficient $\|b_j(U)\|^2$.

Suppose we have a random sample $\{(\mathbf{x}_i, Y_i, U_i)\}, i = 1, \ldots, n$, similar to the setting of additive model, we can estimate $a_j(U)$, $b_j(U)$, and $a_0(U)$ using B-spline technique. Let $\mathbf{B}(U) = (B_1(U), \ldots, B_L(U))^\top$ be a B-spline basis and the coefficients of B-splines can be estimated by the following marginal regression problems:

$$(\widehat{\boldsymbol{\eta}}_j, \widehat{\boldsymbol{\theta}}_j) = \min_{\boldsymbol{\eta}_j, \boldsymbol{\theta}_j} n^{-1} \sum_{i=1}^n (Y_i - \mathbf{B}(U_i)^\top \boldsymbol{\eta}_j - \mathbf{B}(U_i)^\top \boldsymbol{\theta}_j X_{ij})^2,$$

$$\widehat{\boldsymbol{\eta}}_0 = \min_{\boldsymbol{\eta}_0} n^{-1} \sum_{i=1}^n (Y_i - \mathbf{B}(U_i)^\top \boldsymbol{\eta}_0)^2,$$

where $\boldsymbol{\eta}_0 = (\eta_{0_1}, \ldots, \eta_{0_L})^\top$, $\boldsymbol{\eta}_j = (\eta_{j_1}, \ldots, \eta_{j_L})^\top$, and $\boldsymbol{\theta}_j = (\theta_{j1}, \ldots, \theta_{j_L})^\top$ are the B-spline coefficients for $a_0(U)$, $a_j(U)$, and $b_j(U)$, respectively. As a result, $\widehat{a}_j(U), \widehat{b}_j(U)$, and $\widehat{a}_0(U)$ can be estimated by

$$\widehat{a}_j(U) = \mathbf{B}(U)^\top \widehat{\boldsymbol{\eta}}_j, \widehat{b}_j(U) = \mathbf{B}(U)^\top \widehat{\boldsymbol{\theta}}_j, \text{ and } \widehat{a}_0(U) = \mathbf{B}(U)^\top \widehat{\boldsymbol{\eta}}_0.$$

The sample marginal utility for screening is

$$\widehat{\omega}_j = \|\widehat{a}_j(U) + \widehat{b}_j(U)\|_n^2 - \|\widehat{a}_0(U)\|_n^2,$$

where $\|f(U)\|_n^2 = n^{-1}\sum_{i=1}^n f(U_i)^2$. The submodel is selected by

$$\widehat{\mathcal{M}}_{\nu_n} = \{1 \le j \le p : \widehat{\omega}_j \ge \nu_n\}.$$

Under regularity conditions, Fan et al. (2014) established the sure screening property for their proposed screening procedure if the dimensionality satisfies $\log p = o(n^{1-4\kappa}L^{-3})$ for some $0 < \kappa < 1/4$, which is of the same order for the additive model setting. An iterative nonparametric independence screening procedure is also introduced in Fan et al. (2014), which repeatedly applies the feature screening procedure followed by a moderate scale penalized method such as group-SCAD (Wang, Li, & Huang, 2008).

Instead of using the marginal contribution in (10.13) to rank the importance of features, Liu, Li, and Wu (2014) proposed a screening procedure based on conditional correlation for varying-coefficient model. Given U, the conditional correlation between X_j and Y is defined as the conditional Pearson correlation

$$\rho(X_j, Y|U) = \frac{\text{cov}(X_j, Y|U)}{\sqrt{\text{cov}(X_j, X_j|U)\text{cov}(Y, Y|U)}}.$$

Then $E[\rho^2(X_j, Y|U)]$ can be used as a marginal utility to evaluate the importance of X_j at population level. It can be estimated by the kernel regression (Liu et al., 2014). The features with high conditional correlations will be included in the selected submodel. This procedure can be viewed as a natural extension of the SIS by conditioning on the exposure variable U.

10.3.4 Model-Free Feature Screening

In previous sections, we have discussed model-based feature screening procedures for ultra-high dimensional data, which requires us to specify the underlying true model structure. However, it is quite challenging to correctly specify the model structure on the regression function in high dimensional modeling. Mis-specification of the data generation mechanism could lead to large bias. In practice, one may do not know what model to use unless the dimensionality of feature space is reduced to a moderate size. To achieve greater realism, model-free feature screening is necessary for high dimensional modeling. In this section, we review several model-free feature screening procedures.

Recall that under the parametric modeling, the true index set of important features $\mathcal{M}_\star$ is defined as the indices of non-zero elements in $\boldsymbol{\beta}^\star$. Since no assumption is made on the specification of the model, there is no such true parameter $\boldsymbol{\beta}^\star$ and thus we need to redefine the true index set of important features $\mathcal{M}_\star$. Let Y be the response variable and $\mathbf{x} = (X_1, \ldots, X_p)^\top$ be the p-dimensional covariate vector. Define the index set of important features as

$$\mathcal{M}_\star = \{1 \leq j \leq p : F(y|\mathbf{x}) \text{ functionally depends on } X_j \text{ for any } y \in \Psi_y\},$$

where $F(y|\mathbf{x}) = \Pr(Y < y|\mathbf{x})$ is the conditional distribution function of Y given $\mathbf{x}$ and Ψ_y is the support of Y. This indicates that conditional on $\mathbf{x}_{\mathcal{M}_\star}$, Y is statistically independent of $\mathbf{x}_{\mathcal{M}_\star^c}$, where $\mathbf{x}_{\mathcal{M}_\star}$ is a s-dimensional sub-vector of $\mathbf{x}$ consisting of all X_j with $j \in \mathcal{M}_\star$ and $\mathcal{M}_\star^c$ is the complement of $\mathcal{M}_\star$.

Zhu, Li, Li, and Zhu (2011) considered a general model framework under which $F(y|\mathbf{x})$ depends on $\mathbf{x}$ only through $\mathbf{B}^\top \mathbf{x}_{\mathcal{M}_\star}$, where $\mathbf{B}$ is a $s \times K$ unknown parameter matrix. In other words, we assume $F(y|\mathbf{x}) = F(y|\mathbf{B}^\top \mathbf{x}_{\mathcal{M}_\star})$. Note that $\mathbf{B}$ may not be identifiable. What is identifiable is the space spanned by the columns of $\mathbf{B}$. However, the identifiability of $\mathbf{B}$ is of no concern here because our primary goal is to identify important features rather than estimating $\mathbf{B}$ itself. This general framework covers a wide range of existing models including the linear regression model, generalized linear models, the partially linear model (Hardle, Liang, & Gao, 2012), the single-index model (Hardle, Hall, & Ichimura, 1993), and the partially linear single-index model (Carroll, Fan, Gijbels, & Wand, 1997), etc. It also includes the transformation regression model with a general transformation $h(Y)$.

Zhu et al. (2011) proposes a unified screening procedure for this general framework. Without loss of generality, assume $E(X_j) = 0$ and $Var(X_j) = 1$. Define $\boldsymbol{\Omega}(y) = E[\mathbf{x}F(y|\mathbf{x})]$. It then follows by the law of iterated expectations that $\boldsymbol{\Omega}(y) = E[\mathbf{x}E(\mathbf{1}(Y < y|\mathbf{x}))|\mathbf{x}] = \text{cov}(\mathbf{x}, \mathbf{1}(Y < y))$. Let $\Omega_j(y)$ be the jth element of $\boldsymbol{\Omega}(y)$ and define

$$\omega_j(y) = E(\Omega_j^2(y)), \ j = 1, \ldots, p.$$

Under certain conditions, Zhu et al. (2011) showed that

$$\max_{j \in \mathcal{M}_\star^c} \omega_j < \min_{j \in \mathcal{M}_\star} \omega_j \quad \text{uniformly for } p,$$

and $\omega_j = 0$ if and only if $cov(\mathbf{B}^\top \mathbf{x}_{\mathcal{M}_\star}, X_j) = 0$. These results reveal that the quantity ω_j is in fact a measure of the correlation between the marginal covariate X_j and the linear combination $\mathbf{B}^\top \mathbf{x}_{\mathcal{M}_\star}$ and hence can be used as a marginal utility. Here are some insights. If X_j and Y are independent, so are X_j and $\mathbf{1}(Y < y)$. Consequently, $\Omega_j(y) = 0$ for all $y \in \Psi_y$ and $\omega_j = 0$. On the other hand, if X_j and Y are dependent, then there exists some $y \in \Psi_y$ such that $\Omega_j(y) \neq 0$, and hence ω_j must be positive. In practice, one can employ the sample estimate of ω_j to rank the features. Given a random sample $\{(\mathbf{x}_i, Y_i)\}, i = 1, \ldots, n$, and assume the features are standardized in the sense that $n^{-1}\sum_{i=1}^{n} X_{ij} = 0$ and $n^{-1}\sum_{i=1}^{n} X_{ij}^2 = 1$ for all j. A natural estimator for ω_j is

$$\tilde{\omega}_j = \frac{1}{n}\sum_{k=1}^{n}\left\{\frac{1}{n}\sum_{i=1}^{n} X_{ij}\mathbf{1}(Y_i < Y_k)\right\}^2.$$

An equivalent expression of $\tilde{\omega}_j$ is $\widehat{\omega}_j = n^2/(n-1)(n-2)\tilde{\omega}_j$, which is the corresponding U-statistic of $\tilde{\omega}_j$. We use $\widehat{\omega}_j$ as the marginal utility to select important features and the selected submodel is given by

$$\widehat{\mathcal{M}}_{\nu_n} = \{1 \le j \le p : \widehat{\omega}_j > \nu_n\}.$$

This procedure is referred to as sure independent ranking screening (SIRS). Zhu et al. (2011) established the consistency in ranking (CIR) property of the SIRS, which is a stronger result than the sure screening property. It states that if $p = o(\exp(an))$ for some fixed $a > 0$, then there exists some constant $s_\delta \in (0, 4/\delta)$ where $\delta = \min_{j\in\mathcal{M}_\star}\omega_j - \max_{j\in\mathcal{M}_\star^c}\omega_j$ such that

$$\Pr(\max_{j\in\mathcal{M}_\star^c}\widehat{\omega}_j < \min_{j\in\mathcal{M}_\star}\widehat{\omega}_j) \ge 1 - 4p\exp\{n\log(1-\delta s_\delta/4)/3\}. \quad (10.14)$$

Since $p = o(\exp(an))$, the right-hand side of (10.14) approaches to 1 with an exponential rate as $n \to \infty$. Therefore, SIRS ranks all important features above unimportant features with high probability. Provided that an ideal threshold is available, this property would lead to consistency in selection, that is, a proper choice of the threshold can perfectly separate the important and unimportant features. In practice, one can choose the threshold with the help of extra artificial auxiliary variables. The idea of introducing auxiliary variables for thresholding was first proposed by Luo, Stefanski, and Boos (2006) to tune the entry significance level in forward selection, and then extended by Wu, Boos, and Stefanski (2007) to control the false selection rate of forward regression in linear model. Zhu et al. (2011) extended this idea to choose the threshold for feature screening as follows. We generate d auxiliary variables $\mathbf{z} \sim N_d(\mathbf{0}, \mathbf{I}_d)$ such that $\mathbf{z}$ is independent of $\mathbf{x}$ and Y and we regard $(p+d)$-dimensional vector $(\mathbf{x}^\top, \mathbf{z}^\top)^\top$ as the new features. The normality of $\mathbf{z}$ here is not critical here. We know that $\min_{j\in\mathcal{M}_\star}\omega_j > \max_{l=1,\ldots,d}\omega_{p+l}$ since we know $\mathbf{z}$ is truly unimportant features. Given a random sample, we know $\min_{k\in\mathcal{M}_\star}\widehat{\omega}_k > \max_{l=1,\ldots,d}\widehat{\omega}_{p+l}$ holds with high probability according to the consistence in ranking property. Let $C_d = \max_{l=1,\ldots,d}\widehat{\omega}_{p+l}$, the set of selected features is given by

$$\widehat{\mathcal{M}}_{C_d} = \{1 \le k \le p : \widehat{\omega}_k > C_d\}.$$

Li et al. (2012) proposed a model-free feature screening procedure based on the distance correlation. This procedure does not impose any model assumption on

$F(y|\mathbf{x})$. Let $\mathbf{u} \in \mathbb{R}^{d_u}$ and $\mathbf{v} \in \mathbb{R}^{d_v}$ be two random vectors. The distance correlation measures the distance between the joint characteristic function of $(\mathbf{u}, \mathbf{v})$ and the product of marginal characteristic functions of $\mathbf{u}$ and $\mathbf{v}$ (Székely, Rizzo, & Bakirov, 2007). To be precise, let $\phi_{\mathbf{u}}(\mathbf{t})$ and $\phi_{\mathbf{v}}(\mathbf{s})$ be the characteristic functions of $\mathbf{u}$ and $\mathbf{v}$ respectively, and $\phi_{\mathbf{u},\mathbf{v}}(\mathbf{t}, \mathbf{s})$ be the joint characteristic function of $(\mathbf{u}, \mathbf{v})$. The squared distance covariance is defined as

$$\text{dcov}^2(\mathbf{u}, \mathbf{v}) = \int_{\mathbb{R}^{d_u+d_v}} |\phi_{\mathbf{u},\mathbf{v}}(\mathbf{t}, \mathbf{s}) - \phi_{\mathbf{u}}(\mathbf{t})\phi_{\mathbf{v}}(\mathbf{s})|^2 w(\mathbf{t}, \mathbf{s}) d\mathbf{t} d\mathbf{s},$$

where $w(\mathbf{t}, \mathbf{s})$ is some weight function. With a proper choice of the weight function, the squared distance covariance can be expressed in the following closed form:

$$\text{dcov}^2(\mathbf{u}, \mathbf{v}) = S_1 + S_2 - 2S_3,$$

where $S_j, j = 1, 2, 3$ are defined as

$$\begin{aligned} S_1 &= E\{\|\mathbf{u} - \tilde{\mathbf{u}}\|_{d_u} \|\mathbf{v} - \tilde{\mathbf{v}}\|_{d_v}\}, \\ S_2 &= E\{\|\mathbf{u} - \tilde{\mathbf{u}}\|_{d_u}\} E\{\|\mathbf{v} - \tilde{\mathbf{v}}\|_{d_v}\}, \\ S_3 &= E\{E(\|\mathbf{u} - \tilde{\mathbf{u}}\|_{d_u} |\mathbf{u}) E(\|\mathbf{v} - \tilde{\mathbf{v}}\|_{d_v} |\mathbf{v})\}, \end{aligned}$$

where $(\tilde{\mathbf{u}}, \tilde{\mathbf{v}})$ is an independent copy $(\mathbf{u}, \mathbf{v})$ and $\|\mathbf{a}\|_d$ stands for the Euclidean norm of $\mathbf{a} \in \mathbb{R}^d$. The distance correlation (DC) between $\mathbf{u}$ and $\mathbf{v}$ is defined as

$$\text{dcorr}(\mathbf{u}, \mathbf{v}) = \frac{\text{dcov}(\mathbf{u}, \mathbf{v})}{\sqrt{\text{dcov}(\mathbf{u}, \mathbf{u}) \text{dcov}(\mathbf{v}, \mathbf{v})}}.$$

The distance correlation has many appealing properties. The first property is that distance correlation is closely related to the Pearson correlation. If U and V are two univariate normal random variables, the distance correlation dcorr(U, V) is a strictly increasing function of $|\rho|$, where ρ is the Pearson correlation between U and V. This property implies that the DC-based marginal feature screening procedure is equivalent to the SIS in Fan and Lv (2008) for linear regression if features and errors are normally distributed. The second property is that dcorr$(\mathbf{u}, \mathbf{v}) = 0$ if and only if $\mathbf{u}$ and $\mathbf{v}$ are independent (Székely et al., 2007). Note that two univariate random variables U and V are independent if and only if U and $T(V)$, a strictly monotone transformation of V, are independent. This implies that a DC-based feature screening procedure can be more effective than the Pearson correlation based procedure since DC can capture both linear and nonlinear relationship between U and V. In addition, DC is well-defined for multivariate random vectors, thus DC-based screening procedure can be directly used for grouped predictors and multivariate response. These remarkable properties make distance correlation a good candidate for feature screening.

Given a random sample $\{(\mathbf{u}_i, \mathbf{v}_i)\}, i = 1, \ldots, n$ from $(\mathbf{u}, \mathbf{v})$, the squared distance covariance between $\mathbf{u}$ and $\mathbf{v}$ is estimated by $\widehat{\text{dcov}}^2(\mathbf{u}, \mathbf{v}) = \widehat{S}_1 + \widehat{S}_2 - 2\widehat{S}_3$, where

$$\widehat{S}_1 = \frac{1}{n^2} \sum_{i=1}^{n} \sum_{j=1}^{n} \|\mathbf{u}_i - \mathbf{u}_j\|_{d_u} \|\mathbf{v}_i - \mathbf{v}_j\|_{d_v},$$

$$\widehat{S}_2 = \frac{1}{n^2} \sum_{i=1}^{n} \sum_{j=1}^{n} \|\mathbf{u}_i - \mathbf{u}_j\|_{d_u} \frac{1}{n^2} \sum_{i=1}^{n} \sum_{j=1}^{n} \|\mathbf{v}_i - \mathbf{v}_j\|_{d_v},$$

$$\widehat{S}_3 = \frac{1}{n^3} \sum_{i=1}^{n} \sum_{j=1}^{n} \sum_{k=1}^{n} \|\mathbf{u}_i - \mathbf{u}_k\|_{d_u} \|\mathbf{v}_j - \mathbf{v}_k\|_{d_v}.$$

Similarly, we can define the sample distance covariances $\widehat{\text{dcov}}(\mathbf{u}, \mathbf{u})$ and $\widehat{\text{dcov}}(\mathbf{v}, \mathbf{v})$. Accordingly, the sample distance correlation between $\mathbf{u}$ and $\mathbf{v}$ is defined by

$$\widehat{\text{dcorr}}(\mathbf{u}, \mathbf{v}) = \frac{\widehat{\text{dcov}}(\mathbf{u}, \mathbf{v})}{\sqrt{\widehat{\text{dcov}}(\mathbf{u}, \mathbf{u})\widehat{\text{dcov}}(\mathbf{v}, \mathbf{v})}}.$$

Let $\mathbf{y} = (Y_1, \ldots, Y_q)^\top$ be the response vector with support Ψ_y, and $\mathbf{x} = (X_1, \ldots, X_p)^\top$ be the covariate vector. Here we allow the response to be univariate or multivariate and assume q is a fixed number. For each $j = 1, \ldots, p$, we can calculate the sample distance correlation $\widehat{\text{dcorr}}(X_j, \mathbf{y})$. Based on the fact that $\text{dcorr}(X_j, \mathbf{y}) = 0$ if and only if X_j and $\mathbf{y}$ are independent, $\widehat{\text{dcorr}}(X_j, \mathbf{y})$ can be used as a marginal utility to rank the importance of X_j. Therefore, the set of important variables is defined as

$$\widehat{\mathcal{M}}_{\nu_n} = \{1 \leq j \leq p : \widehat{\text{dcorr}}(X_j, \mathbf{y}) > \nu_n\},$$

for some pre-specified threshold ν_n. This model-free feature screening procedure is known as DC-SIS. Under certain moment assumptions and with the choice of $\nu_n = cn^{-\kappa}$ for some constants c and κ, Li et al. (2012) showed that DC-SIS enjoys the sure screening property. This DC-SIS allows for arbitrary regression relationship of Y onto $\mathbf{x}$, regardless of whether it is linear or nonlinear. It also permits univariate and multivariate responses, regardless of whether it is continuous, discrete, or categorical. Note that the SIRS in Zhu et al. (2011) requires that $F(y|\mathbf{x})$ depends on $\mathbf{x}$ through a linear combination $\mathbf{B}^\top \mathbf{x}_{\mathcal{M}_\star}$. Comparing with SIRS, this DC-SIS is completely model-free and it does not require any model assumption on the relationship between features and the response. Another advantage of DC-SIS is that it can be directly utilized for screening grouped variables and multivariate responses while SIRS can only handle univariate response. An iterative version of DC-SIS was proposed in Zhong and Zhu (2015) to address the issues of marginal feature screening.

10.3.5 Feature Screening for Categorical Data

Plenty of feature screening methods have been proposed for models where both the features and the response are continuous. In practice, we are also interested in the situation where features and/or response are categorical data. Fan and Fan (2008) proposed a marginal t-test screening for the linear discriminant analysis and showed that it has the sure screening property. Fan and Song (2010) proposed a maximum marginal likelihood screening for generalized linear models and rank variables according to the magnitudes of coefficient, which can be applied directly to the logistic regression.

Mai and Zou (2012) introduced a nonparametric screening method based on Kolmogorov–Smirnov distance for binary classification. It does not require any modeling assumption and thus is robust and has wide applicability. Let Y be the label and takes value in $\{-1, 1\}$ and let $F_{+j}(x)$ and $F_{-j}(x)$ denote the conditional CDF of X_j given $Y = 1, -1$, respectively. Define

$$K_j = \sup_{-\infty<x<\infty} |F_{+j}(x) - F_{-j}(x)|.$$

The sample version of K_j is defined as

$$K_{nj} = \sup_{-\infty<x<\infty} |\widehat{F}_{+j}(x) - \widehat{F}_{-j}(x)|,$$

where $\widehat{F}_{+j}(x)$ and $\widehat{F}_{-j}(x)$ are the empirical CDF of X_j given $Y = 1, -1$ respectively. This screening procedure is called Kolmogorov filter due to the fact that K_{nj} is actually the Kolmogorov–Smirnov test statistic for testing the equivalence of two distributions. By definition, K_{nj} is invariant under any strictly monotone univariate transformations applied to individual feature. Mai and Zou (2012) recommended using the Kolmogorov filter to select the submodel

$$\widehat{\mathcal{M}}_{d_n} = \{1 \le j \le p : K_{nj} \text{ is among the first } d_n \text{ largest ones}\}.$$

Mai and Zou (2015) extended the idea of Kolmogorov filter to a wide variety of applications including multi-class classification, Poisson regression, and so on by slicing the response. The resulting procedure is a nonparametric model-free feature screening procedure that works with discrete, categorical, or continuous features.

Cui, Li, and Zhong (2015) developed an effective model-free and robust feature screening procedure for ultra-high dimensional discriminant analysis with a possibly diverging number of classes. Without specifying a regression model, define the true index set of important features by

$$\mathcal{M}_\star = \{1 \le j \le p : F(y|\mathbf{x}) \text{ functionally depends on } X_j\}.$$

Let Y be a categorical response with K categories $\{y_1, \ldots, y_K\}$, and assume X is a continuous univariate feature. Let $F(x|Y) = \Pr(X \leq x|Y)$ be the conditional distribution function of X given Y. Denote by $F(x) = \Pr(X \leq x)$ the unconditional distribution function of X and $F_k(x) = \Pr(X \leq x|Y = y_k)$ the conditional distribution function of X given $Y = y_k$. If $F_k(x) = F(x)$ for all x and $k = 1, \ldots, K$, then X and Y are independent. Based on this observation, Cui et al. (2015) proposed the following index

$$\text{MV}(X|Y) = E[Var(F(X|Y))]$$

to measure the dependence between X and Y. Let $p_k = \Pr(Y = y_k) > 0$, then $\text{MV}(X|Y)$ can be written as

$$\text{MV}(X|Y) = \sum_{k=1}^{K} p_k \int (F_k(x) - F(x))^2 dF(x). \tag{10.15}$$

Equation (10.15) implies that $\text{MV}(X|Y)$ can be represented as the weighted average of Cramer–von Mises distances between the conditional distribution of X given $Y = y_k$ and the unconditional distribution function of X. Cui et al. (2015) showed that $\text{MV}(X|Y) = 0$ if and only if X and Y are statistically independent. Another appealing property of $\text{MV}(X|Y)$ is that it characterizes both linear and nonlinear relationships, making it a good marginal utility for ultra-high dimensional discriminant analysis.

Let $\{(X_i, Y_i)\}, i = 1, \ldots, n$ be a random sample from the population (X, Y). Define $\widehat{p}_k = n^{-1}\sum_{i=1}^{n} \mathbf{1}(Y_i = y_k)$, $\widehat{F}(x) = n^{-1}\sum_{i=1}^{n} \mathbf{1}(X_i \leq x)$, and $\widehat{F}_k(x) = n^{-1}\sum_{i=1}^{n} \mathbf{1}(X_i \leq x, Y_i = y_k)/\widehat{p}_k$. Based on the Cramer–von Mises representation (10.15), $\text{MV}(X|Y)$ can be estimated by its sample counterpart

$$\widehat{\text{MV}}(X|Y) = n^{-1} \sum_{k=1}^{K} \sum_{i=1}^{n} \widehat{p}_k (\widehat{F}_k(X_i) - \widehat{F}(X_i))^2.$$

For each of the features $X_j, j = 1 \ldots, p$, we can compute the sample version of the index $\widehat{\text{MV}}(X_j|Y)$ between X_j and Y. We select the submodel by

$$\widehat{\mathcal{M}}_{\nu_n} = \{1 \leq j \leq p : \widehat{\text{MV}}(X_j|Y) > \nu_n\},$$

for some pre-specified threshold ν_n. This MV-based screening procedure is referred to as MV-SIS. The sure screening property holds for MV-SIS under very mild moment conditions of features and it does not require the regression function of Y onto $\mathbf{x}$ to be linear. It is worth noting that MV-SIS is insensitive to heavy tailed distributions of features and potential outliers due to the robustness of conditional distribution function. Furthermore, the sure screening property holds even when number of classes diverges.

In reality, one may also encounter the situation in which both the features and the response are categorical. Huang, Li, and Wang (2014) proposed a chi-square based feature screening procedure for such situation. The idea is to construct a chi-square test statistic for each pair of feature and response. Let $Y_i \in \{1, \ldots, K\}$ be the class label of response, and $\mathbf{x}_i = (X_{i1}, \ldots, X_{ip})^\top$ be the associated categorical features. For simplicity, assume each X_{ij} is binary though the method and theory can be readily applied to multi-class categorical features. Define $\Pr(Y_i = k) = \pi_{yk}, \Pr(X_{ij} = k) = \pi_{jk}$, and $\Pr(Y_i = k_1, X_{ij} = k_2) = \pi_{yj,k_1k_2}$. Those quantities can be estimated by their sample counterparts $\widehat{\pi}_{yk} = n^{-1}\sum_{i=1}^n \mathbf{1}(Y_i = k)$, $\widehat{\pi}_{jk} = n^{-1}\sum_{i=1}^n \mathbf{1}(X_{ij} = k)$ and $\widehat{\pi}_{yj,k_1k_2} = n^{-1}\sum_{i=1}^n \mathbf{1}(Y_i = k_1)\mathbf{1}(X_{ij} = k_2)$, respectively. Subsequently, for each feature, a chi-square type statistic can be constructed as

$$\widehat{\Delta}_j = \sum_{k_1=1}^{K}\sum_{k_2=1}^{2} \frac{(\widehat{\pi}_{yk_1}\widehat{\pi}_{jk_2} - \widehat{\pi}_{yj,k_1k_2})^2}{\widehat{\pi}_{yk_1}\widehat{\pi}_{jk_2}},$$

which is a natural estimator of

$$\Delta_j = \sum_{k_1=1}^{K}\sum_{k_2=1}^{2} \frac{(\pi_{yk_1}\pi_{jk_2} - \pi_{yj,k_1k_2})^2}{\pi_{yk_1}\pi_{jk_2}}.$$

Obviously, features with larger values of $\widehat{\Delta}_j$ are more relevant to the response. As a result, the submodel is selected by

$$\widehat{\mathcal{M}}_{\nu_n} = \{1 \le j \le p : \widehat{\Delta}_j > \nu_n\},$$

where $\nu_n > 0$ is some pre-specified threshold. Note that $n\widehat{\Delta}_j$ has an asymptotic distribution χ^2_{K-1}, where χ^2_{K-1} is the chi-squared distribution with degrees of freedom $K-1$. Then $\widehat{\mathcal{M}}_{\nu_n}$ can be defined in terms of p-value. Let $\widehat{p}_j = \Pr(\chi^2_{K-1} > n\widehat{\Delta}_j)$ and $\widehat{\mathcal{M}}_{\nu_n}$ can be equivalently expressed as $\widehat{\mathcal{M}}_{\nu_n} = \{1 \le j \le p : \widehat{p}_j < p_{\nu_n}\}$ for some $0 < p_{\nu_n} < 1$. When the number of categories of features are different from each other, then features involving more categories are more likely to have larger Δ_j values, regardless of whether the feature is important or not. Based on this observation, it is more appropriate to use the p-values $\widehat{p}_j$ as the marginal utility to select the important features instead of using $\widehat{\Delta}_j$. Assume the jth feature has R_j categories, the p-value $\widehat{p}_j$ can be obtained from the Pearson chi-squared test of independence with degrees of freedom $(K-1)(R_j-1)$.

Huang et al. (2014) suggested using the following maximum ratio criterion to determine how many features should be included in the submodel. Let $\{k_1, \ldots, k_p\}$ be a permutation of $\{1, \ldots, p\}$ such that $\widehat{\Delta}_{k1} \ge \widehat{\Delta}_{k2} \ge \cdots \ge \widehat{\Delta}_{kp}$. Recall that the true model size is $|\mathcal{M}_\star| = s$. As long as $j+1 \le s$, we should have $\widehat{\Delta}_{k_j}/\widehat{\Delta}_{k_{j+1}} \to c_{j,j+1}$ in probability for some $c_{j,j+1} > 0$. On the other hand, if $j > s$, we should have both $\widehat{\Delta}_{k_j}$ and $\widehat{\Delta}_{k_{j+1}}$ converge towards to 0 in probability. If their convergence rates are of the same order, we should have $\widehat{\Delta}_{k_j}/\widehat{\Delta}_{k_{j+1}} = O(1)$. If $j = s$, we expect $\widehat{\Delta}_{k_j} \to c_j > 0$ while $\widehat{\Delta}_{k_{j+1}} \to 0$ in probability. This makes the ratio $\widehat{\Delta}_{k_j}/\widehat{\Delta}_{k_{j+1}} \to \infty$. They suggest selecting the top $\widehat{d}$ features as submodel where

$$\widehat{d} = \arg\max_{0 \le j \le p-1} \widehat{\Delta}_{k_j}/\widehat{\Delta}_{k_{j+1}}$$

and $\widehat{\Delta}_0$ is defined to be 1. That is, we include the $\widehat{d}$ features with the largest $\widehat{\Delta}_j$ in the submodel.

10.4 Time-Dependent Data

10.4.1 Longitudinal Data

Instead of observing independent and identically distributed data, one may observe longitudinal data, that is, the features may change over time. More precisely, longitudinal data, also known as panel data, is a collection of repeated observations of the same subjects over a period of time. Longitudinal data differs from cross-sectional data in that it follows the same subjects over a period of time, while cross-sectional data are collected from different subjects at each time point. Longitudinal data is often seen in economy, finance studies, clinical psychology, etc. For example, longitudinal data is often seen in event studies, which tries to analyze what factors drive abnormal stock returns over time, or how stock prices react to merger and earnings announcements.

Time-varying coefficient model is widely used for modeling longitudinal data. Consider the following time-varying coefficient model,

$$y(t) = \mathbf{x}(t)^\top \boldsymbol{\beta}(t) + \epsilon(t),\ t \in T, \tag{10.16}$$

where $\mathbf{x}(t) = (X_1(t), \ldots, X_p(t))^\top$ are the p-dimensional covariates, $\boldsymbol{\beta}(t) = (\beta_1(t), \ldots, \beta_p(t))^\top$ are the time-varying coefficients, $\epsilon(t)$ is a mean zero stochastic process, and T is the time interval in which the measurements are taken. In model (10.16), t need not to be calendar time, for instance, we can set t to be the age of a subject. In general, it is assumed that T is a closed and bounded interval in $\mathbb{R}$. The goal is to identify the set of true important variables, which is defined as

$$\mathcal{M}_\star = \{1 \le j \le p : \|\beta_j(t)\|_2 \ne 0\},$$

where $\|\beta(t)\|_2 = \frac{1}{|T|}\int_T \beta^2(t)dt$ and $|T|$ is the length of T.

Suppose there is a random sample of n independent subjects $\{\mathbf{x}_i(t), Y_i(t)\}, i = 1, \ldots, n$ from model (10.16). Let t_{ik} and m_i be the time of the kth measurement and the number of repeated measurement for the ith subject. $Y(t_{ik})$ and $\mathbf{x}_i(t_{ik}) =$

$(X_{i1}(t_{ik}), \ldots, X_{ip}(t_{ik}))^\top$ are the ith subject's observed response and covariates at time t_{ik}. Based on the longitudinal observations, the model can be written as

$$Y_i(t_{ik}) = \mathbf{x}_i(t_{ik})^\top \boldsymbol{\beta}(t_{ik}) + \epsilon_i(t_{ik}),$$

where $\boldsymbol{\beta}(t_{ik}) = (\beta_1(t_{ik}), \ldots, \beta_p(t_{ik}))^\top$ is the coefficient at time t_{ik}. Song, Yi, and Zou (2014) considered a marginal time-varying coefficient model for each $j = 1, \ldots, p$,

$$Y_i(t_{ik}) = \beta_j(t_{ik}) X_{ij}(t_{ik}) + \epsilon_i(t_{ik}).$$

Let $\mathbf{B}(t) = (B_1(t), \ldots, B_L(t))^\top$ be a B-spline basis on the time interval T, where L is the dimension of the basis. For the ease of presentation, we use the same B-spline basis for all $\beta_j(t)$. Under smoothness conditions, each $\beta_j(t)$ can be approximated by the linear combination of B-spline basis functions. For each j, consider marginal weighted least square estimation based on B-spline basis

$$\widehat{\boldsymbol{\gamma}}_j = \arg\min_{\gamma_{jl}} \sum_{i=1}^{n} w_i \sum_{k=1}^{m_i} \left(Y_i(t_{ik}) - \sum_{l=1}^{L} X_{ij}(t_{ik}) B_l(t_{ik}) \gamma_{jl} \right)^2,$$

where $\boldsymbol{\gamma}_j = (\gamma_{j1}, \ldots, \gamma_{jL})^\top$ is the unknown parameter and $\widehat{\boldsymbol{\gamma}}_j = (\widehat{\gamma}_{j1}, \ldots, \widehat{\gamma}_{jL})^\top$ is its estimate. Choices of w_i can be 1 or $1/m_i$, that is equal weights to observations or equal weights to subjects. See Song et al. (2014) for more details on how to obtain $\widehat{\boldsymbol{\gamma}}_j$. The B-spline estimator of $\beta_k(t)$ is given by $\widehat{\beta}_j(t) = \widehat{\boldsymbol{\gamma}}_j^\top \mathbf{B}(t)$. The selected set of features is given by

$$\widehat{\mathcal{M}}_{\nu_n} = \{1 \le j \le p : \|\widehat{\beta}_j(t)\|_2 \ge \nu_n\},$$

where ν_n is a pre-specified threshold. To evaluate $\|\widehat{\beta}_j(t)\|_2$, one can take N equally spaced time points $t_1 < \cdots < t_N$ in T, and compute $\|\widehat{\beta}_{Nj}(t)\|_2 = N^{-1} \sum_{i=1}^{N} \widehat{\beta}_j^2(t_i)$. As long as N is large enough, $\|\widehat{\beta}_{Nj}(t)\|_2$ should be close enough to $\|\widehat{\beta}_j(t)\|_2$. This varying-coefficient independence screening is referred to as VIS and enjoys the sure screening property (Song et al., 2014). An iterative VIS (IVIS) was also introduced in Song et al. (2014), which utilizes the additional contribution of unselected features by conditioning on the selected features that survived the previous step.

Cheng, Honda, Li and Peng (2014) proposed a similar nonparametric independence screening method for the time-varying coefficient model. In their setting, they allow some of the important features simply have constant effects, i.e.,

$$Y_i(t_{ik}) = \sum_{j=1}^{q} X_{ij}(t_{ik}) \beta_j + \sum_{j=q+1}^{p} X_{ij}(t_{ik}) \beta_j(t_{ik}) + \epsilon_i(t_{ik}).$$

The first q coefficients $\beta_j, j = 1, \ldots, q$ do not change over time. Cheng et al. (2014) points out that it is very important to identify the non-zero constant coefficients because treating a constant coefficient as time varying will yield a convergence rate that is slower than $\sqrt{n}$.

Both Song et al. (2014) and Cheng et al. (2014) ignore the covariance structure of $\epsilon(t)$ and carry out the feature screening on a working independence structure. Chu, Li, and Reimherr (2016) extended the VIS by incorporating within-subject correlation and dynamic error structure. They also allow baseline variables in their model, which are believed to have impact on the response based on empirical evidence or relevant theories and are not subject to be screened. These baseline features are called Z-variables and the longitudinal features to be screened are called X-variables. Consider the following model,

$$Y_i(t_{ik}) = \sum_{j=1}^{q} \beta_j(t_{ik}) Z_{ij}(t_{ik}) + \sum_{j=1}^{p} \beta_j(t_{ik}) X_{ij}(t_{ik}) + \epsilon_i(t_{ik}), \tag{10.17}$$

where Z-variables are the known important variables by prior knowledge and X-variables are ultra-high dimensional features. It is assumed that $\epsilon_i(t)$ have variances that vary across time, are independent across i (between subjects) and correlated across t (within the same subject). Incorporating the error structure into the model estimation is expected to increase screening accuracy. Chu et al. (2016) proposed a working model without any X-variables to estimate the covariance structure,

$$Y_i(t_{ik}) = \beta_0^w + \sum_{l=1}^{q} \beta_l^w(t_{ik}) Z_{il}(t_{ik}) + \epsilon_i^w(t_{ik}). \tag{10.18}$$

Although model (10.18) is mis-specified, valuable information about the covariance structure can still be gained. Standard ordinary least squares and regression spline technique can be applied to (10.18) (Huang, Wu, & Zhou, 2004), and we can obtain the corresponding residuals $r_i(t_{ik})$. Let $V(t_{ik})$ be a working variance function for $\epsilon(t_{ik})$ and it can be approximated by $V(t_{ik}) \approx \sum_{l=1}^{L} \alpha_l B_l(t_{ik})$, where $B_1(t), \ldots, B_L(t)$ is a B-spline basis. The coefficients $\alpha_l, l = 1, \ldots, L$ can be estimated by minimizing the following least squares

$$\widehat{\boldsymbol{\alpha}} = (\widehat{\alpha}_1, \ldots, \widehat{\alpha}_L)^\top = \min_{\alpha_1, \ldots, \alpha_L} \sum_{i=1}^{n} \sum_{k=1}^{m_i} \left(r_i^2(t_{ik}) - \sum_{l=1}^{L} \alpha_l B_l(t_{ik}) \right)^2.$$

Then define $\widehat{V}(t_{ik}) = \sum_{l=1}^{L} \widehat{\alpha}_l B_l(t_{ik})$. Denote by $\mathbf{R}_i$ the $m_i \times m_i$ working correlation matrix for the ith subject. A parametric model can be used to estimate the working correlation matrix. These models include autoregressive (AR) structure, stationary or non-stationary M-dependent correlation structure, parametric families such as the Matern. Now assume we obtain the working correlation matrix $\mathbf{R}_i$ based on some parametric model, then the weight matrix for ith subject is given by

$$\mathbf{W}_i = \frac{1}{m_i}\widehat{\mathbf{V}}_i^{-1/2}\mathbf{R}_i^{-1}\widehat{\mathbf{V}}_i^{-1/2},$$

where $\widehat{\mathbf{V}}_i$ is the $m_i \times m_i$ diagonal matrix consisting of the time-varying variance

$$\widehat{\mathbf{V}}_i = \begin{pmatrix} \widehat{V}(t_{i1}) & 0 & \dots & 0 \\ 0 & \widehat{V}(t_{i2}) & \dots & 0 \\ \vdots & \vdots & \ddots & \vdots \\ 0 & 0 & \dots & \widehat{V}(t_{im_i}) \end{pmatrix}.$$

For each j, define a marginal time-varying model with the jth X-variable,

$$Y_i(t_{ik}) = \sum_{l=1}^{q} \beta_{lj} Z_{il}(t_{ik}) + \beta_j(t_{ik}) X_{ij}(t_{ik}) + \epsilon_i(t_{ik}). \tag{10.19}$$

Using the B-spline technique and the weight matrix $\mathbf{W}_i$, one can obtain the weighted least squares estimate for model (10.19), and thus the fitted value $\widehat{Y}^{(j)}(t_{ik})$, see Chu et al. (2016) for a detailed description. Then the weighted mean squared errors are given by

$$\widehat{u}_j = \frac{1}{n}\sum_{i=1}^{n}(\mathbf{y}_i - \widehat{\mathbf{y}}_i^{(j)})^\top \mathbf{W}_i(\mathbf{y}_i - \widehat{\mathbf{y}}_i^{(j)}),$$

where $\mathbf{y}_i = (Y(t_{i1}), \dots, Y(t_{im_i}))^\top$ and $\widehat{\mathbf{y}}_i^{(j)} = (\widehat{Y}^{(j)}(t_{i1}), \dots, \widehat{Y}^{(j)}(t_{im_i}))^\top$. Note that a small value of $\widehat{u}_j$ indicates a strong marginal association between the jth feature and the response. Thus, the selected set of important variables is given by

$$\widehat{\mathcal{M}}_{\nu_n} = \{1 \le j \le p : \widehat{u}_j \le \nu_n\}.$$

This procedure has sure screening property, meaning that with probability tending to 1, all important variables will be included in the submodel defined by $\widehat{\mathcal{M}}_{\nu_n}$ provided certain conditions are satisfied. See the supplementary material of Chu et al. (2016).

Different form the B-spline techniques, Xu, Zhu, and Li (2014) proposed a generalized estimating equation (GEE) based sure screening procedure for longitudinal data. Without risk of confusion, we slightly abuse the notations here. Let $\mathbf{y}_i = (Y_{i1}, \dots, Y_{im_i})^\top$ be the response vector for the ith subject, and $\mathbf{X}_i = (\mathbf{x}_{i1}, \dots, \mathbf{x}_{im_i})^\top$ be the corresponding $m_i \times p$ matrix of features. Suppose the conditional mean of Y_{ik} given $\mathbf{x}_{ik}$ is

$$\mu_{ik}(\boldsymbol{\beta}) = E(Y_{ik}|\mathbf{x}_{ik}) = g^{-1}(\mathbf{x}_{ik}^\top\boldsymbol{\beta}),$$

where $g(\cdot)$ is a known link function, and $\boldsymbol{\beta}$ is a p-dimensional unknown parameter vector. Let $\mathbf{A}_i(\boldsymbol{\beta})$ be an $m_i \times m_i$ diagonal matrix with kth diagonal element $\sigma_{ik}^2(\boldsymbol{\beta}) = \mathrm{Var}(Y_{ik}|\mathbf{x}_{ik})$, and $\mathbf{R}_i$ be an $m_i \times m_i$ working correlation matrix. The GEE estimator of $\boldsymbol{\beta}$ is defined to be the solution to

$$\mathbf{G}(\boldsymbol{\beta}) = n^{-1}\sum_{i=1}^{n}\mathbf{X}_i^\top \mathbf{A}_i^{1/2}(\boldsymbol{\beta})\mathbf{R}_i^{-1}\mathbf{A}_i^{1/2}(\boldsymbol{\beta})(\mathbf{y}_i - \boldsymbol{\mu}_i(\boldsymbol{\beta})) = \mathbf{0}, \tag{10.20}$$

where $\boldsymbol{\mu}_i(\boldsymbol{\beta}) = (\mu_{i1}(\boldsymbol{\beta}), \ldots, \mu_{im_i}(\boldsymbol{\beta}))^\top$. Let $\mathbf{g}(\boldsymbol{\beta}) = (g_1(\boldsymbol{\beta}), \ldots, g_p(\boldsymbol{\beta}))^\top = E(\mathbf{G}(\boldsymbol{\beta}))$. Then $g_j(\mathbf{0})$ can be used as a measure of the dependence between the response and the jth feature. Let $\widehat{\mathbf{R}}_i$ be an estimate of $\mathbf{R}_i$. Then $\widehat{\mathbf{G}}(\mathbf{0})$ is defined as

$$\widehat{\mathbf{G}}(\mathbf{0}) = n^{-1}\sum_{i=1}^{n}\mathbf{X}_i^\top \mathbf{A}_i^{1/2}(\mathbf{0})\widehat{\mathbf{R}}_i^{-1}\mathbf{A}_i^{-1/2}(\mathbf{0})(\mathbf{y}_i - \boldsymbol{\mu}_i(\mathbf{0})).$$

Hence, we would select the set of important features using

$$\widehat{\mathcal{M}}_{\nu_n} = \{1 \le j \le p : |\widehat{G}_j(\mathbf{0})| > \nu_n\},$$

where $\widehat{G}_j(\mathbf{0})$ is the jth component of $\widehat{\mathbf{G}}(\mathbf{0})$ and ν_n is a pre-specified threshold. If we consider the linear regression model $Y_i = \mathbf{x}_i^\top\boldsymbol{\beta} + \epsilon_i$, the GEE function in (10.20) reduces to

$$\mathbf{G}(\mathbf{0}) = n^{-1}\sum_{i=1}^{n}\mathbf{x}_i(Y_i - \mathbf{x}_i^\top\boldsymbol{\beta}).$$

Therefore, for any given ν_n, the GEE based screening (GEES) selects the submodel using

$$\widehat{\mathcal{M}}_{\nu_n} = \{1 \le k \le p : n^{-1}|\mathbf{x}_{(j)}^\top\mathbf{y}| > \nu_n\},$$

where $\mathbf{y} = (Y_1, \ldots, Y_n)^\top$ and $\mathbf{x}_{(j)}$ is the jth column of the design matrix $\mathbf{X} = (\mathbf{x}_1, \ldots, \mathbf{x}_n)^\top$, which coincides with the original SIS proposed in Fan and Lv (2008). One desiring property of GEES is that even the working correlation matrix structure of $\widehat{\mathbf{R}}$ is mis-specified, all the important features will be retained by the GEES with probability approaching to 1.

10.4.2 Time Series Data

The analysis of time series data is common in economics and finance. For example, the market model in finance relates the return of an individual stock to the return of a market index or another individual stock. Another example is the term structure of interest rates in which the time evolution of the relationship between interest rates with different maturities is investigated. In this section, we briefly review some feature screening methods in time series. The SIS (Fan & Lv, 2008) was originally proposed for linear regression and assumes the random errors follow normal distribution. Yousuf (2018) analyzes the theoretical properties of SIS for high dimensional linear models with dependent and/or heavy tailed covariates and errors. They also introduced a generalized least squares screening (GLSS) procedure which utilizes the serial correlation present in the data. With proper assumptions on the moment, the strength of dependence in the error and covariate processes, Yousuf (2018) established the sure screening properties for both screening procedures. GLSS is shown to outperform SIS in many cases since GLSS utilizes the serial correlation when estimating the marginal effects.

Yousuf (2018)'s work is limited to the linear model and ignores some unique qualities of time series data. The dependence structure of longitudinal data is too restrictive to cover the type of dependence present in most time series. Yousuf and Feng (2018) studied a more general time series setting. Let $\mathbf{y} = (Y_1, \ldots, Y_n)^\top$ be the response time series, and let $\mathbf{x}_{t-1} = (X_{t-1,1}, \ldots, X_{t-1,m})^\top$ denote the m predictor series at time $t-1$. Given that the lags of these predictor series are possible covariates, let $\mathbf{z}_{t-1} = (\mathbf{x}_{t-1}, \ldots, \mathbf{x}_{t-h}) = (Z_{t-1,1}, \ldots, Z_{t-1,p})$ denote the p-dimensional vector of covariates, where $p = mh$. The set of important covariates is defined as

$$\mathcal{M}_\star = \{1 \le j \le p : F(y_t|Y_{t-1}, \ldots, Y_{t-h}, \mathbf{z}_{t-1}) \text{ functionally depends on } Z_{t-1,j}\},$$

where $F(y_t|\cdot)$ is the conditional distribution function of Y_t. The value h represents the maximum lag order for the response and predictor series. The value of h can be pre-specified by the user, or can be determined by some data driven method. Yousuf and Feng (2018) proposed a model-free feature screening method based on the partial distance correlation (PDC). More specifically, the PDC between $\mathbf{u}$ and $\mathbf{v}$, controlling for $\mathbf{z}$, is defined as

$$\text{pdcor}(\mathbf{u}, \mathbf{v}; \mathbf{z}) = \frac{\text{dcor}^2(\mathbf{u}, \mathbf{v}) - \text{dcor}^2(\mathbf{u}, \mathbf{z})\text{dcor}^2(\mathbf{v}, \mathbf{z})}{\sqrt{1 - \text{dcor}^4(\mathbf{u}, \mathbf{z})}\sqrt{1 - \text{dcor}^4(\mathbf{v}, \mathbf{z})}}, \tag{10.21}$$

if $\text{dcor}(\mathbf{u}, \mathbf{z}), \text{dcor}(\mathbf{v}, \mathbf{z}) \neq 1$, otherwise $\text{pdcor}(\mathbf{u}, \mathbf{v}; \mathbf{z}) = 0$. For more details and interpretation of PDC, see Székely and Rizzo (2014). $\text{pdcor}(\mathbf{u}, \mathbf{v}; \mathbf{z})$ can be estimated by its sample counterpart $\widehat{\text{pdcor}}(\mathbf{u}, \mathbf{v}; \mathbf{z})$ which replaces dcor by $\widehat{\text{dcor}}$ in (10.21).

The corresponding feature screening procedure PDC-SIS was introduced in Yousuf and Feng (2018). They first define the conditioning vector for the lth lag of predictor series k as $\mathcal{S}_{k,l} = (Y_t, \ldots, Y_{t-h}, X_{t-1,k}, \ldots, X_{t-l+1,k})$ with $1 \le l \le h$. Besides that a certain number of lags of Y_t are included in the model, the conditioning vector also includes all lower order lags for each lagged covariate of interest. By including the lower order lags in the conditioning vector, PDC-

SIS tries to shrink towards submodels with lower order lags. For convenience, let $\mathbf{C} = \{\mathcal{S}_{1,1}, \ldots, \mathcal{S}_{m,1}, \mathcal{S}_{1,2}, \ldots, \mathcal{S}_{m,h}\}$ denote the set of conditioning vectors where $C_{k+(l-1)*m} = \mathcal{S}_{k,l}$ is the conditioning vector for covariate $Z_{t-1,(l-1)*m+k}$. The selected submodel is

$$\widehat{\mathcal{M}}_{\nu_n} = \{1 \leq j \leq p : |\widehat{\text{pdcor}}(Y_t, Z_{t-1}; C_j)| \geq \nu_n\}. \quad (10.22)$$

The PDC-SIS attempts to utilize the time series structure by conditioning on previous lags of the covariates. Yousuf and Feng (2018) also proposed a different version, namely PDC-SIS+, to improve the performance of PDC-SIS. Instead of only conditioning on the previous lags of one covariate, PDC-SIS+ also conditions on additional information available from previous lags of other covariates as well. To attempt this, PDC-SIS+ identifies strong conditional signals at each lag level and add them to the conditioning vector for all higher order lag levels. By utilizing this conditioning scheme we can pick up on hidden significant variables in more distant lags, and also shrink toward models with lower order lags by controlling for false positives resulting from high autocorrelation, and cross-correlation.

10.5 Survival Data

10.5.1 Cox Model

Survival analysis is a branch of statistics for analyzing the expected duration of time until one or more events happen, such as death in biological organisms and failure in mechanical systems. This topic is referred to as reliability theory or reliability analysis in engineering, duration analysis or duration modeling in economics, and event history analysis in sociology. It is inevitable to analyze survival data in many scientific studies since the primary outcomes or responses are subject to be censored. The Cox model (Cox, 1972) is the most commonly used regression model for survival data. Let T be the survival time and $\mathbf{x}$ be the p-dimensional covariate vector. Consider the following Cox proportional hazard model

$$h(t|\mathbf{x}) = h_0(t) \exp\{\mathbf{x}^\top \boldsymbol{\beta}\}, \quad (10.23)$$

where $h_0(t)$ is the unknown baseline hazard functions. In survival analysis, survival time T is typically censored by the censoring time C. Denote the observed time by $Z = \min\{T, C\}$ and the event indicator by $\delta = \mathbf{1}(T \leq C)$. For simplicity we assume that T and C are conditionally independent given $\mathbf{x}$ and the censoring mechanism is non-informative. The observed data is an independently and identically distributed random sample $\{(\mathbf{x}_i, z_i, \delta_i)\}, i = 1, \ldots, n$. Let $t_1^0 < \cdots < t_N^0$ be the ordered distinct observed failure times and (k) index its associate covariates $\mathbf{x}_{(k)}$. $\mathcal{R}(t)$ denotes the risk set right before the time t: $\mathcal{R}(t) = \{i : z_i \geq t\}$. Under (10.23), the likelihood function is

$$L(\boldsymbol{\beta}) = \prod_{k=1}^{N} h_0(z_{(k)}) \exp(\mathbf{x}_{(k)}^\top \boldsymbol{\beta}) \prod_{i=1}^{n} \exp\{H_0(z_i) \exp\{\mathbf{x}_i^\top \boldsymbol{\beta}\}\},$$

where $H_0(t) = \int_0^t h_0(s)ds$ is the corresponding cumulative baseline hazard function. Consider the 'least informative' nonparametric modeling for H_0 with the form $H_0(t) = \sum_{k=1}^N h_k \mathbf{1}(t_k^0 \le t)$, then $H_0(z_i) = \sum_{k=1}^N h_k \mathbf{1}(i \in \mathcal{R}(t_k^0))$. Consequently the log-likelihood becomes

$$\ell(\boldsymbol{\beta}) = \sum_{k=1}^{N} \{\log(h_k) + \mathbf{x}_{(k)}^\top \boldsymbol{\beta}\} - \sum_{i=1}^{n} \left\{ \sum_{k=1}^{N} h_k \mathbf{1}(i \in \mathcal{R}(t_k^0)) \exp(\mathbf{x}_i^\top \boldsymbol{\beta}) \right\}. \qquad (10.24)$$

Given $\boldsymbol{\beta}$, the maximizer of (10.24) is given by $\widehat{h}_k = 1/\sum_{i \in \mathcal{R}(t_j^0)} \exp\{\mathbf{x}_i^\top \boldsymbol{\beta}\}$. Plugging in the maximizer, the log-likelihood function can be written as

$$\ell(\boldsymbol{\beta}) = \left(\sum_{i=i}^{n} \delta_i \mathbf{x}_i^\top \boldsymbol{\beta} - \sum_{i=1}^{n} \delta_i \log \left\{ \sum_{k \in \mathcal{R}(t_i)} \exp\{\mathbf{x}_k^\top \boldsymbol{\beta}\} \right\} \right), \qquad (10.25)$$

which is also known as the partial likelihood (Cox, 1972).

10.5.2 Feature Screening for Cox Model

A marginal feature screening procedure is developed in Fan, Feng, and Wu (2010). The marginal utility $\widehat{u}_j$ of the feature X_j is defined as the maximum of the partial likelihood only with respect to X_j,

$$\widehat{u}_j = \max_{\beta_j} \left(\sum_{i=i}^{n} \delta_i X_{ij} \beta_j - \sum_{i=1}^{n} \delta_i \log \left\{ \sum_{k \in \mathcal{R}(t_i)} \exp\{X_{kj} \beta_j\} \right\} \right). \qquad (10.26)$$

Here X_{ij} is the jth element of $\mathbf{x}_i = (X_{i1}, \ldots, X_{ip})^\top$. Intuitively, a larger marginal utility indicates that the associated feature contains more information about the survival outcome. One can rank all features according to the marginal utilities from the largest to the smallest and define the selected submodel

$$\widehat{\mathcal{M}}_{\nu_n} = \{1 \le j \le p : \widehat{u}_j > \nu_n\}.$$

Zhao and Li (2012) proposed to fit a marginal Cox model for each feature, namely the hazard function has the form $h_0(t)\exp\{\beta_j X_j\}$ for feature X_j. Let $N_i(t) = \mathbf{1}(z_i \le t, \delta_i = 1)$ be independent counting process for each subject i and $Y_i(t) = \mathbf{1}(z_i \ge t)$ be the at-risk processes. For $k = 0, 1, \ldots$, define

$$S_j^{(k)}(t) = n^{-1} \sum_{i=1}^{n} X_{ij}^k Y_i(t) h(t|\mathbf{x}_i),$$

$$S_j^{(k)}(\beta, t) = n^{-1} \sum_{i=1}^{n} X_{ij}^k Y_i(t) \exp\{\beta X_{ij}\}.$$

Then the maximum marginal partial likelihood estimator $\widehat{\beta}_j$ is defined as the solution to the following estimating equation

$$U_j(\beta) = \sum_{i=1}^{n} \int_0^C \left\{ X_{ij} - \frac{S_j^{(1)}(\beta, t)}{S_j^{(0)}(\beta, t)} \right\} dN_i(t) = 0. \tag{10.27}$$

Define the information to be $I_j(\beta) = -\partial U_j(\beta)/\partial\beta$. The submodel of selected important feature is given by

$$\widehat{\mathcal{M}}_{\nu_n} = \{1 \le j \le p : I_j(\widehat{\beta}_j)^{1/2} |\widehat{\beta}_j| \ge \nu_n\}.$$

Zhao and Li (2012) also proposed a practical way to choose the threshold $\widehat{\nu}_n$ such that the proposed method has control on the false positive rate, which is the proportion of unimportant features incorrectly selected, i.e., $|\widehat{\mathcal{M}}_{\nu_n} \cap \mathcal{M}_\star^c| / |\mathcal{M}_\star^c|$. The expected false positive rate can be written as

$$E\left(\frac{|\widehat{\mathcal{M}}_{\nu_n} \cap \mathcal{M}_\star^c|}{|\mathcal{M}_\star^c|} \right) = \frac{1}{p - s} \sum_{j \in \mathcal{M}_\star^c} \Pr(I_j(\widehat{\beta}_j)^{1/2} |\widehat{\beta}_j| \ge \nu_n).$$

Zhao and Li (2012) showed that $I_j(\widehat{\beta}_j)^{1/2}\widehat{\beta}_j$ has an asymptotically standard normal distribution. Therefore, the expected false positive rate is $2(1 - \Phi(\nu_n))$, where $\Phi(\cdot)$ is the cumulative distribution function of standard normal. The false positive rate decreases to 0 as p increases with n. In practice, we can first fix the number of false positives f that we are willing to tolerate, which corresponds to a false positive rate of $f/(p-s)$. Because s is unknown, we can be conservative by letting $\nu_n = \Phi^{-1}(1 - f/p)$, so that the expected false positive is less than f. The choice of ν_n is also related to a false discovery rate (FDR). By definition, the FDR is $|\mathcal{M}_\star^c \cap \widehat{\mathcal{M}}_{\nu_n}| / |\widehat{\mathcal{M}}_{\nu_n}|$, which is the false positive rate multiplying by $|\mathcal{M}_\star^c| / |\widehat{\mathcal{M}}_\nu|$. Since $|\mathcal{M}_\star^c| / |\widehat{\mathcal{M}}_\nu| \le p / |\widehat{\mathcal{M}}_\nu|$, in order to control the false positive rate at level $q = f/p$, we can control the FDR at level $f / |\widehat{\mathcal{M}}_{\nu_n}|$. This proposed method is called the principled Cox sure independence screening procedure (PSIS) and we summarize the PSIS as follows.

Step 1. Fit a marginal Cox model for each feature and obtain the parameter estimate $\widehat{\beta}_j$ and variance estimate $I_j(\widehat{\beta}_j)^{-1}$.
Step 2. Fix the false positive rate $q = f/p$ and set $\nu_n = \Phi(1 - q/2)$.
Step 3. Select the feature X_j if $I_j(\widehat{\beta}_j)^{1/2}|\widehat{\beta}_j| \geq \nu_n$.

Zhao and Li (2012) showed that this PSIS enjoys the sure screening property and is able to control the false positive rate. Under regularity conditions (see Appendix in Zhao and Li (2012)), if we choose $\nu_n = \Phi^{-1}(1 - q/2)$, and $\log p = O(n^{1/2-\kappa})$ for some $\kappa < 1/2$, then there exists constants $c_1, c_2 > 0$ such that

$$\Pr(\mathcal{M} \subset \widehat{\mathcal{M}}_{\nu_n}) \geq 1 - s\exp(-c_1 n^{1-2\kappa})$$

and

$$E\left(\frac{|\widehat{\mathcal{M}}_{\nu_n} \cap \mathcal{M}_\star^c|}{|\mathcal{M}_\star^c|}\right) \leq q + c_2 n^{-1/2}.$$

Distinguished from marginal screening procedure in Fan et al. (2010) and Zhao and Li (2012), Yang et al. (2016) proposed a joint screening procedure based on the joint likelihood for the Cox's model. They considered the constrained partial likelihood

$$\widehat{\boldsymbol{\beta}}_m = \arg\max_{\boldsymbol{\beta}} \ell(\boldsymbol{\beta}) \text{ subject to } \|\boldsymbol{\beta}\|_0 \leq m, \tag{10.28}$$

where $\ell(\boldsymbol{\beta})$ is defined in (10.25) and m is some pre-specified integer and is assumed to be greater than the number of non-zero elements in the true parameter $\boldsymbol{\beta}^\star$. The constraint $\|\boldsymbol{\beta}^\star\|_0 \leq m$ guarantees that the solution $\widehat{\boldsymbol{\beta}}_m$ is sparse. However, it is almost impossible to solve the constrained problem (10.28) in the high dimensional setting directly. Alternatively, one can approximate the likelihood function by its Taylor expansion. Let $\boldsymbol{\gamma}$ be in the neighborhood of $\boldsymbol{\beta}$, then

$$\ell(\boldsymbol{\gamma}) \approx \ell(\boldsymbol{\beta}) + (\boldsymbol{\gamma} - \boldsymbol{\beta})^\top \ell'(\boldsymbol{\beta}) + \frac{1}{2}(\boldsymbol{\gamma} - \boldsymbol{\beta})^\top \ell''(\boldsymbol{\beta})(\boldsymbol{\gamma} - \boldsymbol{\beta}), \tag{10.29}$$

where $\ell'(\boldsymbol{\beta})$ and $\ell''(\boldsymbol{\beta})$ are the first and second gradient of $\ell(\boldsymbol{\beta})$, respectively. When $p > n$, the Hessian matrix $\ell''(\boldsymbol{\beta})$ is not invertible. To deal with the singularity of $\ell''(\boldsymbol{\beta})$ and save computational costs, Yang et al. (2016) further approximated $\ell(\boldsymbol{\gamma})$ only including the diagonal elements in $\ell''(\boldsymbol{\beta})$,

$$g(\boldsymbol{\gamma}|\boldsymbol{\beta}) = \ell(\boldsymbol{\beta}) + (\boldsymbol{\gamma} - \boldsymbol{\beta})^\top \ell'(\boldsymbol{\beta}) - \frac{u}{2}(\boldsymbol{\gamma} - \boldsymbol{\beta})^\top \mathbf{W}(\boldsymbol{\gamma} - \boldsymbol{\beta}), \tag{10.30}$$

where u is a scaling constant to be specified and $\mathbf{W}$ is a diagonal matrix with $\mathbf{W} = -\text{diag}\{\ell''(\boldsymbol{\beta})\}$. Then the original problem can be approximated by

$$\max_{\boldsymbol{\gamma}} g(\boldsymbol{\gamma}|\boldsymbol{\beta}) \text{ subject to } \|\boldsymbol{\gamma}\|_0 \leq m. \tag{10.31}$$

Since $\mathbf{W}$ is a diagonal matrix, there is a closed form solution to (10.31) and thus the computational cost is low. In fact, the maximizer of $g(\boldsymbol{\gamma}|\boldsymbol{\beta})$ without the constraint is

$$\tilde{\boldsymbol{\gamma}} = (\tilde{\gamma}_1, \ldots, \tilde{\gamma}_p)^\top = \boldsymbol{\beta} + u^{-1}\mathbf{W}^{-1}\ell'(\boldsymbol{\beta}).$$

Denote the order statistics of $\tilde{\gamma}_j$ by $|\tilde{\gamma}_{(1)}| \geq |\tilde{\gamma}_{(2)}| \geq \cdots \geq |\tilde{\gamma}_{(p)}|$. The solution to (10.31) is given by $\widehat{\boldsymbol{\gamma}} = (\widehat{\gamma}_1, \ldots, \widehat{\gamma}_p)^\top$ with $\widehat{\gamma}_j = \tilde{\gamma}_j \mathbf{1}\{|\tilde{\gamma}_j| > |\tilde{\gamma}_{(m+1)}|\} := H(\tilde{\gamma}_j; m)$. We summarize the joint feature screening as follows.

Step 1. Initialize $\boldsymbol{\beta}^{(0)} = \mathbf{0}$.

Step 2. Set $t = 0, 1, 2, \ldots$ and iteratively conduct Step 2a and Step 2b until the algorithm converges.

Step 2a. Compute $\tilde{\boldsymbol{\gamma}}^{(t)}$ and $\tilde{\boldsymbol{\beta}}^{(t)}$ where $\tilde{\boldsymbol{\gamma}}^{(t)} = \boldsymbol{\beta}^{(t)} + u_t^{-1}\mathbf{W}^{-1}(\boldsymbol{\beta}^{(t)})\ell'(\boldsymbol{\beta}^{(t)})$, and $\tilde{\boldsymbol{\beta}}^{(t)} = (H(\tilde{\gamma}_1^{(t)}; m), \ldots, H(\tilde{\gamma}_p^{(t)}; m))^\top$. Set $\mathcal{M}_t = \{j : \tilde{\beta}_j^{(t)} \neq 0\}$.

Step 2b. Update $\boldsymbol{\beta}$ by $\boldsymbol{\beta}^{(t+1)} = (\beta_1^{(t+1)}, \ldots, \beta_p^{(t+1)})^\top$ as follows. If $j \notin \mathcal{M}_t$, set $\beta_j^{(t+1)} = 0$; otherwise, set $\{\beta_j^{(t+1)} : j \in \mathcal{M}_t\}$ be the maximum partial likelihood estimate of the submodel $\mathcal{M}_t$.

This procedure is referred to as sure joint screening (SJS) procedure. Yang et al. (2016) showed the sure screening property of the SJS under proper regularity conditions. This SJS is expected to perform better than the marginal screening procedure when there are features that are marginally independent of the survival time, but not jointly independent of the survival time. In practical implementation, Yang et al. (2016) suggested setting $m = \lfloor n/\log n \rfloor$ in practice based on their numerical studies.

Acknowledgements This work was supported by a NSF grant DMS 1820702 and NIDA, NIH grant P50 DA039838. The content is solely the responsibility of the authors and does not necessarily represent the official views of NSF, NIH, or NIDA.

References

Candes, E., & Tao, T. (2007). The Dantzig selector: Statistical estimation when p is much larger than n. *The Annals of Statistics, 35*(6), 2313–2351.

Carroll, R. J., Fan, J., Gijbels, I., & Wand, M. P. (1997). Generalized partially linear single-index models. *Journal of the American Statistical Association, 92*(438), 477–489.

Cheng, M.-Y., Honda, T., Li, J., & Peng, H. (2014). Nonparametric independence screening and structure identification for ultra-high dimensional longitudinal data. *The Annals of Statistics, 42*(5), 1819–1849.

Chu, W., Li, R., & Reimherr, M. (2016). Feature screening for time-varying coefficient models with ultrahigh dimensional longitudinal data. *The Annals of Applied Statistics, 10*(2), 596.

Cox, D. (1972). Regression models and life-tables. *Journal of the Royal Statistical Society: Series B (Statistical Methodology), 34*(2), 87–22.

Cui, H., Li, R., & Zhong, W. (2015). Model-free feature screening for ultrahigh dimensional discriminant analysis. *Journal of the American Statistical Association, 110*(510), 630–641.

Fan, J., & Fan, Y. (2008). High dimensional classification using features annealed independence rules. *The Annals of Statistics, 36*(6), 2605.

Fan, J., Feng, Y., & Song, R. (2011). Nonparametric independence screening in sparse ultra-high-dimensional additive models. *Journal of the American Statistical Association, 106*(494), 544–557.

Fan, J., Feng, Y., & Wu, Y. (2010). High-dimensional variable selection for cox's proportional hazards model. In *Borrowing strength: Theory powering applications–a festschrift for lawrence d. brown* (pp. 70–86). Bethesda, MD: Institute of Mathematical Statistics.

Fan, J., & Li, R. (2001). Variable selection via nonconcave penalized likelihood and its oracle properties. *Journal of the American statistical Association, 96*(456), 1348–1360.

Fan, J., & Lv, J. (2008). Sure independence screening for ultrahigh dimensional feature space. *Journal of the Royal Statistical Society: Series B (Statistical Methodology), 70*(5), 849–911.

Fan, J., & Lv, J. (2010). A selective overview of variable selection in high dimensional feature space. *Statistica Sinica, 20*(1), 101.

Fan, J., Ma, Y., & Dai, W. (2014). Nonparametric independence screening in sparse ultra-high-dimensional varying coefficient models. *Journal of the American Statistical Association, 109*(507), 1270–1284.

Fan, J., Samworth, R., & Wu, Y. (2009). Ultrahigh dimensional feature selection: Beyond the linear model. *The Journal of Machine Learning Research, 10*, 2013–2038.

Fan, J., & Song, R. (2010). Sure independence screening in generalized linear models with np-dimensionality. *The Annals of Statistics, 38*(6), 3567–3604.

Fan, J., & Zhang, W. (2008). Statistical methods with varying coefficient models. *Statistics and Its Interface, 1*(1), 179.

Freund, Y., & Schapire, R.E. (1997). A decision-theoretic generalization of on-line learning and an application to boosting. *Journal of Computer and System Sciences, 55*(1), 119–139.

Hardle, W., Hall, P., & Ichimura, H. (1993). Optimal smoothing in single-index models. *The Annals of Statistics, 21*(1), 157–178.

Hardle, W., Liang, H., & Gao, J. (2012). *Partially linear models*. Berlin: Springer Science & Business Media.

Huang, D., Li, R., & Wang, H. (2014). Feature screening for ultrahigh dimensional categorical data with applications. *Journal of Business & Economic Statistics, 32*(2), 237–244.

Huang, J. Z., Wu, C. O., & Zhou, L. (2004). Polynomial spline estimation and inference for varying coefficient models with longitudinal data. *Statistica Sinica, 14*, 763–788.

Huber, P. J. (1964). Robust estimation of a location parameter. *The Annals of Mathematical Statistics, 35*(1), 73–101.

Li, R., Zhong, W., & Zhu, L. (2012). Feature screening via distance correlation learning. *Journal of the American Statistical Association, 107*(499), 1129–1139.

Liu, J., Li, R., & Wu, R. (2014). Feature selection for varying coefficient models with ultrahigh-dimensional covariates. *Journal of the American Statistical Association, 109*(505), 266–274.

Luo, X., Stefanski, L. A., & Boos, D. D. (2006). Tuning variable selection procedures by adding noise. *Technometrics, 48*(2), 165–175.

Mai, Q., & Zou, H. (2012). The Kolmogorov filter for variable screening in high-dimensional binary classification. *Biometrika, 100*(1), 229–234.

Mai, Q., & Zou, H. (2015). The fused Kolmogorov filter: A nonparametric model-free screening method. *The Annals of Statistics, 43*(4), 1471–1497.

Meier, L., Van de Geer, S., & Bühlmann, P. (2009). High-dimensional additive modeling. *The Annals of Statistics, 37*(6B), 3779–3821.

Song, R., Yi, F., & Zou, H. (2014). On varying-coefficient independence screening for high-dimensional varying-coefficient models. *Statistica Sinica, 24*(4), 1735.

Székely, G. J., & Rizzo, M. L. (2014). Partial distance correlation with methods for dissimilarities. *The Annals of Statistics, 42*(6), 2382–2412.

Székely, G. J., Rizzo, M. L., & Bakirov, N. K. (2007). Measuring and testing dependence by correlation of distances. *The Annals of Statistics, 35*(6), 2769– 2794.

Tibshirani, R. (1996). Regression shrinkage and selection via the lasso. *Journal of the Royal Statistical Society. Series B (Methodological), 58*, 267–288.

Vapnik, V. (2013). *The nature of statistical learning theory*. Berlin: Springer science & business media.

Wang, L., Li, H., & Huang, J. Z. (2008). Variable selection in nonparametric varying-coefficient models for analysis of repeated measurements. *Journal of the American Statistical Association, 103*(484), 1556–1569.

Wu, Y., Boos, D. D., & Stefanski, L. A. (2007). Controlling variable selection by the addition of pseudovariables. *Journal of the American Statistical Association, 102*(477), 235–243.

Xu, C., & Chen, J. (2014). The sparse MLE for ultrahigh-dimensional feature screening. *Journal of the American Statistical Association, 109*(507), 1257–1269.

Xu, P., Zhu, L., & Li, Y. (2014). Ultrahigh dimensional time course feature selection. *Biometrics, 70*(2), 356–365.

Yang, G., Yu, Y., Li, R., & Buu, A. (2016). Feature screening in ultrahigh dimensional Cox's model. *Statistica Sinica, 26*, 881.

Yousuf, K. (2018). Variable screening for high dimensional time series. *Electronic Journal of Statistics, 12*(1), 667–702.

Yousuf, K., & Feng, Y. (2018). Partial distance correlation screening for high dimensional time series. *Preprint arXiv:1802.09116*.

Zhang, C.-H. (2010). Nearly unbiased variable selection under minimax concave penalty. *The Annals of Statistics, 38*(2), 894–942.

Zhao, S. D., & Li, Y. (2012). Principled sure independence screening for Cox models with ultra-high-dimensional covariates. *Journal of Multivariate Analysis, 105*(1), 397–411.

Zhong, W., & Zhu, L. (2015). An iterative approach to distance correlation-based sure independence screening. *Journal of Statistical Computation and Simulation, 85*(11), 2331–2345.

Zhu, L., Li, L., Li, R., & Zhu, L. (2011). Model-free feature screening for ultrahigh-dimensional data. *Journal of the American Statistical Association, 106*(496), 1464–1475.

Part IV
Dealing with Model Uncertainty

Chapter 11
Frequentist Averaging

Felix Chan, Laurent Pauwels, and Sylvia Soltyk

11.1 Introduction

> In combining the results of these two methods, one can obtain a result whose probability law of error will be more rapidly decreasing – Laplace (1818)

While Bates and Granger (1969) is often considered to be the seminal work that promoted the benefit of forecast combination, the idea of combining different approaches to improve performance can be traced back to Laplace (1818) as pointed out by Stigler (1973). Today, forecast combination is a standard practice both in academic and policy related work relying on the enormous volume of research in the past 30 years. As the availability of 'big data' becomes more common, so is the importance of forecast combination. In the era of 'big data', a common problem faced by practitioners is the large number of predictors which is often greater than the number of observations. Thus, it is not possible to utilise all predictors in one 'super model', at least not in the traditional econometrics sense, which would involve estimating more parameters than the number of observations. Even if there are sufficient observations to estimate such a 'super model', it often becomes intractable due to the number of predictors. As such, forecasters often seek more parsimonious representations. The two most common approaches are model averaging and forecast combination. As it will become clearer later, the former accommodates model uncertainty while the latter focus on forecast uncertainty. It is certainly true

F. Chan (✉) · S. Soltyk
School of Economics, Finance and Property, Curtin University, Perth, WA, Australia
e-mail: F.Chan@curtin.edu.au; S.Soltyk@curtin.edu.au

L. Pauwels
Discipline of Business Analytics, University of Sydney, Sydney, NSW, Australia
e-mail: laurent.pauwels@sydney.edu.au

P. Fuleky (ed.), *Macroeconomic Forecasting in the Era of Big Data*,
Advanced Studies in Theoretical and Applied Econometrics 52,
https://doi.org/10.1007/978-3-030-31150-6_11

that the two approaches are fundamentally different with different motivations, together they offer a complementary toolbox to macroeconomic forecasters.

As such, the objective of this chapter is to summarise some of the recent approaches of model averaging and optimal forecast combination from the perspective of macroeconomic forecasting. It covers both point forecasts and density forecasts. In addition to discussing the various estimators and formulation, a large part of this chapter is to highlight the current understanding on the foundation of forecast combination. This is particularly important as noted in Chan and Pauwels (2018) and Claeskens, Magnus, Vasnev, and Wang (2016), the theoretical foundation of forecast combination is less than complete and its superior performance is not guaranteed. The presence of the so-called forecast combination puzzle demonstrates the lack of theoretical understanding on the foundation of forecast combination. Thus one of the aims of this chapter is to present the gap in the current theoretical foundation and encourage further research in this area.

This chapter is organised as follows. Section 11.2 provides an overview of the basic principles of model averaging from the frequentist perspective. Section 11.3 presents an overview of the optimal forecast combination problem applied to points forecasts. The asymptotic theory for both the Mean Squared Forecast Error (MSFE) and Mean Absolute Deviation (MAD) optimal combination weights is derived. This is followed by Sect. 11.4 which focuses on their applications to density forecasts. A novel approach of density forecast combination by matching moments is also proposed. Section 11.5 contains some concluding remarks and discussions on possible future directions.

11.2 Background: Model Averaging

This section provides some background on frequentist model averaging. It is important to point out that there exist surveys on the various topics covered in this section. Moral-Benito (2015) provides an excellent survey on model averaging from both Bayesian and frequentist perspectives. See also Elliott (2011), Elliott, Gargano, and Timmermann (2013), Hansen (2007, 2014), Liu and Kuo (2016) for some recent advances.

While the motivation of model averaging originated from minimising model uncertainty, such minimisation can also lead to reducing forecast variance. Consider the following data generating process (DGP)

$$y_t = \mathbf{x}_t'\boldsymbol{\beta} + \mathbf{z}_t'\boldsymbol{\gamma} + \varepsilon_t \qquad t = 1, \ldots, T \tag{11.1}$$

where $\mathbf{x}_t$ and $\mathbf{z}_t$ are $M_1 \times 1$ and $M_2 \times 1$ vectors of predictors. Both M_1 and M_2 are fixed finite constant with $M = M_1 + M_2$. The difference between the two sets of predictors is that some or all of the elements in the parameter vector $\boldsymbol{\gamma}$

can be zeros, whereas all elements in $\boldsymbol{\beta}$ are non-zeros.[1] In other words, this setup allows researchers to nominate variables that are important for their investigation but unclear about the exact specification of the model amongst a large collection of predictors. This is particularly relevant if researchers are interested in conducting statistical inference on $\boldsymbol{\beta}$ where $\mathbf{z}_t$ are essentially a set of *control variables*. Since Least Squares type estimates of $\boldsymbol{\beta}$ are generally biased if the model is misspecified (omitted variable bias), this leads to the problem of selecting appropriate predictors out of all possible M_2 variables in $\mathbf{z}_t$. Note that there are in total 2^{M_2} possible models.

Let $\Omega = \{F_1, \ldots, F_{2^{M_2}}\}$ of all possible models, and let $\hat{\boldsymbol{\beta}}_i$ be the estimator for model F_i then the Frequentist Model Average Estimator can be defined as

$$\hat{\boldsymbol{\beta}}_{FMA} = \sum_{i=1}^{2^{M_2}} w_i \hat{\boldsymbol{\beta}}_i \tag{11.2}$$

where $w_i \in [0, 1]$ with $\sum_{i=1}^{M_2} w_i = 1$. This estimator is shown to be consistent and asymptotically Normal under fairly standard regularity conditions. Specifically,

$$\sqrt{T}\left(\hat{\boldsymbol{\beta}}_{FMA} - \boldsymbol{\beta}_0\right) \xrightarrow{d} \Lambda \tag{11.3}$$

where

$$\boldsymbol{\Lambda} = \sum_{i=1}^{2^{M_2}} w_i \boldsymbol{\Lambda}_i$$

with $\boldsymbol{\Lambda}_i$ being the asymptotic distribution of $\hat{\boldsymbol{\beta}}_i$.

The variance of each element in $\hat{\boldsymbol{\beta}}_{FMA}$ can be modified further to accommodate uncertainty surrounding model selection. Buckland, Burnham, and Augustin (1997) proposed

$$\widehat{SE}\left(\hat{\beta}_{FMA,k}\right) = \sum_{i=1}^{2^{M_2}} w_i \sqrt{\frac{\hat{\sigma}^2_{i,k}}{N} + \left(\hat{\beta}_{j,k} - \hat{\beta}_{FMA,k}\right)^2} \qquad k = 1, \ldots, M_1$$

where $\hat{\sigma}^2_{i,k}$ denotes the variance estimate of $\hat{\beta}_{i,k}$. That is, the kth element in the parameter vector $\hat{\boldsymbol{\beta}}_j$. For details see Hjort and Claeskens (2003), Claeskens and Hjort (2003) and Claeskens and Hjort (2008). It should also note that this result does not hold for perfectly correlated estimators, see Buckland et al. (1997).

[1]Strictly speaking, some elements in $\boldsymbol{\beta}$ can be 0 but variables selection is not being conducted on $\mathbf{x}_t$.

The theoretical results presented above assume the weights, w_i, are fixed. Since the asymptotic properties of the estimator depends on the weights, as shown in Eq. (11.3), the choice of weights is therefore important in terms of efficiency and other statistical properties. While researchers may have some idea about which models would appear to be more important or more likely in practice, it is difficult to determine the exact weighting scheme. A common approach for choosing w_i is based on different information criteria, such as Akaike (AIC), Schwarz-Bayesian (SBIC) and Hannan–Quinn (HQIC). These criteria share a general form which is

$$IC_i = -2\log(L_i) + \gamma_i \tag{11.4}$$

where L_i and γ_i denote the maximum likelihood and penalty functions for model $i = 1, \ldots, 2^{M_2}$, respectively. The penalty functions of the three aforementioned criteria are

$$\begin{aligned}\gamma^{AIC} &= 2N\\ \gamma^{SBIC} &= 2N\log T\\ \gamma^{HQC} &= 2N\log\log T\end{aligned}$$

where N and T denote the number of parameters and observations, respectively. In order to transform the information criteria into a proper weighting scheme, Buckland et al. (1997) proposes a likelihood ratio

$$w_i = \frac{c_i}{\sum_{i=1}^{2^{M_2}} c_i}$$

where $c_i = \exp\left(-IC_i/2\right)$.

Hansen (2007) proposes a slightly different approach to obtain the weights. Instead of considering the model in the form of Eq. (11.1), Hansen (2007) considers the following

$$y_t = \sum_{i=1}^{\infty} \pi_i x_{it} + \varepsilon_t, \quad \varepsilon_t \sim \text{iid}\left(0, \sigma_\varepsilon^2\right) \qquad t = 1, \ldots, T. \tag{11.5}$$

Under this setup, models with any finite subset of predictors, x_{it}, can be interpreted as an approximation of Eq. (11.5). Specifically, let model j be the model with the first N_j predictors in Eq. (11.5), that is

$$y_t = \sum_{i=1}^{N_j} \pi_i x_{it} + u_{jt} \tag{11.6}$$

where

$$u_{jt} = \sum_{i=N_j+1}^{\infty} \pi_i x_{it} + \varepsilon_t. \tag{11.7}$$

Equation (11.6) can be expressed in matrix form

$$\mathbf{y} = \mathbf{X}_j \boldsymbol{\pi}_j + \mathbf{u}_j \tag{11.8}$$

for convenience. Define an estimator

$$\hat{\boldsymbol{\pi}}_J(\mathbf{w}) = \sum_{i=1}^{J} w_i \begin{pmatrix} \hat{\boldsymbol{\pi}}_i \\ \mathbf{0} \end{pmatrix} \tag{11.9}$$

where $\hat{\boldsymbol{\pi}}_j$ denotes the least squares estimator of $\boldsymbol{\pi}_j$. Define the Mallow's criterion as

$$MC(\mathbf{w}) = \left[\mathbf{y} - \mathbf{X}_J \hat{\boldsymbol{\pi}}_J(\mathbf{w})\right]' \left[\mathbf{y} - \mathbf{X}_J \hat{\boldsymbol{\pi}}_J(\mathbf{w})\right] + 2\sigma_\varepsilon^2 \mathbf{w}'\mathbf{N} \tag{11.10}$$

where $\mathbf{N} = (N_1, \ldots, N_J)$. The optimal weight based on minimising the Mallow's criterion is therefore

$$\hat{\mathbf{w}}_{MC} = \arg\min_{\mathbf{w} \in \mathbf{W}} MC(\mathbf{w}). \tag{11.11}$$

Model averaging based on $\hat{\mathbf{w}}_{MC}$ is often called the *Mallows Model Averaging* (Moral-Benito, 2015) which contains certain optimality properties. Specifically, if $\hat{\mathbf{w}}_{MC}$ is constrained to a discrete set in $\mathbf{W}$ then the estimator has the lowest possible squared error. This result, however, does not hold if ε_t is not homoskedastic. Hansen and Racine (2012) proposes the Jackknife Model Averaging which is asymptotically optimal (least squared error) in the presence of heteroskedastic errors.

11.3 Forecast Combination

In standard model averaging as introduced above, the central idea is to estimate (a subset of) the parameter vector by combining the parameter vector estimates from different model specifications. The forecast can then be generated through this 'combined' parameter vector. While this approach would be robust to model specification, or more specifically, robust to different variable selection bias, the forecast performance from this 'combined' estimate is not clear. A more direct approach, is to combine the forecast generated from different models in such a way that would minimise a specific forecast criterion.

Since the seminal work of Bates and Granger (1969), the combination of point forecasts have been widely studied and implemented as a way to improve forecast performance vis-à-vis the forecasts of individual models. Bates and Granger (1969) considered a weighted linear combination of a pair of point forecasts using the relative variance and covariances to construct a combined forecast that minimised the MSFE. The seminal paper by Bates and Granger (1969) resulted numerous studies, including work by Newbold and Granger (1974), Granger and Ramanathan (1984) and Diebold (1989). Clemen (1989)'s survey of the literature, reviewing over 200 papers, up until the time of its publication concluded that when the objective is to minimise MSFE, the simple average of the individual forecasts generally outperforms the forecast combination with optimal weights.

Despite its age, the survey by Timmermann (2006) remains an authority on forecast combination. Some recent advances, however, can be found in Claeskens et al. (2016) and Chan and Pauwels (2018). Other significant works in the literature will also be presented and highlighted throughout the rest of the chapter.

Timmermann (2006) presents further motivation to combine forecasts, adding to the well-known result that forecast combination lead to improved forecast performance (smaller MSFE) compared to individual forecasts. These motivations can be summarised as follows:

1. Pooling the underlying information set from each individual forecast model to construct a full information set may not be possible due to the often unavailable individual model information sets. Using Monte Carlo simulation, Hsiao and Wan (2014) find that when the full information set is available, the combination of information is preferred over forecast combination, otherwise combining methods are preferred over individual forecasts.
2. Different models have varying adaptation rates to structural breaks. Pesaran and Timmermann (2007) show that combinations of forecasts from models with varying levels of adaptability to structural break will outperform forecasts from individual models. For structural break and forecast combination, see amongst others: Diebold and Pauly (1987), Hendry and Clements (2004), Jore, Mitchell, and Vahey (2010), Tian and Anderson (2014) and Hsiao and Wan (2014).
3. Combining forecasts of different models may result in increased robustness regarding both model misspecification biases which may be present in the individual forecast models, and measurement errors in the underlying data sets of the individual forecasts (Stock & Watson, 1998, 2004). This is particularly prominent when forecast combination is conducted jointly with model averaging.

These motivations are driven mostly by empirical observations rather than theoretical investigation. The theoretical underpinning of forecast combination has received less attention in the literature until recently by Smith and Wallis (2009), Elliott (2011), Claeskens et al. (2016) and Chan and Pauwels (2018) to mention a few. Chan and Pauwels (2018) derive the necessary and sufficient conditions for a simple average forecast combination to outperform an individual forecast, and hence provide broad theoretical justification why forecast combination works.

11.3.1 The Problem

This section presents the forecast combination problem. The discussion follows closely to the framework proposed by Chan and Pauwels (2018) and enables the derivation of most of the basic results in the literature on combining point forecasts.

It is often convenient to express the variable of interest as a sum of its forecast and forecast error as shown in Chan and Pauwels (2018). Specifically, consider

$$y_t = \hat{y}_{it} + v_{it} \qquad i = 0, \cdots, K \tag{11.12}$$

where v_{it} are the forecast errors. Define 'best forecast' under a given forecast criterion (loss function), $L\left(y_t, \hat{y}_{it}\right)$, as

Definition 11.1 Let $\hat{y}_{0t}$ be the best forecast for y_t under the loss function $L\left(y_t, \hat{y}_t\right)$ then for any $\varepsilon > 0$,

$$\Pr\left[L\left(y_t, \hat{y}_{0t}\right) < \varepsilon\right] \geq \Pr\left[L\left(y_t, \hat{y}_{it}\right) < \varepsilon\right] \quad \forall i = 1, \ldots, K. \tag{11.13}$$

Following from the standard result in stochastic dominance, Definition 11.1 implies that

$$\mathbb{E}\left[L\left(y_t, \hat{y}_{0t}\right)\right] \leq \mathbb{E}\left[L\left(y_t, \hat{y}_{it}\right)\right] \quad \forall i = 1, \ldots, K. \tag{11.14}$$

The expression in Eq. (11.14) is more common from a practical viewpoint as Definition 11.1 is difficult to verify in practice.

Let $\hat{y}_{0t}$ denotes the best forecast of y_t based on Definition 11.1 and the forecast criterion (loss function) $L\left(y_t, \hat{y}_{it}\right) \geq 0$. Define $u_{it} = v_{0t} - v_{it}$, rearranging Eq. (11.12) yields

$$\hat{y}_{it} = y_t - v_{0t} + u_{it} \qquad i = 1, \cdots, K. \tag{11.15}$$

This framework decomposes the prediction errors v_{it} into two parts. The first part, v_{0t}, represents the prediction error of the best forecast, and the second part, u_{it}, represents the difference in prediction errors between the best forecast and forecast i.

Equation (11.15) can be written in matrix form. Let $\mathbf{Y} = (y_0, \cdots, y_T)'$, $\hat{\mathbf{y}}_t = \left(\hat{y}_{1t}, \cdots, \hat{y}_{Kt}\right)'$, $\hat{\mathbf{Y}} = \left(\hat{\mathbf{y}}_1, \cdots, \hat{\mathbf{y}}_T\right)'$, $\hat{\mathbf{Y}}_0 = \left(\hat{y}_{01}, \cdots, \hat{y}_{0T}\right)$ and $\mathbf{u} = (\mathbf{u}_1, \cdots, \mathbf{u}_T)'$ with $\mathbf{u}_t = (u_{1t}, \cdots, u_{Kt})'$, $\boldsymbol{v}_t = (\boldsymbol{v}_{1t}, \cdots, \boldsymbol{v}_{Kt})'$ with $\boldsymbol{v} = (\boldsymbol{v}_1, \cdots, \boldsymbol{v}_T)'$ and $\boldsymbol{v}_0 = (v_{01}, \cdots, \boldsymbol{v}_{0T})'$. Equation (11.15) is expressed as

$$\hat{\mathbf{Y}} = (\mathbf{Y} - \boldsymbol{v}_0) \otimes \mathbf{i}' + \mathbf{u} \tag{11.16}$$

where $\mathbf{i}$ denotes a $k \times 1$ vector of ones, and $\otimes$ denotes the Kronecker product. Forecasts for $t = 1, \cdots, T$, based on a linear combination of forecasts from the

k models, are therefore

$$\hat{\mathbf{Y}}\mathbf{w} = \mathbf{Y}\mathbf{i}'\mathbf{w} - \boldsymbol{\nu}_0\mathbf{i}'\mathbf{w} + \mathbf{u}\mathbf{w}. \tag{11.17}$$

If $\mathbf{w}$ is an affine combination, i.e., $\mathbf{i}'\mathbf{w} = 1$, then $\boldsymbol{\nu}_0 + \mathbf{u}\mathbf{w}$ is a $T \times 1$ vector containing the forecast errors of the forecast combination. If $\mathbf{w}$ is not an affine combination, then $\hat{\mathbf{Y}}\mathbf{w}$ does not produce unbiased forecasts because $\mathbb{E}\left(\hat{\mathbf{Y}}\mathbf{w}\right) = \mathbb{E}\left(\mathbf{Y}\right)\mathbf{i}'\mathbf{w}$, under the standard assumptions that $\mathbb{E}\left(\nu_{0t}\right) = \mathbb{E}\left(u_{it}\right) = 0$ for all i and t. Hence the affine constraint, $\mathbf{i}'\mathbf{w}$, plays a significant role in avoiding potential bias due to forecast combination.

These forecasts can be based on models with different predictors but it can also be 'subjective' forecasts, i.e., not generated by any particular models or predictors such as professional forecaster surveys. Perhaps more importantly, it can also contain forecasts generated by combinations of other forecasts in $\hat{\mathbf{y}}_t$.

The fundamental problem in forecast combination which can be presented as the following optimization problem for point forecast

$$\mathbf{w}^* = \arg\min \ \mathbb{E}\left[L\left(y_t, \mathbf{w}'\hat{\mathbf{y}}_t\right)\right] \tag{11.18}$$

$$\text{s.t.} \quad \mathbf{i}'\mathbf{w} = 1. \tag{11.19}$$

where $\mathbf{i}$ denotes a column of 1's. While this setup is convenient for theoretical analysis, it is often not straightforward to evaluate the expectation in (11.18) without any assumption on the distribution of forecast errors. As such, the finite sample counterpart of Eq. (11.18) is used to estimate the optimal weight. Specifically, practitioners solve

$$\hat{\mathbf{w}} = \arg\min \ T^{-1}\sum_{t=1}^{T} L\left(y_t, \mathbf{w}'\hat{\mathbf{y}}_t\right) \tag{11.20}$$

$$\text{s.t.} \quad \mathbf{i}'\mathbf{w} = 1. \tag{11.21}$$

11.3.2 Forecast Criteria

An important aspect in forecasting is the criteria with which the forecasts are being evaluated, and hence the specification of $\mathbb{E}\left[L\left(y_t, \mathbf{w}'\hat{\mathbf{y}}_t\right)\right]$ in the optimization problem in Eqs. (11.18) and (11.19). The fundamental problem is to seek a combination of $\hat{\mathbf{y}}_{t|s}$ to produce the 'best' forecast for y_t under some specific loss function or forecast criterion. The following three loss functions are amongst the most popular forecast criteria:

$$L_{MSFE}\left(y_t, \hat{y}_t\right) = \left(y_t - \hat{y}_t\right)^2 \tag{11.22}$$

$$L_{MAD}\left(y_t, \hat{y}_t\right) = \left|y_t - \hat{y}_t\right| \tag{11.23}$$

$$L_{MAPE}\left(y_t, \hat{y}_t\right) = \left|\frac{y_t - \hat{y}_t}{y_t}\right|. \tag{11.24}$$

The sample counterparts of Eqs. (11.22)–(11.24) give the well-known forecast criteria

$$MSFE = T^{-1}\sum_{t=1}^{T}\left(y_t - \hat{y}_t\right)^2 \tag{11.25}$$

$$MAD = T^{-1}\sum_{t=1}^{T}\left|y_t - \hat{y}_t\right| \tag{11.26}$$

$$MAPE = T^{-1}\sum_{t=1}^{T}\left|\frac{y_t - \hat{y}_t}{y_t}\right|. \tag{11.27}$$

where T denotes the number of forecasts generated. Equation (11.25) is the well-known *Mean Squared Forecast Error*, Eq. (11.26) gives the *Mean Absolute Deviation* or *Mean Absolute Error* and Eq. (11.27) is the *Mean Absolute Percentage Error*. Under the assumptions of the *Weak Law of Large Numbers* (WLLN),

$$T^{-1}\sum_{t=1}^{T} L\left(y_t, \hat{y}_{t|s}\right) - \mathbb{E}\left[L\left(y_t, \hat{y}_{t|s}\right)\right] = o_p(1). \tag{11.28}$$

Equation (11.28) gives an important insight about forecast evaluation in practice as pointed out by Chan and Pauwels (2018). The three forecast criteria as defined in Eqs. (11.25)–(11.27) can be interpreted as the estimated mean of the forecast errors, $e_t = y_t - \hat{y}_t$. Like any other estimators, the estimates are subject to finite sampling errors. Therefore, simple ranking based on these criteria is not sufficient to identify the 'best' model. Statistical tests must be conducted in order to determine if a set of forecasts does in fact produce superior performance over another. See for examples, Harvey, Leybourne, and Newbold (1997), Clark and West (2006, 2007) and the references within.

This problem is amplified when the number of forecasts, K, is large. One possible way to reduce the complexity is by combining these forecasts in a way that would minimise a specific forecast criterion. There are at least two possible approaches, namely *model averaging* and *forecast combination*. Note that these approaches are not mutually exclusive. In fact, they can be used jointly in minimising a specific forecast criteria. A more thorough discussion can be found in Sects. 11.2 and 11.3.1.

11.3.3 MSFE

Consider the optimization problem as presented in Eqs. (11.18) and (11.19) with the MSFE loss function in Eq. (11.22). The Lagrangian function associated with the optimization problem is

$$L = \mathbb{E}\left[y_t - \mathbf{w}'\hat{\mathbf{y}}_t\right]^2 + \lambda\left(1 - \mathbf{w}'\mathbf{i}\right). \tag{11.29}$$

Note that $y_t = \hat{y}_{it} + \nu_{it}$ where y_{it} denotes the forecast of y_t from model i with ν_{it} denotes the forecast error of model i. Thus, under the affine constraint, Eq. (11.29) can be written as

$$L = \mathbf{w}'\mathbf{\Omega}\mathbf{w} + \lambda\left(1 - \mathbf{w}'\mathbf{i}\right) \tag{11.30}$$

where $\mathbf{\Omega} = \mathbb{E}(\boldsymbol{\nu}_t\boldsymbol{\nu}_t')$ is the variance-covariance matrix of forecast errors from the K competing models, with $\nu_t = (\nu_{1t}, \ldots, \nu_{kt})'$.

Differentiate Eq. (11.29) with respect to $\mathbf{w}$ and λ and set them to zero gives

$$\left.\frac{\partial L}{\partial \mathbf{w}}\right|_{\mathbf{w}=\mathbf{w}^*,\lambda=\lambda^*} = 2\mathbf{\Omega}\mathbf{w}^* - \mathbf{i}\lambda^* = \mathbf{0} \tag{11.31}$$

$$\left.\frac{\partial L}{\partial \lambda}\right|_{\mathbf{w}=\mathbf{w}^*,\lambda=\lambda^*} = 1 - \mathbf{w}^{*\prime}\mathbf{i} = 0. \tag{11.32}$$

Pre-multiply the first equation above with $\mathbf{w}$ and using the second equation, $\mathbf{wi} = 1$, gives

$$\lambda^* = 2\mathbf{w}^{*\prime}\mathbf{\Omega}\mathbf{w}^* \tag{11.33}$$

and therefore

$$\mathbf{\Omega}\mathbf{w}^* - \mathbf{i}\mathbf{w}^{*\prime}\mathbf{\Omega}\mathbf{w}^* = \mathbf{0}. \tag{11.34}$$

Rearranging gives

$$\mathbf{\Omega}\mathbf{w}^*\left(\mathbf{w}^{*\prime}\mathbf{\Omega}\mathbf{w}^*\right)^{-1} = \mathbf{i}. \tag{11.35}$$

The expression in Eq. (11.35) turns out to be extremely useful in terms of understanding the properties of forecast combination under MSFE. This includes the conditions that leads to the simple average being the optimal weight vector, $\mathbf{w}^* = K^{-1}\mathbf{i}$. For more details, see Chan and Pauwels (2018).

It is, however, also possible to derive a closed form solution for $\mathbf{w}^*$ under the assumption that $\mathbf{\Omega}$ is positive definite with non-zero determinant. Note that the first order conditions as shown in Eqs. (11.31) and (11.32) form a linear system, that is

$$\begin{pmatrix} 2\boldsymbol{\Omega} & -\mathbf{i} \\ \mathbf{i}' & 0 \end{pmatrix} \begin{pmatrix} \mathbf{w}^* \\ \lambda^* \end{pmatrix} = \begin{pmatrix} \mathbf{0} \\ 1 \end{pmatrix}.$$

Since $\boldsymbol{\Omega}^{-1}$ exists, the coefficient matrix has the following inverse

$$\begin{pmatrix} \frac{1}{2}\left[\boldsymbol{\Omega}^{-1} - \boldsymbol{\Omega}^{-1}\mathbf{i}\left(\mathbf{i}'\boldsymbol{\Omega}{-}1\mathbf{i}\right)^{-1}\mathbf{i}'\boldsymbol{\Omega}^{-1}\right] & \boldsymbol{\Omega}^{-1}\mathbf{i}\left(\mathbf{i}'\boldsymbol{\Omega}^{-1}\mathbf{i}\right)^{-1} \\ -\left(\mathbf{i}'\boldsymbol{\Omega}^{-1}\mathbf{i}\right)^{-1}\mathbf{i}'\boldsymbol{\Omega}^{-1} & 2\left(\mathbf{i}'\boldsymbol{\Omega}^{-1}\mathbf{i}\right)^{-1} \end{pmatrix}$$

and solving for the optimal weight $\mathbf{w}^*$ gives

$$\mathbf{w}^* = \boldsymbol{\Omega}^{-1}\mathbf{i}\left(\mathbf{i}'\boldsymbol{\Omega}^{-1}\mathbf{i}\right)^{-1} \tag{11.36}$$

Given convexity of the loss function, this solution is a unique global minimum of the optimization problem.

An interesting observation from the existing literature is that there appears to be a lack of statistical results on the estimated optimal weight. While there exists a closed form solution for minimising MSFE under the affine constraints, as shown in Eq. (11.36), the variance-covariance matrix of the forecast error, $\boldsymbol{\Omega}$, is seldom known in practice. This means the optimal weight must be estimated, rather than calculated, based on the estimated variance-covariance matrix. Also worth noting is that only the sample counterpart of the loss function can be calculated in practice, which leads to finite sampling errors. Thus, the estimation errors from estimating $\boldsymbol{\Omega}$, along with the finite sample errors in computing MSFE, provide two major causes to the 'forecast combination puzzle' as noted by Claeskens et al. (2016) and Chan and Pauwels (2018). Given the stochastic nature of the problem, distributional properties of the optimal weight vector would appear to be useful in practice. Specifically, the asymptotic distribution of the estimated optimal weight can facilitate inferences and provide a mean to test statistically the estimated weight against other weighting schemes, such as simple average.

Let $\hat{\mathbf{w}}_T^{MSFE} = \hat{\boldsymbol{\Omega}}_T^{-1}\mathbf{i}\left(\mathbf{i}'\hat{\boldsymbol{\Omega}}_T^{-1}\mathbf{i}\right)^{-1}$ where $\hat{\boldsymbol{\Omega}}_T$ denotes the sample estimate of $\boldsymbol{\Omega}$ based on T observations. Consider the following assumptions for purpose of deriving asymptotic results for the estimate of the optimal weight vector.

(i) $\boldsymbol{v}_t$ is independently and identically distributed with finite first and second moments.
(ii) $\hat{\boldsymbol{\Omega}}_T$ is a consistent estimator of $\boldsymbol{\Omega}$ such that $\hat{\boldsymbol{\Omega}}_T - \boldsymbol{\Omega} = o_p(1)$ and $\sqrt{T}\text{vec}\left(\hat{\boldsymbol{\Omega}}_T - \boldsymbol{\Omega}\right) \sim N(0, \mathbf{A})$.
(iii) There exists $0 < \eta < \infty$ such that $\left(\mathbf{i}'\hat{\boldsymbol{\Omega}}_T^{-1}\mathbf{i}\right) - \eta = o_p(1)$.

These assumptions are fairly mild and somewhat standard. Assumption (i) imposes existence of moment on the forecast errors. Note that this assumption does not exclude the case where y_t is integrated with order 1. In other words, y_t can have a unit root, as long as all models for y_t induce forecast errors that have finite second moment. Assumption (ii) requires that the variance-covariance matrix of the forecast errors can be estimated consistently. While it can be numerically challenging, it can generally be achieved in practice. Assumption (iii) is to ensure that the closed form solution of the optimal weights exists under MSFE.

The asymptotic distribution of $\hat{\mathbf{w}}_T^{MSFE}$ under $\hat{\boldsymbol{\Omega}}_T$ can be found in Proposition 11.1.

Proposition 11.1 *Under Assumption (iii),* $\sqrt{T}vec\left(\hat{\mathbf{w}}_T^{MSFE} - \mathbf{w}^*\right) \sim N\left(0, \mathbf{B}\right)$ *where*

$$\mathbf{B} = \frac{1}{\eta^2}\left(\mathbf{i}'\boldsymbol{\Omega}^{-1} \otimes \boldsymbol{\Omega}^{-1}\right)\mathbf{A}\left(\boldsymbol{\Omega}^{-1}\mathbf{i} \otimes \boldsymbol{\Omega}^{-1}\right) \tag{11.37}$$

with $\mathbf{A}$ *denotes the variance-covariance matrix of* $vec\left(\hat{\boldsymbol{\Omega}}_T^{-1}\right)$ *and* $\eta = \mathbf{i}'\boldsymbol{\Omega}^{-1}\mathbf{i}$.

Proof See Appendix. □

The literature has investigated various aspects of forecast combinations. One of the main issues is centred around the fact that taking a simple average of forecasts often provides a simple and effective way to combine forecasts. Two questions have been investigated, namely: Why does the simple average often outperform more complex weighting techniques in practice? And, is the simple average optimal? The former question has been named the 'forecast combination puzzle' by Stock and Watson (2004).

The Forecast Combination Puzzle

While the aforementioned literature at the beginning of Sect. 11.3 presents the justification for using forecast combination and the closed form solution derived above in Eq. (11.36) appears to be straightforward to compute in practice, the common empirical finding known as the 'forecast combination puzzle' continues to be of interest in the literature. The puzzle arises from the empirical observation that the simple average generally outperforms more complicated weighting strategies.

Stock and Watson (1998) compare 49 univariate forecasting methods and various forecast combination methods using U.S. macroeconomic data, while Stock and Watson (2004) examine numerous linear and non-linear forecasting models and forecast combination methods for macroeconomic data from seven countries. In both of these papers, the authors find that various combination methods outperformed individual forecasts, with the simple average and median forecast combination being best according to MSFE. Marcellino (2004) extended their analysis to European data and arrived at similar conclusions.

Smith and Wallis (2009) propose an explanation for the 'forecast combination puzzle' by re-examining the Survey of Professional Forecasters (SPF) data used in Stock and Watson (2003) and by using Monte Carlo simulations. The authors explain that the 'forecast combination puzzle' is due, in part, to the parameter estimation effect: the estimation error in the weight estimation process results in finite sample errors; hence, the simple average outperforms the weighted schemes with respect to MSFE. The authors recommend omitting the covariances between forecast errors and calculate the weight estimates on the (inverse) MSFE (in contrast to Bates and Granger, 1969). Elliott (2011) elaborates, showing that not only must there be estimation error, relatively the gains from optimal weight estimation must be small for the 'forecast combination puzzle' to hold.

While the majority of the literature finds that the simple average of forecasts generally outperforms more complicated weighing strategies empirically, there are exceptions. Aiolfi, Capistrán, and Timmermann (2010) discuss how to combine survey forecasts and time series model forecasts, and apply these methods to SPF data. The authors find that the simple average weighted survey forecast outperforms the best individual time series forecast the majority of the time, but combining both survey forecasts and a selection of time series forecast combination, there is an improved performance overall. Genre, Kenny, Meyler, and Timmermann (2013) analyzed various weighting schemes using European Central Bank SPF data, and found that some of the more complicated methods outperformed simple averages occasionally; however, no approach consistently outperformed the simple average over time and over a range of target variables. Pinar, Stengos, and Yazgan (2012) use stochastic dominance efficiency to evaluate a number of time series model forecast combination and find that the simple average does not outperform a number of more complicated weighting strategies where the weights are estimated.

Is the Simple Average Optimal?

Until recently, theoretical investigations in the literature have raised the question: is the simple average combination optimal or is this finding a result of some other mechanism? Some of the ambiguous results suggest that further investigation into the theoretical properties of forecast combination is necessary to clarify whether the simple average is optimal and why the 'forecast combination puzzle' occurs.

Claeskens et al. (2016) offer a theoretical solution to the 'forecast combination puzzle' and conclude that forecast combination is biased when the weights are estimated, and the variance of the combined forecast is larger than the simple average where the weights are not estimated. Hsiao and Wan (2014) propose the use of different geometric approaches and impose a multi-factor structure on forecast errors to combine forecasts with and without the presence of structural break. The authors provide the necessary and sufficient condition when the simple average is an optimal combination for their approaches. Some of the results in Hsiao and Wan (2014) are generalized by Chan and Pauwels (2018). Chan and Pauwels (2018) derive the necessary and sufficient condition for the forecast of an individual model to outperform the simple average forecast combination. Together, these two papers confirm the hypotheses put forward by Elliott (2011) theoretically. These conditions lead to the case where the simple average is the optimal weight, and are represented in the expression in Eq. (11.35) in Sect. 11.3.3.

In the literature, it is common to rank the performance of individual forecasts and their combination using MSFE. Recall Eq. (11.28) in Sect. 11.3.2. This equation gives an important insight about forecast evaluation. The three forecast criteria defined in Eqs. (11.25)–(11.27) can be interpreted as estimated mean of the forecast errors, which means that MSFE, for example, is subject to finite sampling errors. This is because it is an estimator of the forecast errors and may not be a consistent estimator. Therefore, ranking based on MSFE or any of the three forecast criterion is not sufficient to identify the optimal model or combination. Statistical tests must be conducted in order to determine if a set of forecasts or model does in fact produce superior performance over another. The asymptotic result presented in Proposition 11.1 should prove useful in this regard. See also Chan and Pauwels (2018) for some further discussion.

Elliott and Timmermann (2004) observe that the vast majority of studies in the literature base performance of forecast combination on MSFE. Generally, most of these studies have found that the simple average combination is optimal. The authors discuss that this result may not arise when there are asymmetries in the loss function and skewness in the forecast error distribution, which raises the question of optimal forecasts under general loss functions, rather than simply using the symmetric loss function MSFE.

Also noted is the work of Hansen (2008) who suggests that the combination weights should be chosen by minimising the Mallows criterion, a variation of Eq. (11.10) that gives an approximation of MSFE by the sum of squared errors and a penalty term, therefore addressing the trade-off between model complexity and fit.

11.3.4 MAD

The literature on combining point forecasts has focused mainly on MSFE, with some notable exception including Chan and James (2011) and Chan and Pauwels (2019). The former investigates the benefit of forecast combination in forecasting conditional variance and when the forecast criteria is not a well behaved mathematical function, such as the number of Value-at-Risk violations. The latter provides a theoretical exposition on the performance of forecast combination under *Mean Absolute Deviation*. The authors show that, under mild assumptions, the optimal solution under minimising MSFE is the same as minimising MAD. The key to this equivalence result is due to the affine constraint, specifically, Eq. (11.19).

The solution to the optimization problem as presented in Eqs. (11.18) and (11.19) with MAD as defined in Eq. (11.23) is not straightforward to obtain. Due to the non-differentiability nature of the loss function, the derivation of the First Order Necessary Conditions (FONC) is somewhat technical as shown in Chan and Pauwels (2019). The authors show that the necessary and sufficient condition for the optimal solution must satisfy

$$\mathbb{E}\left(\mathbf{u}|\nu_0+\mathbf{u}'\mathbf{w}^*>0\right)\Pr\left(\nu_0+\mathbf{u}'\mathbf{w}^*>0\right) \\ =\mathbb{E}\left(\mathbf{u}|\nu_0+\mathbf{u}'\mathbf{w}^*<0\right)\Pr\left(\nu_0+\mathbf{u}'\mathbf{w}^*<0\right). \tag{11.38}$$

Surprisingly, the MSFE optimal weight is the same as the MAD optimal weight when solving Eqs. (11.18) and (11.19) as shown in Chan and Pauwels (2019). This equivalence has several implications. First, given the result in Proposition 11.1, the asymptotic distribution of the estimated optimal weight under MAD can be derived directly by an application of epi-convergence. The basic idea is to establish the connection between the asymptotic distributions of the solutions from different optimization problem. For a more technical introduction to the concept see Knight (1998, 2001).

Proposition 11.2 *Let* $\hat{\mathbf{w}}_T^{MAD}$ *be the solution to the minimisation problem as defined in Eqs. (11.20) and (11.21). Under the Assumptions (i)–(iii),* $\sqrt{T}vec\left(\hat{\mathbf{w}}_T^{MAD}-\mathbf{w}^*\right)\sim N\left(0,\mathbf{B}\right)$ *where*

$$\mathbf{B}=\frac{1}{\eta^2}\left(\mathbf{i}'\boldsymbol{\Omega}^{-1}\otimes\boldsymbol{\Omega}^{-1}\right)\mathbf{A}\left(\boldsymbol{\Omega}^{-1}\mathbf{i}\otimes\boldsymbol{\Omega}^{-1}\right)$$

with $\mathbf{A}$ *denotes the variance-covariance matrix of* $vec\left(\hat{\boldsymbol{\Omega}}_T^{-1}\right)$ *and* $\eta=\mathbf{i}'\boldsymbol{\Omega}^{-1}\mathbf{i}$.

Proof See Appendix. □

Given the equivalence of the two optimal solutions, it is perhaps not surprising that both share the same asymptotic distribution as implied by Propositions 11.1 and 11.2.

The second implication is that there are two choices for estimating the optimal weight under MSFE and MAD. The first utilises the closed form solution as shown in Eq. (11.36) and the second requires solving the FONC as show in Eq. (11.38) numerically or solving the optimization problem as shown in Eqs. (11.20) and (11.21) with linear programming techniques. The latter would appear to be more computationally intensive but the closed form solution requires estimating the variance-covariance matrix of the forecast errors, which is known to be a difficult problem when K is large. It is also well known that the estimation of large variance-covariance matrix can be extremely sensitive to outliers. The optimal choice between these options remains unclear and could be an area for future research.

11.4 Density Forecasts Combination

The majority of studies in the literature have focused on combination of point forecasts, which generally does not provide any description of the associated uncertainty. A number of authors have attempted to provide some estimate of uncertainty for point forecasts by extending the theory to interval forecasts (see, for example, Chatfield (1993) and Christoffersen (1998)). Moreover, as noted by Diebold and Lopez (1996), forecasts of economic and financial variables are often represented by probabilities. A density forecast is an estimate of the probability distribution of the possible future values of a variable. On density forecasting see, for example, Diebold, Gunther, and Tay (1998), Corradi and Swanson (2006) provide a survey on the evaluation of density and interval forecasts, and Tay and Wallis (2000) for a comprehensive survey on density forecasting. Since the combination of point forecasts generally provides better forecasting results compared to individual forecasts, a natural extension of this framework is density forecast combination.

Wallis (2005) uses equally weighted combined density forecasts to demonstrate the statistical framework for combining density forecasts. To find optimal weights for density forecasts, Mitchell and Hall (2005), Hall and Mitchell (2007) and Geweke and Amisano (2011) propose to maximise the average logarithmic score of the combined density forecasts by minimising the Kullback–Liebler distance between the forecasts and the 'true' densities, to obtain a linear combination of the density forecasts. This approach is summarised as optimization problems in the next section.

Kascha and Ravazzolo (2010) compare linear and logarithmic combinations of density forecasts using equal weights, recursive log score weights and (inverse) MSFE weights. The authors find that while the combinations do not always outperform individual models, they provide insurance against selecting an inappropriate individual model. Kapetanios, Mitchell, Price, and Fawcett (2015) allow the density combination weights to follow a more general scheme by letting the weights depend on the variable of interest using piecewise linear weight functions. The authors find that their generalized density forecast combination outperforms the linear counterparts proposed in the preceding literature. Pauwels and Vasnev (2016) show that the number of forecasting periods must be sufficiently large for the asymptotic properties of the method proposed by Hall and Mitchell (2007) to hold. The authors note that if the number of forecasting periods is small, the optimal weights by Hall and Mitchell (2007) may result in only one density being selected, rather than a combination. In contrast to the preceding authors who use linear and logarithmic combinations of densities, Busetti (2017) proposes a slightly different approach by averaging the quantiles of the individual densities using a method known as quantile aggregation.

11.4.1 Optimal Weights

A popular formulation using the concept of *Kullback–Leibler divergence* leads to the following optimization problem

$$\mathbf{w}^{*\,KLIC} = \arg\min_{\mathbf{w}} \int f(y) \log\left(\frac{f(y)}{\mathbf{w}'\mathbf{p}(y)}\right) dy \tag{11.39}$$

$$\text{s.t.} \quad \mathbf{i}'\mathbf{w} = 1 \tag{11.40}$$

where $f(y)$ denotes the true probability density of the variable y_t and $\mathbf{p}(y) = (p_1(y), \ldots, p_k(y))'$ denotes the vector of K competing forecast densities. Equation (11.39) represents the most common forecast criterion for density forecast, namely, the *Kullback–Liebler Information Criterion* (KLIC). Again, Eqs. (11.39) and (11.40) are convenient for theoretical analysis but it is generally not practical since the underlying distribution is often unknown. In fact, if $f(y)$ is known, then there is no need for density combination!

Similar to point forecast, practitioners are therefore required to solve the finite sample counterpart of the optimization problem as presented in Eqs. (11.39) and (11.40). That is,

$$\hat{\mathbf{w}}^{KLIC} = \arg\min_{\mathbf{w}} T^{-1} \sum_{t=1}^{T} \log f_t(y_t) - \log\left(\mathbf{w}'\mathbf{p}_t(y_t)\right)$$
$$\text{s.t.} \quad \mathbf{i}'\mathbf{w} = 1$$

where $f_t(y_t)$ denotes an empirical estimate of $f(y)$. However, since $f_t(y_t)$ does not involve $\mathbf{w}$ and it can be considered fixed. The optimization problem can be simplified further to

$$\hat{\mathbf{w}} = \arg\max_{\mathbf{w}} T^{-1} \sum_{t=1}^{T} \log\left(\mathbf{w}'\mathbf{p}_t(y_t)\right) \tag{11.41}$$
$$\text{s.t.} \quad \mathbf{i}'\mathbf{w} = 1. \tag{11.42}$$

The optimal weights, $\hat{\mathbf{w}}$, obtained from optimizing (11.41) subject to (11.42) are often called log score weights as introduced by Mitchell and Hall (2005), Hall and Mitchell (2007) and Geweke and Amisano (2011).

The consistency and asymptotic distribution of the optimal log score weights are derived in Pauwels, Radchenko, and Vasnev (2018). Pauwels et al. (2018) show that under some assumptions these log score weights are consistent as the number of observations increase (Theorem 3.1). Furthermore, the weights are shown to be asymptotically normal under mild conditions (Theorem 3.2). The authors also extend the KLIC optimization framework subject to high moment constraints and derive the corresponding asymptotic theory. This is particularly relevant when making Value-at-Risk forecasts especially when the data exhibit asymmetry and fat tails.

11.4.2 Theoretical Discussions

As shown in the results below, it is often convenient to explore the theoretical properties of density forecast combination by using concepts from information theory. Let $f(y)$ and $p(y)$ be two density functions with $f(y)$ being the true density function for the random variable y_t. Define

$$H(f) = -\int f(y)\log f(y)dy \tag{11.43}$$

$$H(p||f) = -\int f(y)\log p(y)dy \tag{11.44}$$

Equations (11.43) and (11.44) define the entropy and cross entropy for the random variable, Y, respectively. Note that $H(f) \geq 0$ and $H(f||p) \geq 0$. The KLIC between f and p is defined as

$$D(p, f) = \int f(y)\log\frac{f(y)}{p(y)}dy. \tag{11.45}$$

Straightforward algebra shows that $D(p, f) = H(p||f) - H(f)$. Gibbs inequality ensures that KLIC is always positive. An implication is that $H(p||f) \geq H(f)$ with equality if and only if $p(y) = f(y)$. Recall the optimization problem as defined in Eqs. (11.39) and (11.40). Under the assumption that $H(f) = \int f(y)\log f(y)dy = \mathbb{E}(\log f(y))$ is well defined, the optimization can be rewritten as

$$\begin{aligned}\mathbf{w}^{*\ KLIC} &= \arg\max_{\mathbf{w}} \int f(y)\log\left(\mathbf{w}'\mathbf{p}(y)\right)dy \\ \text{s.t.}\quad \mathbf{w}'\mathbf{i} &= 1.\end{aligned}$$

Note that the objective function in this case is $-H(\mathbf{w}'\mathbf{p}||f)$ as defined in Eq. (11.44). This is the population version of the objective function in Hall and Mitchell (2007). The Lagrangian is

$$L(\mathbf{w}) = \int f(y)\log\left(\mathbf{w}'\mathbf{p}(y)\right)dy - \lambda\left(\mathbf{w}'\mathbf{i} - 1\right) \tag{11.46}$$

and the first order necessary conditions are therefore

$$\left.\frac{\partial L}{\partial \mathbf{w}}\right|_{\lambda=\lambda^*,\mathbf{w}=\hat{\mathbf{w}}^{KLIC}} = \int f(y)\frac{\mathbf{p}(y)}{\mathbf{w}'\mathbf{p}(y)}dy - \lambda\mathbf{i} = \mathbf{0} \tag{11.47}$$

$$\left.\frac{\partial L}{\partial \lambda}\right|_{\lambda=\lambda^*,\mathbf{w}=\hat{\mathbf{w}}^{KLIC}} = \mathbf{w}'\mathbf{i} - 1 = 0. \tag{11.48}$$

Pre-multiplying Eq. (11.47) with $\mathbf{w}'$ and simplifying using Eq. (11.48) gives $\lambda = 1$ and this implies the optimal weight vector must satisfy

$$\int \frac{\mathbf{p}(y)}{\mathbf{w}'\mathbf{p}(y)}f(y)dy = \mathbf{i} \tag{11.49}$$

or

$$\mathbb{E}\left(\frac{\mathbf{p}(y)}{\mathbf{w}'\mathbf{p}(y)}\right) = \mathbf{i}. \tag{11.50}$$

Several theoretical considerations are warranted here. The first is concerned with the performance of density forecast when the weight vector is the solution to the optimization problem as stated in Eqs. (11.39) and (11.40). The following propositions shed light on this issue.

Proposition 11.3 *Let* $p(y) = wp_1(y) + (1 - w)p_2(y)$ *then* $D(p, f) \leq wD(p_1, f) + (1 - w)D(p_2, f)$.

Proof See Appendix. □

Proposition 11.3 provides the upper bound of divergence for forecast density combination and it is the density forecast combination analogue to Proposition 1 in Chan and Pauwels (2018). An implication of this result is that it is possible for individual density forecast to outperform forecast combination and forms the foundation to demonstrate the relative performance of optimal weight as shown in Proposition 11.4.

Proposition 11.4 *Let* $\mathbf{w}$ *satisfies Eq. (11.49), then* $D(\mathbf{w}'\mathbf{p}, f) \leq D(p_i, f)$ *for all* $i = 1, \ldots N$.

Proof See Appendix. □

Following from Proposition 11.3, Proposition 11.4 shows that forecast combination based on the optimal weight as derived in Eq. (11.49) will improve density forecast in the KLIC sense. This demonstrates the fact that, similar to point forecast, combining density forecast will in general, improve forecast performance, at least in the KLIC sense.

The second theoretical consideration is the density counterpart of the 'forecast combination puzzle'. That is, how well will the simple average of density forecast perform? This is partially answered by the proposition below.

Proposition 11.5 *The simple average will outperform an individual density,* $p_i(y)$, *in the KLIC sense if*

$$\mathbb{E}\left(\frac{p_i(y)}{\mathbf{i}'\mathbf{p}(y)}\right) < \log \frac{1}{N}. \tag{11.51}$$

Proof See Appendix. □

Proposition 11.5 suggests that simple average may still outperform individual density if there are sufficient additional information from combining different densities.[2] In other words, there must be additional information content from the other densities in order for the simple average to be beneficial, even though the simple average is not the optimal weight.

11.4.3 Extension: Method of Moments

The method discussed above uses KLIC as a measure of 'distance' between the two densities. An alternative approach to this is to seek a linear combination of the K densities, so that the resulting random variable shares certain features, such as the same set of moments, with y_t. This is akin to the *generalized method of moments* proposed by Hansen (1982). There are two motivations of this approach, one theoretical and one practical.

The theoretical motivation is that the minimisation of KLIC aims to seek a linear combination so that the two densities are as 'close' to each other as possible. This applies to the density over all, which means certain aspects of the distribution may not be as close as it could be in order for the other features to be 'closer' such that the 'closeness' overall is achieved. However, there are applications where certain features of the distribution are more important than the others. This method allows practitioners, at least in principle, to choose which aspects or features of the distribution are more important. For example, when applied to the forecast of Value-at-Risk, the variance and the kurtosis of the distribution would perhaps be more important than other moments. See, for example, Chan and James (2011) and Pauwels et al. (2018) for further discussion.

The practical motivation for this extension is that there is no closed form solution to Eq. (11.50). Numerical methods must be employed in order to obtain the optimal weight vector for density forecast combination when optimizing KLIC. However, a closed form solutions is often possible for the method of moments as shown below.

Following from the previous section, the basic idea is to stipulate a linear combination of densities, $p(y) = \mathbf{w}'\mathbf{p}(y)$, to produce a combined density which has a set of moments as close to the moments of y_t as possible. Note that the jth moment of $p(y)$ is

$$\begin{aligned}\int y^j p(y)dy =& \int y^j \mathbf{w}'\mathbf{p}(y)dy \\ =&\mathbf{w}' \int y^j \mathbf{p}(y)dy\end{aligned}$$

[2]More specifically, the proposition shows that the additional information from combining different densities must be greater than N nit.

and therefore

$$\mathbf{w}'\left[\int y\mathbf{p}(y)dy, \ldots, \int y^N \mathbf{p}(y)dy\right] = \boldsymbol{\mu}_p \mathbf{w}$$

where

$$\boldsymbol{\mu}_p = \begin{bmatrix} \mu_{1,1} & \cdots & \mu_{K,1} \\ \vdots & \ddots & \vdots \\ \mu_{1,N} & \ddots & \mu_{K,N} \end{bmatrix} \quad (11.52)$$

with $\mu_{i,j}$ denotes the jth moment implied by the i density, $p_i(y)$. Let $\boldsymbol{\mu} = (\mu_1, \ldots, \mu_N)$ be the vector of the raw N moments of y_t, then one can seek the linear combination to minimise the following quadratic form

$$\hat{\mathbf{w}}^{MM} = \arg\min_{\mathbf{w}} \left(\boldsymbol{\mu}_p \mathbf{w} - \boldsymbol{\mu}\right)' \left(\boldsymbol{\mu}_p \mathbf{w} - \boldsymbol{\mu}\right) \quad (11.53)$$

$$\text{s.t.} \qquad \mathbf{i}'\mathbf{w} = 1. \quad (11.54)$$

Under the assumption that the Nth moment of y_t exists and $N \geq K$ such that $\boldsymbol{\mu}_p' \boldsymbol{\mu}_p$ is non-singular, it is straightforward to obtain a closed form solution to the optimization problem defined in Eqs. (11.53) and (11.54).

The Lagrangian associated with the optimization problem is

$$L = \left(\boldsymbol{\mu}_p \mathbf{w} - \boldsymbol{\mu}\right)' \left(\boldsymbol{\mu}_p \mathbf{w} - \boldsymbol{\mu}\right) + \lambda \left(1 - \mathbf{i}'\mathbf{w}\right). \quad (11.55)$$

The FONC is therefore

$$\left.\frac{\partial L}{\partial \mathbf{w}}\right|_{\mathbf{w}=\hat{\mathbf{w}}^{MM}, \lambda=\lambda^*} = 2\boldsymbol{\mu}_p' \left(\boldsymbol{\mu}_p \mathbf{w} - \boldsymbol{\mu}\right) - \lambda \mathbf{i} = \mathbf{0} \quad (11.56)$$

$$\left.\frac{\partial L}{\partial \lambda}\right|_{\mathbf{w}=\hat{\mathbf{w}}^{MM}, \lambda=\lambda^*} = 1 - \mathbf{i}'\mathbf{w}. \quad (11.57)$$

Under the affine constraint and pre-multiplying Eq. (11.56) with $\mathbf{w}'$ implies

$$\lambda^* = 2\hat{\mathbf{w}}^{MM\prime} \boldsymbol{\mu}_p' \left(\boldsymbol{\mu}_p \hat{\mathbf{w}}^{MM} - \boldsymbol{\mu}\right). \quad (11.58)$$

Similar to the optimal weight for combining point forecast under MSFE, the FONC as stated in Eqs. (11.56)–(11.58) forms a linear system. That is

$$\begin{pmatrix} 2\mathbf{U} & -\mathbf{i} \\ \mathbf{i}' & 0 \end{pmatrix} \begin{pmatrix} \hat{\mathbf{w}}^{MM} \\ \lambda^* \end{pmatrix} = \begin{pmatrix} 2\boldsymbol{\mu}_p' \boldsymbol{\mu} \\ 1 \end{pmatrix} \quad (11.59)$$

where $\mathbf{U} = \boldsymbol{\mu}'_p \boldsymbol{\mu}_p$. Under the assumption that $\mathbf{U}$ is non-singular, then the coefficient matrix in Eq. (11.59) has the inverse

$$\begin{pmatrix} \frac{1}{2}\left[\mathbf{U}^{-1} - \mathbf{U}^{-1}\mathbf{i}\left(\mathbf{i}'\mathbf{U}^{-1}\mathbf{i}\right)^{-1}\mathbf{i}'\mathbf{U}^{-1}\right] & -\mathbf{U}^{-1}\mathbf{i}\left(\mathbf{i}'\mathbf{U}^{-1}\mathbf{i}\right)^{-1} \\ -\left(\mathbf{i}'\mathbf{U}^{-1}\mathbf{i}\right)^{-1}\mathbf{i}'\mathbf{U}^{-1} & 2\left(\mathbf{i}'\mathbf{U}^{-1}\mathbf{i}\right)^{-1} \end{pmatrix}.$$

Solving for $\hat{\mathbf{w}}^{MM}$ gives

$$\hat{\mathbf{w}}^{MM} = \mathbf{U}^{-1}\left[\boldsymbol{\mu}'_p\boldsymbol{\mu} - \mathbf{i}\left(\mathbf{i}'\mathbf{U}^{-1}\mathbf{i}\right)^{-1}\left(\mathbf{i}'\mathbf{U}^{-1}\boldsymbol{\mu}'_p\boldsymbol{\mu} - 1\right)\right]. \tag{11.60}$$

One potential drawback of this approach is that $N \geq K$ is a necessary condition for $\mathbf{U}$ to have an inverse. While the existence of all N moments seems restrictive, especially if K large, it is possible to replace the raw moment constraints with a set of *orthogonality conditions*. This is in the same spirit as the *generalized Method of Moments* proposed by Hansen (1982) which extend the method of moments by replacing the moment constraints with a set of orthogonality conditions. In other words, practitioners may be able to propose N distinct features of the distribution that are important without imposing strict existence on higher order moments.

Let $\mathbf{g}(\mathbf{w})$ be a $N \times 1$ vector of twice differentiable functions, then consider the following optimization problem

$$\begin{aligned} \hat{\mathbf{w}}^{GMM} &= \arg\min_{\mathbf{w}} \mathbf{g}'(\mathbf{w})\mathbf{g}(\mathbf{w}) \\ \text{s.t.} \quad \mathbf{i}'\mathbf{w} &= 1. \end{aligned} \tag{11.61}$$

Again, construct the associated Lagrangian function which leads to the following first order condition

$$\begin{aligned} \left.\frac{\partial L}{\partial \mathbf{w}}\right|_{\mathbf{w}=\hat{\mathbf{w}}^{GMM},\lambda=\lambda^*} &= 2\frac{\partial \mathbf{g}'}{\partial \mathbf{w}}\mathbf{g}(\mathbf{w})\lambda\mathbf{i} = \mathbf{0} \\ \left.\frac{\partial L}{\partial \lambda}\right|_{\mathbf{w}=\hat{\mathbf{w}}^{GMM},\lambda=\lambda^*} &= 1 - \mathbf{i}'\mathbf{w} = 0. \end{aligned}$$

Using the similar argument as in the raw moment case, one can obtain

$$\lambda^* = 2\hat{\mathbf{w}}^{GMM\prime}\left.\frac{\partial \mathbf{g}'}{\partial \mathbf{w}}\right|_{\mathbf{w}=\hat{\mathbf{w}}^{GMM\prime}}\mathbf{g}\left(\hat{\mathbf{w}}^{GMM\prime}\right)$$

and therefore $\hat{\mathbf{w}}^{GMM}$ must satisfy

$$\left(\mathbf{I}_K - \mathbf{i}\hat{\mathbf{w}}^{GMM\prime}\right)\left.\frac{\partial \mathbf{g}'}{\partial \mathbf{w}}\right|_{\mathbf{w}=\hat{\mathbf{w}}^{GMM}}\mathbf{g}\left(\hat{\mathbf{w}}^{GMM}\right) = \mathbf{0}. \tag{11.62}$$

The connection between $\hat{\mathbf{w}}^{GMM}$ and $\hat{\mathbf{w}}^{KLIC}$ can be established in several ways. Perhaps the most straightforward approach is to consider the optimization problem as presented in Eqs. (11.39) and (11.40) with additional moment constraints. That is,

$$\begin{aligned} \hat{\mathbf{w}}^{R} &= \arg\min_{\mathbf{w}} - \int f(y) \log \mathbf{w}'\mathbf{p}(y) dy \\ \text{s.t.} \quad \mathbf{i}'\mathbf{w} &= 1 \\ \mathbf{g}(\mathbf{w}) &= 0 \end{aligned} \tag{11.63}$$

and the associated Lagrangian is

$$L(\mathbf{w}) = - \int f(y) \log \mathbf{w}'\mathbf{p}(y) dy + \lambda_0 \left(1 - \mathbf{i}'\mathbf{w}\right) + \boldsymbol{\lambda}'\mathbf{g}(\mathbf{w}). \tag{11.64}$$

$\boldsymbol{\lambda} = (\lambda_1, \ldots, \lambda_N)'$ denotes the vector of Lagrange multipliers associated with the N moment constraints. The FONC in this case are

$$\left.\frac{\partial L}{\partial \mathbf{w}}\right|_{\mathbf{w}=\hat{\mathbf{w}}^R, \boldsymbol{\lambda}=\boldsymbol{\lambda}^*, \lambda_0=\lambda_0^*} = \int f(y) \frac{\mathbf{p}(y)}{\mathbf{w}'\mathbf{p}(y)} dy - \lambda_0 \mathbf{i} - \mu_p' \boldsymbol{\lambda} = \mathbf{0} \tag{11.65}$$

$$\left.\frac{\partial L}{\partial \lambda_0}\right|_{\mathbf{w}=\hat{\mathbf{w}}^R} = 1 - \mathbf{w}'\mathbf{i} = 0 \tag{11.66}$$

$$\left.\frac{\partial L}{\partial \boldsymbol{\lambda}}\right|_{\mathbf{w}=\hat{\mathbf{w}}^R} = \mathbf{g}(\mathbf{w}) = \mathbf{0}. \tag{11.67}$$

Under the affine constraints and pre-multiplying Eq. (11.65) by $\mathbf{w}'$ yields

$$\lambda_0^* = 1 + \hat{\mathbf{w}}^R \frac{\partial \mathbf{g}'}{\partial \mathbf{w}} \boldsymbol{\lambda}^*$$

Substituting this expression into Eq. (11.65) gives

$$\mathbb{E}\left(\frac{\mathbf{p}(y)}{\hat{\mathbf{w}}^{R\prime}\mathbf{p}(y)}\right) - \mathbf{i} = -\left(\mathbf{I}_K - \mathbf{i}\hat{\mathbf{w}}^{R\prime}\right) \left.\frac{\partial \mathbf{g}'}{\partial \mathbf{w}}\right|_{\mathbf{w}=\hat{\mathbf{w}}^R} \boldsymbol{\lambda}. \tag{11.68}$$

This expression is quite insightful for connecting $\hat{\mathbf{w}}^{KLIC}$ and $\hat{\mathbf{w}}^{GMM}$. Note that the left hand side of Eq. (11.68) contains the expression in the FONC for $\hat{\mathbf{w}}^{KLIC}$ as shown in Eq. (11.49). If $\boldsymbol{\lambda}^* = \mathbf{g}(\hat{\mathbf{w}}^{GMM})$ then the right hand side of Eq. (11.68) contains the expression in the FONC for $\hat{\mathbf{w}}^{GMM}$ as shown in Eq. (11.62). In that case, if both sides are identically 0 then $\hat{\mathbf{w}}^R = \hat{\mathbf{w}}^{GMM} = \hat{\mathbf{w}}^{KLIC}$.

The intuition behind this result is as follows. The equivalence between the two optimal weight vectors can occur when the random variable associated with the density $p_{KLIC}(y) = \hat{\mathbf{w}}^{KLIC\prime}\mathbf{p}(y)$ satisfies the orthogonality conditions, that is

$\mathbf{g}\left(\hat{\mathbf{w}}^{KLIC}\right) = \mathbf{0}$. In that case, the constraint is binding and $\boldsymbol{\lambda}^* = \mathbf{0}$ and the FONC of the optimization as defined in Eq. (11.63) coincides with the FONC of the optimization as defined in Eq. (11.61).

Even if the constraints are not binding, equivalence between $\hat{\mathbf{w}}^{GMM}$ and $\hat{\mathbf{w}}^{KLIC}$ can still occur. That is, when $\hat{\mathbf{w}}^{KLIC}$ is also the minima for the quadratic form $\mathbf{g}'(\mathbf{w})\mathbf{g}(\mathbf{w})$, then clearly $\hat{\mathbf{w}}^{KLIC} = \hat{\mathbf{w}}^{GMM}$. At the present moment, the relation between $\mathbf{g}(\mathbf{w})$ and KLIC is unclear. That is, given $\mathbf{p}(y)$ and $p(y)$, is it possible to generate $\mathbf{g}(\mathbf{w})$ so that $\hat{\mathbf{w}}^{KLIC} = \hat{\mathbf{w}}^{GMM}$? Such results will no doubt enhance the current understanding of density forecast combination and it may also aid the computation of the optimal weight given the computation $\hat{\mathbf{w}}^{GMM}$ does not require numerical approximation when $\mathbf{g}(\mathbf{w})$ is a set of linear functions.

Note that it is possible to test if

$$H_0 : \hat{\mathbf{w}}^{GMM} = \hat{\mathbf{w}}^{KLIC}$$

$$H_1 : \hat{\mathbf{w}}^{GMM} \neq \hat{\mathbf{w}}^{KLIC}.$$

Under the null,

$$\mathbf{z} = \mathbb{E}\left(\frac{\mathbf{p}(y)}{\hat{\mathbf{w}}^{GMM\prime}\mathbf{p}(y)}\right) - \mathbf{i} = \mathbf{0}$$

let z_i be the ith element in $\mathbf{z}$ then

$$\sqrt{T}\frac{z_i}{SE(z_i)} \overset{d}{\sim} N(0,1)$$

by Central Limit Theorem under appropriate set of regularity conditions. In practice, the expectation can be estimated by Monte Carlo Integration method. That is

$$\mathbb{E}\left(\frac{\mathbf{p}(y)}{\hat{\mathbf{w}}^{GMM\prime}\mathbf{p}(y)}\right) \approx T^{-1}\sum_{t=1}^{T}\frac{\mathbf{p}(y_t)}{\hat{\mathbf{w}}^{GMM\prime}\mathbf{p}(y_t)}.$$

11.5 Conclusion

This chapter provided an overview on averaging models and forecasts with a focus on the theoretical foundation of forecast combination for both point forecast and density forecast. Since the optimal weight can only be estimated, this chapter provided some statistical properties of optimal weight estimate under both MSFE and MAD. Specifically, this chapter derived the asymptotic distribution of the optimal weight under MSFE. Due to the equivalence of the optimal weight under MSFE and MAD, the chapter also provided the asymptotic distribution of optimal weight under MAD.

In terms of forecast combination for probability density, this chapter provided several theoretical results concerning the performance of combining density forecast. It also proposed a novel method of combining density based on the generalized method of moments approach. It has shown the connection between the proposed GMM method and the conventional approach of minimising KLIC. A formal testing procedure had also been suggested to examine if the estimated weights between the two approaches are equivalent in practice.

While the materials in this chapter were presented from an econometric perspective, some of the central ideas can be found in the machine learning literature. For example, forecast combination via MSFE is closely related to model stacking, a popular form of model ensemble techniques in addition to bagging and boosting. The intimate connection between the theory presented here and those being investigated in the machine learning literature should be an interesting and important area of future research.

Throughout the chapter, it should be clear that the general theory of forecast combination is still developing. Such theory is not only useful in enhancing current understanding of forecast combination, it also has practical implications. The 'forecast combination puzzle' is an excellent example on the usefulness of theory in addressing practical issues, which leads to the improvement of best practice in forecasting.

Technical Proofs

Proof (Proposition 11.1) By definition

$$\hat{\mathbf{w}}_T^{MSE} - \mathbf{w}^* = \hat{\boldsymbol{\Omega}}_T^{-1}\mathbf{1}\left(\mathbf{1}^\top\hat{\boldsymbol{\Omega}}_T^{-1}\mathbf{1}\right)^{-1} - \boldsymbol{\Omega}^{-1}\mathbf{1}\left(\mathbf{1}^\top\boldsymbol{\Omega}^{-1}\mathbf{1}\right)$$

$$\sqrt{T}\left(\hat{\mathbf{w}}_T^{MSE} - \mathbf{w}^*\right) = \sqrt{T}\left\{\hat{\boldsymbol{\Omega}}_T^{-1}\mathbf{1}\left[\left(\mathbf{1}^\top\hat{\boldsymbol{\Omega}}_T^{-1}\mathbf{1}\right)^{-1} - \left(\mathbf{1}^\top\boldsymbol{\Omega}^{-1}\mathbf{1}\right)^{-1}\right] + \left(\hat{\boldsymbol{\Omega}}_T^{-1} - \boldsymbol{\Omega}^{-1}\right)\mathbf{1}\left(\mathbf{1}^\top\boldsymbol{\Omega}\mathbf{1}\right)^{-1}\right\}$$

$$\xrightarrow{D} \frac{\sqrt{T}}{\eta}\left[\left(\hat{\boldsymbol{\Omega}}_T^{-1} - \boldsymbol{\Omega}^{-1}\right)\mathbf{1}\right].$$

This implies

$$\sqrt{T}\operatorname{vec}\left(\hat{\mathbf{w}}_T^{MSE} - \mathbf{w}^*\right) \overset{a}{\sim} \sqrt{T}\eta^{-1}\left(\mathbf{1}'\otimes\mathbf{I}\right)\operatorname{vec}\left(\hat{\boldsymbol{\Omega}}_T^{-1} - \boldsymbol{\Omega}^{-1}\right).$$

Under Assumption ii, $\hat{\boldsymbol{\Omega}}_T$ has an asymptotic normal distribution, see for examples, Cook (1951) and Iwashita and Siotani (1994). Then, by the delta method, it follows that $\sqrt{T}\left(\hat{\mathbf{w}}_T^{MSE} - \mathbf{w}^*\right)$ is normally distributed with the variance-covariance

$$\mathbf{B} = \frac{1}{\eta^2}\left(\mathbf{1}^\top\boldsymbol{\Omega}^{-1}\otimes\boldsymbol{\Omega}^{-1}\right)\mathbf{A}\left(\boldsymbol{\Omega}^{-1}\mathbf{1}\otimes\boldsymbol{\Omega}^{-1}\right).$$

This completes the proof. □

Proof (Proposition 11.2) Corollary 3 in Chan and Pauwels (2019) showed that $\hat{\mathbf{w}}_T^{MAD} = \mathbf{w}^* + o_p(1)$ and $\hat{\mathbf{w}}_T^{MSFE} = \mathbf{w}^* + o_p(1)$, the result then follows directly from an application of Theorem 1 in Knight (2001). This completes the proof. □

Proof (Proposition 11.3) Note that

$$\begin{aligned}\log p(y) =& \log\left(wp_1(y) + (1-w)p_2(y)\right)\\ \geq& w\log p_1(y) + (1-w)\log p_2(y).\end{aligned}$$

Therefore

$$\begin{aligned}-\log p(y) \leq& -w\log p_1(y) - (1-w)\log p_2(y)\\ -f(y)\log p(y) \leq& -wf(y)\log p_1(y) - (1-w)f(y)\log p_2(y)\\ -\int f(y)\log p(y)dy \leq& -w\int f(y)\log p_1(y)dy - (1-w)\int f(y)\log p_2(y)dy\\ H(p||f) \leq& wH(p_1||f) + (1-w)H(p_2||f)\\ D(p,f) \leq& wD(p_1,f) + (1-w)D(p_2,f).\end{aligned}$$

□

Proof (Proposition 11.4) It is sufficient to show that $H(\mathbf{w}'\mathbf{p}||f) - H(p_i||f) \leq 0$ for all i.

$$\begin{aligned}&H(\mathbf{w}'\mathbf{p}||f) - H(p_i||f)\\ =&\int f(y)\log\left(\frac{p_i(y)}{\mathbf{w}'\mathbf{p}(y)}\right)dy\\ =&\mathbb{E}\left[\log\left(\frac{p_i(y)}{\mathbf{w}'\mathbf{p}(y)}\right)\right]\\ \leq&\log\mathbb{E}\left(\frac{p_i(y)}{\mathbf{w}'\mathbf{p}(y)}\right)\\ \leq&0.\end{aligned}$$

Proof (Proposition 11.5) The result is straightforward if the simple average, $\mathbf{i}/N$ satisfied Eq. (11.50). Assuming that it does not, then the simple average outperforms an individual density in the KLIC sense implies

$$H\left(\frac{\mathbf{i}'}{N}\mathbf{p}(y)||f\right) - H\left(p_i(y)||f\right) < 0$$

$$-\int f(y)\log\frac{\mathbf{i}'\mathbf{p}(y)}{N}dy + \int f(y)\log p_i(y)dy < 0$$

$$\int f(y)\log\frac{p_i(y)}{\mathbf{i}'\mathbf{p}(y)}dy < \log\frac{1}{N}.$$

References

Aiolfi, A., Capistrán, C., & Timmermann, A. (2010). *A simple explanation of the forecast combination puzzle*. Aarhus: Center for Research in Econometric Analysis of Time Series (CREATS), Aarhus University.

Bates, J. M., & Granger, C. W. J. (1969). The combination of forecasts. *Operational Research Quarterly, 20*, 451–468.

Buckland, S., Burnham, K., & Augustin, N. (1997). Model selection: An integral part of inference. *Biometrics, 53*, 603–618.

Busetti, F. (2017). Quantile aggregation of density forecasts. *Oxford Bulletin of Economics and Statistics, 79*(4), 495–512. https://doi.org/10.1111/obes.12163.

Chan, F., & James, A. (2011). Application of forecast combination in volatility modelling. In F. Chan, D. Marinova, & R. Anderssen (Eds.), *Modsim2011, 19th international congress on modelling and simulation* (pp. 1610–1616). Canberra: Modelling, Simulation Society of Australia, and New Zealand. Retrieved from http://mssanz.org.au/modsim2011/D10/james.pdf

Chan, F., & Pauwels, L. L. (2018). Some theoretical results on forecast combinations. *International Journal of Forecasting, 34*(1), 64–74.

Chan, F., & Pauwels, L. L. (2019). Equivalence of optimal forecast combinations under affine constraints. Working paper.

Chatfield, C. (1993). Calculating interval forecasts. *Journal of Business and Economic Statistics, 11*, 121–135.

Christoffersen, P. F. (1998). Evaluating interval forecasts. *International Economic Review, 39*, 841–862.

Claeskens, G., & Hjort, N. L. (2003). The focused information criterion. *Journal of the American Statistical Association, 98*(464), 900–916. https://doi.org/10.1198/016214503000000819.

Claeskens, G., & Hjort, N. L. (2008). *Model selection and model averaging*. Cambridge: Cambridge University Press.

Claeskens, G., Magnus, J. R., Vasnev, A. L., & Wang, W. (2016). A simple theoretical explanation of the forecast combination puzzle. *International Journal of Forecasting, 32*(3), 754–62.

Clark, T. E., & West, K. D. (2006). Using out-of-sample mean squared prediction errors to test the martingale difference hypothesis. *Journal of Econometrics, 135*(1–2), 155–186. https://doi.org/10.1016/j.jeconom.2005.07.014

Clark, T. E., & West, K. D. (2007). Approximately normal tests for equal predictive accuracy in nested models. *Journal of Econometrics, 138*(1), 291–311. https://doi.org/10.1016/j.jeconom.2006.05.023

Clemen, R. T. (1989). Combining forecasts: A review and annotated bibliography. *International Journal of Forecasting, 5*, 559–583.

Cook, M. B. (1951). Bi-variate k-statistics and cumulants of their joint sampling distribution. *Biometrika, 38*(1), 179–195.

Corradi, V., & Swanson, N. R. (2006). Predictive density evaluation. In *Handbook of economic forecasting* (Chap. 5, Vol. 1, pp. 135–196). https://doi.org/10.1016/S15740706(05)01004-9.
Diebold, F. X. (1989). Forecast combination and encompassing: Reconciling two divergent literatures. *International Journal of Forecasting, 5*, 589–592.
Diebold, F. X., Gunther, T. A., & Tay, A. S. (1998). Evaluating interval forecasts with applications to financial risk management. *International Economic Review, 39*, 863–883.
Diebold, F. X., & Lopez, J. A. (1996). *Forecast evaluation and combination*. Cambridge: National Bureau of Economic Research (NBER).
Diebold, F. X., & Pauly, P. (1987). Structural change and the combination of forecasts. *Journal of Forecasting, 6*, 21–40.
Elliott, G. (2011). *Averaging and the optimal combination of forecasts*. San Diego: University of California.
Elliott, G., Gargano, A., & Timmermann, A. (2013). Complete subset regressions. *Journal of Econometrics, 177*, 357–373. https://doi.org/10.1016/j.jeconom.2013.04.017.
Elliott, G., & Timmermann, A. (2004). Optimal forecast combinations under general loss functions and forecast error distributions. *Journal of Econometrics, 122*, 47–79.
Genre, V., Kenny, G., Meyler, A., & Timmermann, A. (2013). Combining expert forecasts: Can anything beat the simple average? *International Journal of Forecasting, 29*, 108–121.
Geweke, J., & Amisano, G. (2011). Optimal prediction pools. *Journal of Econometrics, 164*, 130–141. https://doi.org/10.1016/j.jeconom.2011.02.017.
Granger, C. J. W., & Ramanathan, R. (1984). Improved methods of combining forecasts. *Journal of Forecasting, 3*, 197–204.
Hall, S. G., & Mitchell, J. (2007). Combining density forecasts. *International Journal of Forecasting, 23*, 1–13.
Hansen, B. E. (2007). Leasts squares model averaging. *Econometrica, 75*(4), 1175–1189.
Hansen, B. E. (2008). Least squares forecast averaging. *Journal of Econometrics, 1146*, 342–350.
Hansen, B. E. (2014). Model averaging, asymptotic risk, and regressor groups. *Quantitative Economics, 5*(3), 495–530. https://doi.org/10.3982/QE332.
Hansen, B. E., & Racine, J. (2012). Jackknife model averaging. *Journal of Econometrics, 167*, 38–46.
Hansen, L. (1982). Large sample properties of generalized method of moments estimators. *Econometrica, 50*, 1029–1054.
Harvey, D., Leybourne, S., & Newbold, P. (1997). Testing the equality of prediction mean squared errors. *International Journal of Forecasting, 13*(2), 281–291. https://doi.org/10.1016/S0169-2070(96)00719-4.
Hendry, D. F., & Clements, M. P. (2004). Pooling of forecasts. *Econometrics Journal, 7*, 1–31.
Hjort, N. L., & Claeskens, G. (2003). Frequentist model average estimators. *Journal of the American Statistical Association, 98*(464), 879–899. https://doi.org/10.1198/016214503000000828. arXiv:1011.1669v3.
Hsiao, C., & Wan, S. (2014). Is there an optimal forecast combination? *Journal of Econometrics, 178*, 294–309.
Iwashita, T., & Siotani, M. (1994). Asymptotic distributions of functions of a sample covariance matrix under the elliptical distribution. *Canadian Journal of Statistics, 22*(2), 273–283.
Jore, A. S., Mitchell, J., & Vahey, S. P. (2010). Combining forecast densities from VARs with instabilities. *Journal of Applied Econometrics, 25*, 621–634.
Kapetanios, G., Mitchell, J., Price, S., & Fawcett, N. (2015). Generalised density forecast combinations. *Journal of Econometrics, 188*, 150–165. https://doi.org/10.1016/j.jeconom.2015.02.047.
Kascha, C., & Ravazzolo, F. (2010). Combining inflation density forecasts. *Journal of Forecasting, 29*, 231–250.
Knight, K. (1998). Limiting distributions for L1 regression estimators under general conditions. *The Annals of Statistics, 26*(2), 755–770.

Knight, K. (2001). Limiting distributions of linear programming estimators. *Extremes, 4*(2), 87–103. Retrieved from http://link.springer.com/article/10.1023/A%7B%5C%%7D3A1013991808181.

Laplace, P. (1818). *Deuxième supplément a la théorie analytique des probabilitiés*. Paris: Courcier.

Liu, C. A., & Kuo, B. S. (2016). Model averaging in predictive regressions. *The Econometrics Journal, 19*, 203–231. https://doi.org/10.1111/ectj.12063.

Marcellino, M. (2004). Forecast pooling for European macroeconomic variables. *Oxford Bulletin of Economics and Statistics, 66*, 91–112.

Mitchell, J., & Hall, S. G. (2005). Evaluating, comparing and combining density forecasts using the KLIC with an application to the bank of England and NIESR 'fan' charts of inflation. *Oxford Bulletin of Economics and Statistics, 67*, 995–1033.

Moral-Benito, E. (2015). Model averaging in economics: An overview. *Journal of Economic Surveys, 29*(1), 46–75. https://doi.org/10.1111/joes.12044.

Newbold, P., & Granger, C. J. W. (1974). Experience with forecasting univariate time series and the combination of forecasts. *Journal of the Royal Statistical Society A, 137*, 131–165.

Pauwels, L. L., Radchenko, P., & Vasnev, A. (2018). Higher moment constraints for predictive density combinations. *SSRN Electronic Journal*. https://doi.org/10.2139/ssrn.3315025.

Pauwels, L. L., & Vasnev, A. L. (2016). A note on the estimation of optimal weights for density forecast combinations. *International Journal of Forecasting, 32*, 391–397.

Pesaran, M. H., & Timmermann, A. (2007). Selection of estimation window in the presence of breaks. *Journal of Econometrics, 137*, 134–161.

Pinar M., Stengos, T., & Yazgan, M. E. (2012). *Is there an optimal forecast combination? a stochastic dominance approach to forecast combination*. Waterloo, On: The Rimini Centre for Economic Analysis.

Smith, J., & Wallis, K. F. (2009). A simple explanation of the forecast combination puzzle. *Oxford Bulletin of Economics and Statistics, 71*, 331–355.

Stigler, S. M. (1973). Laplace, fisher and the discovery of the concept of sufficiency. *Biometrika, 60*(3), 439–445.

Stock, J. H., & Watson, M. W. (1998). *A comparison of linear and nonlinear univariate models for forecasting macroeconomic time series*. Cambridge: National Bureau of Economic Research (NBER).

Stock, J. H., & Watson, M. W. (2003). How did leading indicator forecasts perform during the 2001 recession? *Quarterly Economic Review – Federal Reserve Bank of Richmond, 89*, 71–90.

Stock, J. H., & Watson, M. W. (2004). Combination forecasts of output growth in a seven-country data set. *Journal of Forecasting, 23*, 405–430.

Tay, A. S., & Wallis, K. F. (2000). Density forecasting: A survey. *Journal of Forecasting, 19*, 235–254.

Tian, J., & Anderson, H. M. (2014). Forecast combination under structural break uncertainty. *International Journal of Forecasting, 30*, 161–175.

Timmermann, A. (2006). Forecast combinations. In *Handbook of economic forecasting* (Chap. 4, Vol. 1, pp. 135–196). https://doi.org/10.1016/S1574-0706(05)01004-9.

Wallis, K. F. (2005). Combining density and interval forecasts: A modest proposal. *Oxford Bulletin of Economics and Statistics, 67*, 983–994.

Chapter 12
Bayesian Model Averaging

Paul Hofmarcher and Bettina Grün

12.1 Introduction

In recent years, macroeconomic forecasting has seen a rise in the application of advanced statistical methods and models to address the challenges and opportunities of "big data." Big data is a relatively new phenomenon to economists, but following Koop (2017), *big data has the potential to revolutionize empirical macroeconomics*, as the information contained in these data sets could improve our forecasts and our understanding of the macroeconomic environment. The era of big data has emerged from the increased ease of collecting data from new sources, including, for example, social media content, and the automatic processing of text files to extract information. This increases the number of variables in the data set and calls for suitable statistical methods which combine the information from traditional sources with the new ones; for an overview on the new data sources being now available in the big data area see Chap. 1.

In many macroeconomic forecasting applications the aim is to predict a variable of interest, such as inflation, unemployment, growth, etc., based on a set of covariates. For a numeric dependent variable linear regression is the standard predictive model. However, the specification of a single linear model with a specific set of covariates requires conclusive theory on which covariates to include. Especially in the era of big data with an enormous number of *new* variables this

---

P. Hofmarcher (✉)
Salzburg Centre of European Union Studies (SCEUS), Department of Business, Economics and Social Theory, Paris Lodron University of Salzburg, Salzburg, Austria
e-mail: paul.hofmarcher@sbg.ac.at

B. Grün
Department of Applied Statistics, Johannes Kepler University Linz, Linz, Austria
e-mail: Bettina.Gruen@jku.at

P. Fuleky (ed.), *Macroeconomic Forecasting in the Era of Big Data*,
Advanced Studies in Theoretical and Applied Econometrics 52,
https://doi.org/10.1007/978-3-030-31150-6_12

might be challenging due to a lack of theory and the availability of a large number of predictors which are potentially not all relevant. In addition to the forecasting context this is also a virulent and challenging problem when aiming at identifying relevant determinants for the dependent variable.

A naïve approach to resolve this problem is to use a data-driven approach to select a single "best" model (e.g., via stepwise regression methods). This "best" model would then be used to describe the relationships between the covariates and the dependent variable and predict the dependent variable of interest. This approach, however, neglects the uncertainty associated with the selection of this specific model. This uncertainty is usually referred to as model uncertainty, i.e., the uncertainty of choosing the appropriate set of covariates to be included in the model. Disregarding this uncertainty in an empirical application has been shown to lead to inferior predictive performance and misleading insights with respect to covariates being determinants for the dependent variable.

One approach to account for model uncertainty is to use a modeling approach which simultaneously includes all models from a set of possible models $\{M_1, M_2, \ldots\}$. In the regression case, usually the models in this set differ with respect to the covariates included in the model. A statistical tool is needed which allows to combine the models in this set to obtain the final model used for prediction and assessing covariate importance. In statistics, Barnard (1963) was the first who proposed a model combination to forecast air passenger data. Bates and Granger (1969) compared a number of methods to combine two forecasts and indicated that changing the weights of those methods might result in better forecasts than constant weights. The modern theory of Bayesian model averaging (henceforth BMA) was finally presented by Leamer (1978). BMA assumes that prior knowledge is available on a set of possible models which contains the true model. Other forecast combination approaches were proposed including frequentist averaging, which is discussed and presented in detail in Chap. 11.

This work gives a coherent overview of BMA and its economic applications. For a policy maker two veins of inference might be of interest: (1) to determine robust determinants of the dependent variable of interest, like inflation or unemployment and (2) to improve forecasting performance. Both veins can be addressed via BMA. BMA represents a well-founded statistical approach to avoid selecting a specific model, but is based on using an ensemble of models. Using an ensemble usually has a better forecasting performance than any single model (Madigan & Raftery, 1994). A tutorial introducing BMA is provided in Hoeting, Madigan, Raftery, and Volinsky (1999); Moral-Benito (2015) and Steel (2018) provide overview papers on BMA and its use in economics.

This work is organized as follows: The next section gives an overview of the use of BMA in economics. In particular, we list a number of relevant and recent applications of BMA in economics. Section 12.3 describes the general methodology of BMA for the linear model class including a description of widely used model and parameter priors. Inference and posterior analysis under these priors is discussed. Section 12.4 applies the presented BMA framework to forecast box office revenues. The data set of this exercise is from Lehrer and Xie (2017). The data set includes not only traditional covariates such as budget and genre classification but also variables derived from social media content. The BMA results are obtained with the open-source package BMS (Zeugner & Feldkircher, 2015) available for the R environment for statistical computing and graphics (R Core Team., 2019). Finally, Sect. 12.5 concludes.

12.2 BMA in Economics

Inference under model uncertainty is a pervasive problem in economics, particularly when modeling or aiming at predicting macroeconomic variables. Model uncertainty is caused by lack of theory or the abundance of competing theories not excluding each other. In this case researchers are not guided by economic theory to select the covariates to be included in the regression model (Brock & Durlauf, 2001). Today there is a huge strand of literature applying BMA methods to account for model uncertainty in economics.

One of the most prominent examples of such a problem in economics is the identification of robust determinants for economic long-term cross-country growth (Fernández, Ley, & Steel, 2001; Sala-i-Martin, Doppelhofer, & Miller, 2004). Modern growth theory has been an area with many potential covariates being suggested and empirical evidence has struggled to resolve the open-endedness of theory. The open-endedness of theory is caused by competing growth theories not ruling out each other (see Brock and Durlauf, 2001). To assess the robustness of empirical determinants of economic long-term cross-country growth, BMA has been applied, among others, by Brock and Durlauf (2001), Danquah, Moral-Benito, and Ouattara (2014), Eicher, Henn, and Papageorgiou (2012), Fernández et al. (2001), Moral-Benito (2012), Sala-i-Martin et al. (2004). In the spirit of Acemoglu, Johnson, and Robinson (2001), who argue that intellectual property rights are a key determinant of long-term growth, Eicher and Newiak (2013) use a two stage least squares BMA approach, developed by Lenkoski, Eicher, and Raftery (2014), to study the influence of intellectual property rights on country development. Arin and Braunfels (2018) study the impact of natural resources on long-term growth using BMA methods.

Other economic applications of BMA include, e.g., the identification of indicators for a financial crisis (Feldkircher, Horvath, & Rusnak, 2014), or the modeling of aggregated default rates for firms (Hofmarcher, Kerbl, Grün, Sigmund, & Hornik, 2014). Wright (2008) uses BMA to forecast exchange rates and finds that BMA results in slightly better out-of-sample forecasts than a random walk benchmark model. Tobias and Li (2004) estimate the returns to education via BMA. Jetter and Parmeter (2018) use BMA methods to identify robust determinants of corruption among cultural, economic, institutional, and geographical factors. Time-varying BMA approaches were developed by, e.g., Raftery, Karny, and Ettler (2010) to model inflation and output forecasts and inter alia applied by Koop and Korobilis (2012).

12.2.1 Jointness

One of the aims of applying BMA is to identify robust determinants for the dependent variable. Following an early discussion of Bayesian measures of variable importance, posterior inclusion probabilities (PIPs) have become a standard measure to achieve this and assess and interpret the results from BMA applications (Hofmarcher, Crespo Cuaresma, Grün, Humer, & Moser, 2018). PIPs provide valuable insight into the overall importance of a covariate. Regardless of which other covariates are included in the regression model the PIP of a covariate reflects its marginal inclusion probability and thus may be interpreted as a robust measure of importance.

The drawback of PIPs is that they do not account for the interdependence of inclusion and exclusion of variables. Using PIPs it is, for example, not possible to determine if the importance of a variable is evenly spread across all potential model specifications or it is specific to a certain combination of explanatory variables. Further on, using solely PIPs we cannot infer whether two covariates tend to appear together in model specifications (complements), are independently from each other included, or are substitutes in terms of model inclusion.

To analyze the joint inclusion/exclusion of covariates so-called bivariate jointness measures were introduced in the economic literature and a vivid discussion emerged about the criteria or properties a bivariate jointness measure should fulfill. Doppelhofer and Weeks (2005)[1] proposed a measure, which was criticized by Ley and Steel (2007) and Strachan (2009) because it does not fulfill a set of desirable properties. These authors then propose additional measures which were again criticized by Doppelhofer and Weeks (2009b) due to non-desirable properties. Doppelhofer and Weeks (2009b) again propose yet another measure. Finally, Hofmarcher et al. (2018) present a rigorous list of properties which should be fulfilled by a suitable bivariate jointness measure by merging two strands of literature: the BMA jointness literature as well as the literature on interestingness measures for association rules in machine learning (Glass, 2013; Wu, Chen, & Han, 2010). Hofmarcher et al. (2018) propose a regularized version of Yule's Q association coefficient as a suitable bivariate measure to assess jointness. This measure meets all criteria in the list of desirous properties for bivariate jointness measures.

Bivariate jointness measures only allow to analyze pairs of covariates instead of groups of explanatories consisting of more than two covariates. An alternative approach to investigate the interdependence structure of covariate inclusion in BMA has been proposed by Crespo Cuaresma, Grün, Hofmarcher, Humer, and Moser (2016) by describing the overall dependency structure of the covariates via latent class analysis (LCA). The main idea of this approach is that the overall dependency structure between the sampled models is driven by a latent discrete variable. This latent variable groups the sampled models in such a way that the single covariates

[1] This working paper was published in a slightly different version as Doppelhofer and Weeks (2009a).

are included independently in the models within each latent group. By applying LCA to the models sampled via BMA, we are able to capture the dependency structure across included covariates through an unobserved latent discrete variable. Conditional on this latent discrete variable covariate inclusion is independent. Such a setting implies that PIPs within groups constitute sufficient information to describe the importance of the variables and the differences of PIPs between groups cause the marginal dependency of inclusion observed for different covariates.

12.2.2 Functional Uncertainty

The standard BMA approach relies on the assumption that the functional form of the relationship between the variables and the mean of the dependent variable is known. Usually, variables included in the model are assumed to have a linear effect on the dependent variable. This implies that a unit change in the variable has the same effect on the dependent variable regardless of the specific value the variable has. Alternatively additional covariates can be derived from the observed variable to model a non-linear functional relationship. For example, Henderson and Parmeter (2016) manually added new covariates via polynomial terms of the original variables and Tobias and Li (2004) defined piecewise linear functions based on the original variables.

Functional misspecification occurs if only linear effects are included, but in fact non-linear effects should be considered. In this case several issues may arise which preclude the correct robust identification of effects (see Malsiner-Walli, Hofmarcher, & Grün, 2019): (1) irrelevant covariates may be included into the model to compensate for the missing non-linear effects and are overestimated in terms of importance, (2) important non-linear covariates may not be identified due to the functional misspecification, and (3) effects of covariates, which vary over the covariate range, are forced to be constant over the entire covariate range. Inference based on such a misspecified model may lead to incorrect conclusions.

To circumvent this, non-parametric as well as semi-parametric methods have been suggested (see Henderson & Parmeter, 2016). For instance, local-linear least square regressions are used in Henderson, Papageorgiou, and Parmeter (2011) to investigate non-linearities of different economic growth determinants and in Delgado, Henderson, and Parmeter (2014) to study the effect of education on economic growth. Non-parametric methods, however, have the disadvantage that they are only applicable for a small number of covariates requiring a pre-selection of the relevant variables.

Semi-parametric methods using model matrix expansions to obtain more flexible regression functions are discussed in Smith and Kohn (1996) to model housing values, and in Koop and Tobias (2006) to study returns to education. Malsiner-Walli et al. (2019) show how Bayesian semi-parametric regression methods in combination with stochastic search variable selection (Scheipl, 2011; Scheipl, Fahrmeir, & Kneib, 2012) can be used to address two model uncertainties simultaneously: firstly, the model uncertainty with respect to the included covariates and secondly, the uncertainty concerning the functional form of the relationship between dependent and independent variables. In contrast to the non-parametric approaches this approach allows to include a large number of variables. The drawback is that different priors need to be employed than usually preferred in BMA applications. This is necessary to be able to simultaneously estimate the smoothness parameters in a computationally feasible way while accounting for model uncertainty.

12.3 Statistical Model and Methods

12.3.1 Model Specification

In the following we consider the standard BMA setting where the aim is to predict or model a dependent variable y given a number of covariates $x_1, \ldots, x_K$ based on a linear model. The assumption is that not all covariates are necessary, but only a subset $S \subset \{1, \ldots, K\}$ such that

$$y = \beta_0 + \sum_{s \in S} x_s \beta_s + \epsilon,$$

with $\epsilon \sim N(0, \sigma^2)$.

The selection of the set S is the model selection step and each S induces a different model $\mathcal{M}$. If there are K covariates and each covariate can either be included or excluded, the total number of different selection sets and thus models equals 2^K. In addition one can associate with each model a 0/1 vector $\boldsymbol{\gamma}$ of length K indicating if a covariate is included in the model or not.

The linear model represents the data model. In a Bayesian setting this is complemented by priors on the parameters. For BMA one needs (1) model priors and (2) regression parameter priors on the coefficients $\boldsymbol{\beta}$ and the error variance σ^2. The posterior is obtained based on a sample with n observations such that $\boldsymbol{y} = (y_1, \ldots, y_n)$ is the vector of the observed values for the dependent variable and $\boldsymbol{x}_k = (x_{1k}, \ldots, x_{nk})$ is the vector of the observed values for the kth covariate. The covariates are combined in the matrix $\boldsymbol{X}$ by column-wise stacking of the covariate vectors.

The model prior gives non-zero prior weights to a specific set of models and zero weight to all other models not included in this set. The Bayesian analysis then is performed under the assumption that the true model is among the models which have non-zero prior model weights. The posterior model probabilities obtained thus need to be interpreted conditional on this assumption. The posterior probability of a model $\mathcal{M}_j$ is then given by

$$p(\mathcal{M}_j | \boldsymbol{y}, \boldsymbol{X}) \propto p(\boldsymbol{y} | \boldsymbol{X}, \mathcal{M}_j) p(\mathcal{M}_j). \tag{12.1}$$

Taking model uncertainty into account the posterior distribution of the parameters given the data corresponds to

$$p(\beta_0, \boldsymbol{\beta}, \sigma^2 | \boldsymbol{y}, \boldsymbol{X}) = \sum_{j=1}^{2^K} p(\beta_0, \boldsymbol{\beta}, \sigma^2 | \boldsymbol{y}, \boldsymbol{X}, \mathcal{M}_j) p(\mathcal{M}_j | \boldsymbol{y}, \boldsymbol{X}).$$

Conditional on fixing the model $\mathcal{M}_j$ with $j \in \{1, \ldots, 2^K\}$ one can use general Bayesian regression analysis methods to obtain the posterior $p(\beta_0, \boldsymbol{\beta}, \sigma^2 | \boldsymbol{y}, \boldsymbol{X}, \mathcal{M}_j)$ with $\boldsymbol{\beta} = (\beta_1, \ldots, \beta_K)$.

12.3.2 Regression Parameter Priors

The prior usually selected for $(\beta_0, \boldsymbol{\beta}, \sigma^2)$ has the following characteristics:

(a) The intercept and the error variance are a-priori independent, while the regression coefficients depend a-priori on σ^2:

$$p(\beta_0, \boldsymbol{\beta}, \sigma^2) = p(\beta_0) p(\boldsymbol{\beta} | \sigma^2) p(\sigma^2).$$

(b) Uninformative, flat priors are used for the intercept and the error variance:

$$p(\beta_0, \sigma^2) \propto \frac{1}{\sigma^2}.$$

(c) Zellner's g prior (Zellner, 1986) is used for the regression coefficients prior. The regression coefficients are assumed to follow a normal distribution and are a-priori correlated as implied by the covariate structure:

$$p(\boldsymbol{\beta} | \sigma^2) \sim N\left(\boldsymbol{0}, g\sigma^2 \left(\boldsymbol{X}'\boldsymbol{X}\right)^{-1}\right).$$

The hyper-parameter g influences how much the ordinary least squares (OLS) estimates are shrunken towards zero. Different fixed values were proposed such as the unit information prior with $g = n$ recommended by Eicher, Papageorgiou, and Raftery (2011). Alternatively also a hyper-prior may be put on g (Ley & Steel, 2012). For $g \to \infty$ a flat prior emerges and this corresponds to using the OLS estimator (Sala-i-Martin et al., 2004).

These priors are in general selected because they are uninformative where possible and lead to computational efficient estimation. In particular the use of Zellner's g prior considerably simplifies estimation and posterior inference. For Zellner's g prior the posterior distribution of the model parameters is available in closed form. The posterior mean and variance of the regression coefficients can be analytically determined. Also the marginal likelihood of the data given a specific model is available in closed form. An alternative to Zellner's g prior would be the independence prior where the regression coefficients are assumed to follow a normal distribution and are a-priori independent. However, estimation and posterior inference are complicated. For a detailed comparison of the independence prior and Zellner's g prior see Malsiner-Walli et al. (2019).

12.3.3 Model Priors

Priors over models $\mathcal{M}_j$, $j = 1, \ldots, 2^K$ are usually constructed by exploiting the one-to-one relationship between a model $\mathcal{M}_j$ and the binary covariate inclusion vector $\boldsymbol{\gamma}_j$ and assigning probabilities to the inclusion of covariates.

The specification of an appropriate model prior builds on two pieces of information: firstly, the expected number of covariates included in the model and secondly, if covariates are assumed to be included independent of the inclusion of other covariates. In the following we will refer to model priors which include covariates independently of the inclusion of other covariates as *independent model priors* and model priors for which the a-priori inclusion of one covariate depends on the inclusion of specific other covariates as *dependent model priors*.

Independent Model Priors

Several independent priors were proposed in the literature for the covariate inclusion vector $\boldsymbol{\gamma}$ which can be used to represent a specific model. In the following we will focus on the two most widely used independent model priors. These are the binomial model prior proposed by Sala-i-Martin et al. (2004) and the extension to the beta-binomial model prior put forward by Ley and Steel (2009).

Binomial Model Prior

A naïve approach to specify a model prior assigns the same probability to each model $\mathcal{M}_j$ in the considered model class represented by the inclusion vector $\boldsymbol{\gamma}_j$. This seems to represent an uninformative prior where a uniform distribution over the model space is used. However, in the regression case where the inclusion of different covariate sets is considered this prior puts unequal mass to models of different size. For K covariates there are K different models of size one, while there are $\binom{K}{k}$ different models of size k. For this naïve model prior specification the expected prior model size is $K/2$. The expected prior model size thus increases with the number of potential explanatories and for many applications this prior would put too much weight to large models containing many covariates.

This naïve prior where all models have the same weight emerges if for each covariate the prior inclusion probability is set to 1/2. Choosing different prior inclusion probabilities generalizes this prior and allows to vary the prior expected model size. This model prior was suggested by Sala-i-Martin et al. (2004). They selected a prior expected model size $\bar{k}$ and assumed that each covariate has a prior inclusion probability of $\theta = \bar{k}/K$. For a chosen inclusion probability θ, the probability of a specific model $\boldsymbol{\gamma}$ is given by

$$p(\boldsymbol{\gamma}|\theta) = \theta^k (1-\theta)^{K-k},$$

where $\boldsymbol{\gamma}$ contains k covariates, i.e., $\boldsymbol{\iota}'\boldsymbol{\gamma} = k$, with $\boldsymbol{\iota}$ a vector of ones of length K.

A simple way to elicit θ is to determine the prior expected model size. The prior expected model size implied by the binomial model prior equals θK with variance $\theta(1-\theta)K$. Sala-i-Martin et al. (2004) argued that in terms of interpretability the expected prior model size should not linearly increase with K. In their empirical application, they choose θ in a way, that $\theta K = 7$, for their benchmark model. In addition they also show that the choice of θ can have a substantial impact on the results.

Beta-Binomial Model Prior

To increase the flexibility of the model prior and to reduce the influence of the prior expected model size $\bar{k}$, Ley and Steel (2009) proposed a hyper-prior on the inclusion probability θ by imposing a beta distribution. This changes the prior for $\boldsymbol{\gamma}$ from the binomial to the beta-binomial prior, with the hierarchical specification:

$$p(\boldsymbol{\gamma}|\theta) = \theta^k(1-\theta)^{K-k},$$
$$\theta \sim \text{Beta}(a, b).$$

This gives

$$\begin{aligned} p(\boldsymbol{\gamma}|a,b) &= \frac{\Gamma(a+b)}{\Gamma(a)\Gamma(b)} \int_{\Omega_\theta} \theta^k(1-\theta)^{(K-k)}\theta^{a-1}(1-\theta)^{b-1}d\theta \\ &= \frac{\Gamma(a+b)}{\Gamma(a)\Gamma(b)\Gamma(a+b+K)}\Gamma(a+k)\Gamma(b+K-k), \end{aligned}$$

which leads to much less informative and therefore more flexible model priors. Again the prior probability of a specific model does only depend on the number of covariates included k, but not the specific covariates. A detailed discussion of this model prior can be found in Ley and Steel (2009). They proposed to set $a = 1$ and $b = (K - \bar{k})/\bar{k}$ to induce a prior expected model size of $\bar{k}$ and provided a rigorous comparison of the binomial and the beta-binomial model priors.

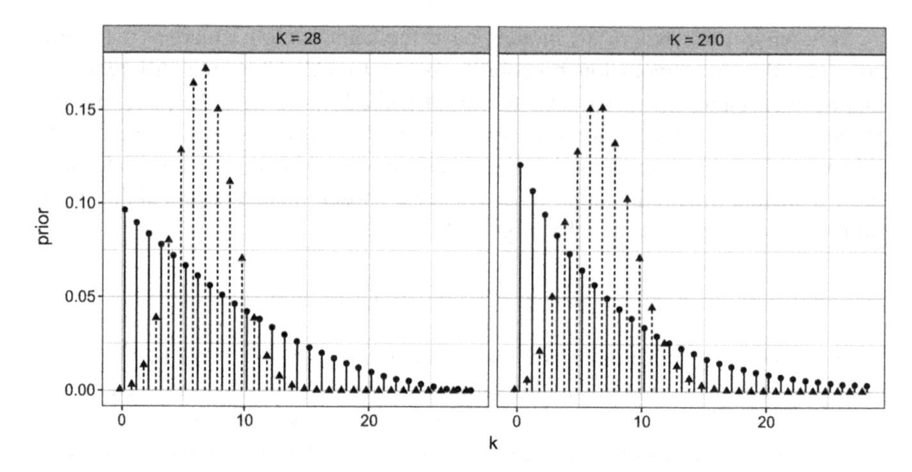

Fig. 12.1 Prior model size distributions for the binomial and the beta-binomial model prior for either $K = 28$ covariates (left) and $K = 210$ covariates (right). The prior expected model size $\bar{k}$ is set to 7. The binomial model prior is drawn with dashed lines and triangles, the beta-binomial prior with $a = 1$ and $b = (K - \bar{k})/\bar{k}$ with full lines and bullets

Figure 12.1 indicates how the induced prior model size distributions differ for the binomial and beta-binomial model priors for different number of covariates K assuming an expected prior model size of $\bar{k} = 7$. The number of covariates are set to 28 and 210 because this corresponds to the number of covariates included in the different model specifications considered in the application in Sect. 12.4. The binomial model prior has a clear mode close to the prior expected model size and assigns most of its mass to model sizes close to the mode. This implies that the choice of the prior expected model size is highly influential for the binomial model prior. The beta-binomial prior has a mode at the smallest model size with decreasing weights assigned when model size increases. This reflects that a-priori smaller models are preferred and indicates that the prior expected model size is less influential for this prior.

Dependent Model Priors

The beta-binomial prior has become a default prior in BMA applications. However, as an independent model prior, this prior does not allow to vary the prior inclusion probabilities for covariates in dependence of other covariates being included. This limitation is particularly virulent in case interaction terms are to be included in the linear model or if the covariates are highly correlated.

Heredity Priors

Under independent model priors, the inclusion of an interaction effect does not depend on the inclusion of the associated main effects. Masanjala and Papageorgiou (2008), for example, treated interaction terms as normal covariates, which was criticized by Crespo Cuaresma (2011), who argued that "the interpretation of an interaction term parameter requires that the effect is exclusive to that particular product of covariates and not driven through the independent effects of the interacted variables." He argued that a possible way to account for this problem is to restrict the models considered to those where an interaction term is only included in case the main effects are also part of the model. This approach corresponds to selecting a model prior where the prior model probabilities are set to zero for models including the interaction term without both main effects.

Priors which take the presence of main effects into account when considering the inclusion of an interaction effect are referred to as *heredity priors*. Chipman (1996) proposed two heredity priors for handling two-way interactions: the *weak* and the *strong heredity prior*. The former prior enforces that at least one main effect of the interaction term is included in the model, while the latter prior requires both main effects to be included, i.e., models which include an interaction term but not the associated main effects are down-weighted to zero. Crespo Cuaresma (2011) argued in favor of the strong heredity prior.

Collinearity Adjusted Dilution Model Prior

Independent model priors do not take the collinearity between the potential covariates into account, but assume two covariates are a-priori equally likely to be included regardless of if they are independent or highly correlated. George (2010) criticized this approach as putting disproportionate mass on parts of the covariate space spanned by correlated covariates and proposed to adjust the prior weight of a model $\boldsymbol{\gamma}$ by taking the value of the determinant of the correlation matrix of $\boldsymbol{X}$ given by $|C_{\boldsymbol{\gamma}}|$ into account. Note that $|C_{\boldsymbol{\gamma}}| = 1$ if the included covariates are uncorrelated and equals zero if they are perfectly collinear. The prior probability of a specific model $\boldsymbol{\gamma}$ is given by

$$p(\boldsymbol{\gamma}|\theta) \propto f(|C_{\boldsymbol{\gamma}}|)\theta^k(1-\theta)^{K-k},$$

for some monotonic function f satisfying $f(0) = 0$ and $f(1) = 1$, e.g., the identity function for f. The function f controls the down-weighting applied to a set of covariates in dependence of their correlation structure compared to the binomial model prior with $\theta = \bar{k}/K$.

Using the data set of Masanjala and Papageorgiou (2008), Moser and Hofmarcher (2013) presented a rigorous analysis of different dilution model priors as well as the heredity model prior. The results for their empirical application indicate that the

PIP of interaction terms is sensitive to the prior selected, while the out-of-sample predictive performance does not change with the model prior choice.

Dirichlet Process Model Priors

Grün and Hofmarcher (2018) merge the strands of literature where different model priors are considered and post-hoc jointness analysis is suggested. They put forward the idea of using a Dirichlet process model prior. This model prior constitutes a natural choice if one aims at a-priori taking into account that different groups of covariates may be relevant for the outcome variable.[2] Using this model prior, the covariate inclusion probability θ is assumed to vary between the groups and within each group for each of the covariates.

12.3.4 *Inference*

For forecasting one is interested in the posterior distribution (or a point estimate thereof) of a new observation $y^{(n)}$ given its associated covariate values $\boldsymbol{x}^{(n)}$ and the observed data $\boldsymbol{y}$ and $\boldsymbol{X}$:

$$p(y^{(n)}|\boldsymbol{x}^{(n)}, \boldsymbol{y}, \boldsymbol{X}).$$

This posterior also includes the model choice, i.e., in BMA the model class considered, and the prior specification. The model class considered is the linear model with constant variance and a specific set of K potential covariates.

If the mean is used as forecast this can be determined using

$$\hat{y}^{(n)} = \sum_{j=1}^{2^K} \left(\mathrm{E}[\beta_0|\boldsymbol{y}, \boldsymbol{X}, \mathcal{M}_j] + (\boldsymbol{x}^{(n)})' \,\mathrm{E}[\boldsymbol{\beta}|\boldsymbol{y}, \boldsymbol{X}, \mathcal{M}_j] \right) p(\mathcal{M}_j|\boldsymbol{y}, \boldsymbol{X}).$$

This is a weighted sum of the predicted mean given the posterior means of the regression coefficients conditional on a specific model. The weights are equal to the posterior model probabilities. For small values of K all models 2^K can be enumerated and this sum can be exactly calculated. Alternatively the sum is approximated by summing over the models with non-negligible posterior model probability.

---

[2]In the context of economic growth, this is in line with the model formulation in Durlauf, Kourtellos, and Tan (2008) where competing groups of explanatory variables emerging from different theories are assumed to be relevant.

If Zellner's g prior is used for the regression coefficients the posterior mean and variance of the regression coefficients conditional on a specific model can be determined in closed form. These quantities are given by

$$\mathrm{E}[\beta_0 | \boldsymbol{y}, \boldsymbol{X}, \mathcal{M}_j] = \bar{y},$$
$$\mathrm{E}[\boldsymbol{\beta} | \boldsymbol{y}, \boldsymbol{X}, \mathcal{M}_j] = \frac{g}{1+g} \hat{\boldsymbol{\beta}}_{\mathrm{OLS},j},$$

where $\bar{y} = \frac{1}{n}\boldsymbol{\iota}'\boldsymbol{y}$ and $\hat{\boldsymbol{\beta}}_{\mathrm{OLS},j}$ are the OLS estimates for model $\mathcal{M}_j$. The posterior model probabilities are available up to a normalization factor using the marginal likelihood given by

$$p(\boldsymbol{y}|\boldsymbol{X}, \mathcal{M}_j) \propto ((\boldsymbol{y} - \bar{y}\boldsymbol{\iota})'(\boldsymbol{y} - \bar{y}\boldsymbol{\iota}))^{-\frac{n-1}{2}} (1+g)^{-\frac{k}{2}} \left(1 - \frac{g}{1+g} R^2_{\mathrm{OLS},j}\right)^{-\frac{n-1}{2}},$$

where k is the number of covariates included in the model and $R^2_{\mathrm{OLS},j}$ is the proportion of explained variance of model $\mathcal{M}_j$ when fitted with OLS. The posterior model probabilities can be obtained by enumerating all possible models, determining the marginal likelihoods for each of them and then calculating the posterior model probabilities proportional to the marginal likelihoods times prior model probability.

If the model space is too large to be enumerated, Markov chain Monte Carlo (MCMC) methods are usually employed to explore the space in a suitable way to concentrate on models with non-negligible posterior model weight. MCMC sampling may be used to either (1) determine a set of models with non-negligible posterior model probability to be considered for determining the posterior model probabilities or (2) use the relative frequencies of how often the different models are visited as estimates for the posterior model probabilities.

Two different MCMC samplers have been suggested and are implemented in standard software, e.g., the R package BMS. The *birth-death sampler* randomly selects one of the K covariates and creates a proposed model by either dropping the covariate from the current model if this covariate is included in the current model or adding the covariate to the current model if this is not the case. The proposed model is accepted with probability equal to the minimum of 1 and the ratio of the posterior model probabilities of the proposed and current model. The *reversible-jump sampler* (Madigan & York, 1995) extends the birth-death sampler by performing half of the time a birth-death move and in the other half a swap move, where one covariate is randomly dropped and another randomly added.

Both samplers perform only local moves and search for promising models in the neighborhood of the current model. This means that—as usual for MCMC sampling—the draws are auto-correlated and the model selected to start the sampler influences how many burn-in iterations are required for the chain to converge to a part of the model space with high posterior model probabilities. After omitting the burn-in iterations the set of visited models I is determined and the point estimate for a new observation n is obtained by using an approximation based on the results from the MCMC sampler

$$\hat{y}^{(n)} = \sum_{k \in I} \left(\mathrm{E}[\beta_0 | \mathbf{y}, \mathbf{X}, \mathcal{M}_j] + (\mathbf{x}^{(n)})' \mathrm{E}[\boldsymbol{\beta} | \mathbf{y}, \mathbf{X}, \mathcal{M}_j] \right) p(\widehat{\mathcal{M}_j | \mathbf{y}, \mathbf{X}}).$$

One can estimate $p(\widehat{\mathcal{M}_j | \mathbf{y}, \mathbf{X}})$ in two different ways. One may either use the analytical formulas or determine the empirical relative frequencies how often the model was visited during MCMC sampling.

12.3.5 Post-Processing

Further insights into the results of a BMA analysis are gained by determining and inspecting the posterior inclusion probabilities (PIP), the posterior means of the regression coefficients (PM), the posterior standard deviations of the regression coefficients (PSD), and the conditional positive sign certainty (PSC). These quantities indicate if a covariate may be a robust determinant of the dependent variable or not.

The posterior inclusion probability (PIP) for covariate k is given by

$$\mathrm{PIP}_k = \sum_{j=1}^{2^K} 1\{\gamma_k = 1 | \mathcal{M}_j\} p(\mathcal{M}_j | \mathbf{y}, \mathbf{X}),$$

where $1\{\cdot\}$ is the indicator function. The PIP is estimated by summing only over the set of visited models I and plugging in suitable estimates $p(\widehat{\mathcal{M}_j | \mathbf{y}, \mathbf{X}})$. These estimates may either be based on the analytical formulas or the empirical relative frequencies how the model was visited during MCMC sampling. PIP is often interpreted as measuring how important it is to include this variable in the model.

The posterior mean (PM) of the regression coefficient for covariate k is determined by

$$\mathrm{PM}_k = \sum_{j=1}^{2^K} \mathrm{E}(\beta_k | \mathbf{y}, \mathbf{X}, \mathcal{M}_j) p(\mathcal{M}_j | \mathbf{y}, \mathbf{X}),$$

and the posterior standard deviation (PSD) of the regression coefficient for covariate k by

$$\mathrm{PSD}_k = \sqrt{\sum_{j=1}^{2^K} (\mathrm{var}(\beta_k | \mathbf{y}, \mathbf{X}, \mathcal{M}_j) + (\mathrm{E}(\beta_k | \mathbf{y}, \mathbf{X}, \mathcal{M}_j) - \mathrm{PM}_k)^2) p(\mathcal{M}_j | \mathbf{y}, \mathbf{X})}.$$

Estimates may be obtained in a similar way to the PIP, i.e., by considering only the set of visited models I and plugging in suitable estimates $\widehat{p(\mathcal{M}_j|\mathbf{y}, \mathbf{X})}$. PM represents the posterior mean effect a unit change in the covariate k has on the dependent variable and PSD is a measure for the uncertainty associated with the estimated regression coefficient.

The conditional positive sign certainty (PSC) for covariate k is given by

$$\mathrm{PSC}_k = \frac{\sum_{j=1}^{2^K} p(\beta_k > 0|\mathcal{M}_j, \mathbf{y}, \mathbf{X}) p(\mathcal{M}_j|\mathbf{y}, \mathbf{X})}{\sum_{j=1}^{2^K} 1\{\gamma_k = 1|\mathcal{M}_j\} p(\mathcal{M}_j|\mathbf{y}, \mathbf{X})}$$

and indicates how certain one is a-posteriori that this covariate has a positive effect on the dependent variable. Again estimates may be obtained in a similar way than for PIP, PM, and PSD.

While PIP is a measure of variable importance, PM indicates the robust estimate of the effect of the regressor across all sampled models with PSD indicating the uncertainty, and PSC is a measure for the posterior confidence in the sign of this robust effect estimate being positive (Sala-i-Martin et al., 2004). For a covariate to be identified as a robust determinant for the dependent variable, PIP would be expected to be high and PSC either close to zero or one.

12.4 Application

The following data analysis is performed in R, an open-source environment for statistical computing and graphics. Package BMS is used to obtain the BMA analysis results.

12.4.1 Data Description

The application of BMA is illustrated using a data set consisting of all movies released in the USA between October, 1 2010 and June, 30 2013 with budgets between \$20 and \$100 million[3] (see Lehrer & Xie, 2017). In total the data set contains 94 movies and 26 potential variables to predict the opening weekend box office revenues. The potential explanatories consist of characteristics of the film, including genre classification and MPAA film ratings, but also information on the budget and the scheduled number of weeks and screens the film will be on air. In addition to this set of "classical" explanatories, covariates constructed from data

[3]Following Lehrer and Xie (2017), this sample selection criterion was proposed by IHS film consulting and accounts for 41.4% of released movies.

sources which have now only become available in the big data area are included in this data set (see Chap. 1). These explanatories are constructed from Twitter social media information. Based on Twitter messages posted in the weeks before release volume and sentiment scores are constructed for specific time windows before release. As pointed out in Chap. 1, forecasting with Twitter data has also been considered in Asur and Huberman (2010) and Arias, Arratia, and Xuriguera (2013).

Each film may be associated with twelve different genres with several genres potentially being assigned. The Motion Picture Association of America (MPAA) provides film ratings to help parents determine what films are appropriate for their children. The following ratings are possible: general audiences (G), parental guidance suggested (PG), parents strongly cautioned (PG-13), and restricted (R).

Twitter messages are transformed into sentiment scores, through an algorithm developed by Hannak et al. (2012). In a message that mentions a specific film title or key word, sentiment is calculated by examining the emotion words and icons that are captured in the same Twitter message (see the supplemental material of Lehrer and Xie 2017). Following Lehrer and Xie (2017) $75,056$ unique emotion words and icons that appeared in at least 20 tweets are given a specific emotional value. The sentiment score is estimated as a weighted average of those emotional values. Sentiment scores are determined for five different time windows before release and the variables are referred to in the following way such that, e.g., the variable T-21/-27 represents the score for weeks 21–27 before release. In addition to the sentiment scores, measures capturing the volume of the tweets are calculated for the same time windows. In total, for the films analyzed, the Twitter message volume is $1,100,439$.

12.4.2 *Exploratory Data Analysis*

A descriptive summary of the data is given in Tables 12.1 and 12.2. Table 12.1 contains information on the dependent variable and the "classical" explanatories, whereas Table 12.2 summarizes the explanatories constructed based on the Twitter social media information.

In Table 12.1, first the metric variables consisting of the dependent variable ("Open box"), the film budget in million $ ("Budget"), the scheduled number of weeks ("Weeks"), and the scheduled number of screens in hundreds ("Screens") are summarized. The mean and standard deviation (SD) are given as well as median and the first quartile (Q1) and the third quartile (Q3). These results indicate that the dependent variable as well as budget are right-skewed with the median being substantially smaller than the mean and the distance between median and third quartile being much larger than between median and first quartile. The variables Weeks and Screens seem to be more symmetric.

Next the MPAA ratings are given. Each film can only have one of these categories assigned and there is only a single film which is classified as G. In the subsequent analysis we thus merge category G with PG to avoid including a categorical variable

Table 12.1 Summary statistics of the dependent variable and covariates (without the covariates constructed from the Twitter data)

Variable	Mean	SD	Median	Q1	Q3	MPAA	%	Genre	%	Genre	%
Open box	19.1	18.5	14.9	9.5	23.0	G	1.1	Action	37.2	Family	6.4
Budget	50.0	20.4	44.8	35.0	65.0	PG	14.9	Adventure	16.0	Fantasy	7.4
Weeks	13.8	5.5	13.0	10.0	16.8	PG13	37.2	Animation	7.4	Mystery	8.5
Screens	30.0	5.2	30.3	27.4	33.4	R	46.8	Comedy	42.6	Romance	12.8
								Crime	26.6	Sci-Fi	9.6
								Drama	34.0	Thriller	24.5

Table 12.2 Summary statistics of the variables based on Twitter social medial information

Sentiment						Volume					
Variable	Mean	SD	Median	Q1	Q3	Variable	Mean	SD	Median	Q1	Q3
T-1/-3	74.3	2.1	74.8	74.0	75.5	T-1/-3	41.3	110.3	17.9	9.0	38.1
T-4/-6	74.3	2.1	74.8	73.8	75.4	T-4/-6	25.2	112.8	8.6	4.3	19.2
T-7/-13	74.3	1.8	74.7	73.8	75.4	T-7/-13	21.5	89.6	6.8	3.1	17.7
T-14/-20	74.1	2.4	74.7	73.8	75.5	T-14/-20	19.1	90.5	5.2	2.1	12.4
T-21/-27	73.7	3.1	74.6	73.4	75.4	T-21/-27	17.7	92.9	4.5	1.5	10.9

where one category occurs only very rarely. Regarding the genre categorizations each film has at least one and up to three genres assigned. Nineteen percent of the films have a single genre category assigned, 29% two genre categories, and the majority of 53% have three genre categories assigned. The genre analysis indicates that the most popular genre is Comedy which is attributed to 42.6% of the films followed by Action (37.2%) and Drama (34%). Furthermore five genres are assigned to less than 10% of the films.

The variables based on the Twitter social media information are summarized in Table 12.2. The distribution of the sentiment scores remains rather stable over time, whereas the volume increases the closer the time period is to the opening date. Also the sentiment distribution is rather symmetric, whereas volume is right-skewed. This suggests to transform the volume by taking the logarithm to obtain a more symmetric distribution and to define the impact of changes in volume through relative rather than absolute changes.

The distributions of sentiment and logarithmized volume are illustrated for the last time period (T-1/-3) in Fig. 12.2. Qualitatively these distributions remain rather similar for the other time periods. For the sentiment values there are always a few films which have quite low numbers and are separated from the remaining observations. For the logarithmized volume the mean value increases and the spread decreases if the opening date gets closer. Both these measures remain for each film rather stable over time with an average pairwise correlation of 0.84 for the sentiment scores and 0.90 for the logarithmized volume values.

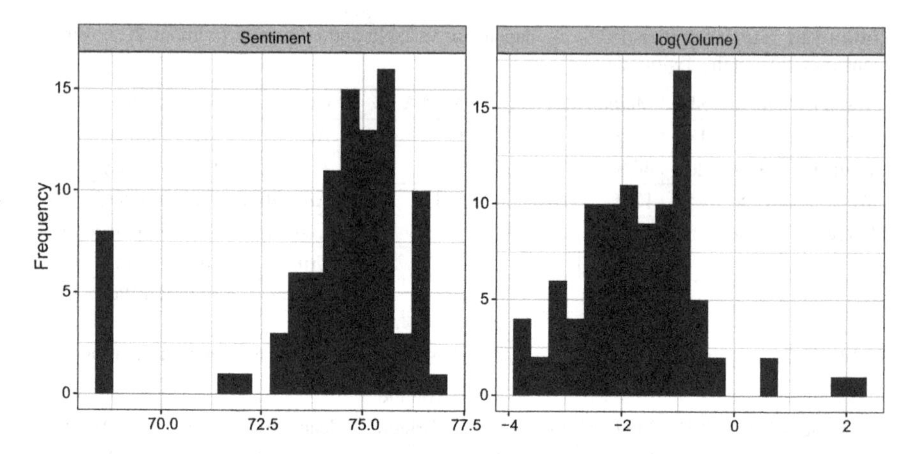

Fig. 12.2 Histograms of sentiment and logarithmized volume values for the Twitter social media information for T-1/-3

12.4.3 BMA Results

In the following the BMA methodology is applied to the open box office data set. For the BMA analysis we employ a standard setting varying a few aspects with respect to the model prior, the data pre-processing and the covariates included in the model to indicate how results depend on these modeling choices.

The model specifications varied are:

- The use of a binomial or a beta-binomial model prior.
- Taking the logarithm of the dependent variable or not.
- Taking the logarithm of the volumes of the tweets or not.
- Including interactions between the dummy variables and the numeric covariates or not.

If interaction terms are considered the data contain 210 potential covariates compared to only 28 covariates if these are excluded. For both model priors the prior expected model size is set to 7. The prior distributions for these settings are shown in Fig. 12.1.

The regression parameters prior employed is the same for all model specifications and consists of a flat prior for intercept and error variance and Zellner's g prior for the remaining regression coefficients with a hyper-prior put on g as suggested in Ley and Steel (2012). In case interactions are included the heredity model prior is used ensuring that interactions are only included in a model if the main effects are also included.

Overall, these settings result in 16 different model specifications for BMA analysis. The BMA analysis for each model specification is based on birth-death MCMC sampling where 1,000,000 iterations are recorded after discarding 500,000 iterations as burn-in.

Table 12.3 Summary of the BMA analysis results for the 16 different BMA specifications

Name	Model prior	log(y)	log(volume)	Interactions	Mean post. model size	# models visited	PMP correlation	Avg. shrinkage (SD)
M-1	B	✓	✓	✓	1.32	2972	0.9999	0.985 (0.019)
M-2	BB	✓	✓	✓	1.08	744	1.0000	0.988 (0.017)
M-3	B	✗	✓	✓	4.95	8358	0.9986	0.978 (0.017)
M-4	BB	✗	✓	✓	4.60	5725	0.9988	0.977 (0.016)
M-5	B	✓	✗	✓	1.32	2972	0.9999	0.985 (0.018)
M-6	BB	✓	✗	✓	1.08	744	1.0000	0.988 (0.017)
M-7	B	✗	✗	✓	4.95	8358	0.9986	0.978 (0.009)
M-8	BB	✗	✗	✓	4.60	5725	0.9988	0.987 (0.009)
M-9	B	✓	✓	✗	3.47	181,311	0.9996	0.960 (0.034)
M-10	BB	✓	✓	✗	1.49	36,400	1.0000	0.978 (0.025)
M-11	B	✗	✓	✗	6.35	308,900	0.9867	0.923 (0.050)
M-12	BB	✗	✓	✗	6.25	214,324	0.9852	0.929 (0.053)
M-13	B	✓	✗	✗	3.47	181,311	0.9996	0.963 (0.034)
M-14	BB	✓	✗	✗	1.49	36,400	1.0000	0.983 (0.022)
M-15	B	✗	✗	✗	6.50	222,544	0.9845	0.982 (0.011)
M-16	BB	✗	✗	✗	6.25	214,324	0.9852	0.983 (0.011)

Table 12.3 lists the specification of the 16 different models and introduces names for each model. These model names will be subsequently used to refer to them. In addition the MCMC output of the BMA analysis is summarized by reporting the mean posterior model size, the number of different models visited by the MCMC sampling procedure, the correlation between the posterior model probabilities determined using the analytic formulas as well as based on the empirical frequencies from the MCMC chain and the posterior distribution of shrinkage factor implied by the Zellner's g prior including a hyper-prior on g.

The first column in the table gives the name for the BMA specification with the following four columns providing the information on how the model prior is set (binomial (B) or beta-binomial (BB) model prior), if the dependent variable is logarithmized or not, if the Twitter volume variables are logarithmized or not and if the interactions between the dummy (Genre, Rating) and the numeric covariates (Budget, Screens, Weeks, Twitter volume, Twitter sentiment) are included.

The remaining four columns summarize the MCMC output. Column *Mean post. model size* reports the average number of regressors included in the visited models. This number varies between 1.08 (for model M-2) and 6.50 (for model M-15). *# models visited* indicates the number of unique sampled models and *PMP correlation* indicates how close the empirical frequencies and analytical posterior model probabilities (PMPs) are measured by the Pearson correlation. These values vary between 0.9845 and 1 indicating high congruence and thus confirming convergence of the MCMC chain. The closer the PMP correlation is to one the better is the approximation of the analytical PMPs by the empirical frequencies. Column *Avg. shrinkage* reports the posterior mean shrinkage factors of the g-prior with the corresponding posterior standard deviations in parentheses. The prior distribution of the shrinkage factor is given by $\frac{g}{1+g} \sim \text{Beta}(1, \frac{a}{2} - 1)$ with $a \in [2, 4]$ such that the prior mean corresponds to the unit information prior. The average shrinkage factors are between 0.923 and 0.988 with rather small standard deviations. This indicates that the shrinkage imposed on the regression coefficients is very small, corresponding to $g \approx 49$.

We can also infer from Table 12.3, that the mean posterior model size is smaller if we take the logarithm of the dependent variable. E.g., model M-1 and M-3 differ only with respect to if the logarithm of the dependent variable is taken or not, but the mean posterior model size of model M-3 (4.95) is nearly 4 times larger than that of M-1 (1.32).

In the following we will mainly focus on three model specifications, namely models M-2, M-4, and M-11. M-2 is the model with the smallest mean number of regressors of all considered specifications. M-4 differs from M-2 solely by not taking the logarithm of the dependent variable. As a consequence the mean number of regressors increases from 1.08 to 4.60, but also the number of visited models increases to 5725. The specification of model M-11 results in both, one of the highest number of mean regressors, and the highest number of visited models. Compared to M-2 and M-4, model M-11 does not include interaction terms, i.e., the number of potential covariates is 28 for this specification compared to 210 in the other two model specifications.

Figure 12.3 displays both the prior and the posterior model size distributions of the mentioned models. From the prior model size distributions we can clearly infer that the beta-binomial prior of models M-2 and M-4 is more flexible in terms of model size than the binomial model prior of model M-11. Table 12.3 shows that not

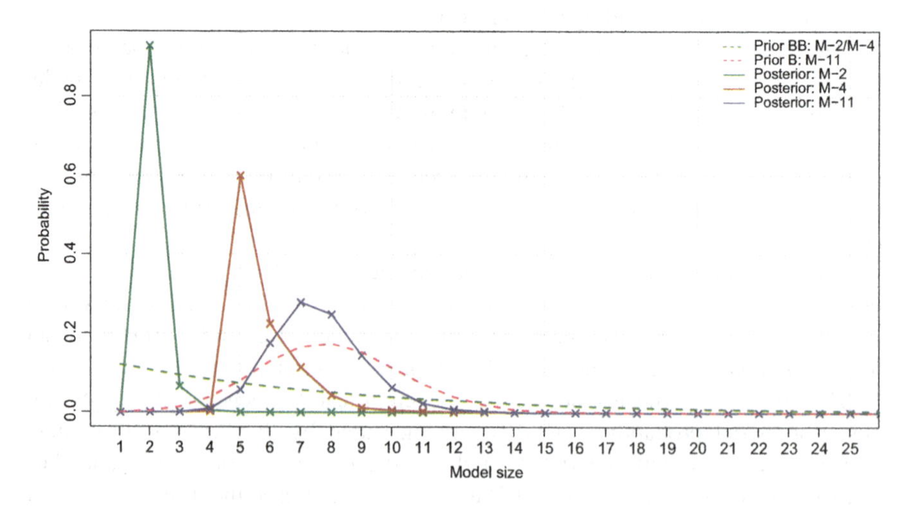

Fig. 12.3 Prior and posterior model size distributions for models M-2, M-4, and M-11

only the choice of the model prior influences the mean posterior model size, but also other model specifications. In addition we find that the mean posterior model size distributions have modes at rather low values and put less weight on high model sizes than the prior. This is because the expected prior model size is set to 7 and the data seem to favor smaller models than proposed by the prior. This is more pronounced for the beta-binomial prior which is less informative.

Next we present the covariate-specific BMA results for models M-2, M-4, and M-11 in Tables 12.4, 12.5, and 12.6. In each of the tables the 20 most important covariates, according to the PIP, are included in the rows. The first column contains the PIPs, which correspond to the fraction of all models where this covariate is included. PIP can be interpreted as the posterior probability that the variable is included in the true model. In the following two columns *PM* and *PSD* present the posterior mean, respectively, posterior standard deviation of the regression coefficients for these 20 most important covariates.

For model M-2 we find that the covariate representing number of screens has a PIP of 1, i.e., this variable is included in all sampled models, followed by Volume T-1/-3 with a PIP of only 0.0210. This indicates that the simple linear regression model with variable Screens as the only explanatory variable seems to already describe the data well. Additionally, the variable Screens has a relative large posterior mean regression coefficient in particular compared to the other covariates displayed in

Table 12.4 BMA coefficient results for model M-2

	PIP	PM	PSD	PSC
Screens	1.0000	0.6916	0.0756	1.0000
Volume T-1/-3	0.0210	0.0039	0.0294	1.0000
Volume T-21/-27	0.0157	0.0028	0.0241	1.0000
Volume T-14/-20	0.0128	0.0023	0.0216	1.0000
Volume T-7/-13	0.0064	0.0009	0.0141	0.9521
Crime	0.0054	0.0011	0.0196	1.0000
Fantasy	0.0028	0.0008	0.0216	1.0000
Sentiment T-1/-3	0.0027	0.0002	0.0060	1.0000
Weeks	0.0023	−0.0001	0.0046	0.0000
Budget	0.0021	0.0003	0.0074	1.0000
Action	0.0016	0.0003	0.0092	1.0000
Mystery	0.0015	0.0000	0.0105	1.0000
Volume T-4/-6	0.0013	0.0002	0.0059	1.0000
Animation	0.0012	−0.0003	0.0126	0.0000
Drama	0.0010	−0.0001	0.0062	0.0000
Sentiment T-21/-27	0.0009	0.0000	0.0023	1.0000
Thriller	0.0008	−0.0001	0.0061	0.0000
Comedy	0.0006	−0.0001	0.0059	0.0000
Adventure	0.0004	0.0001	0.0057	1.0000
Sentiment T-4/-6	0.0004	0.0000	0.0021	1.0000

Top 20 covariates in terms of PIP

Table 12.5 BMA coefficient results for model M-4

	PIP	PM	PSD	PSC
Sci-Fi	0.9996	−0.9064	0.3379	0.0000
Screens	0.9994	0.3567	0.0651	1.0000
Volume T-1/-3	0.9220	0.1869	0.1055	1.0000
Volume T-1/-3#Sci-Fi	0.9036	1.8988	1.0698	1.0000
Volume T-4/-6	0.1195	−0.0080	0.0639	0.0260
Volume T-4/-6#Sci-Fi	0.1127	−0.3401	0.9831	0.0000
Screens#Sci-Fi	0.0999	0.2167	0.6760	0.9855
Budget	0.0493	0.0069	0.0335	1.0000
Sentiment T-1/-3	0.0484	0.0074	0.0354	1.0000
Sentiment T-4/-6	0.0441	0.0064	0.0326	1.0000
Weeks	0.0427	0.0064	0.0329	1.0000
Sentiment T-14/-20	0.0413	0.0058	0.0306	1.0000
Sentiment T-7/-13	0.0286	0.0039	0.0251	1.0000
PG13	0.0200	−0.0044	0.0352	0.0000
Family	0.0179	0.0065	0.0741	0.8526
Volume T-21/-27	0.0179	0.0021	0.0219	0.9867
Volume T-14/-20	0.0166	0.0023	0.0241	0.8833
Crime	0.0144	0.0017	0.0214	1.0000
Volume T-7/-13	0.0121	−0.0002	0.0238	0.3591
Fantasy	0.0106	0.0028	0.0383	0.9618

Top 20 covariates in terms of PIP

Table 12.6 BMA coefficient results for model M-11

	PIP	PM	PSD	PSC
Screens	0.9957	0.3684	0.0933	1.0000
Volume T-1/-3	0.8343	0.3593	0.2416	1.0000
Weeks	0.6117	0.1292	0.1246	1.0000
Adventure	0.4320	0.1965	0.2679	1.0000
Animation	0.2570	−0.1460	0.3106	0.0001
Thriller	0.2420	0.0688	0.1513	1.0000
Sentiment T-1/-3	0.2192	0.0328	0.0924	0.9967
Budget	0.2189	0.0275	0.0656	1.0000
Sentiment T-4/-6	0.2052	0.0270	0.0879	0.9577
Volume T-4/-6	0.1969	−0.0386	0.1610	0.2136
Sentiment T-7/-13	0.1891	0.0235	0.0762	0.9666
Volume T-7/-13	0.1751	−0.0285	0.1435	0.2311
Sci-Fi	0.1698	0.0582	0.1727	1.0000
Sentiment T-14/-20	0.1461	0.0132	0.0583	0.8571
Drama	0.1446	−0.0280	0.0990	0.0017
Volume T-14/-20	0.1353	0.0130	0.0883	0.6175
Volume T-21/-27	0.1310	0.0171	0.0789	0.8523
Family	0.1251	0.0404	0.1673	0.9970
Fantasy	0.1229	0.0335	0.1401	0.9984
Sentiment T-21/-27	0.1163	0.0047	0.0454	0.5997

Top 20 covariates in terms of PIP

Table 12.4. A further indicator to study the effect of a covariate is its positive sign certainty (PSC). For variable Screens we find that this covariate enters all models with a positive coefficient sign.

The only difference between models M-2 and M-4 is that model M-4 uses the original dependent variable instead of the logarithmized dependent variable. Table 12.5 presents the results for this specification. We find that, compared to model M-2 the PIPs are more evenly spread between four covariates. These are Sci-Fi, Screens, Volume T-1/-3, and the interaction term of Volume T-1/-3 and Sci-Fi. All four covariates have a PIP above 0.9. Further, compared to model M-2, the posterior mean of the regressions coefficient of variable Screens halves in size. Volume T-1/-3 which has a PIP of solely 0.021 in model M-2 increases its importance in model M-4 with a PIP of 0.9220. Also its posterior mean regression coefficient changes from 0.0039 to 0.1869, while sign certainty remains close to 1.

Model M-11 differs from models M-2 and M-4 by not taking interaction terms into account. The results are summarized in Table 12.6. We find 13 covariates with a PIP above 0.15, which is a huge increase in the number of important covariates compared to model M-2 (where only one covariate has a PIP above 0.15) and model M-4 (where 4 covariates have a PIP above 0.15). Table 12.6 shows that Screens, the volume of tweets for weeks 1–3 before release (Volume T-1/-3), and Weeks are the three most important covariates for predicting open box office revenues. All those covariates have a positive posterior mean regression coefficient. Compared to model M-4 Sci-Fi looses importance and is now ranked 13th. On the other hand, variable Budget increases its importance with a PIP above 0.21 and a positive mean coefficient. A graphical comparison indicating the differences of models M-2, M-4, and M-11 in terms of PIPs is given in Fig. 12.4.

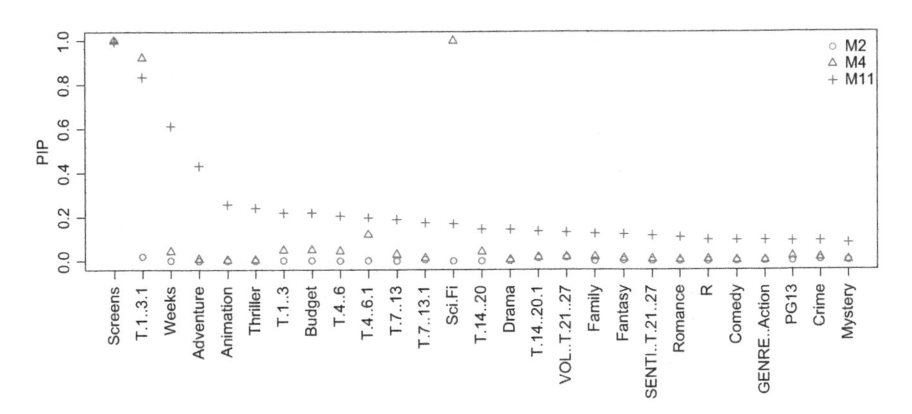

Fig. 12.4 Posterior inclusion probabilities for models M-2, M-4, and M-11 (ordered by PIPs of M-11)

12.4.4 Iterations Matter

The quality of the BMA output depends crucially on the number of MCMC iterations. In particular in situations, where the sampler starts with some model that is not a "good" one, many iterations are already required to reach the part of the model space where models with high posterior model probabilities are located. In this case the first batch of iterations will not draw models with a high posterior model probability and are therefore usually discarded as burn-in iterations. However there is no clear guidance on the number of iterations which should be removed. Table 12.7 displays the estimated PIPs as well as the posterior means (PM) and

Table 12.7 Variation in PIPs for model M-11 when changing the number of burn-in and recorded iterations (B corresponds to the number of discarded burn-in iterations, I to the number of recorded iterations)

	$B = 1000$, $I = 1000$				$B = 1000$, $I = 1{,}000{,}000$				$B = 100{,}000$, $I = 1{,}000{,}000$			
	PIP	PM	PSD	Rank	PIP	PM	PSD	Rank	PIP	PM	PSD	Rank
Screens	1.00	0.36	0.09	1	1.00	0.37	0.09	1	1.00	0.37	0.09	1
Volume T-1/-3	0.95	0.42	0.23	2	0.84	0.36	0.24	2	0.84	0.36	0.24	2
Adventure	0.60	0.26	0.27	3	0.44	0.20	0.27	4	0.44	0.20	0.27	4
Weeks	0.59	0.12	0.12	4	0.61	0.13	0.12	3	0.61	0.13	0.12	3
Budget	0.42	0.05	0.08	5	0.22	0.03	0.07	7	0.22	0.03	0.06	7
Animation	0.30	−0.18	0.34	6	0.25	−0.15	0.31	5	0.25	−0.14	0.31	5
Volume T-4/6	0.30	−0.08	0.20	7	0.19	−0.04	0.16	10	0.19	−0.04	0.16	10
Sentiment T-4/-6	0.27	0.04	0.09	8	0.21	0.03	0.09	9	0.20	0.03	0.09	9
Sentiment T-1/-3	0.26	0.04	0.09	9	0.22	0.03	0.09	8	0.22	0.03	0.09	8
Sentiment T-14/-20	0.20	0.01	0.07	10	0.15	0.01	0.06	14	0.14	0.01	0.06	15
Family	0.19	0.08	0.22	11	0.13	0.04	0.17	19	0.13	0.04	0.17	18
Sci-Fi	0.18	0.05	0.16	12	0.17	0.06	0.17	13	0.17	0.06	0.17	13
Sentiment T-21/-27	0.18	0.00	0.06	13	0.11	0.00	0.04	21	0.12	0.00	0.05	20
Volume T-7/-13	0.16	−0.03	0.14	14	0.18	−0.03	0.14	12	0.18	−0.03	0.14	12
Drama	0.16	−0.02	0.09	15	0.15	−0.03	0.10	15	0.15	−0.03	0.10	14
Thriller	0.15	0.04	0.12	16	0.24	0.07	0.15	6	0.23	0.07	0.15	6
R	0.15	0.01	0.07	17	0.11	0.01	0.07	22	0.10	0.01	0.06	22
Volume T-21/-27	0.15	0.02	0.07	18	0.13	0.02	0.08	17	0.13	0.02	0.08	16
Crime	0.13	0.00	0.07	19	0.09	0.01	0.06	26	0.09	0.01	0.06	26
Mystery	0.11	−0.00	0.09	20	0.09	−0.00	0.08	27	0.09	−0.00	0.08	27
Romance	0.10	−0.02	0.10	21	0.11	−0.02	0.10	20	0.11	−0.02	0.10	21
PG13	0.09	−0.01	0.05	22	0.09	−0.01	0.06	25	0.10	−0.01	0.06	24
Fantasy	0.08	0.02	0.11	23	0.13	0.04	0.14	18	0.12	0.03	0.14	19
Volume T-14/-20	0.06	−0.00	0.05	24	0.13	0.01	0.09	16	0.13	0.01	0.09	17
Sentiment T-7/-13	0.04	0.00	0.04	25	0.19	0.02	0.08	11	0.19	0.02	0.08	11
Comedy	0.01	−0.00	0.02	26	0.10	−0.01	0.06	23	0.10	−0.01	0.06	23
Action	0.01	−0.00	0.02	27	0.09	−0.00	0.06	24	0.10	−0.00	0.06	25

posterior standard deviations (PSD) of the regression coefficients for model M-11 for different burn-in values as well as number of iterations. The first three columns display the results for 1000 burn-in and 1000 recorded iterations. In columns 4–6 the number of burn-in iterations remains constant but the number of recorded iterations is increased to 1,000,000 and the last three columns provide the results for a burn-in period of 1,000,000 iterations.

We can infer that the results are nearly identical for the last two specifications, i.e., essentially the same results are obtained for 1,000,000 recorded iterations regardless of the length of the burn-in period. Comparing the sampling setup ($B = 1000, I = 1000$) with the other two specifications, we find that the importance of some variables changes. For example Thriller is ranked 16th for ($B = 1000$, $I = 1000$) but is on the 6th rank when the number of iterations is increased. This indicates that the MCMC chain has not converged yet when $B = I = 1000$.

To check for convergence of the chain, one might inspect the correlation between the empirical frequencies of the visited models and their analytical PMP. For $B = I = 1000$ we find a *PMP correlation* of 0.2563, while for $B = 1000$, $I = 1{,}000{,}000$, and $B = 100{,}000$, $I = 1{,}000{,}000$ we find *PMP correlation* values above 0.993. While the correlation value for the first setup indicates poor congruence between the results, the values of the other two settings for burn-in/recorded iterations indicate a very high degree of congruence. For a visual inspection of the congruence between analytical and MCMC-based PMPs we can plot both measures. Figure 12.5 displays the best 250 models according to their analytical PMP and plots their MCMC-based empirical frequency counts. On the left we find the results for $B = I = 1000$ and on the right for $B = 1000$, $I = 1{,}000{,}000$. The figure confirms the insights gained from inspecting the PMP correlation values.

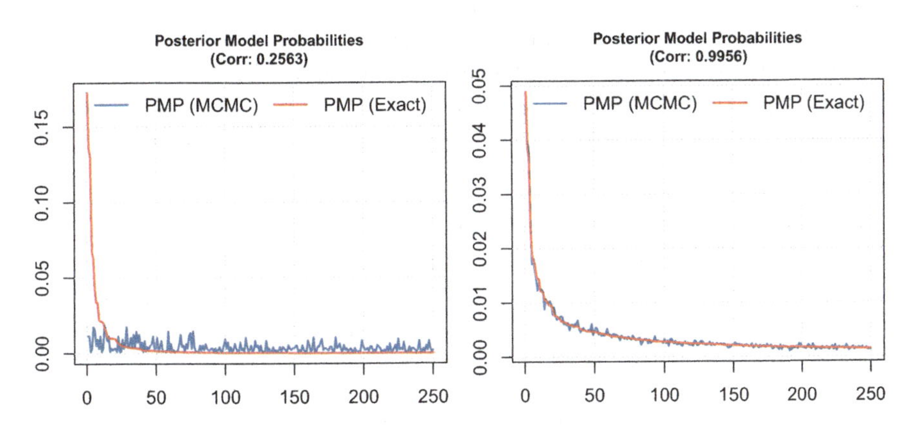

Fig. 12.5 Convergence plot for model specification M-11 for a subset of 250 models. On the left for $I = B = 1000$ and on the right for $B = 1000$, $I = 1{,}000{,}000$

12.4.5 Assessing the Forecasting Performance

In order to evaluate the relative advantages of using different model specifications in the context of open box office forecasting, we compare the forecasting ability of the model specifications described in Table 12.3, making use of an out-of-sample prediction exercise. Using the data set of Lehrer and Xie (2017) we shuffle the data such that the n observations are split into n_T training observations and $n_E = n - n_T$ evaluation observations. For n_E we use a grid of $n_T \in \{10, 20, 30, 40\}$. When using the training set, all models M-1 to M-16 are estimated to evaluate the forecasting performance of the different model specifications. The forecasts from these strategies are then evaluated by calculating mean squared prediction errors (MPE) and mean absolute errors (ABS). The mean squared prediction errors are defined as

$$\text{MPE} = \frac{1}{n_E} \sum_{f \in n_E} (y_f - \text{E}(y_f | \boldsymbol{x}_f, \boldsymbol{y}_{n_T}, \boldsymbol{X}_{n_T}))^2.$$

In an analogous way, the mean absolute errors are defined.

The following five models are compared: the two models determined with BMA using either the relative frequency counts or the marginal likelihoods to determine the posterior model weights, the two models having the highest posterior model weight based on either of the two approaches to calculate the posterior model weight and the full model including all main effects. The full model was also considered in Lehrer and Xie (2017) as a benchmark model in their forecasting evaluation.

Focusing on the three models previously considered in detail, Tables 12.8, 12.9, and 12.10 give the performance evaluation results for the models M-2, M-4, and M-11 for different sizes n_T of the training data sets. The results clearly indicate that model M-2 performs worst. While a statistical model which complies better with the assumptions might be obtained after transforming the dependent variable, the predictive performance measured by equally weighting deviations on the original scale deteriorates. For models M-4 and M-11 using the BMA approach to either

Table 12.8 Mean absolute and squared errors for model M-2

		Relative frequency counts		Marginal likelihoods		
Crit.	n_E	BMA	Top model	BMA	Top model	Full model
ABS	10	1.211 (0.245)	1.212 (0.245)	1.211 (0.245)	1.212 (0.245)	0.690 (0.253)
ABS	20	1.226 (0.208)	1.225 (0.207)	1.225 (0.207)	1.225 (0.207)	0.730 (0.182)
ABS	30	1.223 (0.170)	1.224 (0.168)	1.223 (0.170)	1.224 (0.168)	0.797 (0.178)
ABS	40	1.196 (0.168)	1.195 (0.168)	1.196 (0.168)	1.195 (0.168)	0.858 (0.167)
MPE	10	1.996 (0.805)	1.999 (0.802)	1.997 (0.805)	1.999 (0.802)	1.089 (1.185)
MPE	20	2.034 (0.684)	2.028 (0.680)	2.026 (0.682)	2.028 (0.680)	1.267 (1.148)
MPE	30	2.032 (0.552)	2.031 (0.543)	2.031 (0.552)	2.031 (0.543)	1.524 (1.247)
MPE	40	1.962 (0.559)	1.953 (0.552)	1.962 (0.559)	1.953 (0.552)	1.827 (1.517)

Table 12.9 Mean absolute and squared errors for model M-4

		Relative frequency counts		Marginal likelihoods		
Crit.	n_E	BMA	Top model	BMA	Top model	Full model
ABS	10	0.5410 (0.216)	0.5470 (0.215)	0.5430 (0.215)	0.5470 (0.215)	0.6829 (0.209)
ABS	20	0.5436 (0.137)	0.5528 (0.136)	0.5422 (0.138)	0.5528 (0.136)	0.6809 (0.126)
ABS	30	0.5305 (0.085)	0.5452 (0.090)	0.5315 (0.085)	0.5452 (0.090)	0.7149 (0.122)
ABS	40	0.5602 (0.136)	0.5865 (0.140)	0.5617 (0.136)	0.5865 (0.140)	0.7825 (0.164)
MPE	10	0.9532 (1.328)	0.9791 (1.358)	0.9582 (1.328)	0.9791 (1.358)	1.0523 (0.906)
MPE	20	1.0073 (0.873)	1.0344 (0.884)	1.0070 (0.877)	1.0344 (0.884)	1.0725 (0.560)
MPE	30	0.9559 (0.591)	1.0342 (0.675)	0.9599 (0.589)	1.0342 (0.675)	1.1923 (0.429)
MPE	40	1.1278 (1.088)	1.1868 (1.098)	1.1293 (1.090)	1.1868 (1.098)	1.3854 (0.520)

Table 12.10 Mean absolute and squared errors for model M-11

		Relative frequency counts		Marginal likelihoods		
Crit.	n_E	BMA	Top model	BMA	Top model	Full model
ABS	10	0.5122 (0.202)	0.5528 (0.206)	0.5162 (0.202)	0.5528 (0.206)	0.6884 (0.211)
ABS	20	0.5231 (0.126)	0.5574 (0.133)	0.5279 (0.126)	0.5574 (0.133)	0.6863 (0.131)
ABS	30	0.5251 (0.103)	0.5650 (0.107)	0.5304 (0.102)	0.5650 (0.107)	0.7239 (0.121)
ABS	40	0.5401 (0.091)	0.5886 (0.107)	0.5465 (0.093)	0.5886 (0.107)	0.7827 (0.159)
MPE	10	0.8249 (1.199)	0.8631 (1.143)	0.8266 (1.190)	0.8631 (1.143)	1.0701 (0.922)
MPE	20	0.8621 (0.778)	0.9039 (0.760)	0.8659 (0.770)	0.9039 (0.760)	1.0808 (0.558)
MPE	30	0.8368 (0.570)	0.9002 (0.571)	0.8408 (0.565)	0.9002 (0.571)	1.1726 (0.412)
MPE	40	0.8635 (0.441)	0.9417 (0.437)	0.8705 (0.435)	0.9417 (0.437)	1.3679 (0.473)

obtain a combined model or select the model with the highest posterior model probability leads to an improved predictive performance compared to the full model. In addition a slightly better performance of the BMA based predictions compared to a single model is indicated.

12.5 Summary

The era of big data increases the size of available data sets. Not only the number of observations is increased, but also the number of available covariates. In particular additional information becomes available by extracting data from social media and creating new variables from text content. This aggravates the problem of model uncertainty and makes it even more challenging to specify a suitable model based on theory. In particular if predictive performance is a key criterion, an approach which combines all potential models generally outperforms a single best model. BMA is a principled statistical approach to take model uncertainty into account and estimate predictive models with good forecasting capabilities. Pursuing a Bayesian approach requires the specification of suitable priors as well as approximative estimation methods based on MCMC. In this paper a comprehensive overview on model specification and estimation as well as inference tools for interpreting results in BMA applications is given. For illustration the proposed methods are applied to a data set aiming at predicting the box office revenues of films based on characteristics such as budget and genre, but also information extracted from social media content.

References

Acemoglu, D., Johnson, S., & Robinson, J. (2001). The colonial origins of comparative development: An empirical investigation. *American Economic Review, 91*(5), 1369–1401.

Arias, M., Arratia, A., & Xuriguera, R. (2013). Forecasting with Twitter data. *ACM Transactions on Intelligent Systems and Technology, 5*(1), 8.

Arin, K. P., & Braunfels, E. (2018). The resource curse revisited: A Bayesian model averaging approach. *Energy Economics, 70*(100), 170–178.

Asur, S., & Huberman, B. A. (2010). Predicting the future with social media. In *Proceedings of the 2010 IEEE/WIC/ACM International Conference on Web Intelligence and Intelligent Agent Technology* (Vol. 1, pp. 492–499). Piscataway: IEEE Computer Society.

Barnard, G. A. (1963). New methods of quality control. *Journal of the Royal Statistical Society, Series A, 126*, 255–258.

Bates, J. M., & Granger, C. W. J. (1969). The combination of forecasts. *Operational Research Quarterly, 20*, 451–468.

Brock, W., & Durlauf, S. (2001). Growth empirics and reality. *World Bank Economic Review, 15*, 229–272.

Chipman, H. (1996). Bayesian variable selection with related predictors. *Canadian Journal of Statistics, 24*, 17–36.

Crespo Cuaresma, J. (2011). How different is Africa? A comment on Masanjala and Papageorgiou. *Journal of Applied Econometrics, 26*, 1041–1047.

Crespo Cuaresma, J., Grün, B., Hofmarcher, P., Humer, S., & Moser, M. (2016). Unveiling covariate inclusion structures in economic growth regressions using latent class analysis. *European Economic Review, 81*, 189–202.

Danquah, M., Moral-Benito, E., & Ouattara, B. (2014). TFP growth and its determinants: A model averaging approach. *Empirical Economics, 47*(1), 227–251.

Delgado, M. S., Henderson, D. J., & Parmeter. C. F. (2014). Does education matter for economic growth? *Oxford Bulletin of Economics and Statistics, 76*(3), 334–359.

Doppelhofer, G., & Weeks, M. (2005). *Jointness of growth determinants*. Cambridge: University of Cambridge.

Doppelhofer, G., & Weeks, M. (2009a). Jointness of growth determinants. *Journal of Applied Econometrics, 24*(2), 209–244.

Doppelhofer, G., & Weeks, M. (2009b). Jointness of growth determinants: Reply to comments by Rodney Strachan, Eduardo Ley and Mark FJ Steel. *Journal of Applied Econometrics, 24*(2), 252–256.

Durlauf, S. N., Kourtellos, A., & Tan, C. M. (2008). Are any growth theories robust? *The Economic Journal, 118*(527), 329–346.

Eicher, T. S., Henn, C., & Papageorgiou, C. (2012). Trade creation and diversion revisited: Accounting for model uncertainty and natural trading partner effects. *Journal of Applied Econometrics, 27*(2), 296–321.

Eicher, T. S., & Newiak, M. (2013). Intellectual property rights as development determinants. *Canadian Journal of Economics, 46*(1), 4–22.

Eicher, T. S., Papageorgiou, C., & Raftery, A. E. (2011). Default priors and predictive performance in Bayesian model averaging, with application to growth determinants. *Journal of Applied Econometrics, 26*(1), 30–55.

Feldkircher, M., Horvath, R., & Rusnak, M. (2014). Exchange market pressures during the financial crisis: A Bayesian model averaging evidence. *Journal of International Money and Finance, 40*, 21–41.

Fernández, C., Ley, E., & Steel, M. F. J. (2001). Model uncertainty in cross-country growth regressions. *Journal of Applied Econometrics, 16*(5), 563–576.

George, E. I. (2010). Dilution priors: Compensating for model space redundancy. *IMS Collections Borrowing Strength: Theory Powering Applications – A Festschrift for Lawrence D. Brown, 6*, 158–165.

Glass, D. H. (2013). Confirmation measures of association rule interestingness. *Knowledge-Based Systems, 44*, 65–77.

Grün, B., & Hofmarcher, P. (2018). *Identifying groups of determinants in Bayesian model averaging using Dirichlet process clustering model priors*. Salzburg: University of Salzburg.

Hannak, A., Anderson, E., Barrett, L. F., Lehmann, S., Mislove, A., & Riedewald, M. (2012). Tweetin' in the rain: Exploring societal-scale effects of weather on mood. In *Proceedings of the Sixth International AAAI Conference on Weblogs and Social Media* (pp. 479–482).

Henderson, D. J., Papageorgiou, C., & Parmeter, C. F. (2011). Growth empirics without parameters. *The Economic Journal, 122*(559), 125–154.

Henderson, D. J., & Parmeter, C. F. (2016). Model averaging over nonparametric estimators. In G. González-Rivera, R. C. Hill, & T.-H. Lee (Eds.), *Essays in honor of Aman Ullah*. Advances in Econometrics (Vol. 36, pp. 539–560). Bingley: Emerald Group Publishing Limited.

Hoeting, J. A., Madigan, D., Raftery, A. E., & Volinsky, C. T. (1999). Bayesian model averaging: A tutorial. *Statistical Science, 14*(4), 382–417.

Hofmarcher, P., Crespo Cuaresma, J., Grün, B., Humer, S., & Moser, M. (2018). Bivariate jointness measures in Bayesian model averaging: Solving the conundrum. *Journal of Macroeconomics, 57*, 150–165.

Hofmarcher, P., Kerbl, S., Grün, B., Sigmund, M., & Hornik, K. (2014). Model uncertainty and aggregated default probabilities: New evidence from Austria. *Applied Economics, 46*(8), 871–879.

Jetter, M., & Parmeter, C. F. (2018). Sorting through global corruption determinants: Institutions and education matter – Not culture. *World Development, 109*(100), 279–294.

Koop, G. (2017). Bayesian methods for empirical macroeconomics with big data. *Review of Economic Analysis, 9*, 33–56.

Koop, G., & Korobilis, D. (2012). Forecasting inflation using dynamic model averaging. *International Economic Review, 53*(3), 867–886.

Koop, G., & Tobias, J. L. (2006). Semiparametric Bayesian inference in smooth coefficient models. *Journal of Econometrics, 134*(1), 283–315.

Leamer, E. E. (1978). *Specification searches*. New York: John Wiley and Sons.

Lehrer, S., & Xie, T. (2017). Box office buzz: Does social media data steal the show from model uncertainty when forecasting for Hollywood? *The Review of Economics and Statistics, 99*(5), 749–755.

Lenkoski, A., Eicher, T., & Raftery, A. (2014). Two-stage Bayesian model averaging in endogenous variable models. *Econometric Reviews, 33*, 122–151.

Ley, E., & Steel, M. F. J. (2007). Jointness in Bayesian variable selection with applications to growth regression. *Journal of Macroeconomics, 29*(3), 476–493.

Ley, E., & Steel, M. F. J. (2009). On the effect of prior assumptions in Bayesian model averaging with applications to growth regression. *Journal of Applied Econometrics, 24*(4), 651–674.

Ley, E., & Steel, M. F. J. (2012). Mixtures of g-priors for Bayesian model averaging with economic applications. *Journal of Econometrics, 171*(2), 251–266.

Madigan, D., & Raftery, A. E. (1994). Model selection and accounting for model uncertainty in graphical models using Occam's window. *Journal of the American Statistical Association, 89*(428), 1535–1546.

Madigan, D., & York, J. (1995). Bayesian graphical models for discrete data. *International Statistical Review, 63*(2), 215–232.

Malsiner-Walli, G., Hofmarcher, P., & Grün, B. (2019). Semi-parametric regression under model uncertainty: Economic applications. *Oxford Bulletin of Economics and Statistics, 81*(5), 1117–1143.

Masanjala, W. H., & Papageorgiou, C. (2008). Rough and lonely road to prosperity: A reexamination of the sources of growth in Africa using Bayesian model averaging. *Journal of Applied Econometrics, 23*, 671–682.

Moral-Benito, E. (2012). Determinants of economic growth: A Bayesian panel data approach. *Review of Economics and Statistics, 94*(2), 566–579.

Moral-Benito, E. (2015). Model averaging in economics: An overview. *Journal of Economic Surveys, 29*(1), 46–75.

Moser, M., & Hofmarcher, P. (2013). Model priors revisited: Interaction terms in BMA growth applications. *Journal of Applied Econometrics, 29*(2), 344–347.

R Core Team. (2019). *R: A language and environment for statistical computing*. Vienna: R Foundation for Statistical Computing. Retrieved from https://www.R-project.org/.

Raftery, A., Karny, M., & Ettler, P. (2010). Online prediction under model uncertainty via dynamic model averaging: Application to a cold rolling mill. *Technometrics, 52*, 52–66.

Sala-i-Martin, X., Doppelhofer, G., & Miller, R. I. (2004). Determinants of long-term growth: A Bayesian averaging of classical estimates (BACE) approach. *The American Economic Review, 94*(4), 813–835.

Scheipl, F. (2011). spikeSlabGAM: Bayesian variable selection, model choice and regularization for generalized additive mixed models in R. *Journal of Statistical Software, 43*(14), 1–24.

Scheipl, F., Fahrmeir, L., & Kneib, T. (2012). Spike-and-slab priors for function selection in structured additive regression models. *Journal of the American Statistical Association, 107*(500), 1518–1532.

Smith, M., & Kohn, R. (1996). Nonparametric regression using Bayesian variable selection. *Journal of Econometrics, 75*(2), 317–343.

Steel, M. F. J. (2018). *Bayesian model averaging and its use in economics*. arXiv:1709.08221 [stat.AP]. Retrieved from http://arxiv.org/abs/1709.08221.

Strachan, R. W. (2009). Comment on 'Jointness of growth determinants' by Gernot Doppelhofer and Melvyn Weeks. *Journal of Applied Econometrics, 24*(2), 245–247.

Tobias, J. L., & Li, M. (2004). Returns to schooling and Bayesian model averaging: A union of two literatures. *Journal of Economic Surveys, 18*(2), 153–180.

Wright, J. H. (2008). Bayesian model averaging and exchange rate forecasts. *Journal of Econometrics, 146*(2), 329–341.

Wu, T., Chen, Y., & Han, J. (2010). Re-examination of interestingness measures in pattern mining: A unified framework. *Data Mining and Knowledge Discovery, 21*(3), 371–397.

Zellner, A. (1986). On assessing prior distributions and Bayesian regression analysis with g prior distributions. In P. Goel & A. Zellner (Eds.), *Bayesian inference and decision techniques: Essays in honor of Bruno de Finetti* (Vol. 6, pp. 233–243). New York: Elsevier.

Zeugner, S., & Feldkircher, M. (2015). Bayesian model averaging employing fixed and flexible priors: The BMS package for R. *Journal of Statistical Software, 68*(4), 1–37.

Chapter 13
Bootstrap Aggregating and Random Forest

Tae-Hwy Lee, Aman Ullah, and Ran Wang

13.1 Introduction

The last 30 years witnessed the dramatic developments and applications of Bagging and Random Forests. The core idea of Bagging is model averaging. Instead of choosing one estimator, Bagging considers a set of estimators trained on the bootstrap samples and then takes the average output of them, which is helpful in improving the robustness of an estimator. In Random Forest, we grow a set of Decision Trees to construct a "forest" to balance the accuracy and robustness for forecasting.

This chapter is organized as follows. First, we introduce Bagging and some variants. Second, we discuss Decision Trees in details. Then, we move to Random Forest which is one of the most attractive machine learning algorithms combining Decision Trees and Bagging. Finally, several economic applications of Bagging and Random Forest are discussed. As we mainly focus on the regression problems rather than classification problems, the response y is a real number, unless otherwise mentioned.

13.2 Bootstrap Aggregating and Its Variants

Since the Bagging method combines many base functions in an additive form, there are more than one strategies to construct the aggregating function. In this section, we introduce the Bagging and its two variants, Subagging and Bragging. We also

T.-H. Lee (✉) · A. Ullah · R. Wang
Department of Economics, University of California, Riverside, CA, USA
e-mail: tae.lee@ucr.edu; aman.ullah@ucr.edu; ran.wang@email.ucr.edu

P. Fuleky (ed.), *Macroeconomic Forecasting in the Era of Big Data*,
Advanced Studies in Theoretical and Applied Econometrics 52,
https://doi.org/10.1007/978-3-030-31150-6_13

discuss the Out-of-Bag Error as an important way to measure the out-of-sample error for Bagging methods.

13.2.1 Bootstrap Aggregating (Bagging)

The first Bagging algorithm was proposed in Breiman (1996). Given a sample and an estimating method, he showed that Bagging can decrease the variance of an estimator compared to the estimator running on the original sample only, which provides a way to improve the robustness of a forecast.

Let us consider a sample $\{(y_1, x_1), \ldots, (y_N, x_N)\}$, where $y_i \in \mathbb{R}$ is the dependent variable and $x_i \in \mathbb{R}^p$ are p independent variables. Suppose the data generating process is $y = E(y|x) + u = f(x) + u$ where $E(u|x) = 0$ and $Var(u|x) = \sigma^2$. To estimate the unknown conditional mean function of y given x, $E(y|x) = f(x)$, we choose a function $\hat{f}(x)$ as an approximator, such as linear regression, polynomial regression, or spline, via minimizing the L_2 loss function

$$\min_{\hat{f}} \sum_{i=1}^{N} \left(y_i - \hat{f}(x_i) \right)^2 .$$

A drawback of this method is that, if $\hat{f}(x)$ is a nonlinear function, the estimated function $\hat{f}(x)$ may suffer from the **over-fitting** risk.

Consider the bias–variance decomposition of mean square error (MSE)

$$\begin{aligned} MSE &= E(y - \hat{f}(x))^2 \\ &= \left(E\hat{f}(x) - f(x) \right)^2 + Var(\hat{f}(x)) + Var(u) \\ &= Bias^2 + Variance + \sigma^2. \end{aligned}$$

There are three components included in the MSE: the bias of $\hat{f}(x)$, the variance of $\hat{f}(x)$, and $\sigma^2 = Var(u)$ is the variance of the irreducible error. The bias and the variance are determined by $\hat{f}(x)$. The more complex the forecast $\hat{f}(x)$ is, the lower its bias will be. But a more complex $\hat{f}(x)$ may suffer from a larger variance. By minimizing the L_2 loss function, we often decrease the bias to get the "optimal" $\hat{f}(x)$. As a result, $\hat{f}(x)$ may not be robust as it may result in much larger variance and thus a larger MSE. This is the over-fitting risk. To resolve this problem, the variance of $\hat{f}(x)$ needs to be controlled. There are several ways to control the variance, such as adding regularization term or adding random noise. Bagging is an alternative way to control the variance of $\hat{f}(x)$ via model averaging.

The procedure of Bagging is as follows:

- Based on the sample, we generate bootstrap sample $\{(y_1^b, x_1^b), \ldots, (x_N^b, y_N^b)\}$ via randomly drawing with replacement, with $b = 1, \ldots, B$.
- To each bootstrap sample, estimate $\hat{f}_b(x)$ via minimizing the L_2 loss function

$$\min_{\hat{f}_b(x)} \sum_{i=1}^{N} \left(y_i^b - \hat{f}_b(x_i^b) \right)^2 .$$

- Combine all the estimated forecasts $\hat{f}_1(x), \ldots, \hat{f}_B(x)$ to construct a Bagging estimate

$$\hat{f}(x)_{bagging} = \frac{1}{B} \sum_{b=1}^{B} \hat{f}_b(x).$$

Breiman (1996) proved that Bagging can make prediction more robust. Several other papers have studied why/how Bagging works. Friedman and Hall (2007) showed that Bagging could reduce the variance of the higher order terms but have no effect on the linear term when a smooth estimator is decomposed. Buja and Stuetzle (2000a) showed that Bagging could potentially improve the MSE based on second and higher order asymptotic terms but do not have any effects on the first order linear term. At the same time, Buja and Stuetzle (2000b) also showed that Bagging could even increase the second order MSE terms. Bühlmann and Yu (2002) studied in the Tree-based Bagging, which is a non-smooth and non-differentiable estimator, and found that Bagging does improve the first order dominant variance term in the MSE asymptotic terms. In summary, Bagging works with its main effects on variance and it can make prediction more robust by decreasing the variance term.

13.2.2 Sub-sampling Aggregating (Subagging)

The effectiveness of Bagging method is rooted in the Bootstrap method, the resampling with replacement. Sub-sampling, as another resampling method without replacement, can also be introduced to the same aggregating idea. Compared to the Bootstrap method, the Sub-sampling method often provides a similar outcome without relatively heavy computations and random sampling in Bootstrap. Theoretically, Sub-sampling needs weaker assumptions than the Bootstrap method.

Comparing to the Bootstrap, Sub-sampling method needs extra parameters. Let d be the number of sample points contained in each sub-sample. Since sub-sampling method draws samples without replacement from the original sample, the number of sub-sample is $M = \binom{N}{d}$. Thus, instead of aggregating the base predictors based on Bootstrap, we consider **Sub-sampling aggregating**, or **Subagging**, which combines predictors trained on samples from sub-sampling.

The procedure of Subagging is as follows:

- Based on the sample, construct $M = \binom{N}{d}$ different sub-samples $\{(y_1^m, x_1^m), \ldots, (y_d^m, x_d^m)\}$ via randomly drawing M times without replacement, where $m = 1, \ldots, M$.
- To each sub-sample, estimate $\hat{f}_m(x)$ via minimizing the L_2 loss function

$$\min_{\hat{f}_m(x)} \sum_{i=1}^{d} \left(y_i^m - \hat{f}_m(x_i^m) \right)^2 .$$

- Combine all the estimated models $\hat{f}_1(x), \ldots, \hat{f}_M(x)$ to construct a Subagging estimate

$$\hat{f}(x)_{subagging} = \frac{1}{M} \sum_{m=1}^{M} \hat{f}_m(x).$$

Practically, we choose $d = \alpha \times N$ where $0 < \alpha < 1$. There are several related research papers considered the similar settings for d (Buja and Stuetzle, 2000a,b). Since the d is related to the computational cost, $d = N/2$ is widely used in practice.

13.2.3 Bootstrap Robust Aggregating (Bragging)

In Sects. 13.2.1 and 13.2.2, we have discussed Bagging and Subagging that are based on bootstrap samples and sub-sampling samples, respectively. Although they are shown to improve the robustness of a predictor, both of them are based on the mean for aggregation, which may suffer from the problem of outliers. A common way to resolve the problem of outliers is to use median instead of the mean. To construct an outlier-robust model averaging estimator, a median-based Bagging method is discussed by Bühlmann (2004), which is called **Bootstrap Robust Aggregating** or **Bragging**.

The procedure of Bragging is the following:

- Based on the sample, we generate bootstrap samples $\{(y_1^b, x_1^b), \ldots, (y_N^b, x_N^b)\}$ via random draws with replacement, with $b = 1, \ldots, B$.
- With each bootstrap sample, estimate $\hat{f}_b(x)$ via minimizing the L_2 loss function

$$\min_{\hat{f}_b(x)} \sum_{i=1}^{N} \left(y_i^b - \hat{f}_b(x_i^b) \right)^2 .$$

- Combine all the estimated models $\hat{f}_1(x), \ldots, \hat{f}_B(x)$ to construct a Bragging estimate

$$\hat{f}(x)_{bragging} = \text{median}\left(\hat{f}_b(x);\ b = 1, \ldots, B\right).$$

To sum up, instead of taking the mean (average) on the base predictors in Bagging, Bragging takes the median of the base predictors. According to Bühlmann (2004), there are some other robust estimators, like estimating $\hat{f}_b(x)$ based on Huber's estimator, but Bragging works slightly better in practice.

13.2.4 Out-of-Bag Error for Bagging

In Sects. 13.2.1–13.2.3, we have discussed Bagging and its two variants. In the Bootstrap-based methods like Bagging and Bragging, when we train $\hat{f}_b(x)$ on the bootstrap sample, there are many data points not selected by resampling with replacement with the probability

$$P\left((x_i, y_i) \notin Boot_b\right) = \left(1 - \frac{1}{N}\right)^N \to e^{-1} \approx 37\%,$$

where $Boot_b$ is the bth bootstrap sample. There are roughly 37% of the original sample points not included in the bth bootstrap sample. Actually, this is very useful since it can be treated as a "test" sample for checking the out-of-sample error for $\hat{f}_b(x)$. The sample group containing all the samples not included in the bth bootstrap sample is called the **Out-of-Bag sample** or **OOB sample**. The error that the $\hat{f}_b(x)$ has on the bth out-of-bag sample is called the **Out-of-Bag Error**, which is equivalent to the error generated from the real test set. This is discussed in Breiman (1996) in detail. The bth Out-of-Bag error is calculated by

$$\begin{aligned}\widehat{err}_{OOB,b} &= \frac{\sum_{i=1}^{N} I\left((y_i, x_i) \notin Boot_b\right) \times Loss(y_i, \hat{f}_b(x_i))}{\sum_{i=1}^{N} I\left((y_i, x_i) \notin Boot_b\right)} \\ &= \frac{1}{N_b} \sum_{i=1}^{N_b} Loss\left(y_{i,OOB}^b, \hat{f}_b(x_{i,OOB}^b)\right).\end{aligned}$$

The procedure of implementing the Out-of-Bag Error is the following:

- Based on the sample, we generate B different bootstrap samples $\{(y_1^b, x_1^b), \ldots, (y_N^b, x_N^b)\}$ via randomly drawing with replacement.

- To each bootstrap sample, estimate $\hat{f}_b(x)$ via minimizing the loss function

$$\min_{\hat{f}_b(x)} \sum_{i=1}^{N} Loss\left(y_i^b - \hat{f}_b(x_i^b)\right).$$

- Compare the bth bootstrap sample to the original sample to get the bth Out-of-Bag sample $\{(y_{1,OOB}^b, x_{1,OOB}^b), \ldots, (y_{N_b,OOB}^b, x_{N_b,OOB}^b)\}$, where N_b is the number of data points for the bth Out-of-Bag sample.
- Calculate the Out-of-Bag error of $\hat{f}_b(x)$ among all the Out-of-Bag samples

$$\begin{aligned}\widehat{err}_{OOB} &= \frac{1}{B}\sum_{b=1}^{B}\frac{1}{N_b}\sum_{i=1}^{N_b} Loss\left(y_{i,OOB}^b, \hat{f}_b(x_{i,OOB}^b)\right)\\ &= \frac{1}{B}\sum_{b=1}^{B}\widehat{err}_{OOB,b}.\end{aligned}$$

13.3 Decision Trees

Although many machine learning methods, like spline and neural networks, are introduced as the base predictors in Bagging method, the most popular Bagging-based method is the so-called Random Forest proposed by Breiman (2001). Random Forest has been applied to many studies and becomes an indispensable tool for data mining and knowledge discovery. Intuitively, the main idea behind Random Forest is combining a large number of decision trees into a big forest via Bagging. In this section, we concentrate on how to construct the base learner, **Decision Tree**, for Random Forest. In Sect. 13.4, we discuss the Random Forest in detail. Several effective variants of Random Forest are discussed in detail in Sect. 13.5.

13.3.1 The Structure of a Decision Tree

The basic idea of the decision tree has a long history and has been used in many areas including biology, computer science, and business. Biologists usually introduce a very large tree chart to describe the structure of classes containing animals or plants; in computer science, tree structure is a widely used data type or data structure with a root value and sub-trees of children with a parent node, represented as a set of linked nodes; in business, the decision tree is a usual structure choice for a flowchart that each internal node has a series of questions based on input variables.

Figure 13.1 gives an example of book data with the tree structure. Firstly, in all kinds of books, we have economic books. Then, economic books contain

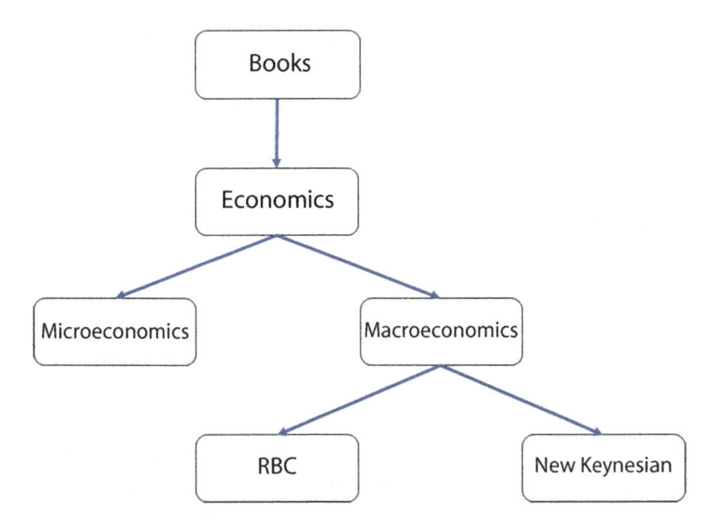

Fig. 13.1 A tree of structured data about economic books

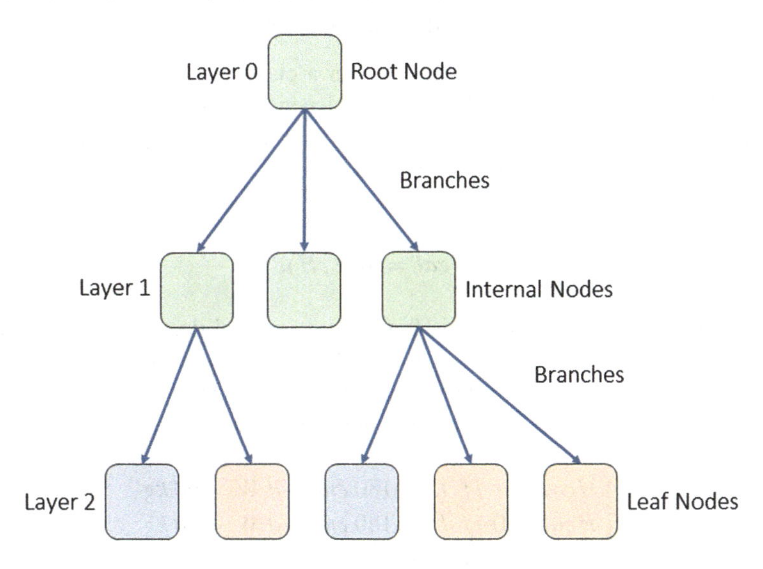

Fig. 13.2 The components in a decision tree

books about macroeconomics, microeconomics, and others. If we concentrate on macroeconomic books, it contains books about real business cycle (RBC) theory, new Keynesian theory, etc.

First of all, let us explore the structure of the decision tree and clarify the names of components in the decision tree. Figure 13.2 illustrates a decision tree with three layers. We can see that there are four components in a decision tree: root nodes, internal nodes, leaf nodes, and branches between every two layers. The root node is the beginning of a decision tree. From the only one root node, there could be two or more branches connecting to the internal nodes in the next layer. Each internal node is also called the parent node to the connected nodes in the next layer. The nodes in the next layer are called child nodes or sub-nodes. Also, every internal node contains a decision rule to decide how to connect to its sub-nodes in the next layer. At the bottom, there are several leaf nodes. They are the end of one decision tree and they represent different outputs for prediction. For example, to a regression problem, each leaf node contains a continuous output. To a classification problem, each leaf node contains a discrete output corresponding to the labels of classes.

Intuitively, all the tree structure methods share the same intuition: the recursive splitting. Given a node, we split it into several branches connecting to its sub-nodes in the next layer. Then, to each sub-node, we split it again to get more sub-nodes in the next layer until the end of the decision tree.

In data mining and machine learning, the decision tree is widely used as a learning algorithm called decision tree learning. We first construct the structure of a decision tree structure. Each node contains a decision rule. To compute the prediction of a decision tree, we feed the input to the root node and then propagate through all the layers to a leaf node, which outputs the final prediction of the decision tree. We discuss this procedure in detail via the following two examples.

Example 1 (People's Health) Let us consider a classification problem about people's health. Suppose a people's health $Heal$ depends on two explanatory variables, weight W and height H. Health is a binary variable with two potential outcomes: $Heal = 1$ means healthy and $Heal = 0$ means not healthy. The function of $Heal$ given H and W is

$$Heal = h(W, H).$$

Now suppose we can represent this function via several decision rules. Based on our experience, to a people with a large height, it is not healthy if this people have a relatively small weight; to a people with a small height, it is not healthy if this people have a large weight. We can write down these rules:

$$\begin{cases} Heal = 1 \ if \ H > 180 \, cm \ and \ W > 60 \, kg \\ Heal = 0 \ if \ H > 180 \, cm \ and \ W < 60 \, kg \\ Heal = 1 \ if \ H < 180 \, cm \ and \ W < 80 \, kg \\ Heal = 0 \ if \ H < 180 \, cm \ and \ W > 80 \, kg. \end{cases}$$

We first consider height H. Based on the outcome of H, there are different decision rules for weight W. Thus, it is straightforward to construct a tree to encode this procedure.

In Fig. 13.3, the node containing H is the root node, which is the beginning of the decision procedure. The node containing W is the internal node in the first layer. In the second layer, there are four leaf nodes that give the final prediction of health. For example, to a sample ($H = 179cm$, $W = 60kg$), according to the decision rule

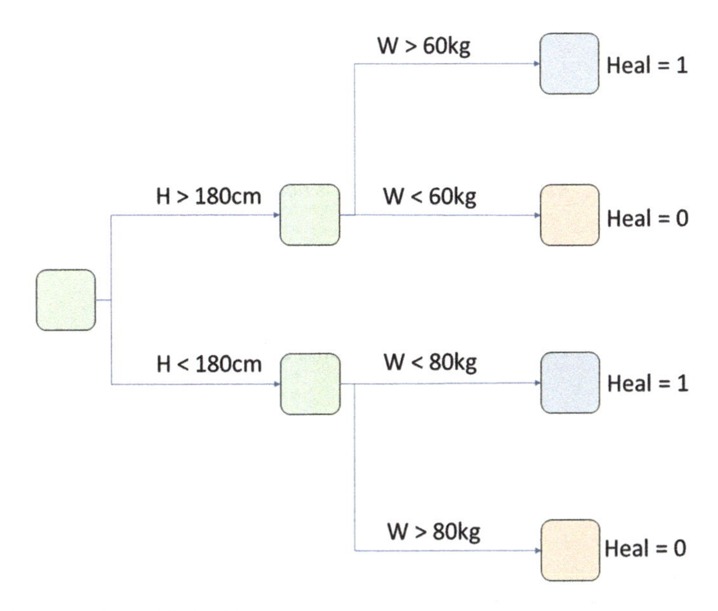

Fig. 13.3 A tree of people's health

in the root node, we choose the lower part of branches since $179 < 180$. Then, since $60 < 80$ based on the decision rule in the internal node, we go to the third leaf node and output $Heal = 1$ as the prediction. This decision tree encodes the four decision rules into a hierarchical decision procedure.

Example 2 (Women's Wage) Another example is about the classic economic research: women's wage. Suppose women's wage depends on two factors: education level Edu and working experience $Expr$. Thus, this is a regression problem. The nonlinear function of women's wage is

$$Wage = g(Edu, Expr).$$

If a woman has higher education level or a longer working experience, it is much possible that woman have higher wage rate. As in Example 1, we suppose the nonlinear function g can be represented by the following rules:

$$\begin{cases} Wage = 50 \ if \ Expr > 10 \ years \ and \ Edu = college \\ Wage = 20 \ if \ Expr > 10 \ years \ and \ Edu \neq college \\ Wage = 10 \ if \ Expr < 10 \ years \ and \ Edu = college \\ Wage = 0 \ if \ Expr < 10 \ years \ and \ Edu \neq college. \end{cases}$$

In this case, we first consider the experience $Expr$. Based on it, we use different decision rules for education Edu. This procedure can also be encoded into a decision tree.

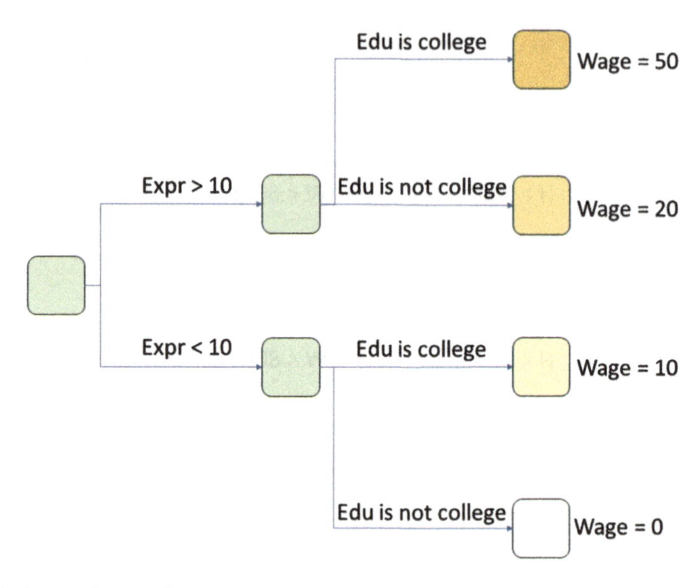

Fig. 13.4 A tree of woman's wage

Figure 13.4 illustrates the decision tree for predicting women's wage. To a woman who has 11 years of working experience with a college degree, it is more likely that she has a higher wage rate. Thus the decision tree outcomes 50; if a woman has 3 years of working experience without a college degree, we expect the woman could have a hard time in searching for her job. Thus, the decision tree reports 0.

13.3.2 Growing a Decision Tree for Classification: ID3 and C4.5

In Sect. 13.3.1, we have discussed how a decision tree works. Given the correct decision rules in the root and internal nodes and the outputs in the leaf nodes, the decision tree can output the prediction we need. The next question is how to decide the decision rules and values for all the nodes in a decision tree. This is related to the learning or growing of a decision tree. There are more than 20 methods to grow a decision tree. In this chapter, we only consider two very important methods. In this section, we discuss ID3 and C4.5 methods for the classification problem. In the Sects. 13.3.3 and 13.3.4, we will introduce the Classification and Regression Tree (CART) method for the classification problem and the regression problem, respectively.

Let us go back to the weight, height, and health example. Since there are two explanatory variables, H and W, we can visualize the input space in a 2D plot.

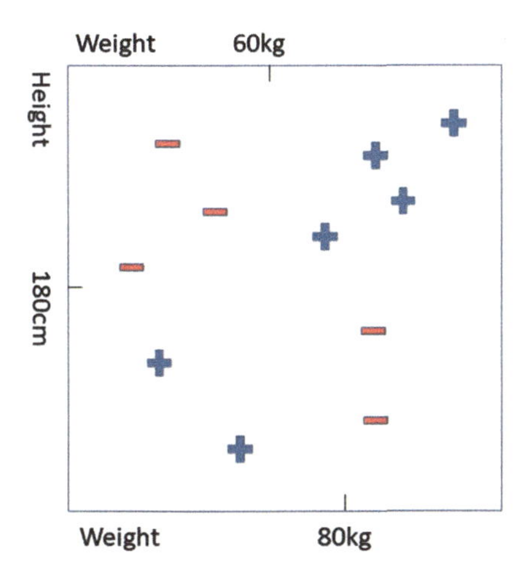

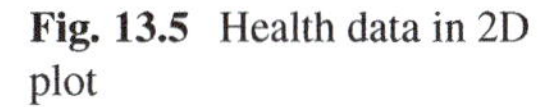
Fig. 13.5 Health data in 2D plot

Figure 13.5 illustrates all the data points $\{(Heal_1, W_1, H_1), \ldots, (Heal_N, W_N, H_N)\}$ in a 2D plot. The horizontal axis is the weight W and the vertical axis represents the height H. The red minus symbol means $Heal = 0$ and the blue plus symbol represents $Heal = 1$.

Figure 13.6 illustrates the implementation of a decision tree in a 2D plot to predict a person's health. First of all, in level 1, the decision rule at the root node is $Height > 180$ or not. In the 2D plot, this rule could be represented as a **decision stump** which is a horizontal line at $H = 180$ cm. The decision stump splits the sample space into two sub-spaces that are corresponding to the two sub-nodes in level 1. The upper space is corresponding to $H > 180$ cm and the lower space represents $H < 180$ cm.

Next, we have two sub-spaces in level two. To the upper spaces, we check the rule at the right internal node, $W > 60$ kg or not. This can be represented as another vertical decision stump at $W = 60$ kg to separate upper space to two sub-spaces. Similarly, to the lower space, we also draw another vertical decision stump, which is corresponding to the decision rule at the left internal node.

Finally, we designate the final output for each of the four sub-spaces that represent the four leaf nodes. In classification problems, given a sub-space corresponding to a leaf node, we consider the number of samples for each class and then choose the class with the most number of samples as the output at this leaf node. For example, the upper left space should predict $Heal = 0$, the upper right space is corresponding to $Heal = 1$. For the regression problems, we often choose the average of all the samples at one sub-space as the output of this leaf node.

To sum up, each node in a decision tree is corresponding to space or a sub-space. The decision rule in each node is corresponding to a decision stump in this space. Then, every leaf node computes its output based on the average outputs belonging to this leaf. To grow a decision tree, there are two kinds of "parameters" need to be figured out: the positions of all the decision stumps corresponding to the non-leaf nodes and the outputs of all the leaf nodes.

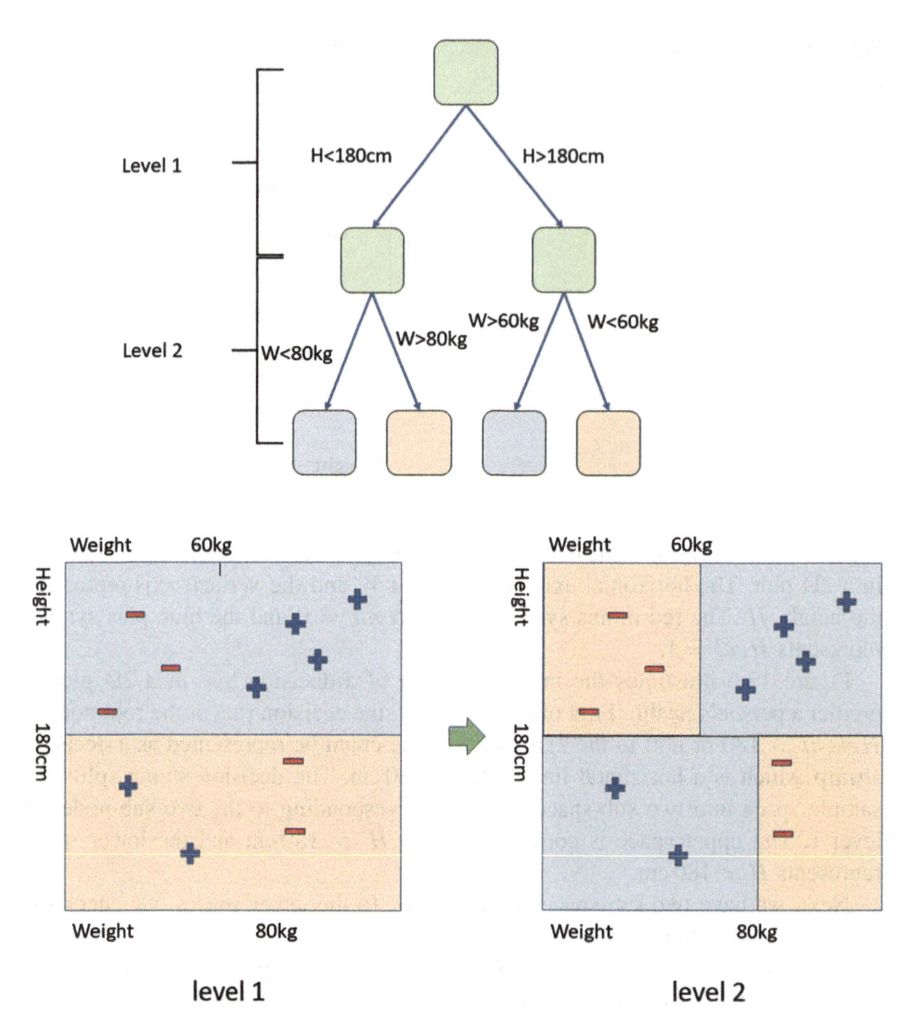

Fig. 13.6 Grow a tree for health data

In decision tree learning, we often grow a decision tree from the root node to leaf nodes. Also in each node, we usually choose only one variable for the decision stump. Thus, the decision stump should be orthogonal to the axis corresponding to the variable we choose. At first, we decide that the optimal decision stump for the root node. Then, to two internal nodes in layer 1, we figure out two optimal decision stumps. Then, we estimate the outputs to four leaf nodes. In other words, decision tree learning is to hierarchically split input space into sub-spaces. Comparing the two plots at the bottom of Fig. 13.6, we can see the procedure of hierarchical splitting for a decision tree learning.

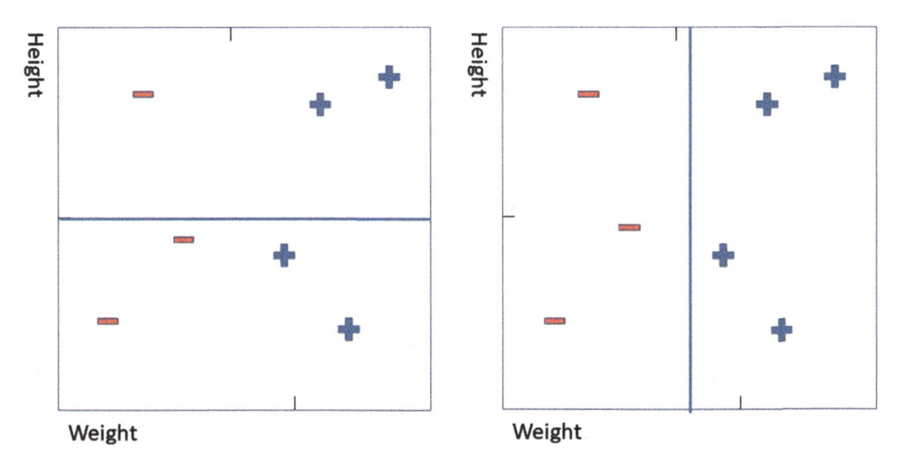

Fig. 13.7 Sub-spaces generated by decision stumps

Thus, the core question is how to measure the goodness of a decision stump to a node. An important measure of this problem is called **impurity**. To understand it, we consider two decision stumps for one sample set.

Figure 13.7 shows the different cases of the sub-spaces split by two decision stumps. To the left panel, H is selected for the decision stump. In two sub-spaces, the samples have two labels. To the right panel, W is selected. The left sub-space only contains samples with label $Heal = 0$ and the right sub-space only contains samples with label $Heal = 1$. Intuitively, we can say that the two sub-spaces in the left panel are impure compared to the sub-spaces in the right panel. The sub-spaces in the right panel should have lower impurity. Obviously, the decision stump in the right panel is better than the left panel since it generates more pure sub-spaces.

Mathematically, the **information entropy** is a great measure of impurity. The more labels of samples are contained in one sub-space, the higher entropy of the sub-space has. To discuss the entropy-based tree growing clearly, we introduce a new definition: **information gain**. The information or entropy for an input space S is

$$Info(S) = -\sum_{c=1}^{C} p_c \log_2(p_c), \tag{13.1}$$

where C is the total number of classes or labels contained in space S. p_c is the frequency of samples for one class in the space S. It can be estimated by

$$p_c = \frac{1}{N_S} \sum_{x_i \in S} I(y_i = c), \tag{13.2}$$

where N_S is the total number of samples in space S. $I(y_i = c)$ is an indicator function measuring the label y_i is the cth class or not.

Suppose we choose D as a decision stump and it separates the space S into two sub-spaces. For example, if we choose D as $x = 5$, the two sub-spaces are corresponding to $x < 5$ and $x > 5$. Then, we calculate the distinct entropies for two sub-spaces. Thus, if the space S is separated into v different sub-spaces, the average entropy of S after splitting is

$$Info_D(S) = \sum_{j=1}^{v} \frac{N_{S_j}}{N_S} \times Info(S_j), \tag{13.3}$$

where v is the number of sub-spaces generated by D. To binary splitting, $v = 2$. S_j is the jth sub-space and it satisfies: $S_i \cap S_j = \emptyset$ if $i \neq j$ and $\bigcup_i S_i = S$. N_{S_j} and N_S are the number of samples contained in S_j and S.

Obviously, the information or entropy for space S changes before and after splitting based on decision stump D. Thus, we define the information gain of D as

$$Gain(D) = Info(S) - Info_D(S). \tag{13.4}$$

Example 3 (Predicting Economic Growth) Consider an example of predicting economic growth G based on two factors: inflation rate I and net export NX. Suppose G is a binary variable where $G = 1$ for expansion and $G = 0$ for recession. Then, the growth G is an unknown function of the inflation rate I and the net export NX

$$G = G(I, NX).$$

From the left panel in Fig. 13.8, we can see the sample distribution of economic growth G. For example, if there is high inflation rate I and high net export NX, we

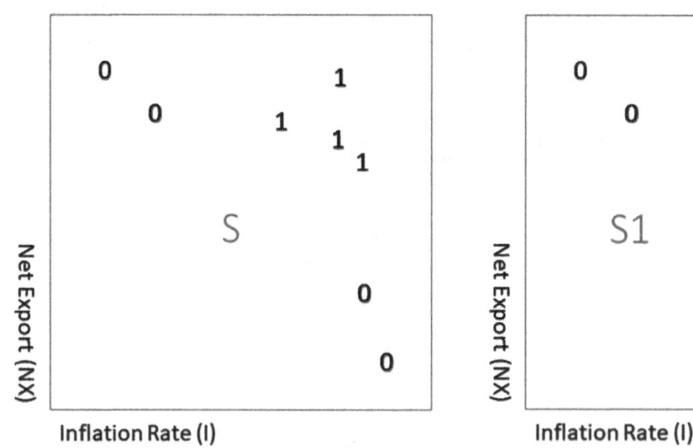

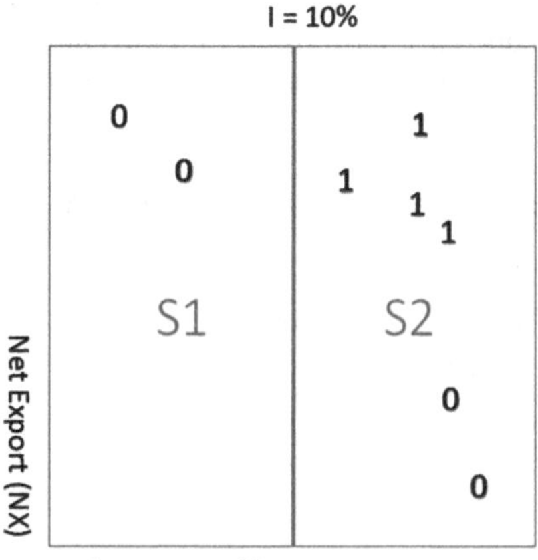

Fig. 13.8 Plots for economic growth data

observe the economic expansion where $G = 1$; if there are high inflation rate I but low net export NX, the economy will be in recession with $G = 0$.

Let us consider a decision tree with only the root node and two leaf nodes to fit the samples. In the right panel, we choose $D : I = 10\%$ as the decision stump in the root node. Thus, the space S is splitted into two sub-spaces S_1 and S_2. According to Eq. (13.1), the information to the original space S is

$$\begin{aligned} Info(S) &= -\sum_{c=1}^{2} p_c \log_2(p_c) \\ &= -(p_1 \log_2(p_1) + p_2 \log_2(p_2)) \\ &= -\left(\frac{4}{8}\log_2\left(\frac{4}{8}\right) + \frac{4}{8}\log_2\left(\frac{4}{8}\right)\right) \\ &= 1, \end{aligned}$$

where class 1 is corresponding to $G = 0$ and class 2 to $G = 1$. And $p_1 = \frac{4}{8}$ means that there are 4 samples with $G = 0$ out of 8 samples.

After splitting, the information to the sub-space S_1 is

$$\begin{aligned} Info(S_1) &= -\sum_{c=1}^{2} p_c \log_2(p_c) \\ &= -p_1 \log_2(p_1) + 0 \\ &= -\left(\frac{2}{2}\right)\log_2\left(\frac{2}{2}\right) = 0. \end{aligned}$$

The information to the sub-space S_2 is

$$\begin{aligned} Info(S_2) &= -\sum_{c=1}^{2} p_c \log_2(p_c) \\ &= -(p_1 \log_2(p_1) + p_1 \log_2(p_1)) \\ &= -\left(\frac{2}{6}\log_2\left(\frac{2}{6}\right) + \frac{4}{6}\log_2\left(\frac{4}{6}\right)\right) \\ &= 0.92. \end{aligned}$$

Based on Eq. (13.3), the average entropy of S after splitting is

$$\begin{aligned} Info_D(S) &= \sum_{j=1}^{v} \frac{N_{S_j}}{N_S} \times Info(S_j) \\ &= \frac{N_{S_1}}{N_S} \times Info(S_1) + \frac{N_{S_2}}{N_S} \times Info(S_2) \\ &= \frac{2}{8} \times 0 + \frac{6}{8} \times 0.92 \\ &= 0.69. \end{aligned}$$

After splitting, the information decreases from 1 to 0.69. According to Eq. (13.3), the information gain of D is

$$Gain(D) = Info(S) - Info_D(S) = 0.31.$$

To sum up, we can find the best decision stump to maximizing the information gain such that the optimal decision stump can be found. From the root node, we repeat finding the best decision stump to each internal node until stopped at the leaf nodes. This method for tree growing is called **ID3** introduced by Quinlan (1986).

Practically, the procedure of implementing the decision tree for classification based on ID3 is the following:

- Suppose the sample is $\{(y_1, x_1), \ldots, (y_N, x_N)\}$ where $y_i \in (0, 1)$ and $x_i \in \mathbb{R}^p$. To the first dimension, gather all the data orderly as $x_{1,(i)}, \ldots, x_{1,(N)}$.
- Search the parameter d_1 respect to $D_1 : x_1 = d_1$ through $x_{1,(i)}$ to $x_{1,(N)}$ such that

$$\max_{D_1} Gain(D_1) = \max_{D_1} \left(Info(S) - Info_{D_1}(S) \right).$$

- Find the best $D_2 : x_2 = d_2, \ldots, D_p : x_p = d_p$ and then choose the optimal D such that

$$\max_{D} Gain(D) = \max_{D} \left(Info(S) - Info_D(S) \right).$$

- Repeatedly run the splitting procedure until every node containing one label of y. Finally, take the label of y from one leaf node as its output.

One problem this method suffer from is related to over-fitting. Suppose we have N data points in space S. According to the rule that maximizing the information gain, we can find that the optimal result is separating one sample into one sub-space such that the entropy is zero in each sub-space. This is not a reasonable choice since it is not robust to noise in the samples. To prevent that, we can introduce a revised version of information gain from **C4.5** method.

C4.5 introduces a measure for information represented via splitting, which is called **Splitting Information**

$$Split\ Info_D(S) = -\sum_{j=1}^{v} \frac{N_{S_j}}{N_S} \times \log_2 \frac{N_{S_j}}{N_S}, \tag{13.5}$$

where $v = 2$ for the binary splitting.

Obviously, this is an entropy based on the number of splitting or the number of sub-spaces. The more the sub-spaces are, the higher the splitting information we will get. To show this conclusion, let us go back to the economic growth case (See Fig. 13.9).

To the left case, the splitting information is computed based on Eq. (13.5) as

$$\begin{aligned} Split\ Info_D(S) &= -\sum_{j=1}^{v} \frac{N_{S_j}}{N_S} \times \log_2 \frac{N_{S_j}}{N_S} \\ &= -\left(\frac{N_{S_1}}{N_S} \times \log_2 \frac{N_{S_1}}{N_S} + \frac{N_{S_2}}{N_S} \times \log_2 \frac{N_{S_2}}{N_S}\right) \\ &= -\left(\frac{2}{8} \times \log_2 \frac{2}{8} + \frac{6}{8} \times \log_2 \frac{6}{8}\right) \\ &= 0.81. \end{aligned}$$

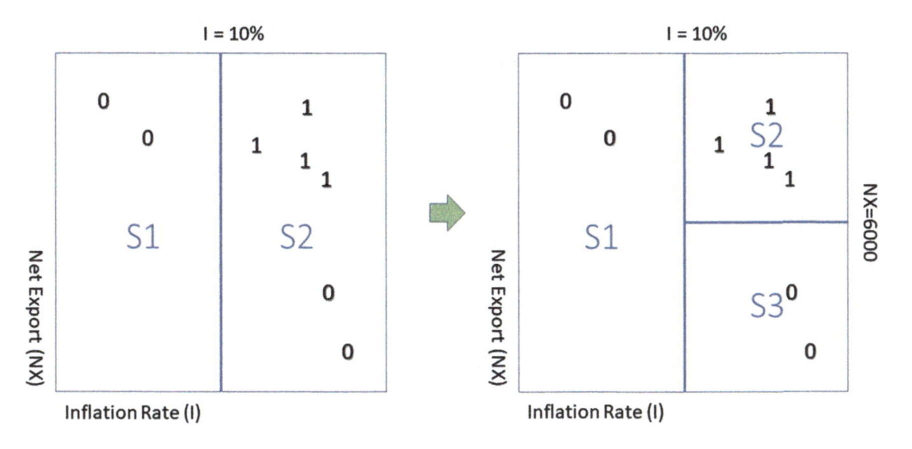

Fig. 13.9 Grow a tree for economic growth prediction

To the right case, the splitting information is

$$
\begin{aligned}
Split\ Info_D(S) &= -\sum_{j=1}^{v} \frac{N_{S_j}}{N_S} \times \log_2 \frac{N_{S_j}}{N_S} \\
&= -\left(\frac{N_{S_1}}{N_S} \times \log_2 \frac{N_{S_1}}{N_S} + \frac{N_{S_2}}{N_S} \times \log_2 \frac{N_{S_2}}{N_S} + \frac{N_{S_3}}{N_S} \times \log_2 \frac{N_{S_3}}{N_S} \right) \\
&= -\left(\frac{2}{8} \times \log_2 \frac{2}{8} + \frac{4}{8} \times \log_2 \frac{4}{8} + \frac{2}{8} \times \log_2 \frac{2}{8} \right) \\
&= 1.5.
\end{aligned}
$$

Thus, when there are more sub-spaces, the splitting information increases. In other words, splitting information is the "cost" for generating sub-spaces. Now, instead of information gain, we can use a new measure called **Gain Ratio(D)**

$$
Gain\ Ratio(D) = \frac{Gain(D)}{Split\ Info(D)}. \tag{13.6}
$$

When we generate more sub-spaces, the information gain increases but splitting information is higher at the same time. Thus, to maximize the Gain Ratio of D, we can make great trade-offs. This is the main idea of **C4.5**, an improved version of ID3 introduced by Quinlan (1994).

Summarizing, the procedure of implementing the decision tree for classification based on C4.5 is the following:

- Suppose the sample is $\{(y_1, x_1), \ldots, (y_N, x_N)\}$ where $y_i \in (0, 1)$ and $x_i \in \mathbb{R}^p$. To the first dimension, gather all the data orderly as $x_{1,(i)}, \ldots, x_{1,(N)}$.
- Search the parameter d_1 respect to $D_1 : x_1 = d_1$ through $x_{1,(i)}$ to $x_{1,(N)}$ such that

$$
\max_{D_1} Gain\ Ratio(D_1) = \max_{D_1} \left(\frac{Gain(D_1)}{Split\ Info(D_1)} \right).
$$

- Find the best $D_2 : x_2 = d_2, \ldots, D_p : x_p = d_p$ and then choose the optimal D such that

$$
\max_{D} Gain\ Ratio(D) = \max_{D} \left(\frac{Gain(D)}{Split\ Info(D)} \right).
$$

- Repeatedly run the splitting procedure until the Gain Ratio is less than 1. Finally, take the most frequency label of y from one leaf node as its output.

13.3.3 Growing a Decision Tree for Classification: CART

In Sect. 13.3.2, we have discussed related methods about how to grow a tree based on ID3 and C4.5 methods. In this section, we introduce another way to construct a decision tree, the **Classification and Regression Tree** (**CART**), which not only features great performance but very easy to implement in practice for both classification and regression tasks.

The main difference between ID3, C4.5, and CART is the measure of information. ID3 and C4.5 choose the entropy to construct the Information Gain and Gain Ratio. In CART, we introduce a new measure for deciding the best decision stump called the **Gini Index** or **Gini Impurity**. The definition of Gini Impurity is

$$Gini(S) = \sum_{j=1}^{M} p_j(1 - p_j) = 1 - \sum_{j=1}^{M} p_j^2, \tag{13.7}$$

where M is the number of classes in node spaces S and p_j is the frequency of class j in node space S. Intuitively, this is the variance of the binary distribution. That is, CART chooses the variance as the impurity measure.

Figure 13.10 illustrates the difference between entropy and Gini Impurity. Given x-axis as the proportion of sample belonging to one class, we can see that two curves are very similar. Then, we have the new Gini Impurity after binary splitting

$$Gini_D(S) = \frac{N_{S_1}}{N_S} Gini(S_1) + \frac{N_{S_2}}{N_S} Gini(S_2). \tag{13.8}$$

where the N_S, N_{S_1}, N_{S_2} are the numbers of sample points in space S, S_1, S_2, respectively. Similarly to the information gain in ID3, we consider the difference of Gini impurity as the measure of goodness of decision stump

$$\Delta Gini_D(S) = Gini(S) - Gini_D(S). \tag{13.9}$$

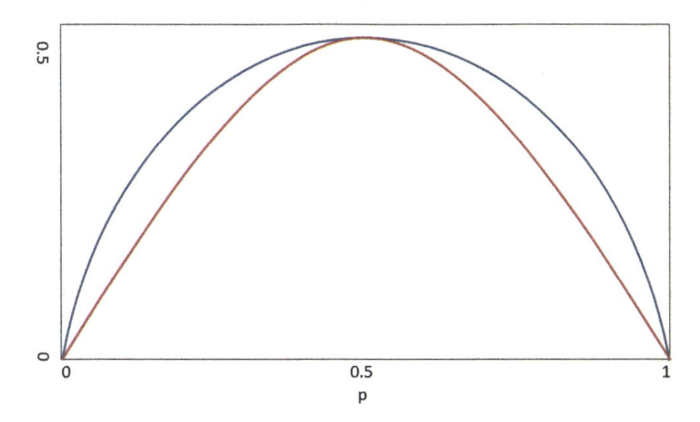

Fig. 13.10 Entropy (blue) and gini impurity (red)

As we discuss in ID3 method, if we grow a decision tree via maximizing the information gain in each node, it is the best choice that we split all the data points in one space such that each sub-space contains one sample point. ID3 and CART may suffer from this risk. C4.5 should be a better choice than ID3 and CART, but it has a fixed rule to prevent over-fitting which cannot be adaptive to data.

To solve this problem, let us consider the total cost of growing a decision tree

$$Total\ Cost = Measure\ of\ Fit + Measure\ of\ Complexity. \tag{13.10}$$

The total cost contains two main parts: the measure of fit is related to the goodness of the model, as the error rate in classification problem; the measure of complexity describes the power of the model. To balance the two measures in growing a decision tree, we often choose the following function as the objective:

$$L = Loss(y_i, x_i; tree) + \lambda\Omega(numbers\ of\ leaf\ nodes).$$

The first term is related to the loss of the decision tree. To classification problem, we can use the error rate on the samples as the loss. The second term is a measure of complexity based on the number of leaf nodes. Ω is an arbitrary function like the absolute function. λ is a tuning parameter balancing the loss and the complexity. Many machine learning and regressions like Lasso and Ridge Regression follow this framework. Also, since the second term penalizes on the number of leaf nodes, this is also called **pruning** a decision tree.

The procedure of implementing the decision tree for classification based on CART is the following:

- Suppose the sample is $\{(y_1, x_1), \ldots, (y_N, x_N)\}$ where $y_i \in \{0, 1\}$ and $x_i \in \mathbb{R}^p$. To the first dimension, gather all the data orderly as $x_{1,(i)}, \ldots, x_{1,(N)}$.
- Search the parameter d_1 respect to $D_1 : x_1 = d_1$ through $x_{1,(i)}$ to $x_{1,(N)}$ such that

$$\max_{D_1} \Delta Gini_{D_1}(S) = \max_{D_1} \left(Gini(S) - Gini_{D_1}(S)\right).$$

- Find the best $D_2 : x_2 = d_2, \ldots, D_p : x_p = d_p$ and then choose the optimal D such that

$$\max_{D} \Delta Gini_{D}(S) = \max_{D} \left(Gini(S) - Gini_{D}(S)\right).$$

- Based on the new decision stump, calculate the error rate for the decision tree and the total loss function

$$L = error\ rate(y_i, x_i; tree) + \lambda\Omega(numbers\ of\ leaf\ nodes).$$

- Repeatedly run the splitting procedure until the total loss function starting to increase. Finally, take the most frequency label of y from one leaf node as its output.

13.3.4 Growing a Decision Tree for Regression: CART

The Information Gain and Gini Impurity are very important measures when we are implementing a classification problem. In economic research, we often consider more regression problems with the continuous response. Thus, instead of the information gain, we choose the variation to measure the goodness of a decision stump

$$Variation(S) = \sum_{i=1}^{N} (y_i - \bar{y})^2, \tag{13.11}$$

where N is the number of data points belong to the space S. After several splitting, the space S is separated into v sub-spaces $S_1, \ldots, S_v$, we can define the average variance after splitting the space S

$$Variation_D(S) = \frac{1}{v} \sum_{j=1}^{v} Variation_j(S), \tag{13.12}$$

where v is the number of the sub-spaces separated by D. Again, to binary splitting, we have $v = 2$. Thus, we have a new information gain for regression method

$$Gain(D) = Variation(S) - Variation_D(S). \tag{13.13}$$

Based on the total cost in Eq. (13.10), we choose the same formula for regression

$$L = Loss(y_i, x_i; tree) + \lambda\Omega(numbers\ of\ leaf\ nodes),$$

where $Loss(y_i, x_i; tree)$ is the L_2 loss function.

Thus, the procedure of implementing the decision tree for regression based on CART is the following:

- Suppose the sample is $\{(y_1, x_1), \ldots, (y_N, x_N)\}$ where $y_i \in \mathbb{R}$ and $x_i \in \mathbb{R}^p$. To the first dimension, gather all the data orderly as $x_{1,(i)}, \ldots, x_{1,(N)}$.
- Search the parameter d_1 respect to $D_1 : x_1 = d_1$ through $x_{1,(i)}$ to $x_{1,(N)}$ such that

$$\max_{D_1} Gain(D) = \max_{D_1} \left(Variation(S) - Variation_{D_1}(S) \right).$$

- Find the best $D_2 : x_2 = d_2, \ldots, D_p : x_p = d_p$ and then choose the optimal D such that

$$\max_D Gain(D) = \max_D (Variation(S) - Variation_D(S)).$$

- Based on the new decision stump, compute the loss for the decision tree and the total loss function

$$L = Loss(y_i, x_i; tree) + \lambda\Omega(numbers\ of\ leaf\ nodes).$$

- Repeatedly run the splitting procedure until the total loss function starting to increase. Finally, take an average of y from one leaf node as its output.

13.3.5 Variable Importance in a Decision Tree

In Sects. 13.3.2–13.3.4, we discussed how to grow a decision tree. In this section, we consider another problem: how to measure the importance of the variable.

In the procedure of growing a decision tree, each time we split one internal node into two child nodes, one variable should be selected based on the information gain or variation gain. Thus, for an important variable, the decision tree should choose it frequently among all the internal nodes. Conversely, the variables may be selected just a few times if the variables are not very important. To the jth variable, Breiman, Friedman, Stone, and Olshen (1984) defined a **relative importance** as

$$I_j^2 = \sum_{t=1}^{T-1} e_t^2 I(v(t) = j), \tag{13.14}$$

where T is the number of internal nodes (non-leaf nodes) in a decision tree, $v(t)$ is the variable selected by node t. e_t is the error improvement based on before and after splitting the space via variable $v(t)$. To regression task, it can be a gain of variation. To classification problem, it is related to information gain of entropy or the difference of Gini Impurity.

For example, let us consider a CART tree to a regression problem. Suppose we split the tth node into two nodes based on variable j selected by the decision stump D. Then, we can calculate the value of the information gain $Gain(D) = Variation(S) - Variation_D(S)$. This is the error improvement e_t. Thus, considering all the internal nodes, we compute all the e_t^2 to get I_j^2.

If variable j is very important, the error improvement should be very large and $I(v(t) = j)$ often equals to 1 since variable j is usually selected. As a result, the measure I_j^2 is relatively large; conversely, if a variable is not very important, the error improvement based on this variable cannot be so large, which leads to a small

I_j^2. After we growing a decision tree on a training set, we often calculate it on the test set.

13.4 Random Forests

Random Forest is a combination of many decision trees based on Bagging. In the first paper about Random Forest, Breiman (2001) discussed the theories behind the Random Forest and compared Random Forest with other ensemble methods. From this section, we start to discuss Random Forests in detail.

13.4.1 Constructing a Random Forest

As we discussed in Sect. 13.2, Bagging method can generate a lot of base learners trained on bootstrap samples and then combine them to predict. If we consider combining a set of unbiased estimators or predictors, Bagging works by decreasing the variances of the predictors but keeping the means unaffected.

For example, let us consider B numbers of unbiased estimators $f_1, f_2, \ldots, f_B$ with same variance σ^2. If they are i.i.d, it is easy to show that the variance of average estimator is

$$Var(g) = Var\left(\frac{1}{B}\sum_{b=1}^{B} f_b\right) = \frac{1}{B}\sigma^2. \tag{13.15}$$

But if the unbiased estimators are correlated, the variance of the average estimator is

$$\begin{aligned} Var(g) &= \frac{1}{B^2} Var\left(\sum_{b=1}^{B} f_b\right) \\ &= \frac{1}{B^2}\left(\sum_{b=1}^{B} Var(f_b) + 2\sum_{b\neq c} cov(f_b, f_c)\right) \\ &= \frac{1}{B^2}(B\sigma^2 + (B^2 - B)\rho\sigma^2) \\ &= \rho\sigma^2 + \frac{(1-\rho)}{B}\sigma^2, \end{aligned} \tag{13.16}$$

where ρ is the correlation coefficient between two estimators.

The variance of average estimator depends on the number of base estimators and the correlation between estimators. Even if we can decrease the second term to zero via adding increasingly large numbers of estimators, the first term remains at the same level if the estimators are not independent. Similarly, in Bagging, even if we can combine a lot of predictors based on Bootstrap, the variance cannot keep decreasing if the predictors are dependent with each other.

In practice, since most of the bootstrap samples are very similar, the decision trees trained on these sample sets are often similar and highly correlated with others. Thus, average estimators of similar decision trees can be more robust but do not perform much better than a single decision tree. That is the reason why Bagging decision trees or other base learners may not work so well in prediction.

Compare to Bagging decision trees, which only combines many trees based on Bootstrap to decrease the second term $\frac{(1-\rho)}{B}\sigma^2$, Random Forest also considers controlling the first term $\rho\sigma^2$. To decrease the correlation between decision trees, Random Forest introduces the so-called **random subset projection** or **random feature projection** during growing a decision tree. That is, instead of applying all the variables in one tree, each decision tree chooses only a subset of variables at each potential split in Random Forest. Also, comparing to the classic decision tree, in Random Forest, decision trees are not necessarily pruned by penalizing the number of leaf nodes but grow all the way to the end. Random subset projection can significantly decrease the correlations between trees since different trees grow on different sets of attributes, which leads to a smaller $\rho\sigma^2$. But it could affect the second term $\frac{(1-\rho)}{B}\sigma^2$ and the unbiasedness of decision trees since they cannot predict dependent variables based on all the attributes. Thus, we need to select the number of variables to select in each split to balance the first and the second term.

The procedure of constructing a Random Forest is the following:

- Generate B number of bootstrap sample sets.
- On each sample set, grow a decision tree all the way to the end.
- During growing a tree, randomly select m variables at each potential split (random feature projection).
- Combine the B decision trees to a Random Forest. To regression, take the average output among all the trees; to classification, consider the vote of all the trees.

We can choose the hyper-parameter m based on cross-validation but this is very time-consuming when B is very large. Thus, to the classification task, m is often chosen as $1 \leq m \leq \sqrt{p}$; to regression task, we choose m as $1 \leq m \leq p/3$, where p is the number of variables. To the node size, for every decision tree, we grow it all the way to the end for the classification task, while we grow to that every leaf node has no more than $n_{min} = 5$ sample inside for regression task.

13.4.2 Variable Importance in a Random Forest

In Sect. 13.3.5, we discussed the relative importance I_j^2 to measure the importance of a variable in a decision tree. Since Random Forest is a linear combination of decision trees, we can introduce an average relative importance

$$I_j^2 = \frac{1}{B}\sum_{b=1}^{B} I_j^2(b), \tag{13.17}$$

where $I_j^2(b)$ is the relative importance for the bth decision tree

$$I_j^2(b) = \sum_{t=1}^{T_b-1} e_t^2 I(v(t)_b = j). \tag{13.18}$$

A drawback for this measure is that we need to check every node in a decision tree. This is not very efficient if there is too many samples or large numbers of the decision tree in a Random Forest.

Surprisingly, Random Forest provides a much simpler but very effective way to measure the importance of variables via **random permutation**. That is, for one variable, we perturb the samples by random permutation. For example, after constructing a Random Forest, to the jth variable along all the samples $x_j = (x_{j,1}, x_{j,2}, \ldots, x_{j,i}, \ldots, x_{j,N})$, we randomly rearrange all the x to generate a new series of samples $x_j^* = (x_{j,1}^*, x_{j,2}^*, \ldots, x_{j,i}^*, \ldots, x_{j,N}^*) = (x_{j,2}, x_{j,10}, \ldots, x_{j,N-4}, \ldots, x_{j,i+5})$, which is the original x_j with random sample order. Then, we test the Random Forest on that to get the error rate or mean square error under random permutation. Intuitively, if one variable is not important, comparing the test error on the original test sample, the test error on permutation test samples should not change a lot since this variable may not usually be selected by the nodes in a decision tree. Given a test set with N_t samples, the variable importance under random permutation is

$$\begin{aligned} VI_j &= \frac{1}{B}\sum_{b=1}^{B}\frac{1}{N_t}\sum_{i=1}^{N_t} Loss(y_i, tree_b(x_{1,i}, \ldots, x_{j,i}^*, \ldots)) - Loss(y_i, tree_b(x_{1,i}, \ldots, x_{j,i}, \ldots)) \\ &= \frac{1}{B}\sum_{b=1}^{B}\frac{1}{N_t}\sum_{i=1}^{N_t} \Delta Loss(y_i, tree_b(x_{1,i}, \ldots, x_{j,i}^*, \ldots)). \end{aligned} \tag{13.19}$$

In practice, one way to estimate the test error is sample splitting. We split one data set into a training set and a validation set and then estimate the test error on the validation set. But this is not efficient because of the loss of samples. When we discussed in Bagging in Sect. 13.2.4, in terms of Bootstrap sampling, all the Bagging methods could leave about one third sample points untouched, that are the Out-of-Bag samples. Since Random Forest is a Bagging method, we can use the OOB error as the test error. This is a very efficient way to implement since each time we add a decision tree based on a new bootstrap sample, we can test the variable importance on the new OOB samples.

Based on the OOB error, we redefine the measure of variable importance as

$$\begin{aligned} VI_j^{OOB} &= \frac{1}{B}\sum_{b=1}^{B}\frac{1}{N_b}\sum_{i=1}^{N_b}(Loss(y_{i,OOB}, tree_b(x_{1,i,OOB}, \ldots, x^*_{j,i,OOB}, \ldots)) \\ &\quad - Loss(y_{i,OOB}, tree_b(x_{1,i,OOB}, \ldots, x_{j,i,OOB}, \ldots; tree_b))) \\ &= \frac{1}{B}\sum_{b=1}^{B}\left(\widehat{err}^*_{OOB,b} - \widehat{err}_{OOB,b}\right) \\ &= \frac{1}{B}\sum_{b=1}^{B}\Delta\widehat{err}_{OOB,b}, \end{aligned} \tag{13.20}$$

where N_b is the sample size of the bth OOB sample.

The implementing procedure is the following:

- To bth bootstrap sample set, grow a decision tree.
- Find the sample point not contained in the sample set and construct the bth Out-of-Bag sample set.
- Compute OOB error for the bth decision tree based on the OOB sample with and without random permutation.
- Calculate VI_j^{OOB} to measure the jth variable importance.

One related topic is about the variable selection in Random Forest. Based on the variable importance, we can compare the importance between two variables. Thus, could we select relevant variables based on this measure? A simple way to implement is designating a threshold value for variable importance and select the variables with high importance only. But there is no theory about how to decide the threshold value such that we can select relevant variables correctly. Recently, Strobl, Boulesteix, Kneib, Augustin, and Zeileis (2008) and Janitza, Celik, and Boulesteix (2016) considered the hypothesis testing to select variables in Random Forest.

The last but not the least, the issue of variable dependence need to be considered when we measure the variable importance via random permutation. For example, if we implement a linear model by regressing the level of health on weight and height, the coefficient on the weight could be very unstable if weight and height are highly correlated. Similarly, in random forest, if two variables are correlated, we cannot get an accurate measure of importance via random permuting on the variable. Strobl, Boulesteix, Zeileis, and Hothorn (2007) discussed the topics about the bias in random forest variable importance.

To resolve this issue, we need to check the dependence among all the variables. Some methods like PCA could be introduced to decorrelate the variables, but they may affect the interpretations of the variables. Strobl et al. (2008) proposed a method called the **conditional variable importance**.

The implementing procedure is the following:

- Given variable x_j, find a group of variables $Z = \{z_1, z_2, \ldots, \}$ that are correlated with x_j.
- To the bth decision tree, find out all the internal nodes containing the variables in Z.
- Extract the cutpoints from the nodes and create a grid by means of bisecting the sample space in each cutpoint.
- In this grid, permute the x_j to compute the OOB accuracy. The OOB error of the bth tree is the difference between OOB accuracy with and without permutation given Z.
- Consider the average of all the trees' OOB error as the forest's OOB error.

13.4.3 *Random Forest as the Adaptive Kernel Functions*

Now we start to discuss some related theories behind Random Forest to uncover why Random Forest works. Basically, Random Forest or decision tree ensemble methods can be seen as a local method. For example, it is easy to find that the predicted value of a given data totally depends on the average of y_i in one of the leaf node. In other words, the predicted value only depends on "neighborhood" samples belong to the leaf node. Similarly, Breiman (2000) showed that Random Forest which is grown using i.i.d random vectors in the tree construction are equivalent to a kernel acting on the true margin.

Without loss of generality, let us consider a Random Forest with B decision trees for a binary classification task. To one decision tree, suppose R as the area of one of leaf node with the responses as $R = +1$ or $R = -1$. We have the labeling rule for $R = +1$ to this leaf node

$$\int_R P(+1|z)P(dz) \geq \int_R P(-1|z)P(dz), \tag{13.21}$$

where z represents all the possible inputs included in the leaf node. Intuitively, by considering all the samples in R, if more samples with the label as $+1$, the response of R is $+1$. Otherwise, we label the response of R as -1.

Based on Eq. (13.21), we have the output $+1$ from a decision tree given an input x when Eq. (13.22) holds

$$\int_{R_x(\theta)} P(1|z)P(dz) \geq \int_{R_x(\theta)} P(-1|z)P(dz), \tag{13.22}$$

where $R_x(\theta)$ is the area of the leaf node containing x and θ is the parameter of the decision tree. Let $D(z) = P(1|z) - P(-1|z)$, then Eq. (13.22) can be written as

$$\int_{R_x(\theta)} D(z)P(dz) \geq 0. \tag{13.23}$$

According to Eq. (13.23), the prediction of the bth decision tree is

$$\widehat{y} = \begin{cases} 1 & if \int_{R_x(\theta_b)} D(z)P(dz) \geq 0 \\ -1 & if \int_{R_x(\theta_b)} D(z)P(dz) \leq 0. \end{cases}$$

Now let us introduce an indicator function $I(x, z \in R(\theta_b))$ to represent the event $z \in R_x(\theta_b)$, we have

$$\widehat{y} = \begin{cases} 1 & if \int I(x, z \in R(\theta_b))D(z)P(dz) \geq 0 \\ -1 & if \int I(x, z \in R(\theta_b))D(z)P(dz) \leq 0. \end{cases}$$

Obviously, the indicator function $I(x, z \in R(\theta_b))$ can be seen as a kernel weighted function $K(x, z)$. Also, this kernel function is not smooth since it only considers the sample in the leaf node $R(\theta_b)$. Intuitively, it means that one decision tree can learn to construct a distribution plot and then works via the "hard" kernel weighting.

Let us consider a Random Forest. Compare to a single decision tree, Random Forest contains B decision trees. Assume in bth decision tree, the number of leaf nodes is T_b. Thus, we can derive a kernel function for Random Forest

$$K_{RF}(x, z) = \frac{1}{B} \sum_{b=1}^{B} \sum_{t=1}^{T_b} I(x, z \in R^t(\theta_b)). \tag{13.24}$$

This is a discrete kernel combining all the leaf nodes from B decision trees. Additionally, when $B \to \infty$, we have

$$\begin{aligned} K_{RF}(x, z) &= \frac{1}{B} \sum_{b=1}^{B} \sum_{t=1}^{T_b} I(x, z \in R^t(\theta_b)) \\ &\to P_\theta(x, z \in A(\theta)), \end{aligned} \tag{13.25}$$

where $A(\theta)$ is the area based on the Random Forest and it contains infinite number of leaf nodes from infinite decision trees. When $B \to \infty$, we can see that the kernel function will converge to a probability measure. That is, the hard kernel function will be a smoother kernel function when we have increasingly number of decision

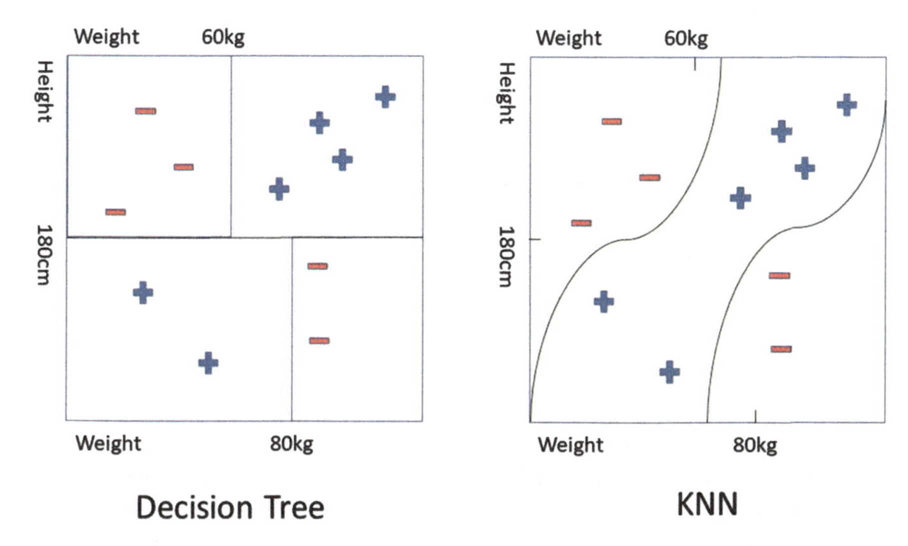

Fig. 13.11 2D plot of decision tree and KNN

trees. Thus, the final output for Random Forest in this case should be

$$\widehat{y}_{RF} = \begin{cases} 1 & if \int K_{RF}(x,z)D(z)P(dz) \geq 0 \\ -1 & if \int K_{RF}(x,z)D(z)P(dz) \leq 0. \end{cases}$$

From another perspective, Lin and Jeon (2006) discussed Random Forest from a point of view of K-nearest neighbor (KNN). To show the connection between Random Forest and KNN, they proposed a new method called potential nearest neighbor (PNN). They also showed that Random Forest could be converted to an adaptive kernel smooth method described by PNN.

To sum up, Random Forest not only combines a large number of decision trees to reduce the variance of prediction like bagging, but also decreases the dependence among decision trees via random feature projection to get a much lower prediction error than Bagging Decision Tree. Theoretically, Random Forest makes prediction via constructing an adaptive kernel function. That is very similar to other local methods such as nonparametric kernel method and KNN. Figure 13.11 illustrates the difference between Decision Tree and KNN.

13.5 Recent Developments of Random Forest

As one of the most effective ensemble method in solving real-world issues, the random forest also has many variants for different modeling tasks in statistics, data mining, and econometrics literature. In this section, we introduce several attractive variants of Random Forest.

13.5.1 Extremely Randomized Trees

For Bagging method, we discussed its effectiveness related to the variance of ensemble model. According to Eq. (13.16)

$$Var(g) = \rho\sigma^2 + \frac{(1-\rho)}{B}\sigma^2,$$

the variance is decomposed into two parts: the first term $\rho\sigma^2$ depends on the correlation among base models and the second term $\frac{(1-\rho)}{B}\sigma^2$ is related to the number of base models.

Since we often combine a large number of base models, we can assume B goes to infinity and the main part of the variance converges to $\rho\sigma^2$. Thus, bagging can largely decrease the second term.

Random Forest, besides controlling the second term via Bagging, also controls the first term by decreasing the ρ via random feature projection simultaneously. Because of the random feature projection, the decision tree suffers from a higher bias. It means that we need to focus on decreasing correlations among decision trees such that the ensemble model becomes more effective.

Random feature projection is not the only way to decrease the correlations. Geurts, Ernst, and Wehenkel (2006) introduced another way to achieve the goal and derived a new technique called the Extremely Randomized Trees (Extra-Trees). Compare to Random Forest, Extra-Trees works on the original samples instead of bootstrap samples. More importantly, Extra-Trees method generates the base decision tree via a more random way to split sample space than the random feature projection in Random Forest.

The Extra-Trees splitting algorithm is the following:

- To a node space in decision tree, first choose K variables $(x_1, x_2, \ldots, x_K)$ among all the p variables.
- To all K attributes, randomly choose a splitting point to each one of them via choosing a uniform number from (x_{min}, x_{max}) belong to this node.
- Compare the criteria among all the random splitting point and choose the attribute x_k giving the best splitting outcome.
- Choose variable x_k and the random splitting point as the final decision stump in this node.
- Stop splitting when the number of sample points $= n_{min}$.

Practically, we set $K = \sqrt{p}$ and $n_{min} = 5$ by default. But we can tune them based on the cross-validation.

Table 13.1 A summary of three ensemble methods

Names	Main part of variance	Bootstrap	Hyper-parameters
Bagging decision trees	$\rho\sigma^2$	Yes	B, n_{min}
Random forest	$\rho\sigma^2$	Yes	B, m, n_{min}
Extremely randomized trees	$\frac{(1-\rho)}{B}\sigma^2$	No	B, K, n_{min}

The key difference of constructing base decision trees between Random Forest and Extra-Trees is the splitting rule for each node. In Random Forest, we choose m variables and then find the optimal decision stump directly. But in Extra-Trees, we choose K variables to randomly generate decision stump and then choose the "optimal" decision stump. As a consequence, randomly growing decision trees in extra-trees will be less dependent than the trees in Random Forest, which leads to lower correlations ρ. Thus, even though Extra-Trees do not introduce Bootstrap, it works well in many data mining and predicting tasks. This idea about being "random" is also used in many other machine learning algorithms such as extreme learning machine proposed by Huang, Zhu, and Siew (2006).

We summarize Bagging Decision Trees, Random Forest, and Extremely Randomized Trees in Table 13.1.

13.5.2 Soft Decision Tree and Forest

Based on the previous discussion, we find that Random Forest and its variants are based on the decision tree. The decision tree is growing via splitting the space into optimal sub-spaces recursively and the function defined by a decision tree is a non-smooth step function. The decision tree is naturally suitable for implementing the classification problem because of the discrete outputs. Since most economic problems are related to the regression problems, we could expect that the decision tree should be so large that it can handle a smooth function "non-smoothly."

To resolve this problem, we can consider a "soft" decision tree instead of the "hard" decision tree. Given a decision tree with only one root node and two leaf nodes, it can have two possible outcomes

$$f(x) = \begin{cases} \mu_1 & if\ g(x) > 0 \\ \mu_2 & if\ g(x) < 0, \end{cases}$$

where μ_1 and μ_2 are correspond to the first and second leaf nodes. $g(x)$ is called gate function. It decides which leaf node should be selected. We can also rewrite the formula based on an indicator function

$$f(x) = \mu_1 \times I(g(x) > 0) + \mu_2 \times (1 - I(g(x) > 0)) .$$

For example, in women wage case we have discussed, the decision stump is D : $Expr = 10$. Given that, we can use a gate function $g(Expr) = Expr - 10$ to represent the decision stump

$$f(Expr) = \mu_1 \times I(g(Expr) > 0) + \mu_2 \times (1 - I(g(Expr) > 0)) . \quad (13.26)$$

That is, if $Expr > 10$, we choose the first leaf node and $Expr < 10$ choose the second one.

Generally, to the mth node, we can use a similar function to represent its output

$$F_m(x) = F_m^L(x) \times I(g_m(x) > 0) + F_m^R(x) \times (1 - I(g_m(x) > 0)). \quad (13.27)$$

If $F_m^L(x)$ and $F_m^R(x)$ are leaf nodes, we have $F_m^L(x) = \mu_L$ and $F_m^R(x) = \mu_R$. If not, they are corresponding to the child nodes in the next layer $F_m^L(x) = F_{m+1}^L(x)$. Because of the indicator function, the $F_m(x)$ is a step function with two outcomes, $F_m^L(x)$ or $F_m^R(x)$. It is a hard decision tree.

In Eq. (13.27), we can use a smooth gate function instead of the identity function such that the decision tree is "soft" and $F_m(x)$ is a smooth function. Let us change the indicator function $I(h)$ to a logistic function $L(h)$, we have

$$F_m(x) = F_m^L(x) \times L(g_m(x)) + F_m^R(x) \times (1 - L(g_m(x))), \quad (13.28)$$

where $L(h) = \frac{1}{1+e^{-h}}$ is a logistic function and $g_m(x) = \beta^T x$ is a linear single index function of input variables. In the soft decision tree, instead of selecting one from two child nodes, a smooth $F_m(x)$ is taking weighed average between $F_m^L(x)$ and $F_m^R(x)$. In Fig. 13.12, we compare the hard decision tree with the soft decision tree.

Back to the women's wage example, we choose $L(g(Expr)) = \frac{1}{1+e^{-(Expr-10)}}$. That is, if $Expr > 10$, we consider the left node more and consider the right node more when $Expr < 10$.

Compared to the hard decision tree, the soft decision tree has many advantages:

- Since soft decision tree can represent any smooth function, it is more suitable to handle the regression problem than the original decision tree. That may be the most important advantage since economic research often cares more about the regression problem, such as economic growth rate prediction and derivative estimation for partial effect analysis.
- Soft decision tree contains a bunch of differentiable gate functions, which means we can train all the parameters via the Expectation-Maximization (EM) method very quickly.

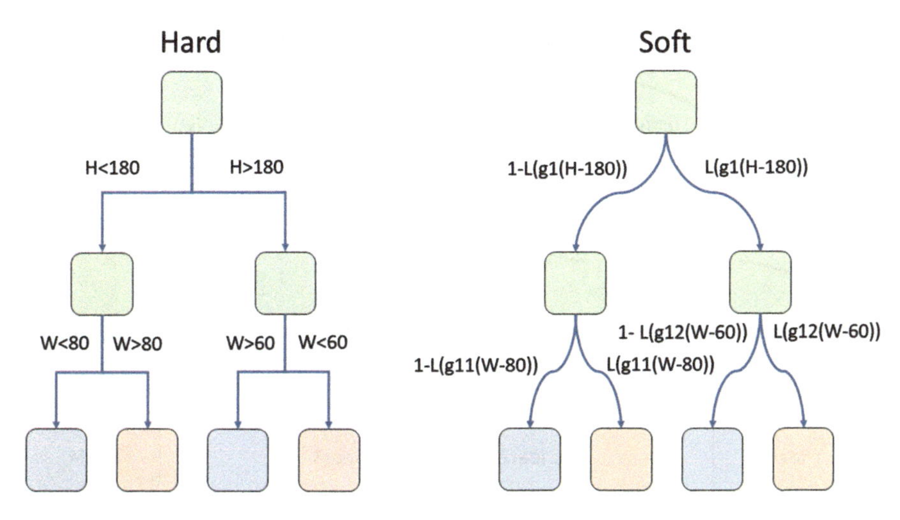

Fig. 13.12 Hard decision tree and soft decision tree

- In all the leaf nodes of a soft decision tree, we could not only choose a constant μ, but consider more flexible methods, like the linear formula or even the neural networks.
- Because of its hierarchical structure, the soft decision tree is a local method as the hard decision tree. Thus, it has similar theories and properties as other local methods like kernel regression.

There are many research papers related to the soft version of the decision tree. This first soft decision tree model is called hierarchical mixtures of experts (HME) discussed by Jordan and Jacob (1994). Instead of growing a decision tree via splitting recursively, in the HME method, we first designate the structure of a soft decision tree, like the number of layers, then optimize all the parameters in this tree.

Consider a soft decision tree with S layers and one split in each node. Thus, the number of total leaf nodes is 2^S and the function of this soft decision tree is

$$
\begin{aligned}
f_{HME}(x) &= \sum_{leaf=1}^{2^S} P_{leaf}(x)\mu_{leaf} \\
&= \sum_{leaf=1}^{2^S} \prod_{p \to leaf} G_p(x)\mu_{leaf},
\end{aligned}
$$

where $p \to leaf$ means all the gate functions contained in the nodes located on the path to the sth leaf node. According to Eq. (13.28), we have

$$
G_p(x) = I(p = left) \times L\left(g_p(x)\right) + I(p = right) \times \left(1 - L\left(g_p(x)\right)\right).
$$

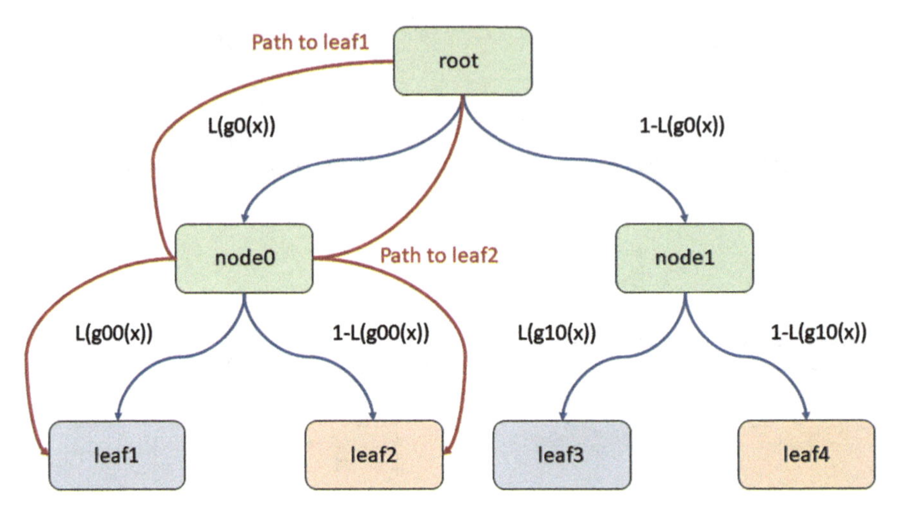

Fig. 13.13 Hierarchical mixtures of experts

It decides the gate function for each node on the path. μ_{leaf} represents the function in each leaf, which could be a constant, a simple linear function, or other nonlinear models.

For example, Fig. 13.13 shows the structure of an HME with two layers. Let us consider the path to the first leaf node $p \to 1$. The path starts from the root node in layer 0. Since the path chooses the left node, the node0, $I(p = left) = 1$ and $G_0(x)$ should be

$$\begin{aligned} G_0(x) &= I(0 = left) \times L(g_0(x)) + (1 - I(0 = left)) \times (1 - L(g_0(x))) \\ &= L(g_0(x)). \end{aligned}$$

Then, the path contains the node1 at layer 1 and then choose the left node, the leaf1. Thus, $G_1(x)$ should be

$$\begin{aligned} G_1(x) &= I(1 = left) \times L(g_{00}(x)) + (1 - I(1 = left)) \times (1 - L(g_{00}(x))) \\ &= L(g_{00}(x)). \end{aligned}$$

Thus, to $P_{leaf=1}(x)$, we have

$$\begin{aligned} P_1(x) &= \prod_{p \to 1} G_p(x) = G_0(x) \times G_1(x) \\ &= L(g_0(x)) \times L(g_{00}(x)). \end{aligned} \tag{13.29}$$

Similarly, to the path to leaf 2, we have

$$\begin{aligned} P_2(x) &= \prod_{p \to 2} G_p(x) = G_0(x) \times G_1(x) \\ &= L(g_0(x)) \times (1 - L(g_{00}(x))). \end{aligned} \tag{13.30}$$

Now we find that HME is similar to a mixture model since $\sum_{leaf} P_{leaf}(x) = 1$. Suppose the μ_{leaf} is a parameter, like mean, of a distribution $P_{leaf}(y|x)$. Then we have the conditional probability of y given x

$$\begin{aligned} P(y|x) &= \sum_{leaf=1}^{2^S} \prod_{p \to leaf} G_p(x) P_{leaf}(y|x) \\ &= \sum_{leaf=1}^{2^S} P_{leaf}(x) P_{leaf}(y|x). \end{aligned}$$

Thus, we can have the log-likelihood function of HME with unknown parameter β

$$\begin{aligned} L(y|x;\beta) &= \sum_{i=1}^{N} \log P(y_i|x_i;\beta) \\ &= \sum_{i=1}^{N} \log \sum_{leaf=1}^{2^S} P_{leaf}(x_i;\beta) P_{leaf}(y_i|x_i;\beta). \end{aligned}$$

To optimize the likelihood function, Jordan and Jacob (1994) considered the Expectation-Maximization (EM) to optimize it. The main idea behind EM is based on the so-called complete log-likelihood function

$$L_c(y|x;\beta) = \sum_{i=1}^{N} \sum_{leaf=1}^{2^S} z_{leaf} \prod_{p \to leaf} G_p(x_i;\beta) P_{leaf}(y_i|x_i;\beta),$$

where z_{leaf} are implicit variables that represent the indicators of leaf nodes. Take the expectation of $L_c(y|x;\beta)$, we have

$$Q(y|x;\beta) = E_z(L_c(y|x;\beta)) = \sum_{i=1}^{N} \sum_{leaf=1}^{2^S} E(z_{leaf}) \prod_{p \to leaf} G_p(x_i;\beta) P_{leaf}(y_i|x_i;\beta).$$

To $E(z_{leaf})$, we have

$$\begin{aligned} E(z_{leaf}) &= P(z_{leaf} = 1|y, x, \beta) \\ &= \frac{P(y|z_{leaf} = 1, x, \beta)P(z_{leaf} = 1|x, \beta)}{P(y|x, \beta)} \\ &= \frac{\prod_{p\to leaf} g_p(x; \beta)P(y|x, \beta)}{\sum_{leaf=1}^{S^2} \prod_{\to leaf} g_p(x; \beta)P(y|x, \beta)}. \end{aligned}$$

We can see that $Q(y|x; \beta)$ is the lower bound of $L(y|x; \beta)$ because of Jensen's inequality. The log-likelihood function $L(y|x)$ is optimized if we can optimize the lower bound $Q(y|x; \beta)$. This is the key to the EM method.

To sum up, the training procedure for HME is as follows:

- Randomly initializes all the parameters β, then propagate forward the input to get the distribution of x_i.
- To each mini-batch, propagate forward all the x to the leaves to get the predicted outputs. Then compute all the $E(z_{i,leaf})$ (E-step).
- Optimize the expectation likelihood function $Q(x, y; \beta)$ (M-step).
- Redo E-step and M-Step until that all the parameters converge.

One possible drawback to the soft decision tree method is that the HME could lead to a long-time training process. More importantly, since HME needs a predetermined structure of a soft decision tree, it is not adaptive to data. To resolve this issue, another way to implement soft decision trees was discussed by Irsoy, Yildiz, and Alpaydin (2012). The authors introduced a new way to grow a soft decision tree. In each node, they used gradient descent to find the optimal splitting line then compare the predicting outcome between the two trees with and without the new splitting line to decide that this new node should be added or not. Thus, this method can adaptively learn the structure of soft decision tree and could be faster. Similarly to Random Forest, Yildiiz, Írsoy, and Alpaydin (2016) constructed an ensemble of soft decision trees via Bagging to explore the ensemble of soft decision trees.

Basically, the soft decision tree method is not so popular as the decision tree since the training process is slower than growing a decision tree. But in many recent research papers, the soft decision tree shows the power for learning hierarchical features adaptively. Kontschieder et al. (2015) proposed deep neural decision forest that combines the convolutional neural network (CNN) feature extractor and tree structure into one differential hierarchical CNN and implements it into the tasks of computer vision. Frosst and Hinton (2017) explored the soft decision tree in distilling the knowledge or features extracted by a neural network based on its hierarchical structure. They found that the soft decision tree method can definitely learn hierarchical features via training.

13.6 Applications of Bagging and Random Forest in Economics

13.6.1 Bagging in Economics

Recently, Bootstrap Aggregating is widely used in macroeconomic analysis and forecasting. Panagiotelis, Athanasopoulos, Hyndman, Jiang, and Vahid (2019) explored the performance of the ensemble a large number of predictors in predicting macroeconomic series data in Australia. Precisely, they compared Bagging LARS with Dynamic Factor Model, Ridge Regression, LARS, and Bayesian VAR, respectively, on GDP growth, CPI inflation, and IBR (the interbank overnight cash rate equivalent to the Federal funds rate in the USA). They found that Bagging method can help in more accurate forecasting.

As discussed in this chapter, Bagging has been proved to be effective to improve on unstable forecast. Theoretical and empirical works using classification, regression trees, variable selection in linear and nonlinear regression have shown that bagging can generate substantial prediction gain. However, most of the existing literature on bagging has been limited to the cross-sectional circumstances with symmetric cost functions. Lee and Yang (2006) extend the application of bagging to time series settings with asymmetric cost functions, particularly for predicting signs and quantiles. They use quantile predictions to construct a binary predictor and the majority-voted bagging binary prediction and show that bagging may improve the binary prediction. For empirical application, they presented results using monthly S&P500 and NASDAQ stock index returns.

Inoue and Kilian (2008) considered the Bagging method in forecasting economic time series of US CPI data. They explored how the Bagging may be adapted to application involving dynamic linear multiple regression for the inflation forecasting. And then they compare several models' performances, including correlated regressor models, factor models, and shrinkage estimation of regressor models (with LASSO) with or without Bagging. Their empirical evidence showed that Bagging can achieve large reductions in prediction mean squared error, even in challenging applications such as inflation forecasting.

Lee, Tu, and Ullah (2014, 2015) and Hillebrand, Lee, and Medeiros (2014) consider parametric, nonparametric, and semiparametric predictive regression models for financial returns subject to various hard-thresholding constraints using indicator functions. The purpose is to incorporate various economic constraints that are implied from economic theory or common priors such as monotonicity or positivity of the regression functions. They use bagging to smooth the hard-thresholding constraints to reduce the variance of the estimators. They show the usefulness of bagging when such economic constraints are imposed in estimation and forecasting, by deriving asymptotic properties of the bagging constrained estimators and forecasts. The advantages of the bagging constrained estimators and forecasts are also demonstrated by extensive Monte Carlo simulations. Applications to predicting financial equity premium are taken for empirical illustrations, which show imposing constraints and bagging can mitigate the chance of making large size forecast errors and bagging can make these constrained forecasts even more robust.

Jin, Su, and Ullah (2014) propose a revised version of bagging as a forecast combination method for the out-of-sample forecasts in time series models. The revised version explicitly takes into account the dependence in time series data and can be used to justify the validity of bagging in the reduction of mean squared forecast error when compared with the unbagged forecasts. Their Monte Carlo simulations show that their method works quite well and outperforms the traditional one-step-ahead linear forecast as well as the nonparametric forecast in general, especially when the in-sample estimation period is small. They also find that the bagging forecasts based on misspecified linear models may work as effectively as those based on nonparametric models, suggesting the robustification property of bagging method in terms of out-of-sample forecasts. They then re-examine forecasting powers of predictive variables suggested in the literature to forecast the excess returns or equity premium and find that, consistent with Welch and Goyal (2008), the historical average excess stock return forecasts may beat other predictor variables in the literature when they apply traditional one-step linear forecast and the nonparametric forecasting methods. However, when using the bagging method or the revised version, which help to improve the mean squared forecast error for unstable predictors, the predictive variables have a better forecasting power than the historical average excess stock return forecasts.

Audrino and Medeiros (2011) proposed a new method called smooth transition tree. They found that the leading indicators for inflation and real activity are the most relevant predictors in characterizing the multiple regimes' structure. They also provided empirical evidence of the model in forecasting the first two conditional moments when it is used in connection with Bagging.

Hirano and Wright (2017) considered forecasting with uncertainty about the choice of predictor variables and compare the performances of model selection methods under Rao–Blackwell theorem and Bagging, respectively. They investigated the distributional properties of a number of different schemes for model choice and parameter estimation: in-sample model selection using the Akaike Information Criterion, out-of-sample model selection, and splitting the data into sub-samples for model selection and parameter estimation. They examined how Bagging affected the local asymptotic risk of the estimators and their associated forecasts. In their numerical study, they found that for many values of the local parameter, the out-of-sample and split-sample schemes performed poorly if implemented in a conventional way. But they performed well if implemented in conjunction with model selection methods under Rao–Blackwell theorem or Bagging.

13.6.2 Random Forest in Economics

To introduce Random Forest into economic research, many economic and statistic researchers studied in extending the theory of random forest not only for forecasting but for inference.

In the literature of economic inference, Strobl et al. (2008) discussed the consistency of Random Forest in the context of additive regression models, which sheds light on the forest-based statistical inference. Wager and Athey (2018) studied in the application of random forest in economic research. They proposed the Causal Forest, an unbiased random forest method for estimating and testing the heterogeneous treatment effect. They first showed that classic Random Forest cannot have unbiasedness because of Bagging. Then, they proposed the Causal Forest which combines a bunch of unbiased Honest Tree based on Sub-sampling aggregating. They also showed that Causal Forest is unbiased and has asymptotic normality under some assumptions. Finally, they discussed the importance and advantage of Causal Forest in applications to economic causal inference.

To the application of economic forecasting, Hothorn and Zeileis (2017) discussed a new Random Forest method, the Transformation Forest. Based on a parametric family of distributions characterized by their transformation function, they proposed a dedicated novel transformation tree and transformation forest as an adaptive local likelihood estimator of conditional distribution functions, which are available for inference procedures. In macroeconomic forecasting, Random Forest is applied in Euro area GDP forecasting (Biau and D'Elia, 2011) and financial volatility forecasting (Luong and Dokuchaev, 2018). Finally, Fischer, Krauss, and Treichel (2018) assess and compare the time series forecasting performance of several machine learning algorithms such as Gradient Boosting Decision Trees, Neural Networks, Logistic Regression, Random Forest, and so on in a simulation study. Nyman and Ormerod (2016) explore the potential of Random Forest for forecasting the economic recession on the quarterly data over 1970Q2 to 1990Q2.

13.7 Summary

In this chapter, we discuss the Bagging method and Random Forest. At first, we begin with introducing Bagging and its variants, the Subagging and Bragging. Next, we introduce Decision Tree, which provides the foundation of the Random Forest. Also, we introduce the related theories about Random Forest and its important variants like Extreme Random Trees and Soft Decision Tree. At last, we discussed many applications of Bagging and Random Forest in macroeconomic forecasting and economic causal inference.

References

Audrino, F., & Medeiros, M. C. (2011). Modeling and forecasting short-term interest rates: The benefits of smooth regimes, macroeconomic variables, and Bagging. *Journal of Applied Econometrics, 26*(6), 999–1022.

Biau, O., & D'Elia, A. (2011). Euro area GDP forecast using large survey dataset - A random forest approach. In *EcoMod 2010*.

Breiman, L. (1996). Bagging predictors. *Machine Learning, 26*(2), 123–140.

Breiman, L. (2000). *Some infinity theory for predictor ensembles*. Berkeley: University of California.

Breiman, L. (2001). Random forests. *Machine Learning, 45*, 5–32.

Breiman, L., Friedman, J., Stone, C., & Olshen, R. (1984). *Classification and regression trees. The Wadsworth and Brooks-Cole Statistics-Probability Series*. Oxfordshire: Taylor & Francis.

Bühlmann, P. (2004). *Bagging, boosting and ensemble methods* (pp. 877–907). *Handbook of Computational Statistics: Concepts and Methods*. Berlin: Springer.

Bühlmann, P., & Yu, B. (2002). Analyzing bagging. *Annals of Statistics, 30*(4), 927–961.

Buja, A., & Stuetzle, W. (2000a), *Bagging does not always decrease mean squared error definitions* (Preprint). Florham Park: AT&T Labs-Research.

Buja, A., & Stuetzle, W. (2000b). *Smoothing effects of bagging* (Preprint). Florham Park: AT&T Labs-Research.

Fischer, T., Krauss, C., & Treichel, A. (2018). *Machine learning for time series forecasting - a simulation study (2018)*. FAU Discussion Papers in Economics, Friedrich-Alexander University Erlangen-Nuremberg, Institute for Economics.

Friedman, J. H., & Hall, P. (2007). On Bagging and nonlinear estimation. *Journal of Statistical Planning and Inference, 137*(3), 669–683.

Frosst, N., & Hinton, G. (2017). Distilling a neural network into a soft decision tree. In *Ceur workshop proceedings*.

Geurts, P., Ernst, D., & Wehenkel, L. (2006). Extremely randomized trees. *Machine Learning, 63*(1), 3–42.

Hillebrand, E., Lee, T.-H., & Medeiros, M. (2014). Bagging constrained equity premium predictors (Chap. 14, pp. 330–356). In *Essays in Nonlinear Time Series Econometrics, Festschrift in Honor of Timo Teräsvirta*. Oxford: Oxford University Press.

Hirano, K., & Wright, J. H. (2017). Forecasting with model uncertainty: Representations and risk reduction. *Econometrica, 85*(2), 617–643.

Hothorn, T., & Zeileis, A. (2017). *Transformation forests*. Technical report. https://arxiv.org/abs/1701.02110.

Huang, G.-B., Zhu, Q.-Y., & Siew, C.-K. (2006). Extreme learning machine: Algorithm, theory and applications. *Neurocomputing, 70*, 489–501.

Inoue, A., & Kilian, L. (2008). How useful is Bagging in forecasting economic time. *Journal of the American Statistical Association, 103*(482), 511–522.

Irsoy, O., Yildiz, O. T., & Alpaydin, E. (2012). A soft decision tree. In *21st International Conference on Pattern Recognition (ICPR 2012)*.

Janitza, S., Celik, E., & Boulesteix, A. L. (2016). A computationally fast variable importance test for Random Forests for high-dimensional data. *Advances in Data Analysis and Classification, 185*, 1–31.

Jin, S., Su, L., & Ullah, A. (2014). Robustify financial time series forecasting with Bagging. *Econometric Reviews, 33*(5-6), 575–605.

Jordan, M., & Jacob, R. (1994). Hierarchical Mixtures of Experts and the EM algorithm. *Neural Computation, 6*, 181–214.

Kontschieder, P., Fiterau, M., Criminisi, A., Bul, S. R., Kessler, F. B., & Bulo', S. R. (2015). Deep Neural Decision Forests. In *The IEEE International Conference on Computer Vision (ICCV)* (pp. 1467–1475).

Lee, T.-H., & Yang, Y. (2006). Bagging binary and quantile predictors for time series. *Journal of Econometrics, 135*(1), 465–497.
Lee, T.-H., Tu, Y., & Ullah, A. (2014). Nonparametric and semiparametric regressions subject to monotonicity constraints: estimation and forecasting. *Journal of Econometrics, 182*(1), 196–210.
Lee, T.-H., Tu, Y., & Ullah, A. (2015). Forecasting equity premium: Global historical average versus local historical average and constraints. *Journal of Business and Economic Statistics, 33*(3), 393–402.
Lin, Y., & Jeon, Y. (2006). Random forests and adaptive nearest neighbors. *Journal of the American Statistical Association, 101*(474), 578–590.
Luong, C., & Dokuchaev, N. (2018). Forecasting of realised volatility with the random forests algorithm. *Journal of Risk and Financial Management, 11*(4), 61.
Nyman, R., & Ormerod, P. (2016). *Predicting economic recessions using machine learning*. arXiv:1701.01428.
Panagiotelis, A., Athanasopoulos, G., Hyndman, R. J., Jiang, B., & Vahid, F. (2019). Macroeconomic forecasting for Australia using a large number of predictors. *International Journal of Forecasting, 35(2)*, 616–633.
Quinlan, J. (1986). Induction of decision trees. *Machine Learning, 1*, 81–106.
Quinlan, J. R. (1994). C4.5: programs for machine learning. *Machine Learning, 16*(3), 235–240.
Strobl, C., Boulesteix, A.-L., Zeileis, A., & Hothorn, T. (2007). Bias in random forest variable importance measures: Illustrations, sources and a solution. *BMC Bioinformatics, 8*, 25.
Strobl, C., Boulesteix, A. L., Kneib, T., Augustin, T., & Zeileis, A. (2008). Conditional variable importance for Random Forests. *BMC Bioinformatics, 9*, 1–11.
Wager, S., & Athey, S. (2018). Estimation and inference of heterogeneous treatment effects using Random Forests. *Journal of the American Statistical Association, 113*(523), 1228–1242.
Welch, I., & Goyal, A. (2008). A comprehensive look at the empirical performance of equity premium prediction. *Review of Financial Studies, 21-4* 1455–1508.
Yildiiz, O. T., Írsoy, O., & Alpaydin, E. (2016). Bagging soft decision trees. In *Machine Learning for Health Informatics* (Vol. 9605, pp. 25–36).

Chapter 14
Boosting

Jianghao Chu, Tae-Hwy Lee, Aman Ullah, and Ran Wang

14.1 Introduction

The term *Boosting* originates from the so-called *hypothesis boosting problem* in the *distribution-free* or *probably approximately correct* model of learning. In this model, the learner produces a classifier based on random samples from an unknown data generating process. Samples are chosen according to a fixed but unknown and arbitrary distribution on the population. The learner's task is to find a classifier that correctly classifies new samples from the data generating process as positive or negative examples. A weak learner produces classifiers that perform only slightly better than random guessing. A strong learner, on the other hand, produces classifiers that can achieve arbitrarily high accuracy given enough samples from the data generating process.

In a seminal paper, Schapire (1990) addresses the problem of improving the accuracy of a class of classifiers that perform only slightly better than random guessing. The paper shows the existence of a weak learner implies the existence of a strong learner and vice versa. A boosting algorithm is then proposed to convert a weak learner into a strong learner. The algorithm uses *filtering* to modify the distribution of samples in such a way as to force the weak learning algorithm to focus on the harder-to-learn parts of the distribution.

Not long after the relation between weak learners and strong learners is revealed, Freund and Schapire (1997) propose the Adaptive Boost (AdaBoost) for binary classification. AdaBoost performs incredibly well in practice and stimulates the invention of boosting algorithms for multi-class classifications. On the other hand,

J. Chu · T.-H. Lee (✉) · A. Ullah · R. Wang
Department of Economics, University of California, Riverside, CA, USA
e-mail: jianghao.chu@email.ucr.edu; tae.lee@ucr.edu; aman.ullah@ucr.edu; ran.wang@email.ucr.edu

P. Fuleky (ed.), *Macroeconomic Forecasting in the Era of Big Data*, Advanced Studies in Theoretical and Applied Econometrics 52,
https://doi.org/10.1007/978-3-030-31150-6_14

researchers try to explain the success of AdaBoost in a more theoretical way, e.g., Friedman, Hastie, and Tibshirani (2000), Bartlett, Jordan, and McAuliffe (2006), and Bartlett and Traskin (2007). Further understanding of the theory behind the success of boosting algorithms in turn triggers a bloom of Boosting algorithm with better statistical properties, e.g., Friedman (2001), Bühlmann (2003), and Mease, Wyner, and Buja (2007).

Boosting is undoubtedly the most popular machine learning algorithm in the online data science platform such as Kaggle. It is efficient and easy to implement. There are numerous packages in Python and R which implement Boosting algorithms in one way or another, e.g., *XBoost*. In the following sections, we will introduce the AdaBoost as well as other Boosting algorithms in detail together with examples to help the readers better understand the algorithms and statistical properties of the Boosting methods.

This chapter is organized as follows. Section 14.1 provides an overview on the origination and development of Boosting. Sections 14.2 and 14.3 are an introduction of AdaBoost which is the first practically feasible Boosting algorithm with its variants. Section 14.4 introduces a Boosting algorithm for linear regressions, namely L_2Boosting. Section 14.5 gives a generalization of the above mentioned algorithms which is called Gradient Boosting Machine. Section 14.6 gives more variants of Boosting, e.g., Boosting for nonlinear models. Section 14.7 provides applications of the Boosting algorithms in macroeconomic studies. In Sect. 14.8 we summarize.

14.2 AdaBoost

The first widely used Boosting algorithm is AdaBoost which solves binary classification problems with great success. A large number of important variables in economics are binary. For example, whether the economy is going into expansion or recession, whether an individual is participating in the labor force, whether a bond is going to default, and etc. Let

$$\pi(\mathbf{x}) \equiv \Pr(y = 1|\mathbf{x})$$

and y takes value 1 with probability $\pi(\mathbf{x})$ and -1 with probability $1 - \pi(\mathbf{x})$. The goal of the researchers is often to predict the unknown value of y given known information on $\mathbf{x}$.

14.2.1 AdaBoost Algorithm

This section introduces the AdaBoost algorithm of Freund and Schapire (1997). The algorithm of AdaBoost is shown in Algorithm 1.

Algorithm 1 Discrete AdaBoost (DAB, Freund & Schapire, 1997)

1. Start with weights $w_i = \frac{1}{n}, i = 1, \ldots, n$.
2. For $m = 1$ to M
   a. For $j = 1$ to k (for each variable)
      i. Fit the classifier $f_{mj}(x_{ij}) \in \{-1, 1\}$ using weights w_i on the training data.
      ii. Compute $err_{mj} = \sum_{i=1}^{n} w_i 1_{(y_i \neq f_{mj}(x_{ji}))}$.
   b. Find $\hat{j}_m = \arg\min_j err_{mj}$
   c. Compute $c_m = \log\left(\frac{1-err_{m,\hat{j}_m}}{err_{m,\hat{j}_m}}\right)$.
   d. Set $w_i \leftarrow w_i \exp[c_m 1_{(y_i \neq f_{m,\hat{j}_m}(x_{\hat{j}_m,i}))}], i = 1, \ldots, n$, and normalize so that $\sum_{i=1}^{n} w_i = 1$.
3. Output the binary classifier $\text{sign}[F_M(\mathbf{x})]$ and the class probability prediction $\hat{\pi}(\mathbf{x}) = \frac{e^{F_M(\mathbf{x})}}{e^{F_M(\mathbf{x})}+e^{-F_M(\mathbf{x})}}$ where $F_M(\mathbf{x}) = \sum_{m=1}^{M} c_m f_{m,\hat{j}_m}(x_{\hat{j}_m})$.

Let y be the binary class taking a value in $\{-1, 1\}$ that we wish to predict. Let $f_m(\mathbf{x})$ be the weak learner (weak classifier) for the binary target y that we fit to predict using the high-dimensional covariates $\mathbf{x}$ in the mth iteration. Let err_m denote the error rate of the weak learner $f_m(\mathbf{x})$, and $E_w(\cdot)$ denote the weighted expectation (to be defined below) of the variable in the parenthesis with weight w. Note that the error rate $E_w\left[1_{(y \neq f_m(\mathbf{x}))}\right]$ is estimated by $err_m = \sum_{i=1}^{n} w_i 1_{(y_i \neq f_m(x_i))}$ with the weight w_i given by step 2(d) from the previous iteration. n is the number of observations. The symbol $1_{(\cdot)}$ is the indicator function which takes the value 1 if a logical condition inside the parenthesis is satisfied and takes the value 0 otherwise. The symbol $\text{sign}(z) = 1$ if $z > 0$, $\text{sign}(z) = -1$ if $z < 0$, and hence $\text{sign}(z) = 1_{(z>0)} - 1_{(z<0)}$.

Remark Note that the presented version of Discrete AdaBoost in Algorithm 1 as well as Real AdaBoost (RAB), LogitBoost (LB), and Gentle AdaBoost (GAB) which will be introduced later in the next section are different from their original version when they were first introduced. The original version of these algorithms only output the class label. In this paper, we follow the idea of Mease et al. (2007) and modified the algorithms to output both the class label and the probability prediction. The probability prediction is attained using

$$\hat{\pi}(\mathbf{x}) = \frac{e^{F_M(\mathbf{x})}}{e^{F_M(\mathbf{x})} + e^{-F_M(\mathbf{x})}},$$

where $F_M(\mathbf{x})$ is the sum of weak learners in the algorithms. □

Remark The only hyperparameter, i.e., the user specified parameter, in the AdaBoost as well as other Boosting algorithms is the number of iterations, M. It is also known as the stopping rule and is commonly chosen by cross-validation as well as information criterion such as AICc (Bühlmann, 2003). The choice of the stopping rule is embedded in most implementation of AdaBoost and should not be a concern for most users. Interesting readers could check Hastie, Tibshirani, and Friedman (2009) for more details of cross-validation. □

The most widely used weak learner is the classification tree. The simplest classification tree, the stump, takes the following functional form

$$f\left(x_j, a\right) = \begin{cases} 1 & x_j > a \\ -1 & x_j < a, \end{cases}$$

where the parameter a is found by minimizing the error rate

$$\min_a \sum_{i=1}^{n} w_i 1\left(y_i \neq f\left(x_{ji}, a\right)\right).$$

The other functional form of the stump can be shown as exchanging the greater and smaller sign in the previous from

$$f\left(x_j, a\right) = \begin{cases} 1 & x_j < a \\ -1 & x_j > a, \end{cases}$$

where the parameter a is found by minimizing the same error rate.

14.2.2 An Example

Now we present an example given by Ng (2014) for predicting the business cycles to help the readers understand the AdaBoost algorithm. Consider classifying whether the 12 months in 2001 is in expansion or recession using 3 months lagged data of the help-wanted index (*HWI*), new orders (*NAPM*), and the 10yr-FF spread (*SPREAD*). The data are listed in Columns 2–4 of Table 14.1. The NBER expansion and recession months are listed in Column 5, where 1 indicates a recession month and -1 indicates an expansion month. We use a stump as the weak learner (f). The stump uses an optimally chosen threshold to split the data into two partitions. This requires setting up a finite number of grid points for *HWI*, *NAPM*, and *SPREAD*, respectively, and evaluating the goodness of fit in each partition.

The algorithm begins by assigning an equal weight of $w_i^{(1)} = \frac{1}{n}$ where $n = 12$ to each observation. For each of the grid points chosen for *HWI*, the sample of y values is partitioned into parts depending on whether HWI_i exceeds the grid point or not. The grid point that minimizes classification error is found to be -0.044. The procedure is repeated with *NAPM* as a splitting variable, and then again with *SPREAD*. A comparison of the three sets of residuals reveals that splitting on the basis of *HWI* gives the smallest weighted error. The first weak learner thus labels Y_i to 1 if $HWI_i < -0.044$. The outcome of the decision is given in Column 6.

Table 14.1 An example

	Data: lagged 3 months				$f_1(\mathbf{x})$	$f_2(\mathbf{x})$	$f_3(\mathbf{x})$	$f_4(\mathbf{x})$	$f_5(\mathbf{x})$
	HWI	*NAPM*	*SPREAD*		*HWI*	*NAPM*	*HWI*	*SPREAD*	*NAPM*
Date	−0.066	48.550	0.244	y	< -0.044	<49.834	< -0.100	> -0.622	<47.062
2001.1	0.014	51.100	−0.770	−1	−1	−1	−1	−1	−1
2001.2	−0.091	50.300	−0.790	−1	1	−1	−1	−1	−1
2001.3	0.082	52.800	−1.160	−1	−1	−1	−1	−1	−1
2001.4	−0.129	49.800	−0.820	1	1	1	1	1	1
2001.5	−0.131	50.200	−0.390	1	1	−1	1	1	1
2001.6	−0.111	47.700	−0.420	1	1	1	1	1	1
2001.7	−0.056	47.200	0.340	1	1	1	1	1	1
2001.8	−0.103	45.400	1.180	1	1	1	1	1	1
2001.9	−0.093	47.100	1.310	1	1	1	1	1	1
2001.10	−0.004	46.800	1.470	1	−1	1	−1	1	1
2001.11	−0.174	46.700	1.320	1	1	1	1	1	1
2001.12	−0.007	47.500	1.660	−1	−1	1	−1	1	−1
c					0.804	1.098	0.710	0.783	0.575
Error rate					0.167	0.100	0.138	0.155	0

Ng (2014)

Compared with the NBER dates in Column 5, we see that months 2 and 10 are mislabeled, giving a misclassification rate of $\frac{2}{12} = 0.167$. This is err_1 of step 2(b). Direct calculations give $c_1 = \log(\frac{1-err_1}{err_1})$ of 0.804. The weights $w_i^{(2)}$ are updated to complete step 2(d). Months 2 and 10 now each have a weight of 0.25, while the remaining 10 observations each have a weight of 0.05. Three thresholds are again computed using weights $w^{(2)}$. Of the three, the *NAPM* split gives the smallest weighted residuals. The weak learner for step 2 is identified. The classification based on the sign of

$$F_2(\mathbf{x}) = 0.804 \cdot 1_{(HWI<-0.044)} + 1.098 \cdot 1_{(NAPM<49.834)}$$

is given in Column 7. Compared with Column 5, we see that months 5 and 12 are mislabeled. The weighted misclassification rate is decreased to 0.100. The new weights $w_t^{(3)}$ are 0.25 for months 5 and 12, 0.138 for months 2 and 10, and 0.027 for the remaining months. Three sets of weighted residuals are again determined using new thresholds. The best predictor is again *HWI* with a threshold of −0.100. Classification based on the sign of $F_3(\mathbf{x})$ is given in Column 8, where

$$F_3(\mathbf{x}) = 0.804 \cdot 1_{(HWI<0.044)} + 1.098 \cdot 1_{(NAPM<48.834)} + 0.710 \cdot 1_{(HWI<-0.100)}.$$

The error rate after three steps actually increases to 0.138. The weak learner in round four is $1_{(SPREAD>-0.622)}$. After $NAPM$ is selected for one more round, all recession dates are correctly classified. The strong learner is an ensemble of five weak learners defined by $\text{sign}(F_5(\mathbf{x}))$, where

$$F_5(\mathbf{x}) = 0.804 \cdot 1_{(HWI<-0.044)} + 1.098 \cdot 1_{(NAPM<49.834)} + 0.710 \cdot 1_{(HWI<-0.100)} + 0.783 \cdot 1_{(SPREAD>-0.622)} + 0.575 \cdot 1_{(NAPM<47.062)}.$$

Note that the same variable can be chosen more than once by AdaBoost which is the key difference from other stage-wise algorithms, e.g., forward stage-wise regression. The weights are adjusted at each step to focus more on the misclassified observations. The final decision is based on an ensemble of models. No single variable can yield the correct classification, which is the premise of an ensemble decision rule.

For more complicated applications, several packages in the statistical programming language `R` provide off-the-shelf implementations of AdaBoost and its variants. For example, *JOUSBoost* gives an implementation of the Discrete AdaBoost algorithm from Freund and Schapire (1997) applied to decision tree classifiers and provides a convenient function to generate test sample of the algorithms.

14.2.3 AdaBoost: Statistical View

After AdaBoost is invented and shown to be successful, numerous papers have attempted to explain the effectiveness of the AdaBoost algorithm. In an influential paper, Friedman et al. (2000) show that AdaBoost builds an additive logistic regression model

$$F_M(\mathbf{x}) = \sum_{m=1}^{M} c_m f_m(\mathbf{x})$$

via Newton-like updates for minimizing the exponential loss

$$J(F) = E\left(e^{-yF(\mathbf{x})} \middle| \mathbf{x}\right).$$

We hereby show the above statement using the greedy method to minimize the exponential loss function iteratively as in Friedman et al. (2000).

After m iterations, the current classifier is denoted as $F_m(\mathbf{x}) = \sum_{s=1}^{m} c_s f_s(\mathbf{x})$. In the next iteration, we are seeking an update $c_{m+1} f_{m+1}(\mathbf{x})$ for the function fitted from previous iterations $F_m(\mathbf{x})$. The updated classifier would take the form

$$F_{m+1}(\mathbf{x}) = F_m(\mathbf{x}) + c_{m+1} f_{m+1}(\mathbf{x}).$$

The loss for $F_{m+1}(\mathbf{x})$ will be

$$\begin{aligned} J\left(F_{m+1}(\mathbf{x})\right) &= J\left(F_m(\mathbf{x}) + c_{m+1} f_{m+1}(\mathbf{x})\right) \\ &= E\left[e^{-y\left(F_m(\mathbf{x}) + c_{m+1} f_{m+1}(\mathbf{x})\right)}\right]. \end{aligned} \tag{14.1}$$

Expand w.r.t. $f_{m+1}(\mathbf{x})$

$$\begin{aligned} J\left(F_{m+1}(\mathbf{x})\right) &\approx E\left[e^{-yF_m(\mathbf{x})}\left(1 - y c_{m+1} f_{m+1}(\mathbf{x}) + \frac{y^2 c_{m+1}^2 f_{m+1}^2(\mathbf{x})}{2}\right)\right] \\ &= E\left[e^{-yF_m(\mathbf{x})}\left(1 - y c_{m+1} f_{m+1}(\mathbf{x}) + \frac{c_{m+1}^2}{2}\right)\right]. \end{aligned}$$

The last equality holds since $y \in \{-1, 1\}$, $f_{m+1}(\mathbf{x}) \in \{-1, 1\}$, and $y^2 = f_{m+1}^2(\mathbf{x}) = 1$. $f_{m+1}(\mathbf{x})$ only appears in the second term in the parenthesis, so minimizing the loss function (14.1) w.r.t. $f_{m+1}(\mathbf{x})$ is equivalent to maximizing the second term in the parenthesis which results in the following conditional expectation

$$\max_f E\left[e^{-yF_m(\mathbf{x})} y c_{m+1} f_{m+1}(\mathbf{x}) \,|\mathbf{x}\right].$$

For any $c > 0$ (we will prove this later), we can omit c_{m+1} in the above objective function

$$\max_f E\left[e^{-yF_m(\mathbf{x})} y f_{m+1}(\mathbf{x}) \,|\mathbf{x}\right].$$

To compare it with the Discrete AdaBoost algorithm, here we define weight $w = w(y, \mathbf{x}) = e^{-yF_m(\mathbf{x})}$. Later we will see that this weight w is equivalent to that shown in the Discrete AdaBoost algorithm. So the above optimization can be seen as maximizing a weighted conditional expectation

$$\max_f E_w\left[y f_{m+1}(\mathbf{x}) \,|\mathbf{x}\right], \tag{14.2}$$

where $E_w(y|\mathbf{x}) := \frac{E(wy|\mathbf{x})}{E(w|\mathbf{x})}$ refers to a weighted conditional expectation. Note that (14.2) can be re-written as

$$\begin{aligned} & E_w\left[y f_{m+1}(\mathbf{x}) \,|\mathbf{x}\right] \\ &= P_w(y = 1|\mathbf{x}) f_{m+1}(\mathbf{x}) - P_w(y = -1|\mathbf{x}) f_{m+1}(\mathbf{x}) \\ &= \left[P_w(y = 1|\mathbf{x}) - P_w(y = -1|\mathbf{x})\right] f_{m+1}(\mathbf{x}) \\ &= E_w(y|\mathbf{x}) f_{m+1}(\mathbf{x}), \end{aligned}$$

where $P_w(y|\mathbf{x}) = \frac{E(w|y,\mathbf{x})P(y|\mathbf{x})}{E(w|\mathbf{x})}$. Solve the maximization problem (14.2). Since $f_{m+1}(\mathbf{x})$ only takes 1 or -1, it should be positive whenever $E_w(y|\mathbf{x})$ is positive and -1 whenever $E_w(y|\mathbf{x})$ is negative. The solution for $f_{m+1}(\mathbf{x})$ is

$$f_{m+1}(\mathbf{x}) = \begin{cases} 1 & E_w(y|\mathbf{x}) > 0 \\ -1 & \text{otherwise.} \end{cases}$$

Next, minimize the loss function (14.1) w.r.t. c_{m+1}

$$c_{m+1} = \arg\min_{c_{m+1}} E_w\left(e^{-c_{m+1}yf_{m+1}(\mathbf{x})}\right)$$

$$E_w\left(e^{-c_{m+1}yf_{m+1}(\mathbf{x})}\right) = P_w(y = f_{m+1}(\mathbf{x}))\,e^{-c_{m+1}} + P_w(y \neq f_{m+1}(\mathbf{x}))\,e^{c_{m+1}}$$

$$\frac{\partial E_w\left(e^{-cyf_{m+1}(\mathbf{x})}\right)}{\partial c_{m+1}} = -P_w(y = f_{m+1}(\mathbf{x}))\,e^{-c_{m+1}} + P_w(y \neq f_{m+1}(\mathbf{x}))\,e^{c_{m+1}}.$$

Let

$$\frac{\partial E_w\left(e^{-c_{m+1}yf_{m+1}(\mathbf{x})}\right)}{\partial c_{m+1}} = 0,$$

and we have

$$P_w(y = f_{m+1}(\mathbf{x}))\,e^{-c_{m+1}} = P_w(y \neq f_{m+1}(\mathbf{x}))\,e^{c_{m+1}}.$$

Solving for c_{m+1}, we obtain

$$c_{m+1} = \frac{1}{2}\log\frac{P_w(y = f_{m+1}(\mathbf{x}))}{P_w(y \neq f_{m+1}(\mathbf{x}))} = \frac{1}{2}\log\left(\frac{1 - err_{m+1}}{err_{m+1}}\right),$$

where $err_{m+1} = P_w(y \neq f_{m+1}(\mathbf{x}))$ is the error rate of $f_{m+1}(\mathbf{x})$. Note that $c_{m+1} > 0$ as long as the error rate is smaller than 50%. Our assumption $c_{m+1} > 0$ holds for any learner that is better than random guessing.

Now we have finished the steps of one iteration and can get our updated classifier by

$$F_{m+1}(\mathbf{x}) \leftarrow F_m(\mathbf{x}) + \left(\frac{1}{2}\log\left(\frac{1 - err_{m+1}}{err_{m+1}}\right)\right) f_{m+1}(\mathbf{x}).$$

Note that in the next iteration, the weight we defined w_{m+1} will be

$$w_{m+1} = e^{-yF_{m+1}(\mathbf{x})} = e^{-y(F_m(\mathbf{x}) + c_{m+1}f_{m+1}(\mathbf{x}))} = w_m \times e^{-c_{m+1}f_{m+1}(\mathbf{x})y}.$$

Since $-yf_{m+1}(\mathbf{x}) = 2 \times 1_{\{y \neq f_{m+1}(\mathbf{x})\}} - 1$, the update is equivalent to

$$w_{m+1} = w_m \times e^{\left(\log\left(\frac{1-err_{m+1}}{err_{m+1}}\right)1_{[y \neq f_{m+1}(\mathbf{x})]}\right)} = w_m \times \left(\frac{1-err_{m+1}}{err_{m+1}}\right)^{1_{[y \neq f_{m+1}(\mathbf{x})]}}.$$

Thus the function and weight update are of an identical form to those used in AdaBoost. AdaBoost could do better than any single weak classifier since it iteratively minimizes the loss function via a Newton-like procedure.

Interestingly, the function $F(\mathbf{x})$ from minimizing the exponential loss is the same as maximizing a logistic log-likelihood. Let

$$\begin{aligned} J(F(\mathbf{x})) &= E\left[E\left(e^{-yF(\mathbf{x})}\middle|\mathbf{x}\right)\right] \\ &= E\left[P(y=1|\mathbf{x})\,e^{-F(\mathbf{x})} + P(y=-1|\mathbf{x})\,e^{F(\mathbf{x})}\right]. \end{aligned}$$

Taking derivative w.r.t. $F(\mathbf{x})$ and making it equal to zero, we obtain

$$\begin{aligned} \frac{\partial E\left(e^{-yF(\mathbf{x})}|\mathbf{x}\right)}{\partial F(\mathbf{x})} &= -P(y=1|\mathbf{x})\,e^{-F(\mathbf{x})} + P(y=-1|\mathbf{x})\,e^{F(\mathbf{x})} = 0 \\ F^*(\mathbf{x}) &= \frac{1}{2}\log\left[\frac{P(y=1|\mathbf{x})}{P(y=-1|\mathbf{x})}\right]. \end{aligned}$$

Moreover, if the true probability is

$$P(y=1|\mathbf{x}) = \frac{e^{2F(\mathbf{x})}}{1+e^{2F(\mathbf{x})}},$$

for $Y = \frac{y+1}{2}$, the log-likelihood is

$$E(\log L|\mathbf{x}) = E\left[2YF(\mathbf{x}) - \log\left(1+e^{2F(\mathbf{x})}\right)\middle|\mathbf{x}\right].$$

The solution $F^*(\mathbf{x})$ that maximizes the log-likelihood must equal the $F(\mathbf{x})$ in the true model $P(y=1|\mathbf{x}) = \frac{e^{2F(\mathbf{x})}}{1+e^{2F(\mathbf{x})}}$. Hence,

$$\begin{aligned} e^{2F^*(\mathbf{x})} &= P(y=1|\mathbf{x})\left(1+e^{2F^*(\mathbf{x})}\right) \\ e^{2F^*(\mathbf{x})} &= \frac{P(y=1|\mathbf{x})}{1-P(y=1|\mathbf{x})} \\ F^*(\mathbf{x}) &= \frac{1}{2}\log\left[\frac{P(y=1|\mathbf{x})}{P(y=-1|\mathbf{x})}\right]. \end{aligned}$$

AdaBoost that minimizes the exponential loss yields the same solution as the logistic regression that maximizes the logistic log-likelihood.

From the above, we can see that AdaBoost gives high weights to and thus, focuses on the samples that are not correctly classified by the previous weak learners. This is exactly what Schapire (1990) referred to as *filtering* in Section 1.

14.3 Extensions to AdaBoost Algorithms

In this section, we introduce three extensions of (Discrete) AdaBoost (DAB) which is shown in Algorithm 1: namely, Real AdaBoost (RAB), LogitBoost (LB), and Gentle AdaBoost (GAB). We discuss how some aspects of the DAB may be modified to yield RAB, LB, and GAB. In the previous section, we learned that Discrete AdaBoost minimizes an exponential loss via iteratively adding a binary weaker learner to the pool of weak learners. The addition of a new weak learner can be seen as taking a step on the direction that loss function descents in the Newton method. There are two major ways to extend the idea of Discrete AdaBoost. One focuses on making the minimization method more efficient by adding a more flexible weak learner. The other is to use different loss functions that may lead to better results. Next, we give an introduction to three extensions of Discrete AdaBoost.

14.3.1 Real AdaBoost

Algorithm 2 Real AdaBoost (RAB, Friedman, Hastie, and Tibshirani, 2000)

1. Start with weights $w_i = \frac{1}{n}, i = 1, \ldots, n$.
2. For $m = 1$ to M

   a. For $j = 1$ to k (for each variable)

      i. Fit the classifier to obtain a class probability estimate $p_m(x_j) = \hat{P}_w(y = 1|x_j) \in [0, 1]$ using weights w_i on the training data.
      ii. Let $f_{mj}(x_j) = \frac{1}{2} \log \frac{p_m(x_j)}{1-p_m(x_j)}$.
      iii. Compute $err_{mj} = \sum_{i=1}^{n} w_i 1_{(y_i \neq \text{sign}(f_{mj}(x_{ji})))}$.

   b. Find $\hat{j}_m = \arg\min_j err_{mj}$.
   c. Set $w_i \leftarrow w_i \exp[-y_i f_{m,\hat{j}_m}(x_{\hat{j}_m,i})], i = 1, \ldots, n$, and normalize so that $\sum_{i=1}^{n} w_i = 1$.

3. Output the classifier $\text{sign}[F_M(\mathbf{x})]$ and the class probability prediction $\hat{\pi}(\mathbf{x}) = \frac{e^{F_M(\mathbf{x})}}{e^{F_M(\mathbf{x})}+e^{-F_M(\mathbf{x})}}$ where $F_M(\mathbf{x}) = \sum_{m=1}^{M} f_m(\mathbf{x})$.

Real AdaBoost that Friedman et al. (2000) propose focuses solely on improving the minimization procedure of Discrete AdaBoost. In Real AdaBoost, the weak learners are continuous comparing to Discrete AdaBoost where the weak learners are binary (discrete). Real AdaBoost is minimizing the exponential loss with continuous updates where Discrete AdaBoost minimizes the exponential loss with discrete updates. Hence, Real AdaBoost is more flexible with the step size and direction of the minimization and minimizes the exponential loss faster and more accurately. However, Real AdaBoost also imposes restriction that the classifier must produce a probability prediction which reduces the flexibility of the model. As pointed out in the numerical examples by Chu, Lee, and Ullah (2018), Real AdaBoost may achieve a larger in-sample training error due to the flexibility of its model. On the other hand, this also reduces the chance of over-fitting and would in the end achieve a smaller out-of-sample test error.

14.3.2 LogitBoost

Friedman et al. (2000) also propose LogitBoost by minimizing the Bernoulli log-likelihood via an adaptive Newton algorithm for fitting an additive logistic regression model. LogitBoost extends Discrete AdaBoost in two ways. First, it uses the Bernoulli log-likelihood instead of the exponential loss function as a loss function. Furthermore, it updates the classifier by adding a linear model instead of a binary weak learner.

Algorithm 3 LogitBoost (LB, Friedman, Hastie, and Tibshirani, 2000)

---

1. Start with weights $w_i = \frac{1}{n}, i = 1, \ldots, n$, $F(\mathbf{x}) = 0$ and probability estimates $p(x_i) = \frac{1}{2}$.
2. For $m = 1$ to M

   a. Compute the working response and weights

   $$z_i = \frac{y_i^* - p(x_i)}{p(x_i)(1 - p(x_i))}$$
   $$w_i = p(x_i)(1 - p(x_i))$$

   b. For $j = 1$ to k (for each variable)

      i. Fit the function $f_{mj}(x_{ji})$ by a weighted least-squares regression of z_i to x_{ji} using weights w_i on the training data.
      ii. Compute $err_{mj} = 1 - R^2_{mj}$ where R^2_{mj} is the coefficient of determination from the weighted least-squares regression.

   c. Find $\hat{j}_m = \arg\min_j err_{mj}$
   d. Update $F(\mathbf{x}) \leftarrow F(\mathbf{x}) + \frac{1}{2} f_{m,\hat{j}}(x_{\hat{j}})$ and $p(\mathbf{x}) \leftarrow \frac{e^{F(\mathbf{x})}}{e^{F(\mathbf{x})} + e^{-F(\mathbf{x})}}$.

3. Output the classifier $\text{sign}[F_M(\mathbf{x})]$ and the class probability prediction $\hat{\pi}(\mathbf{x}) = \frac{e^{F_M(\mathbf{x})}}{e^{F_M(\mathbf{x})} + e^{-F_M(\mathbf{x})}}$ where $F_M(\mathbf{x}) = \sum_{m=1}^{M} f_{m,\hat{j}_m}(x_{\hat{j}_m})$.

In LogitBoost, continuous weak learner is used similarly to Real AdaBoost. However, LogitBoost specifies the use of a linear weak learner while Real AdaBoost allows any weak learner that returns a probability between zero and one. A more fundamental difference here is that LogitBoost uses the Bernoulli log-likelihood as a loss function instead of the exponential loss. Hence, LogitBoost is more similar to logistic regression than Discrete AdaBoost and Real AdaBoost. As pointed out in the numerical examples by Chu et al. (2018), LogitBoost has the smallest in-sample training error but the largest out-of-sample test error. This implies that while LogitBoost is the most flexible of the four, it suffers the most from over-fitting.

14.3.3 Gentle AdaBoost

Algorithm 4 Gentle AdaBoost (GAB, Friedman, Hastie, and Tibshirani, 2000)

1. Start with weights $w_i = \frac{1}{n}, i = 1, \ldots, n$.
2. For $m = 1$ to M
   a. For $j = 1$ to k (for each variable)
      i. Fit the regression function $f_{mj}(x_{ji})$ by weighted least-squares of y_i on x_{ji} using weights w_i on the training data.
      ii. Compute $err_{mj} = 1 - R^2_{mj}$ where R^2_{mj} is the coefficient of determination from the weighted least-squares regression.
   b. Find $\hat{j}_m = \arg\min_j err_{mj}$
   c. Set $w_i \leftarrow w_i \exp[-y_i f_{m,\hat{j}_m}(x_{\hat{j}_m,i})]$, $i = 1, \ldots, n$, and normalize so that $\sum_{i=1}^{n} w_i = 1$.
3. Output the classifier $\text{sign}[F_M(\mathbf{x})]$ and the class probability prediction $\hat{\pi}(\mathbf{x}) = \frac{e^{F_M(\mathbf{x})}}{e^{F_M(\mathbf{x})}+e^{-F_M(\mathbf{x})}}$ where $F_M(\mathbf{x}) = \sum_{m=1}^{M} f_{m,\hat{j}_m}(x_{\hat{j}_m})$.

In Friedman et al. (2000), Gentle AdaBoost extends Discrete AdaBoost in the sense that it allows each weak learner to be a linear model. This is similar to LogitBoost and more flexible than Discrete AdaBoost and Real AdaBoost. However, it is closer to Discrete AdaBoost and Real AdaBoost than LogitBoost in the sense that Gentle AdaBoost, Discrete AdaBoost, and Real AdaBoost all minimize the exponential loss while LogitBoost minimizes the Bernoulli log-likelihood. On the other hand, Gentle AdaBoost is more similar to Real AdaBoost than Discrete AdaBoost since the weak learners are continuous and there is no need to find an optimal step size for each iteration because the weak learner is already optimal. As pointed out in the numerical examples by Chu et al. (2018), Gentle Boost often lies between Real AdaBoost and LogitBoost in terms of in-sample training error and out-of-sample test error.

14.4 L_2Boosting

In addition to classification, the idea of boosting can also be applied to regressions. Bühlmann (2003) propose L_2Boosting that builds a linear model by minimizing the L_2 loss. Bühlmann (2006) further proves the consistency of L_2Boosting in terms of predictions. L_2Boosting is the simplest and perhaps most instructive Boosting algorithm for economists and econometricians. It is very useful for regression, in particular in the presence of high-dimensional explanatory variables.

We consider a simple linear regression

$$y = \mathbf{x}\boldsymbol{\beta} + u,$$

where y is the dependent variable, $\mathbf{x}$ is the independent variable and $u \sim N(0, 1)$. Note that the number of independent variables $\mathbf{x}$ could be high-dimensional, i.e., the number of independent variables in $\mathbf{x}$ can be larger than the number of observations.

This model, in the low dimension case, can be estimated by the ordinary least squares. We minimize the sum of squared errors

$$L = \sum_{i=1}^{n} (y_i - \hat{y}_i)^2,$$

where

$$\hat{y}_i = \mathbf{x}_i \hat{\boldsymbol{\beta}}.$$

The solution to the problem is

$$\hat{\beta} = (\mathbf{X}'\mathbf{X})^{-1}\mathbf{X}'\mathbf{y}.$$

The residual from the previous problem is

$$\hat{u_i} = y_i - \hat{y}_i.$$

In the high-dimension case, the ordinary least-squares method falls down because the matrix $(\mathbf{X}'\mathbf{X})$ is not invertible. Hence, we need to use a modified least-squares method to get over the high-dimension problem.

The basic idea of L_2Boosting is to use only one explanatory variable at a time. Since the number of variables p is larger than the length of the sample period n, the matrix $\mathbf{X}'\mathbf{X}$ is not invertible. However, if we use only one variable in one particular iteration, the matrix $\mathbf{x}_j'\mathbf{x}_j$ is a scalar and thus invertible. In order to exploit the information in the explanatory variables, in the following iterations, we can use other explanatory variables to fit the residuals which are the unexplained part from previous iterations. L_2Boosting can be seen as iteratively using the least-squares technique to explain the residuals from the previous

Algorithm 5 L_2Boosting (Bühlmann, 2003)

1. Start with y_i from the training data.
2. For $m = 1$ to M

   a. For $j = 1$ to k (for each variable)

      i. Fit the regression function $y_i = \beta_{m,0,j} + \beta_{m,j} x_{ji} + u_i$ by least-squares of y_i on x_{ji}.
      ii. Compute $err_{mj} = 1 - R^2_{mj}$ where R^2_{mj} is the coefficient of determination from the least-squares regression.

   b. Find $\hat{j}_m = \arg\min_j err_{mj}$
   c. Set $y_i \leftarrow y_i - \hat{\beta}_{m,0,\hat{j}_m} - \hat{\beta}_{m,\hat{j}_m} x_{\hat{j}_m,i}$, $i = 1, \ldots, n$.

3. Output the final regression model $F_M(\mathbf{x}) = \sum_{m=1}^{M} \beta_{m,0,\hat{j}_m} + \hat{\beta}_{m,\hat{j}_m} x_{\hat{j}_m}$.

least-squares regressions. In the L_2Boosting algorithm, we use the least-squares technique to fit the dependent variable y with only one independent variable $\mathbf{x}_j$. Then, we iteratively take the residual from the previous regression as the new dependent variable y and fit the new dependent variable with, again, only one independent variable $\mathbf{x}_j$. The detailed description of L_2Boosting is listed in Algorithm 5.

The stopping parameter M is the main tuning parameter which can be selected using cross-validation or some information criterion in practice. Bühlmann (2003) proposed to use the corrected AIC to choose the stopping parameter. According to Bühlmann (2003), the square of the bias of the L_2Boosting decays exponentially fast with increasing M, the variance increases exponentially slow with increasing M, and $\lim_{M\to\infty} MSE = \sigma^2$. L_2Boosting is computationally simple and successful if the learner is sufficiently weak. If the learner is too strong, then there will be over-fitting problem as in all the other boosting algorithms. Even though it is straightforward for econometricians to use the simple linear regression as the weak learner, Bühlmann (2003) also suggest using smoothing splines and classification and regression trees as the weak learner.

14.5 Gradient Boosting

This section discusses the Gradient Boosting Machine first introduced by Friedman (2001). Breiman (2004) shows that the AdaBoost algorithm can be represented as a steepest descent algorithm in function space which we call functional gradient descent (FGD). Friedman et al. (2000) and Friedman (2001) then developed a more general, statistical framework which yields a direct interpretation of boosting as a method for function estimation. In their terminology, it is a "stage-wise, additive modeling" approach. Gradient Boosting is a generalization of AdaBoost and L_2Boosting. AdaBoost is a version of Gradient Boosting that uses the exponential loss and L_2Boosting is a version of Gradient Boosting that uses the L_2 loss.

14.5.1 Functional Gradient Descent

Before we introduce the algorithm of Gradient Boosting, let us talk about functional gradient descent in a general way. We consider $F(\mathbf{x})$ to be the function of interest and minimize a risk function $R(F) = E(L(y, F))$ with respect to $F(\mathbf{x})$. For example, in the L_2Boosting, the loss function $L(y, F(\mathbf{x}))$ is the L_2 loss, i.e., $L(y, F(\mathbf{x})) = (y - F(\mathbf{x}))^2$. Notice that we do not impose any parametric assumption on the functional form of $F(\mathbf{x})$, and hence, the solution $F(\mathbf{x})$ is entirely nonparametric.

The Functional Gradient Descent minimizes the risk function $R(F)$ at each $\mathbf{x}$ directly with respect to $F(\mathbf{x})$. In each iteration m, like in gradient descent, we look for a pair of optimal direction $f_m(\mathbf{x})$ and step size c_m. The optimal direction at $\mathbf{x}$ is the direction that the loss function $R(F)$ decreases the fastest. Hence, the optimal direction

$$f_m(\mathbf{x}) = E_y\left[-\frac{\partial L(y, F(\mathbf{x}))}{\partial F(\mathbf{x})}\bigg|\mathbf{x}\right]_{F(\mathbf{x})=F_{m-1}(\mathbf{x})}.$$

The optimal step size c_m can be found given $f_m(\mathbf{x})$ by a line search

$$c_m = \arg\min_{c_m} E_{y,\mathbf{x}} L(y, F_{m-1}(\mathbf{x}) + c_m f_m(\mathbf{x})).$$

Next, we update the estimated function $F(\mathbf{x})$ by

$$F_m(\mathbf{x}) = F_{m-1}(\mathbf{x}) + c_m f_m(\mathbf{x}).$$

Thus, we complete one iteration of Gradient Boosting. In practice, the stopping iteration, which is the main tuning parameter, can be determined via cross-validation or some information criteria. The choice of step size c is of minor importance, as long as it is "small," such as $c = 0.1$. A smaller value of c typically requires a larger number of boosting iterations and thus more computing time, while the predictive accuracy will be better and tend to over-fit less likely.

14.5.2 Gradient Boosting Algorithm

The algorithm of Gradient Boosting is shown in Algorithm 6.

Algorithm 6 Gradient Boosting (GB, Friedman, 2001)

1. Start with $F_0(\mathbf{x}) = \arg\min_{const} \sum_{i=1}^{n} L(y_i, const)$.
2. For $m = 1$ to M
   a. calculate the pseudo-residuals $r_i^m = -\left[\frac{\partial L(y_i, F(\mathbf{x}_i))}{\partial F(\mathbf{x}_i)}\right]_{F(\mathbf{x})=F_{m-1}(\mathbf{x})}, i = 1, \ldots, n$.
   b. $f_m(\mathbf{x}) = \arg\min_{f_m(\mathbf{x})} \sum_{i=1}^{N} (r_i^m - f_m(\mathbf{x}_i))^2$.
   c. $c_m = \arg\min_c \sum_{i=1}^{N} L(y_i, F_{m-1}(\mathbf{x}_i) + c_m f_m(\mathbf{x}_i))$.
   d. $F_m(\mathbf{x}) = F_{m-1}(\mathbf{x}) + c_m f_m(\mathbf{x})$.
3. Output $F_M(\mathbf{x}) = \sum_{m=1}^{M} c_m f_m(\mathbf{x})$.

In theory, any fitting criterion that estimates the conditional expectation could be used to fit the negative gradient at step $1(a)$. In Gradient Boosting, the negative gradient is also called "pseudo-residuals" r_i^m and Gradient Boosting fits this residuals in each iteration. The most popular choice to fit the residuals is the Classification/Regression Tree which we will discuss in detail in Sect. 14.5.3.

14.5.3 Gradient Boosting Decision Tree

Gradient Boosting Decision Tree (GBDT) or Boosting Tree is one of the most important methods for implementing nonlinear models in data mining, statistics, and econometrics. According to the results of data mining tasks at the data mining challenges platform, *Kaggle*, most of the competitors choose Boosting Tree as their basic technique to model the data for predicting tasks.

Obviously, Gradient Boosting Decision Tree combines the decision tree and gradient boosting method. The gradient boosting is the gradient descent in functional space,

$$f_{m+1}(\mathbf{x}) = f_m(\mathbf{x}) + \lambda_m \left(\frac{\partial L}{\partial f}\right)_m,$$

where m is the number of iteration, L is the loss function we need to optimize, λ_m is the learning rate. In each round, we find the best direction $-\left(\frac{\partial L}{\partial f}\right)_m$ to minimize the loss function. In gradient boosting, we can use some simple functions to find out the best direction. That is, we use some functions to fit the "pseudo-residuals" of the loss function. In AdaBoost, we often use the decision stump, a line or hyperplane orthogonal to only one axis, to fit the residual. In the Boosting Tree, we choose a decision tree to handle this task. Also, the decision stump could be seen as a decision tree with one root node and two leaf nodes. Thus, the Boosting Tree is a natural way to generalize Gradient Boosting.

Basically, the Boosting Tree learns an additive function, which is similar to other aggregating methods like Random Forest. But the decision trees are grown

Algorithm 7 Gradient Boosting Decision Tree (Tree Boost, Friedman, 2001)

1. Initially, estimate the first residual via $r_i^0 = -2(y_i - \bar{y}) = -2(y_i - f_1(x_i))$.
2. For $m = 1$ to M

   a. Based on new samples $(r_i^m, x_i), i = 1, \ldots, n$, fit a regression tree $h_m(\mathbf{x})$.
   b. Let $f_{m+1}(\mathbf{x}) = f_m(\mathbf{x}) + \lambda_m h_m(\mathbf{x})$, then, $\lambda_m = \arg\min_\lambda L(y, f_m(\mathbf{x}) + \lambda h_m(\mathbf{x}))$.
   c. Update $f_{m+1}(\mathbf{x})$ via $f_{m+1}(\mathbf{x}) = f_m(\mathbf{x}) + \lambda_m h_m(\mathbf{x})$.
   d. Calculate the new residual $r_i^{m+1} = -2(y_i - f_{m+1}(x_i))$, then update the new samples as $(r_i^{m+1}, x_i), i = 1, \ldots$.

3. Output the Gradient Boosting Decision Tree $F_M(\mathbf{x}) = \sum_{m=1}^{M} \lambda_m f_m(\mathbf{x})$.

very differently among these methods. In the Boosting Tree, a new decision tree is growing based on the "error" from the decision tree which grew in the last iteration. The updating rule comes from Gradient Boosting method and we will dive into the details later.

Suppose we need to implement a regression problem given samples $(y_i, x_i), i = 1, \ldots, n$. If we choose the square loss function, the "pseudo-residual" should be $r_i^m = -\left(\frac{\partial L}{\partial f}\right)_m = -\left(\frac{\partial (y-f)^2}{\partial f}\right)_m = 2(y - f_m)$.

The algorithm of Gradient Boosting Decision Tree is shown in Algorithm 7.

According to Algorithm 7, the main difference between Gradient Boosting and Boosting Tree is at step $1(a)$. In Boosting Tree, we use a decision tree to fit the "residual" or the negative gradient. In other words, Boosting Tree implements the Functional Gradient Descent by following the functional gradient learned by the decision tree.

Additionally, to implement Gradient Boosting Decision Tree, we need to choose several hyperparameters: (1) N, the number of terminal nodes in trees; (2) M, the number of iterations in the boosting procedure.

Firstly, N, the number of terminal nodes in trees, controls the maximum allowed level of interaction between variables in the model. With $N = 2$ (decision stumps), no interaction between variables is allowed. With $N = 3$, the model may include effects of the interaction between up to two variables, and so on. Hastie et al. (2009) comment that typically $4 < N < 8$ work well for boosting and results are fairly insensitive to the choice of N in this range, $N = 2$ is insufficient for many applications, and $N > 10$ is unlikely to be required. Figure 14.1 shows the test error curves corresponding to the different number of nodes in Boosting Tree. We can see that Boosting with decision stumps provides the best test error. When the number of nodes increases, the final test error increases, especially in boosting with trees containing 100 nodes. Thus, practically, we often choose $4 < N < 8$.

Secondly, to the number of iterations M, we will discuss that in Sect. 5.4.1 in detail, which is related to the regularization method in Boosting Tree.

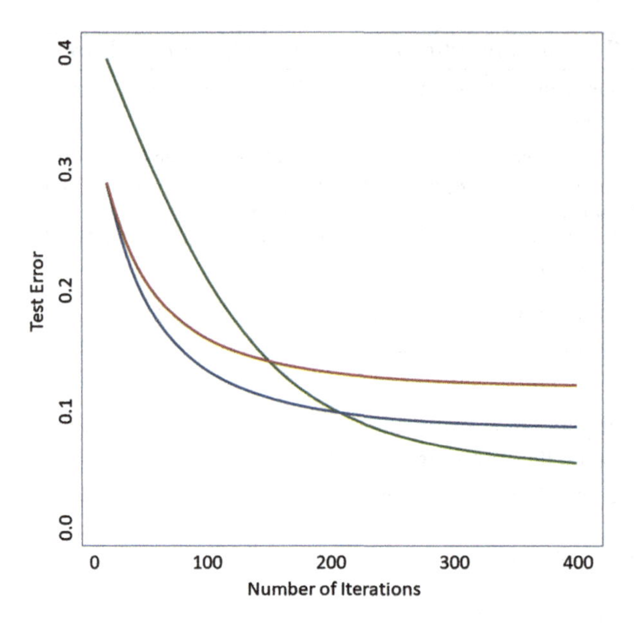

Fig. 14.1 Illustration of Gradient Boosting Decision Trees with different nodes (green: decision stump; red line: tree with 10 leaf nodes; blue: tree with 100 leaf nodes)

14.5.4 Regularization

By following the discussion above, the Gradient Boosting Decision Trees method contains more trees when M is larger. A further issue is related to over-fitting. That is, when there are increasingly large numbers of decision trees, Boosting Tree can fit any data with zero training error, which leads to a bad test error on new samples. To prevent the model from over-fitting, we will introduce two ways to resolve this issue.

Early Stopping

A simple way to resolve this issue is to control the number of iterations M in the Boosting Trees. Basically, we can treat M as a hyperparameter during the training procedure of Boosting Trees. Cross-Validation is an effective way to select hyperparameters including M. Since Boosting Trees method is equivalent to the steepest gradient descent in functional space, selecting the optimal M means that this steepest gradient descent will stop at the Mth iteration.

Shrinkage Method

The second way to resolve the problem of over-fitting is shrinkage. That is, we add a shrinkage parameter during the training process. Let us consider the original formula for updating Boosting Trees:

$$f_{m+1}(\mathbf{x}) = f_m(\mathbf{x}) + \lambda_m h_m(\mathbf{x}). \tag{14.3}$$

In the Boosting Trees, we first fit $h_m(\mathbf{x})$ based on a decision tree. Then, we optimize λ_m for the best step size. Thus, we can shrink the step size by adding a shrinkage parameter ν:

$$f_{m+1}(\mathbf{x}) = f_m(\mathbf{x}) + \nu\lambda_m h_m(\mathbf{x}). \tag{14.4}$$

Obviously, if we set $\nu = 1$, Eq. (14.4) is equivalent to Eq. (14.3). Suppose we set $0 \leq \nu \leq 1$, it can shrink the optimal step size λ_m to $\nu\lambda_m$, which leads to a slower optimization. In other words, compared to the original Boosting Tree, Shrinkage Boosting Tree learns the unknown function slower but more precise in each iteration. As a consequence, to a given $\nu < 1$, we need more steps M to minimize the error. Figure 14.2 shows this consequence. To a binary classification problem, we consider two measures: the test set deviations, which is the negative binomial log-likelihood loss on the test set, and the test set misclassification error. In the left and right panels, we can see that, with the shrinkage parameter less than

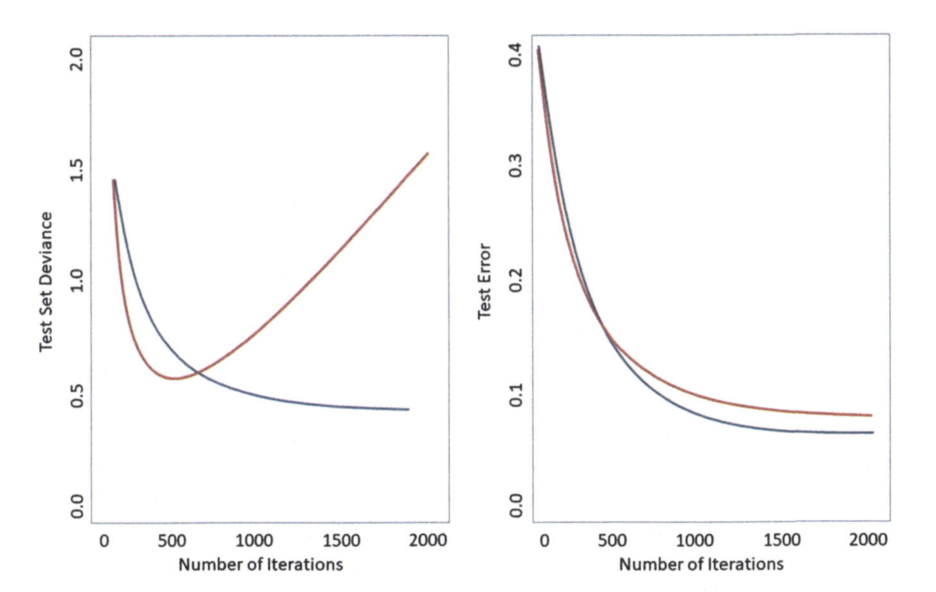

Fig. 14.2 Gradient Boosting Decision Tree (6 leaf nodes) with different shrinkage parameters (blue: shrinkage $\nu = 0.6$; red: no shrinkage)

1, Boosting Tree typically need more iterations to converge but it can hit a better prediction result. Friedman (2001) found that a smaller ν will lead to a larger optimal M but the test errors in the new datasets are often better than the original Boosting Tree. Although large M may need more computational resources, this method may be inexpensive because of the faster computers.

14.5.5 Variable Importance

After training Boosting Tree, the next question is to identify the variable importance. Practically, we often train boosting tree on a dataset with a large number of variables and we are interested in finding important variables for analysis.

Generally, this is also an important topic in tree-based models like Random Forest discussed in Chap. 13. Since Boosting Tree method is also an additive trees aggregating, we can use I_j^2 to measure the importance of a variable j:

$$I_j^2 = \frac{1}{M} \sum_{m=1}^{M} I_j^2(m),$$

and $I_j^2(m)$ is the importance of variable j for the mth decision tree:

$$I_j^2(m) = \sum_{t=1}^{T_m - 1} e_t^2 I(v(t)_m = j),$$

where T_m is the number of internal nodes (non-leaf nodes) in the mth decision tree, $v(t)_m$ is the variable selected by node t, and e_t is the error improvement based on before and after splitting the space via variable $v(t)_m$.

In Random Forest or Bagging Decision Tree method, we can measure the variable importance based on the so-called Out-of-Bag errors. In Boosting Tree, since there are no Out-of-Bag samples, we can only use I_j^2. In practice, OOB-based method and I_j^2 method often provide similar results and I_j^2 works very well especially when M is very large.

Let us consider an example about the relative importance of variables for predicting spam mail via Boosting Trees. The input variable $\mathbf{x}$ could be a vector of counts of the keywords or symbols in one email. The response y is a binary variable $(Spam,\ Not\ Spam)$. We regress y on $\mathbf{x}$ via Boosting Tree and then calculate the variable importance for each word or symbol. On one hand, the most important keywords and symbols may be "!", "$", "free", that is related to money and free; on the other hand, the keywords like "3d", "addresses", and "labs" are not very important since they are relatively neutral. Practically, the variable importance measure often provides a result consistent with common sense.

14.6 Recent Topics in Boosting

In this section, we will focus on four attractive contributions of Boosting in recent years. First of all, we introduce two methods that are related to Boosting in time series and volatility models respectively. They are relevant topics in macroeconomic forecasting. The third method is called Boosting with Momentum (BOOM), which is a generalized version of Gradient Boosting and is more robust than the Gradient Boosting. The fourth method is called Multi-Layered Gradient Boosting Decision Tree, which is a deep learning method via non-differentiable Boosting Tree and shed light on representation learning in tabular data.

14.6.1 Boosting in Nonlinear Time Series Models

In macroeconomic forecasting, nonlinear time series models are widely used in the last 40 years. For example, Tong and Lim (1980) discuss the Threshold Autoregressive (TAR) model to describe the time dependence when the time series is higher or lower than a threshold value. Chan and Tong (1986) propose the Smooth Transition Autoregressive (STAR) model to catch the nonlinear time dependence changing continuously between two states over time. Basically, nonlinear time series models not only perform better than linear time series models but also provide a clear way to analyze the nonlinear dependence among time series data.

Although nonlinear time series models are successful in macroeconomic time series modeling, we also need to consider their assumptions and model settings so that they can work for time series modeling. Unfortunately, in the era of big data, they cannot handle the large datasets since they often contain more complicated time dependence and higher dimensional variables along time that does not satisfy the assumptions. Essentially, the Boosting method provides an effective and consistent way to handle the time series modeling among big datasets especially with relatively fewer assumptions required.

Robinzonov, Tutz, and Hothorn (2012) discuss the details of Boosting for nonlinear time series models. Suppose we have a bunch of time series dataset $z_t = (y_{t-1}, \ldots, y_{t-p}, x_{1,t-1}, \ldots, x_{q,t-1}, \ldots, x_{1,t-p}, \ldots, x_{q,t-p}) = (y_{t-1}, \ldots, y_{t-p}, \mathbf{x}_{t-1}, \ldots, \mathbf{x}_{t-p}) \in \mathbb{R}^{(q+1)p}$, where z_t is the information set at time t, y is a series of endogenous variable with lags of p and $(\mathbf{x}_{t-1}, \ldots, \mathbf{x}_{t-p})$ is a q dimensional vector series with lags of p. Consider a nonlinear time series model for the conditional mean of y_t:

$$E(y_t|z_t) = F(z_t) = F(y_{t-1}, \ldots, y_{t-p}, x_{1,t-1}, \ldots, x_{q,t-1}, \ldots, x_{1,t-p}, \ldots, x_{q,t-p}),$$

where $F(z_t)$ is an unknown nonlinear function. Chen and Tsay (1993) discuss an additive form of $F(z_t)$ for nonlinear time series modeling, which is called Nonlinear

Additive Auto Regressive with exogenous variables (NAARX):

$$
\begin{aligned}
E(y_t|z_t) &= F(z_t) \\
&= \sum_{i=1}^{p} f_i(y_{t-i}) + \sum_{i=1}^{p} f_{1,i}(x_{1,t-i}) + \ldots + \sum_{i=1}^{p} f_{q,i}(x_{q,t-i}) \\
&= \sum_{i=1}^{p} f_i(y_{t-i}) + \sum_{j=1}^{q}\sum_{i=1}^{p} f_{j,i}(x_{j,t-i}).
\end{aligned}
$$

To optimize the best $F(z_t)$ given data, we need to minimize the loss function:

$$
\hat{F}(z_t) = \arg\min_{F(z_t)} \frac{1}{T}\sum_{t=1}^{T} L(y_t, F(z_t)).
$$

For example, we can use L_2 loss function $L(y_t, F(z_t)) = \frac{1}{2}(y_t - F(z_t))^2$. If we consider a parametric function $F(z_t, \beta)$, we can have the following loss function:

$$
\hat{\beta} = \arg\min_{\beta} \frac{1}{T}\sum_{t=1}^{T} L(y_t, F(z_t; \beta)).
$$

Since the true function of $E(y|z)$ has the additive form, the solution to the optimization problem should be represented by a sum over a bunch of estimated functions. In Boosting, we can use M different weak learner to implement:

$$
F(z_t; \hat{\beta}^M) = \sum_{m=0}^{M} \nu h(z_t; \hat{\gamma}^m),
$$

where ν is a shrinkage parameter for preventing over-fitting. Similar to original gradient boosting, in each iteration, we can generate a "pseudo residual" term $r^m(z_t)$ which is

$$
r^m(z_t) = -\left.\frac{\partial L(y_t, F)}{\partial F}\right|_{F=F(z_t;\hat{\beta}^{m-1})}.
$$

Thus, we can optimize $\hat{\gamma}^m$ based on the loss function

$$
\hat{\gamma}^m = \arg\min_{\gamma} \sum_{t=1}^{T} L(r^m(z_t), h(z_t; \gamma)).
$$

Algorithm 8 Component-wise boosting with linear weak learner (Robinzonov et al., 2012)

1. Start with y_t from training data.
2. For $m = 1$ to M

   a. For $j = 1$ to $(1+q)p$ (for each variable)

      i. Fit the regression function $y_t = \beta_{m,0,j} + \beta_{m,j} z_{j,t} + u_t$ by least-squares of y_t on $z_{j,t}$ on the training data.
      ii. Compute $err_{mj} = 1 - R^2_{mj}$ from the weighted least-squares regression.

   b. Find $\hat{j}_m = \arg\min_j err_{mj}$.
   c. Set $y_t \leftarrow y_t - \hat{\beta}_{m,0,\hat{j}_m} - \hat{\beta}_{m,\hat{j}_m} z_{t,\hat{j}_m}$, $t = 1, \ldots, T$.

3. Output the final regression model $F_M(z) = \sum_{m=1}^{M} \beta_{m,0,\hat{j}_m} + \hat{\beta}_{m,\hat{j}_m} z_{\hat{j}_m}$.

After that, we update the $F(z_t; \hat{\beta}^m)$ as

$$F(z_t; \hat{\beta}^m) = F(z_t; \hat{\beta}^{m-1}) + \nu h(z_t; \hat{\gamma}^m).$$

Now go back to the NAARX model. Since each function f only contains one variable, y_{t-i} or $x_{j,t-i}$, we can construct same additive form via L_2Boosting. That is, in each iteration, we only choose one variable from the whole vector $z_t = (y_{t-1}, \ldots, y_{t-p}, \ldots, x_{q,t-1}, \ldots, x_{q,t-p})$ and then fit a weak learner. This is called Component-wise Boosting.

Robinzonov et al. (2012) discussed two methods of component-wise boosting with different weak learners: linear weak learner and P-Spline weak learner. The first method is called component-wise linear weak learner. For this method, we choose a linear function with one variable of z_t as a weak learner in each iteration. The algorithm of Component-wise Boosting with linear weak learner is shown in Algorithm 8.

Obviously, this method only provides a linear solution like an Autoregressive model with exogenous variables (ARX). We can also consider more complicated weak learner such that the nonlinear components could be caught. In the paper, P-Spline with B base learners is considered as the weak learner. The algorithm of Component-wise Boosting with P-Spline weak learner is shown in Algorithm 9.

14.6.2 *Boosting in Volatility Models*

Similarly to Boosting in nonlinear time series models for the mean, it is possible to consider Boosting in volatility models, like GARCH. Audrino and Bühlmann (2016) discussed volatility estimation via functional gradient descent for high-dimensional

Algorithm 9 Component-wise boosting with P-Spline weak learner (Robinzonov et al., 2012)

1. Start with y_t from training data.
2. For $m = 1$ to M
   a. For $j = 1$ to $(1+q)p$ (for each variable)
      i. Fit the P-Spline with B Base learners $\hat{y}_t = Spline_m(z_{j,t})$ by regressing y_t on $z_{j,t}$ on the training data.
      ii. Compute $err_{mj} = 1 - R^2_{mj}$ from the P-Spline regression.
   b. Find $\hat{j}_m = \arg\min_j err_{mj}$.
   c. Set $y_i \leftarrow y_t - \hat{y}_t, t = 1, \ldots, T$.
3. Output the final regression model $F_M(z) = \sum_{m=1}^{M} Spline_m(z_{\hat{j}_m})$.

financial time series. Matías, Febrero-Bande, González-Manteiga, and Reboredo (2010) compare Boost-GARCH with other methods, like neural networks GARCH.

Let us begin with the classic $GARCH(p,q)$ model by Bollerslev (1986):

$$
\begin{aligned}
y_t &= \mu + e_t, t = 1, \ldots, T \\
e_t &\sim N(0, h_t) \\
h_t &= c + \sum_{i=1}^{p} \alpha_i e_{t-i}^2 + \sum_{j=1}^{q} \beta_j h_{t-j}.
\end{aligned}
$$

We can implement a Maximum Likelihood Estimation (MLE) method to estimate all the coefficients. Generally, consider a nonlinear formula of the volatility function h_t:

$$h_t = g(e_{t-1}^2, \ldots, e_{t-p}^2, h_{t-1}, \ldots, h_{t-q}) = g(E_t^2, H_t),$$

where $E_t^2 = (e_{t-1}^2, \ldots, e_{t-p}^2)$ and $H_t = (h_{t-1}, \ldots, h_{t-q})$. Similarly to NAARX model, we can consider an additive form of the function g:

$$h_t = \sum_{m=1}^{M} g_m(E_t^2, H_t).$$

For simplicity, let $p = q = 1$, we have:

$$h_t = \sum_{m=1}^{M} g_m(e_{t-1}^2, h_{t-1}).$$

Algorithm 10 Boost-GARCH (1,1) (Audrino & Bühlmann, 2016)

1. Start with estimating a linear GARCH (1,1) model:

$$y_t = \mu + e_t, t = 1, \dots, T$$
$$e_t \sim N(0, h_t)$$
$$h_t = c + \alpha_1 e_{t-1}^2 + \beta_1 h_{t-1}$$

2. Getting the $\hat{\mu}_0, \hat{h}_{t-1,0}$
3. For $m = 1$ to M

   a. Calculate the residual:

$$e_{t,m}^2 = (y_t - \hat{\mu}_{m-1})^2$$
$$r(\mu)_{t,m} = -\left(\frac{\partial L}{\partial \mu}\right)_m = \frac{y_t - \hat{\mu}_{t,m}}{\hat{h}_{t,m}}$$
$$r(h)_{t,m} = -\left(\frac{\partial L}{\partial h_t}\right)_m = \frac{1}{2}\left(\frac{(y_t - \hat{\mu}_{t,m})^2}{\hat{h}_{t,m}^2} - \frac{1}{\hat{h}_{t,m}}\right)$$

   b. Fit a nonlinear base learner $\hat{y}_t = f_m(e_{t-1}^2, h_{t-1})$ by regressing $r(h)_{t,m}$ on $e_{t-1,m}^2, \hat{h}_{t-1,m}$ on the training data.
   c. Set $\hat{h}_{t,m} \leftarrow \hat{h}_{t,m-1} + f_m(e_{t,m-1}^2, \hat{h}_{t,m-1})$.

4. Output the final regression model $\hat{h}_t = \sum_{m=1}^{M} f_m(e_{t-1}^2, h_{t-1})$.

Thus, we can use L_2Boosting to approximate the formula above. Since we use MLE to estimate the original GARCH model, for Boost-GARCH, we can also introduce the likelihood function for calculating the "pseudo residual" $r_{t,m}$ instead of using the loss function. Finally, Boost-GARCH can fit an additive nonlinear formula as the estimation of h_t:

$$\hat{h}_t = \sum_{m=1}^{M} f_m(e_{t-1}^2, h_{t-1})$$

The algorithm of Boost-GARCH (1, 1) is shown in Algorithm 10.

14.6.3 Boosting with Momentum (BOOM)

In Sect. 14.5 on Gradient Boosting, we show that Gradient Boosting can be represented as a steepest gradient descent in functional space. In the optimization literature, gradient descent is widely discussed on its properties. First, gradient descent is easily revised for many optimization problems. Second, gradient descent often finds out good solutions no matter the optimization problem is convex or non-convex.

But gradient descent also suffers from some drawbacks. Let us consider the plots of loss surface in Fig. 14.4. Suppose the loss surface is convex. Obviously, gradient descent should converge to the global minimum eventually. But what we can see in the plot (a) is that the gradient descent converges very slow and the path of gradient descent is a zig-zag path. Thus, original gradient descent may spend a long time on converging to the optimal solution. Furthermore, the convergence is worse in a non-convex optimization problem.

To resolve this issue, a very practical way is to consider "momentum" term to the gradient descent updating rule:

$$\theta_{m+1} = \theta_m - \lambda V_m,$$

$$V_m = V_{m-1} + \nu \left(\frac{\partial L}{\partial \theta} \right)_m,$$

where θ_m is the parameter we want to optimize at mth iteration and V_m is the momentum term with another corresponding updating rule.

Basically, in original gradient descent method, we have $V_m = \left(\frac{\partial L}{\partial \theta}\right)_m$. In $(m+1)$th iteration, the parameter θ_{m+1} is updated by following the gradient $\left(\frac{\partial L}{\partial \theta}\right)_m$ only. But when we consider momentum term, the parameter θ_m is updated by following the updating direction in previous iteration V_{m-1} and the gradient $\left(\frac{\partial L}{\partial \theta}\right)_m$ together. Intuitively, this is just like the effect of momentum in physics. When a ball is rolling down from the top, even though it comes to a flat surface, it keeps rolling for a while because of momentum.

Plot (b) in Fig. 14.3 illustrates the difference between Gradient Descent without and with Momentum. Compared to the path of convergence in the plot (a), if we consider momentum in gradient descent, the path becomes better and spends less time on moving to the optimal solution which is shown in plot (b).

As the generalized version of gradient descent in function space, gradient boosting may also suffer from the same problem when the loss surface is complicated. Thus, a natural way to improve the gradient boosting method is considering the momentum term in its updating rule. Mukherjee et al. (2013) discuss a general analysis of a fusion of Nesterov's accelerated gradient with parallel coordinate

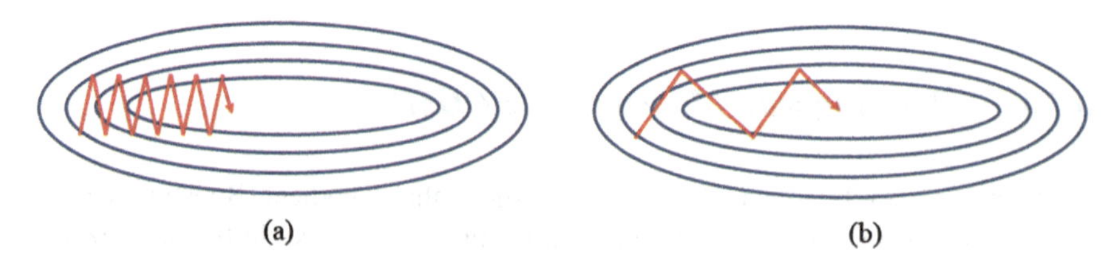

Fig. 14.3 Gradient descent (**a**) without momentum and (**b**) with momentum

descent. The resulting algorithm is called Boosting with Momentum (BOOM). Namely, BOOM retains the momentum and convergence properties of the accelerated gradient method while taking into account the curvature of the objective function. They also show that BOOM is especially effective in large scale learning problems. Algorithm 11 provides the procedure of BOOM via Boosting Tree.

Obviously, the main difference between Boosting with Momentum and ordinary Boosting Tree is a step to update V_m. Also, we have one more hyperparameter to decide ν, which decides the fraction of gradient information saved for next iteration updating of $f_m(\mathbf{x})$. Practically, we set $0.5 < \nu < 0.9$ but it is more reasonable to tune ν via cross-validation.

Basically, this method can be generalized to Stochastic Gradient Boosting discussed by Friedman (2002). Algorithm 12 shows the procedure of BOOM via Stochastic Gradient Boosting Tree.

There some differences between BOOM with Boosting Tree and Stochastic Boosting Tree. In Boosting Tree, we use all the n samples to update the decision

---

Algorithm 11 Gradient Boosting Decision Tree with momentum (Mukherjee et al., 2013; Friedman, 2002)

1. Initially, estimate the first residual via $r_i^0 = -2(y_i - \bar{y}) = -2(y_i - f_1(x_i))$.
2. For $m = 1$ to M
   a. Based on new samples $(r_i^m, x_i), i = 1, \ldots, n$, fit a regression tree $h_m(\mathbf{x})$.
   b. Let $V_m = V_{m-1} + \lambda_m h_m(\mathbf{x})$.
   c. Let $f_{m+1}(\mathbf{x}) = f_m(\mathbf{x}) + \nu V_m$, then optimize λ_m via $\lambda_m = \arg\min_\lambda L(y, f_m(\mathbf{x}) + \nu V_m) = \arg\min_\lambda L(y, f_m(\mathbf{x}) + \nu(V_{m-1} + \lambda h_m(\mathbf{x})))$.
   d. Update $f_{m+1}(\mathbf{x})$ via $f_{m+1}(\mathbf{x}) = f_m(\mathbf{x}) + \nu V_m$.
   e. Calculate the new residual $r_i^{m+1} = -2(y_i - f_{m+1}(x_i))$, then update the new samples as $(r_i^{m+1}, x_i), i = 1, \ldots, n$.
3. Output the Gradient Boosting Decision Tree $F_M(\mathbf{x}) = \sum_{m=1}^{M} \nu V_m$.

Algorithm 12 Stochastic Gradient Boosting Decision Tree with momentum (Mukherjee et al., 2013; Friedman, 2002)

1. Initially, randomly select a subset of the samples $(y_i, x_i), i = 1, \ldots, n_s$, where $0 < n_s < n$.
2. Estimate the first residual via $r_i^0 = -2(y_i - \bar{y}) = -2(y_i - f_1(x_i))$.
3. For $m = 1$ to M.
   a. Based on new samples $(r_i^m, x_i), i = 1, \ldots, n_s$, fit a regression tree $h_m(\mathbf{x})$.
   b. Let $V_m = V_{m-1} + \lambda_m h_m(\mathbf{x})$.
   c. Let $f_{m+1}(\mathbf{x}) = f_m(\mathbf{x}) + \nu V_m$, then optimize λ_m via $\lambda_m = \arg\min_\lambda L(y, f_m(\mathbf{x}) + \nu V_m) = \arg\min_\lambda L(y, f_m(\mathbf{x}) + \nu(V_{m-1} + \lambda h_m(\mathbf{x})))$.
   d. Update $f_{m+1}(\mathbf{x})$ via $f_{m+1}(\mathbf{x}) = f_m(\mathbf{x}) + \nu V_m$.
   e. Calculate the new residual $r_i^{m+1} = -2(y_i - f_{m+1}(x_i))$, then update the new samples as $(r_i^{m+1}, x_i), i = 1, \ldots, n_s$.
4. Output the Gradient Boosting Decision Tree $F_M(\mathbf{x}) = \sum_{m=1}^{M} \nu V_m$.

tree in each iteration. But Stochastic Boosting Tree randomly selects $\frac{n_s}{n}$ fraction of samples to grow a decision tree in each iteration. When the sample size n is increasingly large, selecting a subset of samples could be a better and more efficient way to implement the Boosting Tree algorithm.

14.6.4 Multi-Layered Gradient Boosting Decision Tree

Last 10 years witnessed the dramatic development in the fields about deep learning, which mainly focus on distilling hierarchical features via multi-layered neural networks automatically. From 2006 deep learning methods have changed so many areas like computer vision and natural language processing.

Basically, multi-layered representation is the key ingredient of deep neural networks. Thus, the combination of multi-layered representation and Boosting Tree are expected in handling very complicated tabular data analysis tasks. But there are few research papers exploring multi-layered representation via non-differentiable models, like Boosted Decision Tree. That is, the gradient-based optimization method which is always used in training multi-layered neural networks cannot be introduced in training multi-layered Boosting methods.

Feng, Yu, and Zhou (2018) explored one way to construct Multi-Layered Gradient Boosting Decision Tree (mGBDT) with an explicit emphasis on exploring the ability to learn hierarchical representations by stacking several layers of regression GBDTs. The model can be jointly trained by a variant of target propagation across layers, without the need to derive back-propagation or to require differentiability.

Figure 14.4 provides the structure of a Multi-Layered Gradient Boosting Decision Tree. $F_m, m = 1, \ldots, M$ are the M layers of a mGBDT. Similar to the multi-layered neural networks, the input o_0 is transformed to $o_1, \ldots, o_M$ via $F_1, \ldots, F_M$. Then, the final output o_M is the prediction of the target variable y. But

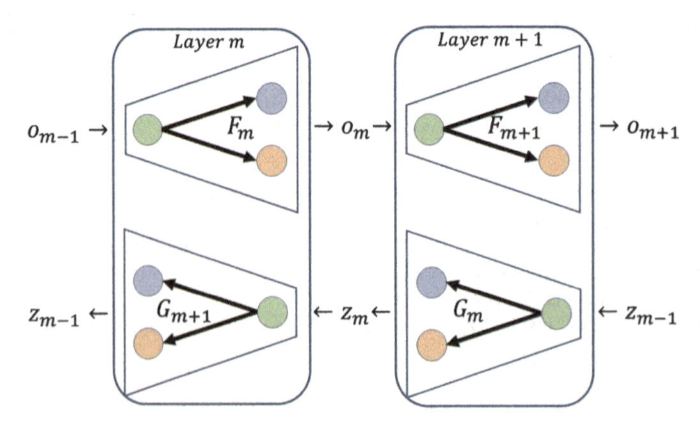

Fig. 14.4 Illustration of multi-layered Gradient Boosting Decision Tree

all the F_m are constructed via Gradient Boosting Decision Tree, we cannot training them via back-propagation method used in training multi-layered neural networks. Feng et al. (2018) introduced another group of functions $G_m, m = 1, \ldots, M$ and corresponding variables $z_m, m = 1, \ldots, M$.

Intuitively, the group of function G_m are introduced for achieving back-propagation algorithm in non-differentiable Boosting Tree. To train Multi-layered Gradient Boosting Decision Tree, firstly, we use "forward propagation" method to calculate all the $o_m, m = 1, \ldots, M$. Secondly, to (o_m), G_m are trained to reconstruct o_m via optimizing the loss function $L(o_m, G_m(F_m(o_m)))$. That is, we train G_m to learn "back-propagation." Then, after training all the $G_m, m = 1, \ldots, M$, we can do "back-propagation" to generate $z_m, m = 1, \ldots, M$ that represents the information to each layer. Next, to the pairs of (z_m, z_{m-1}), we train F_m to optimize another loss function $L(z_m, F_m(z_{m-1}))$. Finally, we can update all the F_m and G_m via Boosting Tree method. Algorithm 13 shows the procedure of Multi-Layered Gradient Boosting Decision Tree.

Algorithm 13 Multi-layered Gradient Boosting Decision Tree (Feng et al., 2018)

1. Input: Number of layers M, layer dimension d_m, samples $(y_i, x_i), i = 1, \ldots, n$. Loss function L. Hyper-parameters α, γ, K_1, K_2, T, σ^2.
2. Initially, set $F_m^0 = Initialize(M, d_m), m = 1, \ldots, M$;
3. For $t = 1$ to T
   a. Propagate the o_0 to calculate $o_m = F(o_{m-1}), m = 1, \ldots, M$
   b. $z_M^t = o_M - \alpha \frac{\partial L(y, o_M)}{\partial o_M}$
   c. For $m = M$ to 2
      i. $G_m^t = G_m^{t-1}$
      ii. $o_{m-1}^{noise} = o_{m-1} + \epsilon, \epsilon \sim N(0, diag(\sigma^2))$
      iii. $L_m^{inv} = L(o_m^{noise}, G_m^t(F_m^{t-1}(o_m^{noise})))$
      iv. for $k = 1$ to K_1
         A. $r_k = -\frac{\partial L_m^{inv}}{\partial G_m^t(F_m^{t-1}(o_m^{noise}))}$
         B. Fit a decision tree h_k to r_k
         C. $G_m^t = G_m^t + \gamma h_k$
      v. $z_{m-1} = G_m^t(z_m)$
   d. For $m = 1$ to M
      i. $F_m^t = F_m^{t-1}$
      ii. $L_m = L(z_m^t, F_m^t(o_{m-1}))$ using gradient boosting decision tree
      iii. for $k = 1$ to K_2
         A. $r_k = -\frac{\partial L_m}{\partial F_j^t(o_m)}$
         B. Fit a decision tree h_k to r_k
         C. $F_m^t = F_m^t + \gamma h_k$
      iv. $o_m = F_m^t(o_{m-1})$
4. Output the trained multi-layered gradient boosting decision tree.

Feng et al. (2018) suggested to optimize $L^{inv} = L(o_m^{noise}, G_m^t(F_m^{t-1}(o_m^{noise})))$ instead of $L_m^{inv} = L(o_m, G_m^t(F_m^{t-1}(o_m)))$ to make the training of G_m more robust. Also, the authors found that the multi-layered Gradient Boosting Decision Tree is very robust to most hyperparameters. Without fine-tuning the parameters, this method can achieve very attractive results.

Furthermore, consider the noisy loss function from the perspective of minimizing the reconstruction error, this process could be seen as an encoding-decoding process. First, in each layer F_m encodes the input via a nonlinear transform. Then, G_m learns how to decode the transformed output back to the original input. This is similar to the Auto Encoder method in deep learning. Thus, we can also use the Multi-layered Gradient Boosting Decision Tree to do encoding-decoding, which shed a light on implementing unsupervised learning tasks in the tabular data in economics.

14.7 Boosting in Macroeconomics and Finance

Boosting methods are widely used in classification and regression. Gradient Boosting implemented in the packages, like *XGBoost* and *LightGBM*, is a very popular algorithm among data science competitions and industrial applications. In this section, we discuss four applications of boosting algorithms in macroeconomics.

14.7.1 Boosting in Predicting Recessions

Ng (2014) uses boosting to predict recessions 3, 6, and 12 months ahead. Boosting is used to screen as many as 1500 potentially relevant predictors consisting of 132 real and financial time series and their lags. The sample period is 1961:1–2011:12. In this application, boosting is used to select relevant predictors from a set of potential predictors as well as probability estimation and prediction of the recessions. In particular, the analysis uses the Bernoulli loss function as implemented in the *GBM* package of Ridgeway (2007). The package returns the class probability instead of classifications. For recession analysis, the probability estimate is interesting in its own right, and the flexibility to choose a threshold other than one-half is convenient.

14.7.2 Boosting Diffusion Indices

Bai (2009) uses boosting to select and estimate the predictors in factor-augmented autoregressions. In their application, boosting is used to make 12 months ahead of forecast on inflation, the change in Federal Funds rate, the growth rate of industrial production, the growth rate of employment, and the unemployment rate. A sample period from 1960:1 to 2003:12 was used for a total of 132 times series. They use two boosting algorithms, namely L_2Boosting and Block Boosting.

14.7.3 Boosting with Markov-Switching

Adam, Mayr, and Kneib (2017) propose a novel class of flexible latent-state time series regression models called Markov-switching generalized additive models for location, scale, and shape. In contrast to conventional Markov-switching regression models, the presented methodology allows users to model different state-dependent parameters of the response distribution—not only the mean, but also variance, skewness, and kurtosis parameters—as potentially smooth functions of a given set of explanatory variables. The authors also propose an estimation approach based on the EM algorithm using the gradient boosting framework to prevent over-fitting while simultaneously performing variable selection. The feasibility of the suggested approach is assessed in simulation experiments and illustrated in a real-data setting, where the authors model the conditional distribution of the daily average price of energy in Spain over time.

14.7.4 Boosting in Financial Modeling

Rossi and Timmermann (2015) construct a new procedure for estimating the covariance risk measure in ICAPM model. First, one or more economic activity indices are extracted from macroeconomic and financial variables for estimating the covariance matrix. Second, given realized covariance matrix as the covariance matrix measure, Boosting Regression Tree is applied in projecting realized covariance matrix on the indices extracted in the first step. Lastly, predictions of the covariance matrix are made based on the nonlinear function approximated by Boosting Regression Tree and applied into the analysis of ICAPM method.

14.8 Summary

In this chapter, we focus on Boosting method. We start with an introduction of the well-known AdaBoost. Several variants of AdaBoost, like Real AdaBoost, LogitBoost, and Gentle AdaBoost are also discussed. Then, we consider in regression problem and introduce L_2Boosting. Next, Gradient Boosting and Gradient Boosting Decision Tree are discussed in theory and practice. Then, we introduce the several variants of Gradient Boosting such as Component-wise Boosting and Boost-GARCH for nonlinear time series modeling, Boosting with Momentum, and multi-layered Boosting Tree. Finally, we discuss several applications of Boosting in macroeconomic forecasting and financial modeling.

References

Adam, T., Mayr, A., & Kneib, T. (2017). *Gradient boosting in Markov-switching generalized additive models for location, scale and shape* (Unpublished paper). https://www.semanticscholar.org/paper/Gradient-boosting-in-Markov-switching-generalized-Adam-Mayr/ee085649e4fe36bb2f58015b6dc29870f34fa45e

Audrino, F., & Bühlmann, P. (2016). Volatility estimation with functional gradient descent for very high-dimensional financial time series. *The Journal of Computational Finance*, *6*(3), 65–89.

Bai, J. (2009). Panel data models with interactive fixed effects. *Econometrica*, *77*(4), 1229–1279.

Bartlett, P. L., Jordan, M. I., & McAuliffe, J. D. (2006). Convexity, classification, and risk bounds. *Journal of the American Statistical Association*, *101*(473), 138–156.

Bartlett, P. L., & Traskin, M. (2007). AdaBoost is consistent. *Journal of Machine Learning Research*, *8*, 2347–2368.

Bollerslev, T. (1986). Generalized autoregressive conditional heteroskedasticity. *Journal of Econometrics*, *31*(3), 307–327.

Breiman, L. (2004). Population theory for Boosting ensembles. *Annals of Statistics*, *32*(1), 1–11.

Bühlmann, P. (2003). *Bagging, subagging and bragging for improving some prediction algorithms*. Zürich: Seminar für Statistik, Eidgenössische Technische Hochschule (ETH).

Bühlmann, P. (2006). Boosting for high-dimensional linear models. *Annals of Statistics*, *34*(2), 559–583.

Chan, K. S., & Tong, H. (1986). On estimating thresholds in autoregressive models. *Journal of Time Series Analysis*, *7*(3), 179–190.

Chen, R., & Tsay, R. S. (1993). Nonlinear additive ARX models. *Journal of the American Statistical Association*, *88*(423), 955–967.

Chu, J., Lee, T.-H., & Ullah, A. (2018). Component-wise AdaBoost algorithms for high-dimensional binary classification and class probability prediction. In *Handbook of statistics*. Amsterdam: Elsevier.

Feng, J., Yu, Y., & Zhou, Z.-H. (2018). Multi-layered gradient boosting decision trees. In *Proceedings of the 32nd Conference on Neural Information Processing Systems*.

Freund, Y., & Schapire, R. E. (1997). A decision-theoretic generalization of on-line learning and an application to Boosting. *Journal of Computer and System Sciences*, *55*, 119–139.

Friedman, J. H. (2001). Greedy function approximation: a Gradient boosting machine. *The Annals of Statistics*, *29*, 1189–1232.

Friedman, J. H. (2002). Stochastic gradient boosting. *Computational Statistics and Data Analysis*, *38*(4), 367–378.

Friedman, J. H., Hastie, T., & Tibshirani, R. (2000). Additive logistic regression: A statistical view of boosting. *Annals of Statistics*, *28*(2), 337–407.

Hastie, T., Tibshirani, R., & Friedman, J. (2009). *The Elements of Statistical Learning*. Berlin: Springer

Matías, J., Febrero-Bande, M., González-Manteiga, W., & Reboredo, J. (2010). Boosting GARCH and neural networks for the prediction of heteroskedastic time series. *Mathematical and Computer Modelling*, *51*(3–4), 256–271.

Mease, D., Wyner, A., & Buja, A. (2007). Cost-weighted boosting with jittering and over/under-sampling: Jous-boost. *Journal of Machine Learning Research*, *8*, 409–439.

Mukherjee, I., Canini, K., Frongillo, R., & Singer, Y. (2013). Parallel boosting with momentum. In Blockeel, H., Kersting, K., Nijssen, S., Železný, F. (Eds.), *Machine learning and knowledge discovery in databases. ECML PKDD 2013*. Lecture notes in computer science (Vol. 8190). Berlin: Springer.

Ng, S. (2014). Viewpoint: Boosting recessions. *Canadian Journal of Economics*, *47*(1), 1–34.
Ridgeway, G. (2007). Generalized boosted models: A guide to the gbm package (Technical Report No. 4).
Robinzonov, N., Tutz, G., & Hothorn, T. (2012). Boosting techniques for nonlinear time series models. *AStA Advances in Statistical Analysis*, *96*(1), 99–122.
Rossi, A. G., & Timmermann, A. (2015). Modeling covariance risk in Merton's ICAPM. *Review of Financial Studies*, *28*(5), 1428–1461.
Schapire, R. E. (1990). The strength of weak learnability. *Machine Learning*, *5*, 197–227.
Tong, H., & Lim, K. S. (1980). Threshold autoregression, limit cycles and cyclical data. *Journal of the Royal Statistical Society. Series B (Methodological)*, *42*(3), 245–292.

Chapter 15
Density Forecasting

Federico Bassetti, Roberto Casarin, and Francesco Ravazzolo

15.1 Introduction

Economic decision in real time is made under a high degree of uncertainty. One of the prominent features of this uncertainty is that relevant information is missing at the moment of the decision. This requires to build forecasts to try to track the future evolution of the economic processes and to inform decision-makers. Researchers recognized the fundamental importance of forecasts a long time ago; but the focus was mainly on point forecasting. Point forecasting is often associated with the mean of a distribution and it is optimal for highly restricted loss functions, such as quadratic loss function. More generally, the value of a point forecast can be increased by supplementing it with some measure of uncertainty and complete probability distributions over outcomes provide information helpful for making economic decisions; see, for example, Anscombe (1968) and Zarnowitz (1969) for early works and the discussions in Granger and Pesaran (2000), Timmermann (2006), and Gneiting (2011). Recently, probabilistic forecasts in the form of predictive

F. Bassetti
Politcenico di Milano, Milan, Italy
e-mail: federico.bassetti@polimi.it

R. Casarin
University Ca' Foscari of Venice, Venice, Italy
e-mail: r.casarin@unive.it

F. Ravazzolo (✉)
Free University of Bozen-Bolzano, Bolzano, Italy

BI Norwegian Business School, Oslo, Norway
e-mail: francesco.ravazzolo@unibz.it

P. Fuleky (ed.), *Macroeconomic Forecasting in the Era of Big Data*,
Advanced Studies in Theoretical and Applied Econometrics 52,
https://doi.org/10.1007/978-3-030-31150-6_15

probability distributions have become prevalent in various fields, including macro economics with routine publications of fancharts from central banks, finance with asset allocation strategies based on higher-order moments, and meteorology with operational ensemble forecasts of future weather Tay and Wallis (2000), Gneiting and Katzfuss (2014). For example in central bank forecasting, the Bank of England, Norges Bank, Sveriges Riksbank publish so-called fancharts for macroeconomic variables such as inflation and GDP growth.

This chapter reviews several methods to construct density forecasts for parametric models. The first method assumes a distribution for the errors and ignore parameter uncertainty; the second method, bootstrapping, accounts for parameter and error uncertainties in a frequentist environment; and the third method relies on Bayesian inference. The three methods rely on different assumptions. We describe them in the case of the simple linear regression models and provide tools to extend the analysis to more complex models. We also discuss density combinations as a tool to deal in the case there are several density forecasts and an *a priori* selection is difficult. This is a challenging case for big data applications, where not all data have predictive power. And we provide some evaluation tools to measure the accuracy of density forecasts, accounting for the fact that the "true" density forecast is never observed, even *ex post*.

Moreover, in order to cope with the fact that relevant information is missing at the moment of decision, several papers (e.g., see Stock and Watson, 1999, 2002, 2005, 2014, and Bańbura, Giannone, and Reichlin, 2010) suggest to forecast with large sets of data. The recent fast growth in (real time) big data allows researchers to forecast variables of interest more accurately (e.g., see Choi and Varian, 2012; Einav and Levin, 2014; Varian, 2014; Varian and Scott, 2014). Stock and Watson (2005, 2014), Bańbura et al. (2010), and Koop and Korobilis (2013) suggest that there are also potential gains from forecasting using a large set of forecasts. However, forecasting with big data sets including many forecasts and high-dimensional models requires new modeling strategies, efficient inference methods, and extra computing power possibly resulting from parallel computing. We refer to Granger (1998) for an early discussion of these issues. In the application, we propose Graphical Processor Units (GPUs) as a tool to reduce computation time based on massively parallel computation and review the GPU computing functions introduced in the MATLAB parallel computing toolbox to reduce the steep learning curve of a dedicated programming language.

The structure of the chapter is organized as follows. Section 15.2 presents the different methods to compute density forecasts. Section 15.3 describes density combinations and Sect. 15.4 proposes different methods for density evaluation. Section 15.5 introduces GPU computing and applies to examples based on Monte Carlo (MC) simulations and an accept–reject algorithm to compute density. Section 15.6 concludes.

15.2 Computing Density Forecasts

This section reviews several methods to construct density forecasts. We discuss methodologies applied to a simple linear regression model:

$$y_t = \mathbf{x}_t'\boldsymbol{\beta} + \epsilon_t,\ t = 1, \ldots, T,\ \epsilon_t \sim i.i.d.(0, \sigma^2) \tag{15.1}$$

where $\theta = (\boldsymbol{\beta}, \sigma^2)$ is a $((m+1) \times 1)$ vector of parameters; $\boldsymbol{\beta}$ a $(m \times 1)$ vector of coefficients; σ^2 the variance of the error term ε_t; and $\mathbf{x}_t$ is a $(m \times 1)$ vector of covariates, which can include exogenous variables $\mathbf{z}_t$ and lagged values of the dependent variable, y_{t-p}, $p > 0$.

We present three methods to deal with constructing density forecasts: assume a distribution for the errors and ignore parameter uncertainty; bootstrapping for accounting for parameter and error uncertainties in a frequentist environment; and Bayesian inference. The three methods rely on different assumptions. The first one requires to specify a distribution for a given model; the second one requires some assumptions and can be applied to any model that respects such assumptions; the third one requires prior information that are usually model dependent.

15.2.1 Distribution Assumption

The easiest method to compute a density forecast is to assume a given distribution for the error term, e.g., $\epsilon_t \sim N(0, \sigma^2)$ in (15.1), and to ignore parameter uncertainty. The $h-$step ahead density prediction, with $h > 1$, conditional to information available up to time T, $\mathcal{D}^T$, results to

$$f(y_{T+h}|\mathcal{D}^T) = N(\mathbf{x}_{T+h}'\mathbf{b}, s^2) \tag{15.2}$$

where $\mathbf{b} = (\mathbf{X}'\mathbf{X})^{-1}\mathbf{X}'\mathbf{y}$, with $\mathbf{y} = (y_1, \cdots, y_T)'$ a $(T \times 1)$ vector, $\mathbf{X} = (\mathbf{x}_1, \cdots, \mathbf{x}_T)'$ a $(T \times m)$ matrix, and $s^2 = \mathbf{e}'\mathbf{e}/(T-m)$, with $\mathbf{e} = (\mathbf{y} - \mathbf{X}\mathbf{b})$. In the linear model (15.1) there is a closed-form solution accounting for parameter uncertainty, see for example Hansen (2006). Simple modifications of that model have also closed-form solution. For example, Clements and Galvao (2014) show how to compute the variance of the MIDAS predictive density to also account for parameter uncertainty.

The expression in (15.2) requires to know $\mathbf{x}_{T+h}$. This is possible only in limited cases where the data generating process of $\mathbf{X}$ is known. In most cases, in particular when $\mathbf{x}_t$ includes also lags of y_t, this condition is not valid. There are several options to deal with it. For example, $\mathbf{x}_{T+h}$ can fixed to include only information up to time T, that is x_T is a function of exogenous $(\mathbf{z}_1, \ldots, \mathbf{z}_T)$ and lagged dependent $(y_T, \ldots, y_{T-p})$ variables. This strategy is often called direct forecasting and regressors in (15.1) should be changed accordingly. Otherwise, the system can be iterated to produce future values, that is $\mathbf{x}_{T+j}$ is computed conditional on $\mathbf{x}_{T+j-1}$ for $j = 1, \ldots, h$.

A special case is when $\mathbf{x}_t$ contains only lags of y_t. The density forecasts changes expression. The variable y_t can be expressed as a function of past errors and initial values as

$$y_t = \sum_{j=0}^{t-1} \phi_j \varepsilon_{t-j} + \pi, \ \varepsilon_t \sim \text{i.i.d.} N(0, \sigma^2)$$

where ϕ_j is the moving average parameter of order j, π summarizes the initial conditions. Assuming that the past errors and coefficients are known, the conditional expectation corresponds to the point forecast

$$y_{T+h} = \sum_{j=h}^{T-1} \phi_j \varepsilon_{T+h-j} + \pi$$

and the forecast error is $\sum_{j=0}^{h-1} \phi_j \varepsilon_{t+h-j}$. It follows that the forecast error variance is given by $s^2(h) = \sigma^2 \sum_{j=0}^{h-1} \phi_j^2$. The predictive density is therefore normally distributed with mean given by the usual point forecast and variance given by the above expression, $N(\mathbf{x}'_{T+h}\mathbf{b}, s^2(h))$.

15.2.2 Bootstrapping

Ignoring parameter and distribution uncertainties can be very costly, in particular for small sample sizes and when the error distribution is not Gaussian, see Pascual, Romo, and Ruiz (2001). A solution to it is to apply a bootstrapping approach. The bootstrapping procedures are distribution-independent and account for parameter uncertainty.

Earlier studies in economics have mostly focused on bootstrapping in linear regressions and univariate autoregressions, see e.g., Berkowitz and Kilian (2000) and Clements and Taylor (2001). More recently, bootstrapping procedures for more advanced models have been proposed. These include models that deal with a large amount of data such as factor models, see e.g., Goncalves and Perron (2014), Djogbenou, Goncalves, and Perron (2015), Djogbenou, Goncalves, and Perron (2017), models with mixed frequency information, see Aastveit, Gerdrup, Jore, and Thorsrud (2014) and Mixed Data Sampling (MIDAS) models, see Aastveit, Foroni, and Ravazzolo (2016).

A Residual-Based Bootstrapping of Density Forecasts

We first consider a parametric residual-based bootstrap to derive forecast densities, accounting for both parameter and shock uncertainty as in Berkowitz and Kilian (2000) and Clements and Taylor (2001). The bootstrap procedure relies on the algorithm in Davison and Hinkley (1997) (Section 7.2.4) for prediction in generalized linear models. The residual-based bootstrap is valid under the following assumptions:

(A1) ε_t are i.i.d. with $E(\varepsilon_t) = 0$, $E(\varepsilon_t^2) = \sigma^2$ with $\sigma^2 < \infty$, and $E(\varepsilon_t^{2(s+1)}) < \infty$ for $s \geq 3$.
(A2) $(\varepsilon_1, \varepsilon_1^2)$ satisfies Cramer's condition, i.e., for every $d > 0$, there exists δ such that $sup_{||t||>d} \exp(\varepsilon_1, \varepsilon_1^2)| \leq \exp(-\delta)$.
(A3) $x_{t_m+w-h_m}^{(m)}$ are exogenous fixed variables.
(A4) The process is stationary.

The steps conducted in the residual-based bootstrap are as follows.

1. Estimate equation (15.1), and obtain $\mathbf{b}$.
2. For $r = 1, \ldots, R$, simulate $\widetilde{y}_{r,t} = \mathbf{x}_t\mathbf{b} + \widetilde{e}_{r,t}$, where $\widetilde{e}_{r,t}$ is resampled from $\check{e}_t \equiv \left(\frac{n}{n-k}\right)^{0.5} e_t$.[1]
3. Re-estimate (15.1) for each $\widetilde{y}_{r,t}$, and obtain $\widetilde{y}_{r,T+h}$, where the shock uncertainty is included by resampling from $\check{e}_t$.

Davison and Hinkley (1997) fix the value of y_T equal to the value of the original series.

In practice, R vectors of pseudo-random numbers are generated to replicate the same properties of the residuals of the model, via the bootstrapping technique. For each $r = 1, \ldots, R$ replications, a new set of simulated data is generated, and a new forecast $\widetilde{y}_{r,T+h}$ is obtained. The empirical distribution of $\left\{\widetilde{y}_{r,T+h}\right\}_{r=1}^{R}$ is then our density.

If the error terms in Eq. (15.1) are independent and identically distributed with common variance, then we can generally make very accurate inferences by using the residual bootstrap. Given the assumptions (A1)–(A4), Davison and Hinkley (1997) discuss how the method is a generalization of the bootstrapping algorithm for linear models and Bose (1988) provides proofs of its convergence.[2]

[1]Davidson and MacKinnon (2006) suggest to rescale the residuals so that they have the correct variance by $\check{e}_t \equiv \left(\frac{n}{n-k}\right)^{0.5} \widehat{e}_t$.

[2]Bose (1988) focuses on linear AR models with imposed stationarity (see assumption (A4) above). For an extension accounting for a possible unit root, see Inoue and Kilian (2002).

Accounting for Autocorrelated or Heteroskedastic Errors

One limitation of the standard residual-based bootstrapping method above is that it treats the errors as i.i.d. The i.i.d. assumption does not follow naturally from economic models, and in many empirical applications the actual data are not well represented by models with i.i.d. errors, see e.g. Goncalves and Kilian (2004) and Davidson and MacKinnon (2006). Typically, economic and financial variables exhibit evidence of autocorrelation and/or conditional heteroskedasticity. In these cases, assumption **(A1)** is violated and the residual-based bootstrap is not valid.

Block bootstrap methods, suggested by Hall (1985) and Kunsch (1989), account for autocorrelated errors. The block bootstrap divides the quantities that are being resampled into blocks of b consecutive observations. The blocks can be either overlapping or non-overlapping, nevertheless Andrews (2002) finds small differences in performance between the two methods.

The wild bootstrap suggested by Wu (1986) and Liu (1988) is specifically designed to handle heteroskedasticity in regression models. Goncalves and Kilian (2004) have also shown that heteroskedasticity is an important feature in many macroeconomic and financial series and apply the wild bootstrap to autoregressive models.

A Block Wild Bootstrapping of Density Forecasts

To account for both autocorrelation and heteroskedasticity at the same time, we suggest using a block wild bootstrap, first proposed by Yeh (1998). Djogbenou et al. (2015, 2017) have recently proposed adapting the block wild bootstrap to the case of factor models and Aastveit et al. (2016) to MIDAS models. Non-overlapping blocks of size n_T of consecutive residuals are formed. Assume that $(T-h)/n_T = k_T$, where k_T is an integer and denotes the number of blocks of size n_T. For $l = 1, \ldots, b_T$ and $j = 1, \ldots, k_T$, we let

$$y^*_{(j-1)n_T+l+h} = \mathbf{x}_{(j-1)n_T+l}\mathbf{b} + e^*_{(j-1)n_T+l+h}$$

where

$$e^*_{(j-1)n_T+l+h} = \check{e}_{(j-1)n_T+l+h} \cdot \nu_j$$

There are various ways to specify the distribution of ν_j. Davidson and Flachaire (2008) assume that ν_j is a Rademacher random variable

$$\nu_j = \begin{cases} 1 \text{ with probability } 1/2 \\ -1 \text{ with probability } 1/2 \end{cases}$$

Davidson and Flachaire (2008) study the wild bootstrap in the context of regression models with heteroskedastic disturbances and find that, among several popular candidates, this has the most desirable properties.

By replacing step 2 and 3 in the residual-based bootstrap above with the block wild bootstrap it is possible to accommodate both serial correlation and heteroskedasticity in $\widetilde{e}_{r,t}$. Djogbenou et al. (2017) set the block size equal to h.

15.2.3 Bayesian Inference

A different approach to construct density forecasts rely on Bayesian inference. Bayesian analysis formulates prior distributions on parameters that multiplied by the likelihood results on parameter posterior distributions. Accounting for the uncertainty on parameter posterior distribution, future probabilistic statements derive without any further assumption. Moreover, prior distributions allow to impose restrictions on the parameters if useful and necessary. However, a user must specify prior statements before to start the analysis.

As example, we present the main derivation for model (15.1). The objective of Bayesian inference is to compute a predictive density

$$f(y_{T+h}|\mathcal{D}^T) = \int p(y_{T+h}, \mathbf{x}_{T+h}, \boldsymbol{\theta}|\mathcal{D}^T)d\boldsymbol{\theta} = \int l(y_{T+h}|\mathbf{x}_{T+h}, \boldsymbol{\theta}, \mathcal{D}^T)p(\boldsymbol{\theta}|\mathcal{D}^T)d\boldsymbol{\theta} \tag{15.3}$$

where $\mathcal{D}^T = (\mathbf{y}, \mathbf{X}, \mathbf{x}_{T+h})$ is the information set, $l(y_{T+h}|\mathbf{x}_{T+h}, \boldsymbol{\theta})$ is the likelihood of the model for time $T+h$, $p(\boldsymbol{\theta}|\mathcal{D}^T)$ is the parameter marginal distribution computed with information up to time T.

Regarding the choice of the prior distribution, if the prior is conjugate then the posterior and the predictive distribution can be computed analytically. If non-conjugate priors are used, then posterior and predictive are in integral form and need to be evaluated by means of numerical methods such as Monte Carlo simulation methods. In the regression model, in practice one usually defines $\tau = 1/\sigma^2$ and assumes a conjugate normal-gamma prior

$$\boldsymbol{\beta}|\tau \sim N(\underline{\boldsymbol{\beta}}, \tau^{-1}\underline{\mathbf{V}}), \quad \tau \sim G(\underline{s}^{-2}, \underline{\nu}), \quad \boldsymbol{\beta}, \tau \sim NG(\underline{\boldsymbol{\beta}}, \underline{\mathbf{V}}, \underline{s}^{-2}, \underline{\nu})$$

where $\underline{\boldsymbol{\beta}}$, $\underline{\mathbf{V}}$, $\underline{s}^{-2}$, and $\underline{\nu}$ are parameters of the normal and gamma prior distributions. Define $\overline{\mathbf{V}} = (\mathbf{V}^{-1} + \mathbf{X}'\mathbf{X})^{-1}$, $\overline{\boldsymbol{\beta}} = \overline{\mathbf{V}}(\underline{\mathbf{V}}^{-1}\underline{\boldsymbol{\beta}} + \mathbf{b}\mathbf{X}'\mathbf{X})$, $\overline{\nu} = \underline{\nu} + T$, $\overline{\nu s}^2 = \underline{\nu s}^2 + \nu s^2 + (\mathbf{b} - \underline{\boldsymbol{\beta}})'(\underline{\mathbf{V}} + (\mathbf{X}'\mathbf{X})^{-1})^{-1}(\mathbf{b} - \underline{\boldsymbol{\beta}})$, $\nu s^2 = (\mathbf{y} - \mathbf{X}\mathbf{b})'(\mathbf{y} - \mathbf{X}\mathbf{b})$, the conditional posteriors of $\boldsymbol{\beta}$ given σ^2 and σ^2 given $\boldsymbol{\beta}$ are

$$p(\boldsymbol{\beta}|\tau, \mathbf{y}) \sim N(\overline{\boldsymbol{\beta}}, \tau^{-1}\overline{\mathbf{V}}), \quad p(\tau|\boldsymbol{\beta}, \mathbf{y}) \sim G(\overline{\nu s}^2, \overline{\nu})$$

See Koop (2003). The target is the marginal posterior distribution, that has a closed-form solution for model (15.1) and Normal-gamma priors

$$\begin{aligned}\boldsymbol{\beta}|\mathcal{D}^T &\sim t(\overline{\boldsymbol{\beta}},\overline{s}^2\overline{\mathbf{V}},\overline{\nu})\\ \tau|\mathcal{D}^T &\sim G(\overline{\nu s}^2,\overline{\nu}-2)\end{aligned}$$

The conditional and marginal predictive densities have also a closed-form solution, see Koop (2003):

$$\begin{aligned}f(y_{T+h}|\boldsymbol{\beta},\sigma,\mathcal{D}^T) &\sim N(\mathbf{x}_{T+h}\overline{\boldsymbol{\beta}},\overline{s}^2\widetilde{\mathbf{x}}'\widetilde{\mathbf{X}})\\ f(y_{T+h}|\mathcal{D}^T) &\sim t(\mathbf{x}_{T+h}\overline{\boldsymbol{\beta}},\overline{s}^2(\mathbf{I}_T+\widetilde{\mathbf{X}}\overline{\mathbf{V}}\widetilde{\mathbf{X}}),\overline{\nu})\end{aligned}$$

As in the normal and bootstrapping cases, computation is more complex when $\mathbf{x}_{T+h}$ is not available at time T and direct forecasting is avoided. The algorithm in (15.3) generalizes to

$$f(\mathbf{y}_{T+h}|\mathcal{D}^T)=\int l(y_{T+h}|\mathbf{x}_{T+h},\theta)p(\mathbf{x}_{T+h}|\mathbf{X},\boldsymbol{\theta})p(\boldsymbol{\theta}|\mathcal{D}^T)d\boldsymbol{\theta} \tag{15.4}$$

Closed-form solutions do not exist for most of economic models, but simulation methods can be used to compute the integral and derive the marginal predictive density. Assume a set of random samples $\boldsymbol{\theta}_r$, $r=1,\ldots,R$ from $p(\boldsymbol{\theta}|\mathcal{D}^T)$ is available, then the predictive density in Eq. (15.4) can be approximated as follows

$$\hat{f}_R(y_{T+h}|\mathcal{D}^T)=\frac{1}{R}\sum_{r=1}^{R}l(y_{T+h}|\mathbf{x}_{T+h},\boldsymbol{\theta}_r)p(\mathbf{x}_{T+h}|\mathbf{X},\boldsymbol{\theta}_r)$$

See Sect. 15.5 for an introduction to simulation methods.

15.3 Density Combinations

When multiple forecasts are available from different models or sources it is possible to combine these in order to make use of all relevant information on the variable to be predicted and, as a consequence, to produce better forecasts. This is particular important when working with large database and selection of relevant information *a priori* is not an easy task. Early papers on forecasting with model combinations are Barnard (1963), who considered air passenger data, and Roberts (1965) who introduced a distribution which includes the predictions from two experts (or models). This latter distribution is essentially a weighted average of the posterior distributions of two models and is similar to the result of a Bayesian Model Averaging (BMA) procedure. See Raftery, Madigan, and Hoeting (1997) for a review on BMA, with a historical perspective and Hofmarcher and Grün (2019) for a recent overview on the statistical methods and foundations of BMA. Raftery, Gneiting, Balabdaoui, and Polakowski (2005) and Sloughter, Gneiting, and Raftery (2010) extend the BMA framework by introducing a method for obtaining probabilistic forecasts from ensembles in the form of predictive densities and apply it to weather forecasting. McAlinn and West (2019) extend it to Bayesian predictive synthesis.

Bates and Granger (1969) deal with the combination of predictions from different forecasting models using descriptive regression. Granger and Ramanathan (1984) extend this and propose to combine forecasts with unrestricted regression coefficients as weights. Terui and van Dijk (2002) generalize the least square weights by representing the dynamic forecast combination as a state space with weights that are assumed to follow a random walk process. Guidolin and Timmermann (2009) introduce Markov-switching weights, and Hoogerheide, Kleijn, Ravazzolo, van Dijk, and Verbeek (2010) propose robust time-varying weights and account for both model and parameter uncertainty in model averaging. Raftery, Karny, and Ettler (2010) derive time-varying weights in "dynamic model averaging," following the spirit of Terui and van Dijk (2002), and speed up computations by applying forgetting factors in the recursive Kalman filter updating.

A different line was started by Ken Wallis in several papers, see for example Wallis (2003, 2005, 2011) and Mitchell and Wallis (2011). Here the use of the full predictive distribution is proposed when forecasting. Benefits and problems related to it are discussed in detail. One focus has been to measure to the importance of density combinations. Hall and Mitchell (2007) introduce the Kullback–Leibler divergence as a unified measure for the evaluation and suggest weights that maximize such a distance, see also Amisano and Geweke (2010) and Geweke and Amisano (2011) for a compressive discussion on how such weights are robust to model incompleteness, that is the true model is not included in the model set. Gneiting and Raftery (2007) recommend strictly proper scoring rules, such as the cumulative rank probability score. Billio, Casarin, Ravazzolo, and van Dijk (2013) develops a general method that can deal with most of issues discussed above, including time-variation in combination weights, learning from past performance, model incompleteness, correlations among weights and joint combined predictions of several variables. See, also Waggoner and Zha (2012), Kapetanios, Mitchell, Price, and Fawcett (2015), Pettenuzzo and Ravazzolo (2016), Aastveit, Ravazzolo, and van Dijk (2018), and Del Negro, Hasegawa, and Schorfheide (2016).

We refer to Aastveit, Mitchell, Ravazzolo, and van Dijk (2019) for a recent survey on the evolution of forecast density combinations in economics. In the following we provide some details on two basic methodologies, the Bayesian Model Averaging (BMA) and the linear opinion pool (LOP), and discuss briefly some extensions.

15.3.1 Bayesian Model Averaging

Let $\mathcal{D}^T$ be the set of information available up to time t, then BMA combines the individual forecast densities $f(y_{T+h}|\mathcal{D}^T, M_j)$, $i = 1, \ldots, N$, into a composite-weighted predictive distribution $f(y_{T+h}|\mathcal{D}^T)$ given by

$$f(y_{T+h}|\mathcal{D}^T) = \sum_{j=1}^{N} P\left(M_j | \mathcal{D}^T\right) f(y_{T+h}|\mathcal{D}^T, M_j)$$

where $P\left(M_j | \mathcal{D}^T\right)$ is the posterior probability of model j, derived by Bayes' rule,

$$P\left(M_j | \mathcal{D}^T\right) = \frac{P\left(\mathcal{D}^T | M_j\right) P\left(M_j\right)}{\sum_{j=1}^{N} P\left(\mathcal{D}^T | M_j\right) P\left(M_j\right)}, \qquad j = 1, \ldots, N$$

and where $P\left(M_j\right)$ is the prior probability of model M_j, with $P\left(\mathcal{D}^T | M_j\right)$ denoting the corresponding marginal likelihood. We shall notice that the model posterior probability can be written in terms of Bayes factors

$$P\left(M_j | \mathcal{D}^T\right) = \frac{\alpha_j b_{1j}}{\sum_{j=2}^{N} \alpha_j b_{1j}}$$

where $\alpha_j = P(M_j)/P(M_1)$ and $b_{1j} = P\left(\mathcal{D}^T | M_j\right) / P\left(\mathcal{D}^T | M_1\right)$, $j = 2, \ldots, N$ are the Bayes factors. An alternative averaging weighting scheme can be define by using the predictive distributions

$$P\left(M_j | \mathcal{D}^T\right) = \frac{P\left(y_T | \mathcal{D}^{T-1}, M_j\right) P\left(M_j\right)}{\sum_{j=1}^{N} P\left(y_T | \mathcal{D}^{T-1}, M_j\right) P\left(M_j\right)}, \qquad j = 1, \ldots, N$$

15.3.2 Linear Opinion Pool

Linear Opinion Pool (LOP) gives a predictive density $f(y_{T+h}|\mathcal{D}^T)$ for the variable of interest to be predicted at horizon $T+h$ with $h > 0$, y_{T+h}, using the information available up to time T, $\mathcal{D}^T$, from a set of predictions generated by the models M_j, $j = 1, \ldots, N$.

$$f(y_{T+h}|\mathcal{D}^T) = \sum_{j=1}^{N} w_{j,T+h} f(y_{T+h}|\mathcal{D}^T, M_j)$$

where $w_{j,T+h}$ is the (0, 1)-valued weight given to model M_j computed at time T and $f(y_{T+h}|\mathcal{D}^T, M_j)$ is the density forecast of y_{T+h} conditional on predictor M_j, and on the information available up to time T. The individual prediction can be model based, parametric or non-parametric, or individual subjective predictions. Each of these predictive densities must be non-negative for all the support of y_{T+h} and their cumulative density functions must add to 1. To guarantee that the combined forecast density $f(y_{T+h})$ also satisfies these features, some restrictions can be imposed to the combination weights $w_{j,T+h}$, $j = 1, \ldots, N$. Sufficient conditions are that weights are non-negative, $w_{j,T+h} \geq 0$, $j = 1, \ldots, N$, and that add to unity, $\sum_{j=1}^{N} w_{j,T+h} = 1$.

Standard practice, see for example Hall and Mitchell (2007), Kascha and Ravazzolo (2010), and Mazzi, Mitchell, and Montana (2014), is to use the cumulative log score, see Eq. (15.5). The combination weights are computed as

$$w_{j,T+h}^{LS} = \frac{\exp(\eta_{j,T}^{LS})}{\sum_{j=1}^{N} \exp(\eta_{j,T}^{LS})}$$

where $\eta_{j,T}^{LS}$ is the cumulative log score for model M_j at time T. We note that at time T when predictions are made, the cumulative log score can be computed up to the same time and therefore weights are based on the statistic $\eta_{j,T}^{LS}$, $j = 1, \ldots, N$. Such statistic contains information on how the predictor M_j associated with prediction $f(y_{T+h}|\mathcal{D}^T, M_j)$ has performed in the past in terms of forecasting. Therefore, the major difference between LOP and BMA is in weights definition. BMA weights depend on model posterior probabilities; LOP weights are computed using distance measures.

15.3.3 Generalized Opinion Pool

Following the notation used in Gneiting and Ranjan (2013), it is possible to define a general pooling method as a parametric family of combination formulas. Let $F_{jT}(y_{T+h}) = F(_{T+h}|\mathcal{D}^T, M_j)$ denote the cdf of the density $f(y_{T+h}|\mathcal{D}^T, M_j)$, a generalized pool is a map

$$H : \begin{bmatrix} \times^N \mathcal{F} \rightarrow \mathcal{F} \\ (F_{1T}(\cdot), \cdots, F_{NT}(\cdot)) \mapsto F(\cdot|\xi, \mathcal{D}^T) = H(F_{1T}(\cdot), \ldots, F_{NT}(\cdot), \xi) \end{bmatrix}$$

indexed by the parameter $\xi \in \Xi$, where Ξ is a parameter space and $\mathcal{F}$ is a suitable space of distributions. Following (see DeGroot, Dawid, & Mortera, 1995; DeGroot & Mortera, 1991) we consider pooling scheme of the form

$$H(F_{1T}(\cdot), \ldots, F_{NT}(\cdot), \xi) = \varphi^{-1}\left(\sum_{j=1}^{N} \omega_j \varphi(F_{jT}(\cdot)\right)$$

where φ is a continuous increasing monotone function with inverse φ^{-1} and $\xi = (\omega_1, \cdots, \omega_N)'$ is a vector of combination weights, with $\omega_1 + \ldots + \omega_N = 1$ and $\omega_j \geq 0$, for all i. If $\varphi(x) = x$ then we obtain the Linear Opinion Pool

$$F(y_{T+h}|\mathcal{D}^T, \xi) = \sum_{j=1}^{N} \omega_j F(y_{T+h}|\mathcal{D}^T, M_j)$$

The harmonic opinion pool is obtained for $\varphi(x) = 1/x$

$$F(y_{T+h}|\mathcal{D}^T, \xi) = \left(\sum_{j=1}^{N} \omega_j F(y_{T+h}|\mathcal{D}^T, M_j)^{-1} \right)^{-1}$$

whereas by choosing $\varphi(x) = \log(x)$ one obtains the logarithmic opinion pool

$$F(y_{T+h}|\mathcal{D}^T, \xi) = \prod_{j=1}^{N} F(y_{T+h}|\mathcal{D}^T, M_j)^{\omega_j}$$

If φ is differentiable then the generalized combination model can be re-written in terms of pdf as follows

$$f(y_{T+h}|\mathcal{D}^T, \xi) = \frac{1}{\varphi'(F(y_{T+h}|\mathcal{D}^T, \xi))} \sum_{j=1}^{N} \omega_j \varphi'(F(y_{T+h}|\mathcal{D}^T, M_j)) f({}_{T+h}y|\mathcal{D}^T, M_j)$$

where φ' denotes the first derivative of φ. The related density functions are

$$f(y_{T+h}|\mathcal{D}^T, \xi) = \sum_{j=1}^{N} \omega_j f(y_{T+h}|\mathcal{D}^T, M_j)$$

$$f(y_{T+h}|\mathcal{D}^T, \xi) = F(y_{T+h}|\mathcal{D}^T, \xi)^2 \sum_{j=1}^{N} \omega_j F(y_{T+h}|\mathcal{D}^T, M_j)^{-2} f(y_{T+h}|\mathcal{D}^T, M_j)$$

$$f(y_{T+h}|\mathcal{D}^T, \xi) = F(y_{T+h}|\mathcal{D}^T, \xi) \sum_{j=1}^{N} \omega_j F(y_{T+h}|\mathcal{D}^T, M_j)^{-1} f(y_{T+h}|\mathcal{D}^T, M_j)$$

for the linear opinion pool, harmonic opinion pool and logarithmic opinion pool, respectively. Generalized combination schemes have developed further in Kapetanios et al. (2015) and Bassetti, Casarin, and Ravazzolo (2018). We illustrate the three combination methods by assuming that two density forecasts are available, $F(y_{T+h}|\mathcal{D}^T, M_1) \sim N(4, 1)$ and $F(y_{T+h}|\mathcal{D}^T, M_2) \sim N(0, 2)$, and an equally

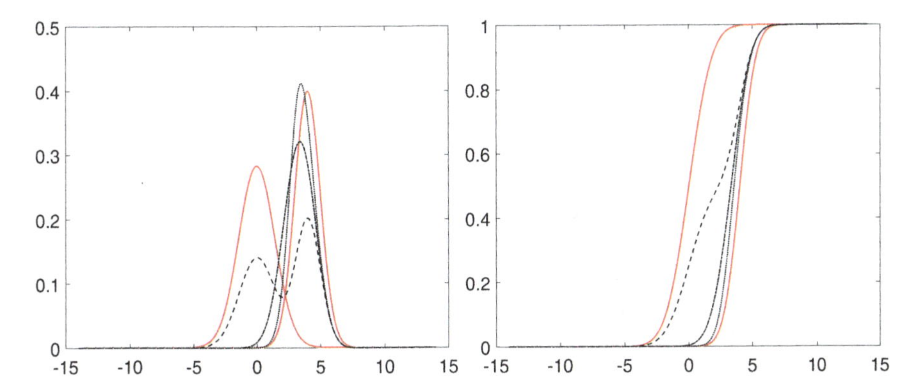

Fig. 15.1 Pdfs (left) and cdfs (right) of the two forecasting models $F(y|\mathcal{D}^T, M_1) \sim N(4, 1)$ and $F(y|\mathcal{D}^T, M_2) \sim N(0, 2)$ (red solid lines) and of their linear (dashed), harmonic (dotted), and logarithmic (dotted-dashed) combination

weighted pooling is used ($\omega_1 = \omega_2 = 0.5$). From Fig. 15.1 one can see that harmonic and logarithmic pools concentrate the probability mass on one of the model in the pool.

15.4 Density Forecast Evaluation

The density of the variable of interest y_{T+h} at given time $T + h$ is never observed. This complicates the evaluation of density forecasts. In economics, there are two main approaches to evaluate density forecasts. The first one is based on properties of a density and refers to absolute accuracy. The second one is based on comparison of different forecasts and refers to relative accuracy.

15.4.1 Absolute Accuracy

The absolute accuracy can be studied by testing forecast accuracy relative to the "true" but unobserved density. Dawid (1982) introduced the criterion of *complete calibration* for comparing prequential probabilities with binary random outcomes. This criterion requires that the averages of the prequential probabilities and of the binary outcomes converges to the same limit. For continuous random variables Dawid (1982) exploited the concept of probability integral transform (PIT) that is the value that a predictive cdf attains at the observations. The PITs summarize the properties of the densities and may help us judge whether the densities are biased in a particular direction and whether the width of the densities has been roughly correct on average Diebold, Gunther, and Tay (1998). More precisely, the PITs represents the ex-ante inverse predictive cumulative distributions, evaluated at the ex-post actual observations. The PIT at time T are

$$PIT_{T+h} = \int_{-\infty}^{y_{T+h}} f(y_{T+h}|\mathcal{D}^T)dy$$

and should be uniformly, independently, and identically distributed if the h-step-ahead forecast densities $f(y_{T+h}|\mathcal{D}^T)$ conditional on the information set available at time T, are correctly calibrated.

As an example assume that a set of observations are generated from a standard normal, $Y_t \sim N(0, 1)$, i.i.d. $t = T+1, \ldots, T+1000$ and that four predictive cdfs are used:

$$\begin{array}{ll} F(y_{T+h}|\mathcal{D}^T, M_1) \sim N(0.5, 1), & F(y_{T+h}|\mathcal{D}^T, M_2) \sim N(0, 2) \\ F(y_{T+h}|\mathcal{D}^T, M_3) \sim N(-0.5, 1), & F(y_{T+h}|\mathcal{D}^T, M_4) \sim N(0, 0.5) \end{array}$$

The first model is wrong in predicting the mean of the distribution, the second one is wrong in predicting the variance. In Fig. 15.2, which show the cdfs of PITs. In each plot the red line indicates the PITs of the true model. Errors in mean induce a cdf that overestimate (left plot) or underestimate (right plot), depending on error sign, the "true" cumulative density function. Variance overestimation appears as an underestimate in the left side of the distribution, and an overestimate in the right side, whereas variance underestimation appears as an overestimate in the left side of the distribution, and an underestimate in the right side. In both cases, the discontinuity point corresponds at the mean, in which the two line intersect.

Calibration can be gauge by testing jointly for uniformity and (for one-step ahead forecasts) independence of the PITs, applying the tests proposed by Berkowitz

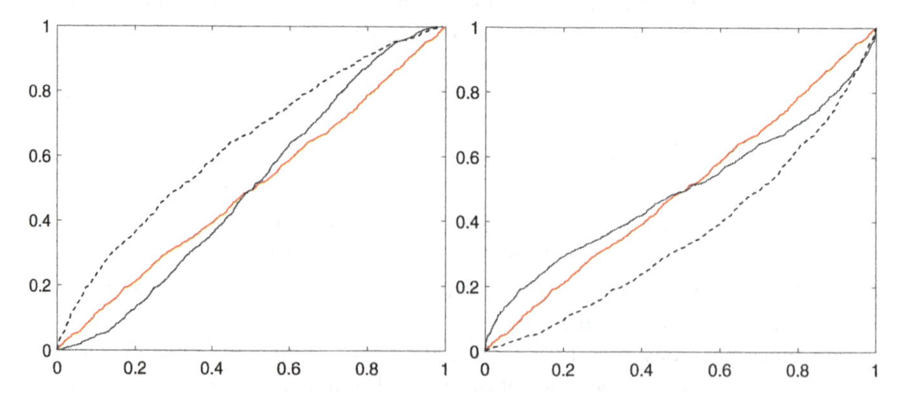

Fig. 15.2 Empirical cdfs of the PITs. Left: PITs generated by $F(y_{T+h}|\mathcal{D}^T, M_1) \sim N(0.5, 1)$ (dashed line), $F(y_{T+h}|\mathcal{D}^T, M_2) \sim N(0, 2)$ (dotted line). Right: PITs generated by $F(y_{T+h}|\mathcal{D}^T, M_3) \sim N(-0.5, 1)$ (dashed line), $F(y_{T+h}|\mathcal{D}^T, M_4) \sim N(0, 0.5)$ (dotted line). In each plot the red solid line indicates the PITS of the true model ($N(0, 1)$)

(2001) and Knuppel (2015).[3] Rossi and Sekhposyan (2013) extend the evaluation in the presence of instabilities; Rossi and Sekhposyan (2014) apply to large database and Rossi and Sekhposyan (2019) compare alternative tests for correct specification of density forecasts.

15.4.2 Relative Accuracy

When moving to relative comparison, density forecasts can be evaluated by the Kullback–Leibler Information Criterion (KLIC)-based measure, utilizing the expected difference in the Logarithmic Scores of the candidate forecast densities; see, for example, Mitchell and Hall (2005), Hall and Mitchell (2007), Kascha and Ravazzolo (2010), and Billio et al. (2013). The KLIC is the distance between the true density $p(y_{T+h}|\mathcal{D}^T)$ of a random variable y_{T+h} and some candidate density $f(y_{T+h}|\mathcal{D}^T, M_j)$ obtained from the model M_j and chooses the model that on average gives the higher probability to events that actually occurred. An estimate of it can be obtained from the average of the sample information, $y_{\underline{T}+1}, \ldots, y_{\overline{T}+1}$, on $p(y_{T+h})$ and $f(y_{T+h}|\mathcal{D}^T, M_j)$:

$$\overline{KLIC}_{j,h} = \frac{1}{T^*} \sum_{T=\underline{T}}^{\overline{T}} [\ln p(y_{T+h}|\mathcal{D}^T) - \ln f(y_{T+h}|\mathcal{D}^T, M_j)]$$

where $T^* = (\overline{T} - \underline{T} + 1)$. Although we do not know the true density, we can still compare different densities, $f(y_{T+h}|M_j)$. For the comparison of two competing models, it is sufficient to consider the Logarithmic Score (LS) given as

$$LS_{j,h} = -\frac{1}{T^*} \sum_{T=\underline{t}}^{\overline{T}} \ln f(y_{T+h}|\mathcal{D}^T, M_j) \tag{15.5}$$

for all j and choose the model for which this score is minimal.

Alternative, density forecasts can be evaluated on the continuous rank probability score (CRPS); see, for example, Gneiting and Raftery (2007), Gneiting and Ranjan (2013), Groen, Paap, and Ravazzolo (2013), and Ravazzolo and Vahey (2014). The CRPS for the model j measures the average absolute distance between the empirical cumulative distribution function (CDF) of y_{T+h}, which is simply a step function in y_{T+h}, and the empirical CDF that is associated with model j's predictive density:

$$\text{CRPS}_{j,T+h} = \int_{-\infty}^{+\infty} \left(F(y_{T+h}|\mathcal{D}^T, M_j) - \mathbb{I}_{[y_{T+h},+\infty)}(y_{T+h}) \right)^2 \mathrm{d}y_{T+h}$$

[3]For longer horizons, test for independence is skipped.

where F is the CDF from the predictive density $f(y_{t+h}|\mathcal{D}^T, M_j)$ of model j. The sample average CRPS is computed as

$$\text{CRPS}_{j,h} = -\frac{1}{T^*}\sum_{T=\underline{t}}^{\overline{T}} \text{CRPS}_{j,T+h}$$

Smaller CRPS values imply higher precisions.

Finally, the Diebold and Mariano (1995) and West (1996) t-tests for equality of the average loss (with loss defined as log score, or CRPS) can be applied.

15.4.3 Forecast Calibration

An expert is well-calibrated if the subjective predictive distribution (or density function) agrees with the sample distribution of the realizations of the unknown variable in the long run. When a predictive density $F(y|\mathcal{D}^T)$ is not well-calibrated, a calibration procedure can be applied, by introducing a monotone non-decreasing map

$$\psi : \begin{bmatrix} [0,1] \rightarrow [0,1] \\ F(\cdot|\mathcal{D}^T, \xi) \mapsto F(\cdot|\mathcal{D}^T, \xi) = \psi(F(\cdot|\mathcal{D}^T)) \end{bmatrix}$$

such that $F(y_{T+h}|\mathcal{D}^T, \xi)$ is well-calibrated. Bassetti et al. (2018) propose to use the cdf of a mixture of Beta II distributions as calibration functional, that is

$$F(y_{T+h}|\mathcal{D}^T, \xi) = \sum_{j=1}^{J} \omega_j B_{\alpha_j,\beta_j}(F(y_{T+h}|\mathcal{D}^T))$$

with $\xi = (\alpha_1, \ldots, \alpha_J, \beta_1, \ldots, \beta_J, \omega_1, \ldots, \omega_J)$ and $\alpha_j, \beta_j > 0$, $\omega_1 + \ldots + \omega_J = 1$, $\omega_j \geq 0$, and $B_{\alpha,\beta}(u)$ the cdf of the Beta II distribution. This calibration functional has the beta calibration scheme of Ranjan and Gneiting (2010) and Gneiting and Ranjan (2013) as special case for $J = 1$ and allows for more flexibility in calibrating in presence of fat tails, skewness, and multiple-modes.

As an example assume that a set of observations are generated from a standard normal, $y_t \sim N(0, 1)$, i.i.d. $t = T + 1, \ldots, T + n$, $n = 1000$ and that the predictive density results from the following linear pooling:

$$F(y_{T+h}|\mathcal{D}^T) \sim \frac{1}{3}N(0.5, 1) + \frac{1}{3}N(0, 2) + \frac{1}{3}N(-0.5, 1)$$

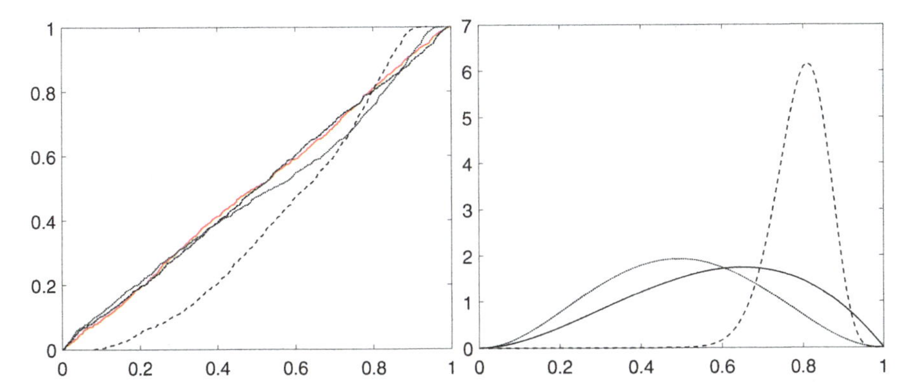

Fig. 15.3 PITs calibration exercise. Left: PITs generated by the true model (red solid), the forecasting model $F(y|\mathcal{D}^T)$ (dashed), the beta calibrated model (dotted) and the beta mixture calibrated model (dashed-dotted). Right: beta calibration function (solid) and the first (dashed) and second (dotted) component of the beta mixture calibration function

Since the PITs of the density forecasts are not well-calibrated (dashed line, in the left panel of Fig. 15.3), we apply the following calibration functions:

$$F(y_{T+h}|\mathcal{D}^T,\xi) = B_{\alpha,\beta}(F(y_{T+h}|\mathcal{D}^T))$$
$$F(y_{T+h}|\mathcal{D}^T,\xi) = \omega B_{\alpha_1,\beta_1}(F(y_{T+h}|\mathcal{D}^T)) + (1-\omega)B_{\alpha_2,\beta_2}(F(y_{T+h}|\mathcal{D}^T))$$

where the parameters $\alpha = 2.81$ and $\beta = 2.01$ and $\alpha_1 = 23.13$, $\beta_1 = 6.61$, $\alpha_2 = 2.95$, $\beta_2 = 3.19$ and $\omega = 0.36$ have been optimally chosen by maximizing the likelihood function

$$l(y_{T+n}|\xi) = \prod_{t=T+1}^{T+n} \left(\sum_{j=1}^{J} \omega_j B_{\alpha_j,\beta_j}(F(y_t|\mathcal{D}^T)) \right)$$

with respect to ξ. For a Bayesian approach to the estimation of the calibration function see Bassetti et al. (2018). The dashed line in the left panel suggests that the beta calibration model is not able to produce well-calibrated PITs, whereas the 2-component beta mixture functional (dotted-dashed line) allows for a better calibration. The first mixture component $B_{\alpha_1,\beta_1}(u)$ for $u \in (0, 1)$ (dotted line in the right plot) is calibrating all the PITs, whereas the second component $B_{\alpha_2,\beta_2}(u)$ (dashed line) is reducing the value of the PITs below the 60%. A Bayesian approach to inference on the calibration functional can carried out by eliciting a prior on the parameter ξ and then using Markov-chain Monte Carlo methods for posterior simulation (e.g., see Robert & Casella, 2004). As an example consider the beta calibration exercise of this section. We assume $\alpha, \beta \sim Ga(2, 4)$ where $Ga(c, d)$ is a gamma distribution with shape and scale parameters c and d, respectively and pdf

$$p(z) = \frac{1}{\Gamma(c)} c^{-d} \exp\left(-\frac{1}{d}z\right) z^{c-1}, \quad z > 0$$

Let $p(\xi) = p(\alpha)p(\beta)$ be the joint prior with $\xi = (\alpha, \beta)$. The joint posterior distribution

$$p(\xi|\mathcal{D}^{T+n}) \propto l(y_{T+n}|\xi)p(\xi)$$

is not tractable, thus we apply a Metropolis-Hastings simulation algorithm (see Sect. 15.5) which generates at the iteration r a candidate ξ^* from the random-walk proposal $\log \xi^* = \log \xi + \eta_{r-1}$, $\eta_t \sim N_2(0, \text{diag}\{0.05, 0.05\})$, where η_{r-1} is the previous iteration random sample from the simulation algorithm. The MH samples are used to estimate the posterior distribution of the calibrated PITs (left plot in Fig. 15.4) and the calibration parameters (right plot).

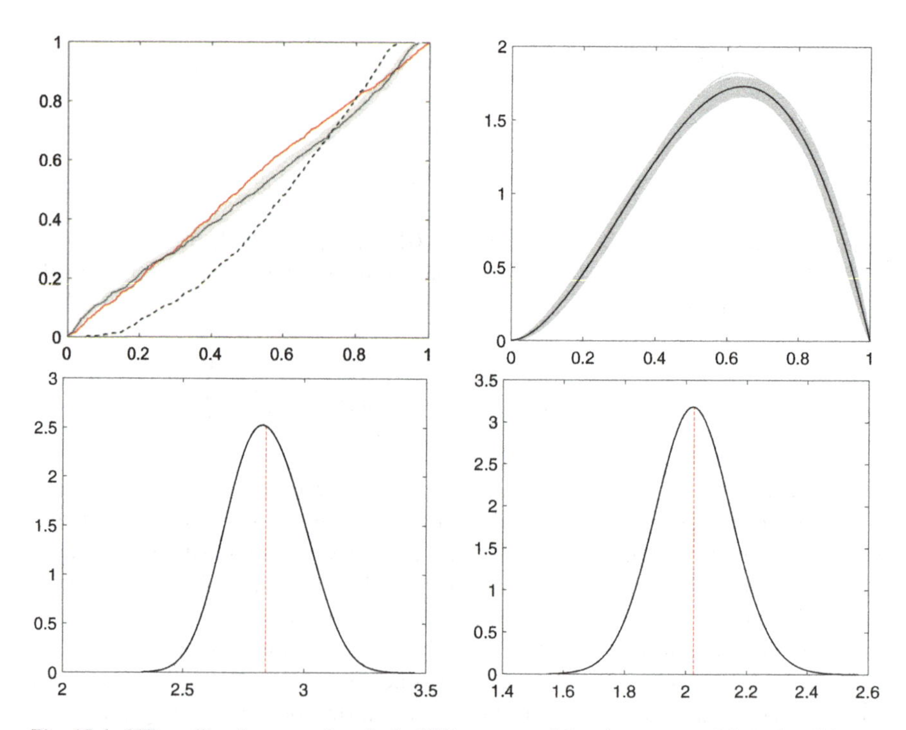

Fig. 15.4 PITs calibration exercise. Left: PITs generated by the true model (red solid), the forecasting model $F(y|\mathcal{D}^T)$ (dashed), the Bayesian beta calibrated model (dotted) and the MCMC posterior coverage (light gray lines). Right: beta calibration function (solid), he MCMC posterior coverage (light gray lines), posterior mean (vertical dashed)

15.5 Monte Carlo Methods for Predictive Approximation

In the next sections, we report some Monte Carlo (MC) simulation methods which can be used for approximating predictive densities expressed in integral form. MC simulation is an approximation method to solve numerically several optimization and integration problems, and already found widespread application in economics and business, e.g., see Kloek and van Dijk, 1978 and Geweke, 1989.

15.5.1 Accept–Reject

The Accept–Reject (AR) algorithm (Robert & Casella, 2004) is used to generate samples from a density $f(y)$ (called target density) by using an density $g(y)$ (called instrumental density). The AR algorithm iterates the following steps for $r = 1, \ldots, R$

1. Generate x_r from g and a uniform u_r from $U_{[0,1]}$,
2. Accept and set $y_r = x_r$ if $u_r \leq f(x_r)/g(x_r)$

As an example consider the target density

$$f(x) \propto \exp(-x^2/2)(\sin(6x)^2 + 3\cos(x)^2 \sin(4x)^2 + 1),$$

which is not easy to simulate, and assume the following instrumental density

$$g(x) \propto \exp(-x^2/2)/\sqrt{2\pi}$$

which is the density of a standard normal distribution and is easier to simulate. The top panel in Fig. 15.5 reports a graphical comparison of the two densities. The

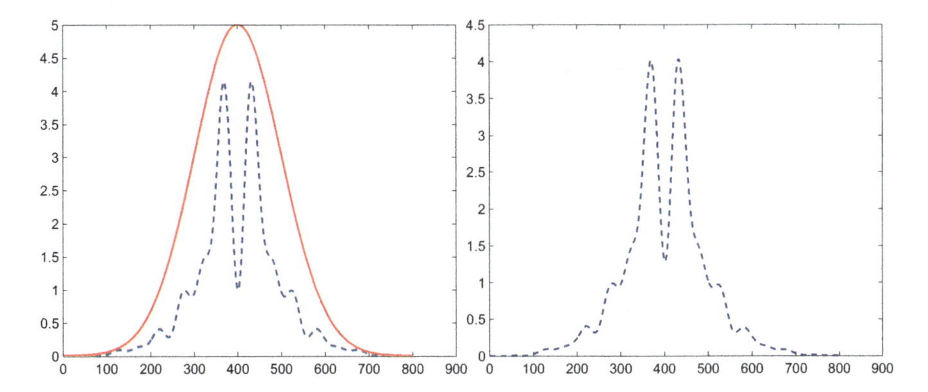

Fig. 15.5 Accept–Reject example. Left: target (dashed) and instrumental (solid) density. Right: target histogram approximated with 1,000,000 draws

bottom panel of Fig. 15.5 shows the simulated target density using the AR algorithm based on 1,000,000 draws. See Listing 15.3 in the Appendix for the MATLAB code.

15.5.2 Importance Sampling

Let $f(y)$ be a target density function, h a measurable function and

$$\Im = \int h(y)f(y)dy$$

the integral of interest. In importance sampling (IS) (see Robert & Casella, 2004, chapter 3) a distribution g (called importance distribution or instrumental distribution) is used to apply a change of measure

$$\Im = \int \frac{f(y)}{g(y)}h(y)g(y)dy$$

The resulting integral is then evaluated numerically by using i.i.d. samples $Y_1, \ldots, Y_R$ from g, that is

$$\Im_R^{IS} = \frac{1}{R}\sum_{r=1}^{R} w(y_r)h(y_r)$$

where $w(y_r) = f(y_r)/g(y_r)$, $r = 1, \ldots, R$ are called importance weights. A set of sufficient conditions for the IS estimators to have finite variance is the following:

(B1) $f(y)/g(y) < M\ \forall y \in \dagger$ and $\mathbb{V}_f(h) < \infty$
(B1) $\mathcal{Y}$ is compact, $f(y) < c$ and $g(y) > \varepsilon\ \forall y \in \mathcal{Y}$

The condition **(B1)** implies that the distribution g has thicker tails than f. If the tails of the importance density are lighter than those of the target then the importance weight $w(y)$ is not a.e. bounded and the variance of the estimator will be infinite for many functions h. A way to address this issue is to consider the self-normalized importance sampling (SNIS) estimator

$$\Im_R^{SNIS} = \frac{\sum_{r=1}^{R} w(y_r)h(y_r)}{\sum_{r=1}^{R} w(y_r)}$$

It is biased on a finite sample, but it converges to $\Im$ by the strong law of large number.

As an example let $h(y) = \sqrt{|y/(1-y)|}$ and y follow a Student-t distribution $\mathcal{T}(\nu, \theta, \sigma^2)$ with density

$$f(y) = \frac{\Gamma((\nu+1)/2)}{\sigma\sqrt{\nu\pi}\Gamma(\nu/2)}\left(1+\frac{(y-\theta)^2}{\nu\sigma^2}\right)^{-(\nu+1)/2} \mathbb{I}_{\mathbb{R}}(y)$$

We study the performance of the importance sampling estimator when the following instrumental distributions are used:

1. Student-t, $t(\nu^*, 0, 1)$ with $\nu^* < \nu$ (e.g., $\nu^* = 7$);
2. Cauchy, $C(0, 1)$.

We shall recall that the Cauchy distribution $C(\alpha, \beta)$ has density function

$$g(y) = \frac{1}{\pi\beta(1+((y-\alpha)/\beta)^2)} \mathbb{I}_{\mathbb{R}}(y)$$

where $-\infty < \alpha < +\infty$ and $\beta > 0$ and cumulative distribution function

$$G(y) = \left(\frac{1}{2} + \frac{1}{\pi}\text{arc tan}\,\frac{y-\alpha}{\beta}\right) \mathbb{I}_{\mathbb{R}}(y)$$

The inverse cdf method can be applied in order to generate from the Cauchy: if $Y = G^{-1}(U)$, where $U \sim \mathcal{U}_{[0,1]}$, then $Y \sim C(\alpha, \beta)$. See Listing 15.4 in Appendix for a MATLAB code. We generate 10000 draws from the instrumental distributions. Figure 15.6 shows that the importance weights for Student-t and Cauchy are stable (left panel), but the Cauchy proposal seems to converge faster than the Student-t (right panel).

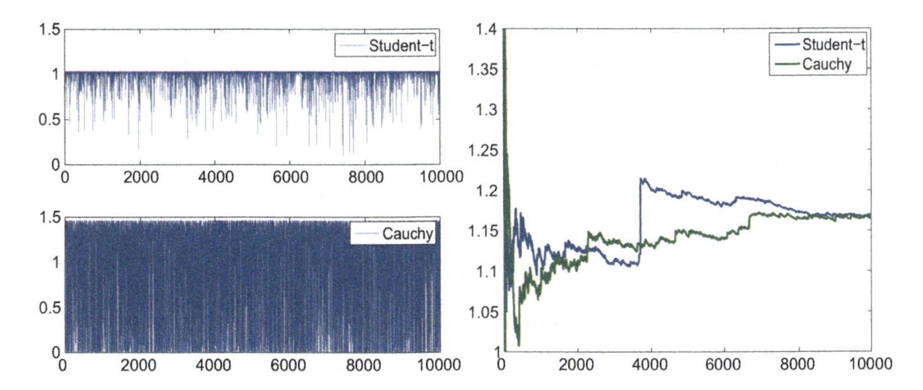

Fig. 15.6 Importance sampling draws for the two different instrumental distributions. Left: importance sampling weights $w(Y_j)$. Right: importance sampling estimator

15.5.3 Metropolis-Hastings

In IS and AR samples from a target distribution can be generated by using a different distribution. A similar idea motivates the use of Markov chain Monte Carlo methods, where samples are generated from an ergodic Markov chain process with the target as a stationary distribution. A general MCMC method is the Metropolis-Hastings (MH) algorithm. Let $f(y)$ be the target distribution and $q(x|y)$ a proposal distribution. The MH algorithm (see Ch. 6–10 in Robert and Casella, 2004) generates a sequence of samples $y_1, \ldots, Y_R$ by iterating the following steps. At the r-th iteration, given y_{r-1} from the previous iteration

1. generate $x^* \sim q(x|y_{r-1})$;
2. set

$$y_r = \begin{cases} x^* & \text{with probability} \quad \alpha(x^*, y_{r-1}) \\ y_{r-1} & \text{with probability } 1 - \alpha(x^*, y_{r-1}) \end{cases}$$

   where

$$\alpha(x, y) = \min\left\{\frac{f(y)}{f(x)}\frac{q(x|y)}{q(y|x)}, 1\right\}$$

The generality of the MH relies on the assumption that the target density is known up to a normalizing constant, which is common in many Bayesian inference problems. A drawback of the MH method is that the sequence of samples is not independent and the degree of dependence depends on the choice of the proposal distribution. In order to illustrate this aspect, we consider a toy example. Assume the target distribution is a bivariate normal mixture $1/3N_2(-\iota, I_2) + 2/3N_2(\iota, I_2)$ where $\iota = (1, 1)'$ and I_2 is the two-dimensional identity matrix and design a random-walk MH algorithm with candidate samples X^* generated from $N_2(y_{r-1}, \tau^2 I_2)$.

Figure 15.7 shows the output of 500 iterations of the MH sampler for different values of the scale parameter τ (different panels). In each plot, the two-dimensional random vectors $y_r, r = 1, \ldots, 500$ (red dots), the trajectory of the M.-H. chain (red line connecting the dots), the initial value of the algorithm (blue dot) and the level sets of the target distribution (solid black lines).

Left plot shows an example of missing mass problem. The scale of the proposal is too small ($\tau^2 = 0.01$), thus the M.-H. chain gets trapped by one of the mode and is not able to visit the other mode. In this case one expects that the results of the approximated inference procedure are sensitive to the choice of the initial condition of the MH chain. The MH chain in the right plot has a better mixing and is able to generate samples from the two components of the mixture.

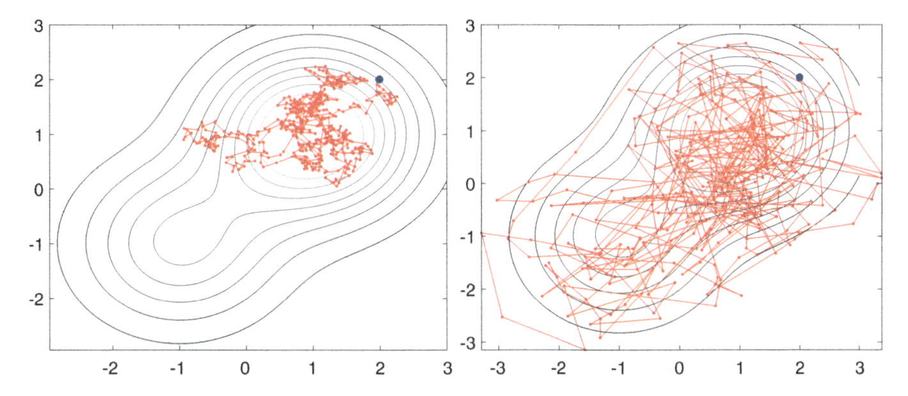

Fig. 15.7 Output of the Metropolis-Hastings for different choices of the random-walk scale parameter, $\tau^2 = 0.01$ (left) and $\tau = 1$ (right). In each plot: the trajectory of the M.-H. chain (red line), the initial value of the algorithm (blue dot) and the level sets of the target distribution (solid black lines)

15.5.4 Constructing Density Forecasting Using GPU

There is a recent trend in using Graphical Processor Units (GPUs) for general, non-graphics, applications (prominently featuring those in scientific computing) the so-called General-Purpose computing on Graphics Processing Units (GPGPU). GPGPU has been applied successfully in different fields such as astrophysics, biology, engineering, and finance, where quantitative analysts started to use this technology ahead of academic economists, see Morozov and Mathur (2011) for a literature review. To date, the adoption of GPU computing technology in economics and econometrics has been relatively slow compared to other fields. There are just a few papers that deal with this interesting topic (e.g., see Casarin, Craiu, & Leisen, 2016; Casarin, Grassi, Ravazzolo, & van Dijk, 2015; Durham & Geweke, 2014; Geweke & Durham, 2012; Morozov & Mathur, 2011; Vergé, Dubarry, Del Moral, & Moulines, 2015). This is odd given the fact that parallel computing in economics has a long history and specifically for this chapter computing density forecasts based on bootstrapping or Bayesian inference requires extensive computation that can be paralleled. The low diffusion of this technology in the economics and econometrics literature, according to Creel (2005), is related to the steep learning curve of a dedicated programming language and expensive hardware. Modern GPUs can easily solve the second problem (hardware costs are relatively low), but the first issue still remains open. Among the popular software used in econometrics (e.g., see LeSage, 1998), MATLAB has introduced from the version R2010b the support to GPU computing in its parallel computing toolbox. This allows for using raw CUDA code within a MATLAB code and MATLAB functions that are executed on the GPU. See Geweke and Durham (2012) for a discussion about CUDA programming in econometrics. As showed in the Appendix, using the build-in functions, GPGPU can be almost effortless where the only knowledge required is a decent programming skill in MATLAB.

15.6 Conclusion

This chapter reviews different methods to construct density forecasts based on error assumptions, bootstrapping, and Bayesian inference. We describe different assumptions of the three methods in the case of the simple linear regression models and provide tools to extend the analysis to more complex models. We also discuss density combinations as a tool to deal in the case there are several density forecasts and an *a priori* selection is difficult. This is particular challenging in big data applications where all the data are not useful. And we provide some evaluation tools to measure the accuracy of density forecasts, accounting for the fact that the "true" density forecast is never observed, even *ex post*.

As example, we present how to use GPU computing almost effortless with MATLAB. The only knowledge required is a decent programming skill and a knowledge of the GPU computing functions introduced in the MATLAB parallel computing toolbox. We generate random numbers, estimate a linear regression model and present a Monte Carlo simulation based on accept/rejection algorithm. We expect large benefits in computational time when dealing with big database with GPU computing.

Appendix

There is little difference between a CPU and a GPU MATLAB code as Listings 15.1 and 15.2, for example, show. The pseudo code, reported in the listings, generates random variables Y and X and estimates the linear regression model $Y = X\beta + \epsilon$, on CPU and GPU, respectively.

The GPU code, Listing 15.2, uses the command gpuArray.randn to generate a matrix of normal random numbers. The build-in function is handled by the NVIDIA plug-in that generates the random number with an underline raw CUDA code. Once the variables vY and mX are created and saved in the GPU memory all the related calculations are automatically executed on the GPU, e.g., *inv* is executed directly on the GPU. This is completely transparent to the user.

If further calculations are needed on the CPU then the command *gather* transfers the data from GPU to the CPU, see line 5 of Listing 15.2. There exist already a lot of supported functions and this number continuously increases with new releases.[4]

```
1 iRows = 1000; iColumns = 5; % number of rows and columns
2 mX = randn(iRows, iColumns); % generate random numbers
3 vY = randn(iRows, 1);
4 vBeta = inv(mX ' * mX) * mX' * vY;
```

Listing 15.1 MATLAB CPU code that generate random numbers and estimate a linear regression model

```
1 iRows = 1000; iColumns = 5; % number of rows and columns
2 mX = gpuArray.randn(iRows, iColumns); % generate random
      numbers
3 vY = gpuArray.randn(iRows, 1);
4 vBeta = inv(mX ' * mX) * mX' * vY;
5 vBeta = gather (vBeta); % transfer data to CPU
```

Listing 15.2 MATLAB GPU code that generate random numbers and estimate a linear regression model

As further examples in Listings 15.3 and 15.4 we show the GPU implementation of the accept/reject and the importance sampling algorithms presented in Sect. 15.5.

```
1 sampsize = 1000000;        % sample size to use for examples
2 sig = 1;                   % standard deviation of the
     instrumental density
3 samp = gpuArray.randn(sampsize, 1) .* sig; % step 1 in the A/R
      algorithm
4 ys = exp((-samp.^2)/2) .* (sin(6 * samp).^2 + 3 *((cos(samp)
     .^2).*(sin(4*samp).^2)) + 1);
5 wts = (1/sqrt(2*pi)) .* exp(-samp.^2/2);
6 samp2 = gpuArray.rand(sampsize, 1);
7 dens = samp(samp2<=(ys)./wts);    % step 2 in the A/R algorithm
8 target = gather(dens);            % step 3 in the A/R algorithm
```

Listing 15.3 Accept/reject MATLAB GPU code

[4]See for the complete list of functions http://www.mathworks.com/help/distcomp/using-gpuarray.html.

```
nIS = 10000; nu = gpuArray(12); nustar = gpuArray(7);% number
    of simulations; degree of freedom of the target density;
    degree of freedom of the proposal
muIS = gpuArray.nan(nIS, 2);
wIS = gpuArray.nan(nIS, 2);
x1 = rant_GPU(nIS, nustar);                    % Student t
    proposals
x2 =  tan((gpuArray.rand(nIS, 1) - 0.5) * pi); % Cauchy
    proposals
wIS(:, 1) = w1_GPU(x1, nu, nustar);            % Importance
    weights
wIS(:, 2) = w3_GPU(x2, nu);                    % Importance
    weights
muIS(:, 1) = sqrt(abs(x1./(1-x1)));
muIS(:, 2) = sqrt(abs(x2./(1-x2)));
muIScum(:,1)=cumsum(muIS(:,1).*wIS(:,1))./(1:nIS)';
muIScum(:,2)=cumsum(muIS(:,2).*wIS(:,2))./(1:nIS)';
%
% Additional functions
function w = w1_GPU(x,nu,nustar)      % Student's t weights
   w = tpdf_GPU(x, nu)./tpdf_GPU(x, nustar);
end
function w=w3_GPU(x,nu)               % Cauchy weights
     w = tpdf_GPU(x, nu)./ pdfcauchy_GPU(x, 0, 1);
end
function f = tpdf_GPU(x,v)            % Student's t GPU pdf
k = find(v>0 & v<Inf);
    if any(k)
        term = exp(gammaln((v(k) + 1) / 2) - gammaln(v(k)/2));
        f(k) = term ./ (sqrt(v(k)*pi) .* (1 + (x(k) .^ 2) ./ v
            (k)) .^ ((v(k) + 1)/2));
    end
end
function f = pdfcauchy_GPU(x, a, b)  % Cauchy GPU pdf
            f = 1./(pi .* b .* (1 + ((x - a)./b).^2));
end
```

Listing 15.4 Importance sampling GPU code

References

Aastveit, K., Mitchell, J., Ravazzolo, F., & van Dijk, H. (2019). The evolution of forecast density combinations in economics. In J. H. Hamilton (Ed.), *Oxford research encyclopedia of economics and finance*. Amsterdam: North-Holland.

Aastveit, K., Ravazzolo, F., & van Dijk, H. (2018). Combined density nowcasting in an uncertain economic environment. *Journal of Business and Economic Statistics*, *36*(1), 131–145.

Aastveit, K. A., Foroni, C., & Ravazzolo, F. (2016). Density forecasts with MIDAS models. *Journal of Applied Econometrics*, *32*(4), 783–801.

Aastveit, K. A., Gerdrup, K., Jore, A. S., & Thorsrud, L. A. (2014). Nowcasting GDP in real time: A density combination approach. *Journal of Business and Economic Statistics*, *32*(1), 48–68.
Amisano, G., & Geweke, J. (2010). Comparing and evaluating Bayesian predictive distributions of asset returns. *International Journal of Forecasting*, *26*(2), 216–230.
Andrews, D. W. K. (2002). Higher-order improvements of a computationally attractive "k"-step bootstrap for extremum estimators. *Econometrica*, *70*(1), 119–162.
Anscombe, F. (1968). Topics in the investigation of linear relations fitted by the method of least squares. *Journal of the Royal Statistical Society B*, *29*, 1–52.
Bańbura, M., Giannone, D., & Reichlin, L. (2010). Large Bayesian vector auto regressions. *Journal of Applied Econometrics*, *25*, 71–92.
Barnard, G. A. (1963). New methods of quality control. *Journal of the Royal Statistical Society, Series A*, *126*, 255–259.
Bassetti, F., Casarin, R., & Ravazzolo, F. (2018). Bayesian nonparametric calibration and combination of predictive distributions. *Journal of the American Statistical Association*, *113*(522), 675–685.
Bates, J., & Granger, C. (1969). The combination of forecasts. *Operations Research Quarterly*, *20*(4), 451–468.
Berkowitz, J. (2001). Testing density forecasts, with applications to risk management. *Journal of Business and Economic Statistics*, *19*(4), 465–474.
Berkowitz, J., & Kilian, L. (2000). Recent developments in bootstrapping time series. *Econometric Reviews*, *19*(1), 1–48.
Billio, M., Casarin, R., Ravazzolo, F., & van Dijk, H. K. (2013). Time-varying combinations of predictive densities using nonlinear filtering. *Journal of Econometrics*, *177*, 213–232.
Bose, A. (1988). Edgeworth corrections by bootstrap in autoregressions. *Annals of Statistics*, *16*(4), 1709–1722.
Casarin, R., Craiu, R. V., & Leisen, F. (2016). Embarrassingly parallel sequential Markov-chain Monte Carlo for large sets of time series. *Statistics and Its Interface*, *9*(4), 497–508.
Casarin, R., Grassi, S., Ravazzolo, F., & van Dijk, H. K. (2015). Parallel sequential Monte Carlo for efficient density combination: The deco MATLAB toolbox. *Journal of Statistical Software*, *68*(3), 1–30.
Choi, H., & Varian, H. (2012). Predicting the Present with Google Trends. *Economic Record*, *88*, 2–9.
Clements, M. P., & Galvao, A. B. (2014). *Measuring macroeconomic uncertainty: US inflation and output growth* (ICMA Centre Discussion Papers in Finance No. 2014/04). Henley-on-Thames: Henley Business School, Reading University.
Clements, M. P., & Taylor, N. (2001). Bootstrapping prediction intervals for autoregressive models. *International Journal of Forecasting*, *17*(2), 247–267.
Creel, M. (2005). User-friendly parallel computations with econometric examples. *Computational Economics*, *26*, 107–128.
Davidson, R., & Flachaire, E. (2008). The wild bootstrap, tamed at last. *Journal of Econometrics*, *146*(1), 162–169.
Davidson, R., & MacKinnon, J. G. (2006). Bootstrap methods in econometrics. In *Palgrave handbooks of econometrics: Volume 1 econometric theory* (pp. 812–838). Basingstoke: Palgrave Macmillan.
Davison, A., & Hinkley, D. (1997). *Bootstrap methods and their applications*. Cambridge: Cambridge University Press.
Dawid, A. P. (1982). Intersubjective statistical models. *Exchangeability in Probability and Statistics*, 217–232.
DeGroot, M. H., Dawid, A. P., & Mortera, J. (1995). Coherent combination of experts' opinions. *Test*, *4*, 263–313.
DeGroot, M. H., & Mortera, J. (1991). Optimal linear opinion pools. *Management Science*, *37*(5), 546–558.

Del Negro, M., Hasegawa, B. R., & Schorfheide, F. (2016). Dynamic prediction pools: An investigation of financial frictions and forecasting performance. *Journal of Econometrics, 192*(2), 391–405.

Diebold, F., & Mariano, R. (1995). Comparing Predictive Accuracy. *Journal of Business and Economic Statistics, 13*, 253–263.

Diebold, F. X., Gunther, T. A., & Tay, A. S. (1998). Evaluating density forecasts with applications to financial risk management. *International Economic Review, 39*(4), 863–83.

Djogbenou, A., Goncalves, S., & Perron, B. (2015). Bootstrap inference in regressions with estimated factors and serial correlation. *Journal of Time Series Analysis, 36*(3), 481–502.

Djogbenou, A., Goncalves, S., & Perron, B. (2017). Bootstrap prediction intervals for factor models. *Journal of Business and Economic Statistics, 35*(1), 53–69.

Durham, G., & Geweke, J. (2014). Adaptive sequential posterior simulators for massively parallel computing environments. In I. Jeliazkov & D. J. Poirier (Eds.), *Bayesian model comparison (advances in econometrics* (Chap. 34). Bingley: Emerald Group Publishing Limited.

Einav, L., & Levin, J. (2014). Economics in the age of big data. *Science, 346*(6210), 715–718.

Geweke, J. (1989). Bayesian inference in econometric models using Monte Carlo integration. *Econometrica, 57*, 1317–1340.

Geweke, J., & Amisano, G. (2011). Optimal prediction pools. *Journal of Econometrics, 164*(1), 130–141.

Geweke, J., & Durham, G. (2012). *Massively parallel sequential Monte Carlo for Bayesian inference*. Sydney: University of Technology Sydney.

Gneiting, T. (2011). Making and evaluating point forecasts. *Journal of the American Statistical Association, 106*, 746–762.

Gneiting, T., & Katzfuss, M. (2014). Probabilistic forecasting. *Annual Review of Statistics and Its Application, 1*, 125–151.

Gneiting, T., & Raftery, A. E. (2007). Strictly proper scoring rules, prediction, and estimation. *Journal of the American Statistical Association, 102*, 359–378.

Gneiting, T., & Ranjan, R. (2013). Combining predicitve distributions. *Electronic Journal of Statistics, 7*, 1747–1782.

Goncalves, S., & Kilian, L. (2004). Bootstrapping autoregressions with conditional heteroskedasticity of unknown form. *Journal of Econometrics, 123*(1), 89–120.

Goncalves, S., & Perron, B. (2014). Bootstrapping factor-augmented regression models. *Journal of Econometrics, 182*(1), 156–173.

Granger, C. W. J. (1998). Extracting information from mega-panels and high-frequency data. *Statistica Neerlandica, 52*, 258–272.

Granger, C. W. J., & Pesaran, M. H. (2000). Economic and statistical measures of forecast accuracy. *Journal of Forecasting, 19*, 537–560.

Granger, C. W. J., & Ramanathan, R. (1984). Improved methods of combining forecasts. *Journal of Forecasting, 3*, 197–204.

Groen, J. J. J., Paap, R., & Ravazzolo, F. (2013). Real-time inflation forecasting in a changing world. *Journal of Business & Economic Stastistics, 31*, 29–44.

Guidolin, M., & Timmermann, A. (2009). Forecasts of US short-term interest rates: A flexible forecast combination approach. *Journal of Econometrics, 150*, 297–311.

Hall, P. (1985). Resampling a coverage process. *Stochastic Processes and Their Applications, 20*(2), 231–246.

Hall, S. G., & Mitchell, J. (2007). Combining density forecasts. *International Journal of Forecasting, 23*(1), 1–13.

Hansen, B. (2006). Interval forecasts and parameter uncertainty. *Journal of Econometrics, 135*, 377–398.

Hofmarcher, P., & Grün, B. (2019). Bayesian model averaging. In P. Fuleky (Ed.), *Macroeconomic forecasting in the era of big data. Springer series "Advanced studies in theoretical and applied econometrics"*.

Hoogerheide, L., Kleijn, R., Ravazzolo, R., van Dijk, H. K., & Verbeek, M. (2010). Forecast accuracy and economic gains from Bayesian model averaging using time varying weights. *Journal of Forecasting*, *29*(1–2), 251–269.

Inoue, A., & Kilian, L. (2002). Bootstrapping autoregressive processes with possible unit roots. *Econometrica*, 377–391.

Kapetanios, G., Mitchell, J., Price, S., & Fawcett, N. (2015). Generalised density forecast combinations. *Journal of Econometrics*, *188*, 150–165.

Kascha, C., & Ravazzolo, F. (2010). Combining inflation density forecasts. *Journal of Forecasting*, *29*(1–2), 231–250.

Kloek, T., & van Dijk, H. (1978). Bayesian estimates of equation system parameters: An application of integration by Monte Carlo. *Econometrica*, *46*, 1–19.

Knuppel, M. (2015). Evaluating the calibration of multi-step-ahead density forecasts using raw moments. *Journal of Business & Economic Statistics*, *33*(2), 270–281.

Koop, G. (2003). *Bayesian econometrics*. Hoboken: John Wiley and Sons.

Koop, G., & Korobilis, D. (2013). Large time-varying parameter VARs. *Journal of Econometrics*, *177*, 185–198.

Kunsch, H. R. (1989). The jackknife and the bootstrap for general stationary observations. *The Annals of Statistics*, *17*(3), 1217–1241.

LeSage, J. P. (1998). Econometrics: MATLAB toolbox of econometrics functions. Statistical Software Components, Boston College Department of Economics.

Liu, R. (1988). Bootstrap procedures under some non-i.i.d. models. *Annals of Statistics*, *16*, 1696–1708.

Mazzi, G., Mitchell, J., & Montana, G. (2014). Density nowcasts and model combination: Nowcasting Euro-area GDP growth over the 2008-9 recession. *Oxford Bulletin of Economics and Statistics*, *76*(2), 233–256.

McAlinn, K., & West, M. (2019). Dynamic Bayesian predictive synthesis in time series forecasting. *Journal of Econometrics*, *210*(1), 155–169.

Mitchell, J., & Hall, S. G. (2005). Evaluating, comparing and combining density forecasts using the KLIC with an application to the Bank of England and NIESR 'fan' charts of inflation. *Oxford Bulletin of Economics and Statistics*, *67*(s1), 995–1033.

Mitchell, J., & Wallis, K. (2011). Evaluating density forecasts: Forecast combinations, model mixtures, calibration and sharpness. *Journal of Applied Econometrics*, *26*(6), 1023–1040.

Morozov, S., & Mathur, S. (2011). Massively parallel computation using graphics processors with application to optimal experimentation in dynamic control. *Computational Economics*, 1–32.

Pascual, L., Romo, J., & Ruiz, E. (2001). Effects of parameter estimation on prediction densities: a bootstrap approach. *International Journal of Forecasting*, *17*(1), 83–103.

Pettenuzzo, D., & Ravazzolo, F. (2016). Optimal portfolio choice under decision-based model combinations. *Journal of Applied Econometrics*, *31*(7), 1312–1332.

Raftery, A., Karny, M., & Ettler, P. (2010). Online prediction under model uncertainty via dynamic model averaging: Application to a cold rolling mill. *Technometrics*, *52*, 52–66.

Raftery, A. E., Gneiting, T., Balabdaoui, F., & Polakowski, M. (2005). Using Bayesian model averaging to calibrate forecast ensembles. *Monthly Weather Review*, *133*, 1155–1174.

Raftery, A. E., Madigan, D., & Hoeting, J. A. (1997). Bayesian model averaging for linear regression models. *Journal of the American Statistical Association*, *92*(437), 179–91.

Ranjan, R., & Gneiting, T. (2010). Combining probability forecasts. *Journal of the Royal Statistical Society: Series B (Statistical Methodology)*, *72*(1), 71–91.

Ravazzolo, F., & Vahey, S. V. (2014). Forecast densities for economic aggregates from disaggregate ensembles. *Studies in Nonlinear Dynamics and Econometrics*, *18*, 367–381.

Robert, C. P., & Casella, G. (2004). *Monte Carlo statistical methods* (2nd ed.), Berlin: Springer.

Roberts, H. V. (1965). Probabilistic prediction. *Journal of American Statistical Association*, *60*, 50–62.

Rossi, B., & Sekhposyan, T. (2013). Conditional predictive density evaluation in the presence of instabilities. *Journal of Econometrics*, *177*(2), 199–212.

Rossi, B., & Sekhposyan, T. (2014). Evaluating predictive densities of us output growth and inflation in a large macroeconomic data set. *International Journal of Forecasting*, *30*(3), 662–682.

Rossi, B., & Sekhposyan, T. (2019). Alternative tests for correct specification of conditional predictive densities. *Journal of Econometrics*, *208*, 638–657.

Sloughter, J., Gneiting, T., & Raftery, A. E. (2010). Probabilistic wind speed forecasting using ensembles and Bayesian model averaging. *Journal of the American Statistical Association*, *105*, 25–35.

Stock, J. H., & Watson, W. M. (1999). Forecasting inflation. *Journal of Monetary Economics*, *44*, 293–335.

Stock, J. H., & Watson, W. M. (2002). Forecasting using principal components from a large number of predictors. *Journal of American Statistical Association*, *97*, 1167–1179.

Stock, J. H., & Watson, W. M. (2005). Implications of dynamic factor models for VAR analysis. NBER Working Paper No. 11467.

Stock, J. H., & Watson, W. M. (2014). Estimating turning points using large data sets. *Journal of Econometris*, *178*, 368–381.

Tay, A., & Wallis, K. F. (2000). Density forecasting: A survey. *Journal of Forecasting*, *19*, 235–254.

Terui, N., & van Dijk, H. K. (2002). Combined forecasts from linear and nonlinear time series models. *International Journal of Forecasting*, *18*, 421–438.

Timmermann, A. (2006). Forecast combinations. In *Handbook of economic fore-casting* (Chap. 4, Vol. 1, pp. 135–196). https://doi.org/10.1016/S1574-0706(05)01004-9

Varian, H. (2014). Machine learning: New tricks for econometrics. *Journal of Economics Perspectives*, *28*, 3–28.

Varian, H., & Scott, S. (2014). Predicting the present with Bayesian structural time series. *International Journal of Mathematical Modelling and Numerical Optimisation*, *5*, 4–23.

Vergé, C., Dubarry, C., Del Moral, P., & Moulines, E. (2015). On parallel implementation of sequential Monte Carlo methods: The island particle model. *Statistics and Computing*, *25*(2), 243–260.

Waggoner, D. F., & Zha, T. (2012). Confronting model misspecification in macroeconomics. *Journal of Econometrics*, *171*, 167–184.

Wallis, K. F. (2003). Chi-squared tests of interval and density forecasts, and the Bank of England's fan charts. *International Journal of Forecasting*, *19*(3), 165–175.

Wallis, K. F. (2005). Combining density and interval forecasts: A modest proposal. *Oxford Bulletin of Economics and Statistics*, *67*(s1), 983–994.

Wallis, K. F. (2011). Combining forecasts – forty years later. *Applied Financial Economics*, *21*(1–2), 33–41.

West, K. (1996). Asymptotic inference about predictive ability. *Econometrica*, *64*, 1067–1084.

Wu, C. (1986). Jackknife, bootstrap and other resampling methods in regression analysis. *Annals of Statistics*, *14*, 1261–1295.

Yeh, A. B. (1998). A bootstrap procedure in linear regression with nonstationary errors. *The Canadian Journal of Statistical Association*, *26*(1), 149–160.

Zarnowitz, V. (1969). Topics in the investigation of linear relations fitted by the method of least squares. *American Statistician*, *23*, 12–16.

Chapter 16
Forecast Evaluation

Mingmian Cheng, Norman R. Swanson, and Chun Yao

16.1 Forecast Evaluation Using Point Predictive Accuracy Tests

In this section, our objective is to review various commonly used statistical tests for comparing the relative accuracy of point predictions from different econometric models. Four main groups of tests are outlined: (1) tests for comparing two non-nested models, (2) tests for comparing two nested models, (3) tests for comparing multiple models, where at least one model is non-nested, and (4) tests that are consistent against generic alternative models. The papers cited in this section (and in subsequent sections) contain references to a large number of papers that develop alternative related tests (interested readers may also refer to Corradi and Swanson (2006b) for details).

Of note is that the tests that we discuss in the sequel assume that all competing models are approximations to some unknown underlying data generating process, and are thus potentially misspecified. The objective is to select the "best" model from among multiple alternatives, where "best" refers to a given loss function, say.

M. Cheng (✉)
Department of Finance, Lingnan (University) College, Sun Yat-sen University, Guangzhou, China
e-mail: chengmm3@mail.sysu.edu.cn

N. R. Swanson · C. Yao
Department of Economics, School of Arts and Sciences, Rutgers University, New Brunswick, NJ, USA
e-mail: nswanson@economics.rutgers.edu; cyao@economics.rutgers.edu

P. Fuleky (ed.), *Macroeconomic Forecasting in the Era of Big Data*,
Advanced Studies in Theoretical and Applied Econometrics 52,
https://doi.org/10.1007/978-3-030-31150-6_16

16.1.1 Comparison of Two Non-nested Models

The starting point of our discussion is the Diebold-Mariano (DM: Diebold and Mariano, 2002) test for the null hypothesis of equal predictive accuracy between two competing models, given a pre-specified loss function. This test sets the groundwork for many subsequent predictive accuracy tests. The DM test assumes that parameter estimation error is asymptotically negligible by positing that the number of observations used for in-sample model estimation grows faster than the number of observations used in out-of-sample forecast evaluation. Parameter estimation error in DM tests, which are often also called DM-West tests, is explicitly taken into account of in West (1996), although at the cost of requiring that the loss function is differentiable.

To fix ideas and notation, let $u_{i,t+h} = y_{t+h} - f_i(Z_i^t, \theta_i^\dagger)$ be the h-step ahead forecast error associated with the i-th model, $f_i(\cdot, \theta_i^\dagger)$, where the benchmark model is always denoted as "model 0", i.e., $f_0(\cdot, \theta_0^\dagger)$. As $\theta_i^\dagger$ and thus $u_{i,t+h}$ are unknown, we construct test statistics using $\widehat{\theta}_{i,t}$ and $\widehat{u}_{i,t+h} = y_{t+h} - f_i(Z_i^t, \widehat{\theta}_{i,t})$, where $\widehat{\theta}_{i,t}$ is an estimator of $\theta_i^\dagger$ constructed using information in Z_i^t from time periods 1 to t, under a recursive estimation scheme, or from $t - R + 1$ to t, under a rolling-window estimation scheme. Hereafter, for notational simplicity, we only consider the recursive estimation scheme, and the rolling-window estimation scheme can be treated in an analogous manner. To do this, split the total sample of T observations into two subsamples of length R and n, i.e., $T = R + n$, where only the last n observations are used for forecast evaluation. At each step, we first estimate the model parameters as follows:

$$\widehat{\theta}_{i,t} = \arg\min_{\theta_i} \frac{1}{t} \sum_{j=1}^{t} q\Big(y_j - f_i(Z_i^{j-1}, \theta_i)\Big), \quad t \geq R \tag{16.1}$$

These parameters are used to parameterize the prediction model, and an h-step-ahead prediction (and prediction error) is constructed. This procedure is repeated by adding one new observation to the original sample, yielding a new h-step-ahead prediction (and prediction error). In such a manner, we can construct a sequence of $(n - h + 1)$ h-step-ahead prediction errors. For a given loss function, $g(\cdot)$, the null hypothesis of DM test is specified as,

$$H_0 : E\Big(g(u_{0,t+h}) - g(u_{1,t+h})\Big) = 0$$

against

$$H_A : E\Big(g(u_{0,t+h}) - g(u_{1,t+h})\Big) \neq 0$$

Of particular note here is that the loss function $g(\cdot)$ used for forecast evaluation may not be the same as the loss function $q(\cdot)$ used for model estimation in Eq. (16.1). However, if they are the same (e.g., models are estimated by ordinary least square (OLS) and forecasts are evaluated by a quadratic loss function, say), parameter estimation error is asymptotically negligible, regardless of the limiting ratio of n/R, as $T \to \infty$.

Define the following statistic:

$$\widehat{S}_n(0,1) = \frac{1}{\sqrt{n}} \sum_{t=R-h+1}^{T-h} \Big(g(\widehat{u}_{0,t+h}) - g(\widehat{u}_{1,t+h}) \Big)$$

then,

$$\begin{aligned} \widehat{S}_n(0,1) - S_n(0,1) = & E\Big(\nabla_{\theta_0} g(u_{0,t+h})\Big) \frac{1}{\sqrt{n}} \sum_{t=R-h+1}^{T-h} (\widehat{\theta}_{0,t+h} - \theta_0^{\dagger}) \\ & - E\Big(\nabla_{\theta_1} g(u_{1,t+h})\Big) \frac{1}{\sqrt{n}} \sum_{t=R-h+1}^{T-h} (\widehat{\theta}_{1,t+h} - \theta_1^{\dagger}) + o_p(1) \end{aligned} \tag{16.2}$$

The limiting distribution of the right-hand side of Eq. (16.2) is given by Lemma 4.1 and Theorem 4.1 in West (1996). From Eq. (16.2), we can immediately see that if $g(\cdot) = q(\cdot)$, then $E\big(\nabla_{\theta_i} g(u_{i,t+h})\big) = 0$ by the first order conditions, and parameter estimation error is asymptotically negligible. Another situation in which parameter estimation error vanishes asymptotically is when $n/R \to 0$, as $T \to \infty$.

Without loss of generality, consider the case of $h = 1$. All results carry over to the case when $h > 1$. The DM test statistic is given by,

$$\widehat{DM}_n = \frac{1}{\sqrt{n}} \frac{1}{\widehat{\sigma}_n} \sum_{t=R}^{T-1} \Big(g(\widehat{u}_{0,t+1}) - g(\widehat{u}_{1,t+1}) \Big)$$

with

$$\begin{aligned} \widehat{\sigma}_n = & \widehat{S}_{gg} + 2\Pi \widehat{F}_0' \widehat{A}_0 \widehat{S}_{h_0 h_0} + 2\Pi \widehat{F}_1' \widehat{A}_1 \widehat{S}_{h_1 h_1} \widehat{A}_1 \widehat{F}_1 \\ & - 2\Pi (\widehat{F}_1' \widehat{A}_1 \widehat{S}_{h_1 h_0} \widehat{A}_0 \widehat{F}_0 + \widehat{F}_0' \widehat{A}_0 \widehat{S}_{h_0 h_1} \widehat{A}_1 \widehat{F}_1) \\ & + \Pi (\widehat{S}_{g h_1}' \widehat{A}_1 \widehat{F}_1 + \widehat{F}_1' \widehat{A}_1 \widehat{S}_{g h_1}) \end{aligned}$$

where for $i, j = 0, 1$, $\Pi = 1 - \frac{\ln(1+\pi)}{\pi}$, and $q_t(\widehat{\theta}_{i,t}) = q\big(y_t - f_i(Z_i^{t-1}, \widehat{\theta}_{i,t})\big)$,

$$\widehat{S}_{h_i h_j} = \frac{1}{n} \sum_{\tau=-l_n}^{l_n} w_\tau \sum_{t=R+l_n}^{T-l_n} \nabla_\theta q_t(\widehat{\theta}_{i,t}) \nabla_\theta q_{t+\tau}(\widehat{\theta}_{j,t})'$$

$$\widehat{S}_{gh_i} = \frac{1}{n} \sum_{\tau=-l_n}^{l_n} w_\tau \sum_{t=R+l_n}^{T-l_n} \Big(g(\widehat{u}_{0,t}) - g(\widehat{u}_{1,t}) - \frac{1}{n} \sum_{t=R}^{T-1} \big(g(\widehat{u}_{0,t+1}) - g(\widehat{u}_{1,t+1})\big)\Big) \times \nabla_\theta q_{t+\tau}(\widehat{\theta}_{i,t})'$$

$$\widehat{S}_{gg} = \frac{1}{n} \sum_{\tau=-l_n}^{l_n} w_\tau \sum_{t=R+l_n}^{T-l_n} \Big(g(\widehat{u}_{0,t}) - g(\widehat{u}_{1,t}) - \frac{1}{n} \sum_{t=R}^{T-1} \big(g(\widehat{u}_{0,t+1}) - g(\widehat{u}_{1,t+1})\big)\Big) \times \Big(g(\widehat{u}_{0,t+\tau}) - g(\widehat{u}_{1,t+\tau}) - \frac{1}{n} \sum_{t=R}^{T-1} \big(g(\widehat{u}_{0,t+1}) - g(\widehat{u}_{1,t+1})\big)\Big)$$

with $w_\tau = 1 - \frac{\tau}{l_n - 1}$, and

$$\widehat{F}_i = \frac{1}{n} \sum_{t=R}^{T-1} \nabla_{\theta_i} g(\widehat{u}_{i,t+1}), \qquad \widehat{A}_i = \Big(-\frac{1}{n} \sum_{t=R}^{T-1} \nabla^2_{\theta_i} q(\widehat{\theta}_{i,t}) \Big)^{-1}$$

Assumption 16.1 (y_t, Z^{t-1}), with y_t scalar and Z^{t-1} an $\Re^\zeta$-valued $(0 < \zeta < \infty)$ vector, is a strictly stationary and absolutely regular β-mixing process with size $-4(4+\psi)/\psi$, $\psi > 0$.

Assumption 16.2 (i) $\theta^\dagger$ is uniquely identified (i.e., $E(q(y_t, Z^{t-1}, \theta_i))) > E(q(y_t, Z^{t-1}, \theta_i^\dagger)))$ for any $\theta_i \neq \theta_i^\dagger$); (ii) $q(\cdot)$ is twice continuously differentiable on the interior of Θ, and for Θ a compact subset of $\Re^\varrho$; (iii) the elements of $\nabla_\theta q$ and $\nabla^2_\theta q$ are p-dominated on Θ, with $p > 2(2+\psi)$, where ψ is the same positive constant as defined in Assumption 16.1; and (iv) $E(-\nabla^2_\theta q)$ is negatively definite uniformly on Θ.

Proposition 16.1 (From Theorem 4.1 in West, 1996) *With Assumptions 16.1 and 16.2, also, assume that $g(\cdot)$ is continuously differentiable, then, if as $n \to \infty$, $l_n \to \infty$ and $l_n/n^{1/4} \to 0$, then as $n, R \to \infty$, under H_0,*

$$\widehat{DM}_n \xrightarrow{d} N(0,1)$$

Under H_A,

$$\Pr\Big(n^{-1/2} |\widehat{DM}_n| > \epsilon\Big) \to 1, \quad \forall \epsilon > 0$$

It is immediate to see that if either $g(\cdot) = q(\cdot)$ or $n/R \to 0$, as $T \to \infty$, the estimator $\widehat{\sigma}_n$ collapses to $\widehat{S}_{gg}$. Note that the limiting distribution of DM test obtains only for the case of short-memory series. Corradi, Swanson, and Olivetti (2001) extends the DM test to the case of co-integrated variables and Rossi (2005) to the case of series with high persistence. Finally, note that the two competing models are assumed to be non-nested. If they are nested, then $u_{0,t+h} = u_{1,t+h}$ under the null, and both $\sum_{t=R-h+1}^{T-h} \left(g(\widehat{u}_{0,t+h}) - g(\widehat{u}_{1,t+h})\right)$ and $\widehat{\sigma}_n$ converge in probability to zero at the same rate if $n/R \to 0$. Therefore the DM test statistic does not converge in distribution to a standard normal variable under the null. Comparison of nested models is introduced in the next section.

16.1.2 Comparison of Two Nested Models

There are situations in which we may be interested in comparing forecasts from nested models. For instance, one of the driving forces behind the literature on out-of-sample comparison of nested models is the seminal paper by Meese and Rogoff (1983), who find that no models driven by economic fundamentals can beat a simple random walk model, in terms of out-of-sample predictive accuracy, when forecasting exchange rates. The models studied in this paper are nested, in the sense that parameter restrictions can be placed on the more general models that reduce these models to the random walk benchmark studied by these authors. When testing out-of-sample Granger causality, alternative models are also nested. Since the DM test discussed above is valid only when the competing models are non-nested, we introduce alternative tests that address testing among nested models.

Clark and McCracken Tests for Nested Models

Clark and McCracken (2001) (CMa) and Clark and McCracken (2003) (CMb) propose several tests for nested linear models, under the assumption that prediction errors follow martingale difference sequences (this rules out the possibility of dynamic misspecification under the null for these particular tests), where CMa tests are tailored for the case of one-step-ahead forecasts, and CMb tests for the case of multi-step-ahead forecasts.

Consider the following two nested models. The restricted model is,

$$y_t = \sum_{j=1}^{q} \beta_j y_{t-j} + \epsilon_t$$

and the unrestricted model is,

$$y_t = \sum_{j=1}^{q} \beta_j y_{t-j} + \sum_{j=1}^{k} \alpha_j x_{t-j} + u_t \tag{16.3}$$

The null hypothesis of CMa tests is formulated as,

$$H_0 : E(\epsilon_t^2) - E(u_t^2) = 0$$

against

$$H_A : E(\epsilon_t^2) - E(u_t^2) > 0$$

We can immediately see from the null and the alternative hypotheses that CMa tests implicitly assume that the restricted model cannot beat the unrestricted model. This is the case when the models are estimated by OLS and the quadratic loss function is employed for evaluation.

CMa proposes the following three different test statistics:

$$ENC - T = (n-1)^{1/2} \frac{\overline{c}}{(n^{-1}\sum_{t=R}^{T-1}(c_{t+1} - \overline{c}))^{1/2}}$$

$$ENC - REG = (n-1)^{1/2} \frac{n^{-1}\sum_{t=R}^{T-1}(\widehat{\epsilon}_{t+1}(\widehat{\epsilon}_{t+1} - \widehat{u}_{t+1}))}{(n^{-1}\sum_{t=R}^{T-1}(\widehat{\epsilon}_{t+1} - \widehat{u}_{t+1})^2 n^{-1}\sum_{t=R}^{T-1}\widehat{\epsilon}_{t+1}^2 - \overline{c})^{1/2}}$$

$$ENC - NEW = n \frac{\overline{c}}{n^{-1}\sum_{t=1}\widehat{u}_{t+1}^2}$$

where $c_{t+1} = \widehat{\epsilon}_{t+1}(\widehat{\epsilon}_{t+1} - \widehat{u}_{t+1})$, $\overline{c} = n^{-1}\sum_{t=R}^{T-1} c_{t+1}$, and $\widehat{\epsilon}_{t+1}$ and $\widehat{u}_{t+1}$ are OLS residuals.

Assumption 16.3 (y_t, x_t) are strictly stationary and strong mixing processes, with size $\frac{-4(4+\delta)}{\delta}$, for some $\delta > 0$, and $E(y_t^8)$ and $E(x_t^8)$ are both finite.

Assumption 16.4 Let $z_t = (y_{t-1}, \ldots, y_{t-q}, x_{t-1}, \ldots, x_{t-q})$ and $E(z_t u_t | \mathcal{F}_{t-1}) = 0$, where $\mathcal{F}_{t-1}$ is the σ-field up to time $t-1$, generated by $(y_{t-1}, y_{t-2}, \ldots, x_{t-1}, x_{t-2}, \ldots)$. Also, $E(u_t^2 | \mathcal{F}_{t-1}) = \sigma_u^2$.

Note that Assumption 16.4 assumes that the unrestricted model is dynamically correct and that u_t is conditionally homoskedastic.

Proposition 16.2 (From Theorem 3.1–3.3 in Clark and McCracken, 2001) *With Assumptions 16.3 and 16.4, under the null, (i) if as* $T \rightarrow \infty$, $n/R \rightarrow \pi > 0$, *then* $ENC - T$ *and* $ENC - REG$ *converge in distribution to* Γ_1/Γ_2 *where* $\Gamma_1 = \int_{(1+\pi)^{-1}}^{1} s^{-1} W_s' dW_s$ *and* $\Gamma_2 = \int_{(1+\pi)^{-1}}^{1} W_s' W_s ds$, *with* W_s *a k-dimensional standard Brownian motion (here k is the number of restrictions or the number of extra regressors in the unrestricted model).* $ENC - NEW$ *converges in distribution to* Γ_1. *(ii) If as* $T \rightarrow \infty$, $n/R \rightarrow 0$, *then* $ENC - T$ *and* $ENC - REG$ *converge in distribution to* $N(0, 1)$. $ENC - NEW$ *converges in probability to 0.*

Therefore, as $T \to \infty$ and $n/R \to \pi > 0$, all three test statistics have non-standard limiting distributions. Critical values are tabulated for different k and π in CMa. Also note that the above proposition is valid only when $h = 1$, i.e., the case of one-step-ahead forecasts, since Assumption 16.4 is violated when $h > 1$. For this case, CMb propose a modified test statistic for which $MA(h-1)$ errors are allowed. Namely, they propose using the following statistic:

$$ENC-T^{'} = (n-h+1)^{1/2} \times \frac{(n-h+1)^{-1}\sum_{t=R}^{T-h}\widehat{c}_{t+h}}{\left((n-h+1)^{-1}\sum_{j=-\overline{j}}^{\overline{j}}\sum_{t=R+j}^{T-h}K(\frac{j}{M})(\widehat{c}_{t+h}-\overline{c})(\widehat{c}_{t+h-j}-\overline{c})\right)^{1/2}},$$

where $K(\cdot)$ is a kernel and $0 \leq K(\frac{j}{M}) \leq 1$, with $K(0) = 1$ and $M = o(n^{1/2})$, and $\overline{j}$ does not grow with the sample size. Therefore, the denominator of $ENC-T^{'}$ is a consistent estimator of the long-run variance when $E(c_t c_{t+|k|}) = 0$ for all $|k| > h$. Of particular note is that although $ENC-T^{'}$ allows for $MA(h-1)$ errors, dynamic misspecification under the null is still not allowed. Also note that, when $h = 1$, $ENC-T^{'}$ is equivalent to $ENC-T$.

Another test statistic suggested in CMb is a DM-type test with non-standard critical values that are needed in order to modify the DM test in order to allow for the comparison of nested models. The test statistic is:

$$MSE-T^{'} = (n-h+1)^{1/2} \times \frac{(n-h+1)^{-1}\sum_{t=R}^{T-h}\widehat{d}_{t+h}}{\left((n-h+1)^{-1}\sum_{j=-\overline{j}}^{\overline{j}}\sum_{t=R+j}^{T-h}K(\frac{j}{M})(\widehat{d}_{t+h}-\overline{d})(\widehat{d}_{t+h-j}-\overline{d})\right)^{1/2}}$$

where $\widehat{d}_{t+h} = \widehat{u}_{t+h}^2 - \widehat{\epsilon}_{t+h}^2$ and $\overline{d} = (n-h+1)^{-1}\sum_{t=R}^{T-h}\widehat{d}_{t+h}$.

Evidently, this test is a standard DM test, although it should be stressed that the critical values used in the application of this variant of the test are different. The limiting distributions of the $ENC-T^{'}$ and $MSE-T^{'}$ are provided in CMb, and are non-standard. Moreover, for the case of $h > 1$, the limiting distributions contain nuisance parameters, so that critical values cannot be tabulated directly. Instead, CMb suggest a modified version of the bootstrap method in Kilian (1999) to carry out statistical inference. For this test, the block bootstrap can also be used to carry out inference. (see Corradi and Swanson (2007) for details.)

Out-of-Sample Tests for Granger Causality

CMa and CMb tests do not take dynamic misspecification into account under the null. Chao, Corradi, and Swanson (2001) (CCS) propose out-of-sample tests for Granger causality allowing for possible dynamic misspecification and conditional heteroskedasticity. The idea is very simple. If the coefficients $\alpha_j, j = 1, \ldots, k$ in Eq. (16.3) are all zeros, then residuals ϵ_{t+1} are uncorrelated with lags of x. As a result, including regressors $x_{t-j}, j = 1, \ldots, k$ does not help improve predictive accuracy, and the unrestricted model does not outperform the restricted model.

Hereafter, for notational simplicity, we only consider the case of $h = 1$. All results can be generalized to the case of $h > 1$. Formally, the test statistic is

$$m_n = n^{-1/2} \sum_{t=R}^{T-1} \widehat{\epsilon}_{t+1} X_t$$

where $X_t = (x_t, x_{t-1}, \ldots, x_{t-k-1})^{'}$. The null hypothesis and the alternative hypothesis are formulated as,

$$H_0 : E(\epsilon_{t+1} x_{t-j}) = 0, \quad j = 0, 1, \ldots, k-1$$

$$H_A : E(\epsilon_{t+1} x_{t-j}) \neq 0, \quad \text{for some } j$$

Assumption 16.5 (y_t, x_t) are strictly stationary and strong mixing processes, with size $\frac{-4(4+\delta)}{\delta}$, for some $\delta > 0$, and $E(y_t^8)$ and $E(x_t^8)$ are both finite. $E(\epsilon_t y_{t-j}) = 0, j = 1, 2, \ldots, q$.

Proposition 16.3 (From Theorem 1 in Chao et al., 2001) *With Assumption 16.5, as* $T \to \infty$, $n/R \to \pi, 0 \leq \pi < \infty$, *(i) under the null, for* $0 < \pi < \infty$,

$$m_n \xrightarrow{d} N(0, \Xi)$$

with

$$\begin{aligned} \Xi =& S_{11} + 2(1 - \pi^{-1} ln(1+\pi)) F^{'} M S_{22} M F - \\ & (1 - \pi^{-1} ln(1+\pi))(F^{'} M S_{12} + S_{12}^{'} M F) \end{aligned}$$

where $F = E(Y_t X_t^{'})$, $M = plim(\frac{1}{t} \sum_{j=q}^{t} Y_j Y_j^{'})^{-1}$, *and* $Y_j = (y_{j-1}, \ldots, y_{j-q})^{'}$. *Furthermore,*

$$S_{11} = \sum_{j=-\infty}^{\infty} E((X_t \epsilon_{t+1} - \mu)(X_{t-j} \epsilon_{t-j+1} - \mu)^{'})$$

$$S_{22} = \sum_{j=-\infty}^{\infty} E((Y_{t-1} \epsilon_t)(Y_{t-j-1} \epsilon_{t-j})^{'})$$

$$S_{12} = \sum_{j=-\infty}^{\infty} E((\epsilon_{t+1} X_t - \mu)(Y_{t-j-1} \epsilon_{t-j})^{'})$$

where $\mu = E(X_t\epsilon_{t+1})$. *In addition, for* $\pi = 0$,

$$m_n \xrightarrow{d} N(0, S_{11})$$

(ii) Under the alternative,

$$\lim_{n\to\infty} \Pr(|n^{-1/2}m_n| > 0) = 1$$

Corollary 16.1 (From Corollary 2 in Chao et al., 2001) *With Assumption 16.5, as* $T \to \infty$, $n/R \to \pi, 0 \leq \pi < \infty, l_T \to \infty, l_T/T^{1/4} \to 0$, *(i) under the null, for* $0 < \pi < \infty$,

$$m_n^{'}\widehat{\Xi}^{-1}m_n \xrightarrow{d} \chi_k^2$$

with

$$\begin{aligned}\widehat{\Xi} = &\widehat{S}_{11} + 2(1-\pi^{-1}ln(1+\pi))\widehat{F}^{'}\widehat{M}\widehat{S}_{22}\widehat{M}\widehat{F}\\ &-(1-\pi^{-1}ln(1+\pi))(\widehat{F}^{'}\widehat{M}\widehat{S}_{12} + \widehat{S}_{12}^{'}\widehat{M}\widehat{F})\end{aligned}$$

where $\widehat{F} = n^{-1}\sum_{t=R}^{T} Y_t X_t^{'}$, $\widehat{M} = (n^{-1}\sum_{t=R}^{T-1} Y_t Y_t^{'})r^{-1}$, *and*

$$\begin{aligned}\widehat{S}_{11} =&\frac{1}{n}\sum_{t=R}^{T-1}(\widehat{\epsilon}_{t+1}X_t - \widehat{\mu}_1)(\widehat{\epsilon}_{t+1}X_t - \widehat{\mu}_1)^{'}\\ &+\frac{1}{n}\sum_{t=\tau}^{l_T} w_\tau \sum_{t=R+\tau}^{T-1}(\widehat{\epsilon}_{t+1}X_t - \widehat{\mu}_1)(\widehat{\epsilon}_{t+1-\tau}X_{t-\tau} - \widehat{\mu}_1)^{'}\\ &+\frac{1}{n}\sum_{t=\tau}^{l_T} w_\tau \sum_{t=R+\tau}^{T-1}(\widehat{\epsilon}_{t+1-\tau}X_{t-\tau} - \widehat{\mu}_1)(\widehat{\epsilon}_{t+1}X_t - \widehat{\mu}_1)^{'}\end{aligned}$$

$$\begin{aligned}\widehat{S}_{12} =&\frac{1}{n}\sum_{\tau=0}^{l_T} w_\tau \sum_{t=R+\tau}^{T-1}(\widehat{\epsilon}_{t+1-\tau}X_{t-\tau} - \widehat{\mu}_1)(Y_{t-1}\widehat{\epsilon}_t)^{'}\\ &+\frac{1}{n}\sum_{\tau=1}^{l_T} w_\tau \sum_{t=R+\tau}^{T-1}(\widehat{\epsilon}_{t+1}X_t - \widehat{\mu}_1)(Y_{t-1-\tau}\widehat{\epsilon}_{t-\tau})^{'}\end{aligned}$$

$$\widehat{S}_{22} = \frac{1}{n}\sum_{t=R}^{T-1}(Y_{t-1}\widehat{\epsilon}_t)(Y_{t-1}\widehat{\epsilon}_t)'$$
$$+\frac{1}{n}\sum_{\tau=1}^{l_T} w_\tau \sum_{t=R+\tau}^{T-1}(Y_{t-1}\widehat{\epsilon}_t)(Y_{t-1-\tau}\widehat{\epsilon}_{t-\tau})'$$
$$+\frac{1}{n}\sum_{\tau=1}^{l_T} w_\tau \sum_{t=R+\tau}^{T-1}(Y_{t-1-\tau}\widehat{\epsilon}_{t-\tau})(Y_{t-1}\widehat{\epsilon}_t)'$$

with $w_\tau = 1 - \frac{\tau}{l_T+1}$. *In addition, for* $\pi = 0$,

$$m_n'\widehat{S}_{11}^{-1}m_n \xrightarrow{d} \chi_k^2$$

(ii) Under the alternative, $m_n'\widehat{S}_{11}^{-1}m_n$ *diverges at rate n.*

Note that a "nonlinear" variant of the above CCS test has also been developed by the same authors. In this generic form of the test, one can test for nonlinear Granger causality, for example, where the alternative hypothesis is that some (unknown) function of the x_t can be added to the benchmark linear model that contains no x_t in order to improve predictive accuracy. This alternative test is thus consistent against generic nonlinear alternatives. Complete details of this test are given in the next section.

16.1.3 A Predictive Accuracy Test that is Consistent Against Generic Alternatives

The test discussed in the previous subsection is designed to have power against a given (linear) alternative; and while it may have power against other alternatives, it is not designed to do so. Thus, it is not consistent against generic alternatives. Tests that are consistent against generic alternatives are sometimes called portmanteau tests, and it is this sort of extension of the out-of-sample Granger causality test discussed above that we now turn our attention to. Broadly speaking, the above consistency has been studied in the consistent specification testing literature (see Bierens (1990), Bierens and Ploberger (1997), De Jong (1996), Hansen (1996a), Lee, White, and Granger (1993) and Stinchcombe and White (1998)).

Corradi and Swanson (2002) draw on both the integrated conditional moment (ICM) testing literature of Bierens (1990) and Bierens and Ploberger (1997) and on the predictive accuracy testing literature; and propose an out-of-sample version of the ICM test that is consistent against generic nonlinear alternatives. This test is designed to examine whether there exists an unknown (possibly nonlinear) alternative model with better predictive power than the benchmark model, for a given loss function. A typical example is the case in which the benchmark model is a simple autoregressive model and we want to know whether including some unknown functions of the past information can produce more accurate forecasts. This is the case of nonlinear Granger causality testing discussed above. Needless to say, this test can be applied to many other cases. One important feature of this test is that the same loss function is used for in-sample model estimation and out-of-sample predictive evaluation (see Granger (1993) and Weiss (1996)).

Consider the following benchmark model:

$$y_t = \theta_1^{\dagger} y_{t-1} + u_t$$

where $\theta_1^{\dagger} = \arg\min_{\theta_1 \in \Theta_1} E(q(y_t - \theta_1 y_{t-1}))$. The generic alternative model is,

$$y_t = \theta_{2,1}^{\dagger}(\gamma) y_{t-1} + \theta_{2,2}^{\dagger}(\gamma)\omega(Z^{t-1}, \gamma) + \upsilon_t$$

where

$$\theta_2^{\dagger}(\gamma) = (\theta_{2,1}^{\dagger}(\gamma), \theta_{2,2}^{\dagger}(\gamma))' = \arg\min_{\theta_2 \in \Theta_2} E\Big(q\big(y_t - \theta_{2,1}(\gamma) y_{t-1} - \theta_{2,2}(\gamma)\omega(Z^{t-1}, \gamma)\big)\Big)$$

The alternative model is "generic" due to the term $\omega(Z^{t-1}, \gamma)$, where the function $\omega(\cdot)$ is a generically comprehensive function, as defined in Bierens (1990) and Bierens and Ploberger (1997). The test hypotheses are

$$H_0 : E(g(u_t) - g(\upsilon_t)) = 0$$

$$H_A : E(g(u_t) - g(\upsilon_t)) > 0$$

By definition, it is clear that the benchmark model is nested within the alternative model. Thus the former model can never outperform the latter. Equivalently, the hypotheses can be restated as

$$H_0 : \theta_{2,2}^{\dagger}(\gamma) = 0$$

$$H_A : \theta_{2,2}^{\dagger}(\gamma) \neq 0$$

Note that, given the definition of $\theta_2^{\dagger}(\gamma)$, we have that

$$E\left(g'(\upsilon_t) \times \left(-y_t, -\omega(Z^{t-1}, \gamma)\right)'\right) = 0$$

Hence, under the null, we have that $\theta_{2,2}^{\dagger}(\gamma) = 0$, $\theta_{2,1}^{\dagger}(\gamma) = \theta_1^{\dagger}$, and $E(g'(u_t)\omega(Z^{t-1}, \gamma)) = 0$. As a result, the hypotheses can be once again be restated as,

$$H_0 : E(g^{'}(u_t)\omega(Z^{t-1}, \gamma)) = 0$$

$$H_A : E(g^{'}(u_t)\omega(Z^{t-1}, \gamma)) \neq 0$$

The test statistic is given by

$$M_n = \int m_n(\gamma)^2 \phi(\gamma) d\gamma$$

with

$$m_n(\gamma) = n^{-1/2} \sum_{t=R}^{T-1} g^{'}(\widehat{u}_t + 1)\omega(Z^t, \gamma)$$

where $\int \phi(\gamma)d\gamma = 1, \phi(\gamma) \geq 0$, and $\phi(\gamma)$ is absolutely continuous with respect to Lebesgue measure.

Assumption 16.6

(i) (y_t, Z^t) is a strictly stationary and absolutely regular strong mixing sequence with size $-4(4+\psi)/\psi, \psi > 0$; (ii) $g(\cdot)$ is three times continuously differentiable in θ, over the interior of Θ, and $\nabla_\theta g$, $\nabla_\theta^2 g$, $\nabla_\theta g^{'}$, $\nabla_\theta^2 g^{'}$ are $2r$-dominated uniformly in Θ, with $r \geq 2(2+\psi)$; (iii) $E(-\nabla_\theta^2 g(\theta))$ is negative definite, uniformly in Θ; (iv) $\omega(\cdot)$ is a bounded, twice continuously differentiable function on the interior of Γ and $\nabla_\gamma \omega(Z^t, \gamma)$ is bounded uniformly in Γ; (iv) $\nabla_\gamma \nabla_\theta g^{'}(\theta)\omega(Z^t, \gamma)$ is continuous on $\Theta \times \Gamma$, Γ a compact subset of $\Re^d$ and is $2r$-dominated uniformly in $\Theta \times \Gamma$, with $r \geq 2(2+\psi)$.

Assumption 16.7

(i) $E(g^{'}(y_t - \theta_1^\dagger y_{t-1})) < E(g^{'}(y_t - \theta_1 y_{t-1})), \forall \theta \neq \theta^\dagger$; (ii) $\inf_\gamma E(g^{'}(y_t - \theta_{2,1}^\dagger(\gamma)y_{t-1} + \theta_{2,2}^\dagger(\gamma)\omega(Z^{t-1}, \gamma))) < E(g^{'}(y_t - \theta_{2,1}(\gamma)y_{t-1} + \theta_{2,2}(\gamma)\omega (Z^{t-1}, \gamma))), \forall \theta \neq \theta^\dagger(\gamma)$.

Assumption 16.8 $T = R + n$, and as $T \to \infty, n/R \to \pi$, with $0 \leq \pi < \infty$.

Proposition 16.4 (From Theorem 1 in Corradi and Swanson, 2002) *With Assumptions 16.6–16.8, the following results hold: (i) Under the null,*

$$M_n \xrightarrow{d} \int Z(\gamma)^2 \phi(\gamma) d\gamma$$

where Z is a Gaussian process with covariance structure

$$K(\gamma_1,\gamma_2) = S_{gg}(\gamma_1,\gamma_2) + 2\Pi\mu_{\gamma_1}A^{\dagger}S_{hh}A^{\dagger}\mu_{\gamma_2} + \Pi\mu_{\gamma_1}^{'}A^{\dagger}S_{gh}(\gamma_2) + \Pi\mu_{\gamma_2}^{'}A^{\dagger}S_{gh}(\gamma_1)$$

with $\mu_{\gamma_1} = E(\nabla_{\theta_1}(g^{'}(u_t)\omega(Z^t,\gamma_1)))$, $A^{\dagger} = (-E(\nabla^2_{\theta_1}q(u_t)))^{-1}$, *and*

$$S_{gg}(\gamma_1,\gamma_2) = \sum_j E(g^{'}(u_{s+1})\omega(Z^s,\gamma_1)g^{'}(u_{s+j+1})\omega(Z^{s+j},\gamma_1))$$

$$S_{hh} = \sum_j E(\nabla_{\theta_1}q(u_s)\nabla_{\theta_1}q(u_{s+j})^{'})$$

$$S_{gh}(\gamma_1) = \sum_j E(g^{'}(u_{s+1})\omega(Z^s,\gamma_1)\nabla_{\theta_1}q(u_{s+j})^{'})$$

and γ, γ_1, *and* γ_2 *are generic elements of* Γ.
(ii) Under the alternative, for $\epsilon > 0$ *and* $\delta < 1$,

$$\lim_{n\to\infty}\Pr\left(n^{-\delta}\int m_n(\gamma)^2\phi(\gamma)d\gamma > \epsilon\right) = 1$$

The limiting distribution under the null is a Gaussian process with a covariance structure that reflects both the time dependence and the parameter estimation error. Therefore the critical values cannot be tabulated. Valid asymptotic critical values can be constructed by using the block bootstrap for recursive estimation schemes, as detailed in Corradi and Swanson (2007). In particular, define,

$$\widetilde{\theta}^*_{1,t} = arg\min_{\theta_1}\frac{1}{t}\sum_{j=2}^{t}[g(y^*_j - \theta_1 y^*_{j-1}) - \theta_1^{'}\frac{1}{T}\sum_{i=2}^{T}\nabla_{\theta}g(y_i - \widehat{\theta}_1 y_{i-1})]$$

Then the bootstrap statistic is,

$$M^*_n = \int m^*_n(\gamma)^2\phi(\gamma)d\gamma$$

where

$$m^*_n(\gamma) = n^{-1/2}\sum_{t=R}^{T-1}\left(g^{'}(u^*_t)\omega(Z^{*,t},\gamma) - T^{-1}\sum_{i=1}^{T-1}g^{'}(\widehat{u}_t)\omega(Z^i,\gamma)\right)$$

Assumption 16.9 For any t, s and $\forall i, j, k = 1, 2$, and for $\Delta < \infty$,

$$\text{(i)}\quad E(\sup_{\theta,\gamma,\gamma^+} |g'(\theta)\omega(Z^{t-1},\gamma)\nabla_\theta^k g'(\theta)\omega(Z^{s-1},\gamma^+)|^4) < \Delta$$

where $\nabla_\theta^k(\cdot)$ denotes the k-th element of the derivative of its argument with respect to θ.

$$\text{(ii)}\quad E(\sup_\theta |\nabla_\theta^k(\nabla_\theta^i g(\theta))\nabla_\theta^j g(\theta)|^4) < \Delta$$

and

$$\text{(iii)}\quad E(\sup_{\theta,\gamma} |g'(\theta)\omega(Z^{t-1},\gamma)\nabla_\theta^k(\nabla_\theta^j g(\theta))|^4) < \Delta$$

Proposition 16.5 (From Proposition 5 in Corradi and Swanson, 2007) *With Assumptions 16.6–16.9, also assume that as $T \to \infty$, $l \to \infty$, and $l/T^{1/4} \to 0$, then as $T, n, R \to \infty$,*

$$\Pr\left(\sup_\delta \left|\overset{*}{\Pr}\left(\int m_n^*(\gamma)^2\phi(\gamma)d\gamma \le \delta\right) - Pr\left(\int m_n(\gamma)^2\phi(\gamma)d\gamma \le \delta\right)\right| > \epsilon\right) \to 0$$

The above proposition justifies the bootstrap procedure. For all samples except a set with probability measure approaching zero, M_n^* mimics the limiting distribution of M_n under the null, ensuring asymptotic size equal to α. Under the alternative, M_n^* still has a well defined limiting distribution, while M_n explodes, ensuring unit asymptotic power.

In closing, note that $\widetilde{\theta}_{1,t}^*$ can be replaced with $\theta_{1,t}^*$ if parameter estimation error is assumed to be asymptotically negligible. In this case, critical values are constructed via standard application of the block bootstrap.

16.1.4 Comparison of Multiple Models

The predictive accuracy tests that we have introduced to this point are all used to choose between two competing models. However, an even more common situation is when multiple (more than two) competing models are available, and the objective is to assess whether there exists at least one model that outperforms a given "benchmark" model. If we sequentially compare each of the alternative models with the benchmark, we induce the so-called "data snooping" problem, where sequential test bias results in the size of our test increasing to unity, so that the null hypothesis is rejected with probability one, even when the null is true. In this subsection, we review several tests for comparing multiple models and addressing the issue of data snooping.

A Reality Check for Data Snooping

White (2000) proposes a test called the "reality check," which is suitable for comparing multiple models. We use the same notation as that used when discussing the DM test, except that there are now multiple alternative models, i.e., model $i = 0, 1, 2, \ldots, m$. Recall that $i = 0$ denotes the benchmark model. Define the following test statistic:

$$\widehat{S}_n = \max_{i=1,\ldots,m} \widehat{S}_n(0,i) \tag{16.4}$$

where

$$\widehat{S}_n(0,i) = \frac{1}{\sqrt{n}} \sum_{t=R}^{T-1} (g(\widehat{u}_{0,t+1}) - g(\widehat{u}_{i,t+1})), \quad i = 1, \ldots, m$$

The reality check tests the following null hypothesis:

$$H_0 : \max_{i=1,\ldots,m} E(g(u_{0,t+1}) - g(u_{i,t+1})) \leq 0$$

against

$$H_A : \max_{i=1,\ldots,m} E(g(u_{0,t+1}) - g(u_{i,t+1})) > 0$$

The null hypothesis states that no competing model among the set of m alternatives yields more accurate forecasts than the benchmark model, for a given loss function; while the alternative hypothesis states that there is at least one alternative model that outperforms the benchmark model. By jointly considering all alternative models, the reality check controls the family wise error rate (FWER), thus circumventing the issue of data snooping, i.e., sequential test bias.

Assumption 16.10 (i) $f_i(\cdot, \theta_i^\dagger)$ is twice continuously differentiable on the interior of Θ_i and the elements of $\nabla_{\theta_i} f_i(Z^t, \theta_i)$ and $\nabla^2_{\theta_i} f_i(Z^t, \theta_i)$ are p-dominated on Θ_i, for $i = 1, \ldots, m$, with $p > 2(2 + \psi)$, where ψ is the same positive constant defined in Assumption 16.1; (ii) $g(\cdot)$ is positively valued, twice continuously differentiable on Θ_i, and $g(\cdot)$, $g'(\cdot)$, and $g''(\cdot)$ are p-dominated on Θ_i, with p defined in (i); and (iii) let $c_{ii} = \lim_{T\to\infty} \mathrm{Var}\left(T^{-1/2} \sum_{t=1}^{T} (g(u_{0,t+1}) - g(u_{i,t+1}))\right)$, $i = 1, \ldots, m$, define analogous covariance terms, c_{ji}, $j, i = 1, \ldots, m$, and assume that c_{ji} is positive semi-definite.

Proposition 16.6 (Parts (i) and (iii) are from Proposition 2.2 in White, 2000) *With Assumptions 16.1, 16.2, and 16.10, then under the null,*

$$\max_{i=1,\ldots,m} \left(\widehat{S}_n(0,i) - \sqrt{n} E\left(g(u_{0,t+1}) - g(u_{i,t+1})\right)\right) \xrightarrow{d} \max_{i=1,\ldots,m} S(0,i)$$

where $S = (S(0,1), \ldots, S(0,m))'$ *is a zero mean Gaussian process with covariance matrix given by* V*, with* V *an* $m \times m$ *matrix, and: (i) If parameter estimation error vanishes, then for* $i = 0, \ldots, m$*,*

$$V = S_{g_i g_i} = \sum_{\tau=-\infty}^{\infty} E\left(g(u_{0,1}) - g(u_{i,1})\right)\left(g(u_{0,1+\tau}) - g(u_{i,1+\tau})\right)$$

(ii) If parameter estimation error does not vanish, then

$$\begin{aligned} V =& S_{g_i g_i} + 2\Pi \mu_0' A_0^{\dagger} C_{00} A_0^{\dagger} \mu_0 + 2\Pi \mu_i' A_i^{\dagger} C_{ii} A_i^{\dagger} \mu_i \\ & - 4\Pi \mu_0' A_0^{\dagger} C_{0i} A_i^{\dagger} \mu_i + 2\Pi S_{g_i q_0} A_0^{\dagger} \mu_0 - 2\Pi S_{g_i q_i} A_i^{\dagger} \mu_i \end{aligned}$$

where

$$C_{ii} = \sum_{\tau=-\infty}^{\infty} E(\nabla_{\theta_i} q_i(y_{1+s}, Z^s, \theta_i^{\dagger}))(\nabla_{\theta_i} q_i(y_{1+s+\tau}, Z^{s+\tau}, \theta_i^{\dagger}))'$$

$$S_{g_i q_i} = \sum_{\tau=-\infty}^{\infty} E\left((g(u_{0,1}) - g(u_{i,1}))\right)(\nabla_{\theta_i} q_i(y_{1+s+\tau}, Z^{s+\tau}, \theta_i^{\dagger}))'$$

$A_i^{\dagger} = (E(-\nabla_{\theta_i}^2 q_i(y_t, Z^{t-1}, \theta_i^{\dagger})))^{-1}$, $\mu_i = E(\nabla_{\theta_i} g(u_{i,t+1}))$, *and* $\Pi = 1 - \pi^{-1} ln(1+\pi)$. *(iii) Under the alternative,* $\Pr(n^{-1/2}|S_n| > \epsilon) \to 1$ *as* $n \to \infty$.

Of particular note is that since the maximum of a Gaussian process is not Gaussian, in general, the construction of critical values for inference is not straightforward. White (2000) proposes two alternatives. The first is a simulation-based approach starting from a consistent estimator of V, say $\widehat{V}$. With $\widehat{V}$, for each simulation $s = 1, \ldots, S$, one realization is drawn from m-dimensional $N(0, \widehat{V})$ and the maximum value over $i = 1, \ldots, m$ is recorded. Repeat this procedure for S times, with a large S, and use the $(1-\alpha)$-percentile of the empirical distribution of the maximum values. A main drawback to this approach is that we need to first estimate the covariance structure V. However, if m is large and the prediction errors exhibit a high degree of heteroskedasticity and time dependence, the estimator of V becomes imprecise and thus the inference unreliable, especially in finite samples. The second approach relies on bootstrap procedures to construct critical values, which overcomes the problem of the first approach. We resample blocks of $g(\widehat{u}_{0,t+1}) - g(\widehat{u}_{i,t+1})$, and for each bootstrap replication $b = 1, \ldots, B$, we calculate

$$\widehat{S}_n^{*(b)}(0,i) = n^{-1/2} \sum_{t=R}^{T-1} (g^*(\widehat{u}_{0,t+1}) - g^*(\widehat{u}_{i,t+1})) \tag{16.5}$$

and the bootstrap statistic is given by

$$S_n^* = \max_{i=1,\ldots,m} |\widehat{S}_n^{*(b)}(0,i) - \widehat{S}_n(0,i)|$$

the $(1-\alpha)$-percentile of the empirical distribution of B bootstrap statistics is then used for inference. Note that in White (2000), parameter estimation error is assumed to be asymptotically negligible. In light of this, Corradi and Swanson (2007) suggest a "re-centering" bootstrap procedure in order to explicitly handle the issue of non-vanishing parameter estimation error, when constructing critical values for this test. The new bootstrap statistic is defined as,

$$S_n^{**} = \max_{i=1,\ldots,m} S_n^{**}(0,i)$$

where

$$S_n^{**}(0,i) = n^{-1/2}\sum_{t=R}^{T-1}[(g(y_{t+1}^* - f_0(Z^{*,t},\widetilde{\theta}_{0,t}^*)) - g(y_{t+1}^* - f_i(Z^{*,t},\widetilde{\theta}_{i,t}^*)))$$
$$- \frac{1}{T}\sum_{j=1}^{T-1}(g(y_{j+1} - f_0(Z^j,\widehat{\theta}_{0,t})) - g(y_{j+1} - f_i(Z^j,\widehat{\theta}_{i,t})))]$$

Note that $S_n^{**}(0,i)$ is different from the standard bootstrap statistic in Eq. (16.5), which is defined as the difference between the statistic constructed using original samples and that using bootstrap samples. The $(1-\alpha)$-percentile of the empirical distribution of S_n^{**} can be used to construct valid critical values for inference in the case of non-vanishing parameter estimation error. Proposition 2 in Corradi and Swanson (2007) establishes the first order validity for the recursive estimation scheme and Corradi and Swanson (2006a) outline the approach to constructing valid bootstrap critical values for the rolling-window estimation scheme. Finally, note that Corradi and Swanson (2007) explain how to use the simple block bootstrap for constructing critical values when parameter estimation error is assumed to be asymptotically negligible. This procedure is perhaps the most obvious method to use for constructing critical values as it involves simply resampling the original data, carrying out the same forecasting procedures as used using the original data, and then constructing bootstrap statistics. These bootstrap statistics can be used (after subtracting the original test statistic from each of them) to form an empirical distribution which mimics the distribution of the test statistic under the null hypothesis. Finally, the empirical distribution can be used to construct critical values, which are the $(1-\alpha)$-quantiles of said distribution.

From Eq. (16.4) and Proposition 16.6, it is immediate to see that the reality check can be rather conservative when many alternative models are strictly dominated by the benchmark model. This is because those "bad" models do not contribute to the test statistic, simply because they are ruled out by the maximum, but contribute to the bootstrap statistics. Therefore, when many inferior models are included, the probability of rejecting the null hypothesis is actually smaller than α. Indeed, it is only for the least favorable case, in which $E(g(u_{0,t+1}) - g(u_{i,t+1})) = 0, \forall i$, that the distribution of $\widehat{S}_n$ coincides with that of

$$\max_{i=1,\ldots,m} \left(\widehat{S}_n(0,i) - \sqrt{n}E\left(g(u_{0,t+1}) - g(u_{i,t+1})\right)\right)$$

We introduce two approaches for addressing the conservative nature of this test below.

A Test for Superior Predictive Ability

Hansen (2005) proposes a modified reality check called the superior predictive ability (SPA) test that controls the FWER and addresses the inclusion of inferior models. The SPA test statistic is defined as,

$$T_n = \max\left\{0, \max_{i=1,\ldots,m} \frac{\widehat{S}_n(0,i)}{\sqrt{\widehat{v}_{i,i}}}\right\}$$

where $\widehat{v}_{i,i} = \frac{1}{B}\sum_{b=1}^{B}\left(\frac{1}{n}\sum_{t=R}^{T-1}((g(\widehat{u}_{0,t+1}) - g(\widehat{u}_{i,t+1})) - (g(\widehat{u}^*_{0,t+1}) - g(\widehat{u}^*_{i,t+1})))^2\right)$.

The bootstrap statistic is then defined as,

$$T_n^{*(b)} = \max\left\{0, \max_{i=1,\ldots,m}\left\{\frac{n^{-1/2}\sum_{t=R}^{T-1}(\widehat{d}_{i,t}^{*(b)} - \widehat{d}_{i,t}\mathbf{1}_{\{\widehat{d}_{i,t}\geq -A_{T,i}\}})}{\sqrt{\widehat{v}_{i,i}}}\right\}\right\}$$

where $\widehat{d}_{i,t}^{*(b)} = g(\widehat{u}^*_{0,t+1}) - g(\widehat{u}^*_{i,t+1})$, $\widehat{d}_{i,t} = g(\widehat{u}_{0,t+1}) - g(\widehat{u}_{i,t+1})$, and $A_{T,i} = \frac{1}{4}T^{-1/4}\sqrt{\widehat{v}_{i,i}}$.

The idea behind the construction of SPA bootstrap critical values is that when a competing model is too slack, the corresponding bootstrap moment condition is not re-centered, and the bootstrap statistic is not affected by this model. Therefore, the SPA test is less conservative than the reality check. Corradi and Distaso (2011) derive a general class of SPA tests using the generalized moment selection approach of Andrews and Soares (2010) and show that Hansen's SPA test belongs to this class. Romano and Wolf (2005) propose a multiple step extension of the reality check which ensures tighter control of irrelevant models.

A Test Based on Sub-Sampling

The conservative property of the reality check can be alleviated by using the sub-sampling approach to constructing critical values, at the cost of sacrificing power in finite samples. Critical values are obtained from the empirical distribution of a sequence of statistics constructed using subsamples of size $\widetilde{b}$, where $\widetilde{b}$ grows with the sample size, but at a slower rate (see Politis, Romano, and Wolf, 1999).

In the context of the reality check, as $n \to \infty, \widetilde{b} \to \infty$, and $\widetilde{b}/n \to 0$, define

$$S_{n,a,\widetilde{b}} = \max_{i=1,\dots,m} S_{n,a,\widetilde{b}}(0,i), \quad a = R, \dots, T - \widetilde{b} - 1$$

where

$$S_{n,a,\widetilde{b}}(0,i) = \widetilde{b}^{-1/2} \sum_{t=a}^{a+\widetilde{b}-1} \left(g(\widehat{u}_{0,t+1}) - g(\widehat{u}_{i,t+1}) \right)$$

We obtain the empirical distribution of $T - \widetilde{b} - 1$ statistics, $S_{n,a,\widetilde{b}}$, and reject the null if the test statistic $\widehat{S}_n$ is greater than the $(1-\alpha)$-quantile of the empirical distribution. The advantage of the sub-sampling approach over the bootstrap is that the test has correct size when $\max_{i=1,\dots,m} E(g(\widehat{u}_{0,t+1}) - g(\widehat{u}_{i,t+1})) < 0$ for some i, while the bootstrap approach delivers a conservative test in this case. However, although the sub-sampling approach ensures that the test has unit asymptotic power, the finite sample power may be rather low, since $S_{n,a,\widetilde{b}}$ diverges at rate $\sqrt{\widetilde{b}}$ instead of $\sqrt{n}$, under the alternative. Finally, note that the sub-sampling approach is also valid in the case of non-vanishing parameter estimation error because each statistic constructed using subsamples properly mimics the distribution of actual statistic.

16.2 Forecast Evaluation Using Density-Based Predictive Accuracy Tests

In Sect. 16.1, we introduced a variety of tests designed for comparing models based on point forecast accuracy. However, there are many practical situations in which economic decision making crucially depends not only on conditional mean forecasts (e.g., point forecasts) but also on predictive confidence intervals or predictive conditional distributions (also called predictive densities). One such case, for instance, is when value at risk (VaR) measures are used in risk management for assessment of the amount of projected financial losses due to extreme tail behavior, e.g., catastrophic events. Another common case is when economic agents are undertaking to optimize their portfolio allocations, in which case the joint distribution of multiple assets is required to be modeled and fully understood. The purpose of this section is to discuss recent tests for comparing (potentially misspecified) conditional distribution models.

16.2.1 The Kullback–Leibler Information Criterion Approach

A well-known measure of distributional accuracy is the Kullback–Leibler Information Criterion (KLIC). Using the KLIC involves simply choosing the model which minimizes the KLIC (see, e.g., White (1982), Vuong (1989), Gianni and Giacomini (2007), Kitamura (2002)). Of note is that White (1982) shows that quasi maximum likelihood estimators minimize the KLIC, under mild conditions. In order to implement the KLIC, one might choose model 0 over model 1, if

$$E(\ln f_0(y_t|Z^t, \theta_0^\dagger) - \ln f_1(y_t|Z^t, \theta_1^\dagger)) > 0$$

For the i.i.d case, Vuong (1989) suggests using a likelihood ratio test for choosing the conditional density model that is closer to the "true" conditional density, in terms of the KLIC. Gianni and Giacomini (2007) suggests using a weighted version of the likelihood ratio test proposed in Vuong (1989) for the case of dependent observations, while Kitamura (2002) employs a KLIC-based approach to select among misspecified conditional models that satisfy given moment conditions. Furthermore, the KLIC approach has recently been employed for the evaluation of dynamic stochastic general equilibrium models (see, e.g., Schorfheide (2010), Fernández-Villaverde and Rubio-Ramírez (2004), and Chang, Gomes, and Schorfheide (2002)). For example, Fernández-Villaverde and Rubio-Ramírez (2004) show that the KLIC-best model is also the model with the highest posterior probability.

The KLIC is a sensible measure of accuracy, as it chooses the model which on average gives higher probability to events which have actually occurred. Also, it leads to simple likelihood ratio type tests which have a standard limiting distribution and are not affected by problems associated with accounting for parameter estimation error. However, it should be noted that if one is interested in measuring accuracy over a specific region, or in measuring accuracy for a given conditional confidence interval, say, this cannot be done in as straightforward manner using the KLIC. For example, if we want to evaluate the accuracy of different models for approximating the probability that the rate of inflation tomorrow, given the rate of inflation today, will be between 0.5 and 1.5%, say, we can do so quite easily using the square error criterion, but not using the KLIC.

16.2.2 A Predictive Density Accuracy Test for Comparing Multiple Misspecified Models

Corradi and Swanson (2005) (CSa) and Corradi and Swanson (2006a) (CSb) introduce a measure of distributional accuracy, which can be interpreted as a distributional generalization of mean square error. In addition, Corradi and Swanson

(2005) apply this measure to the problem of selecting among multiple misspecified predictive density models. In this section we discuss these contributions to the literature.

Consider forming parametric conditional distributions for a scalar random variable, y_t, given Z^t, where $Z^t = (y_{t-1}, \ldots, y_{t-s_1}, X_t, \ldots, X_{t-s_2+1})$, with s_1, s_2 finite. With a little abuse of notation, now we define the group of conditional distribution models, from which one wishes to select a "best" model, as

$$\{F_i(u|Z^t, \theta_i^{\dagger})\}_{i=1,\ldots,m}$$

and define the true conditional distribution as

$$F_0(u|Z^t, \theta_0) = \Pr(y_{t+1} \leq u|Z^t)$$

Assume that $\theta_i^{\dagger} \in \Theta_i$, where Θ_i is a compact set in a finite dimensional Euclidean space, and let $\theta_i^{\dagger}$ be the probability limit of a quasi-maximum likelihood estimator (QMLE) of the parameters of the conditional distribution under model i. If model i is correctly specified, then $\theta_i^{\dagger} = \theta_0$. If $m > 2$, follow White (2000). Namely, choose a particular conditional distribution model as the "benchmark" and test the null hypothesis that no competing model can provide a more accurate approximation of the "true" conditional distribution, against the alternative that at least one competitor outperforms the benchmark model. Needless to say, pairwise comparison of alternative models, in which no benchmark need be specified, follows as a special case.

In this context, measure accuracy using the above distributional analog of mean square error. More precisely, define the mean square (approximation) error associated with model i, in terms of the average over U of $E\left((F_i(u|Z^t, \theta_i^{\dagger}) - F_0(u|Z^t, \theta_0))^2\right)$, where $u \in U$, and U is a possibly unbounded set on the real line, and the expectation is taken with respect to the conditioning variables. In particular, model 1 is more accurate than model 2, if

$$\int_U E((F_1(u|Z^t, \theta_1^{\dagger}) - F_0(u|Z^t, \theta_0))^2 - (F_2(u|Z^t, \theta_2^{\dagger}) - F_0(u|Z^t, \theta_0))^2)\phi(u)du < 0$$

where $\int_U \phi(u)du = 1$ and $\phi(u)du \geq 0, \forall u \in U \in \Re$.

This measure integrates over different quantiles of the conditional distribution. For any given evaluation point, this measure defines a norm and it implies a standard goodness of fit measure. Note that this measure of accuracy leads to straightforward evaluation of distributional accuracy over a given region of interest, as well as to straightforward evaluation of specific quantiles. A conditional confidence interval version of the above condition which is more natural to use in applications involving predictive interval comparison follows immediately, and can be written as

$$E\Big(((F_1(\bar{u}|Z^t,\theta_1^\dagger)-F_1(\underline{u}|Z^t,\theta_1^\dagger))-(F_1(\bar{u}|Z^t,\theta_0)-F_1(\underline{u}|Z^t,\theta_0)))^2$$

$$-((F_2(\bar{u}|Z^t,\theta_2^\dagger)-F_2(\underline{u}|Z^t,\theta_2^\dagger))-(F_1(\bar{u}|Z^t,\theta_0)-F_1(\underline{u}|Z^t,\theta_0)))^2\Big)\leq 0$$

Hereafter, $F_1(\cdot|\cdot,\theta_1^\dagger)$ is taken as the benchmark model, and the objective is to test whether some competitor model can provide a more accurate approximation of $F_0(\cdot|\cdot,\theta_0)$ than the benchmark. The null and the alternative hypotheses are

$$H_0:\max_{i=2,\ldots,m}\int_U E((F_1(u|Z^t,\theta_1^\dagger)-F_0(u|Z^t,\theta_0))^2$$

$$-(F_i(u|Z^t,\theta_i^\dagger)-F_0(u|Z^t,\theta_0))^2)\phi(u)du\leq 0$$

versus

$$H_A:\max_{i=2,\ldots,m}\int_U E((F_1(u|Z^t,\theta_1^\dagger)-F_0(u|Z^t,\theta_0))^2$$

$$-(F_i(u|Z^t,\theta_i^\dagger)-F_0(u|Z^t,\theta_0))^2)\phi(u)du>0$$

where $\phi(u)\geq 0$ and $\int_U \phi(u)=1$, $u\in U\in\Re$, U possibly unbounded. Note that for a given u, we compare conditional distributions in terms of their (mean square) distance from the true distribution. We then average over U. As discussed above, a possibly more natural version of the above hypotheses is in terms of conditional confidence intervals evaluation, so that the objective is to "approximate" $\Pr(\underline{u}\leq Y_{t+1}\leq\bar{u}|Z^t)$, and hence to evaluate a region of the predictive density. In that case, the null and alternative hypotheses can be stated as

$$H_0':\max_{i=2,\ldots,m}E(((F_1(\overline{u}|Z^t\ ,\ \theta_1^\dagger)-F_1(\underline{u}|Z^t\ ,\ \theta_1^{\text{f}}))$$

$$-(F_0(\overline{u}|Z^t,\ \theta_0)-F_0(\underline{u}|Z^t,\ \theta_0)))^2$$

$$-((F_i(\overline{u}|Z^t,\ \theta_i^\dagger)-F_i(\underline{u}|Z^t,\ \theta_i^\dagger))$$

$$-(F_0(\overline{u}|Z^t,\ \theta_0)-F_0(\underline{u}|Z^t,\ \theta_0)))^2)\ \leq 0$$

versus

$$H'_A : \max_{i=2,\ldots,m} E(((F_1(\overline{u}|Z^t\ , \theta_1^\dagger) - F_1(\underline{u}|Z^t\ , \theta_1^\text{f}))$$

$$-(F_0(\overline{u}|Z^t, \theta_0) - F_0(\underline{u}|Z^t, \theta_0)))^2$$

$$-((F_k(\overline{u}|Z^t, \theta_i^\dagger) - F_i(\underline{u}|Z^t, \theta_i^\dagger))$$

$$-(F_0(\overline{u}|Z^t, \theta_0) - F_0(\underline{u}|Z^t, \theta_0)))^2) > 0$$

Alternatively, if interest focuses on testing the null of equal accuracy of two conditional distribution models, say F_1 and F_i, we can simply state the hypotheses as

$$H''_0 : \int_U E((F_1(u|Z^t, \theta_1^\dagger) - F_0(u|Z^t, \theta_0))^2$$

$$- (F_i\ (u|Z^t, \theta_i^\dagger) - F_0(u|Z^t, \theta_0))^2)\phi(u)\mathrm{d}u = 0$$

versus

$$H''_A : \int_U E((F_1(u|Z^t, \theta_1^\dagger) - F_0(u|Z^t, \theta_0))^2$$

$$- (F_i\ (u|Z^t, \theta_i^\dagger) - F_0(u|Z^t, \theta_0))^2)\phi(u)\mathrm{d}u \neq 0,$$

or we can write the predictive density (interval) version of these hypotheses.

Of course, we do not know $F_0(u|Z^t)$. However, it is easy to see that

$$\begin{aligned} E((F_1(u|Z^t, \theta_1^\dagger)-&F_0(u|Z^t, \theta_0))^2 - (F_i(u|Z^t, \theta_i^\dagger) - F_0(u|Z^t, \theta_0))^2) \\ &= E((1\{y_{t+1}\ \leq u\} - F_1(u|Z^t, \theta_1^\dagger))^2) \\ &- E\ ((1\{y_{t+1}\ \leq u\} - F_i\ (u|Z^t, \theta_i^\dagger))^2) \end{aligned} \tag{16.6}$$

where the right-hand side of Eq. (16.6) does not require any knowledge of the true conditional distribution.

The intuition behind Eq. (16.6) is very simple. First, note that for any given u, $E(1\{y_{t+1} \leq u\}|Z^t) = \Pr(y_{t+1} \leq u|Z^t) = F_0(u|Z^t, \theta_0)$. Thus, $1\{y_{t+1} \leq u\} - F_i(u|Z^t, \theta_i^\dagger)$ can be interpreted as an "error" term associated with computation of the conditional expectation under F_i. Now, for $i = 1, \ldots, m$

$$\mu_i^2(u) = E(\ (1\{y_{t+1}\ \leq u\} - F_i\ (u|Z^t,\ \theta_i^{\dagger}))^2)$$

$$= E(((1\{y_{t+1}\ \leq u\} - F_0(u|Z^t,\ \theta_0)) - (F_i(u|Z^t,\ \theta_i^{\dagger}) - F_0(u|Z^t,\ \theta_0)))^2)$$

$$= E((1\{y_{t+1}\ \leq u\} - F_0(u|Z^t,\ \theta_0))^2) + E\ ((F_i\ (u|Z^t,\ \theta_i^{\dagger}) - F_0(u|Z^t,\ \theta_0))^2)$$

given that the expectation of the cross product is zero (which follows because 1 $\{y_{t+1}\ \leq\ u\} - F_0(u|Z^t,\ \theta_0)$ is uncorrelated with any measurable function of Z^t). Therefore,

$$\begin{aligned}\mu_1^2(u) - \mu_i^2(u) &= E(\ (F_1\ (u|Z^t,\ \theta_1^{\dagger}) - F_0(u|Z^t,\ \theta_0))^2) \\ &\quad - E\ ((F_i\ (u|Z^t,\ \theta_i^{\dagger}) - F_0(u|Z^t,\ \theta_0))^2)\end{aligned} \tag{16.7}$$

The statistic of interest is

$$Z_{n,j} = \max_{i=2,\ldots m} \int_U Z_{n,u,j}(1,\ i)\phi(u)\mathrm{d}u, \quad j = 1, 2,$$

where for $j = 1$ (rolling estimation scheme),

$$\begin{aligned}Z_{n,u,1}(1,\ i) = \frac{1}{\sqrt{n}} \sum_{t=R}^{T-1} &((1\{y_{t+1}\ \leq u\} - F_1(u|Z^t,\ \widehat{\theta}_{1,t,\mathrm{rol}}))^2 \\ &- (1\{y_{t+1}\ \leq u\} - F_i(u|Z^t,\ \widehat{\theta}_{i,t,\mathrm{rol}}))^2)\end{aligned}$$

and for $j = 2$ (recursive estimation scheme),

$$\begin{aligned}Z_{n,u,2}(1,\ i) = \frac{1}{\sqrt{n}} \sum_{t=R}^{T-1} &((1\{y_{t+1}\ \leq u\} - F_1(u|Z^t,\ \widehat{\theta}_{1,t,\mathrm{rec}}))^2 \\ &- (1\{y_{t+1}\ \leq u\} - F_i(u|Z^t,\ \widehat{\theta}_{i,t,\mathrm{rec}}))^2)\end{aligned}$$

where $\widehat{\theta}_{i,t,\mathrm{rol}}$ and $\widehat{\theta}_{i,t,\mathrm{rec}}$ are defined as

$$\widehat{\theta}_{i,t,\mathrm{rol}} = \arg\min_{\theta\in\Theta} \frac{1}{R} \sum_{j=t-R+1}^{t} q(y_j, Z^{j-1},\theta),\ R \leq t \leq T-1$$

and

$$\widehat{\theta}_{i,t,\text{rec}} = \arg\min_{\theta \in \Theta} \frac{1}{t} \sum_{j=1}^{t} q(y_j, Z^{j-1}, \theta),\ t = R, R+1, R+n-1$$

As shown above and in Corradi and Swanson (2005), the hypotheses of interest can be restated as

$$H_0 : \max_{i=2,\ldots,m} \int_U (\mu_1^2(u) - \mu_i^2(u))\phi(u)\mathrm{d}u \leq 0$$

versus

$$H_A : \max_{i=2,\ldots,m} \int_U (\mu_1^2(u) - \mu_i^2(u))\phi(u)\mathrm{d}u > 0$$

where $\mu_i^2(u) = E(\ (1\{y_{t+1}\ \leq u\} - F_i\ (u|Z^t,\ \theta_i^\dagger))^2)$

Assumption 16.11 (i) $\theta_i^\dagger$ is uniquely defined,

$$E(\ln(f_i(y_t, Z^{t-1}, \theta_i))) < E(\ln(f_i(y_t, Z^{t-1}, \theta_i^\dagger))),$$

for any $\theta_i \neq \theta_i^\dagger$; (ii) $\ln f_i$ is twice continuously differentiable on the interior of Θ_i, and $\forall \Theta_i$ a compact subset of $\Re^{\varrho(i)}$; (iii) the elements of $\nabla_{\theta_i} \ln f_i$ and $\nabla^2_{\theta_i} \ln f_i$ are p-dominated on Θ_i, with $p > 2(2+\psi)$, where ψ is the same positive constant as defined in Assumption 16.1; and (iv) $E(-\nabla^2_{\theta_i} \ln f_i)$ is negatively definite uniformly on Θ_i.

Assumption 16.12 $T = R + n$, and as $T \to \infty$, $n/R \to \pi$, with $0 < \pi < \infty$.

Assumption 16.13 (i) $F_i(u|Z^t, \theta_i)$ is continuously differentiable on the interior of Θ_i and $\nabla_{\theta_i} F_i(u|Z^t, \theta_i^\dagger)$ is $2r$-dominated on Θ_i, uniformly in u, $r > 2$, $\forall i$[1]; and (ii) let

$$v_{ii}(u) = \text{plim}_{T\to\infty} \text{Var}\Big(\frac{1}{\sqrt{T}} \sum_{t=s}^{T} (((1\{y_{t+1} \leq u\} - F_1(u|Z^t, \theta_1^\dagger))^2 - \mu_1^2(u))$$

$$-((1\{y_{t+1} \leq u\} - F_i(u|Z^t, \theta_i^\dagger))^2 - \mu_i^2(u))\Big),\ \forall i$$

define analogous covariance terms, $v_{j,i}(u)$, $j, i = 2, \ldots, m$, and assume that $[v_{j,i}(u)]$ is positive semi-definite, uniformly in u.

[1]We require that for $j = 1, \ldots, p_i$, $E(\nabla_\theta F_i(u|Z^t, \theta_i^\dagger))_j \geq D_t(u)$, with $\sup_t \sup_{u\in\Re} E(D_t(u)^{2r}) < \infty$.

Proposition 16.7 (From Proposition 1 in Corradi and Swanson, 2006a) *With Assumptions 16.1, 16.11–16.13, then*

$$\max_{i=2,\ldots,m} \int_U (Z_{n,u,j}\ (1\ ,\ i) - \sqrt{n}(\mu_1^2(u) - \mu_i^2(u)))\phi_U(u)\mathrm{d}u$$

$$\xrightarrow{d} \max_{i=2,\ldots,m} \int_U Z_{1,i,j}(u)\phi_U(u)\mathrm{d}u$$

where $Z_{1,i,j}(u)$ is a zero mean Gaussian process with covariance $C_{i,j}(u,\ u')(j = 1$ corresponds to rolling and $j = 2$ to recursive estimation schemes), equal to

$$E(\sum_{j=-\infty}^{\infty} ((1\{y_{s+1}\ \le u\} - F_1(u|Z^s,\ \theta_1^\dagger))^2 - \mu_1^2(u)) \times\ ((1\{y_{s+j+1}\ \le u'\}$$

$$-F_1(u'|Z^{s+j},\ \theta_1^\dagger))^2 - \mu_1^2(u'))) + E(\sum_{j=-\infty}^{\infty} ((1\{y_{s+1}\ \le u\} - F_i(u|Z^s,\ \theta_i^\dagger))^2 - \mu_i^2(u))$$

$$\times\ ((1\{y_{s+j+1}\ \le u'\} - F_i(u'|Z^{s+j},\ \theta_i^\dagger))^2 - \mu_i^2(u'))) - 2E(\sum_{j=-\infty}^{\infty} ((1\{y_{s+1}\ \le u\}$$

$$-F_1(u|Z^s,\ \theta_1^\dagger))^2 - \mu_1^2(u)) \times\ ((1\{y_{s+j+1}\ \le u'\} - F_i(u'|Z^{s+j},\ \theta_i^\dagger))^2 - \mu_i^2(u')))$$

$$+4\Pi_j m_{\theta_1^\dagger}(u)'A(\theta_1^\dagger) \times E(\sum_{j=-\infty}^{\infty} \nabla_{\theta_1} \ln f_1(y_{s+1}|Z^s,\ \theta_1^\dagger)\nabla_{\theta_1} \ln f_1(y_{s+j+1}|Z^{s+j},\ \theta_1^\dagger)')$$

$$\times A(\theta_1^\dagger)m_{\theta_1^\dagger}(u') + 4\Pi_j m_{\theta_i^\dagger}(u)'A(\theta_i^\dagger) \times E(\sum_{j=-\infty}^{\infty} \nabla_{\theta_i} \ln f_i(y_{s+1}|Z^s,\ \theta_i^\dagger)$$

$$\times \nabla_{\theta_i} \ln f_i(y_{s+j+1}|Z^{s+j},\ \theta_i^\dagger)') \times A(\theta_i^\dagger)m_{\theta_i^\dagger}(u') - 4\Pi_j m_{\theta_1^\dagger}(u,\ )'A(\theta_1^\dagger)$$

$$\times E(\sum_{j=-\infty}^{\infty} \nabla_{\theta_1} \ln f_1(y_{s+1}|Z^s,\ \theta_1^\dagger)\nabla_{\theta_i} \ln f_i(y_{s+j+1}|Z^{s+j} \times A(\theta_i^\dagger)m_{\theta_i^\dagger}(u')$$

$$-4C\Pi_j m_{\theta_1^\dagger}(u)'A(\theta_1^\dagger) \times E(\sum_{j=-\infty}^{\infty} \nabla_{\theta_1} \ln f_1(y_{s+1}|Z^s,\ \theta_1^\dagger) \times ((1\{y_{s+j+1}\ \le u\}$$

$$-F_1\,(u|Z^{s+j},\,\theta_1^{\dagger}))^2-\mu_1^2(u)))+4C\Pi_j m_{\theta_1^{\dagger}}(u)'A(\theta_1^{\dagger})\times E(\sum_{j=-\infty}^{\infty}\nabla_{\theta_1}\ln f_1(y_{s+1}|Z^s,\,\theta_1^{\dagger})$$

$$\times((1\{y_{s+j+1}\ \leq u\}-F_i(u|Z^{s+j},\,\theta_i^{\dagger}))^2-\mu_i^2(u)))-4C\Pi_j m_{\theta_i^{\dagger}}(u)'A(\theta_i^{\dagger})$$

$$\times E(\sum_{j=-\infty}^{\infty}\nabla_{\theta_i}\ln f_i(y_{s+1}|Z^s,\theta_i^{\dagger})'\times((1\{y_{s+j+1}\ \leq u\}-F_i(u|Z^{s+j},\,\theta_i^{\dagger}))^2-\mu_i^2(u)))$$

$$+4C\Pi_j m_{\theta_i^{\dagger}}(u)'A(\theta_i^{\dagger})\times E(\sum_{j=-\infty}^{\infty}\nabla_{\theta_i}\ln f_i(y_{s+1}|Z^s,\,\theta_i^{\dagger})'\times\ ((1\{y_{s+j+1}\ \leq u\}$$

$$-F_1(u|Z^{s+j},\,\theta_1^{\dagger}))^2-\mu_1^2(u)))$$

with

$$m_{\theta_i^{\dagger}}(u)'=E(\nabla_{\theta_i}F_i(u|Z^t,\,\theta_i^{\dagger})'(1\{y_{t+1}\ \leq u\}-F_i(u|Z^t,\,\theta_i^{\dagger})))$$

and

$$A(\theta_i^{\dagger})\ =A_i^{\dagger}\ =\ (E(-\nabla_{\theta_i}^2\ln f_i(y_{t+1}|Z^t,\,\theta_i^{\dagger})))^{-1}$$

and for $j=1$ *and* $n\leq R$, $\Pi_1=(\pi-\frac{\pi^2}{3})$, $C\Pi_1=\frac{\pi}{2}$, *and for* $n>R$, $\Pi_1=(1-\frac{1}{3\pi})$ *and* $C\Pi_1=(1-\frac{1}{2\pi})$. *Finally, for* $j=2$, $\Pi_2=2(1-\pi^{-1}\ln(1+\pi))$ *and* $C\Pi_2=0.5\Pi_2$.

From this proposition, note that when all competing models provide an approximation to the true conditional distribution that is as (mean square) accurate as that provided by the benchmark (i.e., when $\int_U(\mu_1^2(u)-\mu_i^2(u))\phi(u)du\ =\ 0,\forall i)$, then the limiting distribution is a zero mean Gaussian process with a covariance kernel which is not nuisance parameter free. Additionally, when all competitor models are worse than the benchmark, the statistic diverges to minus infinity at rate $\sqrt{n}$. Finally, when only some competitor models are worse than the benchmark, the limiting distribution provides a conservative test, as Z_P will always be smaller than $\max_{i=2,\ldots,m}\int_U(Z_{n,u}\ (1,\ i)-\sqrt{n}(\mu_1^2(u)-\mu_i^2(u)))\phi(u)du$, asymptotically. Of course, when H_A holds, the statistic diverges to plus infinity at rate $\sqrt{n}$.

For the case of evaluation of multiple conditional confidence intervals, consider the statistic

$$V_{n,\tau}\ =\ \max_{i=2,\ldots,m}\ V_{n,\underline{u},\overline{u},\tau}(1,\ i)$$

where

$$V_{n,\underline{u},\overline{u},\tau}(1,\ i) = \frac{1}{\sqrt{n}}\sum_{t=R}^{T-1}((1\{\underline{u} \leq y_{t+1}\ \leq \overline{u}\} - (F_1(\overline{u}|Z^t,\ \widehat{\theta}_{1,t,\tau})$$

$$-F_1(\underline{u}|Z^t,\ \widehat{\theta}_{1,t,\tau})))^2 - (1\{\underline{u} \leq y_{t+1}\ \leq \overline{u}\} - (F_i(\overline{u}|Z^t,\ \widehat{\theta}_{i,t,\tau}) - F_i(\underline{u}|Z^t,\ \widehat{\theta}_{i,t,\tau})))^2)$$

where $s = \max\{s1,\ s2\}, \tau = 1, 2$, and $\widehat{\theta}_{i,t,\tau} = \widehat{\theta}_{i,t,\text{rol}}$ for $\tau = 1$, and $\widehat{\theta}_{i,t,\tau} = \widehat{\theta}_{k,t,\text{rec}}$ for $\tau = 2$.

We then have the following result.

Proposition 16.8 (From Proposition lb in Corradi and Swanson, 2006a) *With Assumptions 16.1, 16.11–16.13, then for* $\tau = 1$,

$$\max_{i=2,\ldots m}\ (V_{n,\underline{u},\overline{u},\tau}\ (1\ ,\ i)\ -\ \sqrt{n}(\mu_1^2 - \mu_i^2)) \xrightarrow{d} \max_{i=2,\ldots m}\ V_{n,i,\tau}(\underline{u},\ \overline{u})$$

where $V_{n,i,\tau}(\underline{u},\ \overline{u})$ *is a zero mean normal random variable with covariance* $c_{ii} = v_{ii} + p_{ii} + cp_{ii}$, *where* v_{ii} *denotes the component of the long-run variance matrix we would have in absence of parameter estimation error,* p_{ii} *denotes the contribution of parameter estimation error and* cp_{ii} *denotes the covariance across the two components. In particular*

$$v_{ii} = E\sum_{j=-\infty}^{\infty}(((1\{\underline{u} \leq y_{s+1}\ \leq \overline{u}\} - (F_1(\overline{u}|Z^s,\ \theta_1^{\dagger}) - F_1(\underline{u}|Z^s,\ \theta_1^{\dagger})))^2 - \mu_1^2)$$

$$\times\ ((1\{\underline{u} \leq y_{s+1+j}\ \leq \overline{u}\} - (F_1(\overline{u}|Z^{s+j},\ \theta_1^{\dagger}) - F_1(\underline{u}|Z^{s+j},\ \theta_1^{\dagger})))^2 - \mu_1^2))$$

$$+E\sum_{j=-\infty}^{\infty}(((1\{\underline{u} \leq y_{s+1}\ \leq \overline{u}\} - (F_i(\overline{u}|Z^s,\ \theta_i^{\dagger}) - F_i(\underline{u}|Z^s,\ \theta_i^{\dagger})))^2 - \mu_i^2)$$

$$\times\ ((1\{\underline{u} \leq y_{s+1+j}\ \leq \overline{u}\} - (F_i(\overline{u}|Z^{s+j},\ \theta_i^{\dagger}) - F_i(\underline{u}|Z^{s+j},\ \theta_i^{\dagger})))^2 - \mu_i^2))$$

$$-2E\sum_{j=-\infty}^{\infty}(((1\{\underline{u} \leq y_{s+1}\ \leq \overline{u}\} - (F_1(\overline{u}|Z^s,\ \theta_1^{\dagger}) - F_1(\underline{u}|Z^s,\ \theta_1^{\dagger})))^2 - \mu_1^2)$$

$$\times\ ((1\{\underline{u} \leq y_{s+1+j}\ \leq \overline{u}\} - (F_i(\overline{u}|Z^{s+j},\ \theta_i^{\dagger}) - F_i(\underline{u}|Z^{s+j},\theta_i^{\dagger})))^2 - \mu_i^2))$$

Also,

$$p_{ii} = 4m'_{\theta_1^\dagger} A(\theta_1^\dagger) E(\sum_{j=-\infty}^{\infty} \nabla_{\theta_1} \ln f_i(y_{s+1}|Z^s, \theta_1^\dagger) \nabla_{\theta_1} \ln f_i(y_{s+1+j}|Z^{s+j}, \theta_1^\dagger)') \times A(\theta_1^\dagger) m_{\theta_1^\dagger}$$

$$+4m'_{\theta_i^\dagger} A(\theta_i^\dagger) E(\sum_{j=-\infty}^{\infty} \nabla_{\theta_i} \ln f_i(y_{s+1}|Z^s, \theta_i^\dagger) \nabla_{\theta_i} \ln f_i(y_{s+1+j}|Z^{s+j}, \theta_i^\dagger)') \times A(\theta_i^\dagger) m_{\theta_i^\dagger}$$

$$-8m'_{\theta_1^\dagger} A(\theta_1^\dagger) E(\nabla_{\theta_1} \ln f_1(y_{s+1}|Z^s, \theta_1^\dagger) \nabla_{\theta_i} \ln f_i(y_{s+1+j}|Z^{s+j}, \theta_i^\dagger)') \times A(\theta_i^\dagger) m_{\theta_i^\dagger}$$

Finally,

$$cp_{ii} = -4m'_{\theta_1^\dagger} A(\theta_1^\dagger) E(\sum_{j=-\infty}^{\infty} \nabla_{\theta_1} \ln f_1(y_{s+1}|Z^s, \theta_1^\dagger)$$

$$\times ((1\{\underline{u} \leq y_{s+j} \leq \overline{u}\} - (F_1(\overline{u}|Z^{s+j}, \theta_1^\dagger) - F_1(\underline{u}|Z^{s+j}, \theta_1^\dagger)))^2 - \mu_1^2)$$

$$+8m'_{\theta_1^\dagger} A(\theta_1^\dagger) E(\sum_{j=-\infty}^{\infty} \nabla_{\theta_1} \ln f_1(y_s|Z^s, \theta_1^\dagger)$$

$$\times ((1\{\underline{u} \leq y_{s+1+j} \leq \overline{u}\} - (F_i(\overline{u}|Z^{s+j}, \theta_i^\dagger) - F_i(\underline{u}|Z^s, \theta_i)))^2 - \mu_i^2))$$

$$-4m'_{\theta_i^\dagger} A(\theta_i^\dagger) E(\sum_{j=-\infty}^{\infty} \nabla_{\theta_i} \ln f_i(y_{s+1}|Z^s, \theta_i^\dagger)$$

$$\times ((1\{\underline{u} \leq y_{s+j} \leq \overline{u}\} - (F_i(\overline{u}|Z^{s+j}, \theta_i^\dagger) - F_i(\underline{u}|Z^{s+j}, \theta_i^\dagger)))^2 - \mu_i^2))$$

with

$$m'_{\theta_i^\dagger} = E(\nabla_{\theta_i}(F_i(\overline{u}|Z^t, \theta_i^\dagger) - F_i(\overline{u}|Z^t, \theta_i^\dagger))$$

$$\times (1\{\underline{u} \leq y_t \leq \overline{u}\} - (F_i(\overline{u}|Z^t, \theta_i^\dagger) - F_i(\overline{u}|Z^t, \theta_i^\dagger))))$$

and

$$A(\theta_i^\dagger) = (E(-\ln \nabla_{\theta_i}^2 f_i(y_t|Z^t, \theta_i^\dagger)))^{-1}$$

An analogous result holds for the case where $\tau = 2$, and is omitted for the sake of brevity.

Due to the contribution of parameter estimation error, simulation error, and the time series dynamics to the covariance kernel (see Proposition 16.7), critical values cannot be directly tabulated. As a result, block bootstrap techniques are used to construct valid critical values for statistical inference. In order to show the first order validity of the bootstrap, the authors derive the limiting distribution of appropriately formed bootstrap statistics and show that they coincide with the limiting distribution given in Proposition 16.7. Recalling that as all candidate models are potentially misspecified under both hypotheses, the parametric bootstrap is not generally applicable in our context. Instead, we must begin by resampling b blocks of length $l, bl = T-1$. Let $Y_t^* = (\Delta \log X_t^*, \ \Delta \log X_{t-1}^*)$ be the resampled series, such that $Y_2^*,\ldots, \ Y_{l+1}^*, \ Y_{l+2}^*,\ldots, \ Y_{T-l+2}^*,\ldots, \ Y_T^*$ equals $Y_{I_1+1},\ldots, \ Y_{I_1+l}, \ Y_{I_2+1},\ldots, \ Y_{I_b+1},\ldots, \ Y_{I_b+T}$, where $I_j, \ i = 1,\ldots, \ b$ are independent, discrete uniform random variates on $1,\ldots, \ T-1+1$. That is, $I_j = i, i = 1,\ldots, \ T-l$ with probability $1/(T-l)$. Then, use Y_t^* to compute $\widehat{\theta}_{j,T}^*$ and plug in $\widehat{\theta}_{j,T}^*$ in order to simulate a sample under model $j, j = 1,\ldots, m$. Let $Y_{j,n}(\widehat{\theta}_{j,T}^*), n = 2,\ldots, S$ denote the series simulated in this manner. At this point, we need to distinguish between the case where $\delta = 0$ (vanishing simulation error) and $\delta > 0$ (non-vanishing simulation error). In the former case, we do not need to resample the simulated series, as there is no need to mimic the contribution of simulation error to the covariance kernel. On the other hand, in the latter case we draw $\widetilde{b}$ blocks of length $\widetilde{l}$ with $\widetilde{bl} = S-1$, and let $Y_{j,n}^*(\widehat{\theta}_{j,T}^*), \ j = 1,\ldots, m, n = 2,\ldots, S$ denote the resampled series under model j. Notice that $Y_{j,2}^*(\widehat{\theta}_{j,T}^*),\ldots, \ Y_{j,l+1}^*(\widehat{\theta}_{j,T}^*),\ldots, \ Y_{j,S}^*(\widehat{\theta}_{j,T}^*)$ is equal to $Y_{j,\widetilde{I}_1}(\widehat{\theta}_{j,T}^*),\ldots, \ Y_{j,\widetilde{I}_1+l}(\widehat{\theta}_{j,T}^*) \ \ldots, \ Y_{j,\widetilde{I}b_1+l}(\widehat{\theta}_{j,T}^*)$ where $\widetilde{I}_i, i = 1,\ldots, \widetilde{b}$ are independent discrete uniform random variates on $1,\ldots, \ S-\widetilde{l}$. Also, note that for each of the m models, and for each bootstrap replication, we draw $\widetilde{b}$ discrete uniform random variates (the $\widetilde{I}_i$) on $1, \ldots, S-\widetilde{l}$, and that draws are independent across models. Thus, in our use of notation, we have suppressed the dependence of $\widetilde{I}_i$ on j.

Thereafter, form bootstrap statistics as follows:

$$Z_{n,\tau}^* = \max_{i=2,\ldots m} \int_U Z_{n,u,\tau}^*(1, \ i)\phi(u)du$$

where for $\tau = 1$ (rolling estimation scheme), and for $\tau = 2$ (recursive estimation scheme)

$$Z_{n,u,\tau}^*(1, \ i) = \frac{1}{\sqrt{n}} \sum_{t=R}^{T-1} \Big((1\{y_{t+1}^* \le u\} - F_1(u|Z^{*,t}\widetilde{\theta}_{1,t,\tau}^*))^2$$

$$- (1\{y_{t+1}^* \le u\} - F_i(u|Z^{*,t}\widetilde{\theta}_{i,t,\tau}^*))^2 \Big)$$

$$-\frac{1}{T} \sum_{j=s+1}^{T-1} \Big((1\{y_{j+1} \le u\} - F_1(u|Z^i, \ \widehat{\theta}_{1,t,\tau}))^2 - (1\{y_{j+1} \le u\} - F_i(u|Z^j, \ \widehat{\theta}_{i,t,\tau}))^2 \Big)$$

Note that each bootstrap term, say $1\{y^*_{t+1} \leq u\} - F_i(u|Z^{*,t}, \widetilde{\theta}^*_{i,t,\tau})$, $t \geq R$, is re-centered around the (full) sample mean $\frac{1}{T}\sum_{j=s+1}^{T-1}(1\{y_{j+1} \leq u\} - F_i(u|Z^j, \widehat{\theta}_{i,t,\tau}))^2$. This is necessary as the bootstrap statistic is constructed using the last n resampled observations, which in turn have been resampled from the full sample. In particular, this is necessary regardless of the ratio n/R. If $n/R \to 0$, then we do not need to mimic parameter estimation error, and so could simply use $\widehat{\theta}_{1,t,\tau}$ instead of $\widetilde{\theta}^*_{1,t,\tau}$, but we still need to re-center any bootstrap term around the (full) sample mean.

Note that re-centering is necessary, even for first order validity of the bootstrap, in the case of over-identified generalized method of moments (GMM) estimators (see, e.g., Hall and Horowitz (1996), Andrews (2002), Andrews (2004), Inoue and Shintani (2006)). This is due to the fact that, in the over-identified case, the bootstrap moment conditions are not equal to zero, even if the population moment conditions are. However, in the context of m-estimators using the full sample, re-centering is needed only for higher order asymptotics, but not for first order validity, in the sense that the bias term is of smaller order than $T^{-1/2}$. Namely, in the case of recursive m-estimators the bias term is instead of order $T^{-1/2}$ and so it does contribute to the limiting distribution. This points to a need for re-centering when using recursive estimation schemes.

For the confidence interval case, define

$$V^*_{n,\tau} = \max_{i=2,\ldots m}, V^*_{n_{\underline{u},\overline{u}},\tau}(1, i)$$

and

$$V^*_{n,\underline{u},\overline{u},\tau}(1, i) = \frac{1}{\sqrt{n}}\sum_{t=R}^{T-1}\Big((1\{\underline{u} \leq y^*_{t+1} \leq \overline{u}\} - (F_1(\overline{u}|Z^{*t},\widetilde{\theta}^*_{1,t,\tau}) - F_1(\underline{u}|Z^{*t},\widetilde{\theta}^*_{1,t,\tau}))^2$$

$$- (1\{\underline{u} \leq y^*_{t+1} \leq \overline{u}\} - (F_i(\overline{u}|Z^{*t},\widetilde{\theta}^*_{i,t,\tau}) - F_1(\underline{u}|Z^{*t},\widetilde{\theta}^*_{i,t,\tau})))^2\Big)$$

$$-\frac{1}{T}\sum_{j=s+1}^{T-1}\Big((1\{\underline{u} \leq y_{i+1} \leq \overline{u}\} - (F_1(\overline{u}|Z^j, \widehat{\theta}_{1,t,\tau}) - F_1(\underline{u}|Z^j, \widehat{\theta}_{1,t,\tau})))^2$$

$$- (1\{\underline{u} \leq y_{j+1} \leq \overline{u}\} - (F_i(\overline{u}|Z^j, \widehat{\theta}_{i,t,\tau}) - F_1(\underline{u}|Z^j, \widehat{\theta}_{i,t,\tau})))^2\Big)$$

where, as usual, $\tau = 1, 2$. The following results then hold.

Proposition 16.9 (From Proposition 6 in Corradi and Swanson, 2006a) *With Assumptions 16.1, 16.11–16.13, also, assume that as* $T \to \infty, l \to \infty$, *and that*

$\frac{l}{T^{1/4}} \to 0$. *Then, as* T, n *and* $R \to \infty$, *for* $\tau = 1, 2$

$$\Pr\left(\sup_{v \in \Re} \left| \overset{*}{\Pr_T}\left(\max_{i=2,\dots m} \int_U Z^*_{n,u,\tau}(1\,,\,i)\phi(u)du \leq v\right)\right.\right.$$

$$\left.\left.- \Pr\left(\max_{i=2,\dots,m} \int_U Z^{\mu}_{n,u,\tau}\,(1\,,\,i)\phi(u)du \leq v\right)\right| > \epsilon\right) \to 0$$

where $Z^{\mu}_{n,u,\tau}(1, i) = Z_{n,u,\tau}(1, i) - \sqrt{n}\left(\mu_1^2(u) - \mu_i^2(u)\right)$, *and where* $\mu_1^2(u) - \mu_i^2(u)$ *is defined as in Eq. (16.7).*

Proposition 16.10 (From Proposition 7 in Corradi and Swanson, 2006a) *With Assumptions 16.1, 16.11–16.13, also assume that as* $T \to \infty, l \to \infty$, *and that* $\frac{l}{T^{1/4}} \to 0$. *Then, as* T, n *and* $R \to \infty$, *for* $\tau = 1, 2$

$$\Pr\left(\sup_{v \in \Re} \left| \overset{*}{\Pr_T}\left(\max_{i=2,\dots m},\, V^*_{n,\underline{u},\overline{u},\tau}(1,\,i)\ \leq v\right)\right.\right.$$

$$\left.\left.- \Pr\left(\max_{i=2,\dots m},\, V^{\mu}_{n,\underline{u},\overline{u},\tau}\,(1,\,i) \leq v\right)\right|\ > \epsilon\right) \to 0$$

where $V^{\mu}_{n,\underline{u},\overline{u},\tau}(1, i) = V_{n,\underline{u},\overline{u},\tau}(1\,,\,i) - \sqrt{n}\left(\mu_1^2(u) - \mu_i^2(u)\right)$.

The above results suggest proceeding in the following manner. For brevity, consider the case of $Z^*_{n,\tau}$. For any bootstrap replication, compute the bootstrap statistic, $Z^*_{n,\tau}$. Perform B bootstrap replications (B large) and compute the quantiles of the empirical distribution of the B bootstrap statistics. Reject H_0, if $Z_{n,\tau}$ is greater than the $(1 - \alpha)$th-percentile. Otherwise, do not reject. Now, for all samples except a set with probability measure approaching zero, $Z_{n,\tau}$ has the same limiting distribution as the corresponding bootstrapped statistic when $E(\mu_1^2(u) - \mu_i^2(u)) = 0, \forall i$, ensuring asymptotic size equal to α. On the other hand, when one or more competitor models are strictly dominated by the benchmark, the rule provides a test with asymptotic size between 0 and α. Under the alternative, $Z_{n,\tau}$ diverges to (plus) infinity, while the corresponding bootstrap statistic has a well defined limiting distribution, ensuring unit asymptotic power.

From the above discussion, we see that the bootstrap distribution provides correct asymptotic critical values only for the least favorable case under the null hypothesis, that is, when all competitor models are as good as the benchmark model. When $\max_{i=2,\dots,m} \int_U (\mu_1^2(u) - \mu_i^2(u))\phi(u)du = 0$, but $\int_U (\mu_1^2(u) - \mu_i^2(u))\phi(u)du < 0$ for some i, then the bootstrap critical values lead to conservative inference. An alternative to our bootstrap critical values in this case is the construction of critical values based on sub-sampling, which is briefly discussed in Sect. 16.1.4. Heuristically, construct $T - 2b_T$ statistics using subsamples of length b_T, where $b_T/T \to 0$. The empirical distribution of these statistics computed over the various subsamples properly mimics the distribution of the statistic. Thus, sub-sampling provides valid critical values even for the case where $\max_{i=2,\ldots,m} \int_U (\mu_1^2(u) - \mu_i^2(u))\phi(u)\mathrm{d}u = 0$, but $\int_U (\mu_1^2(u) - \mu_i^2(u))\phi(u)\mathrm{d}u < 0$ for some i. This is the approach used by Linton, Maasoumi, and Whang (2002), for example, in the context of testing for stochastic dominance. Needless to say, one problem with sub-sampling is that unless the sample is very large, the empirical distribution of the subsampled statistics may yield a poor approximation of the limiting distribution of the statistic. Another alternative approach for addressing the conservative nature of our bootstrap critical values is the Hansen's SPA approach (see Sect. 16.1.4 and Hansen (2005)). Hansen's idea is to re-center the bootstrap statistics using the sample mean, whenever the latter is larger than (minus) a bound of order $\sqrt{2T \log\log T}$. Otherwise, do not re-center the bootstrap statistics. In the current context, his approach leads to correctly sized inference when $\max_{i=2,\ldots,m} \int_U (\mu_1^2(u) - \mu_i^2(u))\phi(u)\mathrm{d}u = 0$, but $\int_U (\mu_1^2(u) - \mu_i^2(u))\phi(u)\mathrm{d}u < 0$ for some i. Additionally, his approach has the feature that if all models are characterized by a sample mean below the bound, the null is "accepted" and no bootstrap statistic is constructed.

16.3 Forecast Evaluation Using Density-Based Predictive Accuracy Tests that are not Loss Function Dependent: The Case of Stochastic Dominance

All predictive accuracy tests outlined in previous two parts of this chapter are loss functions dependent, i.e., loss functions such as mean squared forecast error (MSFE) and mean absolute forecast error (MAFE) must be specified prior to test construction. Evidently, given possible misspecification, model rankings may change under different loss functions. In the following section, we introduce a novel criterion for forecast evaluation that utilizes the entire distribution of forecast errors, is robust to the choice of loss function, and ranks distributions of forecast errors via stochastic dominance type tests.

16.3.1 Robust Forecast Comparison

Jin, Corradi, and Swanson (2017) (JCS) introduce the concepts of general-loss (GL) forecast superiority and convex-loss (CL) forecast superiority and develop tests for GL (CL) superiority that are based on an out-of-sample generalization of the tests introduced by Linton, Maassoumi, and Whang (2005). The JCS tests evaluate the entire forecast error distribution and do not require knowledge or specification of a loss function, i.e., tests are robust to the choice of loss function. In addition, parameter estimation error and data dependence are taken into account, and heterogeneity that is induced by distributional change over time is allowed for.

The concepts of general-loss (GL) forecast superiority and convex-loss (CL) forecast superiority are defined as follows:

(1) For any two sequences of forecast errors $u_{1,t}$ and $u_{2,t}$, $u_{1,t}$ general-loss (GL) outperforms $u_{2,t}$, denoted as $u_1 \succeq_G u_2$, if and only if $E(g(u_{1,t})) \leq E(g(u_{2,t}))$, $\forall g(\cdot) \in GL(\cdot)$, where $GL(\cdot)$ are the set of general-loss functions with properties specified in Granger (1999); and
(2) $u_{1,t}$ convex-loss (CL) outperforms $u_{2,t}$, denoted as $u_1 \succeq_C u_2$, if and only if $E(g(u_{1,t})) \leq E(g(u_{2,t}))$, $\forall g(\cdot) \in CL(\cdot)$, where $CL(\cdot)$ are the set of general-loss functions which in addition are convex.

These authors also establish linkages between GL(CL) forecast superiority and first(second) order stochastic dominance, allowing for the construction of direct tests for GL(CL) forecast superiority. Define

$$G(x) = \Big(F_2(x) - F_1(x)\Big)sgn(x),$$

where $sgn(x) = 1$, if $x \geq 0$, and $sgn(x) = -1$, if $x < 0$. Here, $F_i(x)$ denotes the cumulative distribution function (CDF) of u_i, and

$$C(x) = \int_{-\infty}^{x} \Big(F_1(t) - F_2(t)\Big)dt 1_{\{x<0\}} + \int_{x}^{\infty} \Big(F_2(t) - F_1(t)\Big)dt 1_{\{x\geq 0\}}$$

Assumption 16.14 $g(\cdot) : \Re \to \Re^+$ is continuously differentiable, except for finitely many points, with derivative $\nabla g(\cdot)$, such that $\nabla g(z) \leq 0$, $\forall z \leq 0$ and $\nabla g(z) \geq 0$, $\forall z \geq 0$.

Proposition 16.11 (From Propositions 2.2 and 2.3 in Jin et al., 2017) *With Assumption 16.14, $E(g(u_{1,t})) \leq E(g(u_{2,t}))$, $\forall g(\cdot) \in GL(\cdot)$, if and only if $G(x) \leq 0$, $\forall x \in \mathcal{X}$, where $\mathcal{X}$ is the union of the supports of all forecast errors. Further, if $\int_{-\infty}^{x}(F_1(t) - F_2(t))dt 1_{\{x<0\}}$ and $\int_{x}^{\infty}(F_2(t) - F_1(t))dt 1_{\{x\geq 0\}}$ are well defined for each $x \in \mathcal{X}$, then $E(g(u_{1,t})) \leq E(g(u_{2,t}))$, $\forall g(\cdot) \in CL(\cdot)$ if and only if $C(x) \leq 0$, $\forall x \in \mathcal{X}$.*

The above proposition establishes a clear mapping between GL (CL) forecast superiority and first (second) order stochastic dominance. Intuitively, if we construct a graph that contains a plot of $G(x)$ against x. When $u_1 \succeq_G u_2$, we expect all points lie below or on the zero line. Similarly, if we construct a graph that contains a plot of $C(x)$ against x. When $u_1 \succeq_C u_2$, we expect all points lie below or on the zero line as well.

The hypotheses tested in JCS are

$$H_0 : \max_{i=1,\ldots,m} E\Big(g(u_{0,t+1}) - g(u_{i,t+1})\Big) \leq 0$$

versus

$$H_A : \max_{i=1,\ldots,m} E\Big(g(u_{0,t+1}) - g(u_{i,t+1})\Big) > 0$$

Given Proposition 16.11, the above hypotheses can be restated as

$$H_0^{TG} = H_0^{TG-} \cap H_0^{TG+} : \Big(\max_{i=1,\ldots,m} (F_0(x) - F_i(x)) \le 0,\ \forall x \le 0\Big)$$
$$\cap \Big(\max_{i=1,\ldots,m} (F_i(x) - F_0(x)) \le 0,\ \forall x > 0\Big)$$

versus

$$H_A^{TG} = H_A^{TG-} \cup H_A^{TG+} : \Big(\max_{i=1,\ldots,m} (F_0(x) - F_i(x)) > 0,\ \text{for some } x \le 0\Big)$$
$$\cup \Big(\max_{i=1,\ldots,m} (F_i(x) - F_0(x)) > 0,\ \text{for some } x > 0\Big)$$

for the case of GL forecast superiority. Similarly, for the case of CL forecast superiority, we have that

$$H_0^{TC} = H_0^{TC-} \cap H_0^{TC+} : \Big(\max_{i=1,\ldots,m} \int_{-\infty}^{x} (F_0(x) - F_i(x)) \le 0,\ \forall x \le 0\Big)$$
$$\cap \Big(\max_{i=1,\ldots,m} \int_{x}^{\infty} (F_i(x) - F_0(x)) \le 0,\ \forall x > 0\Big)$$

versus

$$H_A^{TC} = H_A^{TC-} \cup H_A^{TC+} : \Big(\max_{i=1,\ldots,m} \int_{-\infty}^{x} (F_0(x) - F_i(x)) > 0,\ \text{for some } x \le 0\Big)$$
$$\cup \Big(\max_{i=1,\ldots,m} \int_{x}^{\infty} (F_i(x) - F_0(x)) > 0,\ \text{for some } x > 0\Big)$$

Of note is that the above null (alternative) is the intersection (union) of two different null (alternative) hypotheses because of a discontinuity at zero. The test statistics for GL forecast superiority are constructed as follows:

$$TG_n^+ = \max_{i=1,\ldots,k} \sup_{x \in \mathcal{X}^+} \sqrt{n}\widehat{G}_{i,n}(x)$$

and

$$TG_n^- = \max_{i=1,\ldots,k} \sup_{x \in \mathcal{X}^-} \sqrt{n}\widehat{G}_{i,n}(x)$$

with

$$\widehat{G}_{i,n}(x) = \big(\widehat{F}_{0,n}(x) - \widehat{F}_{i,n}(x)\big)sgn(x)$$

where $\widehat{F}_{i,n}(x)$ denotes the empirical CDF of u_i, with

$$\widehat{F}_{i,n}(x) = n^{-1}\sum_{t=R}^{T} 1_{\{u_{i,t} \leq x\}}$$

Similarly, the test statistics for CL forecast superiority are constructed as follows:

$$TC_n^+ = \max_{i=1,\ldots,k} \sup_{x \in \mathcal{X}^+} \sqrt{n}\widehat{C}_{i,n}(x)$$

and

$$TC_n^- = \max_{i=1,\ldots,k} \sup_{x \in \mathcal{X}^-} \sqrt{n}\widehat{C}_{i,n}(x)$$

with

$$\begin{aligned}\widehat{C}_{i,n}(x) &= \int_{-\infty}^{x} \big(\widehat{F}_{0,n}(x) - \widehat{F}_{i,n}(x)\big)dx 1_{\{x<0\}} - \int_{x}^{\infty} \big(\widehat{F}_{i,n}(x) - \widehat{F}_{0,n}(x)\big)dx 1_{\{x\geq 0\}} \\ &= \frac{1}{n}\sum_{t=1}^{n}\Big\{[(u_{0,t} - x)sgn(x)]_+ - [(u_{i,t} - x)sgn(x)]_+\Big\},\end{aligned}$$

where $[z]_+ = \max\{0, z\}$.

Note that in order to reduce computation time, it may be preferable to construct approximations to the suprema in statistics TG^+, TG^-, TC^+, and TC^- by taking maxima over some smaller grid of points, $\mathcal{X}_N = \{x_1, \ldots, x_N\}$, where $N < n$. Theoretically, the distribution theory is unaffected by using this approximation, as the set of evaluation points becomes dense in the joint support. We now require the following assumptions.

Assumption 16.15

(i) $\{(y_t, Z_i^t)'\}$ is a strictly stationary and α-mixing sequence with mixing coefficient $\alpha(l) = O(l^{-C_0})$, for some $C_0 > \max\{(q-1)(q+1), 1 + 2/\delta\}$, with $i = 0, \ldots, m$, where q is an even integer that satisfies $q > 3(g_{\max} + 1)/2$. Here, $g_{\max} = \max\{g_0, \ldots, g_m\}$ and δ is a positive constant;
(ii) For $i = 0, \ldots, m$, $f_i(Z_i^t, \theta_i)$ is differentiable a.s. with respect to θ_i in the neighborhood $\Theta_i^\dagger$ of $\theta_i^\dagger$, with $\sup_{\theta \in \Theta_0^\dagger} ||\nabla_\theta f_i(Z_i^t, \theta_i)||_2 < \infty$;
(iii) The conditional distribution of $u_{i,t}$ given Z_i^t has bounded density with respect to the Lebesgue measure a.s., and $||u_{i,t}||_{2+\delta} < \infty$, $\forall i$.

Assumption 16.16*

(i) $\{(y_t, Z_i^t)'\}$ is a strictly stationary and α-mixing sequence with mixing coefficient $\alpha(l) = O(l^{-C_0})$, for some $C_0 > \max\{rq/(r-q), 1 + 2/\delta\}$, with $i = 0, \ldots, m$, and $r > q > g_{\max} + 1$;
(ii) For $i = 0, \ldots, m$, $f_i(Z_i^t, \theta_i)$ is differentiable a.s. with respect to θ_i in the neighborhood $\Theta_i^\dagger$ of $\theta_i^\dagger$, with $\sup_{\theta \in \Theta_0^\dagger} ||\nabla_\theta f_i(Z_i^t, \theta_i)||_r < \infty$;
(iii) $||u_{i,t}||_r < \infty$, $\forall i$.

Assumption 16.17 $\forall\, i$ and t, $\widehat{\theta}_{i,t}$ satisfies $\widehat{\theta}_{i,t} - \theta_i^\dagger = B_i(t)H_i(t)$, where $B_i(t)$ is a $n_i \times L_i$ matrix and $H_i(t)$ is $L_i \times 1$, with the following:

(i) $B_i(t) \to B_i$ a.s., where B_i is a matrix of rank n_i;
(ii) $H_i(t) = t^{-1}\sum_{s=1}^t h_{i,s}$, $R^{-1}\sum_{s=t-R+1}^t h_{i,s}$, and $R^{-1}\sum_{s=1}^R h_{i,s}$ for the recursive, rolling, and fixed schemes, respectively, where $h_{i,s} = h_{i,s}(\theta_i^\dagger)$;
(iii) $E(h_{i,s}(\theta_i^\dagger)) = 0$; and
(iv) $||h_{i,s}(\theta_i^\dagger)||_{2+\delta} < \infty$, for some $\delta > 0$.

Assumption 16.18

(i) The distribution function of forecast errors, $F_i(x, \theta_i)$ is differentiable with respect to θ_i in a neighborhood $\Theta_i^\dagger$ of $\theta_i^\dagger$, $\forall i$;
(ii) $\forall i$, and $\forall$ sequences of positive constants $\{\xi_n : n \geq 1\}$, such that $\xi_n \to 0$, $\sup_{x \in \mathcal{X}} \sup_{\theta:\, ||\theta - \theta_i^\dagger|| \leq \xi_n} ||\nabla_\theta F_i(x, \theta) sgn(x) - \Delta_i^\dagger(x)|| = O(\xi_n^\eta)$, for some $\eta > 0$, where $\Delta_i^\dagger(x) = \nabla_\theta F_i(x, \theta_i^\dagger) sgn(x)$;
(iii) $\sup_{x \in \mathcal{X}} ||\Delta_i^\dagger(x)|| < \infty, \forall i$.

Assumption 16.19*

(i) Assumption 16.8 (i) holds;
(ii) $\forall i$, and $\forall$ sequences of positive constants $\{\xi_n : n \geq 1\}$, such that $\xi_n \to 0$, $\sup_{x \in \mathcal{X}} \sup_{\theta:\, ||\theta - \theta_i^\dagger|| \leq \xi_n} ||\nabla_\theta \{\int_{-\infty}^x F_i(t, \theta) dt 1_{\{x<0\}} + \int_x^\infty (1 - F_i(x, \theta)) dt 1_{\{x \geq 0\}}\} - \Lambda_i^\dagger(x)|| = O(\xi_n^\eta)$, for some $\eta > 0$, where

$$\Lambda_i^\dagger(x) = \nabla_\theta \left\{ \int_{-\infty}^x F_i(t, \theta_i^\dagger) dt 1_{\{x<0\}} + \int_x^\infty (1 - F_i(x, \theta_i^\dagger)) dt 1_{\{x \geq 0\}} \right\};$$

(iii) $\sup_{x \in \mathcal{X}} ||\Lambda_i^\dagger(x)|| < \infty, \forall i$.

Assumptions 16.16* and 16.19* are needed for testing H_0^{TC}. Note that the first and third assumptions parallel those imposed by Linton et al. (2005), with the uniform continuity conditions in Assumptions 16.18 and 16.19* strengthened. Assumption 16.15 is needed in order to verify the stochastic equicontinuity of the empirical process, for a class of bounded functions that appears in the TG_n test. Assumption 16.16* introduces a trade-off between mixing sizes and moment conditions, and is used to verify the stochastic equicontinuity result for the possibly unbounded functions that appear in the TC_n test. For further details, see Hansen (1996b). Assumptions 16.18 and 16.19* differ in the amount of smoothness required. For the CL forecast superiority test, less smoothness is required. Finally, it is worth stressing that Assumptions 16.8 and 16.17 are identical to Assumptions 1 and 2 in McCracken (2000), respectively.

Proposition 16.12 (From Theorem 3.1 in Jin et al., 2017)

(i) With Assumptions 16.8, 16.15–16.18, under $H_0^{TG^-}$,

$$TG_n^- \xrightarrow{d} \max_{i=1,\ldots,m} \sup_{x \in \mathcal{B}_i^{g-}} [\widetilde{g}_i(x) + \Delta_{i0}(x)' B_i \upsilon_{i0} - \Delta_{10}(x)' B_1 \upsilon_{10}], \text{ if } TG^- = 0$$

$$\to -\infty, \text{ if } TG^- < 0$$

Under H_0^{TG+},

$$TG_n^+ \xrightarrow{d} \max_{i=1,\ldots,m} \sup_{x \in \mathcal{B}_i^{g+}} [\widetilde{g}_i(x) + \Delta_{i0}(x)' B_i \upsilon_{i0} - \Delta_{10}(x)' B_1 \upsilon_{10}], \text{ if } TG^+ = 0$$

$$\to -\infty, \text{ if } TG^+ < 0$$

where $\mathcal{B}_i^{g-} = \{x \in \mathcal{X}^- : F_0(x) = F_i(x)\}$ *and* $\mathcal{B}_i^{g+} = \{x \in \mathcal{X}^+ : F_0(x) = F_i(x)\}$, *and* $\left(\widetilde{g}_i(\cdot), \upsilon_{i0}, \upsilon_{10}\right)'$ *is a mean zero Gaussian process with covariance function given by*

$$\Omega_i^g(x_1, x_2) = \lim_{T\to\infty} E \begin{pmatrix} \upsilon_{i,n}^g(x_1, \theta_i^\dagger) - \upsilon_{0,n}^g(x_1, \theta_0^\dagger) \\ \sqrt{n}\overline{H}_{i,n} \\ \sqrt{n}\overline{H}_{0,n} \end{pmatrix} \begin{pmatrix} \upsilon_{i,n}^g(x_2, \theta_i^\dagger) - \upsilon_{0,n}^g(x_2, \theta_0^\dagger) \\ \sqrt{n}\overline{H}_{i,n} \\ \sqrt{n}\overline{H}_{0,n} \end{pmatrix}'$$

with $\overline{H}_{i,n} = n^{-1} \sum_{t=R}^{T} H_i(t)$, *and* $\upsilon_{i,n}^g(x, \theta)$ *is an empirical process defined as*

$$\upsilon_{i,n}^g(x, \theta) = \frac{1}{\sqrt{n}} \sum_{t=R}^{T} \{1_{\{u_{i,t+\tau}(\theta) \le x\}} - F_i(x, \theta)\} sgn(x)$$

(ii) With Assumptions 16.16, 16.17, 16.19* and 7.5, under* $H_0^{TC^-}$,

$$TC_n^- \xrightarrow{d} \max_{i=1,\ldots,m} \sup_{x \in \mathcal{B}_i^{c-}} [\widetilde{c}_i(x) + \Lambda_{i0}(x)' B_i \upsilon_{i0} - \Lambda_{10}(x)' B_1 \upsilon_{10}], \text{ if } TC^- = 0$$

$$\to -\infty, \text{ if } TC^- < 0$$

Under $H_0^{TC^+}$,

$$TC_n^+ \xrightarrow{d} \max_{i=1,\ldots,m} \sup_{x \in \mathcal{B}_i^{c+}} [\widetilde{c}_i(x) + \Lambda_{i0}(x)' B_i \upsilon_{i0} - \Lambda_{10}(x)' B_1 \upsilon_{10}], \text{ if } TC^+ = 0$$

$$\to -\infty, \text{ if } TC^+ < 0$$

where $\mathcal{B}_i^{c-} = \{x \in \mathcal{X}^- : \int_{-\infty}^{x} (F_i(x) - F_0(x)) dx 1_{\{x<0\}}\}$ *and* $\mathcal{B}_i^{c+} = \{x \in \mathcal{X}^+ : \int_{x}^{\infty} (F_0(x) - F_i(x)) dx 1_{\{x \geq 0\}}\}$. *Similarly,* $\left(\widetilde{c}_i(\cdot), \upsilon_{i0}, \upsilon_{10}\right)'$ *is a mean zero Gaussian process with covariance function given by*

$$\Omega_i^c(x_1, x_2) = \lim_{T\to\infty} E \begin{pmatrix} \upsilon_{i,n}^c(x_1, \theta_i^\dagger) - \upsilon_{0,n}^c(x_1, \theta_0^\dagger) \\ \sqrt{n} H_{i,n} \\ \sqrt{n} H_{0,n} \end{pmatrix} \begin{pmatrix} \upsilon_{i,n}^c(x_2, \theta_i^\dagger) - \upsilon_{0,n}^c(x_2, \theta_0^\dagger) \\ \sqrt{n} H_{i,n} \\ \sqrt{n} H_{0,n} \end{pmatrix}'$$

where $\upsilon_{i,n}^c(x, \theta)$ *is an empirical process defined as*

$$\upsilon_{i,n}^c(x,\theta) = \frac{1}{\sqrt{n}} \sum_{t=R}^{T} \left\{ \int_{-\infty}^{x} [1_{\{u_{i,t+\tau}(\theta) \leq s\}} - F_i(s,\theta)] ds 1_{\{x<0\}} \right.$$
$$\left. - \int_{x}^{\infty} [1_{\{u_{i,t+\tau}(\theta) \leq s\}} - F_i(s,\theta)] ds 1_{\{x \geq 0\}} \right\}$$

The asymptotic null distributions of TG_n^+ (TG_n^-) *and* TC_n^+ (TC_n^-) *depend on the true model parameters and the distribution functions,* $F_i(\cdot)$, $i = 1, \ldots, m$, *which implies that the asymptotic critical values for* TG_n^+ (TG_n^-) *and* TC_n^+ (TC_n^-) *cannot be tabulated. Therefore, the stationary bootstrap is used to approximate the asymptotic null distributions of our test statistics. (Note that the block bootstrap can also be used, as discussed in subsequent research by Corradi, Sin, and Swanson.) The objective is to utilize bootstrap procedure that mimics the asymptotic null distribution in the least favorable case, where* $F_0(x) = \ldots = F_m(x)$, $\forall x \in \mathcal{X}$.

Define the bootstrap statistic as

$$TG_n^{*+} = \max_{i=1,\ldots,k} \sup_{x \in \mathcal{X}^+} \sqrt{n} \left(\widehat{G}_{i,n}^*(x) - \widehat{G}_{i,n}(x) \right)$$

with

$$\widehat{G}_{i,n}^*(x) = \left(\widehat{F}_{0,n}^*(x) - \widehat{F}_{i,n}^*(x) \right) sgn(x)$$

where $\widehat{F}_{i,n}^*(x)$ denotes the empirical CDF of resampled u_i, i.e., u_i^*. TG_n^{*-}, TC_n^{*+} and TC_n^{*-} can be defined analogously.

Assumption 16.20 The smoothing parameter, S_n, determining the mean block length in stationary bootstrap satisfies $0 < S_n < 1$, $S_n \to 0$ and $nS_n^2 \to \infty$, as $n \to \infty$.

Assumption 16.21 For any arbitrary $n_i \times 1$ vector, λ_i, with $\lambda_i^{'}\lambda_i = 1$, and $\forall i$, we have (i)

$$\Pr\Big[\limsup_{t \geq R} n^{1/2} \frac{|\lambda_i^{'}(\widehat{\theta}_{i,t} - \theta_i^{\dagger})|}{(\lambda_i^{'}\Sigma_i\lambda_i \mathrm{loglog}(\lambda_i^{'}\Sigma_i\lambda_i)n)^{1/2}} = 1\Big] = 1$$

for the recursive scheme, where $\Sigma_i = B_i[\lim_{T\to\infty} \mathrm{Var}(n^{-1/2}\sum_{t=R+1}^{T} H_i(t))]B_i^{'}$. (ii)

$$\Pr\Big[\limsup_{t \geq R} R^{1/2} \frac{|\lambda_i^{'}(\widehat{\theta}_{i,t} - \theta_i^{\dagger})|}{(\lambda_i^{'}\Sigma_i\lambda_i \mathrm{loglog}(\lambda_i^{'}\Sigma_i\lambda_i)R)^{1/2}} = 1\Big] = 1$$

for the rolling scheme, where $\Sigma_i = B_i[\lim_{T\to\infty} \mathrm{Var}(R^{-1/2}\sum_{t=R+1}^{T} H_i(t))]B_i^{'}$.

Proposition 16.13 (From Corollary 3.3 in Jin et al., 2017) *With Assumptions 16.15–16.18, 16.20, and 16.21, and that $(n/R)loglogR \to 0$, as $T \to \infty$, then*

$$\rho\Big(L[\max_{i=1,\ldots,m} \sup_{x\in\mathcal{X}^+} \sqrt{n}(\widehat{G}_{i,n}^{*}(x) - \widehat{G}_{i,n}(x))|U_1,\ldots,U_{T+\tau}],$$
$$L[\max_{i=1,\ldots,m} \sup_{x\in\mathcal{X}^+} \sqrt{n}(\widehat{G}_{i,n}(x) - G_i(x))]\Big) \overset{n}{\to} 0$$

and

$$\rho\Big(L[\max_{i=1,\ldots,m} \sup_{x\in\mathcal{X}^-} \sqrt{n}(\widehat{G}_{i,n}^{*}(x) - \widehat{G}_{i,n}(x))|U_1,\ldots,U_{T+\tau}],$$
$$L[\max_{i=1,\ldots,m} \sup_{x\in\mathcal{X}^-} \sqrt{n}(\widehat{G}_{i,n}(x) - G_i(x))]\Big) \overset{n}{\to} 0$$

where ρ is any metric metrizing weak convergence, $L[\cdot]$ denotes the probability law of the corresponding Hilbert space valued random variable, and $U_t = (y_t, Z_0^t, \ldots, Z_m^t)'$.

Therefore, the asymptotic null distribution of TG_n^+ (TG_n^-) can be approximated by $TG_n^{*+} - TG_n^+$ ($TG_n^{*-} - TG_n^-$). Arguments in favor of using the stationary bootstrap with TC_n^+ and TC_n^- are similar.

To conduct inference, use the following approach due to Holm (1979). Define $q_{n,S_n}^{G+}(1-\alpha)$ and $q_{n,S_n}^{G-}(1-\alpha)$ to be the $(1-\alpha)$-th sample quantile of

TG_n^{*+} and TG_n^{*-}, respectively. Then, estimate bootstrap p-values, $p_{B,n,S_n}^{G^+} = \frac{1}{B}\sum_{s=1}^{B}(TG_n^{*+} \geq TG_n^{+})$, and finally use the following rules:

Rule TG: Reject H_0^{TG} at level α, if $\min\left\{p_{B,n,S_n}^{G^+},\ p_{B,n,S_n}^{G^-}\right\} \leq \alpha/2$;

Rule TC: Reject H_0^{TG} at level α, if $\min\left\{p_{B,n,S_n}^{C^+},\ p_{B,n,S_n}^{C^-}\right\} \leq \alpha/2$;

Note that Holm bounds are equivalent to Bonferroni bounds when there are only two hypotheses. From Proposition 16.13, it follows immediately that this test, when implemented using the stationary bootstrap, has asymptotically correct size only in the least favorable case, under the null, and is asymptotically biased towards certain local alternatives.

Proposition 16.14 (From Theorem 4.1 in Jin et al., 2017) *With Assumptions 16.8, 16.15–16.18, under H_A^{TG},*

$$Pr\left(TG_n^{+} > q_{n,S_n}^{G^+}(1-\alpha)\right) \rightarrow 1, \text{ as } T \rightarrow \infty$$

and

$$Pr\left(TG_n^{-} > q_{n,S_n}^{G^-}(1-\alpha)\right) \rightarrow 1, \text{ as } T \rightarrow \infty$$

The above proposition ensures unit asymptotic power under the alternative. Similar arguments apply to TC_n^{+} and TC_n^{-} as well. For details of the power of TG_n^{+} (TG_n^{-}) and TC_n^{+} (TC_n^{-}) tests against a sequence of contiguous local alternatives converging to the null, at rate $n^{-1/2}$, see Jin et al. (2017).

References

Andrews, D. W., & Soares, G. (2010). Inference for parameters defined by moment inequalities using generalized moment selection. *Econometrica, 78*(1), 119–157.

Andrews, D. W. K. (2002). Higher-order improvements of a computationally attractive "k"-step bootstrap for extremum estimators. *Econometrica, 70*(1), 119–162.

Andrews, D. W. K. (2004). The block–block bootstrap: Improved asymptotic refinements. *Econometrica, 72*(3), 673–700.

Bierens, H. J. (1990). A consistent conditional moment test of functional form. *Econometrica, 58*, 1443–1458.

Bierens, H. J., & Ploberger, W. (1997). Asymptotic theory of integrated conditional moment tests. *Econometrica, 65*, 1129–1151.

Chang, Y., Gomes, J. F., & Schorfheide, F. (2002). Learning-by-doing as a propagation mechanism. *American Economic Review, 92*(5), 1498–1520.

Chao, J., Corradi, V., & Swanson, N. R. (2001). Out-of-sample tests for granger causality. *Macroeconomic Dynamics, 5*(4), 598–620.

Clark, T. E., & McCracken, M. W. (2001). Tests of equal forecast accuracy and encompassing for nested models. *Journal of Econometrics, 105*(1), 85–110.

Clark, T. E., & McCracken, M. W. (2003). Evaluating long horizon forecasts. Working Paper, University of Missouri-Columbia.

Corradi, V., & Distaso, W. (2011). Multiple forecast model evaluation. In *The handbook of economic forecasting* (pp. 391–414). Oxford: Oxford University Press.

Corradi, V., & Swanson, N. R. (2002). A consistent test for out of sample nonlinear predictive ability. *Journal of Econometrics*, *110*, 353–381.

Corradi, V., & Swanson, N. R. (2005). A test for comparing multiple misspecified conditional interval models. *Econometric Theory*, *21*(5), 991–1016.

Corradi, V., & Swanson, N. R. (2006a). Predictive density and conditional confidence interval accuracy tests. *Journal of Econometrics*, *135*(1), 187–228.

Corradi, V., & Swanson, N. R. (2006b). Predictive density evaluation. *Handbook of Economic Forecasting*, *1*, 197–284.

Corradi, V., & Swanson, N. R. (2007). Nonparametric bootstrap procedures for predictive inference based on recursive estimation schemes. *International Economic Review*, *48*(1), 67–109.

Corradi, V., Swanson, N. R., & Olivetti, C. (2001). Predictive ability with cointegrated variables. *Journal of Econometrics*, *104*(2), 315–358.

De Jong, R. M. (1996). The bierens test under data dependence. *Journal of Econometrics*, *72*(1), 1–32.

Diebold, F. X., & Mariano, R. S. (2002). Comparing predictive accuracy. *Journal of Business & Economic Statistics*, *20*(1), 134–144.

Fernández-Villaverde, J., & Rubio-Ramírez, J. F. (2004). Comparing dynamic equilibrium models to data: A Bayesian approach. *Journal of Econometrics*, *123*(1), 153–187.

Gianni, A., & Giacomini, R. (2007). Comparing density forecasts via weighted likelihood ratio tests. *Journal of Business & Economic Statistics*, *25*(2), 177–190.

Granger, C. W. J. (1999). Outline of forecast theory using generalized cost function. *Spanish Economic Review*, *1*, 161–173.

Granger, C. W. J. (1993). On the limitations of comparing mean square forecast errors: A comment. *Journal of Forecasting*, *12*(8), 651–652.

Hall, P., & Horowitz, J. L. (1996). Bootstrap critical values for tests based on generalized-method-of-moments estimators. *Econometrica*, *64*(4), 891–916.

Hansen, B. E. (1996a). Inference when a nuisance parameter is not identified under the null hypothesis. *Econometrica*, *64*, 413–430.

Hansen, B. E. (1996b). Stochastic equicontinuity for unbounded dependent heterogeneous arrays. *Econometric Theory*, *12*, 347–359.

Hansen, R. P. (2005). A test for superior predictive ability. *Journal of Business & Economic Statistics*, *23*(4), 365–380.

Holm, S. (1979). A simple sequentially rejective multiple test procedure. *Scandinavian Journal of Statistics*, *6*, 65–70.

Inoue, A., & Shintani, M. (2006). Bootstrapping GMM estimators for time series. *Journal of Econometrics*, *133*(2), 531–555.

Jin, S., Corradi, V., & Swanson, N. R. (2017). Robust forecast comparison. *Econometric Theory*, *33*(6), 1306–1351.

Kilian, L. (1999). Exchange rates and monetary fundamentals: What do we learn from long-horizon regressions? *Journal of Applied Econometrics*, *14*(5), 491–510.

Kitamura, Y. (2002). Econometric comparisons of conditional models. Working Paper, University of Pennsylvania.

Lee, T. H., White, H., & Granger, C. W. J. (1993). Testing for neglected nonlinearity in time series models: A comparison of neural network methods and alternative tests. *Journal of Econometrics*, *56*(3), 269–290.

Linton, O. B., Maasoumi, E., & Whang, Y. J. (2002). Consistent testing for stochastic dominance: A subsampling approach. *Social Science Electronic Publishing*, *72*(3), 735–765.

Linton, O., Maassoumi, E., & Whang, Y. J. (2005). Consistent testing for stochastic dominance: A subsampling approach. *Review of Economic Studies*, *72*, 735–765.

McCracken, M. W. (2000). Robust out-of-sample inference. *Journal of Econometrics*, *99*, 195–223.
Meese, R. A., & Rogoff, K. (1983). Empirical exchange rate models of the seventies: Do they fit out-of-sample? *Journal of International Economics*, *14*, 3–24.
Politis, D. N., Romano, J. P., & Wolf, M. (1999). *Subsampling*. *Springer Series in Statistics*. New York: Springer.
Romano, J. P., & Wolf, M. (2005). Stepwise multiple testing as formalized data snooping. *Econometrica*, *73*(4), 1237–1282.
Rossi, B. (2005). Testing long-horizon predictive ability with high persistence, and the meese–rogoff puzzle. *International Economic Review*, *46*(1), 61–92.
Schorfheide, F. (2010). Loss function-based evaluation of DSGE models. *Journal of Applied Econometrics*, *15*(6), 645–670.
Stinchcombe, M. B., & White, H. (1998). Consistent specification testing with nuisance parameters present only under the alternative. *Econometric Theory*, *14*(3), 295–325.
Vuong, Q. H. (1989). Likelihood ratio tests for model selection and non-nested hypotheses. *Econometrica*, *57*(2), 307–333.
Weiss, A. (1996). Estimating time series models using the relevant cost function. *Journal of Applied Econometrics*, *11*(5), 539–560.
West, K. D. (1996). Asymptotic inference about predictive ability. *Econometrica*, *64*, 1067–1084.
White, H. (1982). Maximum likelihood estimation of misspecified models. *Econometrica*, *50*(1), 1–25.
White, H. (2000). A reality check for data snooping. *Econometrica*, *68*(5), 1097–1126.

Part V
Further Issues

Chapter 17
Unit Roots and Cointegration

Stephan Smeekes and Etienne Wijler

17.1 Introduction

In this chapter we investigate forecasting with Big Data when the series in the dataset may contain unit roots and be cointegrated. As most macroeoconomic time series are at least very persistent, and may contain unit roots, a proper handling of unit roots and cointegration is of paramount importance in macroeconomic forecasting. The theory of unit roots and cointegration in small systems is well-developed and numerous reference works exist to guide the practitioner, see, for example, Enders (2008) or Hamilton (1994) for comprehensive treatments.

In this chapter, we discuss the problems that arise when extending the analysis to high-dimensional data and consider solutions that have been proposed in the literature. In particular, we discuss the applicability of the proposed methods for macroeconomic forecasting, reviewing relevant theoretical properties and practical issues. Moreover, by considering two Big Data applications—that are very different in spirit—we illustrate the issues and analyse the performance of the various methods in practically relevant situations.

The empirical literature dealing with unit roots and cointegration can essentially be split into two different philosophies. The first approach is to apply an appropriate transformation to each series such that one can work with stationary time series, with the most common transformation taking first differences of a series with a unit root. This is the most common approach in high-dimensional forecasting, as it only involves 'straightforward' unit root or stationarity testing on each series. Indeed, commonly used Big Data such as the FRED-MD and -QD datasets (McCracken

S. Smeekes (✉) · E. Wijler
Department of Quantitative Economics, School of Business and Economics, Maastricht University, Maastricht, The Netherlands
e-mail: s.smeekes@maastrichtuniversity.nl; e.wijler@maastrichtuniversity.nl

P. Fuleky (ed.), *Macroeconomic Forecasting in the Era of Big Data*,
Advanced Studies in Theoretical and Applied Econometrics 52,
https://doi.org/10.1007/978-3-030-31150-6_17

& Ng, 2016) already come with pre-determined transformation codes to achieve stationarity. While this approach appears to be conceptually simple, we will argue in this chapter that there are apparently minor issues that are often ignored in practice, but which can have a big impact on the performance of consequent forecasts, in particular when working with less established datasets.

The second approach is to model unit root and cointegration properties directly. In small systems, this is commonly done through vector error correction models (VECM), often using the popular maximum likelihood methodology developed by Johansen (1995b). The rationale for this seemingly more complicated approach is that ignoring long-run relations between the variables, as is done in the first approach, means not incorporating all information into the forecaster's model, which may have a detrimental effect on the forecast quality. Extending these techniques for modelling cointegration to high-dimensional settings requires a careful rethink of how cointegration can be viewed in high dimensions, and is an ongoing area of research. We will discuss recent contributions in this area and analyse the respective merits and drawbacks of each method.

While the importance of the concept of cointegration for macroeconometric analysis cannot be understated, one might argue that for the specific goal of forecasting it is not crucial. In the low-dimensional time series literature a large body of literature exists which compares the relative merits of the two philosophical approaches for forecasting, see, for instance, Clements and Hendry (1995), Christoffersen and Diebold (1998), Diebold and Kilian (2000) and the references therein. Generally, the conclusion is mixed, with the performance of each approach varying depending on forecast horizon, dimensions of the models, estimation accuracy and even specific applications and datasets. As this is no different in a high-dimensional context, we make no attempt to classify one of these approaches as superior. Instead, we aim to provide the practitioner with an overview of tools available to follow either line of thought.

One could discern a third approach to unit roots and cointegration, which is to ignore unit roots all together and estimate all forecasting models in levels. While this approach is at first glance close to the first approach and one might have valid reasons to prefer this approach, we do not recommend this in high-dimensional problems. If cointegration is not present in (parts of) the data, these methods may be very sensitive to spurious regression. The higher the dimensions of the data, the more likely that spurious regression becomes an issue. In particular, given that many methods discussed in this book perform some sort of dimensionality reduction or variable selection, this may actually increase the likelihood of obtaining spurious results. For instance, Smeekes and Wijler (2018b) investigate the sensitivity of penalized regression methods to spurious results, and find that their variable selection mechanisms cannot properly distinguish between cointegrated and spurious regressors. Low-dimensional solutions such as always including lagged levels to avoid spurious regression are not possible in high-dimensional systems, as it would require including too many variables, and the applied dimensionality reduction or variable selection techniques might not be able to retain the lagged levels in the model. As such, we do not consider the approach of estimating everything in levels further in this chapter.[1]

We also illustrate the discussed methods by two empirical applications. In the first we forecast several US macroeconomic variables using the FRED-MD database. This application tests the methods in a known macroeconomic context, thus serving as a benchmark. In our second application, we consider nowcasting unemployment using a dataset constructed from Google Trends with frequencies of unemployment-related search terms. This second application not only serves to highlight the potential of 'modern' Big Data sources for macroeconomic forecasting, but also illustrates that in such Big Data applications, we have little theoretical guidance to decide on unit root and cointegration properties, and proper data-driven methods are needed.

Note that, as is common in the related high-dimensional literature, we focus explicitly on point forecasts. As distributional theory changes when unit roots are present, performing interval forecasts in the presence of unit roots and cointegration is a much more challenging—and largely unresolved—issue in the high-dimensional setting, especially as it adds to the complications of performing inference in high dimensions already present without unit roots. Given the scarcity of literature on this topic, we do not consider interval prediction in this chapter. This is clearly a very important avenue for future research.

The remainder of this chapter is organized as follows: Section 17.2 describes the general setup and introduces the cointegration model, along with some useful representations for later use. We discuss how to transform high-dimensional datasets to stationarity in Sect. 17.3, while Sect. 17.4 introduces high-dimensional approaches for modelling cointegration. In Sect. 17.5 we apply the discussed methods to our two empirical forecasting exercises. Finally, Sect. 17.6 concludes.

17.2 General Setup

In this section we describe a general model for cointegration to be used throughout this chapter. Next to defining the model in the classical error correction form, we also consider alternative representations that will be useful later in this chapter. As is common in the literature, we denote a time series as $I(d)$ if it has to be differenced d times to achieve stationarity, and we will use $I(1)$ interchangeably with a unit root process, and $I(0)$ with a stationary process.[2]

[1]Obviously, this caveat does not mean that forecasting in levels does not yield good results for specific applications. The applied researcher is free to apply any of the methods discussed in this book directly to (suspected) unit root series, but should simply be wary of the results.

[2]In fact, $I(0)$ processes can be non-stationary, for example, through having time-varying unconditional variance. For ease of explanation we still use 'stationary' to describe $I(0)$ processes though.

Let z_t denote an N-dimensional time series observed at time $t = 1, \ldots, T$. Assume that we can represent the series as

$$z_t = \boldsymbol{\mu} + \boldsymbol{\tau} t + \boldsymbol{\zeta}_t, \tag{17.1}$$

where $\boldsymbol{\mu}$ is an N-dimensional vector of intercepts, $\boldsymbol{\tau}$ is an n-dimensional vector of trend slopes, and $\boldsymbol{\zeta}_t$ is the N-dimensional purely stochastic time series. This stochastic component is given by

$$\Delta\boldsymbol{\zeta}_t = \boldsymbol{A}\boldsymbol{B}'\boldsymbol{\zeta}_{t-1} + \sum_{j=1}^{p} \boldsymbol{\Phi}_j \Delta\boldsymbol{\zeta}_{t-j} + \boldsymbol{\varepsilon}_t, \tag{17.2}$$

where $\boldsymbol{\varepsilon}_t$ is the N-dimensional innovation vector. Generally the innovations $\boldsymbol{\varepsilon}_t$ will be a martingale difference sequence, although we abstract from making too specific assumptions at this point.

We can obtain the classical vector error correction model (VECM) for z_t by substituting (17.1) into (17.2):

$$\Delta z_t = \boldsymbol{A}\boldsymbol{B}' \left(z_{t-1} - \boldsymbol{\mu} - \boldsymbol{\tau}(t-1)\right) + \boldsymbol{\tau}^* + \sum_{j=1}^{p} \boldsymbol{\Phi}_j \Delta z_{t-j} + \boldsymbol{\varepsilon}_t, \tag{17.3}$$

where $\boldsymbol{\tau}^* = (\boldsymbol{I}_N - \sum_{j=1}^{p} \boldsymbol{\Phi}_j)\boldsymbol{\tau}$. The long-run relations are contained in the $N \times r$-matrix $\boldsymbol{B}$, while the $N \times r$ matrix $\boldsymbol{A}$ contains the corresponding loadings. Here the variable r describes the number of cointegrating relations in the systems. If $r = 0$, we adopt the convention that $\boldsymbol{A}\boldsymbol{B}' = 0$; in this case z_t is a pure N-dimensional unit root process. If $r = N$, all series are $I(0)$. To ensure that z_t is at most an $I(1)$ process, the lag polynomial $\boldsymbol{C}(z) := (1-z) - \boldsymbol{A}\boldsymbol{B}'z - \sum_{j=1}^{p} \boldsymbol{\Phi}_j(1-z)z^j$ and matrices $\boldsymbol{A}$ and $\boldsymbol{B}$ should satisfy standard conditions that can be found in, inter alia, Johansen (1995b). Under these assumptions, exactly $N - r$ roots of the lag polynomial $\boldsymbol{C}(z)$ are equal to unity, while the remaining r roots lie outside the unit circle.

The typical interpretation of the VECM is that all series are $I(1)$, but r linear combinations of the series are $I(0)$. However, it may also be the case that some individual series within the VECM are actually $I(0)$; these define 'trivial' cointegration relations as any linear combination of these series remain $I(0)$. Thus the setup allows for observing a dataset with a mix of $I(0)$ and $I(1)$ series.

From the Granger Representation Theorem (cf. Johansen, 1995b, p. 49), we can obtain the *common trend representation* of (17.3), which is given by

$$z_t = \boldsymbol{\mu} + \boldsymbol{\tau} t + \boldsymbol{C}\boldsymbol{s}_t + \boldsymbol{u}_t, \tag{17.4}$$

where $\boldsymbol{C}$ is an $N \times N$ matrix of rank $N-r$,[3] $\boldsymbol{s}_t = \sum_{i=1}^{t} \boldsymbol{\varepsilon}_t$ are the stochastic trends and $\boldsymbol{u}_t$ is a stationary process. This representation shows that z_t can be decomposed in a deterministic process, an $I(1)$ part of common trends, $\boldsymbol{C}\boldsymbol{s}_t$, and a stationary part $\boldsymbol{u}_t$.

To see the commonality of the trends, note that as $\boldsymbol{C}$ is of reduced rank, we can define $N \times (N-r)$ matrices $\boldsymbol{\Lambda}$ and $\boldsymbol{\Gamma}$ such that $\boldsymbol{C} = \boldsymbol{\Lambda}\boldsymbol{\Gamma}'$. Then defining the $N-r \times 1$-vector $\boldsymbol{f}_t = \boldsymbol{\Gamma}'\boldsymbol{s}_t$, we can write (17.4) as

$$z_t = \boldsymbol{\mu} + \boldsymbol{\tau} t + \boldsymbol{\Lambda} \boldsymbol{f}_t + \boldsymbol{u}_t. \tag{17.5}$$

We can now see the common trends as *common factors*, which provides a convenient way to think about cointegration in high dimensions.

This brings us to an alternative way to represent cointegration through a common factor structure from the outset. This form was considered by Bai and Ng (2004) among others to investigate different sources of nonstationarity in a panel data context. In this case we start from (17.5), assuming that the elements of both $\boldsymbol{f}_t$ and $\boldsymbol{u}_t$ can be $I(0)$ or $I(1)$. The combination of the two then determines the properties of the series z_t. Consider a single series $z_{i,t}$, which can be represented as

$$z_{i,t} = \mu_i + \tau_i t + \boldsymbol{\lambda}_i' \boldsymbol{f}_t + u_{i,t},$$

where $\boldsymbol{\lambda}_i'$ denotes the i-th row of $\boldsymbol{\Lambda}$. Note that $z_{i,t}$ is $I(0)$ only if both $\boldsymbol{u}_{i,t}$ and $\boldsymbol{\lambda}_i' \boldsymbol{f}_t$ are $I(0)$, where the latter occurs if either all factors $\boldsymbol{f}_t$ are $I(0)$, or no $I(1)$ factors load on series i. Similarly, cointegration between series i and j requires that both $u_{i,t}$ and $u_{j,t}$ are $I(0)$.

Remark 17.1 For expositional simplicity we do not consider $I(2)$ variables here. While the VECM can be extended to allow for $I(2)$ series, see, e.g. Johansen (1995a), in practice most cointegration analyses are performed on $I(1)$ series. If the data contains (suspected) $I(2)$ series, these are generally differenced before commencing the cointegration analysis.

Similarly, one could think of the data generating process (DGP) as being of infinite lag order, rather than fixed order p. In this case the VECM with fixed order can be thought of as an approximation to the infinite order model, where p should be large enough to capture 'enough' of the serial correlation. Either way, in applications p is generally not known and has to be estimated.

[3]If $r = 0$, we set $\boldsymbol{C} = \boldsymbol{0}$.

17.3 Transformations to Stationarity and Unit Root Pre-testing

In this section we discuss how to determine the appropriate transformations—in particular how often the series need to be differenced—in order to obtain only stationary time series in our dataset. While established datasets, such as the FRED-MD, come with an overview of the appropriate transformation for each series, this is generally not the case and data-driven methods are needed. Thus, one normally has to apply unit root or stationarity tests to determine the order of integration, and the corresponding transformation. In this section we investigate how to approach this pre-testing problem.

First, we investigate unit root tests in more detail, and highlight some of their characteristics that one should take into account when considering high-dimensional macroeconomic forecasting. Second, we discuss how to deal with the multiple testing problem that arises from the fact that we need to combine unit root tests on many time series.

17.3.1 Unit Root Test Characteristics

Even though the literature on unit root testing has grown exponentially since the seminal paper of Dickey and Fuller (1979), discussing at length the characteristics of various unit root tests, unit root pre-testing is often done in an automatic, routine-like, way by considering classical tests such as augmented Dickey–Fuller (ADF) tests. However, these tests have various problematic characteristics which may accumulate when applied in high-dimensional problems. While we cannot discuss all of these here, let us briefly mention some of particular relevance for macroeconomic forecasting. An extensive overview of unit root testing is provided by Choi (2015).[4]

Size Distortions

Standard unit root tests are very prone to size distortions. One source is neglected serial correlation (cf. Schwert, 1989), while another is time-varying volatility (Cavaliere, 2005). For both sources, bootstrap methods have proven a successful means to counteract the size distortions; however, while for serial correlation any 'off-the-shelf' time series bootstrap method can be used (see Palm, Smeekes, and Urbain, 2008, for an overview and comparison), dealing with general forms of

[4]Given the greater popularity of tests where the null hypothesis is a unit root over tests with stationarity as the null, we focus exclusively on unit root tests here. However, most of the discussion applies to stationarity tests as well.

heteroskedasticity requires a unit root test based on the wild bootstrap (Cavaliere & Taylor, 2008, 2009).

It should be noted that unconditional volatility changes pose a particular concern for macroeconomic time series. Many datasets such as FRED-MD span the period of the Great Moderation, which has significantly affected the volatility of macroeconomic time series (Justiniano & Primiceri, 2008; Stock & Watson, 2003). It would therefore appear wise to take potential volatility changes into account when selecting an appropriate unit root test.

Power and Specification Considerations

The power properties of the different unit root tests proposed vary considerably, and generally optimal tests do not exist. One particular source of variation is the magnitude of the initial condition, where, for instance, the DF-GLS test of Elliott, Rothenberg, and Stock (1996) is optimal when the initial condition is zero, but the ADF test is much more powerful when the initial condition is large (Müller & Elliott, 2003). An even larger source of variation is the presence or absence of a deterministic trend. Unit root tests with a trend included (or, equivalently, unit root tests performed on detrended data) are considerably less powerful than without trend (performed on demeaned data). On the other hand, if a trend is not included when the data do contain one, the unit root test is not correctly sized anymore (Harvey, Leybourne, & Taylor, 2009).

While dealing with such issues is manageable in unit root testing for a single series, this changes when considering large datasets. For instance, deciding whether to include a trend in the unit root test can be based on a combination of theory, visual inspection, pre-testing and comparing outcomes of different tests with or without a trend. However, such an analysis has to be done manually for each series involved, which quickly becomes problematic if the dimension of the dataset increases. This is even more problematic for modern Big Datasets, such as Google Trends, for which no theory exists to guide the practitioner, and where the dimension can become arbitrarily large.

As such one would like to have an automatic way of choosing good specifications for the unit root tests that may differ across series. One easy way is provided by the union of unit root tests principle proposed by Harvey et al. (2009), Harvey, Leybourne, & Taylor (2012), in which several unit root tests are performed, and the unit root null hypothesis is rejected if one of the tests rejects (when corrected for multiple testing). In particular, Harvey et al. (2012) consider a union of the ADF and DF-GLS tests, both with and without linear trend, to cover uncertainty about both trend and initial condition. Smeekes and Taylor (2012) consider a wild bootstrap version of this test that is robust to time-varying volatility. The test statistic for series i takes the form

$$UR_i = \min\left(\left(\frac{x_i}{c_{i,GLS}^{\mu*}(\alpha)}\right) GLS_i^{\mu}, \left(\frac{x_i}{c_{i,GLS}^{\tau*}(\alpha)}\right) GLS_i^{\tau}, \left(\frac{x_i}{c_{i,ADF}^{\mu*}(\alpha)}\right) ADF_i^{\mu}, \left(\frac{x_i}{c_{i,ADF}^{\tau*}(\alpha)}\right) ADF_i^{\tau}\right), \tag{17.6}$$

where ADF_i and GLS_i are the ADF and DF-GLS test performed on series i, while superscript μ and τ indicate whether the series are demeaned or detrended, respectively. The bootstrap critical values such as $c_{i,GLS}^{\mu*}(\alpha)$ used in the scaling factors are determined in a preliminary bootstrap step as the individual level α critical values of the four tests. The variable x_i is a scaling factor to which the statistics are scaled. Any $x_i < 0$ suffices to preserve the left-tail rejection region; if one additionally takes x_i the same value for all series i, test statistics become comparable across series, which facilitates the multiple comparisons discussed in the next subsection.

17.3.2 Multiple Unit Root Tests

Performing a unit root test for every series separately raises issues associated with multiple testing. In particular, the probability of incorrect classifications rises with the number of tests performed. If each test has a significance level of 5%, we may also expect roughly 5% of the $I(1)$ series to be incorrectly classified as $I(0)$. In a high-dimensional dataset this can quickly lead to a significant number of incorrectly classified series. It will of course depend on the specific application whether this is problematic—a priori we cannot say whether the 'important' series will be correctly classified or not—but to avoid such issues one can formally account for multiple testing.

There is a huge statistical literature about multiple testing; Romano, Shaikh, and Wolf (2008b) provide an overview with a focus on econometric applications. Here we briefly discuss the most prominent methods developed for the purposes of unit root testing. Before discussing the different methods to control for multiple testing, let us set up the general framework. Let $UR_1, \ldots, UR_N$ denote the unit root test statistics for series 1 up to N, assuming they reject for small values of the statistics.[5] It is important to choose the test statistics such that they are directly comparable, in the sense that their marginal distributions are the same. If this is the case, then the

[5]We can assume this without loss of generality as any test statistic can be modified to indeed do so.

ranking

$$UR_{(1)} \leq \ldots \leq UR_{(R)} \leq UR_{(R+1)} \leq \ldots \leq UR_{(N)}, \tag{17.7}$$

where $UR_{(i)}$ denotes the i-th order statistic of $UR_1, \ldots, UR_N$, corresponds to a ranking from 'most significant' to 'least significant'. To ensure the comparability of the test statistics, one needs to eliminate nuisance parameters from their distribution. Hence, simply using the bootstrap to absorb nuisance parameters is not sufficient; instead, one often needs to transform (for instance, to p-values) or scale the statistics appropriately. In the union tests of (17.6), the scaling is done automatically by setting $x_i = -1$ for all units.

Given the ranking in (17.7), the objective is to find an appropriate cut-off point R such that for all statistics less than or equal $UR_{(R)}$ the unit root hypothesis is rejected, and for all statistics larger it is not rejected. How this threshold is determined depends on how multiple testing is controlled for.

Controlling Generalized Error Rates

Generalized error rates provide multivariate extensions of the standard Type I error. The most common is the *familywise error rate (FWE)*, which is defined as the probability of making at least one false rejection of the null hypothesis. This can easily be controlled by the popular Bonferroni correction. However, this is very conservative as it is valid under any form of dependence. On the contrary, if the bootstrap is used to capture the actual dependence structure among the tests, one can control for multiple testing without the need for being conservative. This approach is followed by Hanck (2009), who controls FWE in unit root testing by applying the bootstrap algorithm proposed by Romano and Wolf (2005).

While controlling FWE makes sense when N is small, in typical high-dimensional datasets FWE becomes too conservative. Instead, one can control the *false discovery rate (FDR)* originally proposed by Benjamini and Hochberg (1995), which is defined as

$$FDR = \mathbb{E}\left[\frac{F}{R}\mathbb{1}(R > 0)\right],$$

where R denotes the total number of rejections, and F the number of false rejections. The advantage of the FDR is that it scales with increasing N, and thus is more appropriate for large datasets. However, most non-bootstrap methods are either not valid under arbitrary dependence or overly conservative. Moon and Perron (2012) compare several methods to control FDR and find that the bootstrap method of Romano, Shaikh, and Wolf (2008a), hereafter denoted as BFDR, does not share these disadvantages and clearly outperforms the other methods. A downside of this method however is that the algorithm is rather complicated and time-consuming to implement. Globally, the algorithm proceeds in a sequential way by starting to test the 'most significant' series, that is, the smallest unit root test statistic. This statistic is then compared to an appropriate critical values obtained from the bootstrap algorithm, where the bootstrap evaluates all scenarios possible in terms of false and true rejections given the current progression of the algorithm. If the null hypothesis can be rejected for the current series, the algorithm proceeds to the next most significant statistic and the procedure is repeated. Once a non-rejection is observed, the algorithm stops. For details we refer to Romano et al. (2008a). This makes the bootstrap FDR method a *step-down* method, contrary to the original Benjamini and Hochberg (1995) approach which is a step-up method starting from the least significant statistic.

Sequential Testing

Smeekes (2015) proposes an alternative bootstrap method for multiple unit root testing based on sequential testing. In a first step, the null hypothesis that all N series are $I(1)$—hence $p_1 = 0$ series are $I(0)$—is tested against the alternative that (at least) p_2 series are $I(0)$. If the null hypothesis is rejected, the p_2 most significant statistics in (17.7) are deemed $I(0)$ and removed from consideration. Then the null hypothesis that all remaining $N - p_2$ series are $I(1)$ is tested against the alternative that at least p_2 of them are $I(0)$, and so on. If no rejections are observed, the final rounds test p_K $I(0)$ series against the alternative of N $I(0)$ series. The numbers $p_2, \ldots, p_K$ as well as the number of tests K are chosen by the practitioner based on the specific application at hand. By choosing the numbers as $p_k = [q_k N]$, where $q_1, \ldots, q_K$ are desired quantiles, the method automatically scales with N.

Unlike the BFDR method, this Bootstrap Sequential Quantile Test (BSQT) is straightforward and fast to implement. However, it is dependent on the choice of numbers p_k to be tested; its 'error allowance' is therefore of a different nature to error rates like FDR. Smeekes (2015) shows that, when p_J units are found to be $I(0)$, the probability that the true number of $I(0)$ series lies outside the interval $[p_{J-1}, p_{J+1}]$ is at most the chosen significance level of the test. As such, there is some uncertainty around the cut-off point.

It might therefore be tempting to choose $p_k = k - 1$ for all $k = 1, \ldots, N$, such that this uncertainty disappears. However, as discussed in Smeekes (2015), applying the sequential method to each series individually hurts power if N is large as it amounts to controlling FWE. Instead, a better approach is to iterate the BSQT method; that is, it can be applied in a second stage just to the interval $[p_{J-1}, p_{J+1}]$ to reduce the uncertainty. This can be iterated until few enough series remain to be tested individually in a sequential manner. On the other hand, if $p_1, \ldots, p_K$ are chosen sensibly and not spaced too far apart, the uncertainty is limited to a narrow range around the 'marginally significant' unit root tests. These series are at risk of missclassification anyway, and the practical consequences of incorrect classification for these series on the boundary of a unit root are likely small.

Smeekes (2015) performs a Monte Carlo comparison of the BSQT and BFDR methods, as well as several methods proposed in the panel data literature such as Ng (2008) and Chortareas and Kapetanios (2009). Globally BSQT and BFDR clearly outperform the other methods, where BFDR is somewhat more accurate than BSQT when the time dimension T is at least of equal magnitude as the number of series N. On the other hand, when T is much smaller than N BFDR suffers from a lack of power and BSQT is clearly preferable. In our empirical applications we will therefore consider both BFDR and BSQT, as well as the strategy of performing individual tests without controlling for multiple testing.

Remark 17.2 An interesting non-bootstrap alternative is the panel method proposed by Pedroni, Vogelsang, Wagner, and Westerlund (2015), which has excellent performance in finite samples. However, implementation of this method requires that T is strictly larger than N, thus severely limiting its potential for analysing Big Data. Another alternative would be to apply the model selection approach through the adaptive lasso by Kock (2016) which avoids testing all together. However, this has only been proposed in a univariate context and its properties are unknown for the type of application considered here.

Multivariate Bootstrap Methods

All multiple testing methods described above require a bootstrap method that can not only account for dependence within a single time series, but can also capture the dependence structures between series. Accurately modelling the dependence between the individual test statistics is crucial for proper functioning of the multiple testing corrections. Capturing the strong and complex dynamic dependencies between macroeconomic series requires flexible bootstrap methods that can handle general forms of dependence.

Moon and Perron (2012) and Smeekes (2015) use the moving-blocks bootstrap (MBB) based on the results of Palm, Smeekes, and Urbain (2011) who prove validity for mixed $I(1)/I(0)$ panel datasets under general forms of dependence. However the MBB has two disadvantages. First, it can only be applied to balanced datasets where each time series is observed over the same period. This makes application to datasets such as FRED-MD difficult, at least without deleting observations for series that have been observed for a longer period. Second, the MBB is sensitive to unconditional heteroskedasticity, which makes its application problematic for series affected by the Great Moderation.

Dependent wild bootstrap (DWB) methods address both issues while still being able to capture complex dependence structure. Originally proposed by Shao (2010) for univariate time series, they were extended to unit root testing by Smeekes and Urbain (2014a) and Rho and Shao (2019), where the former paper considers the multivariate setup needed here. A general wild bootstrap algorithm for multivariate unit root testing looks as follows:

1. Detrend the series $\{z_t\}$ by OLS; that is, let $\hat{\boldsymbol{\zeta}}_t = (\widehat{\zeta}_{1,t}, \ldots, \widehat{\zeta}_N)'$, where

$$\widehat{\zeta}_{i,t} = z_{i,t} - \widehat{\mu}_i - \widehat{\tau}_i t, \qquad i = 1, \ldots, N, \quad t = 1, \ldots, T$$

and $(\widehat{\mu}_i, \widehat{\tau}_i)'$ are the OLS estimators of $(\mu_i, \tau_i)'$.

2. Transform $\widehat{\boldsymbol{\zeta}}_t$ to a multivariate $I(0)$ series $\widehat{\boldsymbol{u}}_t = (\widehat{u}_{1,t}, \ldots, \widehat{u}_{N,t})'$ by setting

$$\widehat{u}_{i,t} = \widehat{\zeta}_{i,t} - \widehat{\rho}_i \widehat{\zeta}_{i,t-1}, \qquad i = 1, \ldots, N, \quad t = 1, \ldots, T,$$

   where $\widehat{\rho}_i$ is either an estimator of the largest autoregressive root of $\{\widehat{\zeta}_{i,t}\}$ using, for instance, an (A)DF regression, or $\widehat{\rho}_i = 1$.
3. Generate a univariate sequence of *dependent* random variables $\xi_1^*, \ldots, \xi_N^*$ with the properties that $\mathbb{E}^* \xi_t^* = 0$ and $\mathbb{E}^* \xi_t^{*2} = 1$ for all t. Then construct bootstrap errors $\boldsymbol{u}_t^* = (u_{1,t}^*, \ldots, u_{N,t}^*)'$ as

$$u_{i,t}^* = \xi_t^* \hat{u}_{i,t}, \qquad i = 1, \ldots, N, \quad t = 1, \ldots, T. \tag{17.8}$$

4. Let $z_t^* = \sum_{s=1}^t \boldsymbol{u}_s^*$ and calculate the desired unit root test statistics $UR_1^*, \ldots, UR_N^*$ from $\{z_t^*\}$. Use these bootstrap test statistics in an appropriate algorithm for controlling multiple testing.

Note that, unlike for the MBB, in (17.8) no resampling takes place, and as such missing values 'stay in their place' without creating new 'holes' in the bootstrap samples. This makes the method applicable to unbalanced panels. Moreover, heteroskedasticity is automatically taken into account by virtue of the wild bootstrap principle. Serial dependence is captured through the dependence of $\{\xi_t^*\}$, while dependence across series is captured directly by using the same, univariate, ξ_t^* for each series i. Smeekes and Urbain (2014a) provide theoretical results on the bootstrap validity under general forms of dependence and heteroskedasticity.

There are various options to draw the dependent $\{\xi_t^*\}$; Shao (2010) proposes to draw these from a multivariate normal distributions, where the covariance between ξ_s^* and ξ_t^* is determined by a kernel function with as input the scaled distance $|s - t| / \ell$. The tuning parameter ℓ serves as a similar parameter as the block length in the MBB; the larger it is, the more serial dependence is captured. Smeekes and Urbain (2014a) and Friedrich, Smeekes, and Urbain (2018) propose generating $\{\xi_t^*\}$ through an AR(1) process with normally distributed innovations and AR parameter γ, where γ is again a tuning parameter that determines how much serial dependence is captured. They label this approach the autoregressive wild bootstrap (AWB), and show that the AWB generally performs at least as well as Shao's (2010) DWB in simulations.

Finally, one might consider the sieve wild bootstrap used in Cavaliere and Taylor (2009) and Smeekes and Taylor (2012), where the series $\{\widehat{\boldsymbol{u}}_t\}$ are first filtered through individual AR processes, and the wild bootstrap is applied afterwards to the residuals. However, as Smeekes and Urbain (2014b) show that this method cannot capture complex dynamic dependencies across series, it should not be used in this multivariate context. If common factors are believed to be the primary source of dependence across series, factor bootstrap methods such as those considered by Trapani (2013) or Gonçalves and Perron (2014) could be used as well.

17.4 High-Dimensional Cointegration

In this section, we discuss various recently proposed methods to model high-dimensional (co)integrated datasets. Similar to the high-dimensional modelling of stationary datasets, two main modelling approaches can be distinguished. One approach is to summarize the complete data into a much smaller and more manageable set through the extraction of common factors and their associated loadings, thereby casting the problem into the framework represented by (17.5). Another approach is to consider direct estimation of a system that is fully specified on the observable data as in (17.3), under the implicit assumption that the true DGP governing the long- and short-run dynamics is sparse, i.e. the number of non-zero coefficients in said relationships is small. These two approaches, however, rely on fundamentally different philosophies and estimation procedures, which constitute the topic of this section.[6]

17.4.1 Modelling Cointegration Through Factor Structures

Factor models are based on the intuitive notion that all variables in an economic system are driven by a small number of common shocks, which are often thought of as representing broad economic phenomena such as the unobserved business cycle. On (transformed) stationary macroeconomic datasets, the extracted factors have been successfully applied for the purpose of forecasting by incorporating them in dynamic factor models (Forni, Hallin, Lippi, & Reichlin, 2005), factor-augmented vector autoregressive (FAVAR) models (Bernanke, Boivin, & Eliasz, 2005) or single-equation models (Stock & Watson, 2002a,b). We refer to Chaps. 2 and 3 of this book for further details on these methods. Recent proposals are brought forward in the literature that allow for application of these techniques on non-stationary and possibly cointegrated datasets. In Sect. 17.4.1 the dynamic factor model proposed by Barigozzi, Lippi, and Luciani (2017, 2018) is discussed and Sect. 17.4.1 details the factor-augmented error correction model by Banerjee, Marcellino, and Masten (2014, 2016). As both approaches require an a priori choice on the number of common factors, we briefly discuss estimation of the factor dimension in Sect. 17.4.1.

[6] Some recent papers such as Onatski and Wang (2018) and Zhang, Robinson, and Yao (2018) have taken different, novel approaches to high-dimensional cointegration analysis. However, these methods do not directly lend themselves to forecasting and are therefore not discussed in this chapter.

Dynamic Factor Models

A popular starting point for econometric modelling involving common shocks is the specification of a dynamic factor model. Recall our representation of an individual time series by

$$z_{i,t} = \mu_i + \tau_i t + \boldsymbol{\lambda}_i' \boldsymbol{f}_t + u_{i,t}, \tag{17.9}$$

where $\boldsymbol{f}_t$ is the $N - r \times 1$ dimensional vector of common factors. Given a set of estimates for the unobserved factors, say $\hat{\boldsymbol{f}}_t$ for $t = 1, \ldots, T$, one may directly obtain estimates for the remaining parameters in (17.9) by solving the least-squares regression problem[7]

$$\left(\hat{\boldsymbol{\mu}}, \hat{\boldsymbol{\tau}}, \hat{\boldsymbol{\Lambda}}\right) = \underset{\boldsymbol{\mu},\boldsymbol{\tau},\boldsymbol{\Lambda}}{\arg\min} \sum_{t=1}^{T} \left(z_t - \boldsymbol{\mu} - \boldsymbol{\tau} t - \boldsymbol{\Lambda} \hat{\boldsymbol{f}}_t\right)^2. \tag{17.10}$$

The forecast for the realization of an observable time series at time period $T + h$ can then be constructed as

$$\hat{z}_{i,T+h|T} = \hat{\mu}_i + \hat{\tau}_i (T + h) + \hat{\boldsymbol{\lambda}}_i \hat{\boldsymbol{f}}_{T+h|T}. \tag{17.11}$$

This, however, requires the additional estimate $\hat{\boldsymbol{f}}_{T+h|T}$, which may be obtained through an explicit dynamic specification of the factors.

Barigozzi et al. (2018) assume that the differenced factors admit a reduced-rank vector autoregressive (VAR) representation, given by

$$\boldsymbol{S}(L) \Delta \boldsymbol{f}_t = \boldsymbol{C}(L) \boldsymbol{v}_t, \tag{17.12}$$

where $\boldsymbol{S}(L)$ is an invertible $N - r \times N - r$ matrix polynomial and $\boldsymbol{C}(L)$ is a finite degree $N - r \times q$ matrix polynomial. Furthermore, $\boldsymbol{v}_t$ is a $q \times 1$ vector of white noise common shocks with $N - r > q$. Inverting the left-hand side matrix polynomial and summing both sides gives rise to the specification

$$\boldsymbol{f}_t = \boldsymbol{S}^{-1}(L) \boldsymbol{C}(L) \sum_{s=1}^{t} \boldsymbol{v}_s = \boldsymbol{U}(L) \sum_{s=1}^{t} \boldsymbol{v}_s = \boldsymbol{U}(1) \sum_{s=1}^{t} \boldsymbol{v}_s + \boldsymbol{U}^*(L) \left(\boldsymbol{v}_t - \boldsymbol{v}_0\right), \tag{17.13}$$

where the last equation follows from application of the Beveridge–Nelson decomposition to $\boldsymbol{U}(L) = \boldsymbol{U}(1) + \boldsymbol{U}^*(L)(1 - L)$. Thus, (17.13) reveals that the factors are driven by a set of common trends and stationary linear processes. Crucially,

[7]Typically, the estimation procedure for $\hat{\boldsymbol{f}}_t$ provides the estimates $\hat{\boldsymbol{\Lambda}}$ as well, such that only the coefficients regulating the deterministic specification ought to be estimated.

the assumption that the number of common shocks is strictly smaller than the number of integrated factors, i.e. $\boldsymbol{f}_t$ is a singular stochastic vector, implies that rank $(\boldsymbol{U}(1)) = q - d$ for $0 \leq d < q$. Consequently, there exists a full column rank matrix $\boldsymbol{B}_f$ of dimension $N - r \times N - r - q + d$ with the property that $\boldsymbol{B}'_f \boldsymbol{f}_t$ is stationary. Then, under the general assumption that the entries of $\boldsymbol{U}(L)$ are rational functions of L, Barigozzi et al. (2017) show that $\boldsymbol{f}_t$ admits a VECM representation of the form

$$\Delta \boldsymbol{f}_t = \boldsymbol{A}_f \boldsymbol{B}'_f \boldsymbol{f}_{t-1} + \sum_{j=1}^{p} \boldsymbol{G}_j \Delta \boldsymbol{f}_{t-j} + \boldsymbol{K} \boldsymbol{v}_t, \tag{17.14}$$

where $\boldsymbol{K}$ is a constant matrix of dimension $N - r \times q$.

Since the factors in (17.14) are unobserved, estimation of the system requires the use of a consistent estimate of the space spanned by $\boldsymbol{f}_t$. Allowing idiosyncratic components $v_{i,t}$ in (17.9) to be either $I(1)$ or $I(0)$, and allowing for the presence of a non-zero constant μ_i and linear trend τ_i, Barigozzi et al. (2018) propose an intuitive procedure that enables estimation of the factor space by the method of principal components. First, the data is detrended with the use of a regression estimate:

$$\tilde{z}_{i,t} = z_{i,t} - \hat{\tau}_i t,$$

where $\hat{\tau}_i$ is the OLS estimator of the trend in the regression of $z_{i,t}$ on an intercept and linear trend. Then, similar to the procedure originally proposed by Bai and Ng (2004), the factor loadings are estimated as $\hat{\boldsymbol{\Lambda}} = \sqrt{N}\hat{\boldsymbol{W}}$, where $\hat{\boldsymbol{W}}$ is the $N \times (N - r)$ matrix with normalized right eigenvectors of $T^{-1} \sum_{t=1}^{T} \Delta \tilde{z}_t \Delta \tilde{z}_t'$ corresponding to the $N - r$ largest eigenvalues. The estimates for the factors are given by $\hat{\boldsymbol{f}}_t = \frac{1}{N} \hat{\boldsymbol{\Lambda}}' \tilde{z}_t$.

Plugging $\hat{\boldsymbol{f}}_t$ into (17.14) results in

$$\Delta \hat{\boldsymbol{f}}_t = \boldsymbol{A}_f \boldsymbol{B}'_f \hat{\boldsymbol{f}}_{t-1} + \sum_{j=1}^{p} \boldsymbol{G}_j \Delta \hat{\boldsymbol{f}}_{t-j} + \hat{\boldsymbol{v}}_t, \tag{17.15}$$

which can be estimated using standard approaches, such as the maximum likelihood procedure proposed by Johansen (1995b). Afterwards, the iterated one-step-ahead forecasts $\Delta \hat{\boldsymbol{f}}_{T+1|T}, \ldots, \Delta \hat{\boldsymbol{f}}_{T+h|T}$ are calculated from the estimated system, based on which the desired forecast $\hat{\boldsymbol{f}}_{T+h|T} = \hat{\boldsymbol{f}}_T + \sum_{k=1}^{h} \Delta \hat{\boldsymbol{f}}_{T+k|T}$ is obtained. The final forecast for $\hat{z}_{i,T+h|T}$ is then easily derived from (17.11).

Remark 17.3 Since the idiosyncratic components are allowed to be serially dependent or even $I(1)$, a possible extension is to explicitly model these dynamics. As a simple example, each $u_{i,t}$ could be modelled with a simple autoregressive model, from which the prediction $\hat{u}_{i,T+h|T}$ can be obtained following standard procedures (e.g. Hamilton, 1994, Ch. 4). This prediction is then added to (17.11), leading to the final forecast

$$\hat{z}_{i,T+h|T} = \hat{\mu}_i + \hat{\tau}_i(T+h) + \hat{\boldsymbol{\lambda}}_i \hat{\boldsymbol{f}}_{T+h|T} + \hat{u}_{i,T+h|T}.$$

This extension leads to substantial improvements in forecast performance in the macroeconomic forecast application presented in Sect. 17.5.

Factor-Augmented Error Correction Model

It frequently occurs that the variables of direct interest constitute only a small subset of the collection of observed variables. In this scenario, Banerjee et al. (2014, 2016), Banerjee, Marcellino, and Masten (2017), henceforth referred to as BMM, propose to model only the series of interest in a VECM system while including factors extracted from the full dataset to proxy for the missing information from the excluded observed time series.

The approach of BMM can be motivated starting from the common trend representation in (17.4). Partition the observed time series $z_t = (z'_{A,t}, z'_{B,t})'$, where $z_{A,t}$ is an $N_A \times 1$ vector containing the variables of interest. Then, we may rewrite (17.4) as

$$\begin{bmatrix} z_{A,t} \\ z_{B,t} \end{bmatrix} = \begin{bmatrix} \boldsymbol{\mu}_A \\ \boldsymbol{\mu}_B \end{bmatrix} + \begin{bmatrix} \boldsymbol{\tau}_A \\ \boldsymbol{\tau}_B \end{bmatrix} t + \begin{bmatrix} \boldsymbol{\Lambda}_A \\ \boldsymbol{\Lambda}_B \end{bmatrix} \boldsymbol{f}_t + \begin{bmatrix} \boldsymbol{u}_{A,t} \\ \boldsymbol{u}_{B,t}. \end{bmatrix} \tag{17.16}$$

The idiosyncratic components in (17.16) are assumed to be $I(0)$.[8] Furthermore, both non-stationary $I(1)$ factors and stationary factors are admitted in the above representation. Contrary to Barigozzi et al. (2017), BMM do not require the factors in (17.16) to be singular.

To derive a dynamic representation better suited to forecasting the variables of interest, Banerjee et al. (2014, 2017) use the fact that when the subset of variables is of a larger dimension than the factors, i.e. $N_A > N - r$, $z_{A,t}$ and $\boldsymbol{f}_t$ cointegrate. As a result, the Granger Representation Theorem implies the existence of an error correction representation of the form

$$\begin{bmatrix} \Delta z_{A,t} \\ \boldsymbol{f}_t \end{bmatrix} = \begin{bmatrix} \boldsymbol{\mu}_A \\ \boldsymbol{\mu}_f \end{bmatrix} + \begin{bmatrix} \boldsymbol{\tau}_A \\ \boldsymbol{\tau}_f \end{bmatrix} t + \begin{bmatrix} \boldsymbol{A}_A \\ \boldsymbol{A}_B \end{bmatrix} \boldsymbol{B}' \begin{bmatrix} z_{A,t-1} \\ \boldsymbol{f}_{t-1} \end{bmatrix} + \begin{bmatrix} \boldsymbol{e}_{A,t} \\ \boldsymbol{e}_{f,t} \end{bmatrix}. \tag{17.17}$$

[8] In principle, the proposed estimation procedure remains feasible in the presence of $I(1)$ idiosyncratic components. The theoretical motivation, however, relies on the concept of cointegration between the observable time series and a set of common factors. This only occurs when the idiosyncratic components are stationary.

To account for serial dependence in (17.17), Banerjee et al. (2014) propose the approximating model

$$\begin{bmatrix} \Delta z_{A,t} \\ f_t \end{bmatrix} = \begin{bmatrix} \mu_A \\ \mu_f \end{bmatrix} + \begin{bmatrix} \tau_A \\ \tau_f \end{bmatrix} t + \begin{bmatrix} A_A \\ A_B \end{bmatrix} B' \begin{bmatrix} z_{A,t-1} \\ f_{t-1} \end{bmatrix} + \sum_{j=1}^{p} \Phi_j \begin{bmatrix} \Delta z_{A,t-j} \\ \Delta f_{t-j} \end{bmatrix} + \begin{bmatrix} \epsilon_{A,t} \\ \epsilon_{f,t} \end{bmatrix}, \tag{17.18}$$

where the errors $\left(\epsilon'_{A,t}, \epsilon'_{f,t}\right)'$ are assumed i.i.d.

Similar to the case of the dynamic factor model in Sect. 17.4.1, the factors in the approximating model (17.18) are unobserved and need to be replaced with their corresponding estimates $\hat{f}_t$. Under a set of mild assumptions, Bai (2004) shows that the space spanned by f_t can be consistently estimated using the method of principal components applied to the levels of the data. Assume that $f_t = \left(f'_{ns,t}, f'_{s,t}\right)'$ where $f_{ns,t}$ and $f_{s,t}$ contain r_{ns} non-stationary and r_s stationary factors, respectively. Let $Z = (z_1, \ldots, z_T)$ be the $(N \times T)$ matrix of observed time series. Then, Bai (2004) shows that $f_{ns,t}$ is consistently estimated by $\hat{f}_{ns,t}$, representing the eigenvectors corresponding to the r_{ns} largest eigenvalues of $Z'Z$, normalized such that $\frac{1}{T^2}\sum_{t=1}^{T} \hat{f}_{ns,t}\hat{f}'_{ns,t} = I$. Similarly, $f_{s,t}$ is consistently estimated by $\hat{f}_{s,t}$, representing the eigenvectors corresponding to the next r_s largest eigenvalues of $Z'Z$, normalized such that $\frac{1}{T}\sum_{t=1}^{T} \hat{f}_{s,t}\hat{f}'_{s,t} = I$.

The final step in the forecast exercise consists of plugging in $\hat{f}_t = \left(\hat{f}'_{ns,t}, \hat{f}'_{s,t}\right)'$ into (17.18), leading to

$$\begin{bmatrix} \Delta z_{A,t} \\ \hat{f}_t \end{bmatrix} = \begin{bmatrix} \mu_A \\ \mu_f \end{bmatrix} + \begin{bmatrix} \tau_A \\ \tau_f \end{bmatrix} t + \begin{bmatrix} A_A \\ A_B \end{bmatrix} B' \begin{bmatrix} z_{A,t-1} \\ \hat{f}_{t-1} \end{bmatrix} + \sum_{j=1}^{p} \Phi_j \begin{bmatrix} \Delta z_{A,t-j} \\ \Delta \hat{f}_{t-j} \end{bmatrix} + \begin{bmatrix} \epsilon_{A,t} \\ \epsilon_{f,t} \end{bmatrix}. \tag{17.19}$$

Since in typical macroeconomic applications the number of factors is relatively small, feasible estimates for (17.19) can be obtained from the maximum likelihood procedure of Johansen (1995b). The iterated one-step-ahead forecasts $\Delta\hat{z}_{A,T+1|T}$ to $\Delta\hat{z}_{A,T+h|T}$ are calculated from the estimated system, which are then integrated to obtain the desired forecast $\hat{z}_{A,T+h|T}$.

Estimating the Number of Factors

Implementation of the factor models discussed in this section requires an a priori choice regarding the number of factors. A wide variety of methods to estimate the dimension of the factors is available. The dynamic factor model of Barigozzi et al. (2017, 2018) adopts the estimation strategy proposed by Bai and Ng (2004), which relies on first differencing the data. Since, under the assumed absence of

$I(2)$ variables, all variables in this transformed dataset are stationary, the standard tools to determine the number of factors in the stationary setting are applicable. A non-exhaustive list is given by Bai and Ng (2002), Hallin and Liška (2007), Alessi, Barigozzi, and Capasso (2010), Onatski (2010) and Ahn and Horenstein (2013).

The factor-augmented error correction model of Banerjee et al. (2014, 2016) adopts the estimation strategy proposed by Bai and Ng (2004), which extracts the factors from the data in levels. While the number of factors may still be determined based on the differenced dataset, Bai (2004) proposes a set of information criteria that allows for estimation of the number of non-stationary factors without differencing the data.

Conveniently, it is possible to combine factor selection procedures to separately determine the number of non-stationary and stationary factors. For example, the total number of factors, say $r_{ns} + r_s$, can be found based on the differenced dataset and one of the information criteria in Bai and Ng (2002). Afterwards, the number of non-stationary factors, r_{ns}, is determined based on the data in levels using one of the criteria from Bai (2004). The number of stationary factors follows from the difference between the two criteria. Recently, Barigozzi and Trapani (2018) propose a novel approach to discern the number of $I(0)$ factors, zero-mean $I(1)$ factors and factors with a linear trend. Their method however requires that all idiosyncratic components are $I(0)$.

17.4.2 Sparse Models

Rather than extracting common factors, an alternative approach to forecasting with macroeconomic data is full-system estimation with the use of shrinkage estimators (e.g. De Mol, Giannone, & Reichlin 2008; Stock & Watson 2012; Callot & Kock 2014) as discussed in Chap. 7 of this book. The general premise of shrinkage estimators is the so-called bias–variance trade-off, i.e. the idea that, by allowing a relatively small amount of bias in the estimation procedure, a larger reduction in variance may be attained. A number of shrinkage estimators, among which the lasso originally proposed by Tibshirani (1996), simultaneously perform variable selection and model estimation. Such methods are natural considerations when it is believed that the data generating process is sparse, i.e. only a small subset of variables among the candidate set is responsible for the variation in the variables of interest. Obviously, such a viewpoint is in sharp contrast with the philosophy underlying the common factor framework. However, even in cases where a sparse data generating process is deemed unrealistic, shrinkage estimators can remain attractive due to their aforementioned bias–variance trade-off (Smeekes & Wijler, 2018b).

For expositional convenience, we assume in this section that either $\boldsymbol{\mu}$ and $\boldsymbol{\tau}$ are zero or that z_t is demeaned and detrended. Defining $\boldsymbol{\Pi} = \boldsymbol{A}\boldsymbol{B}'$, model (17.3) is then given by

$$\Delta z_t = \boldsymbol{\Pi} z_{t-1} + \sum_{j=1}^{p} \boldsymbol{\Phi}_j \Delta z_{t-j} + \epsilon_t,$$

which in matrix notation reads as

$$\Delta \boldsymbol{Z} = \boldsymbol{\Pi} \boldsymbol{Z}_{-1} + \boldsymbol{\Phi} \Delta \boldsymbol{X} + \boldsymbol{E}, \tag{17.20}$$

where $\Delta \boldsymbol{Z} = (\Delta z_1, \ldots, \Delta z_T)$, $\boldsymbol{Z}_{-1} = (z_0, \ldots, z_{T-1})$, $\boldsymbol{\Phi} = \left(\boldsymbol{\Phi}_1, \ldots, \boldsymbol{\Phi}_p\right)$ and $\Delta \boldsymbol{X} = (\Delta \boldsymbol{x}_0, \ldots, \Delta \boldsymbol{x}_{T-1})$, with $\boldsymbol{x}_t = \left(z_t', \ldots, z_{t-p+1}'\right)'$.

Full-System Estimation

Several proposals to estimate (17.20) with the use of shrinkage estimators are brought forward in recent literature. Liao and Phillips (2015) propose an automated approach that simultaneously enables sparse estimation of the coefficient matrices $(\boldsymbol{\Pi}, \boldsymbol{\Phi})$, including the cointegrating rank of $\boldsymbol{\Pi}$ and the short-run dynamic lag order in $\boldsymbol{\Phi}$. However, while the method has attractive theoretical properties, the estimation procedure involves non-standard optimization over the complex plane and is difficult to implement even in low dimensions, as also noted by Liang and Schienle (2019). Accordingly, we do not further elaborate on their proposed method, but refer the interested reader to the original paper.

Liang and Schienle (2019) develop an automated estimation procedure that makes use of a QR-decomposition of the long-run coefficient matrix. They propose to first regress out the short-run dynamics, by post-multiplying (17.20) with $\boldsymbol{M} = \boldsymbol{I}_T - \Delta \boldsymbol{X}' \left(\Delta \boldsymbol{X}' \Delta \boldsymbol{X}\right)^{-1} \Delta \boldsymbol{X}$, resulting in

$$\Delta \tilde{\boldsymbol{Z}} = \boldsymbol{\Pi} \tilde{\boldsymbol{Z}}_{-1} + \tilde{\boldsymbol{E}}, \tag{17.21}$$

with $\Delta \tilde{\boldsymbol{Z}} = \Delta \boldsymbol{Z} M$, $\tilde{\boldsymbol{Z}}_{-1} = \boldsymbol{Z}_{-1} M$ and $\tilde{\boldsymbol{E}} = \boldsymbol{E} M$. The key idea behind the method proposed by Liang & Schienle is to decompose the long-run coefficient matrix into

$$\boldsymbol{\Pi}' = \boldsymbol{Q} \boldsymbol{R},$$

where $\boldsymbol{Q}' \boldsymbol{Q} = \boldsymbol{I}_N$ and $\boldsymbol{R}$ is an upper-triangular matrix. Such a representation can be calculated from the QR-decomposition of $\boldsymbol{\Pi}$ with column pivoting.

The column pivoting orders the columns in $\boldsymbol{R}$ according to size, such that zero elements occur at the ends of the rows. As a result, the rank of $\boldsymbol{\Pi}$ corresponds to the number of non-zero columns in $\boldsymbol{R}$. Exploiting this rank property requires an initial estimator for the long-run coefficient matrix, such as the OLS estimator

$$\hat{\boldsymbol{\Pi}}_{OLS} = \left(\Delta \tilde{\boldsymbol{Z}} \tilde{\boldsymbol{Z}}_{-1}'\right) \left(\tilde{\boldsymbol{Z}}_{-1} \tilde{\boldsymbol{Z}}_{-1}'\right)^{-1},$$

proposed by Liang and Schienle (2019). The QR-decomposition with column pivoting is then calculated from $\hat{\boldsymbol{\Pi}}_{OLS}'$, resulting in the representation $\hat{\boldsymbol{\Pi}}_{OLS} =$

$\hat{\boldsymbol{R}}'_{OLS}\hat{\boldsymbol{Q}}'_{OLS}$.[9] Since the unrestricted estimator $\hat{\boldsymbol{\Pi}}_{OLS}$ will be full-rank, $\hat{\boldsymbol{R}}_{OLS}$ is a full-rank matrix as well. However, by the consistency of $\hat{\boldsymbol{\Pi}}_{OLS}$ and the ordering induced by the column-pivoting step, the last $N - r$ columns are expected to contain elements that are small in magnitude. Accordingly, a well-chosen shrinkage estimator that penalizes the columns of $\boldsymbol{R}$ may be able to separate the relevant from the irrelevant columns.

The shrinkage estimator proposed by Liang and Schienle (2019) is given by

$$\hat{\boldsymbol{R}} = \arg\min_{\boldsymbol{R}} \left\| \Delta \boldsymbol{Z} - \boldsymbol{R}'\hat{\boldsymbol{Q}}'\boldsymbol{Z}_{-1} \right\|_2^2 + \lambda \sum_{j=1}^{N} \frac{\|\hat{\boldsymbol{r}}_j\|_2}{\hat{\mu}_j}, \tag{17.22}$$

where $\hat{\boldsymbol{R}} = (\hat{\boldsymbol{r}}_1, \ldots, \hat{\boldsymbol{r}}_N)$, with $\hat{\boldsymbol{r}}_j = (\hat{r}_{1,j}, \ldots, \hat{r}_{N,j})'$, $\|\hat{\boldsymbol{r}}_j\|_2 = \sqrt{\sum_{i=1}^N \hat{r}_{i,j}^2}$ and $\hat{\mu}_k = \sqrt{\sum_{i=k}^N \hat{r}_{k,i}^2}$. Furthermore, λ is a tuning parameter that controls the degree of regularization, with larger values resulting in more shrinkage. Weighting the penalty for each group by $\hat{\mu}_j$ puts a relatively higher penalty on groups for which the initial OLS estimates are small. The estimator clearly penalizes a set of predefined groups of coefficients, i.e. the columns of $\boldsymbol{R}$, and, therefore, is a variant of the group lasso for which numerous algorithms are available (e.g. Friedman, Hastie, & Tibshirani, 2010a; Meier, Van De Geer, & Bühlmann, 2008; Simon, Friedman, Hastie, & Tibshirani, 2013). The final estimate for the long-run coefficient matrix is obtained as $\hat{\boldsymbol{\Pi}} = \hat{\boldsymbol{R}}'\hat{\boldsymbol{Q}}'_{OLS}$.

The procedure detailed thus far focuses solely on estimation of the long-run relationships and requires an a priori choice of the lag order p. Furthermore, a necessary assumption is that initial OLS estimates are available, thereby restricting the admissible dimension of the system to $N(p+1) < T$. Within this restricted dimension, the short-run coefficient matrix $\boldsymbol{\Phi}$ can be consistently estimated by OLS and the corresponding lag order may be determined by standard information criteria such as the BIC. Alternatively, a second adaptive group lasso can be employed to obtain the regularized estimates $\hat{\boldsymbol{\Phi}} = \left(\hat{\boldsymbol{\Phi}}_1, \ldots, \hat{\boldsymbol{\Phi}}_p\right)$, see Liang and Schienle (2019, p. 425) for details. The lag order is then determined by the number of non-zero matrices $\hat{\boldsymbol{\Phi}}_i$ for $i \in \{1, \ldots, p\}$.

Wilms and Croux (2016) propose a penalized maximum likelihood estimator to estimate sparse VECMs. Instead of estimating the cointegrating rank and coefficient matrices for a fixed lag order, the method of Wilms & Croux enables joint estimation of the lag order and coefficient matrices for a given cointegrating rank. Additionally, the penalized maximum likelihood procedure does not require the availability of initial OLS estimates and, therefore, notwithstanding computational constraints, can

[9] As part of their theoretical contributions, Liang and Schienle (2019) show that the first r columns of $\hat{\boldsymbol{Q}}$ consistently estimate the space spanned by the cointegrating vectors $\boldsymbol{B}$ in (17.3), in an asymptotic framework where the dimension N is allowed to grow at rate $T^{1/4-\nu}$ for $\nu > 0$.

be applied to datasets of arbitrary dimension. Under the assumption of multivariate normality of the errors, i.e. $\boldsymbol{\epsilon}_t \sim \mathbb{N}(\mathbf{0}, \boldsymbol{\Sigma})$, the penalized negative log-likelihood is given by

$$\mathcal{L}(\boldsymbol{A}, \boldsymbol{B}, \boldsymbol{\Phi}, \boldsymbol{\Omega}) = \frac{1}{T}\text{tr}\left((\Delta \boldsymbol{Z} - \boldsymbol{A}\boldsymbol{B}'\boldsymbol{Z}_{-1} - \boldsymbol{\Phi}\Delta \boldsymbol{X})'\boldsymbol{\Omega}(\Delta \boldsymbol{Z} - \boldsymbol{A}\boldsymbol{B}'\boldsymbol{Z}_{-1} - \boldsymbol{\Phi}\Delta \boldsymbol{X})\right) \\ - \log|\boldsymbol{\Omega}| + \lambda_1 P_1(\boldsymbol{B}) + \lambda_2 P_2(\boldsymbol{\Phi}) + \lambda_3 P_3(\boldsymbol{\Omega}), \tag{17.23}$$

where $\boldsymbol{\Omega} = \boldsymbol{\Sigma}^{-1}$, and P_1, P_2 and P_3 being three penalty functions. The cointegrating vectors, short-run dynamics and covariance matrix are penalized as

$$P_1(\boldsymbol{B}) = \sum_{i=1}^{N}\sum_{j=1}^{r}|\beta_{i,j}|, \quad P_2(\boldsymbol{\Phi}) = \sum_{i=1}^{N}\sum_{j=1}^{Np}|\phi_{i,j}|, \quad P_3(\boldsymbol{\Omega}) = \sum_{i,j=1,i\neq j}^{N}|\omega_{i,j}|,$$

respectively. The use of L_1-penalization enables some elements to be estimated as exactly zero. The solution that minimizes (17.23) is obtained through an iterative updating scheme, where the solution for a coefficient matrix is obtained by minimizing the objective function conditional on the remaining coefficient matrices. The full algorithm is described in detail in Wilms and Croux (2016, pp. 1527–1528) and R code is provided by the authors online.[10]

Single-Equation Estimation

Frequently, the forecast exercise is aimed at forecasting a small number of time series based on a large number of potentially relevant variables. The means of data reduction thus far considered utilize either data aggregation or subset selection. However, in cases where the set of target variables is small, a substantial reduction in dimension can be obtained through the choice of appropriate single-equation representations for each variable separately.

Smeekes and Wijler (2018a) propose the penalized error correction selector (SPECS) as an automated single-equation modelling procedure on high-dimensional (co)integrated datasets. Assume that the N-dimensional observed time series admits the decomposition $z_t = (y_t, \boldsymbol{x}_t')'$, where y_t is the variable of interest and $\boldsymbol{x}_t$ are variables that are considered as potentially relevant in explaining the variation in y_t. Starting from the VECM system (17.20), a single-equation representation for Δy_t can be obtained by conditioning on the contemporaneous differences $\Delta \boldsymbol{x}_t$. This results in

$$\Delta y_t = \boldsymbol{\delta}' z_{t-1} + \boldsymbol{\pi}' \boldsymbol{w}_t + \epsilon_{y,t}, \tag{17.24}$$

[10] https://feb.kuleuven.be/public/u0070413/SparseCointegration/.

where $\boldsymbol{w}_t = (\Delta \boldsymbol{x}_t', \Delta \boldsymbol{z}_{t-1}', \ldots, \Delta \boldsymbol{z}_{t-p}')'$.[11] The number of parameters to be estimated in the single-equation model (17.24) is reduced to $N(p+2)-1$ as opposed to the original $N^2(p+1)$ parameters in (17.20). Nonetheless, for large N the total number of parameters may still be too large to estimate precisely by ordinary least squares, if possible at all. Therefore, Smeekes and Wijler propose a shrinkage procedure defined as

$$\hat{\boldsymbol{\delta}}, \hat{\boldsymbol{\pi}} = \arg\min_{\boldsymbol{\delta}, \boldsymbol{\pi}} \sum_{t=1}^{T} \left(\Delta y_t - \boldsymbol{\delta}' \boldsymbol{z}_{t-1} + \boldsymbol{\pi}' \boldsymbol{w}_t \right)^2 + P_\lambda(\boldsymbol{\delta}, \boldsymbol{\pi}). \tag{17.25}$$

The penalty function takes on the form

$$P_\lambda(\boldsymbol{\delta}, \boldsymbol{\pi}) = \lambda_G \|\boldsymbol{\delta}\| + \lambda_\delta \sum_{i=1}^{N} \omega_{\delta,i}^{k_\delta} |\delta_i| + \lambda_\pi \sum_{j=1}^{N(p+1)-1} \omega_{\pi,j}^{k_\pi} |\pi_j|, \tag{17.26}$$

where $\omega_{\delta,i}^{k_\delta} = 1/\left|\hat{\delta}_{Init,i}\right|^{k_\delta}$ and $\omega_{\pi,j}^{k_\pi} = 1/\left|\hat{\pi}_{Init,j}\right|^{k_\pi}$, with $\hat{\boldsymbol{\delta}}_{Init}$ and $\hat{\boldsymbol{\pi}}_{Init}$ being some consistent initial estimates, such as OLS or ridge estimates. The tuning parameters k_δ and k_π regulate the degree to which the initial estimates affect the penalty weights.

SPECS simultaneously employs individual penalties on all coefficients and a group penalty on $\boldsymbol{\delta}$, the implied cointegrating vector. Absent of cointegration, this cointegrating vector is equal to zero, in which case the group penalty promotes the removal of the lagged levels as a group.[12] In the presence of cointegration, however, the implied cointegrating vector may still contain many zero elements. The addition of the individual penalties allows for correct recovery of this sparsity pattern. This combination of penalties is commonly referred to as the sparse group lasso and R code is provided by the authors.[13]

In the single-equation model, the variation in y_t is explained by contemporaneous realizations of the conditioning variables $\boldsymbol{x}_t$. Therefore, forecasting the variable of interest requires forecasts for the latter as well, unless their realizations become available to the researcher prior to the realizations of y_t. SPECS is therefore highly suited to nowcasting applications. While not originally developed for the purpose of forecasting, direct forecasts with SPECS can be obtained by modifying the objective function to

$$\sum_{t=1}^{T} \left(\Delta_h y_t - \boldsymbol{\delta}' \boldsymbol{z}_{t-1} + \boldsymbol{\pi}' \boldsymbol{w}_t \right)^2 + P_\lambda(\boldsymbol{\delta}, \boldsymbol{\pi}),$$

[11] Details regarding the relationship between the components of the single-equation model (17.24) and the full system (17.3) are provided in Smeekes and Wijler (2018a, p. 5).

[12] From a theoretical point of view, the group penalty is not required for consistent selection and estimation of the non-zero coefficients.

[13] https://sites.google.com/view/etiennewijler/code?authuser=0.

where $\Delta_h y_t = y_{t+h} - y_t$. The direct h-step ahead forecast is then simply obtained as $\hat{y}_{T+h|T} = y_T + \hat{\boldsymbol{\delta}}' z_{T-1} + \hat{\boldsymbol{\pi}}' \boldsymbol{w}_T$.

17.5 Empirical Applications

In this section we evaluate the methods discussed in Sects. 17.3 and 17.4 in two empirical applications. First we forecast several US macroeconomic variables using the FRED-MD dataset of McCracken and Ng (2016). The FRED-MD dataset is a well-established and popular source for macroeconomic forecasting, and allows us to evaluate the methods in an almost controlled environment. Second we consider nowcasting Dutch unemployment using Google Trends data on frequencies of unemployment-related queries. This application not only highlights the potential of novel Big Datasets for macroeconomic purposes, but also puts the methods to the test in a more difficult environment where less theoretical guidance is available on the properties of the data.

17.5.1 Macroeconomic Forecasting Using the FRED-MD Dataset

We consider forecasting eight US macroeconomic variables from the FRED-MD dataset at 1, 6 and 12 months forecast horizons. We first focus on the strategy discussed in Sect. 17.3 where we first transform all series to I(0) before estimating the forecasting models. We illustrate the unit root testing methods, and show the empirical consequences of specification changes in the orders of integration. Next, we analyse the methods discussed in Sect. 17.4, and compare their forecast accuracy.

Transformations to Stationarity

As the FRED-MD series have already been classified by McCracken and Ng (2016), we have a benchmark for our own classification using the unit root testing methodology discussed in Sect. 17.3. We consider the autoregressive wild bootstrap as described in Sect. 17.3.2 in combination with the union test in (17.6). We set the AWB parameter γ equal to 0.85, which implies that over a year of serial dependence is captured by the bootstrap. Lag lengths in the ADF regressions are selected by the rescaled MAIC criterion of Cavaliere, Phillips, Smeekes, and Taylor (2015), which is robust to heteroskedasticity. To account for multiple testing, we control the false discovery rate at 5% using the bootstrap method of Romano et al. (2008a) (labelled as 'BFDR') and apply the sequential test procedure of Smeekes (2015) (labelled as 'BSQT') with a significance level of 5% and evenly spaced 0.05 quantiles such that

$p_k = [0.05(k-1)]$ for $k = 1, \ldots, 20$. We also perform the unit root tests on each series individually (labelled as 'iADF') with a significance level of 5%.

As some series in the FRED-MD are likely $I(2)$, we need to extend the methodology to detect these as well. We consider two ways to do so. First, we borrow information about the $I(2)$ series from the official FRED-MD classification, and take first differences of the series deemed to be $I(2)$. We then put these first differences together with the other series in levels and test for unit roots. This strategy ensures that the $I(2)$ series are classified at least as $I(1)$, and we only need to perform a single round of unit root testing. Our second approach is fully data-driven and follows a multivariate extension of the 'Pantula principle' (Pantula, 1989), where we first test for a unit root in the first difference of all series. The series for which the null cannot be rejected are classified as $I(2)$ and removed from the sample. The remaining series are then tested in levels and consequently classified as $I(1)$ or $I(0)$. In the results we append an acronym with a 1 if the first strategy is followed, and with a 2 if the second strategy is followed.[14]

As a final method, we include a 'naive' unit root testing approach that we believe is representative of casual unit root testing applied by many practitioners who, understandably, may not pay too much detailed attention to the unit root testing. In particular, we use the `adf.test` function from the popular R package 'tseries' (Trapletti & Hornik, 2018), and apply it with its default options, which implies performing individual ADF tests with a trend and setting a fixed lag length as a function of the sample size.[15] Our goal is not to discuss the merits of this particular unit root test procedure, but instead to highlight the consequences of casually using a 'standard' unit root test procedure that does not address the issues described in Sect. 17.3.

Figure 17.1 presents the found orders of integration. Globally the classification appears to agree among the different methods, which is comforting, although some important differences can be noted. First, none of the data-driven methods finds as many $I(2)$ series as the FRED classification does. Indeed, this may not be such a surprising result, as it remains a debated issue among practitioners whether these series should be modelled as $I(1)$ or as $I(2)$, see, for example, the discussion in Marcellino, Stock, and Watson (2006).

Second, although most methods yield fairly similar classifications, the clear outlier is BFDR2, which finds all series but one to be $I(1)$. The FDR controlling algorithm may, by construction, be too conservative in the early stages of the

[14]We take logarithmic transformations of the series before differencing when indicated by the official FRED-MD classification. Determining when a logarithmic transformation is appropriate is a daunting task for such a high-dimensional system as it seems difficult to automatize, especially as it cannot be seen separately from the determination of the order of integration (Franses & McAleer, 1998; Kramer & Davies, 2002). Klaassen, Kueck, and Spindler (2017) propose a high-dimensional method to determine an appropriate transformation model, but it is not trivial how to combine their method with unit root testing. Therefore we apply the 'true' transformations such that we can abstract from this issue.

[15]The lag length is set equal to $\lfloor (T-1)^{1/3} \rfloor$.

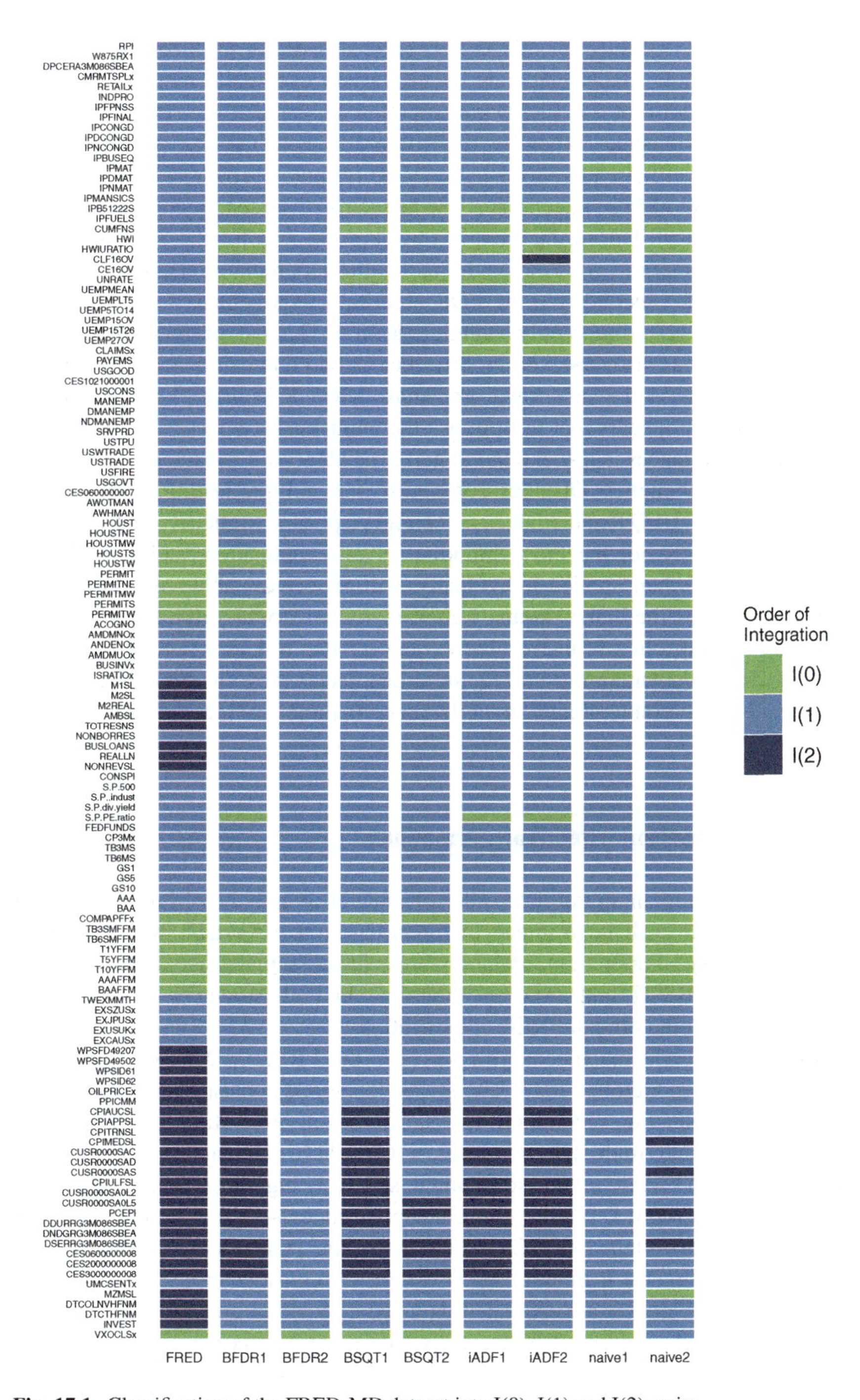

Fig. 17.1 Classification of the FRED-MD dataset into I(0), I(1) and I(2) series

algorithm when few rejections R have been recorded, yet too liberal in the final stages upon finding many rejections. Indeed, when testing the first differences of all series for a unit root, the FRED classification tells us that for most of the series the null can be rejected. When the algorithm arrives at the $I(2)$ series, the unit root hypothesis will already have been rejected for many series. With R being that large, the number of false rejections F can be relatively large too without increasing the FDR too much. Hence, incorrectly rejecting the null for the $I(2)$ series will fall within the 'margins of error' and thus lead to a complete rejection of all null hypotheses. In the second step the FDR algorithm then appears to get 'stuck' in the early stages, resulting in only a single rejection. This risk of the method getting stuck early on was also observed by Smeekes (2015) and can be explained by the fact that early on in the step-down procedure, when R is small, FDR is about as strict as FWE. It appears that in this case the inclusion of the $I(2)$ series in levels rather than differences is just enough to make the algorithm get stuck.

Third, even though iADF does not control for multiple testing, its results are fairly similar to BSQT and FDR1. It therefore appears explicitly controlling for multiple testing is not the most important in this application, and sensible unit root tests, even when applied individually, will give reasonable answers. On first glance even using the 'naive' strategy appears not be very harmful. However, upon more careful inspection of the results, we can see that it does differ from the other methods. In particular, almost no $I(2)$ series are detected by this strategy, and given that there is no reason to prefer it over the other methods, we recommend against its use.

Forecast Comparison After Transformations

While determining an appropriate order of integration may be of interest in itself, our goal here is to evaluate its impact on forecast accuracy. As such, we next evaluate if, and how, the chosen transformation impacts the actual forecast performance of the BFDR, BSQT and iADF methods, all in both strategies considered, in comparison with the official FRED classification.

We forecast eight macroeconomic series in the FRED-MD dataset using data from July 1972 to October 2018. The series of interest consist of four real series, namely real production income (RPI), total industrial production (INDPRO), real manufacturing and trade industries sales (CMRMTSPLx) and non-agricultural employees (PAYEMS), and four nominal series, being the producer index for finished goods (WPSFD49207), consumer price index—total (CPIAUCSL), consumer price index—less food (CPIULFSL) and the PCE price deflator (PCEPI). Each series is forecast h months ahead, where we consider the forecast horizons $h = 1, 6, 12$. All models are estimated on a rolling window spanning 10 years, i.e. containing 120 observations. Within each window, we regress every time series on a constant and linear trend and obtain the corresponding residuals. For the stationary methods, these residuals are transformed to stationarity according to the results of the unit root testing procedure. Each model is fitted to these transformed residuals, after which the h-step ahead forecast is constructed as an iterated one-step-ahead forecast, when possible, and transformed to levels, if needed. The final forecast is obtained by adding the level forecast of the transformed residuals to the forecast of the deterministic components. We briefly describe the implementation of each method below.

We consider four methods here. The first method is a standard vector autoregressive (VAR) model, fit on the eight variables of interest. Considering only the eight series of interest, however, may result in a substantial loss of relevant information contained in the remaining variables in the complete dataset. Therefore, we also consider a factor-augmented vector autoregressive model (FAVAR) in the spirit of Bernanke et al. (2005), which includes factors as proxies for this missing information. We extract four factors from the complete and transformed dataset and fit two separate FAVAR models containing these four factors, in addition to either the four real or the four nominal series. Rather than focusing on the estimation of heavily parameterized full systems, one may attempt to reduce the dimensionality by considering single-equation models, as discussed in Sect. 17.4.2. Conditioning the variable of interest on the remaining variables in the dataset results in an autoregressive distributed lag model with $M = N(p+1) - 1$ parameters. For large N, shrinkage may still be desirable. Therefore, we include a penalized autoregressive distributed lag model (PADL) in the comparison, which is based on the minimization of

$$\sum_{t=1}^{T} \left(y_t^h - \boldsymbol{\pi}' \boldsymbol{w}_t \right)^2 + \lambda \sum_{j=1}^{M} \omega_{\pi,j}^{k_\pi} \left| \pi_j \right|, \tag{17.27}$$

where

$$y_t^h = \begin{cases} y_{t+h} - y_t & \text{if } y_t \sim I(1), \\ y_{t+h} - y_t - \Delta y_t & \text{if } y_t \sim I(2). \end{cases} \tag{17.28}$$

Furthermore, $\boldsymbol{w}_t$ contains contemporaneous values of all transformed time series except y_t, and three lags of all transformed time series. The weights $\omega_{\pi,j}^{k_\pi}$ are as defined in Sect. 17.4.2. In essence, this can be seen as an implementation of SPECS with the built-in restriction that $\boldsymbol{\delta} = \mathbf{0}$, thereby ignoring cointegration. Finally, the concept of using factors as proxies for missing information remains equally useful for single-equation models. Accordingly, we include a factor-augmented penalized autoregressive distributed model (FAPADL) which is a single-equation model derived from a FAVAR. We estimate eight factors on the complete dataset, which are added to the eight variables of interest in the single-equation model. This is then estimated in accordance to (17.27), with $\boldsymbol{w}_t$ now containing contemporaneous values and three lags of the eight time series of interest and the eight factors. The PADL and FAPADL are variants of the adaptive lasso and we implement these in R based on the popular 'glmnet' package (Friedman, Hastie, & Tibshirani, 2010b).

The lag order for the VAR and FAVAR is chosen by the BIC criterion, with a maximum lag order of three.

Our goal is not to be exhaustive, but we believe these four methods cover a wide enough range of available high-dimensional forecast methods such that our results cannot be attributed to the choice of a particular forecasting method and instead genuinely reflect the effect of different transformations to stationarity. For the sake of space, we only report the results based on the FAVAR here for 1 month and 12 months ahead forecasts, as these are representative for the full set of results (which are available upon request). Generally, we find the same patterns within each method as we observe for the FAVAR, though they may be more or less pronounced. Overall the FAVAR is the most accurate of the four methods considered, which is why we choose to focus on it.

We compare the methods through their relative Mean Squared Forecast Errors (MSFEs), where the AR model is taken as benchmark. To attach a measure of statistical significance to these MSFEs, we obtain 90% Model Confidence Sets (MCS) of the best performing model. We obtain the MCS using the autoregressive wild bootstrap as in Smeekes and Wijler (2018b). For the full details on the MCS implementation we refer to that paper.

The results are given in Figs. 17.2 and 17.3. For the 1-month-ahead forecast the results are close for the different transformation methods, but for the 12-months-ahead forecasts, we clearly see big differences for the nominal series. Inspection of the classifications in Fig. 17.1 shows that the decisive factor is the classification of the dependent variable. For the three price series, the methods that classify these as $I(1)$ rather than $I(2)$ obtain substantial gains in forecast accuracy. Interestingly, the FRED classification finds these series to be $I(2)$, and thus deviating from the official classification can lead to substantial gains. These results are in line with the results of Marcellino et al. (2006), who also find that modelling price series as $I(1)$ rather than $I(2)$ results in better forecast accuracy.

As the outlying BFDR2 classification also classifies these series as I(1), this 'lucky shot' eclipses any losses from the missclassification of the other series. However, for the real series it can be observed that BFDR2 does indeed always perform somewhat worse than the other methods, although the MCS does not find it to be significant everywhere.

Concluding, missclassification of the order of integration can have an effect on the performance of high-dimensional forecasting methods. However, unless the dependent variable is missclassified, the high-dimensional nature of the data also ensures that this effect is smoothed out. On the other hand, correct classification of the dependent variable appears to be crucial, in particular regarding the classification as $I(1)$ versus $I(2)$.

Forecast Comparisons for Cointegration Methods

The forecast exercise for the methods that are able to take into account the cointegrating properties of the data proceeds along the same lines as in Sect. 17.5.1.

Fig. 17.2 MCS and relative MSFEs for 1-month horizon. Methods that are included in the MCS are depicted in blue and methods that are excluded from the MCS are depicted in red

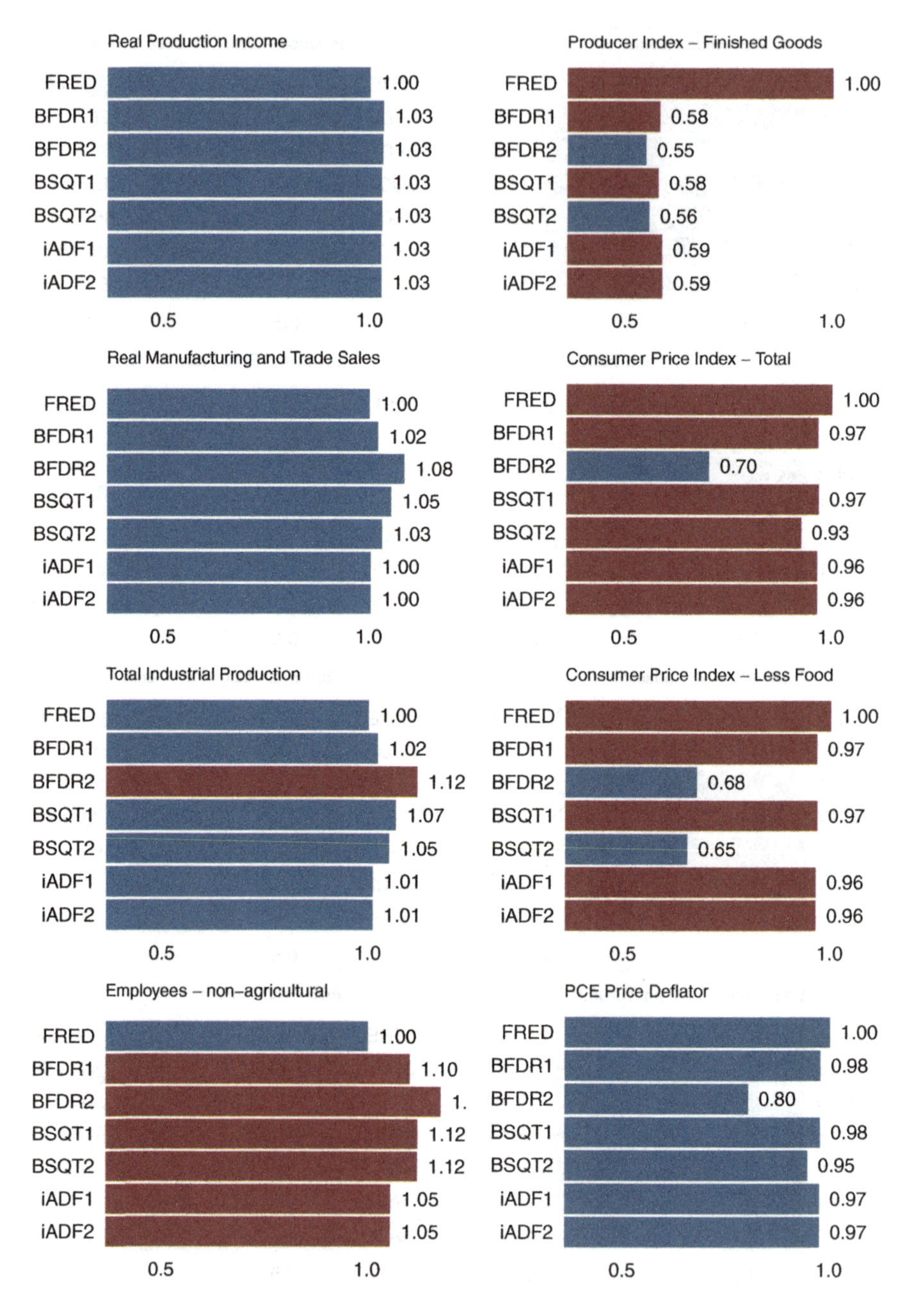

Fig. 17.3 MCS and relative MSFEs for 12-month horizon. Methods that are included in the MCS are depicted as blue and methods that are excluded from the MCS are depicted in red

A noteworthy exception is that the time series that are considered I(1) in the FRED-MD classification are now kept in levels, whereas those that are considered as I(2) are differenced once. The methods included in the comparison are: (i) the factor error correction model (FECM) by Banerjee et al. (2014, 2016, 2017), (ii) the non-stationary dynamic factor model (N-DFM) by Barigozzi et al. (2017, 2018), (iii) the maximum likelihood procedure (ML) by Johansen (1995b), (iv) the QR-decomposed VECM (QR-VECM) by Liang and Schienle (2019), (v) the penalized maximum likelihood (PML) by Wilms and Croux (2016), (vi) the single-equation penalized error correction selector (SPECS) by Smeekes and Wijler (2018a) and (vii) a factor-augmented SPECS (FASPECS). The latter method is simply the single-equation model derived from the FECM, based on the same principles as the FAPADL from the previous section. It is worth noting that the majority of these non-stationary methods have natural counterparts in the stationary world; the ML procedure compares directly to the VAR model, FECM compares to FAVAR and SPECS and FASPECS to PADL and FAPADL, respectively. Finally, all methods are compared against an AR model fit on the dependent variable, the latter being transformed according to the original FRED codes.

We briefly discuss some additional implementation choices for the non-stationary methods. For all procedures that require an estimate of the cointegrating rank, we use the information criteria proposed by Cheng and Phillips (2009). The only exception is the PML method, for which the cointegrating rank is determined by the procedure advocated in Wilms and Croux (2016). Similar to Banerjee et al. (2014), we do not rely on information criteria to select the number of factors, but rather fix the number of factors in the implementation of the FECM and N-DFM methods to four.[16] In the N-DFM approach, we model the idiosyncratic components of the target variables as simple AR models. The ML procedure estimates a VECM system on the eight variables of interest. In congruence with the implementation of the stationary methods, the lag order for FECM, N-DFM and ML is chosen by the BIC criterion, with a maximum lag order of three. The QR-VECM and PML methods are estimated on a dataset containing the eight series of interest and an additional 17 variables, informally selected based on their unique information within each economic category. Details are provided in Table 17.1. We incorporate only a single lag in the QR-VECM implementation, necessitated by the requirement of initial OLS estimates. SPECS estimates the model

$$y_t^h = \boldsymbol{\delta}' \boldsymbol{z}_{t-1} + \boldsymbol{\pi}' \boldsymbol{w}_t + \epsilon_{y,t},$$

where y_t^h is defined in (17.28), with the order of integration based on the original FRED codes. Note that the variables included in $\boldsymbol{z}_t$ are either the complete set of

[16] In untabulated results, we find that the forecast performance does not improve when the number of factors is selected by the information criteria by Bai (2004). Neither does the addition of a stationary factor computed from the estimated idiosyncratic component, in the spirit of Banerjee et al. (2014). Both strategies are therefore omitted from the analysis.

Table 17.1 Overview of the variables included for QR-VECM and PML

	FRED code	Description
Real	RPI	Real personal income
	CMRMTSPLx	Real manufacturing and trade industries sale
	INDPRO	IP index
	PAYEMS	All employees: total nonfarm
Nominal	WPSFD49207	PPI: finished goods
	CPIAUCSL	CPI: all items
	CPIULFSL	CPI: all items less food
	PCEPI	Personal cons. expend.: chain index
Additional	CUMFNS	Capacity utilization: manufacturing
	HWI	Help-Wanted Index for United States
	UNRATE	Civilian unemployment rate
	UEMPMEAN	Average duration of unemployment (weeks)
	HOUST	Housing starts: total new privately owned
	PERMIT	New private housing permits (SAAR)
	BUSINVx	Total business inventories
	M1SL	M1 money stock
	M2SL	M2 money stock
	FEDFUNDS	Effective federal funds rate
	TB3MS	3-month treasury bill
	GS5	5-year treasury rate
	GS10	10-year treasury rate
	EXJPUSx	Japan/US foreign exchange rate
	EXUSUKx	US/UK foreign exchange rate
	EXCAUSx	Canada/US foreign exchange rate
	S.P.500	S&P common stock price index: composite

124 time series or the eight time series of interest plus an additional eight estimated factors, depending on whether the implementation concerns SPECS or FASPECS, respectively. Finally, all parameters that regulate the degree of shrinkage are chosen by time series cross-validation, proposed by Hyndman and Athanasopoulos (2018) and discussed in a context similar to the current analysis in Smeekes and Wijler (2018b, p. 411).

Results are given in Figs. 17.4, 17.5 and 17.6. Considering first the 1-month ahead predictions, we observe similar forecasting performance on the first three real series (RPI, CMRMTSPLx, INDPRO) with almost none of the methods being excluded from the 90% model confidence set. The employment forecasts of the AR benchmark and the FAVAR approach are considered superior to those of the other methods. On the four nominal series, the sparse high-dimensional methods display relatively poor performance, regardless of whether they take into account potential cointegration in the data. Overall, no clear distinction is visible between the non-stationary and stationary methods, although this may not come as a surprise given

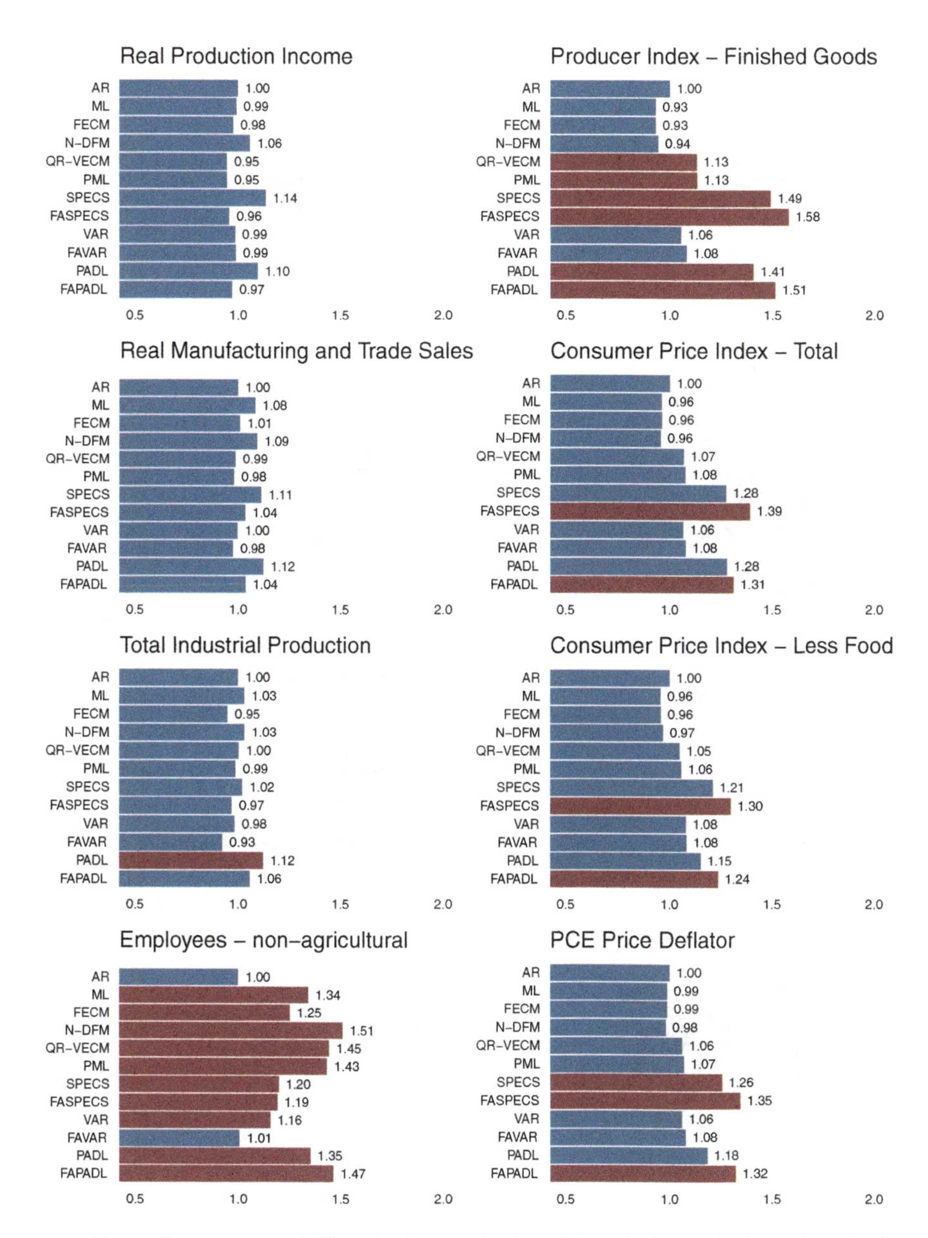

Fig. 17.4 MCS and relative MSFEs for 1-month horizon. Methods that are included in the MCS are depicted in blue and methods that are excluded from the MCS are depicted in red

the short forecast horizon. As usual, the AR benchmark appears hard to beat and is not excluded from any of the model confidence sets here.

The forecast comparisons for longer forecast horizons display stronger differentiation across methods. Our findings are qualitatively similar for the 6-month

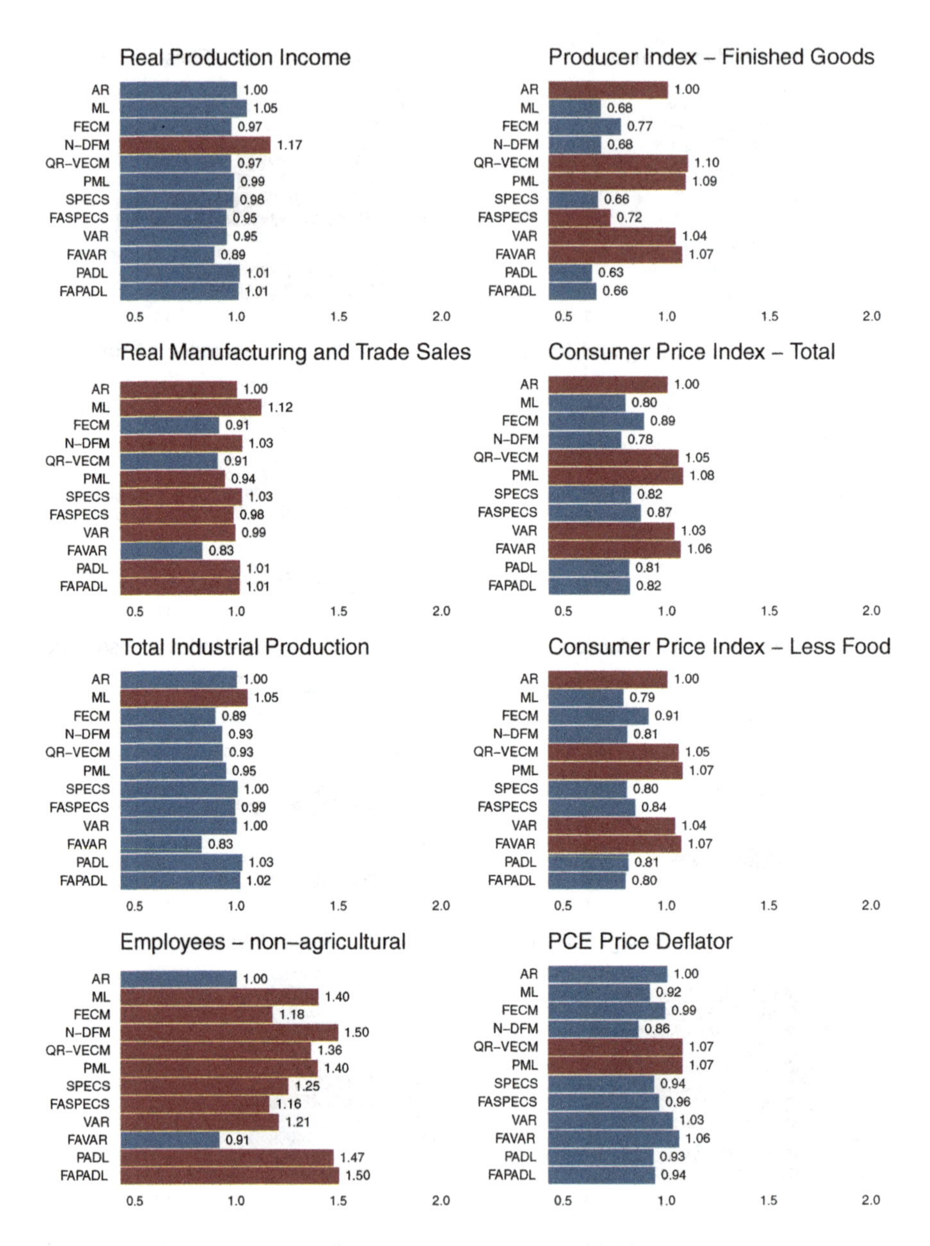

Fig. 17.5 MCS and relative MSFEs for 6-month horizon. Methods that are included in the MCS are depicted in blue and methods that are excluded from the MCS are depicted in red

and 12-month horizons, and, for the sake of brevity, we comment here on the 12-month horizon only. The results for the first three real series again do not portray a preference for taking into account cointegration versus transforming the data. Comparing ML to FECM and VAR to FAVAR, incorporating information across

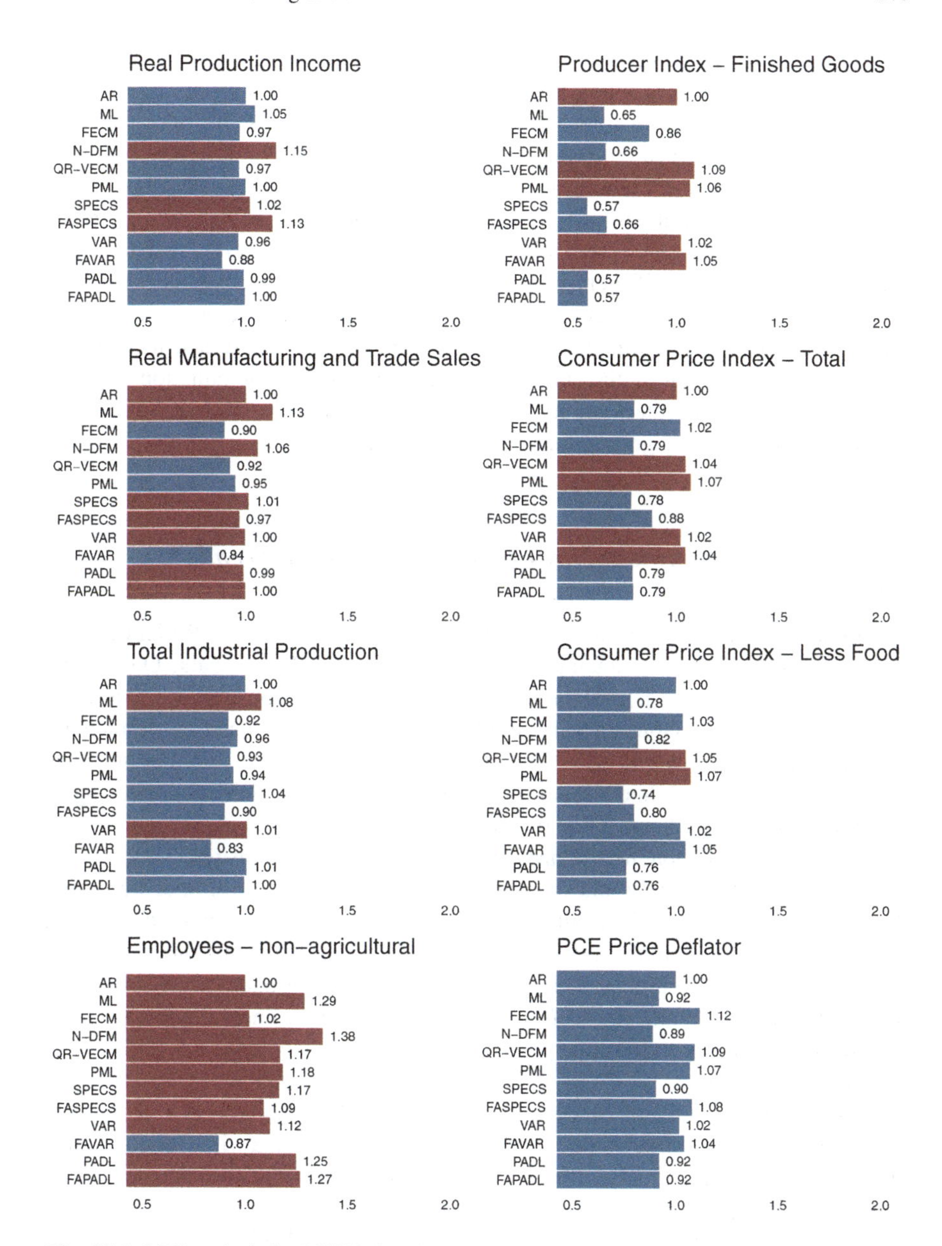

Fig. 17.6 MCS and relative MSFEs for 12-month horizon. Methods that are included in the MCS are depicted in blue and methods that are excluded from the MCS are depicted in red

the whole dataset seems to positively affect forecast performance, a finding that is additionally confirmed by the favourable performance of the penalized VECM estimators. The FAVAR substantially outperforms on the employment series, being the only method included in the model confidence set. On the nominal series, the single-equation methods perform well, again not showing any gain or loss in predictive power by accounting for cointegration. The ML and N-DFM procedure methods show favourable forecast accuracy as well, whereas the two penalized VECM estimators appear inferior on the nominal series. The AR benchmark is excluded for four out of eight series.

In summary, the comparative performance is strongly dependent on the choice of dependent variable and forecast horizon. For short forecast horizons, hardly any statistically significant differences in forecast accuracy are observed. However, for longer horizons the differences are more pronounced, with factor-augmented or penalized full system estimators performing well on the real series, the FAVAR strongly outperforming on the employment series and the single-equation methods appearing superior on the nominal series. The findings do not provide conclusive evidence whether cointegration matters for forecasting.

17.5.2 Unemployment Nowcasting with Google Trends

In this section we revisit the nowcasting application of Smeekes and Wijler (2018a), who consider nowcasting unemployment using Google Trends data. One of the advantages of modern Big Datasets is that information obtained from internet activity is often available on very short notice, and can be used to supplement official statistics produced by statistical offices. For instance, internet searches about unemployment-related issues may contain information about people being or becoming unemployed, and could be used to obtain unemployment estimates before statistical offices are able to produce official unemployment statistics.

Google records weekly and monthly data on the popularity of specific search terms through its publicly available Google Trends service,[17] with data being available only days after a period ends. On the other hand, national statistical offices need weeks to process surveys and produce official unemployment figures for the preceding month. As such, Google Trends data on unemployment-related queries would appear to have the potential to produce timely nowcasts of the latest unemployment figures.

Indeed, Schiavoni, Palm, Smeekes, and van den Brakel (2019) propose a dynamic factor model within a state space context to combine survey data with Google Trends data to produce more timely official unemployment statistics. They illustrate their method using a dataset of about one hundred unemployment-related queries in the Netherlands obtained from Google Trends. Smeekes and Wijler (2018a) consider a similar setup with the same Google Trends data, but consider the conceptually simpler setup where the dependent variable to be nowcasted is the official published

[17]https://trends.google.com/trends.

unemployment by Statistics Netherlands.[18] Moreover, they exclusively focus on penalized regression methods. In this section we revisit their application in the context of the methods discussed in this chapter. For full details on the dataset, which is available on the authors' websites, we refer to Smeekes and Wijler (2018a).

Transformations to Stationarity

As for the FRED-MD dataset, we first consider the different ways to classify the series into I(0), I(1) and I(2) series. However, unlike for the FRED data, here we do not have a pre-set classification available, and therefore unit root testing is a necessity before continuing the analysis. Moreover, as the dataset could easily be extended to an arbitrarily high dimension by simply adding other relevant queries, an automated fully data driven method is required.

This lack of a known classification also means that our first strategy as used in Sect. 17.5.1 has to be adapted, as we cannot differ I(2) series a priori. In particular, for our first strategy we assume that the series can be at most I(1), and hence we perform only a single unit root test on the levels of all series. Our second strategy is again the Pantula principle as in Sect. 17.5.1. Within each strategy we consider the same four tests as before.[19]

The classification results are given in Fig. 17.7. Generally they provide strong evidence that nearly all series are I(1), with most methods only finding very few I(0) and I(2) series. Interestingly, one of the few series that the methods disagree about is the unemployment series, which receives all three possible classifications. From our previous results we may expect this series, our dependent variable, to be the major determinant of forecast accuracy. Aside from this result, the most striking result is the performance of the naive tests that find many more I(0) variables than the other methods. One possible explanation for this result may be the nature of the Google Trends data that can exhibit large changes in volatility. As standard unit root tests are not robust to such changes, a naive strategy might seriously be affected, as appears to be the case here.

Forecast Comparison

We now compare the nowcasting performance of the high-dimensional methods. Given our focus on forecasting the present, that is, $h = 0$, for a single variable, there is little benefit in considering the system estimators we used before. Therefore we only consider the subset of single-equation models that allow for nowcasting.

[18] Additionally, this means the application does not require the use of the private survey data and is based on publicly available data only.

[19] As Google reports the search frequencies in relative terms (both to the past and other searches), we do not take logs anywhere.

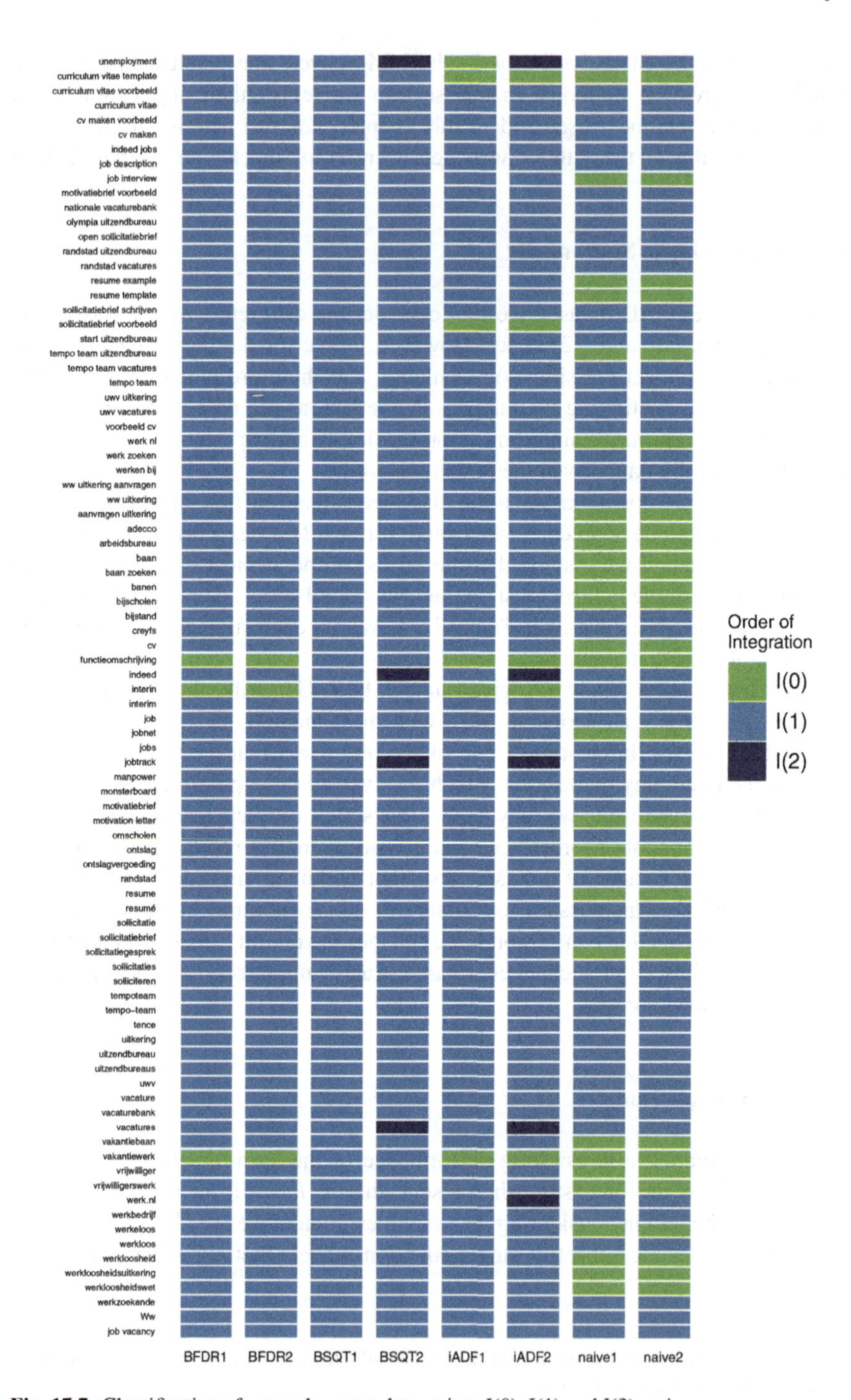

Fig. 17.7 Classification of unemployment dataset into I(0), I(1) and I(2) series

Specifically, we include SPECS as described in Sect. 17.4.2 as well as its modification FASPECS described in Sect. 17.5.1 as methods that explicitly account for unit roots and cointegration. Furthermore, we include PADL and FAPADL as described in Sect. 17.5.1. For all methods, the modification for nowcasting is done by setting $h = 0$, where we implicitly assume that at time t the values for the explanatory variables are available, but that for unemployment is not. This corresponds to the real-life situation.

For SPECS we model unemployment as (at most) $I(1)$, given that this is its predominant classification in Fig. 17.7. Additionally, we include all regressors in levels, thereby implicitly assuming these are at most $I(1)$ as well, which is again justified by the preceding unit root tests. For PADL and FAPADL we transform the series to stationarity according to the obtained classifications. Again we consider an AR model as benchmark, while all other implementational details are the same as in Sect. 17.5.1.

Our dataset covers monthly data from January 2004 until December 2017 for unemployment obtained from Statistics Netherlands, and 87 Google Trends series. We estimate the models on a rolling window of 100 observations each, leaving 64 time periods for obtaining nowcasts. We compare the nowcast accuracy through relative Mean Squared Nowcast Error (MSNE), with the AR model as benchmark, and obtain 90% Model Confidence Sets containing the best models in the same way as in Sect. 17.5.1.

Figure 17.8 presents the results. We see that, with the exception of the PADL—iADF1 method, all methods outperform the AR benchmark, although the 90% MCS does not find the differences to be significant. Factor augmentation generally leads to slightly more accurate forecasts than the full penalization approaches,

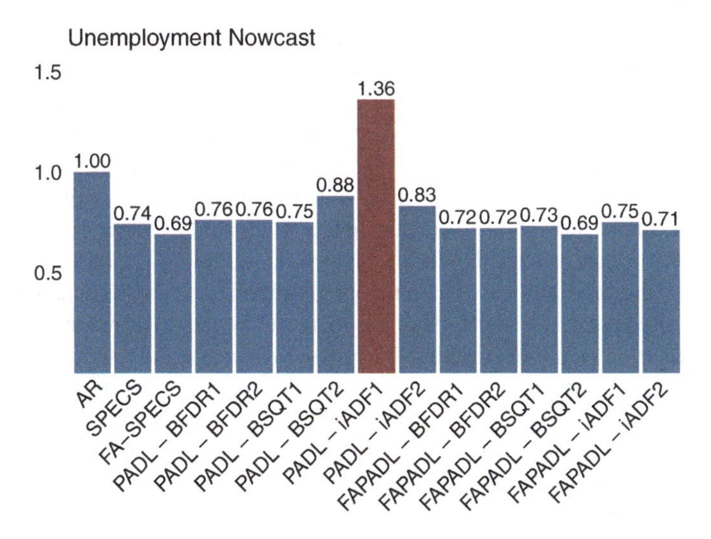

Fig. 17.8 MCS and relative MSNEs for the unemployment nowcasts. Methods that are included in the MCS are depicted in blue and methods that are excluded from the MCS are depicted in red

but differences are marginal. Interestingly, the classification of unemployment appears to only have a minor effect on the accuracy, with $I(0)$, $I(1)$ and $I(2)$ classifications all performing similarly. This does not necessarily contradict the results in Sect. 17.5.1, as differences were only pronounced there for longer forecast horizons, whereas the forecast horizon here is immediate. Finally, we observe that the SPECS methods are always at least as accurate as their counterparts that do not take cointegration into account. It therefore seems to pay off to allow for cointegration, even though differences are again marginal.

17.6 Conclusion

In this chapter we investigated how the potential presence of unit roots and cointegration impacts macroeconomic forecasting in the presence of Big Data. We considered both the strategies of transforming all data to stationarity, and of explicitly modelling any unit roots and cointegrating relationships.

The strategy of transforming to stationarity is commonly thought of as allowing one to bypass the unit root issue. However, this strategy is not innocuous as often thought, as it still relies on a correct classification of the orders of integration of all series. Given that this needs to be done for a large number of series, there is a lot of room for errors, and naive unit root testing is not advised. We discussed potential pitfalls for this classification, and evaluated methods designed to deal with issues of poor size and power of unit root tests, as well as controlling appropriate error rates in multiple testing.

Next we considered modelling unit roots and cointegration directly in a high-dimensional framework. We reviewed methods approaching the problem from two different philosophies, namely that of factor models and that of penalized regression. Within these philosophies we also highlighted differences among the proposed methods both in terms of underlying assumptions and implementation issues.

We illustrated these methods in two empirical applications: the first considered forecasting macroeconomic variables using the well-established FRED-MD dataset, while the second considered nowcasting unemployment using Google Trends data. Both applications showed that transforming to stationarity requires careful considerations of the methods used. While the specific method used for accounting for multiple testing generally only led to marginal differences, a correct classification of the variable to be forecasted is critically important. We therefore recommend paying specific attention to these variables. Moreover, as occasionally some methods can deliver strange results, in general it is advisable to perform the classification using multiple approaches, to ensure that the classification found is credible.

The applications also demonstrated that there is no general way to model cointegration that is clearly superior. Indeed, the results do not show a clear conclusion on whether cointegration should be taken into account. This result, perhaps unsurprisingly, mirrors the literature on low-dimensional time series. It therefore remains up to the practitioner to decide for their specific application if, and if yes how, cointegration should be modelled for forecasting purposes. Overall, the methods we consider in this chapter provide reliable tools to do so, should the practitioner wish to do so.

Concluding, several reliable tools are available for dealing with unit roots and cointegration in a high-dimensional forecasting setting. However, there is no panacea; a single best approach that is applicable in all settings does not exist. Instead, dealing with unit roots and cointegration in practice requires careful consideration and investigation which methods are most applicable in a given particular application. We also note that the field is rapidly developing, and major innovations are still to be expected in the near future. For instance, interval or density forecasting in high-dimensional systems with unit roots remains an entirely open issue. As high-dimensional inference is already complicated by issues such as post-selection bias, extending this to the unit root setting is very challenging indeed. Such tools however will be indispensable for the macroeconomic practitioner, and therefore constitute an exciting avenue for future research.

References

Ahn, S. C., & Horenstein, A. R. (2013). Eigenvalue ratio test for the number of factors. *Econometrica, 81*(3), 1203–1227.

Alessi, L., Barigozzi, M., & Capasso, M. (2010). Improved penalization for determining the number of factors in approximate factor models. *Statistics & Probability Letters, 80*(23), 1806–1813.

Bai, J. (2004). Estimating cross-section common stochastic trends in nonstationary panel data. *Journal of Econometrics, 122*(1), 137–183.

Bai, J., & Ng, S. (2002). Determining the number of factors in approximate factor models. *Econometrica, 70*(1), 191–221.

Bai, J., & Ng, S. (2004). A panic attack on unit roots and cointegration. *Econometrica, 72*(4), 1127–1177.

Banerjee, A., Marcellino, M., & Masten, I. (2014). Forecasting with factor-augmented error correction models. *International Journal of Forecasting, 30*(3), 589–612.

Banerjee, A., Marcellino, M., & Masten, I. (2016). An overview of the factor augmented error-correction model. In E. Hillebrand & S. J. Koopman (Eds.), *Dynamic factor models* (Chap. 1, Vol. 35, pp. 3–41). Advances in Econometrics. Bingley: Emerald Group Publishing Limited.

Banerjee, A., Marcellino, M., & Masten, I. (2017). Structural FECM: Cointegration in large-scale structural FAVAR models. *Journal of Applied Econometrics, 32*(6), 1069–1086.

Barigozzi, M., Lippi, M., & Luciani, M. (2017). *Dynamic factor models, cointegration, and error correction mechanisms* (arXiv e-prints No. 1510.02399).

Barigozzi, M., Lippi, M., & Luciani, M. (2018). *Non-stationary dynamic factor models for large datasets* (arXiv e-prints No. 1602.02398).

Barigozzi, M., & Trapani, L. (2018). *Determining the dimension of factor structures in non-stationary large datasets* (arXiv e-prints No. 1806.03647).

Benjamini, Y., & Hochberg, Y. (1995). Controlling the false discovery rate: A practical and powerful approach to multiple testing. *Journal of the Royal Statistical Society: Series B, 57*(1), 289–300.

Bernanke, B., Boivin, J., & Eliasz, P. S. (2005). Measuring the effects of monetary policy: A factor-augmented vector autoregressive (FAVAR) approach. *The Quarterly Journal of Economics, 120*(1), 387–422.

Callot, L. A., & Kock, A. B. (2014). Oracle efficient estimation and forecasting with the adaptive lasso and the adaptive group lasso in vector autoregressions. *Essays in Nonlinear Time Series Econometrics*, 238–268.

Cavaliere, G. (2005). Unit root tests under time-varying variances. *Econometric Reviews, 23*(3), 259–292.

Cavaliere, G., Phillips, P. C. B., Smeekes, S., & Taylor, A. M. R. (2015). Lag length selection for unit root tests in the presence of nonstationary volatility. *Econometric Reviews, 34*(4), 512–536.

Cavaliere, G., & Taylor, A. M. R. (2008). Bootstrap unit root tests for time series with nonstationary volatility. *Econometric Theory, 24*(1), 43–71.

Cavaliere, G., & Taylor, A. M. R. (2009). Bootstrap *M* unit root tests. *Econometric Reviews, 28*(5), 393–421.

Cheng, X., & Phillips, P. C. B. (2009). Semiparametric cointegrating rank selection. *Econometrics Journal, 12*(suppl1), S83–S104.

Choi, I. (2015). *Almost all about unit roots: Foundations, developments, and applications*. Cambridge: Cambridge University Press.

Chortareas, G., & Kapetanios, G. (2009). Getting PPP right: Identifying mean-reverting real exchange rates in panels. *Journal of Banking and Finance, 33*(2), 390–404.

Christoffersen, P. F., & Diebold, F. X. (1998). Cointegration and long-horizon fore-casting. *Journal of Business & Economic Statistics, 16*(4), 450–456.

Clements, M. P., & Hendry, D. F. (1995). Forecasting in cointegrated systems. *Journal of Applied Econometrics, 10*(2), 127–146.

De Mol, C., Giannone, D., & Reichlin, L. (2008). Forecasting using a large number of predictors: Is Bayesian shrinkage a valid alternative to principal components? *Journal of Econometrics, 146*, 318–328.

Dickey, D. A., & Fuller, W. A. (1979). Distribution of estimators for autoregressive time series with a unit root. *Journal of the American Statistical Association, 74*(366a), 427–431.

Diebold, F. X., & Kilian, L. (2000). Unit-root tests are useful for selecting forecasting models. *Journal of Business & Economic Statistics, 18*(3), 265–273.

Elliott, G., Rothenberg, T. J., & Stock, J. H. (1996). Efficient tests for an autoregressive unit root. *Econometrica, 64*(4), 813–836.

Enders, W. (2008). *Applied econometric time series* (4th ed.). New Delhi: Wiley.

Forni, M., Hallin, M., Lippi, M., & Reichlin, L. (2005). The generalized dynamic factor model: One-sided estimation and forecasting. *Journal of the American Statistical Association, 100*(471), 830–840.

Franses, P. H., & McAleer, M. (1998). Testing for unit roots and non-linear transformations. *Journal of Time Series Analysis, 19*(2), 147–164.

Friedman, J., Hastie, T., & Tibshirani, R. (2010a). *A note on the group lasso and a sparse group lasso* (arXiv e-prints No. 1001.0736).

Friedman, J., Hastie, T., & Tibshirani, R. (2010b). Regularization paths for generalized linear models via coordinate descent. *Journal of Statistical Software, 33*(1), 1–22.

Friedrich, M., Smeekes, S., & Urbain, J.-P. (2018). *Autoregressive wild bootstrap inference for nonparametric trends* (arXiv e-prints No. 1807.02357).

Gonçalves, S., & Perron, B. (2014). Bootstrapping factor-augmented regression models. *Journal of Econometrics, 182*(1), 156–173.

Hallin, M., & Liška, R. (2007). Determining the number of factors in the general dynamic factor model. *Journal of the American Statistical Association, 102*(478), 603–617.

Hamilton, J. D. (1994). *Time series analysis*. Princeton: Princeton University Press.

Hanck, C. (2009). For which countries did PPP hold? A multiple testing approach. *Empirical Economics, 37*(1), 93–103.

Harvey, D. I., Leybourne, S. J., & Taylor, A. M. R. (2009). Unit root testing in practice: Dealing with uncertainty over the trend and initial condition. *Econometric Theory, 25*(3), 587–636.

Harvey, D. I., Leybourne, S. J., & Taylor, A. M. R. (2012). Testing for unit roots in the presence of uncertainty over both the trend and initial condition. *Journal of Econometrics, 169*(2), 188–195.

Hyndman, R. J., & Athanasopoulos, G. (2018). *Forecasting: Principles and practice*. OTexts.

Johansen, S. (1995a). A statistical analysis of cointegration for i(2) variables. *Econometric Theory, 11*(1), 25–59.

Johansen, S. (1995b). *Likelihood-based inference in cointegrated vector autoregressive models*. Oxford: Oxford University Press.

Justiniano, A., & Primiceri, G. (2008). The time-varying volatility of macroeconomic fluctuations. *American Economic Review, 98*(3), 604–641.

Klaassen, S., Kueck, J., & Spindler, M. (2017). *Transformation models in high-dimensions* (arXiv e-prints No. 1712.07364).

Kock, A. B. (2016). Consistent and conservative model selection with the adaptive lasso in stationary and nonstationary autoregressions. *Econometric Theory, 32*, 243–259.

Kramer, W., & Davies, L. (2002). Testing for unit roots in the context of misspecified logarithmic random walks. *Economics Letters, 74*(3), 313–319.

Liang, C., & Schienle, M. (2019). Determination of vector error correction models in high dimensions. *Journal of Econometrics, 208*(2), 418–441.

Liao, Z., & Phillips, P. C. B. (2015). Automated estimation of vector error correction models. *Econometric Theory, 31*(3), 581–646.

Marcellino, M., Stock, J. H., & Watson, M. W. (2006). A comparison of direct and iterated multistep AR methods for forecasting macroeconomic time series. *Journal of Econometrics, 135*(2), 499–526.

McCracken, M. W., & Ng, S. (2016). FRED-MD: A monthly database for macroeconomic research. *Journal of Business & Economic Statistics, 34*(4), 574–589.

Meier, L., Van De Geer, S., & Bühlmann, P. (2008). The group lasso for logistic regression. *Journal of the Royal Statistical Society: Series B, 70*(1), 53–71.

Moon, H. R., & Perron, B. (2012). Beyond panel unit root tests: Using multiple testing to determine the non stationarity properties of individual series in a panel. *Journal of Econometrics, 169*(1), 29–33.

Müller, U. K., & Elliott, G. (2003). Tests for unit roots and the initial condition. *Econometrica, 71*(4), 1269–1286.

Ng, S. (2008). A simple test for nonstationarity in mixed panels. *Journal of Business and Economic Statistics, 26*(1), 113–127.

Onatski, A. (2010). Determining the number of factors from empirical distribution of eigenvalues. *The Review of Economics and Statistics, 92*(4), 1004–1016.

Onatski, A., & Wang, C. (2018). Alternative asymptotics for cointegration tests in large VARs. *Econometrica, 86*(4), 1465–1478.

Palm, F. C., Smeekes, S., & Urbain, J.-P. (2008). Bootstrap unit root tests: Comparison and extensions. *Journal of Time Series Analysis, 29*(1), 371–401.

Palm, F. C., Smeekes, S., & Urbain, J.-P. (2011). Cross-sectional dependence robust block bootstrap panel unit root tests. *Journal of Econometrics, 163*(1), 85–104.

Pantula, S. G. (1989). Testing for unit roots in time series data. *Econometric Theory, 5*(2), 256–271.

Pedroni, P., Vogelsang, T. J., Wagner, M., & Westerlund, J. (2015). Nonparametric rank tests for non-stationary panels. *Journal of Econometrics, 185*(2), 378–391.

Rho, Y., & Shao, X. (2019). Bootstrap-assisted unit root testing with piecewise locally stationary errors. *Econometric Theory, 35*(1), 142–166.

Romano, J. P., Shaikh, A. M., & Wolf, M. (2008a). Control of the false discovery rate under dependence using the bootstrap and subsampling. *Test, 17*(3), 417–442.

Romano, J. P., Shaikh, A. M., & Wolf, M. (2008b). Formalized data snooping based on generalized error rates. *Econometric Theory, 24*(2), 404–447.

Romano, J. P., & Wolf, M. (2005). Stepwise multiple testing as formalized data snooping. *Econometrica, 73*(4), 1237–1282.

Schiavoni, C., Palm, F., Smeekes, S., & van den Brakel, J. (2019). *A dynamic factor model approach to incorporate big data in state space models for official statistics* (arXiv e-print No. 1901.11355).

Schwert, G. W. (1989). Tests for unit roots: A Monte Carlo investigation. *Journal of Business and Economic Statistics, 7*(1), 147–159.

Shao, X. (2010). The dependent wild bootstrap. *Journal of the American Statistical Association, 105*(489), 218–235.

Simon, N., Friedman, J., Hastie, T., & Tibshirani, R. (2013). A sparse-group lasso. *Journal of Computational and Graphical Statistics, 22*(2), 231–245.

Smeekes, S. (2015). Bootstrap sequential tests to determine the order of integration of individual units in a time series panel. *Journal of Time Series Analysis, 36*(3), 398–415.

Smeekes, S., & Taylor, A. M. R. (2012). Bootstrap union tests for unit roots in the presence of nonstationary volatility. *Econometric Theory, 28*(2), 422–456.

Smeekes, S., & Urbain, J.-P. (2014a). *A multivariate invariance principle for modified wild bootstrap methods with an application to unit root testing* (GSBE Research Memorandum No. RM/14/008). Maastricht University.

Smeekes, S., & Urbain, J.-P. (2014b). On the applicability of the sieve bootstrap in time series panels. *Oxford Bulletin of Economics and Statistics, 76*(1), 139–151.

Smeekes, S., & Wijler, E. (2018a). *An automated approach towards sparse single-equation cointegration modelling* (arXiv e-print No. 1809.08889).

Smeekes, S., & Wijler, E. (2018b). Macroeconomic forecasting using penalized regression methods. *International Journal of Forecasting, 34*(3), 408–430.

Stock, J. H., & Watson, M. W. (2002a). Forecasting using principal components from a large number of predictors. *Journal of the American Statistical Association, 97*(460), 1167–1179.

Stock, J. H., & Watson, M. W. (2002b). Macroeconomic forecasting using diffusion indexes. *Journal of Business & Economic Statistics, 20*(2), 147–162.

Stock, J. H., & Watson, M. W. (2003). Has the business cycle changed and why? In M. Gertler & K. Rogoff (Eds.), *NBER macroeconomics annual 2002* (Chap. 4, Vol. 17, pp. 159–230). Cambridge: MIT Press.

Stock, J. H., & Watson, M. W. (2012). Generalized shrinkage methods for forecasting using many predictors. *Journal of Business & Economic Statistics, 30*, 481–493.

Tibshirani, R. (1996). Regression shrinkage and selection via the Lasso. *Journal of the Royal Statistical Society (Series B), 58*(1), 267–288.

Trapani, L. (2013). On bootstrapping panel factor series. *Journal of Econometrics, 172*(1), 127–141.

Trapletti, A., & Hornik, K. (2018). *Tseries: Time series analysis and computational finance*. R package version 0.10-46. Retrieved from https://CRAN.R-project.org/package=tseries

Wilms, I., & Croux, C. (2016). Forecasting using sparse cointegration. *International Journal of Forecasting, 32*(4), 1256–1267.

Zhang, R., Robinson, P., & Yao, Q. (2018). Identifying cointegration by eigenanalysis. *Journal of the American Statistical Association, 114*, 916–927.

Chapter 18
Turning Points and Classification

Jeremy Piger

18.1 Introduction

It is common in studies of economic time series for each calendar time period to be categorized as belonging to one of some fixed number of recurrent regimes. For example, months and quarters of macroeconomic time series are separated into periods of recession and expansion, and time series of equity returns are divided into bull vs. bear market regimes. Other examples include time series measuring the banking sector, which can be categorized as belonging to normal vs. crises regimes, and time series of housing prices, for which some periods might be labeled as arising from a "bubble" regime. A key feature of these regimes in most economic settings is that they are thought to be persistent, meaning the probability of each regime occurring increases once the regime has become active.

In many cases of interest, the regimes are never explicitly observed. Instead, the historical timing of regimes is inferred from time series of historical economic data. For example, in the USA, the National Bureau of Economic Research (NBER) Business Cycle Dating Committee provides a chronology of business cycle expansion and recession dates developed from the study of local minima and maxima of many individual time series. Because the NBER methodology is not explicitly formalized, a literature has worked to develop and evaluate formal statistical methods for establishing the historical dates of economic recessions and expansions in both US and international data. Examples include Hamilton (1989), Vishwakarma (1994), Chauvet (1998), Harding and Pagan (2006), Fushing, Chen, Berge, and Jordá (2010), Berge and Jordá (2011), and Stock and Watson (2014).

J. Piger
Department of Economics, University of Oregon, Eugene, OR, USA
e-mail: jpiger@uoregon.edu

P. Fuleky (ed.), *Macroeconomic Forecasting in the Era of Big Data*,
Advanced Studies in Theoretical and Applied Econometrics 52,
https://doi.org/10.1007/978-3-030-31150-6_18

In this chapter I am also interested in determining which regime is active based on information from economic time series. However, the focus is on real-time prediction rather than historical classification. Specifically, the ultimate goal will be to identify the active regime toward the end of the observed sample period (nowcasting) or after the end of the observed sample period (forecasting). Most of the prediction techniques I consider will take as given a historical categorization of regimes, and will use this categorization to learn the relationship between predictor variables and the occurrence of alternative regimes. This learned relationship will then be exploited in order to classify time periods that have not yet been assigned to a regime. I will also be particularly interested in the ability of alternative prediction techniques to identify turning points, which mark the transition from one regime to another. When regimes are persistent, so that there are relatively few turning points in a sample, it is quite possible for a prediction technique to have good average performance for identifying regimes, but consistently make errors in identifying regimes around turning points.

Consistent with the topic of this book, I will place particular emphasis in this chapter on the case where regime predictions are formed in a data-rich environment. In our specific context, this environment will be characterized by the availability of a large number of time-series predictor variables from which we can infer regimes. In typical language, our predictor dataset will be a "wide" dataset. Such datasets create issues when building predictive models, since it is difficult to exploit the information in the dataset without overfitting, which will ultimately lead to poor out-of-sample predictions.

The problem of regime identification discussed above is an example of a statistical classification problem, for which there is a substantial literature outside of economics. I will follow the tradition of that literature and refer to the regimes as "classes," the task of inferring classes from economic indicators as "classification," and a particular approach to classification as a "classifier." Inside of economics, there is a long history of using parametric models, such as a logit or probit, as classifiers. For example, many studies have used logit and probit models to predict US recessions, where the model is estimated over a period for which the NBER business cycle chronology is known. A small set of examples from this literature include Estrella and Mishkin (1998), Estrella, Rodrigues, and Schich (2003), Kauppi and Saikkonen (2008), Rudebusch and Williams (2009), and Fossati (2016). Because they use an available historical indicator of the class to estimate the parameters of the model, such approaches are an example of what is called a "supervised" classifier in the statistical classification literature. This is in contrast to "unsupervised classifiers," which endogenously determine clustering of the data, and thus endogenously determine the classes. Unsupervised classifiers have also been used for providing real-time nowcasts and forecasts of US recessions, with the primary example being the Markov-switching framework of Hamilton (1989). Chauvet (1998) proposes a dynamic factor model with Markov-switching (DFMS) to identify expansion and recession phases from a group of coincident indicators, and Chauvet and Hamilton (2006), Chauvet and Piger (2008), and Camacho, Perez-Quiros, and Poncela (2018) evaluate the performance of variants of this DFMS model to identify NBER turning points in real time. An important feature of Markov-switching models is that they explicitly model the persistence of the regime, by assuming the regime indicator follows a Markov process.

Recently, a number of authors have applied machine learning techniques commonly used outside of economics to classification problems involving time series of economic data. As an example, Qi (2001), Ng (2014), Berge (2015), Davig and Smalter Hall (2016), Garbellano (2016), and Giusto and Piger (2017) have applied machine learning techniques such as artificial neural networks, boosting, Naïve Bayes, and learning vector quantization to forecasting and nowcasting US recessions, while Ward (2017) used random forests to identify periods of financial crises. These studies have generally found improvements from the use of the machine learning algorithms over commonly used alternatives. For example, Berge (2015) finds that the performance of boosting algorithms improves on equal weight model averages of recession forecasts produced by logistic models, while Ward (2017) finds a similar result for forecasts of financial crises produced by a random forest.

Machine learning algorithms are particularly attractive in data-rich settings. Such algorithms typically have one or more "regularization" mechanisms that trade off in-sample fit against model complexity, which can help prevent overfitting. These algorithms are generally also fit using techniques that explicitly take into account out-of-sample performance, most typically using cross-validation. This aligns the model fitting stage with the ultimate goal of out-of-sample prediction, which again can help prevent overfitting. A number of these methods also have built-in mechanisms to conduct model selection jointly with estimation in a fully automated procedure. This provides a means to target relevant predictors from among a large set of possible predictors. Finally, these methods are computationally tractable, making them relatively easy to apply to large datasets.

In this chapter, I survey a variety of popular off-the-shelf supervised machine learning classification algorithms for the purpose of classifying economic regimes in real time using time-series data. Each classification technique will be presented in detail, and its implementation in the R programming language will be discussed.[1] Particular emphasis will be placed on the use of these classifiers in data-rich environments. I will also present DFMS models as an alternative to the machine learning classifiers in some settings. Finally, each of the various classifiers will be evaluated for their real-time performance in identifying US business cycle turning points from 2000 to 2018.

As discussed above, an existing literature in economics uses parametric logit and probit models to predict economic regimes. A subset of this literature has utilized these models in data-rich environments. For example, Owyang, Piger, and Wall (2015) and Berge (2015) use model averaging techniques with probit and logit models to utilize the information in wide datasets, while Fossati (2016) uses principal components to extract factors from wide datasets to use as predictor

[1] http://www.R-project.org/.

variables in probit models. I will not cover these techniques here, instead opting to provide a resource for machine learning methods, which hold great promise, but have received less attention in the literature to date.

The remainder of this chapter proceeds as follows: Section 18.2 will formalize the forecasting problem we are interested in and describe metrics for evaluating class forecasts. Section 18.3 will then survey the details of a variety of supervised machine learning classifiers and their implementation, while Sect. 18.4 will present details of DFMS models for the purpose of classification. Section 18.5 will present an application to real-time nowcasting of US business cycle turning points. Section 18.6 concludes.

18.2 The Forecasting Problem

In this section I lay out the forecasting problem of interest, as well as establish notation used for the remainder of this chapter. I also describe some features of economic data that should be recognized when conducting a classification exercise. Finally, I detail common approaches to evaluating the quality of class predictions.

18.2.1 Real-Time Classification

Our task is to develop a prediction of whether an economic entity in period $t+h$ is (or will be) in each of a discrete number (C) of classes. Define a discrete variable $S_{t+h} \in \{1, \ldots, C\}$ that indicates the active class in period $t+h$. It will also be useful to define C binary variables $S^c_{t+h} = I\left(S_{t+h} = c\right)$, where $I\left(\right) \in \{0, 1\}$ is an indicator function, and $c = 1, \ldots, C$.

Assume that we have a set of N predictors to draw inference on S_{t+h}. Collect these predictors measured at time t in the vector X_t, with an individual variable inside this vector labeled $X_{j,t}$, $j = 1, \ldots, N$. Note that X_t can include both contemporaneous values of variables as well as lags. I define a classifier as $\widehat{S^c_{t+h}}\left(X_t\right)$, where this classifier produces a prediction of S^c_{t+h} conditional on X_t. These predictions will take the form of conditional probabilities of the form $\Pr\left(S_{t+h} = c|X_t\right)$. Note that a user of these predictions may additionally be interested in binary predictions of S^c_{t+h}. To generate a binary prediction we would combine our classifier $\widehat{S^c_{t+h}}\left(X_t\right)$ with a rule, $L\left(\right)$, such that $L\left(\widehat{S^c_{t+h}}\left(X_t\right)\right) \in \{0, 1\}$. Finally, assume we have available T observations on X_t and S_{t+h}, denoted as $\{X_t, S_{t+h}\}_{t=1}^{T}$. I will refer to this in-sample period as the "training sample." This training sample is used to determine the parameters of the classifier, and I refer to this process as "training" the classifier. Once trained, a classifier can then be used to forecast S^c_{t+h} outside of the training sample. Specifically, given an X_{T+q}, we can predict S^c_{T+q+h} using $\widehat{S^c_{T+q+h}}\left(X_{T+q}\right)$ or $L\left(\widehat{S^c_{T+q+h}}\left(X_{T+q}\right)\right)$.

I will also be interested in the prediction of turning points, which mark the transition from one class to another. The timely identification of turning points in economic applications is often of great importance, as knowledge that a turning point has already occurred, or will in the future, can lead to changes in behavior on the part of firms, consumers, and policy makers. As an example, more timely information suggesting the macroeconomy has entered a recession phase should lead to quicker action on the part of monetary and fiscal policymakers, and ultimately increased mitigation of the effects of the recession. In order to predict turning points we will require another rule to convert sequences of $\widehat{S^c_{t+h}}(X_t)$ into turning point predictions. Of course, how cautious one is in converting probabilities to turning point predictions is determined by the user's loss function, and in particular the relative aversion to false positives. In the application presented in Sect. 18.5, I will explore the real-time performance of a specific strategy for nowcasting turning points between expansion and recession phases in the macroeconomy.

18.2.2 Classification and Economic Data

As is clear from the discussion above, we are interested in this chapter in classification in the context of time-series data. This is in contrast to much of the broader classification literature, which is primarily focused on classification in cross-sectional datasets, where the class is reasonably thought of as independent across observations. In most economic time series, the relevant class is instead characterized by time-series persistence, such that a class is more likely to continue if it is already operational than if it is not. In this chapter, I survey a variety of off-the-shelf machine learning classifiers, most of which do not explicitly model persistence in the class indicator. In cases where the forecast horizon h is reasonably long, ignoring persistence of the class is not likely to be an issue, as the dependence of the future class on the current class will have dissipated. However, in short horizon cases, such as that considered in the application presented in Sect. 18.5, this persistence is more likely to be important. To incorporate persistence into the machine learning classifiers' predictions, I follow Giusto and Piger (2017) and allow for lagged values to enter the X_t vector. Lagged predictor variables will provide information about lagged classes, which should improve classification of future classes.[2]

Economic data is often characterized by "ragged edges," meaning that some values of the predictor variables are missing in the out-of-sample period (Camacho et al., 2018). This generally occurs due to differences in the timing of release dates for different indicators, which can leave us with only incomplete observation of

[2]When converting $\widehat{S^c_{t+h}}(X_t)$ into turning point predictions, one might also consider conversion rules that acknowledge class persistence. For example, multiple periods of elevated class probabilities could be required before a turning point into that class is predicted.

the vector X_{T+j}. There are a variety of approaches that one can take to deal with these missing observations when producing out-of-sample predictions. A simple, yet effective, approach is to use k nearest neighbors (kNN) imputation to impute the missing observations. This approach imputes the missing variables based on fully observed vectors from the training sample that are similar on the dimension of the non-missing observations. kNN imputation is discussed in more detail in Sect. 18.3.3.

18.2.3 Metrics for Evaluating Class Forecasts

In this section I discuss metrics for evaluating the performance of classifiers. Suppose we have a set of $\widetilde{T}$ class indicators, S_{t+h}^c, and associated classifier predictions, $\widehat{S_{t+h}^c}(X_t)$, where $t \in \Theta$. Since $\widehat{S_{t+h}^c}(X_t)$ is interpreted as a probability, an obvious metric to evaluate these predictions is Brier's quadratic probability score (QPS), which is the analogue of the traditional mean squared error for discrete data:

$$QPS = \frac{1}{\widetilde{T}} \sum_{t \in \Theta} \sum_{c=1}^{C} (S_{t+h}^c - \widehat{S_{t+h}^c}(X_t))^2$$

The QPS is bounded between 0 and 2, with smaller values indicating better classification ability.

As discussed above, in addition to predictions that are in the form of probabilities, we are often interested in binary predictions produced as $L\left(\widehat{S_{t+h}^c}(X_t)\right)$. In this case, there are a variety of commonly used metrics to assess the accuracy of classifiers. In describing these, it is useful to restrict our discussion to the two class case, so that $c \in \{1, 2\}$.[3] Also, without loss of generality, label $c = 1$ as the "positive" class and $c = 2$ as the "negative" class. We can then define a confusion matrix:
where TP is the number of true positives, defined as the number of instances of $c = 1$ that were classified correctly as $c = 1$, and FP indicates the number of false positives, defined as the instances of $c = 2$ that were classified incorrectly as $c = 1$. FN and TN are defined similarly.

A number of metrics of classifier performance can then be constructed from this confusion matrix. The first is *Accuracy*, which simply measures the proportion of periods that are classified correctly:

$$Accuracy = \frac{TP + TN}{TP + TN + FP + FN}$$

[3]Generalizations of these metrics to the multi-class case generally proceed by considering each class against all other classes in order to mimic a two class problem.

		Actual	
		positive	negative
Predicted	positive	TP	FP
	negative	FN	TN

Of course, *Accuracy* is strongly affected by the extent to which classes are balanced in the sample period. If one class dominates the period under consideration, then it is easy to have very high accuracy by simply always forecasting that class with high probability. In many economic applications, classes are strongly unbalanced, and as a result the use of *Accuracy* alone to validate a classifier would not be recommended. Using the confusion matrix we can instead define accuracy metrics for each class. Specifically, the *true positive rate*, or TPR, gives us the proportion of instances of $c = 1$ that were classified correctly:

$$TPR = \frac{TP}{TP + FN}$$

while the *true negative rate*, or TNR, gives us the proportion of instances of $c = 2$ that were classified correctly:

$$TNR = \frac{TN}{TN + FP}$$

It is common to express the information in TNR as the *false positive rate*, which is given by $FPR = 1 - TNR$.[4]

It is clear that which of these metrics is of primary focus depends on the relative preferences of the user for true positives vs. false positives. Also, it should be remembered that the confusion matrix and the metrics defined from its contents are dependent not just on the classifier $\widehat{S^c_{t+h}}(X_t)$, but also on the rule L used to

[4]In the classification literature, *TPR* is referred to as the *sensitivity* and *TNR* as the *specificity*.

convert this classifier to binary outcomes. In many cases, these rules are simply of the form:

$$L\left(\widehat{S^c_{t+h}}(X_t)\right) = \begin{cases} 1 & \text{if } \widehat{S^c_{t+h}}(X_t) > d \\ 2 & \text{if } \widehat{S^c_{t+h}}(X_t) \leq d \end{cases}$$

such that $c = 1$ is predicted if $\widehat{S^c_{t+h}}(X_t)$ rises above the threshold d, and $0 \leq d \leq 1$ since our classifier is a probability. A useful summary of the performance of a classifier is provided by the "receiver operator characteristic" (ROC) curve, which is a plot of combinations of TPR (y-axis) and FPR (x-axis), where the value of d is varied to generate the plot. When $d = 1$ both TPR and FPR are zero, since both TP and FP are zero if class $c = 1$ is never predicted. Also, $d = 0$ will generate TPR and FPR that are both one, since FN and TN will be zero if class $c = 1$ is always predicted. Thus, the ROC curve will always rise from the origin to the (1,1) ordinate. A classifier for which X_t provides no information regarding S^c_{t+h}, and is thus constant, will have $TPR = FPR$, and the ROC curve will lie along the 45 degree line. Classifiers for which X_t does provide useful information will have an ROC curve that lies above the 45 degree line. Figure 18.1 provides an example of such a ROC curve. Finally, suppose we have a perfect classifier, such that there exists a value of $d = d^*$ where $TPR = 1$ and $FPR = 0$. For all values of $d \geq d^*$, the ROC curve will be a vertical line on the y-axis from (0,0) to (0,1), where for values of $d \leq d^*$, the ROC curve will lie on a horizontal line from (0,1) to (1,1).

As discussed in Berge and Jordá (2011), the area under the ROC curve (AUROC) can be a useful measure of the classification ability of a classifier. The AUROC

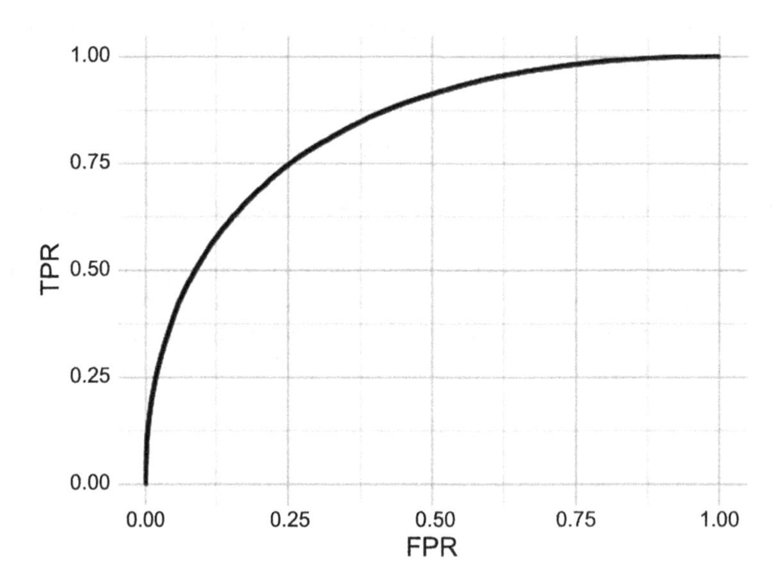

Fig. 18.1 Example receiver operator characteristic (ROC) curve

for the 45 degree line, which is the ROC curve for the classifier when X_t has no predictive ability, is 0.5. The AUROC for a perfect classifier is 1. In practice, the AUROC will lie in between these extremes, with larger values indicating better classification ability.

In the application presented in Sect. 18.5, I consider both the QPS and the AUROC to assess out-of-sample classification ability. I will also evaluate the ability of these classifiers to forecast turning points. In this case, given the relative rarity of turning points in most economic applications, it seems fruitful to evaluate performance through case studies of the individual turning points. Examples of such a case study will be provided in Sect. 18.5.

18.3 Machine Learning Approaches to Supervised Classification

In this section I will survey a variety of off-the-shelf supervised machine learning classifiers. The classifiers I survey have varying levels of suitability for data-rich environments. For all the classifiers considered, application to datasets with many predictors is computationally feasible. Some of the classifiers also have built-in mechanisms to identify relevant predictors, while others use all predictor variables equally. Where relevant, I will discuss the wide-data attributes of each classifier below.

All of the classifiers that I discuss in this section involve various specification choices that must be set in order to implement the classifier. In some cases, these choices involve setting the value of a parameter, while in others they may involve the choice between two or more variants of the classifier. Following the machine learning literature, I refer to these various choices as tuning parameters. While these tuning parameters can be set a priori, in this chapter I instead implement the commonly used procedure of cross-validation to set the tuning parameters automatically in a data-based way. In the following subsection I briefly describe cross-validation, before moving to discussions of the individual classifiers.

18.3.1 Cross-Validation

The central idea of cross-validation is to randomly partition the full training sample into a new training sample and a (non-overlapping) evaluation sample. For specific values of the tuning parameters, the classifier is trained on the partition of data labeled the training sample, and is then used to classify the partition labeled the evaluation sample. The performance of the classifier on the evaluation sample is recorded for each point in a grid for the tuning parameters, and the values of the tuning parameters with the best performance classifying the evaluation sample are selected. These optimal values for the tuning parameters are then used to train the classifier over the full training sample. Performance on the evaluation sample is generally evaluated using a scalar metric for classification performance. For example, in the application presented in Sect. 18.5, I use the AUROC for this purpose.

In k-fold cross-validation, k of these partitions (or "folds") are randomly generated, and the performance of the classifier for specific tuning parameter values is averaged across the k evaluation samples. A common value for k, which is the default in many software implementations of k-fold cross-validation, is $k = 10$. One can increase robustness by repeating k-fold cross-validation a number of times, and averaging performance across these repeats. This is known as repeated k-fold cross-validation. Finally, in settings with strongly unbalanced classes, which is common in economic applications, it is typical to sample the k partitions such that they reflect the percentage of classes in the full training sample. In this case, the procedure is labeled stratified k-fold cross-validation.

Cross-validation is an attractive approach for setting tuning parameters because it aligns the final objective, good out-of-sample forecasts, with the objective used to determine tuning parameters. In other words, tuning parameters are given a value based on the ability of the classifier to produce good out-of-sample forecasts. This is in contrast to traditional estimation, which sets parameters based on the in-sample fit of the model. Cross-validation is overwhelmingly used in machine learning algorithms for classification, and it can be easily implemented for a wide variety of classifiers using the **`caret`** package in `R`.

18.3.2 Naïve Bayes

We begin our survey of machine learning approaches to supervised classification with the **Naïve Bayes** (NB) classifier. NB is a supervised classification approach that produces a posterior probability for each class based on application of Bayes rule. NB simplifies the classification problem considerably by assuming that inside of each class, the individual variables in the vector X_t are independent of each other. This conditional independence is a strong assumption, and would be expected to be routinely violated in economic datasets. Indeed, it would be expected to be violated in most datasets, which explains the "naïve" moniker. However, despite this strong assumption, the NB algorithm works surprisingly well in practice. This is primarily because what is generally needed for classification is not exact posterior probabilities of the class, but only reasonably accurate approximate rank orderings of probabilities. Two recent applications in economics include Garbellano (2016), who used a NB classifier to nowcast US recessions and expansions in real time, and Davig and Smalter Hall (2016), who used the NB classifier, including some extensions, to predict US business cycle turning points.

To describe the NB classifier I begin with Bayes rule:

$$\Pr(S_{t+h} = c|X_t) \propto f(X_t|S_{t+h} = c) \Pr(S_{t+h} = c) \tag{18.1}$$

In words, Bayes rule tells us that the posterior probability that S_{t+h} is in phase c is proportional to the probability density for X_t conditional on S_{t+h} being in phase c multiplied by the unconditional (prior) probability that S_{t+h} is in phase c.

The primary difficulty in operationalizing (18.1) is specifying a model for X_t to produce $f(X_t|S_{t+h} = c)$. The NB approach simplifies this task considerably by assuming that each variable in X_t is independent of each other variable in X_t, conditional on $S_{t+h} = c$. This implies that the conditional data density can be factored as follows:

$$f(X_t|S_{t+h} = c) = \prod_{j=1}^{N} f_j\left(X_{j,t}|S_{t+h} = c\right)$$

where $X_{j,t}$ is one of the variables in X_t. Equation (18.1) then becomes

$$\Pr(S_{t+h} = c|X_t) \propto \left[\prod_{j=1}^{N} f_j\left(X_{j,t}|S_{t+h} = c\right)\right] \Pr(S_{t+h} = c) \tag{18.2}$$

How do we set $f_j\left(X_{j,t}|S_{t+h} = c\right)$? One approach is to assume a parametric distribution, where a typical choice in the case of continuous X_t is the normal distribution:

$$X_{j,t}|S_{t+h} = c \sim N\left(\mu_{j,c}, \sigma_{j,c}^2\right)$$

where $\mu_{j,c}$ and $\sigma_{j,c}^2$ are estimated from the training sample. Alternatively, we could evaluate $f_j\left(X_{j,t}|S_{t+h} = c\right)$ non-parametrically using a kernel density estimator fit to the training sample. In our application of NB presented in Sect. 18.5, I treat the choice of whether to use a normal distribution or a kernel density estimate as a tuning parameter.

Finally, Eq. (18.2) produces an object that is proportional to the conditional probability $\Pr(S_{t+h} = c|X_t)$. We can recover this conditional probability exactly as:

$$\Pr(S_{t+h} = c|X_t) = \frac{\left[\prod\limits_{j=1}^{N} f_j\left(X_{j,t}|S_{t+h} = c\right)\right] \Pr(S_{t+h} = c)}{\sum\limits_{c=1}^{C}\left(\left[\prod\limits_{j=1}^{N} f_j\left(X_{j,t}|S_{t+h} = c\right)\right] \Pr(S_{t+h} = c)\right)}$$

Our NB classifier is then $\widehat{S}_{t+h}^{c}(X_t) = \Pr(S_{t+h} = c|X_t)$.

The NB classifier has a number of advantages. First, it is intuitive and interpretable, directly producing posterior probabilities of each class. Second, it is easily scalable to large numbers of predictor variables, requiring a number of parameters linear to the number of predictors. Third, since only univariate properties of a predictor in each class are estimated, the classifier can be implemented with relatively small amounts of training data. Finally, ignoring cross-variable relationships guards against overfitting the training sample data.

A primary drawback of the NB approach is that it ignores cross-variable relationships potentially valuable for classification. Of course, this drawback is also the source of the advantages mentioned above. Another drawback relates to the application of NB in data-rich settings. As Eq. (18.2) makes clear, all predictors are given equal weight in determining the posterior probability of a class. As a result, the performance of the classifier can deteriorate if there are a large number of irrelevant predictors, the probability of which will increase in data-rich settings.

Naïve Bayes classification can be implemented in `R` via the **`caret`** package, using the **`naive_bayes`** method. Implementation involves two tuning parameters. The first is **`usekernel`**, which indicates whether a Gaussian density or a kernel density estimator is used to approximate $f_j \left(X_{j,t}|S_{t+h} = c\right)$. The second is **`adjust`**, which is a parameter indicating the size of the bandwidth in the kernel density estimator.

18.3.3 k-Nearest Neighbors

The k-nearest neighbor (kNN) algorithm is among the simplest of supervised classification techniques. Suppose that for predictor data X_t, we define a neighborhood, labeled $R_k(X_t)$, that consists of the k closest points to X_t in the training sample (not including X_t itself). Our class prediction for $S_{t+h} = c$ is then simply the proportion of points in the region belonging to class c:

$$\widehat{S}^c_{t+h}(X_t) = \frac{1}{k} \sum_{X_i \in R_k(X_t)} I(S_{i+h} = c)$$

In other words, to predict S^c_{t+h}, we find the k values of X that are closest to X_t, and compute the proportion of these that correspond to class c.

To complete this classifier we must specify a measure of "closeness." The most commonly used metric is Euclidean distance:

$$d(X_t, X_i) = \sqrt{\sum_{j=1}^{N} (X_{j,t} - X_{j,i})^2}$$

Other distance metrics are of course possible. The region $R_k(X_t)$ can also be defined continuously, so that training sample observations are not simply in vs. out of $R_k(X_t)$, but have declining influence as they move farther away from X_t.

kNN classifiers are simple to understand, and can often provide a powerful classification tool, particularly in cases where X_t is of low dimension. However, kNN is adversely affected in cases where the X_t vector contains many irrelevant predictors, as the metric defining closeness is affected by all predictors, irregardless of their classification ability. Of course, the likelihood of containing many irrelevant predictors increases with larger datasets, and as a result kNN classification is not a commonly used classifier when there are a large number of predictors. Other approaches, such as the tree-based methods described below, are preferred in data-rich settings, in that they can automatically identify relevant predictors.

As was discussed in Sect. 18.2.2, in many applications some values of the predictor variables will be missing in the out-of-sample period. It is also possible to have missing values in the training sample. A simple and effective approach to handle missing values is to apply a kNN type procedure to impute the missing values. Specifically, suppose we have a vector X_t that is only partially observed. Segment this vector into X_t^* and $\widetilde{X}_t$, where X_t^* holds the N^* variables that are observed and $\widetilde{X}_t$ holds the $N - N^*$ variables that are not observed. Define a neighborhood, labeled $R_k\left(X_t^*\right)$, that consists of the k closest points to X_t^* over the portion of the training sample for which there are no missing values. Closeness can again be defined in terms of the Euclidean metric:

$$d(X_t^*, X_i^*) = \sqrt{\sum_{j=1}^{N^*}(X_{j,t}^* - X_{j,i}^*)^2}$$

We then impute the missing variable values contained in $\widetilde{X}_t$ using the mean of those same variables in $R_k\left(X_t^*\right)$:

$$\widetilde{X}_t^{imputed} = \frac{1}{k} \sum_{X_i \in R_k(X_t^*)} \widetilde{X}_i$$

kNN classification can be implemented in `R` via the **`caret`** package, using the **`knn`** method. The **`knn`** method involves a single tuning parameter, **`k`**, which indicates the value of k. Also, when implementing any machine learning classification method in `R` using the **`caret`** package, kNN imputation can be used to replace missing values via the **`preProc`** argument to the **`train`** function.

18.3.4 Learning Vector Quantization

Learning Vector Quantization (LVQ) is a classifier that forms predictions on the basis of the closeness of X_t to some key points in the predictor space. In this sense it is like kNN, in that it generates predictions based on a nearest-neighbor strategy. However, unlike kNN, these key points are not collections of training sample data points, but instead are endogenously learned from the training sample data. LVQ is widely used in real-time classification problems in a number of fields and applications, and was used in the economics literature by Giusto and Piger (2017) to identify US business cycle turning points in real time. LVQ methods were developed by Teuvo Kohonen and are described in detail in Kohonen (2001).

To describe LVQ it is useful to begin with vector quantization (VQ). A VQ classifier relies on the definition of certain key points, called *codebook vectors*, defined in the predictor space. Each codebook vector is assigned to a class, and there can be more than one codebook vector per class. We would generally have far fewer codebook vectors than data vectors, implying that a codebook vector provides representation for a group of training sample data vectors. In other words, the codebook vectors *quantize* the salient features of the predictor data. Once these codebook vectors are singled out, data is classified via a majority vote of the nearest group of k codebook vectors in the Euclidean metric.

How is the location of each codebook vector established? An LVQ algorithm is an adaptive learning algorithm in which the locations of the codebook vectors are determined through adjustments of decreasing magnitude. Denote our codebook vectors as $v_i \in R^N, i = 1, \ldots, V$, let $g = 1, 2, \ldots, G$ denote iterations of the algorithm, and let α^g be a decreasing geometric sequence where $0 < \alpha < 1$. Given the initial location of the codebook vectors, the LVQ algorithm makes adjustments to their location as described in Algorithm 15:

This LVQ algorithm is very simple. A data vector is considered, and its nearest codebook vector is identified. If the class attached to this codebook vector agrees with the actual classification of the data vector, its location is moved closer to

---

Algorithm 15 Learning vector quantization (Kohonen 2001)

Initialize $v_i^0, i = 1, \ldots, V$
for $g = 1$ to G **do**.
  for $t = 1$ to T **do**.
    Identify the single codebook vector v_*^{g-1} closest to X_t in the Euclidean metric.
    Adjust the location of v_*^g according to:

$$v_*^g = v_*^{g-1} + \alpha^g (X_t - v_*^{g-1}) \qquad \text{if } X_t \text{ and } v_*^{g-1} \text{ belong to the same class}$$

$$v_*^g = v_*^{g-1} - \alpha^g (X_t - v_*^{g-1}) \qquad \text{otherwise}$$

  end for
end for

---

the data vector. If the selected codebook vector does not classify the data vector correctly, then it is moved farther from the data vector. These adjustments are made in a simple linear fashion. These calculations are repeated for each data vector in the dataset. When the data has all been used, a new iteration is started where the weight α^g, which controls the size of the adjustment to the codebook vectors, is decreased. This continues for G iterations, with the final codebook vectors given by $v_i^G, i = 1, \ldots, V$.

Once the final codebook vectors are established, the LVQ classifier produces a prediction, $\widehat{S}_{t+h}(X_t)$, via a majority voting strategy. First, we identify the k closest codebook vectors to X_t in the Euclidean metric. Second, the predicted class for X_t is set equal to the majority class of these k codebook vectors. Denote this majority class as c^*. Then:

$$\widehat{S}_{t+h}^c(X_t) = 1, \quad \text{if } c = c^*$$

$$\widehat{S}_{t+h}^c(X_t) = 0, \quad \text{otherwise}$$

Here I have laid out the basic LVQ algorithm, which has been shown to work well in many practical applications. Various modifications to this algorithm have been proposed, which may improve classification ability in some contexts. These include LVQ with nonlinear updating rules, as in the generalized LVQ algorithm of Sato and Yamada (1995), as well as LVQ employed with alternatives to the Euclidean measure of distance, such as the generalized relevance LVQ of Hammer and Villmann (2002). The latter allows for adaptive weighting of data series in the dimensions most helpful for classification, and may be particularly useful when applying LVQ to large datasets.

LVQ classification can be implemented in `R` via the **`caret`** package using the **`lvq`** method. The **`lvq`** method has two tuning parameters. The first is the number of codebook vectors to use in creating the final classification, k, and is labeled **`k`**. The second is the total number of codebook vectors V, and is labeled **`size`**. To implement the classifier, one must also set the values of G and α. In the **`lvq`** method, G is set endogenously to ensure convergence of the codebook vectors. The default value of α in the **`lvq`** method is 0.3. Kohonen (2001) argues that classification results from LVQ should be largely invariant to the choice of alternative values of α provided that $\alpha^g \to 0$ as $g \to \infty$, which ensures that the size of codebook vector updates eventually converges to zero. Giusto and Piger (2017) verified this insensitivity in their application of LVQ to identifying business cycle turning points.

The LVQ algorithm requires an initialization of the codebook vectors. This initialization can have effects on the resulting class prediction, as the final placement of the codebook vectors in an LVQ algorithm is not invariant to initialization. A simple approach, which I follow in the application, is to allow all classes to have the same number of codebook vectors, and initialize the codebook vectors attached to each class with random draws of X_t vectors from training sample observations corresponding to each class.

18.3.5 Classification Trees

A number of commonly used classification techniques are based on **classification trees**. I will discuss several of these approaches in subsequent sections, each of which uses aggregations of multiple classification trees to generate predictions. Before delving into these techniques, in this section I describe the single classification trees upon which they are built.

A classification tree is an intuitive, non-parametric, procedure that approaches classification by partitioning the predictor variable space into non-overlapping regions. These regions are created according to a conjunction of binary conditions. As an example, in a case with two predictor variables, one of the regions might be of the form $\{X_t | X_{1,t} \geq \tau_1, X_{2,t} < \tau_2\}$. The partitions are established in such a way so as to effectively isolate alternative classes in the training sample. For example, the region mentioned above might have been chosen because it corresponds to cases where S_{t+h} is usually equal to c. To generate predictions, the classification tree would then place a high probability on $S_{t+h} = c$ if X_t fell in this region.

How specifically are the partitions established using a training sample? Here I will describe a popular training algorithm for a classification tree, namely the classification and regression tree (CART). CART proceeds by recursively partitioning the training sample through a series of binary splits of the predictor data. Each new partition, or split, segments a region that was previously defined by the earlier splits. The new split is determined by one of the predictor variables, labeled the "split variable," based on a binary condition of the form $X_{j,t} < \tau$ and $X_{j,t} \geq \tau$, where both j and τ can differ across splits. The totality of these recursive splits partition the sample space into M non-overlapping regions, labeled A^*_m, $m = 1, \ldots, M$, where there are T^*_m training sample observations in each region. For all X_t that are in region A^*_m, the prediction for S_{t+h} is a constant equal to the within-region sample proportion of class c:

$$P^c_{A^*_m} = \frac{1}{T^*_m} \sum_{X_t \in A^*_m} I(S_{t+h} = c) \tag{18.3}$$

The CART classifier is then:

$$\widehat{S}^c_{t+h}(X_t) = \sum_{m=1}^{M} P^c_{A^*_m} \, I\left(X_t \in A^*_m\right) \tag{18.4}$$

How is the recursive partitioning implemented to arrive at the regions A^*_m? Suppose that at a given step in the recursion, we have a region defined by the totality of the previous splits. In the language of decision trees, this region is called a "node." Further, assume this node has not itself yet been segmented into subregions. I refer to such a node as an "unsplit node" and label this unsplit node generically as A. For a given j and τ^j we then segment the data in this region according to

$\{X_t | X_{j,t} < \tau^j, X_t \in A\}$ and $\{X_t | X_{j,t} \geq \tau^j, X_t \in A\}$, which splits the data in this node into two non-overlapping regions, labeled A^L and A^R, respectively. In these regions, there are T^L and T^R training sample observations. In order to determine the splitting variable, j, and the split threshold, τ^j, we scan through all pairs $\{j, \tau^j\}$, where $j = 1, \ldots N$ and $\tau^j \in \mathcal{T}^{A,j}$, to find the values that maximize a measure of the homogeneity of class outcomes inside of A^L and A^R.[5] Although a variety of measures of homogeneity are possible, a common choice is the Gini impurity:

$$G_L = \sum_{c=1}^{C} P^c_{A^L}(1 - P^c_{A^L})$$

$$G_R = \sum_{c=1}^{C} P^c_{A^R}(1 - P^c_{A^R})$$

The Gini impurity is bounded between zero and one, where a value of zero indicates a "pure" region where only one class is present. Higher values of the Gini impurity indicate greater class diversity. The average Gini impurity for the two new proposed regions is

$$\overline{G} = \frac{T^L}{T^L + T^R} G_L + \frac{T^R}{T^L + T^R} G_R \tag{18.5}$$

The split point j and split threshold τ^j are then chosen to create regions A^L and A^R that minimize the average Gini impurity.

This procedure is repeated for other unsplit nodes of the tree, with each additional split creating two new unsplit nodes. By continuing this process recursively, the sample space is divided into smaller and smaller regions. One could allow the recursion to run until we are left with only pure unsplit nodes. A more common choice in practice is to stop splitting nodes when any newly created region would contain a number of observations below some predefined minimum. This minimum number of observations per region is a tuning parameter of the classification tree. Whatever approach is taken, when we arrive at an unsplit node that is not going to be split further, this node becomes one of our final regions A^*_m. In the language of decision trees, this final unsplit node is referred to as a "leaf." Algorithm 16 provides a description of the CART classification tree.

Classification trees have many advantages. They are simple to understand, require no parametric modeling assumptions, and are flexible enough to capture complicated nonlinear and discontinuous relationships between the predictor variables and class indicator. Also, the recursive partitioning algorithm described above

[5] In a CART classification tree, $\mathcal{T}^{A,j}$ is a discrete set of all non-equivalent values for τ^j, which is simply the set of midpoints of the ordered values for $X_{j,t}$ in the training sample observations relevant for node A.

Algorithm 16 A single CART classification tree (Breiman, Friedman, Olshen, and Stone 1984)

1: Initialize a single unsplit node to contain the full training sample
2: **for** All unsplit nodes A_u with total observations > threshold **do**
3: **for** $j = 1$ to N and $\tau^j \in \mathcal{T}^{A_u,j}$ **do**
4: Create non-overlapping regions $A_u^L = \{X_t | X_{j,t} < \tau^j, X_t \in A_u\}$ and $A_u^R = \{X_t | X_{j,t} \geq \tau^j, X_t \in A_u\}$ and calculate $\bar{G}$ as in (18.5).
5: **end for**
6: Select j and τ^j to minimize $\bar{G}$ and create the associated nodes A_u^L and A_u^R.
7: Update the set of unsplit nodes to include A_u^L and A_u^R.
8: **end for**
9: For final leaf nodes, A_m^*, form $P_{A_m^*}^c$ as in (18.3), for $c = 1, \ldots, C$ and $m = 1, \ldots M$
10: Form the CART classification tree classifier: $\widehat{S_{t+h}^c}(X_t)$ as in (18.4).

scales easily to large datasets, making classification trees attractive in this setting. Finally, unlike the classifiers we have encountered to this point, a CART classification tree automatically conducts model selection in the process of producing a prediction. Specifically, a variable may never be used as a splitting variable, which leaves it unused in producing a prediction by the classifier. Likewise, another variable may be used multiple times as a splitting variable. These differences result in varying levels of variable importance in a CART classifier. Hastie, Tibshirani, and Friedman (2009) detail a measure of variable importance that can be produced from a CART tree.

CART trees have one significant disadvantage. The sequence of binary splits, and the path dependence this produces, generally produces a high variance forecast. That is, small changes in the training sample can produce very different classification trees and associated predictions. As a result, a number of procedures exist that attempt to retain the benefits of classification trees while reducing variance. We turn to these procedures next.

18.3.6 Bagging, Random Forests, and Extremely Randomized Trees

In this section we describe **bagged classification trees**, their close variant, the **random forest**, and a modification to the random forest known as **extremely randomized trees**. Each of these approaches averages the predicted classification coming from many classification trees. This allows us to harness the advantages of tree-based methods while at the same time reducing the variance of the tree-based predictions through averaging. Random forests have been used to identify turning points in economic data by Ward (2017), who uses random forests to identify episodes of financial crises, and Garbellano (2016), who uses random forests to nowcast US recession episodes. Bagging and random forests are discussed in more detail in Chap. 13 of this book.

We begin with bootstrap aggregated (bagged) classification trees, which were introduced in Breiman (1996). Bagged classification trees work by training a large number, B, of CART classification trees and then averaging the class predictions from each of these trees to arrive at a final class prediction. The trees are different because each is trained on a bootstrap training sample of size T, which is created by sampling $\{X_t, S_{t+h}\}$ with replacement from the full training sample. Each tree produces a class prediction, which I label $\widehat{S}^c_{b,t+h}(X_t)$, $b = 1, \ldots, B$. The bagged classifier is then the simple average of the B CART class predictions:

$$\widehat{S}^c_{t+h}(X_t) = \frac{1}{B}\sum_{b=1}^{B}\widehat{S}^c_{b,t+h}(X_t) \tag{18.6}$$

Bagged classification trees are a variance reduction technique that can give substantial gains in accuracy over individual classification trees. As discussed in Breiman (1996), a key determinant of the potential benefits of bagging is the variance of the individual classification trees across alternative training samples. All else equal, the higher is this variance, the more potential benefit there is from bagging.

As discussed in Breiman (2001), the extent of the variance improvement also depends on the amount of correlation across the individual classification trees that constitute the bagged classification tree, with higher correlation generating lower variance improvements. This correlation could be significant, as the single classification trees used in bagging are trained on overlapping bootstrap training samples. As a result, modifications to the bagged classification tree have been developed that attempt to reduce the correlation across trees, without substantially increasing the bias of the individual classification trees. The most well-known of these is the random forest (RF), originally developed in Breiman (2001).[6]

An RF classifier attempts to lower the correlation across trees by adding another layer of randomness into the training of individual classification trees. As with a bagged classification tree, each single classification tree is trained on a bootstrap training sample. However, when training this tree, rather than search over the entire set of predictors $j = 1, \ldots, N$ for the best splitting variable at each node, we instead randomly choose $Q << N$ predictor variables at each node, and search only over these variables for the splitting variable. This procedure is repeated to train B individual classification trees, and the random forest classifier is produced by averaging these classification trees as in Eq. (18.6). Algorithm 17 provides a description of the RF classifier.

When implementing an RF classifier, the individual trees are usually allowed to grow to maximum size, meaning that nodes are split until only pure nodes remain. While such a tree should produce a classifier with low bias, it is likely to be unduly influenced by peculiarities of the specific training sample, and thus will

[6]Other papers that were influential in the development of random forest methods include Amit and Geman (1997) and Ho (1998).

Algorithm 17 Random forest classifier (Breiman 2001)

1: **for** $b = 1$ to B **do**
2: Form a bootstrap training sample by sampling $\{X_t, S_{t+h}\}$ with replacement T times from the training sample observations.
3: Initialize a single unsplit node to contain the full bootstrap training sample
4: **for** All unsplit nodes A_u with total observations > threshold **do**
5: Randomly select Q predictor variables as possible splitting variables. Denote these predictor variables at time t as $\widetilde{X}_t$
6: **for** $X_{j,t} \in \widetilde{X}_t$ and $\tau^j \in \mathcal{T}^{A_u,j}$ **do**
7: Create two non-overlapping regions $A_u^L = \{X_t | X_{j,t} < \tau^j, X_t \in A_u\}$ and $A_R = \{X_t | X_{j,t} \geq \tau^j, X_t \in A_u\}$ and calculate $\bar{G}$ as in (18.5).
8: **end for**
9: Select j and τ^j to minimize $\bar{G}$ and create the associated nodes A_u^L and A_u^R.
10: Update the set of unsplit nodes to include A_u^L and A_u^R
11: **end for**
12: For final leaf nodes, A_m^*, form $P_{A_m^*}^c$ as in (18.3), for $c = 1, \ldots, C$ and $m = 1, \ldots M$
13: Form the single tree classifier: $\widehat{S}_{b,t+h}^c(X_t)$ as in (18.4).
14: **end for**
15: Form the Random Forest classifier: $\widehat{S}_{t+h}^c(X_t) = \frac{1}{B}\sum_{b=1}^{B} \widehat{S}_{b,t+h}^c(X_t)$.

have high variance. The averaging of the individual trees lowers this variance, while the additional randomness injected by the random selection of predictor variables helps maximize the variance reduction benefits of averaging. It is worth noting that by searching only over a small random subset of predictors at each node for the optimal splitting variable, the RF classifier also has computational advantages over bagging.

Extremely Randomized Trees, or "ExtraTrees," is another approach to reduce correlation among individual classification trees. Unlike both bagged classification trees and RF, ExtraTrees trains each individual classification tree on the entire training sample rather than bootstrapped samples. As does a random forest, ExtraTrees randomizes the subset of predictor variables considered as possible splitting variables at each node. The innovation with ExtraTrees is that when training the tree, for each possible split variable, only a single value of τ^j is considered as the possible split threshold. For each j, this value is randomly chosen from the uniform interval $[min(X_{j,t}|X_{j,t} \in A), max(X_{j,t}|X_{j,t} \in A)]$. Thus, ExtraTrees randomizes across both the split variable and the split threshold dimension. ExtraTrees was introduced by Geurts, Ernst, and Wehenkel (2006), who argue that the additional randomization introduced by ExtraTrees should reduce variance more strongly than weaker randomization schemes. Also, the lack of a search over all possible τ^j for each split variable at each node provides additional computational advantages over the RF classifier. Algorithm 18 provides a description of the ExtraTrees classifier.

RF and ExtraTrees classifiers have enjoyed a substantial amount of success in empirical applications. They are also particularly well suited for data-rich environments. Application to wide datasets of predictor variables is computationally tractable, requiring a number of scans that at most increase linearly with the number

Algorithm 18 Extremely randomized trees (Geurts et al. 2006)

1: **for** $b = 1$ to B **do**
2: Initialize a single unsplit node to contain the full training sample
3: **for** All unsplit nodes A_u with total observations > threshold **do**
4: Randomly select Q predictor variables as possible splitting variables. Denote these predictor variables at time t as $\widetilde{X}_t$
5: **for** $X_{j,t} \in \widetilde{X}_t$ **do**
6: Randomly select a single τ^j from the uniform interval: $[min(X_{j,t}|X_{j,t} \in A_u), max(X_{j,t}|X_{j,t} \in A_u)]$
7: Create two non-overlapping regions $A_u^L = \{X_t|X_{j,t} < \tau^j, X_t \in A_u\}$ and $A_R = \{X_t|X_{j,t} \geq \tau^j, X_t \in A_u\}$ and calculate $\bar{G}$ as in (18.5).
8: **end for**
9: Select j to minimize $\bar{G}$ and create the associated nodes A_u^L and A_u^R.
10: Update the set of unsplit nodes to include A_u^L and A_u^R
11: **end for**
12: For final leaf nodes, A_m^*, form $P_{A_m^*}^c$ as in (18.3), for $c = 1, \ldots, C$ and $m = 1, \ldots M$
13: Form the single tree classifier: $\widehat{S}_{b,t+h}^c(X_t)$ as in (18.4).
14: **end for**
15: Form the ExtraTrees classifier: $\widehat{S}_{t+h}^c(X_t) = \frac{1}{B}\sum_{b=1}^{B} \widehat{S}_{b,t+h}^c(X_t)$.

of predictors. Also, because these algorithms are based on classification trees, they automatically conduct model selection as the classifier is trained.

Both RF and ExtraTrees classifiers can be implemented in `R` via the **`caret`** package, using the **`ranger`** method. Implementation involves three tuning parameters. The first is the value of Q and is denoted **`mtry`** in the caret package. The second is **`splitrule`** and indicates whether a random forest (**`splitrule`**= 0) or an extremely randomized tree (**`splitrule`**= 1) is trained. Finally, **`min.node.size`** indicates the minimum number of observations allowed in the final regions established for each individual classification tree. As discussed above, it is common with random forests and extremely randomized trees to allow trees to be trained until all regions are pure. This can be accomplished by setting **`min.node.size`**= 1.

18.3.7 Boosting

Boosting has been described by Hastie et al. (2009) as "one of the most powerful learning ideas introduced in the last twenty years" and by Breiman (1996) as "the best off-the-shelf classifier in the world." Many alternative descriptions and interpretations of boosting exist, and a recent survey and historical perspective is provided in Mayr, Binder, Gefeller, and Schmid (2014). In economics, several recent papers have used boosting to predict expansion and recession episodes, including Ng (2014), Berge (2015), and D opke, Fritsche, and Pierdzioch (2017). Boosting is described in more detail in Chap. 14 of this book.

The central idea of boosting is to recursively apply simple "base" learners to a training sample, and then combine these base learners to form a strong classifier. In each step of the recursive boosting procedure, the base learner is trained on a weighted sample of the data, where the weighting is done so as to emphasize observations in the sample that to that point had been classified *incorrectly*. The final classifier is formed by combining the sequence of base learners, with better performing base learners getting more weight in this combination.

The first boosting algorithms are credited to Schapire (1990), Freund (1995), and Freund and Schapire (1996) and are referred to as **AdaBoost**. Later work by Friedman, Hastie, and Tibshirani (2000) interpreted AdaBoost as a forward stagewise procedure to fit an additive logistic regression model, while Friedman (2001) showed that boosting algorithms can be interpreted generally as non-parametric function estimation using gradient descent. In the following I will describe boosting in more detail using these later interpretations.

For notational simplicity, and to provide a working example, consider a two class case, where I define the two classes as $S_{t+h} = -1$ and $S_{t+h} = 1$. Define a function $F(X_t) \in \mathbb{R}$ that is meant to model the relationship between our predictor variables, X_t, and S_{t+h}. Larger values of $F(X_t)$ signal increased evidence for $S_{t+h} = 1$, while smaller values indicate increased evidence for $S_{t+h} = -1$. Finally, define a loss function, $C(S_{t+h}, F(X_t))$, and suppose our goal is to choose $F(X_t)$ such that we minimize the expected loss:

$$E_{S,X}C(S_{t+h}, F(X_t)) \tag{18.7}$$

A common loss function for classification is exponential loss:

$$C(S_{t+h}, F(X_t)) = \exp(-S_{t+h}F(X_t))$$

The exponential loss function is smaller if the signs of S_{t+h} and $F(X_t)$ match than if they do not. Also, this loss function rewards (penalizes) larger absolute values of $F(X_t)$ when it is correct (incorrect).

For exponential loss, it is straightforward to show Friedman et al. (2000) that the $F(X_t)$ that minimizes Eq. (18.7) is

$$F(X_t) = \frac{1}{2}\ln\left[\frac{\Pr(S_{t+h} = 1|X_t)}{\Pr(S_{t+h} = -1|X_t)}\right]$$

which is simply one-half the log odds ratio. A traditional approach commonly found in economic studies is to assume an approximating parametric model for the log odds ratio. For example, a parametric logistic regression model would specify:

$$\ln\left[\frac{\Pr(S_{t+h} = 1|X_t)}{\Pr(S_{t+h} = -1|X_t)}\right] = X_t'\beta$$

A boosting algorithm alternatively models $F(X_t)$ as an additive model (Friedman et al. 2000):

$$F(X_t) = \sum_{j=1}^{J} \alpha_j T_j\left(X_t; \beta_j\right) \tag{18.8}$$

where each $T_j\left(X_t; \beta_j\right)$ is a base learner with parameters β_j. $T_j\left(X_t; \beta_j\right)$ is usually chosen as a simple model or algorithm with only a small number of associated parameters. A very common choice for $T_j\left(X_t; \beta_j\right)$ is a CART regression tree with a small number of splits.

Boosting algorithms fit Eq. (18.8) to the training sample in a forward stagewise manner. An additive model fit via forward stagewise iteratively solves for the loss minimizing $\alpha_j T_j\left(X_t; \beta_j\right)$, conditional on the sum of previously fit terms, labeled $F_{j-1}(X_t) = \sum_{i=1}^{j-1} \alpha_i T_i(X_t; \beta_i)$. Specifically, conditional on an initial $F_0(X_t)$, we iteratively solve the following for $j = 1, \ldots, J$:

$$\{\alpha_j, \beta_j\} = \min_{\alpha_j, \beta_j} \sum_{t=1}^{T} C\left(S_{t+h}, \left[F_{j-1}(X_t) + \alpha_j T_j\left(X_t; \beta_j\right)\right]\right) \tag{18.9}$$

$$F_j(X_t) = F_{j-1}(X_t) + \alpha_j T_j\left(X_t; \beta_j\right) \tag{18.10}$$

Gradient boosting finds an approximate solution to equation (18.9)–(18.10) via a two-step procedure. First, for each j, compute the "pseudo-residuals" as the negative gradient of the loss function evaluated at $F_{j-1}(X_t)$:

$$e_{j,t+h} = -\left[\frac{\partial C(S_{t+h}, F(X_t))}{\partial F(X_t)}\right]_{F(X_t)=F_{j-1}(X_t)}$$

Next, a CART regression tree $T_j\left(X_t; \beta_j\right)$ is fit to the pseudo-residuals. Specifically, a tree is trained on a training sample made up of $\{e_{j,t+h}, X_t\}_{t=1}^{T}$, with the final tree containing M non-overlapping leaves, $A_{m,j}^{*}$, $m = 1, \ldots, M$. The CART regression tree predicts a constant in each region:

$$T_j\left(X_t; \beta_j\right) = \sum_{m=1}^{M} \bar{e}_{m,j} I(X_t \in A_{m,j}^{*})$$

where $\bar{e}_{m,j}$ is the simple average of $e_{j,t+h}$ inside the leaf $A_{m,j}^{*}$, and β_j represents the parameters of this tree, which would include details such as the split locations and splitting variables. These are chosen as described in Sect. (18.3.5), but as the pseudo-residuals are continuous, a least squares criterion is minimized to choose β_j rather than the Gini impurity. Notice that because the tree predicts a constant in each regime, the solution to equation (18.9) involves simply a single parameter optimization in each of the $A^*_{m,j}$ regions. Each of these optimizations takes the form:

$$\gamma_{m,j} = \min_{\gamma} \sum_{X_t \in A^*_{m,j}} C\left(S_{t+h}, F_{j-1}(X_t) + \gamma\right), \quad m = 1, \ldots, M$$

Given this solution, Eq. (18.10) becomes

$$F_j(X_t) = F_{j-1}(X_t) + \sum_{m=1}^{M} \gamma_{m,j} I(X_t \in A^*_{m,j}) \tag{18.11}$$

As discussed in Friedman (2001), gradient boosting is analogous to AdaBoost when the loss function is exponential. However, gradient boosting is more general, and can be implemented for any differentiable loss function. Gradient boosting also helps expose the intuition of boosting. The gradient boosting algorithm approximates the optimal $F(X_t)$ through a series of Newton steps, and in this sense boosting can be interpreted as a numerical minimization of the empirical loss function in the space of the function $F(X_t)$. Each of these Newton steps moves $F_j(X_t)$ in the direction of the negative gradient of the loss function, which is the direction of greatest descent for the loss function in $F(X_t)$ space. Loosely speaking, the negative gradient, or pseudo-residuals, provides us with the residuals from applying $F_{j-1}(X_t)$ to classify S_{t+h}. In this sense, at each step, the boosting algorithm focuses on observations that were classified incorrectly in the previous step.

Finally, Friedman (2001) suggests a modification of Eq. (18.11) to introduce a shrinkage parameter:

$$F_j(X_t) = F_{j-1}(X_t) + \eta \sum_{m=1}^{M} \gamma_{m,j} I(X_t \in A^*_{m,j})$$

where $0 < \eta \leq 1$ controls the size of the function steps in the gradient based numerical optimization. In practice, η is a tuning parameter for the gradient boosting algorithm.

Gradient boosting with trees as the base learners is referred to under a variety of names, including a gradient boosting machine, MART (multiple additive regression trees), TreeBoost, and a boosted regression tree. The boosting algorithm for our two class example is shown in Algorithm 19.

Upon completion of this algorithm we have $F_J(X_t)$, although in many applications this function is further converted into a more recognizable class prediction. For example, AdaBoost uses the classifier $sign(F_J(X_t))$, which for the two class example with exponential loss classifies S_{t+h} according to its highest probability class. In our application, we will instead convert $F_J(X_t)$ to a class probability by inverting the assumed exponential cost function, and use these probabilities as our

Algorithm 19 Gradient boosting with trees (Friedman 2001)

1: Initialize $F^0(X_t) = \min_{\gamma} \sum_{t=1}^{T} C(S_{t+h}, \gamma)$
2: **for** $j = 1$ to J **do**
3: $e_{j,t+h} = -\left[\frac{\partial C(S_{t+h}, F(X_t))}{\partial F(X_t)}\right]_{F(X_t)=F_{j-1}(X_t)}, t = 1 \ldots T$
4: Fit $T(X_t; \beta_j)$ to $\{e_{j,t+h}, X_t\}_{t=1}^{T}$ to determine regions $A^*_{m,j}$, $m = 1, \ldots M$
5: $\gamma_{m,j} = \min_{\gamma} \sum_{X_t \in A^*_{m,j}} C\left(S_{t+h}, F_{j-1}(X_t) + \gamma\right)$, $m = 1, \ldots M$
6: $F_j(X_t) = F_{j-1}(X_t) + \eta \sum_{m=1}^{M} \gamma_{m,j} I\left(X_t \in A^*_{m,j}\right)$
7: **end for**

classifier, $\widehat{S}_{t+h}$. Again, for the two class case with exponential loss:

$$\widehat{S}^{c=1}_{t+h}(X_t) = \frac{exp(2F_J(X_t))}{1 + exp(2F_J(X_t))}$$

$$\widehat{S}^{c=-1}_{t+h}(X_t) = \frac{1}{1 + exp(2F_J(X_t))}$$

Gradient boosting with trees scales very well to data-rich environments. The forward-stagewise gradient boosting algorithms simplify optimization considerably. Further, gradient boosting is commonly implemented with small trees, in part to avoid overfitting. Indeed, a common choice is to use the so-called stumps, which are trees with only a single split. This makes implementation with large sets of predictors very fast, as at each step in the boosting algorithm, only a small number of scans through the predictor variables is required.

Two final aspects of gradient boosting bear further comment. First, as discussed in Hastie et al. (2009), the algorithm above can be modified to incorporate $K > 2$ classes by assuming a negative multinomial log likelihood cost function. Second, Friedman (2001) suggests a modified version of Algorithm 19 in which, at each step j, a random subsample of the observations is chosen. This modification, known as "stochastic gradient boosting," can help prevent overfitting while also improving computational efficiency.

Gradient boosting can be implemented in `R` via the **`caret`** package, using the **`gbm`** method. Implementation of gbm where regression trees are the base learners involves four tuning parameters. The first is **`n.trees`**, which is the stopping point J for the additive model in Eq. (18.8). The second is **`interaction.depth`**, which is the depth (maximum number of consecutive splits) of the regression trees used as weak learners. **`shrinkage`** is the shrinkage parameter, η in the updating rule Eq. (18.3.7). Finally, **`n.minobsinnode`** is the minimum terminal node size for the regression trees.

18.4 Markov-Switching Models

In this section we describe Markov-switching (MS) models, which are a popular approach for both historical and real-time classification of economic data. In contrast to the machine learning algorithms presented in the previous section, MS models are unsupervised, meaning that a historical time series indicating the class is not required. Instead, MS models assume a parametric structure for the evolution of the class, as well as for the interaction of the class with observed data. This structure allows for statistical inference on which class is, or will be, active. In data-rich environments, Markov-switching can be combined with dynamic factor models to capture the information contained in datasets with many predictors. Obviously, MS models are particularly attractive when a historical class indicator is not available, and thus supervised approaches cannot be implemented. However, MS models have also been used quite effectively for real-time classification in settings where a historical indicator is available. We will see an example of this in Sect. 18.5.

MS models are parametric time-series models in which parameters are allowed to take on different values in each of C regimes, which for our purposes correspond to the classes of interest. A fundamental difference from the supervised approaches we have already discussed is that these regimes are not assumed to be observed in the training sample. Instead, a stochastic process assumed to have generated the regime shifts is included as part of the model, which allows for both in-sample historical inference on which regime is active, as well as out-of-sample forecasts of regimes. In the MS model, introduced to econometrics by Goldfeld and Quandt (1973), Cosslett and Lee (1985), and Hamilton (1989), the stochastic process assumed is a C-state Markov process. Also, and in contrast to the non-parametric approaches we have already seen, a specific parametric structure is assumed to link the observed X_t to the regimes. Following Hamilton (1989), this linking model is usually an autoregressive time-series model with parameters that differ in the C regimes. The primary use of these models in the applied economics literature has been to describe changes in the dynamic behavior of macroeconomic and financial time series.

The parametric structure of MS models comes with some benefits for classification. First, by specifying a stochastic process for the regimes, one can allow for dynamic features that may help with both historical and out-of-sample classification. For example, most economic regimes of interest display substantial levels of persistence. In an MS model, this persistence is captured by the assumed Markov process for the regimes. Second, by assuming a parametric model linking X_t to the classes, the model allows the researcher to focus the classification exercise on the object of interest. For example, if one is interested in identifying high and low volatility regimes, a model that allows for switching in only conditional variance of an AR model could be specified.[7]

[7] MS models generally require a normalization in order to properly define the regimes. For example, in a two regime example where the regimes are high and low volatility, we could specify that $S_{t+h} = 1$ is the low variance regime and $S_{t+h} = 2$ is the high variance regime. In practice this is

Since the seminal work of Hamilton (1989), MS models have become a very popular modeling tool for applied work in economics. Of particular note are regime-switching models of measures of economic output, such as real gross domestic product (GDP), which have been used to model and identify the phases of the business cycle. Examples of such models include Hamilton (1989), Chauvet (1998), Kim and Nelson (1999a), Kim and Nelson (1999b), and Kim, Morley, and Piger (2005). A sampling of other applications includes modeling regime shifts in time series of inflation and interest rates (Ang and Bekaert, 2002; Evans and Wachtel, 1993; Garcia and Perron, 1996), high and low volatility regimes in equity returns (Dueker, 1997; Guidolin and Timmermann, 2005; Hamilton and Lin, 1996; Hamilton and Susmel, 1994; Turner, Startz, and Nelson, 1989), shifts in the Federal Reserve's policy "rule" (Kim, 2004; Sims and Zha, 2006), and time variation in the response of economic output to monetary policy actions (Garcia and Schaller, 2002; Kaufmann, 2002; Lo and Piger, 2005; Ravn and Sola, 2004). Hamilton and Raj (2002), Hamilton (2008), and Piger (2009) provide surveys of MS models, while Hamilton (1994) and Kim and Nelson (1999c) provide textbook treatments.

Following Hamilton (1989), early work on MS models focused on univariate models. In this case, X_t is scalar, and a common modeling choice is a pth-order autoregressive model with Markov-switching parameters:

$$X_t = \mu_{S_{t+h}} + \phi_{1,S_{t+h}} \left(X_{t-1} - \mu_{S_{t+h-1}}\right) + \cdots + \phi_{p,S_{t+h}} \left(X_{t-p} - \mu_{S_{t+h-p}}\right) + \varepsilon_t$$
$$\varepsilon_t \sim N\left(0, \sigma^2_{S_{t+h}}\right) \tag{18.12}$$

where $S_{t+h} \in \{1, \ldots, C\}$ indicates the regime and is assumed to be unobserved, even in the training sample. In this model, each of the mean, autoregressive parameters and conditional variance parameters are allowed to change in each of the C different regimes. Hamilton (1989) develops a recursive filter that can be used to construct the likelihood function for this MS autoregressive model, and thus estimate the parameters of the model via maximum likelihood.

A subsequent literature explored Markov-switching in multivariate settings. In the context of identifying business cycle regimes, Diebold and Rudebusch (1996) argue that considering multivariate information in the form of a factor structure can drastically improve statistical identification of the regimes. Chauvet (1998) operationalizes this idea by developing a statistical model that incorporates both a dynamic factor model and Markov-switching, now commonly called a dynamic factor Markov-switching (DFMS) model. Specifically, if X_t is multivariate, we assume that X_t is driven by a single-index dynamic factor structure, where the dynamic factor is itself driven by a Markov-switching process. A typical example

enforced by restricting the variance in $S_{t+h} = 2$ to be larger than that in $S_{t+h} = 1$. See Hamilton, Waggoner, and Zha (2007) for an extensive discussion of normalization in the MS model.

of such a model is as follows:

$$X_t^{std} = \begin{bmatrix} \lambda_1(L) \\ \lambda_2(L) \\ \vdots \\ \lambda_N(L) \end{bmatrix} F_t + v_t$$

where X_t^{std} is the demeaned and standardized vector of predictor variables, $\lambda_i(L)$ is a lag polynomial, and $v_t = \left(v_{1,t}, v_{2,t}, \ldots, v_{N,t}\right)'$ is a zero-mean disturbance vector meant to capture idiosyncratic variation in the series. v_t is allowed to be serially correlated, but its cross-correlations are limited. In the so-called exact factor model we assume that $E(v_{i,t}v_{j,t}) = 0$, while in the "approximate" factor model v_t is allowed to have weak cross-correlations. Finally, F_t is the unobserved, scalar, "dynamic factor." We assume that F_t follows a Markov-switching autoregressive process as in Eq. (18.12), with X_t replaced by F_t.

Chauvet (1998) specifies a version of this DFMS model where the number of predictors is $N = 4$ and shows how the parameters of both the dynamic factor process and the MS process can be estimated jointly via the approximate maximum likelihood estimator developed in Kim (1994). Kim and Nelson (1998) develop a Bayesian Gibbs-sampling approach to estimate a similar model. Finally, Camacho et al. (2018) develop modifications of the DFMS framework that are useful for real-time monitoring of economic activity, including mixed-frequency data and unbalanced panels. Chapter 2 of this book presents additional discussion of the DFMS model.

As discussed in Camacho, Perez-Quiros, and Poncela (2015), in data-rich environments the joint estimation of the DFMS model can become computationally unwieldy. In these cases, an alternative, two-step, approach to estimation of model parameters can provide significant computational savings. Specifically, in the first step, the dynamic factor F_t is estimated using the non-parametric principal components estimator of Stock and Watson (2002). Specifically, $\widehat{F}_t$ is set equal to the first principal component of X_t^{std}. In a second step, $\widehat{F}_t$ is fit to a univariate MS model as in Eq. (18.12). The performance of this two-step approach relative to the one-step approach was evaluated by Camacho et al. (2015), and the two-step approach was used by Fossati (2016) for the task of identifying US business cycle phases in real time.

For the purposes of this chapter, we are primarily interested in the ability of MS models to produce a class prediction, $\widehat{S}_{t+h}^c$. In an MS model, this prediction comes in the form of a "smoothed" conditional probability: $\widehat{S}_{t+h}^c = \Pr\left(S_{t+h} = c|\widetilde{X}_T\right), c = 1, \ldots, C$, where $\widetilde{X}_T$ denotes the entire training sample, $\widetilde{X}_T = \{X_t\}_{t=1}^T$. Bayesian estimation approaches of MS models are particularly useful here, as they produce this conditional probability while integrating out uncertainty regarding model parameters, rather than conditioning on estimates of these parameters.

In the application presented in Sect. 18.5 I will consider two versions of the DFMS model for classification. First, for cases with a small number of predictor variables, we estimate the parameters of the DFMS model jointly using the Bayesian sampler of Kim & Nelson (1998). Second, for cases where the number of predictor variables is large, we use the two-step approach described above, where we estimate the univariate MS model for $\widehat{F}_t$ via Bayesian techniques. Both of these approaches produce a classifier in the form of the smoothed conditional probability of the class. A complete description of the Bayesian samplers used in estimation is beyond the scope of this chapter. I refer the interested reader to Kim and Nelson (1999c), where detailed descriptions of Bayesian samplers for MS models can be found.

18.5 Application

In this section I present an application of the classification techniques presented above to nowcasting US expansion and recession phases at the monthly frequency. In this case, $S_{t+h} \in \{1, 2\}$, where $S_{t+h} = 1$ indicates a month that is a recession and $S_{t+h} = 2$ indicates a month that is an expansion. As I am interested in nowcasting, I set $h = 0$. As the measure of expansion and recession regimes, I use the NBER business cycle dates, which are determined by the NBER's Business Cycle Dating Committee. I will evaluate the ability of the alternative classifiers to accurately classify out-of-sample months that have not yet been classified by the NBER, and also to provide timely identification of turning points between expansion and recession (peaks) and recession and expansion (troughs) in real time.

Providing improved nowcasts of business cycle phases and associated turning points is of significant importance because there are many examples of turning points that were not predicted ex ante. This leaves policymakers, financial markets, firms, and individuals to try to determine if a new business cycle phase has already begun. Even this is a difficult task, with new turning points usually not identified until many months after they occur. For example, the NBER has historically announced new turning points with a lag of between 4 and 21 months. Statistical models improve on the NBER's timeliness considerably, with little difference in the timing of the turning point dates established.[8] However, these models still generally identify turning points only after several months have passed. For example, Hamilton (2011) surveys a wide range of statistical models that were in place to identify business cycle turning points in real time, and finds that such models did not send a definitive signal regarding the December 2007 NBER peak until late 2008.

There have been a number of existing studies that evaluate the performance of individual classifiers to nowcast US business cycle dates. In this chapter I contribute to this literature by providing a comparison of a broad range of classifiers, including

[8] See, e.g., Chauvet and Piger (2008), Chauvet and Hamilton (2006), and Giusto and Piger (2017).

several that have not yet been evaluated for the purpose of nowcasting business cycles. In doing so, I also evaluate the ability of these classifiers to provide improved nowcasts using large N vs. small N datasets. Most of the existing literature has focused on small N datasets, usually consisting of four coincident monthly series highlighted by the NBER's Business Cycle Dating Committee as important in their decisions. Notable exceptions are Fossati (2016), Davig and Smalter Hall (2016), and Berge (2015), each of which uses larger datasets to classify NBER recessions in real time.

For predictor variables, I begin with the FRED-MD dataset, which is a monthly dataset on a large number of macroeconomic and financial variables maintained by the Federal Reserve Bank of St. Louis. The development of FRED-MD is described in McCracken and Ng (2015). I use the most recent version of this dataset available, which as of the writing of this chapter was the vintage released at the end of November 2018. This vintage provides data for 128 monthly series covering months from a maximum of January 1959 through October 2018. I then delete six series that are not regularly available over the sample period, and add seven series on manufacturing activity from the National Association of Purchasing Managers, obtained from Quandl (www.quandl.com). I also add seven indices of "news implied volatility" as constructed in Manela and Moreira (2017). The addition of these series is motivated by Karnizova and Li (2014), who show that uncertainty measures have predictive power for forecasting US recessions. Finally, I restrict all series to begin in January 1960, which eliminates missing values during 1959 for a number of series.

For all series that are from the original FRED-MD dataset, I transform the series to be stationary using the transformation suggested in McCracken and Ng (2015), implemented using the Matlab code available from Michael McCracken's website. For the seven NAPM series and NVIX series, I leave the series without transformation. In some cases, the transformation involves differencing, which uses up the initial observation. To have a common starting point for our sample, I begin measuring all series in February 1960. The final raw dataset then consists of 136 series, where all series begin in February 1960 and extend to a maximum of October 2018.

In the analysis I consider three alternative datasets. The first, labeled *Narrow* in the tables below, is a small dataset that uses four coincident indicators that have been the focus of much of the US business cycle dating literature. These series are the growth rate of non-farm payroll employment, the growth rate of the industrial production index, the growth rate of real personal income excluding transfer receipts, and the growth rate of real manufacturing and trade sales. The second dataset, labeled *Real Activity*, is a larger dataset consisting of the 70 variables that are in the groupings "output and income," "labor market," "housing," and "consumption," orders and inventories' as defined in McCracken and Ng (2015). These variables define the real activity variables in the dataset, and as such target the most obvious variables with which to date turning points in real economic activity. The third dataset, labeled *Broad*, consists of all the variables in the dataset.

To evaluate the performance of the classifiers using these datasets, I conduct a pseudo out-of-sample nowcasting exercise that covers the last two decades. Specifically, consider an analyst applying a classification technique in real time at the end of each month from January 2000 to November 2018. For each of these months, I assume the analyst has a dataset covering the time period that would have been available in real time. That is, I accurately replicate the real-time data reporting lags the analyst would face. The reason this is a pseudo out-of-sample exercise is that I do not use the vintage of each dataset that would have been available in real time, as such vintage data is not readily available for all the variables in our dataset. Chauvet and Piger (2008) show that data revisions do not cause significant inaccuracies in real time dating of turning points. That said, interesting future work would replicate this analysis with a fully vintage dataset.

In each month, the analyst uses the available dataset to train the supervised classifiers over a period for which the NBER classification of S_t is assumed known. At each month, the lag with which the NBER classification is assumed known is allowed to vary, and is set using the approach taken in Giusto and Piger (2017). Specifically, I assume that: (1) The date of a new peak or trough is assumed to be known once it is announced by the NBER. (2) If the NBER does not announce a new peak within twelve months of a date, then it is assumed that a new peak did not occur at that date. Twelve months is the longest historical lag the NBER has taken in announcing a new business cycle peak. (3) Once the date of a new turning point is announced by the NBER, the new NBER business cycle phase (expansion or recession) is assumed to last at least 6 months. Since the unsupervised DFMS classifier does not require knowledge of the NBER classification, I estimate the parameters of this model over the full period for which the predictor data is available to the analyst. After training, the analyst then uses the classifier to classify those months through the end of the relevant sample of predictor variables for which the NBER dates are not known.

Somewhat more formally, suppose the data sample of predictor variables available to the analyst ends in time H, and the NBER dates are known through time $H - J$. Then the supervised classifiers would be trained on data through $H - J$, and the DFMS model would be estimated on data through H. After training and estimation, all classifiers will be used to classify the unknown NBER dates from month $H - J + 1$ through H, and the accuracy of these monthly classifications will be evaluated. I will also use these out-of-sample classifications to identify new business cycle turning points in real time.

Before discussing the results, there are several details of the implementation to discuss. First, when the analyst applies the classifiers to predict the NBER date out of sample, the most recent data to be classified will be incomplete due to differential reporting lags across series. In general, I handle these "ragged edges" by filling in the missing values using kNN imputation, as discussed in Sect. 18.3.3, prior to performing any subsequent analysis. Second, for the supervised classifiers, I classify S_t on the basis of the contemporaneous values of the predictor variables (month t) and the first lag (month $t - 1$). That is, the X_t vector contains both the contemporaneous and first lag of all variables in the relevant sample. For the unsupervised DFMS approach, all available values of X_t are used in forming the smoothed posterior probability for each class. Third, for each dataset, I replace outliers, defined as data points that are greater than four standard deviations from the mean, with the median of a six quarter window on either side of the outlier. Finally, as is typical in the classification literature, I standardize and demean all variables prior to analysis.[9]

For each of the supervised classifiers, I use the **caret** package in R to train and form predictions. In each case, repeated, stratified k-fold cross-validation is used for tuning parameters, with k set equal to 10, and the number of repeats also set equal to 10.[10] The objective function used in the cross-validation exercise was AUROC. Default ranges from caret were used in tuning parameters. Note that both the kNN imputation for missing values, as well as outlier detection, were done separately on each fold of the cross-validation exercise, which prevents data from outside the fold from informing the within-fold training.

For the unsupervised DFMS model, we must specify a specific version of the Markov-switching equation to apply to the factor F_t. In order to provide a comparison to the existing literature, I use specifications that most closely match those in existing studies. Specifically, for the model applied to the narrow dataset, for which the DFMS model is estimated jointly, I follow Chauvet (1998) and Kim and Nelson (1999b) and allow the factor to follow an AR(2) process with regime switching in mean:

$$F_t = \mu_{S_t} + \phi_1 \left(X_{t-1} - \mu_{S_{t-1}}\right) + \phi_2 \left(X_{t-2} - \mu_{S_{t-2}}\right) + \varepsilon_t$$
$$\varepsilon_t \sim N\left(0, \sigma^2\right)$$

For the DFMS model applied to the real activity dataset, for which the DFMS model is estimated via a two-step procedure, I follow Camacho et al. (2015) and use a simple AR(0) process with a switching mean:

$$F_t = \mu_{S_t} + \varepsilon_t$$
$$\varepsilon_t \sim N\left(0, \sigma^2\right)$$

I do not consider the broad dataset for the DFMS model, as the diversity of series in this dataset is likely not well described by only a single factor as is assumed by the DFMS model.

[9]In unreported results, I also considered a version of each supervised classifier that classified based on predictor variables formed as principal components from the relevant dataset. The performance of this version of the classifier was similar in all cases to the results applied to the full dataset of individual predictors.

[10]This is a relatively small number of repeats, and was chosen to reduce the computational burden of the recursive out-of-sample nowcasting exercise. In unreported results, I confirmed the robustness of several randomly chosen reported results to a larger number of repeats (100).

Table 18.1 presents the QPS and AUROC statistics for each classifier applied to each dataset, calculated over all the out-of-sample observations in the recursive nowcasting exercise. There are several conclusions that can be drawn from these results. First of all, in general, the AUROC statistics are very high, suggesting that each of these classifiers has substantial ability to classify expansion and recession months out of sample. Second, there are only relatively small differences in these statistics across classifiers. The DFMS model applied to the narrow dataset provides the highest AUROC at 0.997, which is very close to perfect classification ability,

Table 18.1 Out-of-sample evaluation metrics for alternative classifiers

Classifier	QPS	AUROC
Naïve Bayes		
Narrow	0.058	0.990
Real activity	0.064	0.974
Broad	0.074	0.968
kNN		
Narrow	0.030	0.989
Real activity	0.033	0.978
Broad	0.055	0.990
Random forest/extra trees		
Narrow	0.034	0.988
Real activity	0.032	0.988
Broad	0.036	0.989
Boosting		
Narrow	0.043	0.980
Real activity	0.037	0.978
Broad	0.039	0.982
LVQ		
Narrow	0.043	0.938
Real activity	0.046	0.930
Broad	0.038	0.952
DFMS		
Narrow	0.041	0.997
Real activity	0.047	0.992
Ensemble		
Narrow	0.034	0.992
Real activity	0.029	0.993
Broad	0.031	0.993

Notes: This table shows the quadratic probability score (QPS) and the area under the ROC curve (AUROC) for out-of-sample nowcasts produced from January 2000 to October 2018 by the supervised and unsupervised classifiers discussed in Sects. 18.3 and 18.4

while the kNN classifier applied to the narrow dataset produces the lowest QPS. However, most classifiers produce AUROCs and QPS values that are reasonably close to these best performing values.

Third, Table 18.1 suggests that the out-of-sample evaluation metrics are only moderately improved, it at all, by considering predictor variables beyond the narrow dataset. The differences in the evaluation statistics that result from changing dataset size are small, and in many cases these changes are not in a consistent direction across the QPS vs. AUROC. Overall, these results do not suggest that considering a larger number of predictors over the narrow dataset is clearly advantageous for nowcasting business cycle phases in US data. Note that some of this result likely comes because there is limited room for improvement over the narrow dataset.

Table 18.1 also presents results for a simple ensemble classifier, which is formed as the average of the alternative classifiers. The ensemble classifier averages the classification from all six classifiers for the narrow and real activity dataset, and averages the classification of the five supervised classifiers for the broad dataset. By averaging across approximately unbiased classifier that is not perfectly correlated, an ensemble classifier holds out the possibility of lower variance forecasts than is produced by the individual classifiers. Interestingly, the ensemble classifiers perform well in this setting, with the ensemble classifier applied to the real activity dataset having the lowest QPS and second highest AUROC of any classifier/dataset combination in the table.

The results in Table 18.1 do not speak directly to the question of identifying turning points (peaks and troughs) in real time. To evaluate the ability of the classifiers to identify turning points, we require a rule to transform the classifier output into turning point predictions. Here I employ a simple rule to identify a new turning point, which can be described as follows: If the most recent known NBER classified month is an expansion month, a business cycle peak is established if $\widehat{S}_t^1(X_t) \geq 0.5$ for the final month in the out-of-sample period. Similarly, if the most recent known NBER classified month is a recession month, a business cycle trough is established if $\widehat{S}_t^1(X_t) < 0.5$ for the final month in the out-of-sample period. This is a rather aggressive rule, which puts a high value on speed of detection. As such, we will be particularly interested in the tendency of this rule to identify false turning points.

Table 18.2 shows the performance of each classifier for identifying the four US business cycle turning points over the 2000–2018 time period. In the table, the dates shown are the first month in which the analyst would have been able to identify a turning point in the vicinity of the relevant turning point. For example, consider the column for the December 2007 business cycle peak. An entry of "Mar 2008" means that an analyst applying the classifiers at the end of March 2008 would have detected a business cycle peak in the vicinity of the December 2007 peak. Because there is a minimum 1 month lag in data reporting for all series in the FRED-MD dataset, the analyst would have been using a dataset that extended through February 2008 to identify this turning point. An entry of "NA" means that the relevant turning point was not identified prior to the NBER Business Cycle Dating Committee making the announcement of a new business cycle turning point.

Table 18.2 Out-of-sample turning point identification for alternative classifiers

Classifier	Peaks		Troughs		False turning points
	Mar 2001	Dec 2007	Nov 2001	Jun 2009	
Naïve Bayes					
Narrow	Feb 2001	May 2008	Jan 2002	Sep 2009	False Peak: Sep 2005
Real activity	Feb 2001	Mar 2008	Mar 2002	Feb 2010	False Peak: Nov 2010
Broad	Oct 2001	Mar 2008	Jan 2002	Feb 2010	None
kNN					
Narrow	May 2001	May 2008	Jan 2002	Aug 2009	None
Real activity	NA	Sep 2008	Jan 2002	Aug 2009	None
Broad	NA	NA	Nov 2001	Jul 2009	None
Random forest/extra trees					
Narrow	May 2001	May 2008	Jan 2002	Aug 2009	None
Real activity	NA	May 2008	Mar 2002	Aug 2009	None
Broad	NA	Oct 2008	Jan 2002	Aug 2009	None
Boosting					
Narrow	May 2001	May 2008	Jan 2002	Aug 2009	False Trough: Nov 2008
Real activity	May 2001	Nov 2008	Dec 2001	Sep 2009	None
Broad	May 2001	Nov 2008	Dec 2001	May 2009	None
LVQ					
Narrow	May 2001	May 2008	Feb 2002	Aug 2009	False Peak: Sep 2005
Real activity	Sep 2001	Mar 2008	Mar 2002	Aug 2009	False Peak: Oct 2010
Broad	May 2001	Mar 2008	Feb 2002	Aug 2009	None
DFMS					
Narrow	May 2001	May 2008	Jun 2002	Sep 2009	False Peak: Sep 2005
Real activity	Apr 2001	April 2008	Apr 2002	May 2010	None
Ensemble					
Narrow	May 2001	May 2008	Jan 2002	Aug 2009	False Peak: Sep 2005
Real activity	Jul 2001	May 2008	Mar 2002	Sep 2009	None
Broad	Mar 2001	May 2008	Mar 2002	Sep 2009	None

Notes: This table shows the earliest month that an analyst would have identified the four NBER business cycle turning points over the January 2000 to October 2018 out-of-sample period using the supervised and unsupervised classifiers discussed in Sects. 18.3 and 18.4. An entry of "NA" means the relevant turning point was not identified prior to the NBER Business Cycle Dating Committee making the announcement of a new business cycle turning point

I begin with the two business cycle peaks in the out-of-sample period. If we focus on the narrow dataset, we see that most of the classifiers identify the March 2001 business cycle peak by the end of May 2001, and the December 2007 peak by the end of May 2008. This is a very timely identification of both of these turning points. For the March 2001 peak, identification at the end of May 2001 means that the classifiers identified this recession using data through April 2001, which was the very first month of the recession. For the December 2007 peak, Hamilton (2011) reports that other real-time approaches in use at the time did not identify a business cycle peak until late 2008 or early 2009, and the NBER business cycle dating committee announced the December 2007 peak in December 2008. Thus, identification at the end of May 2008 is relatively very fast. This timely performance does come with a single false business cycle peak being called for several, although not all, of the classifiers. The date of this false peak was in September 2005 for most classifiers.

If we move to the larger real activity and broad datasets, in most cases the performance of the classifiers for identifying business cycle peaks deteriorates. Two of the classifiers, kNN and random forests, fail to identify the 2001 business cycle peak, while several classifiers identify peaks more slowly when using the larger datasets than the narrow dataset. There are some cases where moving from the narrow to the real activity dataset does improve with detection. The primary example is the DFMS model, where three of the four turning points are identified more quickly when using the real activity dataset, and the false peak that occurs under the narrow dataset is eliminated. Overall, a reasonable conclusion is that there are limited gains from using larger datasets to identify business cycle peaks in real time, with the gains that do occur coming from the use of the real activity dataset with certain classifiers.

Moving to troughs, the five supervised classification techniques identify troughs very quickly in real time when applied to the narrow dataset. For the November 2001 trough, these classifiers identify the trough by January or February of 2002, while the June 2009 trough is identified by August or September of 2009. This is impressive considering the very slow nature of the recovery following these recessions. As an example, the NBER business cycle dating committee did not identify the November 2001 troughs until July 2003 and the June 2009 trough until September 2010. This performance was achieved with only a single false trough identified by one algorithm, Boosted Classification Trees. The DFMS classifier was somewhat slower to detect these troughs than the other classifiers, although still substantially faster than the NBER announcement. Finally, consistent with the results for peaks, larger datasets did not substantially improve the timeliness with which troughs were identified on average.

Given the small number of turning points in the out-of-sample period, it is hard to distinguish definitively between the performance of the individual classifiers. If one were forced to choose a single classifier for the purpose of identifying turning points, both the kNN classifier and the random forest classifier applied to the narrow dataset were quick to identify turning points while producing no false positives. The other classifiers had similar performance, but produced a single false positive. That said, the kNN classifier and random forest classifier both failed to identify business cycle peaks when applied to the larger datasets, which may give us some pause as to the robustness of these classifiers. If one looks more holistically across the various datasets, the boosting algorithm emerges as a possible favorite, as it identifies all four turning points for all four datasets, and does so with speed comparable to the top set of performers for each dataset. Finally, the ensemble classifier has overall performance similar to the boosting algorithm. We might also expect, given the potential effect of averaging on reducing classifier variance, that the ensemble classifier will be the most robust classifier across alternative out-of-sample periods.

18.6 Conclusion

In this chapter I have surveyed a variety of approaches for real-time classification of economic time-series data. Special attention was paid to the case where classification is conducted in a data-rich environment. Much of the discussion was focused on machine learning supervised classification techniques that are common to the statistical classification literature, but have only recently begun to be widely used in economics. I also presented a review of Markov-switching models, which is an unsupervised classification approach that has been commonly used in economics for both historical and real-time classification. Finally, I presented an application to real-time identification of US business cycle turning points based on a wide dataset of 136 macroeconomic and financial time series.

References

Amit, Y., & Geman, D. (1997). Shape quantization and recognition with randomized trees. *Neural Computation, 9*, 1545–1588.

Ang, A., & Bekaert, G. (2002). Regime switches in interest rates. *Journal of Business and Economic Statistics, 20*, 163–182.

Berge, T. (2015). Predicting recessions with leading indicators: Model averaging and selection over the business cycle. *Journal of Forecasting, 34*(6), 455–471.

Berge, T., & Jordá, O. (2011). The classification of economic activity into expansions and recessions. *American Economic Journal: Macroeconomics, 3*(2), 246–277.

Breiman, L. (1996). Bagging predictors. *Machine Learning, 26*(2), 123–140.

Breiman, L. (2001). Random forests. *Machine Learning, 45*(1), 5–32.

Breiman, L., Friedman, J., Olshen, R., & Stone, C. (1984). *Classification and regression trees*. Belmont, CA: Wadsworth.

Camacho, M., Perez-Quiros, G., & Poncela, P. (2015). Extracting nonlinear signals from several economic indicators. *Journal of Applied Econometrics, 30*(7), 1073–1089.

Camacho, M., Perez-Quiros, G., & Poncela, P. (2018). Markov-switching dynamic factor models in real time. *International Journal of Forecasting, 34*, 598–611.

Chauvet, M. (1998). An econometric characterization of business cycle dynamics with factor structure and regime switching. *International Economic Review, 39*, 969–996.

Chauvet, M., & Hamilton, J. D. (2006). Dating business cycle turning points. In P. R. Costas Milas & D. van Dijk (Eds.), *Nonlinear time series analysis of business cycles* (pp. 1–53). North Holland: Elsevier.

Chauvet, M., & Piger, J. (2008). A comparison of the real-time performance of business cycle dating methods. *Journal of Business and Economic Statistics, 26*(1), 42–49.

Cosslett, S., & Lee, L.-F. (1985). Serial correlation in discrete variable models. *Journal of Econometrics, 27*, 79–97.

D opke, J., Fritsche, U., & Pierdzioch, C. (2017). Predicting recessions with boosted regression trees. *International Journal of Forecasting, 33*, 745–759.

Davig, T., & Smalter Hall, A. (2016). *Recession forecasting using Bayesian classification*. Federal Reserve Bank of Kansas City Research Working paper no. 1606.

Diebold, F. X., & Rudebusch, G. D. (1996). Measuring business cycles: A modern perspective. *The Review of Economics and Statistics, 78*(1), 67–77.

Dueker, M. (1997). Markov switching in GARCH processes and mean-reverting stock market volatility. *Journal of Business and Economic Statistics, 15*, 26–34.

Estrella, A., & Mishkin, F. S. (1998). Predicting U.S. recessions: Financial variables as leading indicators. *The Review of Economics and Statistics, 80*(1), 45–61.

Estrella, A., Rodrigues, A. P., & Schich, S. (2003). How stable is the predictive power of the yield curve? Evidence from Germany and the United States. *The Review of Economics and Statistics, 85*(3), 629–644.

Evans, M., & Wachtel, P. (1993). Inflation regimes and the sources of inflation uncertainty. *Journal of Money, Credit and Banking, 25*, 475–511.

Fossati, S. (2016). Dating U.S. business cycles with macro factors. *Studies in Non-linear Dynamics and Econometrics, 20*, 529–547.

Freund, Y. (1995). Boosting a weak learning algorithm by majority. *Information and Computation, 121*(2), 256–285.

Freund, Y., & Schapire, R. (1996). Experiments with a new boosting algorithm. *Proceedings of ICML, 13*, 148–156.

Friedman, J. (2001). Greedy function approximation: A gradient boosting machine. *The Annals of Statistics, 29*(5), 1189–1232.

Friedman, J., Hastie, T., & Tibshirani, R. (2000). Additive logistic regression: A statistical view of boosting. *The Annals of Statistics, 28*(2), 337–407.

Fushing, H., Chen, S.-C., Berge, T., & Jordá, O. (2010). *A chronology of international business cycles through non-parametric decoding*. SSRN Working Paper no: 1705758.

Garbellano, J. (2016). *Nowcasting recessions with machine learning: New tools for predicting the business cycle* (Bachelor's Thesis, University of Oregon).

Garcia, R., & Perron, P. (1996). An analysis of the real interest rate under regime shifts. *Review of Economics and Statistics, 78*(1), 111–125.

Garcia, R., & Schaller, H. (2002). Are the effects of monetary policy asymmetric? *Economic Inquiry, 40*, 102–119.

Geurts, P., Ernst, D., & Wehenkel, L. (2006). Extremely randomized trees. *Machine Learning, 63*(1), 3–42.

Giusto, A., & Piger, J. (2017). Identifying business cycle turning points in real time with vector quantization. *International Journal of Forecasting, 33*, 174–184.

Goldfeld, S. M., & Quandt, R. E. (1973). A Markov model for switching regressions. *Journal of Econometrics, 1*(1), 3–16.

Guidolin, M., & Timmermann, A. (2005). Economic implications of bull and bear regimes in UK stock and bond returns. *Economic Journal, 115*(500), 111–143.

Hamilton, J. D. (1989). A new approach to the economic analysis of nonstationary time series and the business cycle. *Econometrica, 57*(2), 357–384.

Hamilton, J. D. (1994). *Time series analysis*. Princeton, NJ: Princeton University Press.

Hamilton, J. D. (2008). Regime switching models. In S. N. Durlauf & L. E. Blume (Eds.), *New Palgrave dictionary of economics* (2nd ed.). London: Palgrave MacMillan.

Hamilton, J. D. (2011). Calling recessions in real time. *International Journal of Forecasting, 27*(4), 1006–1026.

Hamilton, J. D., & Lin, G. (1996). Stock market volatility and the business cycle. *Journal of Applied Econometrics, 11*, 573–593.

Hamilton, J. D., & Raj, B. (2002). New directions in business cycle research and financial analysis. *Empirical Economics, 27*(2), 149–162.

Hamilton, J. D., & Susmel, R. (1994). Autoregressive conditional heteroskedasticity and changes in regime. *Journal of Econometrics, 64*, 307–333.

Hamilton, J. D., Waggoner, D. F., & Zha, T. (2007). Normalization in econometrics. *Econometric Reviews, 26*(2–4), 221–252.

Hammer, B., & Villmann, T. (2002). Generalized relevance learning vector quantization. *Neural Networks, 15*(8–9), 1059–1068.

Harding, D., & Pagan, A. (2006). Synchronization of cycles. *Journal of Econometrics, 132*(1), 59–79.

Hastie, T., Tibshirani, R., & Friedman, J. (2009). *The elements of statistical learning. data mining, inference and prediction*. New York, NY: Springer.

Ho, T. K. (1998). The random subspace method for constructing decision forests. *IEEE Transactions on Pattern Analysis and Machine Intelligence, 20*(8), 832–844.

Karnizova, L., & Li, J. (2014). Economic policy uncertainty, financial markets and probability of U.S. recessions. *Economics Letters, 125*(2), 261–265.

Kaufmann, S. (2002). Is there an asymmetric effect of monetary policy over time? A Bayesian analysis using Austrian data. *Empirical Economics, 27*, 277–297.

Kauppi, H., & Saikkonen, P. (2008). Predicting U.S. recessions with dynamic binary response models. *The Review of Economics and Statistics, 90*(4), 777–791.

Kim, C.-J. (1994). Dynamic linear models with Markov switching. *Journal of Econometrics, 60*(1–2), 1–22.

Kim, C.-J. (2004). Markov-switching models with endogenous explanatory variables. *Journal of Econometrics, 122*, 127–136.

Kim, C.-J., Morley, J., & Piger, J. (2005). Nonlinearity and the permanent effects of recessions. *Journal of Applied Econometrics, 20*(2), 291–309.

Kim, C.-J., & Nelson, C. R. (1998). Business cycle turning points, a new coincident index, and tests of duration dependence based on a dynamic factor model with regime switching. *Review of Economics and Statistics, 80*(2), 188–201.

Kim, C.-J., & Nelson, C. R. (1999a). Friedman's plucking model of business fluctuations: Tests and estimates of permanent and transitory components. *Journal of Money, Credit and Banking, 31*, 317–334.

Kim, C.-J., & Nelson, C. R. (1999b). Has the U.S. economy become more stable? A Bayesian approach based on a Markov-switching model of the business cycle. *Review of Economics and Statistics, 81*(4), 608–616.

Kim, C.-J., & Nelson, C. R. (1999c). *State-space models with regime switching*. Cambridge, MA: The MIT Press.

Kohonen, T. (2001). *Self-organizing maps*. Berlin: Springer.

Lo, M. C., & Piger, J. (2005). Is the response of output to monetary policy asymmetric? Evidence from a regime-switching coefficients model. *Journal of Money, Credit and Banking, 37*, 865–887.

Manela, A., & Moreira, A. (2017). News implied volatility and disaster concerns. *Journal of Financial Economics, 123*(1), 137–162.

Mayr, A., Binder, H., Gefeller, O., & Schmid, M. (2014). The evolution of boosting algorithms: From machine learning to statistical modelling. *Methods of Information in Medicine, 6*(1), 419–427.

McCracken, M. W., & Ng, S. (2015). *Fred-md: A monthly database for macroeconomic research*. St. Louis Federal Reserve Bank Working Paper no. 2015-012B.

Ng, S. (2014). Boosting recessions. *Canadian Journal of Economics, 47*(1), 1–34.

Owyang, M. T., Piger, J., & Wall, H. J. (2015). Forecasting national recessions using state-level data. *Journal of Money, Credit and Banking, 47*(5), 847–866.

Piger, J. (2009). Econometrics: Models of regime changes. In R. A. Meyers (Ed.), *Encyclopedia of complexity and system science* (pp. 2744–2757). Berlin: Springer.

Qi, M. (2001). Predicting U.S. recessions with leading indicators via neural network models. *International Journal of Forecasting, 17*, 383–401.

Ravn, M., & Sola, M. (2004). Asymmetric effects of monetary policy in the united states. *Federal Reserve Bank of St. Louis Review, 86*, 41–60.

Rudebusch, G., & Williams, J. (2009). Forecasting recessions: The puzzle of the enduring power of the yield curve. *Journal of Business and Economic Statistics, 27*(4), 492–503.

Sato, A., & Yamada, K. (1995). Generalized learning vector quantization. In D. T. G. Tesauro & T. Leen (Eds.), *Advances in neural information processing systems* (pp. 423–429). North Holland: Elsevier.

Schapire, R. E. (1990). The strength of weak learnability. *Machine Learning, 5*, 197–227.

Sims, C. A., & Zha, T. (2006). Were there regime switches in U.S. monetary policy? *American Economic Review, 96*(1), 54–81.

Stock, J. H., & Watson, M. W. (2002). Forecasting using principal components from a large number of predictors. *Journal of the American Statistical Association, 97*, 1167–1179.

Stock, J. H., & Watson, M. W. (2014). Estimating turning points using large data sets. *Journal of Econometrics, 178*(1), 368–381.

Turner, C. M., Startz, R., & Nelson, C. R. (1989). A Markov model of heteroskedasticity, risk, and learning in the stock market. *Journal of Financial Economics, 25*(1), 3–22.

Vishwakarma, K. (1994). Recognizing business cycle turning points by means of a neural network. *Computational Economics, 7*, 175–185.

Ward, F. (2017). Spotting the danger zone: Forecasting financial crises with classification tree ensembles and many predictors. *Journal of Applied Econometrics, 32*, 359–378.

Chapter 19
Robust Methods for High-Dimensional Regression and Covariance Matrix Estimation

Marco Avella-Medina

19.1 Introduction

Statistical models are used in practice as approximations to reality. They are stylized mathematical constructions built under certain assumptions and justified from a purely theoretical point of view, by the statistical properties that they enjoy. One should naturally wonder how a statistical procedure behaves if the assumptions upon which the model is constructed fail to hold. This question has arguably become even more important with the large sizes of modern data sets being analyzed as more complex models inevitably lead to more assumptions. This motivates the development robust procedures that are less sensitive towards stochastic deviations from the assumed models.

Many modern scientific works analyze high-dimensional data sets using statistical models where the number of unknown parameters of interest p is very large relative to the sample size n. While many of the first motivating examples came from successful work genomics and biology (Alizadeh et al., 2000; Golub et al., 1999; Perou et al., 2000), the power of these techniques and the increasing availability of electronic data have also motivated their study in the social sciences. Two notorious examples of active research areas that have benefited from recent work in this direction are the literature in causal inference (Athey & Imbens, 2017; Athey, Imbens, & Wager, 2018; Chernozhukov et al., 2017) and finance (Ait-Sahalia & Xiu, 2017; Bai & Wang, 2016; Fan, Furger, & Xiu, 2016).

A ubiquitous theme in high-dimensional statistics is the need to assume appropriate low dimensional structure, as this is the key condition that ensures the existence of valid statistical methods in high-dimensional regimes. In particular,

M. Avella-Medina
Department of Statistics, Columbia University, New York, NY, USA
e-mail: marco.avella@columbia.edu

P. Fuleky (ed.), *Macroeconomic Forecasting in the Era of Big Data*,
Advanced Studies in Theoretical and Applied Econometrics 52,
https://doi.org/10.1007/978-3-030-31150-6_19

many regularization approaches have been studied in this context. Sparsity inducing regularization techniques have proved to be particularly useful when the ambient dimensional space is very large, but one assumes an intrinsic sparse dimensional parameter space. In regression settings this could mean that there is a very large number of covariates that one would like to include as predictors of some response variable, but it is believed that only a handful of those unknown regressors have real predictive power. In the context of covariance matrix estimation, a sparsity condition could mean that the population covariance matrix is banded or that its off diagonal elements decay very fast to zero.

In this chapter we review two ways of weakening distributional assumptions that are commonly made in the context of high-dimensional regression and covariance matrix estimation. The first approach that we describe builds on the theory of robust statistics pioneered by Hampel (1971, 1974) and Huber (1964). Book length exposition of this area include Hampel, Ronchetti, Rousseeuw, and Stahel (1986), Huber (1981), and Maronna, Martin, and Yohai (2006). Here contamination neighborhoods are defined to account for small, but arbitrary deviations from a targeted generative process that explains the majority of the data. The second approach that we will describe extends popular high-dimensional techniques that were developed under strong sub-Gaussian assumptions to heavy tailed scenarios. For this we build on robust statistics ideas to construct mean estimators that achieve finite sample Gaussian-type deviation errors in the presence of heavy tails. This particular property turns out to be a key one for successful sparse covariance matrix estimation and many other high-dimensional matrix estimation problems.

The outline of this chapter is as follows. In Sect. 19.2 we review some of the basic ideas of robust statistics by introducing small neighborhood contamination models, M-estimators, and influence functions. In Sect. 19.3 we show how robust statistics ideas can be extended to high-dimensional regression settings by considering penalized M-estimators for generalized linear models. In Sect. 19.4 we discuss the problem of covariance matrix estimation in high dimensions and explain how one can obtain optimal rates of convergence for this problem by regularizing a suitable initial estimator that builds on M-estimators. In Sect. 19.5 we discuss how the ideas of Sects. 19.3 and 19.4 can be extended to other models, before we conclude in Sect. 19.6.

19.2 Robust Statistics Tools

19.2.1 Huber Contamination Models

Let us start by reviewing how Huber formalized the robustness problem in Huber (1964). Let F_θ be the assumed parametric model and consider the ε-neighborhood

$P_\varepsilon(F_\theta)$ of F_θ:

$$P_\varepsilon(F_\theta) = \{G | G = (1-\varepsilon)F_\theta + \varepsilon H,\ H \text{ an arbitrary distribution }\}.$$

The basic idea is to study the behavior of statistical procedures such as estimators and tests in $P_\varepsilon(F_\theta)$. This captures the idea that the assumed model F_θ might only be approximately correct. Huber formalized a minimax approach that views the robustness problem as a game between the Nature, which chooses a distribution G in the neighborhood $P_\varepsilon(F_\theta)$ of the model F_θ, and the Statistician, who chooses an estimator for θ in a given class $\{\psi\}$ of estimators. The payoff of the game is the asymptotic variance $V(\psi, G)$ of the estimator under a given distribution in $P_\varepsilon(F_\theta)$. The statistician solves this problem by choosing a minimax strategy that minimizes the asymptotic variance at the least favorable distribution in the neighborhood, leading to a robust estimator.

In the simple normal location model with mean parameter μ one wishes to estimate the unknown mean parameter μ using a given iid sample of normally distributed observations $z_1, \ldots, z_n$ drawn from the distribution Φ_μ. The solution to the minimax problem in $P_\varepsilon(\Phi_\mu)$ is the Huber estimator, i.e., the estimator T_n that solves the estimating equation

$$\sum_{i=1}^{n} \psi_c(z_i - T_n) = 0, \tag{19.1}$$

where $\psi_c(r) = \min\{c \max(-c, r)\}$ is the so-called Huber function displayed in Fig. 19.1. The constant c is a tuning parameter that controls the trade-off between efficiency at the model Φ_μ and robustness. The extreme cases $c = \infty$ and $c = 0$ lead

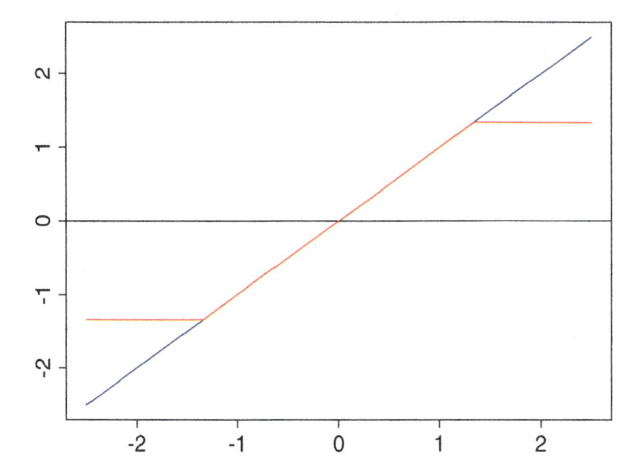

Fig. 19.1 Huber's function is displayed in red. The blue line corresponds to the case where $c = \infty$ in which the Huber estimator becomes the sample mean

respectively to the efficient but non-robust mean estimator and the highly robust but inefficient median estimator. The Huber estimator can be interpreted as an iterative reweighted mean estimator since (19.1) can be written as a

$$T_n = \frac{\sum_{i=1} w_c(r_i) z_i}{\sum_{i=1} w_c(r_i)}$$

where $r_i = z_i - T_n$ are the residuals and the weights are defined by $w_c(r_i) = \psi_c(r_i)/r_i$. Therefore observations with residuals smaller than c will get weight 1, while those large residuals receive smaller weights. The Huber function is of central importance in robust statistics and the idea of "huberizing" is a general and simple way of devising new robust estimators. Huber's minimax theory is a fundamental building block of robust statistics and leads to elegant and exact finite sample results. However, it has also been difficult to extend it to general parametric models. An alternative approach to robustness that has proven to be more easily extended to more complicated models is discussed next. A more extended overview can be found in Avella-Medina and Ronchetti (2015).

19.2.2 Influence Function and M-Estimators

Hampel (1974) opened one of the main lines of research in the robustness literature by formalizing the notion of local robustness, i.e., the stability of statistical procedures under moderate distributional deviations from ideal models. In this setting the quantities of interest are viewed as functionals of the underlying generating distribution. Typically, their linear approximation is studied to assess the behavior of estimators in a neighborhood of the model. Here the influence function plays a crucial role in describing the local stability of the functional analyzed. For a statistical functional $T(F)$, it is defined as

$$IF(z; T, F) = \lim_{\varepsilon \to 0} \frac{T\left((1-\varepsilon)F + \varepsilon \Delta_z\right) - T(F)}{\varepsilon},$$

where Δ_z is the distribution which puts mass 1 at any point z (Hampel, 1974; Hampel et al., 1986).

The influence function allows for an easy assessment of the relative influence of individual observations on the value of an estimate. If it is unbounded, a single outlier could cause trouble. Furthermore, if a statistical functional $T(F)$ is sufficiently regular, a first order von Mises expansion (von Mises, 1947) leads to the approximation

$$T(G) \approx T(F) + \int IF(z; F, T) d(G - F)(z), \tag{19.2}$$

where $IF(z; F, T)$ denotes the influence function of the functional T at the distribution F. Considering the approximation (19.2) over an ϵ neighborhood of the model $\mathcal{F}_\epsilon = \{G | G = (1-\epsilon)F + \epsilon H, H \text{ arbitrary}\}$, we see that the influence function can be used to linearize the asymptotic bias of $T(\cdot)$ in a neighborhood of the ideal model F. Consequently a bounded influence function implies a bounded approximate bias.

A particularly useful class of estimators can be defined as follows. Let $z_1, \ldots, z_n$ be a sample of iid m-dimensional observations, where $Z_i \sim F$ and define the functional $T(F)$ through the implicit equation

$$T(F) : \quad E_F\left[\psi(Z_i; T(F))\right] = 0, \tag{19.3}$$

where $\Psi : \mathbb{R}^m \times \mathbb{R}^p \rightarrow \mathbb{R}^p$. The sample version obtained by plugging in the empirical distribution $F = \hat{F}$ in (19.3) defines the *M-estimator* $T_n = T(\hat{F})$ as the solution of the equation

$$\sum_{i=1}^{n} \psi(z_i; T_n) = 0. \tag{19.4}$$

M-estimators enjoy several nice and useful properties:

- M-estimators generalize regular maximum likelihood estimators, which can be obtained by choosing the score function $\psi(z; \theta) = \frac{\partial}{\partial\theta} \log f_\theta(z)$ in (19.4), where $f_\theta(\cdot)$ is the density of the assumed parametric model.
- For a given parametric model F_θ, the condition $E_{F_\theta}\left[\psi(Z_i; \theta)\right] = 0$ implies $T(F_\theta) = \theta$, i.e., Fisher consistency for the M-estimator.
- Under general conditions (Huber, 1967, 1981), M-estimators are asymptotically normal:

$$\sqrt{n}(T_n - T(F)) \xrightarrow{\mathcal{D}} \mathcal{N}(0, V(\psi, F)),$$

  where

$$\begin{aligned} V(\psi, F) &= M(\psi, F)^{-1} Q(\psi, F) M(\psi, F)^{-T} \\ M(\psi, F) &= -\tfrac{\partial}{\partial\theta} E_F[\psi(Z; T(F))] \\ Q(\psi, F) &= E_F[\psi(Z; T(F)) \cdot \psi(Z; T(F))^T]. \end{aligned}$$

- The influence function of the function T is

$$IF(z; \psi, F) = M(\psi, F)^{-1} \psi(z; T(F)),$$

  i.e., it is proportional to $\psi(\cdot; T(F))$ and bounded if the latter is bounded.

In the following sections we will use M-estimators as building blocks for constructing robust high-dimensional estimators of generalized linear models and sparse covariance matrices.

19.3 Robust Regression in High Dimensions

In this section we review some of the results of Avella-Medina (2017) and Avella-Medina and Ronchetti (2018) on penalized M-estimation.

19.3.1 A Class Robust M-Estimators for Generalized Linear Models

Generalized linear models (McCullagh & Nelder, 1989) include standard linear models and allow the modeling of both discrete and continuous responses belonging to the exponential family. The response variables $Y_1, \ldots, Y_n$ are drawn independently from the densities $f(y_i; \theta_i) = \exp\left[\{y_i\theta_i - b(\theta_i)\}/\phi + c(y_i, \phi)\right]$, where $a(\cdot), b(\cdot)$, and $c(\cdot)$ are specific functions and ϕ a nuisance parameter. Thus $E(Y_i) = \mu_i = b'(\theta_i)$ and var(Y_i)= $v(\mu_i) = \phi b''(\theta_i)$ and $g(\mu_i) = \eta_i = x_i^T\beta_0$, where $\beta_0 \in R^d$ is the vector of parameters, $x_i \in R^d$ is the set of explanatory variables and $g(\cdot)$ the link function.

In this context we will construct penalized M-estimators by penalizing the class of loss functions proposed by Cantoni and Ronchetti (2001). These losses can be viewed as a natural robustification of the quasilikelihood loss of Wedderburn (1974), leading to the robust quasilikelihood

$$\rho_n(\beta) = \frac{1}{n}\sum_{i=1}^{n} Q_M(y_i, x_i^T\beta), \tag{19.5}$$

where the functions $Q_M(y_i, x_i^T\beta)$ can be written as

$$Q_M(y_i, x_i^T\beta) = \int_{\tilde{s}}^{\mu_i} \nu(y_i, t)w(x_i)\mathrm{d}t - \frac{1}{n}\sum_{j=1}^{n}\int_{\tilde{t}}^{\mu_j} E\{\nu(y_i, t)\}w(x_j)\mathrm{d}t$$

with $\nu(y_i, t) = \psi\{(y_i - t)/\sqrt{v(t)}\}/\sqrt{v(t)}$, $\tilde{s}$ such that $\psi\{(y_i - \tilde{s})/\sqrt{v(\tilde{s})}\} = 0$ and $\tilde{t}$ such that $E[\psi\{(y_i - \tilde{s})/\sqrt{v(\tilde{s})}\}] = 0$. The function $\psi(\cdot)$ is bounded and protects against large outliers in the responses, and $w(\cdot)$ downweights leverage points in the covariates. The estimator of $\hat{\beta}$ of β_0 derived from the minimization of this loss function is the solution of the estimating equation

$$\frac{1}{n}\sum_{i=1}^{n}\left\{\psi(r_i)\frac{1}{\sqrt{v(\mu_i)}}w(x_i)\frac{\partial\mu_i}{\partial\beta}-a(\beta)\right\}=0, \qquad (19.6)$$

where $r_i = (y_i - \mu_i)/\sqrt{v(\mu_i)}$ and $a(\beta) = n^{-1}\sum_{i=1}^{n} E\{\psi(r_i)/\sqrt{v(\mu_i)}\}w(x_i)\partial\mu_i/\partial\beta$ ensures Fisher consistency and can be computed as in Cantoni and Ronchetti (2001). Although here we focus the robust quasilikelihood loss, the results presented in this section are more general. Indeed, as explained in Avella-Medina and Ronchetti (2018), they also apply for the class of bounded deviance losses introduced in Bianco and Yohai (1996) and for the class of robust Bregman divergences of Zhang, Guo, Cheng, and Zhang (2014).

19.3.2 Oracle Estimators and Robustness

Sparsity is one of the central assumptions needed in high-dimensional regression. Specifically, one supposes that the true underlying parameter vector has many zero components and without loss of generality we write $\beta_0 = (\beta_1^T, \beta_2^T)^T$, where $\beta_1 \in R^k$, $\beta_2 = 0 \in R^{d-k}$, $k < n$ and $k < d$. In particular, this allows to consider scenarios where the number of unknown parameters p is larger than the sample size n.

Oracle estimators have played an important role in the theoretical analysis of many high-dimensional procedures. They are defined as the ideal estimator we would use if we knew the support $\mathcal{A} = \{j : \beta_{0j} \neq 0\}$ of the true parameter β_0, say maximum likelihood for the set of nonzero parameters. Such estimators can also be used for a simple robustness assessment of more complicated penalized estimators. Indeed, an oracle estimator, whose robustness properties can be easily assessed with standard tools such as the ones reviewed in Sect. 19.2 serves as a benchmark for a best possible procedure that is unfortunately unattainable. For example, since likelihood-based estimators in general do not have a bounded score function, they do not have a bounded influence function and are not robust in this sense. It follows that we could expect that in a neighborhood of the model, a penalized M-estimator based on a loss function with a bounded derivative could behave even better than classical non-robust oracle estimators. Clearly an appropriate benchmark estimator under contamination will only be given by a robust estimator that remains stable in $P_\epsilon(F)$.

19.3.3 Penalized M-Estimator

Penalized methods have proved to be a good alternative to traditional approaches for variable selection, particularly in high-dimensional problems. By allowing estimation and variable selection simultaneously, they overcome the high computational cost of variable selection when the number of covariates is large. Since their introduction for the linear model (Breiman, 1995; Frank & Friedman, 1993; Tibshirani, 1996), many extensions have been proposed (Efron, Hastie, Johnstone, & Tibshirani, 2004; Tibshirani, 2011; Yuan & Lin, 2006; Zou & Hastie, 2005). Their asymptotic properties have been studied when the number of parameters is fixed (Fan & Li, 2001; Knight & Fu, 2000; Zou, 2006) and a large literature treats the high-dimensional case, where the number of parameters is allowed to grow as the sample size increases (Bühlmann & van de Geer, 2011). These results provide strong arguments in favor of such procedures.

A first natural candidate for obtaining a robust high-dimensional estimator based on the quasilikelihood loss (19.5) is to include a ℓ_1 regularization penalty, thus defining the new objective function

$$\rho_n(\beta) + \lambda_n \sum_{j=1}^{d} |\beta_j|, \tag{19.7}$$

where $\lambda > 0$ is a tuning parameter that controls the amount of sparsity induced by the penalty term. The estimator $\hat{\beta}$ obtained by minimizing (19.7) is a natural extension of the lasso estimator (Tibshirani, 1996). Given the extensive literature showing the desirable properties of the lasso for sparse high-dimensional problems (Loh & Wainwright, 2015; Negahban, Ravikumar, Wainwright, & Yu, 2012), one can also expect our robust lasso to inherit those good properties of lasso estimators. The following theorem confirms this intuition. The main technical challenge that needs to be addressed for this problem is that robust quasilikelihood function is not guaranteed to be convex, even in low dimensions. The key observation that allows us to overcome this obstacle is that one can show that the nonconvexity of this problem is not too severe in the sense that for large n, with high probability, (19.5) satisfies the restricted strong convexity condition of Loh and Wainwright (2015), which in turn suffices to consistency. The result also holds if we replace the lasso penalty by a decomposable penalty as defined in Loh and Wainwright (2015). The following result appeared in Avella-Medina and Ronchetti (2018, Theorem 4).

Theorem 19.1 *Denote by $\hat{\beta}$ the robust lasso obtained by solving* (19.7) *with* $\lambda_n = O\{(n^{-1}\log d)^{1/2}\}$. *Further let* $k = o(n^{1/2})$ *and* $\log d = o(n^{1/2})$. *Then, under Conditions 4–6 in Avella-Medina and Ronchetti (2018) we have*

$$\|\hat{\beta} - \beta_0\|_2 = O\left\{\left(\frac{k \log d}{n}\right)^{1/2}\right\},$$

with probability at least $1 - 4e^{-\gamma\kappa_n}$, *where* γ *is some positive constant and* $\kappa_n = \min(n/k^2, \log d)$.

Theorem 19.1 tells us that consistency is achievable even when $d > n$, but it tells us nothing about the quality of the support recovery. In particular, can we guarantee that we found all the 0 components of β_0? In order to address this question one needs to consider more sophisticated penalty functions and assume minimal signal conditions on the nonzero components of β_0. Indeed, it has been shown that in general the lasso cannot be consistent for variable selection (Fan & Lv, 2011; Meinshausen & Bühlmann, 2006; Yuan & Lin, 2006; Zhao & Yu, 2006; Zou, 2006). One valid alternative is to consider a two stage procedure where one first fits an initial lasso estimator and uses it for the construction of a second stage adaptive lasso estimator. More precisely, given the initial lasso estimates $\tilde{\beta}$, we define the weights of the adaptive lasso estimator as

$$\hat{w}_j = \begin{cases} 1/|\tilde{\beta}_j|, & |\tilde{\beta}_j| > 0, \\ \infty, & |\tilde{\beta}_j| = 0, \end{cases} \quad j = 1, \ldots, d. \tag{19.8}$$

As a result of the above weights, the slope parameters that are shrunk to zero by the initial estimator are not included in the robust adaptive lasso minimization of the problem

$$\rho_n(\beta) + \lambda_n \sum_{j=1}^{d} \hat{w}_j |\beta_j|, \tag{19.9}$$

where we define $\hat{w}_j|\beta_j| = 0$ whenever $\hat{w}_j = \infty$ and $\beta_j = 0$. Let $\{(x_i, y_i)\}_{i=1}^n$ denote independent pairs, each having the same distribution.

Theorem 19.2 below appeared in Avella-Medina and Ronchetti (2018, Theorem 3). It shows that given an initial consistent estimate, the robust adaptive lasso enjoys oracle properties provided there is an appropriate scaling of the sparsity and ambient dimension of the problem. In particular, it requires that $k \ll n$ and $\log d = O(n^\alpha)$ for some $\alpha \in (0, 1/2)$. An additional minimum signal strength condition is needed in order to guarantee variable selection consistency, i.e., $s_n = \{|\beta_{0j}| : \beta_{0j} \neq 0\}$ is such that $s_n \gg \lambda_n$.

Theorem 19.2 *Assume Conditions 4–6 in Avella-Medina and Ronchetti (2018) and let* $\tilde{\beta}$ *be a consistent initial estimator with rate* $r_n = \{(k \log d)/n\}^{1/2}$ *in* ℓ_2*-norm defining weights* (19.8). *Let* $\log d = O(n^\alpha)$ *for* $\alpha \in (0, 1/2)$. *Further let the number of nonzero parameters be of order* $k = o(n^{1/3})$ *and assume the minimum signal is such that* $s_n \gg \lambda_n$ *with* $\lambda_n (nk)^{1/2} \to 0$ *and* $\lambda_n r_n \gg \max\{(k/n)^{1/2}, n^{-\alpha}(\log n)^{1/2}\}$. *Finally, let* v *be a k-dimensional vector with* $\|\mathrm{v}\|_2 = 1$. *Then, there exists a minimizer* $\hat{\beta} = (\hat{\beta}_1, \hat{\beta}_2)^T$ *of* (19.9) *such that as* $n \to \infty$,

(*a*) *sparsity:* $pr(\hat{\beta}_2 = 0) \to 1$;
(*b*) *asymptotic normality:* $n^{1/2} \mathrm{v}^T M_{11} Q_{11}^{-1/2} (\hat{\beta}_1 - \beta_1) \to N(0, 1)$ *in distribution.*

As in Zou (2006), the existence of an initial consistent estimator is the key for obtaining variable selection consistency in Theorem 19.2. Combined with Theorem 19.1, it implies that using the robust lasso as initial estimator for its adaptive counterpart yields an estimator that satisfies the oracle properties. One should however be aware of the limitations of estimators satisfying the oracle properties as stated in Theorem 19.2. Indeed, unlike the consistency result for the lasso, the adaptive lasso requires an additional minimal signal strength condition in order to establish variable selection consistency. In the fixed-parameter asymptotic scenario considered above, it requires that the nonzero coefficients will be asymptotically larger than $O(n^{-1/2})$. As shown by Leeb and Pötscher (2005, 2008, 2009), in the presence of weaker signals the distribution of estimators satisfying the oracle properties can be highly non-normal regardless of the sample size. Some recent proposals for uniform post-selection inference do not require minimal signal conditions on the nonzero coefficients. Representative work in this direction includes Belloni, Chen, Chernozhukov, and Hansen (2012), Belloni, Chernozhukov, and Kato (2015), Javanmard and Montanari (2014), Lee, Sun, Sun, and Taylor (2016), Zhang and Zhang (2014). This is a promising direction for future research.

19.3.4 Computational Aspects

There are some non-trivial aspects to be considered for the computation of the estimator introduced in the above section. In particular, the optimization of the objective function is more complicated than in standard non-penalized setups due to the presence of a non-differentiable penalty function and a tuning parameter.

Fisher Scoring Coordinate Descent

The sparsity inducing penalty function given by the weighted ℓ_1 norm in (19.9) requires the implementation of non-standard gradient descent algorithms. For this it suffices to consider the lasso penalty where $\hat{w}_j = 1$ $(j = 1, \ldots, d)$. Indeed, the adaptive lasso estimator can be computed with the same algorithm after one reparametrizes the adaptive lasso problem as discussed in Zou (2006, Section 3.5).

The main idea of our algorithm is to consider a coordinate-descent-type algorithm based on successive expected quadratic approximations of the quasilikelihood about the current estimates. In this way the optimization of the penalized robust quasilikelihood boils down to iteratively solving weighted least squares lasso problems. Specifically, for a given value of the tuning parameter, we successively minimize via coordinate descent the penalized weighted least squares loss

$$\|W(z - X\beta)\|_2^2 + \lambda \|\beta\|_1 \tag{19.10}$$

where $W = \text{diag}(W_1, \ldots, W_n)$ is a weight matrix and $z = (z_1, \ldots, z_n)^T$ a vector of pseudo-data with components

$$W_i^2 = E\{\psi(r_i)r_i\}v(\mu_i)^{-1}w(x_i)\left(\frac{\partial \mu_i}{\partial \eta_i}\right)^2, \quad z_i = \eta_i + \frac{\psi(r_i) - E\{\psi(r_i)\}}{E\{\psi(r_i)r_i\}}v(\mu_i)^{1/2}\frac{\partial \eta_i}{\partial \mu_i}.$$

These are the robust counterparts of the usual expressions appearing in the iterative reweighted least squares algorithm for GLM; cf. Appendix E.3 in Heritier, Cantoni, Copt, and Victoria-Feser (2009). This coordinate descent algorithm is therefore a sequence of three nested loops: (a) outer loop: decrease λ; (b) middle loop: update W and z in (19.10) using the current parameters $\hat{\beta}_\lambda$; (c) inner loop: run the coordinate descent algorithm on the weighted least squares problem (19.10). Interestingly, contrary to non-sparse and differentiable problems, coordinate descent type algorithms have empirically been shown to run fast provided one incorporates some tricks that significantly speed up computations (Friedman, Hastie, & Tibshirani, 2010). Our algorithm differs from the one of Friedman et al. (2010) in the quadratic approximation step, where we compute expected weights. This step assures that W has only positive components in Poisson and binomial regression when using Huberized residuals, which guarantees the convergence of the inner loop. For classical penalized log-likelihood regression with canonical link, the two algorithms coincide. The initial value of the tuning parameter in our algorithm is $\lambda_0 = n^{-1}\|WX^T z\|_\infty$, with W and z computed with $\beta = 0$. This guarantees that the initial solution is the zero vector. We then run our coordinate descent algorithm and solve our lasso problem along a grid of decreasing values of tuning parameters that defines a solution path for the slope parameters. The middle loop uses current parameters as warm starts. In the inner loop, after a complete cycle through all the variables we iterate only on the current active set, i.e., the set of nonzero coefficients. If another complete cycle does not change this set, the inner loop has converged, otherwise the process is repeated. The use of warm starts and active set cycling speeds up computations as discussed in Friedman et al. (2010, p. 7).

Tuning Parameter Selection

One can choose the tuning parameter λ_n based on a robust extended Bayesian information criterion. Specifically we select the parameter λ_n that minimizes

$$\text{EBIC}(\lambda_n) = \rho_n(\hat{\beta}_{\lambda_n}) + \Big(\frac{\log n}{n} + \gamma \frac{\log d}{n}\Big)|\text{supp}\hat{\beta}_{\lambda_n}|, \tag{19.11}$$

where $|\text{supp}\hat{\beta}_{\lambda_n}|$ denotes the cardinality of the support of $\hat{\beta}_{\lambda_n}$ and $0 \leq \gamma \leq 1$ is a constant. We use $\gamma = 0.5$. We write $\hat{\beta}_{\lambda_n}$ to stress the dependence of the minimizer of (19.9) on the tuning parameter. In an unpenalized setup, a Schwartz information criterion was considered by Machado (1993), who provided theoretical justification for it by proving model selection consistency and robustness. In the penalized sparse regression literature, Lambert-Lacroix and Zwald (2011) and Li, Peng, and Zhu (2011) used this criterion to select the tuning parameter. In high dimensions Chen and Chen (2008, 2012), and Fan and Tang (2013) showed the benefits of minimizing (19.11) in a penalized likelihood framework.

19.3.5 Robustness Properties

The theoretical results of Sect. 19.3.3 showed that if there is no contamination, the robust estimators obtained by penalizing the robust quasilikelihood preserve the good properties of its likelihood-based counterparts. The hope is that the resulting estimator is more stable when a small fraction of the data is contaminated, at best as stable as a robust oracle estimator as discussed in Sect. 19.3.2. In this subsection we will see in what sense that intuition is correct and illustrate it in simulated examples.

Finite Sample Bias

The asymptotic results given in Sect. 19.3.3 generalize to a shrinking contamination neighborhood where $\epsilon \to 0$ as $n \to \infty$, as long as $\epsilon = o(\lambda_n)$. Furthermore, if the contamination neighborhood does not shrink, but instead produces a small nonasymptotic bias on the estimated nonzero coefficients and the minimum signal is large enough, Avella-Medina and Ronchetti (2018, Theorem 6) showed that one can obtain correct support recovery and bounded bias. In particular, with high probability, one can obtain robust estimators $\hat{\beta}$ satisfying

(a) sparsity: $\hat{\beta}_2 = 0$;
(b) ℓ_∞-norm: $\|\hat{\beta}_1 - \beta_1\|_\infty = O(n^{-\zeta} \log n + \epsilon)$.

where ζ is such that $s_n \geq n^{-\zeta} \log n$. This statement is in the spirit of the infinitesimal robustness approach to robust statistics discussed in Sect. 19.2.2 and can be viewed as an extension, to a contaminated neighborhood, of the weak oracle properties derived in Fan and Lv (2011, Theorem 2).

Let us illustrate numerically the performance of classical and robust versions of both the lasso and adaptive lasso in a contaminated generalized linear model. For the robust estimators we use the loss function given by (19.6) with $\psi(r) = \psi_c(r)$, the Huber function. We take as the target model a Poisson regression model with canonical link $g(\mu_i) = \log \mu_i = x_i^T \beta$, where $\beta = (1.8, 1, 0, 0, 1.5, 0, \cdots, 0)^T$ and the covariates x_{ij} were generated from standard uniforms with correlation $\mathrm{cor}(x_{ij}, x_{ik}) = \rho^{|j-k|}$ and $\rho = 0.5$ for $j, k = 1, \ldots, d$. This setup is reminiscent of example 1 in Tibshirani (1996). The response variables Y_i were generated according to the Poisson distribution $\mathcal{P}(\mu_i)$ and a perturbed distribution of the form $(1 - b)\mathcal{P}(\mu_i) + b\mathcal{P}(\nu\mu_i)$, where $b \sim Bin(1, \epsilon)$. The latter represents a situation where the distribution of the data lies in a small neighborhood of the model that can produce for instance overdispersion. We set $c = 1.5$, $\nu = 1, 5, 10$, and $\epsilon = 0.05$. The sample sizes and dimensionality were respectively $n = 50, 100, 200$ and $d = 100, 400, 1600$. We implemented the Fisher scoring coordinate descent algorithm as discussed in Sect. 19.3.4 with a grid of values λ of length 100 decreasing on the log scale. We stopped the algorithm when the tuning parameter gave models of size greater than or equal to 20, 30, and 40 respectively when n was 50, 100, and 200. The tuning parameters were selected by BIC and the simulation size was 100. We

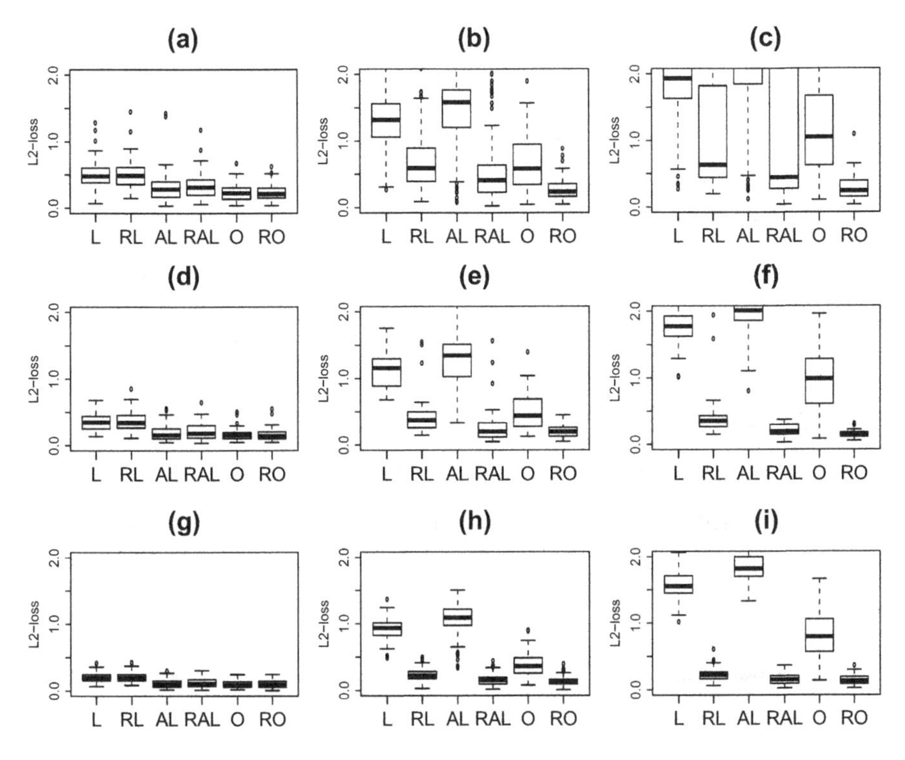

Fig. 19.2 The plots show the $L2$ error of the classical and robust versions of the lasso, adaptive lasso, and oracle. The severity of the contamination $\nu = 1, 5, 10$ increases from left to right and the sample size and dimensionality of the problems $(n, d) = \{(50, 100), (100, 400), (200, 1600)\}$ from top to bottom

show the performance of the classical lasso (L) and adaptive lasso (AL), the robust lasso (RL) and adaptive robust lasso (RAL) as well as the classical oracle (O) and the robust oracle (RO) estimators. Figure 19.2 illustrates the estimation error of the different estimators while Fig. 19.3 illustrates the model selection properties.

It is clear that without contamination the classical procedures and their robust counterparts have a very similar performance. As expected from our theoretical results the estimation error of all the penalized estimators seems to converge to zero as the sample size increases despite of the fact that the parameter dimension increases at an even faster rate. The picture changes drastically under contamination for the classical estimators as they give poor estimation errors and keep too many variables. On the other hand, the robust estimators remain stable under contamination. In particular the robust adaptive lasso performs almost as well as the robust oracle in terms of L_2-loss and is successful in recovering the true support when $n = 100, 200$ even under contamination. In this example the robust adaptive lasso outperforms the classical oracle estimator under contamination. The poor L_2 error of the classical penalized estimators under contamination is

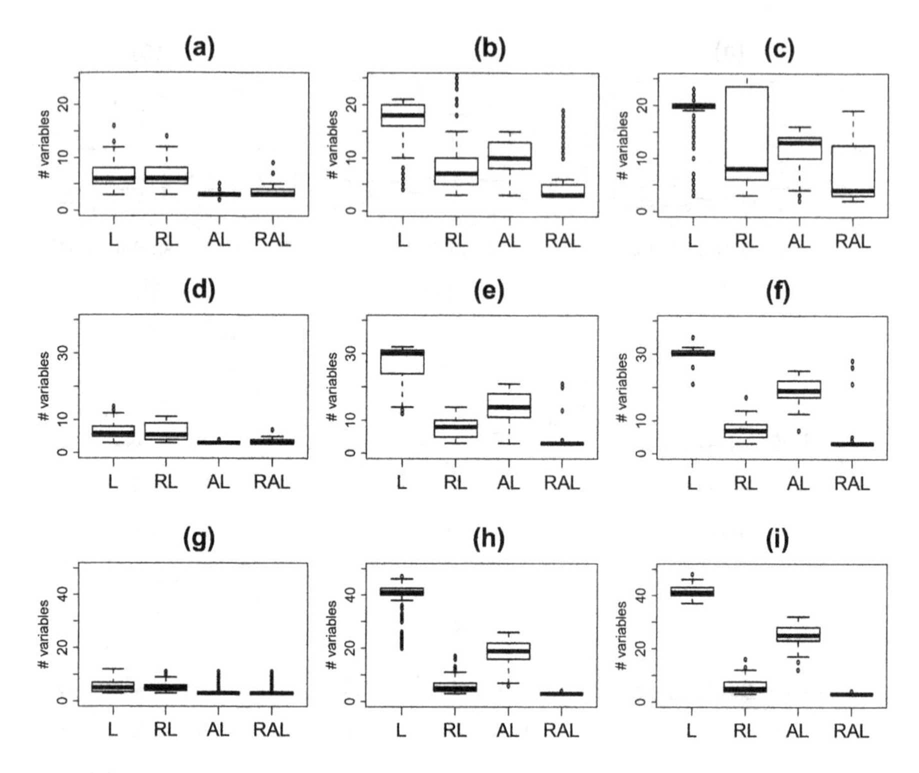

Fig. 19.3 The plots show the cardinality of the support of the classical and robust versions of the lasso and adaptive lasso. The severity of the contamination $\nu = 1, 5, 10$ increases from left to right and the sample size and dimensionality of the problems $(n, d) = \{(50, 100), (100, 400), (200, 1600)\}$ from top to bottom

likely to stem from the large number of noise variables they tend to select in this situation.

Influence Function

Given the discussion in Sect. 19.2.2, it is very natural to seek to assess the robustness properties of penalized M-estimators using the influence function. One should expect that loss functions leading to bounded-influence M-estimators can also be used to construct bounded-influence penalized M-estimators. In particular, intuitively a loss function with a bounded gradient function Ψ should guarantee robustness. This intuition turns out to be correct, but requires a new notion of influence function established in Avella-Medina (2017). Indeed, the typical tools used to derive the influence function of M-estimators suffer from a major problem when considering penalized M-estimators: they cannot handle non-differentiable penalty functions which are necessary for achieving sparsity (Fan & Li, 2001).

The usual argument used to obtain the influence function of M-estimators is as follows. One can look at the M-functional $T(F_\varepsilon)$ defined by (19.3) evaluated at F_ε, as the solution to an implicit equation depending on two arguments $T(F_\varepsilon)$ and ε. Taking this perspective (19.3) becomes $g(T(F_\varepsilon), \varepsilon) = 0$ and the derivative of $T(F_\varepsilon)$ with respect to ε can be computed using the implicit function theorem. Because nondifferentiable penalties do not lead to estimating equations of the form (19.3), an alternative approach is required.

One possible way to circumvent the technical difficulties entailed by a nondifferentiable penalty functions $P(\cdot)$ is to define a sequence of smooth penalty functions $\{P_m(\cdot)\}$ such that $\lim_{m\to\infty} P_m(\cdot) = P(\cdot)$. This trick can be used to define a limiting form of the influence function of penalized M-estimators obtained using smooth penalty functions P_m., denoted by $T(F; P_m)$. These estimators are defined as the minimizers of

$$\Lambda_\lambda(\theta; F, P_m) = E_F[L(Z, \theta)] + P_m(\theta; \lambda).$$

We let $\text{IF}_{P_m}(z; T, F)$ be the influence function of $T(F; P_m)$ and define the influence function of $T(F)$ as

$$\text{IF}(z; F, T) := \lim_{m\to\infty} \text{IF}_{P_m}(z; F, T).$$

A natural question that arises from this definition is whether the limit depends on the sequence $\{P_m\}$ chosen. In order to answer this question we first show that the limiting M-functional $\lim_{m\to\infty} T(F; P_m)$ is unique as well as its influence function. Furthermore, the resulting influence function will also be bounded if and only if Ψ is bounded. Interestingly the limiting influence function can still be viewed as a derivative but now in the sense of distribution theory of Schwartz (Schwartz, 1959). See Avella-Medina (2017) for rigorous statements. Figure 19.4 complements the simulation results shown above by considering a uniform grid of values of $\varepsilon \in [0, 0.1]$ and $\nu = 5$. It clearly shows that the robust lasso and adaptive lasso remain stable under moderate contamination whereas their classical counterparts are not. Indeed, even very small amounts of contamination completely ruin the performance of the classical procedures. It is interesting to note that although our robust estimators clearly outperform the likelihood-based methods under contamination, their performance start to deteriorate as the amount of contamination approaches 10%. This reflects the local nature of the robustness of bounded-influence estimators. Indeed, the robustness properties of our quasilikelihood loss stem from the boundedness of its derivative. This particular feature guarantees that just as the M-estimators analogues, our robust penalized quasilikelihood estimators will have a bounded influence function. Therefore they are only expected to have a bounded bias in a small contamination neighborhood of the model as shown in our simulations.

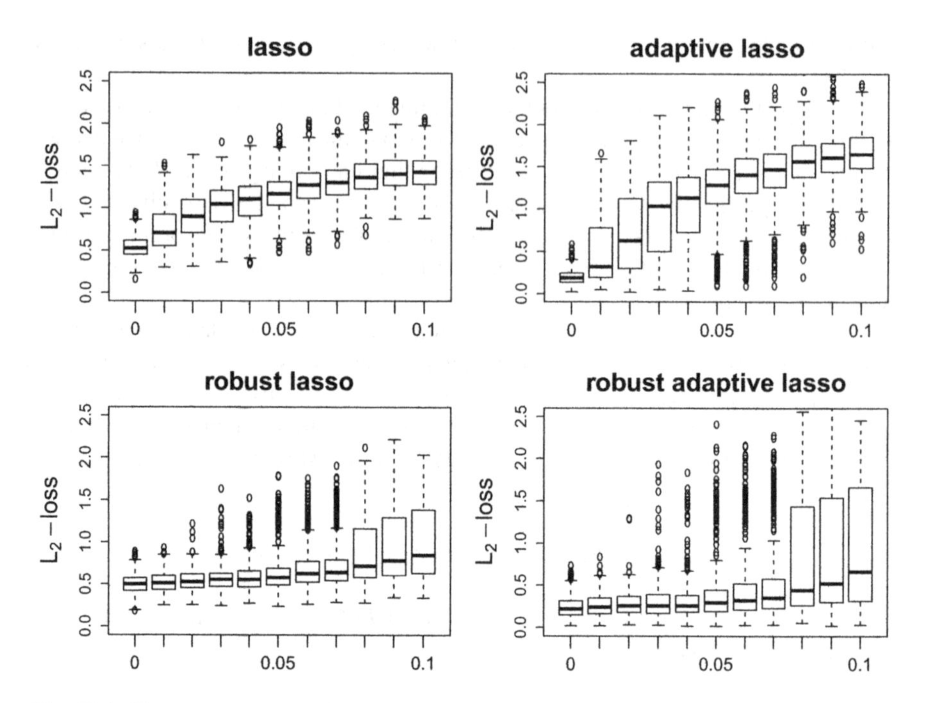

Fig. 19.4 The boxplots show the performance measured by the L2-loss of classical and robust counterparts of the lasso and adaptive lasso as the contamination increases

19.4 Robust Estimation of High-Dimensional Covariance Matrices

In this section we introduce some key ideas arising in the presence of high-dimensional data when the object of study is the unknown population variance matrix. We will then discuss how to introduce robustness considerations by reviewing some results obtained in Avella-Medina, Battey, Fan, and Li (2018).

19.4.1 Sparse Covariance Matrix Estimation

The range of open statistical challenges posed by high-dimensional data, in which the number of variables p is larger than the number of observations n, is still intimidating. Among the most relevant challenges is high-dimensional covariance matrix and inverse covariance (precision) matrix estimation. In fact, almost every procedure from classical multivariate analysis relies on an estimator of at least one of these objects. The difficulty in achieving consistency in certain meaningful matrix norms leaves classical multivariate analysis unable to adequately respond to the needs of the data-rich sciences.

Consistent covariance matrix estimation in high-dimensional settings is achievable under suitable structural assumptions and regularity conditions. For instance, under the assumption that all rows or columns of the covariance matrix belong to a sufficiently small ℓ_q-ball around zero, thresholding (Bickel & Levina, 2008; Rothman, Bickel, Levina, & Zhu, 2008) or its adaptive counterpart (Cai & Liu, 2011) give consistent estimators of the covariance matrix in spectral norm for data drawn from a sub-Gaussian distribution. Here we will focus on a subset of the class $\mathcal{S}^+(\mathbb{R}, p)$ of positive definite symmetric matrices with elements in $\mathbb{R}$. In particular, we assume that $\Sigma^* = (\sigma^*_{uv})$ belongs to the class sparse matrices

$$\mathcal{U}_q\{s_0(p)\} = \left\{\Sigma : \Sigma \in \mathcal{S}^+(\mathbb{R}, p),\ \max_u \sum_{v=1}^{p} (\sigma^*_{uu}\sigma^*_{vv})^{(1-q)/2} |\sigma^*_{uv}|^q \leq s_0(p)\right\}, \tag{19.12}$$

where each column is assumed to be weakly sparse in the sense that they are all required to lie in a ℓ_q ball of size $s_0(p)$. The columns of a covariance matrix in $\mathcal{U}_q\{s_0(p)\}$ are required to lie in a weighted ℓ_q-ball, where the weights are determined by the variance of the entries of the population covariance. The class $\mathcal{U}_q\{s_0(p)\}$ was introduced by Cai and Liu (2011), slightly generalizing the class of matrices considered in Bickel and Levina (2008) and Rothman, Levina, and Zhu (2009).

Assuming the population covariance matrix lies in (19.12), a valid estimation approach is to regularize the sample covariance matrix

$$\widehat{\Sigma} = (\widehat{\sigma}_{uv}) = \frac{1}{n} \sum_{i=1}^{n} (X_i - \overline{X})(X_i - \overline{X})^T$$

with $X_1, \ldots, X_n$ independent and identically distributed copies of $X \in \mathbb{R}^p$ and $\overline{X} = n^{-1} \sum_{i=1}^{n} X_i$. Clearly $\widehat{\Sigma}$ is a very poor estimate of Σ when p is large relative to the sample size n and will not even be positive definite if $p > n$. One can however show that simple elementwise shrinkage and thresholding of the sample covariance can lead to consistent estimation. More precisely, we can let $\tau_\lambda(\cdot)$ be a general thresholding function for which

(i) $|\tau_\lambda(z)| \leq |y|$ for all z, y that satisfy $|z - y| \leq \lambda$;
(ii) $\tau_\lambda(z) = 0$ for $|z| \leq \lambda$;
(iii) $|\tau_\lambda(z) - z| \leq \lambda$, for all $z \in \mathbb{R}$.

Similar thresholding functions were set forth in Antoniadis and Fan (2001) and proposed in the context of covariance estimation via thresholding in Rothman et al. (2009) and Cai and Liu (2011). Some examples of thresholding functions satisfying these three conditions are the soft thresholding rule $\tau_\lambda(z) = \text{sgn}(z)(z - \lambda)_+$, the adaptive lasso rule $\tau_\lambda(z) = z(1 - |\lambda/z|^\eta)_+$ with $\eta \geq 1$ and the smoothly clipped absolute deviation thresholding (SCAD) rule (Rothman et al., 2009). The hard thresholding rule $\tau_\lambda(z) = z1(|z| > \lambda)$ does not satisfy (i), but leads to similar results. The adaptive thresholding estimator is defined as

$$\widehat{\Sigma}^{\mathcal{T}} = (\widehat{\sigma}_{uv}^{\mathcal{T}}) = \left\{\tau_{\lambda_{uv}}(\widehat{\sigma}_{uv})\right\}$$

where $\widehat{\sigma}_{uv}$ is (u, v)th the entry of $\widehat{\Sigma}$ and the threshold λ_{uv} is entry-dependent. Equipped with these adaptive thresholds, Cai & Liu (2011) establish optimal rates of convergence of the resulting estimator under sub-Gaussianity of X. The latter condition requires the existence of $b > 0$ such that $E(\exp[t\{X_{1,u} - E(X_{1,u})\}]) \leq \exp(b^2t^2/2)$ for every $t \in \mathbb{R}$ and every $u = [p]$, and $[p]$ denotes $\{1, \ldots, p\}$. Such a light-tail distributional condition is crucial in high dimensions as discussed in next subsection.

19.4.2 The Challenge of Heavy Tails

The intuition behind the success of thresholding covariance matrix estimators is fairly simple. Even though the sample covariance is a poor high-dimensional estimator, each of the entries of this matrix are consistent estimators of their respective population pairwise covariance. Hence elementwise shrinkage approaches should lead to valid consistent estimation as long as one can guarantee a uniform control over the convergence of all the pairwise sample covariances. Concentration inequalities provide a way to guarantee such a uniform control and are in fact one of the cornerstones of the theory of high-dimensional statistics. Assuming that the distribution of the data generating process is sub-Gaussian, typically allows to derive exponential concentration inequalities (Boucheron, Lugosi, & Massart, 2013). This is critical for instance for the empirical mean, since this estimator cannot concentrate exponentially unless we assume that the underlying data generating process is sub-Gaussian (Bubeck, Cesa-Bianchi, & Lugosi, 2013). Proposition 1 in Avella-Medina et al. (2018) provides a similar negative result for the sample covariance matrix. This has non-trivial consequences for the problem of high-dimensional covariance matrix estimation as it rules out regularizing the sample covariance. Indeed, it cannot satisfy (19.13) in the presence of heavy tails, i.e., if one only wants to assume a few finite moments.

Since sub-Gaussianity seems to be too restrictive in practice, Avella-Medina et al. (2018) suggested new procedures that can achieve the same minimax optimality when data are leptokurtic. Inspection of the proofs of Bickel and Levina (2008) and Cai and Liu (2011) reveals that sub-Gaussianity is needed merely because those thresholding estimators regularize the sample covariance matrix. Indeed, minimax optimal rates of convergence are achievable within a larger class of distributions if the sample covariance matrix is replaced by some other pilot estimator that exhibits better concentration properties.

More precisely, for a p-dimensional random vector X with mean μ, let $\Sigma^* = E\{(X-\mu)(X-\mu)^T\}$ and let $\widetilde{\Sigma} = (\widetilde{\sigma}_{uv})$ denote an arbitrary pilot estimator of $\Sigma^* = (\sigma^*_{uv})$ where $u, v \in [p]$. The key requirement for adaptive elementwise shrinkage on $\widetilde{\Sigma}$ to give optimal covariance estimation is that

$$\text{pr}[\max_{u,v} |\widetilde{\sigma}_{uv} - \sigma^*_{uv}| \leq C_0\{(\log p)/n\}^{1/2}] \geq 1 - \epsilon_{n,p}, \tag{19.13}$$

where C_0 is a positive constant and $\epsilon_{n,p}$ is a deterministic sequence converging to zero as $n, p \to \infty$ such that $n^{-1}\log p \to 0$. It is important to notice that even though the estimation error that (19.13) controls is uniform over $p(p-1)/2$ different elements, we only want to pay a logarithmic price in the dimension p in the high probability error bound. Provided this condition holds true, adaptive thresholding will deliver rates of convergence that match the optimal minimax rates of Cai and Liu (2011) even under violations of their sub-Gaussianity condition.

To accommodate data drawn from distributions violating sub-Gaussianity, one could replace the sample covariance matrix $\widehat{\Sigma}$ by another pilot estimator $\widetilde{\Sigma}$ satisfying Eq. (19.13) under weaker moment conditions. The resulting adaptive thresholding estimator $\widetilde{\Sigma}^{\mathcal{T}}$ is defined as

$$\widetilde{\Sigma}^{\mathcal{T}} = (\widetilde{\sigma}^{\mathcal{T}}_{uv}) = \{\tau_{\lambda_{uv}}(\widetilde{\sigma}_{uv})\}, \tag{19.14}$$

where $\widetilde{\sigma}_{uv}$ is (u, v)th the entry of $\widetilde{\Sigma}$ and the threshold λ_{uv} is entry-dependent. As suggested by Fan, Liao, and Mincheva (2013), the entry-dependent threshold

$$\lambda_{uv} = \lambda\left(\frac{\widetilde{\sigma}_{uu}\widetilde{\sigma}_{vv}\log p}{n}\right)^{1/2} \tag{19.15}$$

is used, where $\lambda > 0$ is a constant. This is simpler than the threshold used by Cai and Liu (2011), as it does not require estimation of var$(\widetilde{\sigma}_{uv})$ and achieves the same optimality. In the next subsection we show how to use tools from robust statistics to construct pilot estimators that have the required elementwise convergence rates of Eq. (19.13).

19.4.3 Revisting Tools from Robust Statistics

Robust statistics, as described in Sect. 19.2, provides a formal framework for understanding how deviations from an assumed model affect statistical methods. In particular, it provides an effective mathematical formulation to the problem of the presence of outliers through a framework that allows us to understand the impact of stochastic deviations from model assumptions on statistics of interest. In this paradigm one is worried about the presence of a small fraction of observations that is not generated from the idealized posited statistical model. One can however think of other ways of ensuring some degree of robustness of a statistical procedure. In particular, it is fairly natural to seek for methods that extend the validity of some classical methods to more flexible classes of models. One can for instance seek to relax light tail distributional assumptions by assuming instead that few number of population moments are finite. Although such an approach is not always guaranteed to give robust estimators in the sense of Huber and Ronchetti (1981), the resulting estimates are natural appealing alternatives to classical estimators.

Here we will describe a Huber-type M-estimator as an alternative to the sample covariance. Crucially, the resulting pilot estimator achieves the same elementwise deviations as the sample covariance matrix under a Gaussian random sample, even when in the presence of heavy tails. Let $Y_{uv} = (X_u - \mu_u^*)(X_v - \mu_v^*)$. Then $\sigma_{uv}^* = E(Y_{uv}) = \mu_{uv}^* - \mu_u^*\mu_v^*$ where $\mu_u^* = E(X_u)$, $\mu_v^* = E(X_v)$, and $\mu_{uv}^* = E(X_u X_v)$. We propose to estimate σ_{uv}^* using robust estimators of μ_u^*, μ_v^*, and μ_{uv}^*. In particular, replacing Z_i in Eq. (19.1) by $X_{i,u}$, $X_{i,u}$ and $X_{i,u}X_{i,v}$ gives Huber estimators, $\widetilde{\mu}_u^c$, $\widetilde{\mu}_v^c$, and $\widetilde{\mu}_{uv}^c$, of μ_u^*, μ_v^*, and μ_{uv}^* respectively, from which the Huber-type estimator of Σ^* is defined as $\widetilde{\Sigma}_c = (\widetilde{\sigma}_{uv}^c) = (\widetilde{\mu}_{uv}^c - \widetilde{\mu}_u^c\widetilde{\mu}_v^c)$.

We depart from the ideas discussed in Sect. 19.2 by allowing c to grow to infinity as n increases. The reason is that our goal here differs from those of Huber and the theory of robust statistics (Huber, 1964; Huber & Ronchetti, 1981). There, the distribution generating the data is assumed to be a contaminated version of a given parametric model, where the contamination level is small, and the objective is to estimate model parameters of the uncontaminated model. The tuning parameter c is therefore typically fixed at a value guaranteeing a given level of efficiency if the underlying distribution is indeed the uncontaminated model. For instance, in the location model, choosing $c = 1.345$ guarantees 95% efficiency if the data generating distribution is Gaussian. Our goal is instead to estimate the mean of the underlying distribution, allowing departures from sub-Gaussianity. In related work, Fan, Liu, and Wang (2015) show that when c is allowed to diverge at an appropriate rate, the Huber estimator of the mean concentrates exponentially fast around the true mean when only a finite second moment exists. In a similar spirit, we allow c to grow with n in order to alleviate the bias. An appropriate choice of c trades off bias and robustness. We build on Fan et al. (2015) and Catoni (2012), showing that our proposed Huber-type estimator satisfies Condition (19.13).

The following proposition lays the foundations for our analysis of high-dimensional covariance matrix estimators under infinite kurtosis. It extends Theorem 5 in Fan et al. (2015), and gives rates of convergence for Huber's estimator of $E(X_u)$ assuming a bounded $1+\varepsilon$ moment for $\varepsilon \in (0, 1]$. The result is optimal in the sense that our rates match the minimax lower bound given in Theorem 3.1 of Devroye, Lerasle, Lugosi, and Oliveira (2017). The rates depend on ε, and when $\varepsilon = 1$ they match those of Catoni (2012) and Fan et al. (2015).

Proposition 19.1 *Let $\delta \in (0, 1)$, $\varepsilon \in (0, 1]$, $n > 12\log(2\delta^{-1})$, and $Z_1, \ldots, Z_n$ be independent and identically distributed random variables with mean μ and bounded $1+\varepsilon$ moment, i.e., $E(|Z_1 - \mu|^{1+\varepsilon}) = v < \infty$. Take $c = \{vn/\log(2\delta^{-1})\}^{1/(1+\varepsilon)}$.*

Then with probability at least $1 - 2\delta$,

$$|\widetilde{\mu}^c - \mu| \le \frac{7+\sqrt{2}}{2} v^{1/(1+\varepsilon)} \left\{ \frac{\log(2\delta^{-1})}{n} \right\}^{\varepsilon/(1+\varepsilon)},$$

where $\widetilde{\mu}^c$ *is as defined in* (19.1).

Proposition 19.1 allows us to show that the elementwise errors of the Huber-type estimator are uniformly controlled in the sense of (19.13). This is the building block of the following result (Avella-Medina et al., 2018, Theorem 3) that establishes the rates of convergence for the adaptive thresholding estimators.

Theorem 19.3 *Suppose* $\Sigma^* \in \mathcal{U}_q\{s_0(p)\}$ *and assume* $\max_{1\le u\le p} E(|X_u|^{2+\varepsilon}) \le \kappa_\varepsilon^2$. *Let* $\widehat{\Sigma}_c^{\mathcal{T}}$ *be the adaptive thresholding estimator* (19.14) *and* (19.15) *based on the Huber pilot estimator* $\widetilde{\Sigma}_c$ *with* $c = K(n/\log p)^{1/(2+\varepsilon)}$ *for* $K \ge 2^{-1}(7+\sqrt{2})\kappa_\varepsilon(2+L)^{\varepsilon/(2+\varepsilon)}$ *and* $L > 0$. *Under the scaling condition* $\log p = O(n^{1/2})$ *and choosing* $\lambda_{uv} = \lambda\{\widetilde{\sigma}_{uu}^c\widetilde{\sigma}_{vv}^c(\log p)/n\}^{\varepsilon/(2+\varepsilon)}$ *for some* $\lambda > 0$, *we have, for sufficiently large n,*

$$\inf_{\Sigma^*\in\mathcal{U}_q\{s_0(p)\}} \mathrm{pr}\left\{ \|\widehat{\Sigma}_c^{\mathcal{T}} - \Sigma^*\|_2 \le C s_0(p) \Big(\frac{\log p}{n}\Big)^{\frac{\varepsilon(1-q)}{(2+\varepsilon)}} \right\} \ge 1 - \epsilon_{n,p},$$

where $\epsilon_{n,p} \le C_0 p^{-L}$ *for positive constants* C_0 *and* L.

The rates match the minimax optimal rates established in Cai and Liu (2011) as long as the underlying distributions have bounded fourth moments, i.e., when $\varepsilon = 2$. It does not prove that these rates are minimax optimal under $2+\varepsilon$ finite moments. However, the proof expands on the elementwise max norm convergence of the pilot estimator, which is optimal by Theorem 3.1 of Devroye et al. (2017). This is a strong indication that our rates are sharp.

19.4.4 On the Robustness Properties of the Pilot Estimators

The construction in the previous subsection will lead to an unbounded influence function when n goes to infinity since we need a diverging constant c. Consequently, the resulting estimators will not be asymptotically robust in the sense of Hampel et al. (1986). This does not contradict the fact that the estimators give optimal mean deviation errors in small samples even under heavy tails. In fact, it is unclear whether the two goals are compatible, i.e., whether one can construct estimator with such finite sample guarantees while remaining asymptotically robust in the sense of the influence function.

As shown in Avella-Medina et al. (2018), one can also construct rank-based pilot estimators using Kendall's tau and obtain optimal rates of convergence provided the underlying distribution is elliptical. Such a result only requires the existence of finite second moments, but imposes additional structure on the data generating process through the ellipticity assumption. From a robust statistics such rank-based pilot estimators and variants such as Ma and Genton (2001) can be expected to lead to good robustness properties in the spirit of Sect. 19.2 given the results of Loh and Tan (2018).

19.5 Further Extensions

In this section we discuss how to extend the ideas described in the previous two sections in related regression and sparse matrix estimation problems.

19.5.1 Generalized Additive Models

Generalized additive models (GAM) are a natural nonparametric extension of the GLM discussed in Sect. 19.3. They relax the linear dependence of the transformed mean and assume the more flexible formulation

$$g(\mu_i) = \eta_0(X_i) = \beta_0 + \sum_{j=1}^{d} f_j(X_{ij}),$$

where now each f_j is a smooth and integrable unknown function of the scalar variable x_j. In other words the transformed mean $m(x) = \beta_0 + \sum_{j=1}^{d} f_j(x_j)$ has an additive form. The ideal population optimization problem that one would like to solve in a perfectly specified GAM is

$$\max_f \left(E\left[\ell\left\{ Y, \beta_0 + \sum_{j=1}^{d} f_j(X_j) \right\} \right] \right),$$

where ℓ denotes the log-likelihood of the random pair (Y, X) and f lies in some Hilbert space $\mathcal{H}$. In practice we can estimate the unknown smooth functions f_j using polynomial spline approximations of the form $f_j(\cdot) \approx \sum_{l=1}^{K_n} \beta_{jl} B_{jl}(\cdot) = B_j^T(\cdot)\beta_{[j]}$, where the $B_{jl}(\cdot)$ are some known spline basis functions for all $j = 1, \ldots, d$. One can combine this spline approach with the robust quasilikelihood introduced in Sect. 19.3 in order to gain some robustness. One can also introduce sparsity inducing penalty functions $p(\|f_j\|; \lambda) = \lambda w_j \|f_j\|$ for $j = 1, \ldots, d$, where w_j is some weight function. Hence we can consider a robust group lasso estimator for GAM defined by

$$\tilde{\beta} = \underset{\beta}{\operatorname{argmax}} \left[E_n \Big\{ Q_M(Y, g^{-1}\Big(\beta_0 + \sum_{j=1}^{d} B_j^T(X_j)\beta_{[j]}\Big)\Big\} - \lambda_n \sum_{j=1}^{d} \sqrt{d_j} \|\beta_{[j]}\|_2 \right],$$

where E_n denotes the expectation with respect to the empirical distribution and d_j is the dimension of parameter vector used in the approximation of the f_j. This robust group lasso approach has been recently studied in Avella-Medina and Ronchetti (2019) where it is shown that under appropriate regularity conditions the above estimator leads to consistent estimation and variable selection.

19.5.2 *Sure Independence Screening*

When the number of covariates is very large compared to the sample size, one can complement lasso-type variable selection methods with simple screening rules that help finding potential variables of interest in a first stage. The basic idea that we describe here is an extension of the sure independence screening idea of Fan and Lv (2008) to the GLM/GAM setup described above. One of the simplest versions of sure independence screening for the linear regression model is based on estimated parameters $\hat{\beta}_j^M$ obtained by running marginal univariate regressions using each of the standardized covariates. One can then define the set of screened covariates as

$$\bar{S}_{\tau_n} = \{1 \leq j \leq d : |\hat{\beta}_j^M| \geq \tau_n\},$$

for a given threshold τ_n. Conditions under which the above screening rule keeps the subset of important variables with high probability have been studied in Fan and Lv (2008). A non-exhaustive list of extensions of this methodology to more complicated models includes GLM, additive models, and varying coefficients proposed respectively in Fan and Song (2010), Fan, Feng, and Song (2011), and Fan, Ma, and Dai (2014). Following this line of work, Avella-Medina and Ronchetti (2019) considered a natural extension of sure independence screening to a robust GAM setup by considering marginal nonparametric fits $\hat{f}_j^M$ obtained by solving

$$\underset{f_0 \in \mathbb{R}, f_j \in \varphi_n}{\operatorname{argmin}} \; E_n \Big\{ Q_M(Y, g^{-1}\Big(f_0 + f_j(X_j)\Big)\Big\},$$

where φ_n is a space of spline functions. These marginal fits can then be used to select a set of variables by the screening rule

$$\hat{S}_{\tau_n} = \{1 \leq j \leq d : \|\hat{f}_j^M\|_n^2 \geq \tau_n\}. \tag{19.16}$$

Once we have reduced the dimension of the initial problem based on the screening rule (19.16), we can apply more refined group lasso methods as discussed in the previous subsection. The success of this operation relies on whether the screening procedure does not mistakenly delete some important variables. In other words the procedure should have the sure screening property of Fan and Lv (2008). Sufficient conditions for establishing such sure screening properties for (19.16) can be found in Avella-Medina and Ronchetti (2019).

19.5.3 Precision Matrix Estimation

The inverse covariance matrix or precision matrix is of great interest in many procedures from classical multivariate analysis and is central for understanding Gaussian graphical models. Similar to the high-dimensional covariance matrix estimation problem, one cannot expect to consistently estimate a very high-dimensional precision matrix without structural assumptions. Consequently, sparsity assumptions on the columns of the precision matrix motivates the use of penalized likelihood approaches such as the graphical lasso (Friedman, Hastie, & Tibshirani, 2008; Yuan & Lin, 2007) and the constrained ℓ_1-minimizer of Cai, Liu, and Luo (2011), Cai, Liu, and Zhou (2016) Yuan (2010), both of which lead to consistent estimation under regularity condition. In fact, under appropriate sub-Gaussian distributional assumptions, Cai et al. (2016) establish that their adaptive constrained ℓ_1-minimizer is minimax optimal within the class sparse precision matrices

$$\mathcal{G}_q(c_{n,p}, M_{n,p}) = \Big\{\Omega \in \mathcal{S}^+(\mathbb{R}, p) : \max_v \sum_{u=1}^{p} |\omega_{uv}|^q \leq c_{n,p}, \|\Omega\|_1 \leq M_{n,p},$$
$$\frac{1}{M_1} \leq \lambda_{\min}(\Omega) \leq \lambda_{\max}(\Omega) \leq M_1\Big\},$$

where $0 \leq q \leq 1$, $M_1 > 0$ is a constant and $M_{n,p}$ and $c_{n,p}$ are positive deterministic sequences, bounded away from zero and allowed to diverge as n and p grow. In this class of precision matrices, sparsity is imposed by restricting the columns of Ω^* to lie in an ℓ_q-ball of radius $c_{n,p}$ $(0 \leq q < 1)$.

Analogously to the covariance matrix estimation problem discussed Sect. 19.4, Avella-Medina et al. (2018) also show that the optimal rates of estimation of the precision matrix $\Omega^* = (\Sigma^*)^{-1}$ can be attained without sub-Gaussian assumptions. Indeed, the adaptive constrained ℓ_1-minimization estimator of Cai et al. (2016) gives optimal rates of convergence as long as it is applied to a pilot estimator satisfying

$$\mathrm{pr}\big[\max_{u,v} \big|(\widetilde{\Sigma}\Omega^* - I_p)_{uv}\big| \leq C_0\{(\log p)/n\}^{1/2}\big] \geq 1 - \epsilon_{n,p}, \tag{19.17}$$

where C_0 and $\epsilon_{n,p}$ are analogous to (19.13), and I_p denotes the $p \times p$ identity matrix. While Eq. (19.17) holds with $\widetilde{\Sigma} = \widehat{\Sigma}$ under sub-Gaussianity of X, it fails otherwise. More precisely, it follows from Theorem 4 and Lemma 10 in Avella-Medina et al. (2018) that if $\max_{1\le u\le p} E(|X_u|^{2+\varepsilon}) \le \kappa_\varepsilon^2$ one can construct an estimator with optimal rates of convergence as long as we have the scaling conditions $c_{n,p} = O\{n^{(1-q)/2}/(\log p)^{(3-q)/2}\}$, $M_{n,p} = O\{n/\log p)^{\epsilon/2}\}$, and $(\log p)/n^{1/3} = O(1)$. In particular one can construct an adaptive estimator $\widehat{\Omega}_H$ using similar ideas to those of Cai et al. (2016) but regularizing the Huber-type estimator introduced in Sect. 19.4.3, instead of the sample covariance matrix. Choosing $c = K(n/\log p)^{1/(2+\varepsilon)}$ for some $K > 0$ and a sufficiently large n one can show that

$$\inf_{\Omega^*\in\mathcal{G}_q(c_{n,p},M_{n,p})} \text{pr}\left\{\|\widetilde{\Omega}_H - \Omega^*\|_2 \le CM_{n,p}^{1-q}c_{n,p}\left(\frac{\log p}{n}\right)^{\frac{\varepsilon(1-q)}{(2+\varepsilon)}}\right\} \ge 1-\epsilon_{n,p}, \tag{19.18}$$

where $\epsilon_{n,p} \le C_0 p^{-L}$ for positive constants C_0 and L. The rates in (19.18) match the minimax optimal ones of Cai et al. (2016) when $\varepsilon = 2$.

19.5.4 Factor Models and High-Frequency Data

For many economics applications the covariance matrix estimation techniques described in Sect. 19.4 are too stylized. For example the class of sparse matrices (19.12) might not be appropriate to model covariance structure of stock returns. In this case however, factor models such as the celebrated Fama-French model (Fama & French, 1993) can capture an alternative low dimensional structure in the data. In other applications an iid framework is also too simplistic. For instance high-frequency financial data cannot be directly tackled by the techniques developed in Sect. 19.4 since in this setting the observations are typically neither identically distributed nor independent.

Despite the obvious limitations of the setting presented in Sect. 19.4, the same idea of constructing an initial estimator with good elementwise convergence properties under heavy tails is more broadly applicable. Indeed, combining this idea with appropriate regularization techniques it can be successfully exploited to deliver more robust estimators of approximate factor models in high dimensions and high-frequency settings (Fan & Kim, 2018; Fan, Wang, & Zhong, 2019).

19.6 Conclusion

We showed that old ideas from robust statistics can provide an interesting set of tools to effectively tackle high-dimensional problems. In particular, we showed that they can be used to obtain procedures that are less sensitive to the presence of outliers and heavy tails. Indeed, some ideas from M-estimation can be extended to high-dimensional regression problems by including appropriate sparsity inducing regularization techniques. In regression settings we showed that when robust losses are used, the resulting penalized M-estimators allow to deal with data deviating from the assumed model and give more reliable inference even in the presence of a small fraction of outliers. Furthermore M-estimators can also be used to construct estimators that exhibit good finite sample deviations errors from a target parameter. This is of paramount importance in the context of covariance matrix estimation where it is critically important to have access to good initial estimators with good elementwise convergence properties in the presence of heavy tails.

References

Ait-Sahalia, Y., & Xiu, D. (2017). Using principal component analysis to estimate a high dimensional factor model with high-frequency data. *Journal of Econometrics, 201*(2), 384–399.

Alizadeh, A. A., Eisen, M. B., Davis, R. E., Ma, C., Lossos, I. S., Rosenwald, A., ... Powell, J. I. (2000). Distinct types of diffuse large b-cell lymphoma identified by gene expression profiling. *Nature, 403*, 6769.

Antoniadis, A., & Fan, J. (2001). Regularization of wavelet approximations. *Journal of the American Statistical Association, 96*, 939–957.

Athey, S., & Imbens, G. W. (2017). The state of applied econometrics: Causality and policy evaluation. *Journal of Economic Perspectives, 31*(2), 3–32.

Athey, S., Imbens, G. W., & Wager, S. (2018). Approximate residual balancing: Debiased inference of average treatment effects in high dimensions. *Journal of the Royal Statistical Society: Series B, 80*(4), 597–623.

Avella-Medina, M. (2017). Influence functions for penalized M-estimators. *Bernoulli, 23*, 3778–96.

Avella-Medina, M., Battey, H., Fan, J., & Li, Q. (2018). Robust estimation of high-dimensional covariance and precision matrices. *Biometrika, 105*(2), 271–284.

Avella-Medina, M., & Ronchetti, E. (2015). Robust statistics: A selective overview and new directions. *Wiley Interdisciplinary Reviews: Computational Statistics, 7*(6), 372–393.

Avella-Medina, M., & Ronchetti, E. (2018). Robust and consistent variable selection in high-dimensional generalized linear models. *Biometrika, 105*(1), 31–44.

Avella-Medina, M., & Ronchetti, E. (2019). *Robust variable selection for generalized additive models*. Working paper.

Bai, J., & Wang, P. (2016). Econometric analysis of large factor models. *Annual Review of Economics, 8*, 53–80.

Belloni, A., Chen, D., Chernozhukov, V., & Hansen, C. (2012). Sparse models and methods for optimal instruments with an application to eminent domain. *Econometrica, 80*, 2369–2429.

Belloni, A., Chernozhukov, V., & Kato, K. (2015). Uniform post-selection inference for least absolute deviation regression and other z-estimation problems. *Biometrika, 102*, 77–94.

Bianco, A. M., & Yohai, V. J. (1996). Robust estimation in the logistic regression model. In H. Rieder (Ed.), *Robust statistics, data analysis and computer intensive methods: In honor of peter Huber's 60th birthday*. New York: Springer.
Bickel, P. J., & Levina, E. (2008). Covariance regularization by thresholding. *The Annals of Statistics, 36*, 2577–2604.
Boucheron, S., Lugosi, G., & Massart, P. (2013). *Concentration inequalities: A nonasymptotic theory of independence*. Oxford: Oxford University Press.
Breiman, L. (1995). Better subset regression using the nonnegative garrote. *Techno-metrics, 37*, 373–384.
Bubeck, S., Cesa-Bianchi, N., & Lugosi, G. (2013). Bandits with heavy tail. *IEEE Transactions on Information Theory, 59*, 7711–7717.
Bühlmann, P., & van de Geer, S. (2011). *Statistics for high-dimensional data: Methods, theory and applications*. Heidelberg: Springer.
Cai, T. T., & Liu, W. (2011). Adaptive thresholding for sparse covariance matrix estimation. *Journal of the American Statistical Association, 106*, 672–674.
Cai, T. T., Liu, W., & Luo, X. (2011). A constrained ℓ_1-minimization approach to sparse precision matrix estimation. *Journal of the American Statistical Association, 106*, 594–607.
Cai, T. T., Liu, W., & Zhou, H. (2016). Estimating sparse precision matrix: Optimal rates of convergence and adaptive estimation. *The Annals of Statistics, 44*, 455–488.
Cantoni, E., & Ronchetti, E. (2001). Robust inference for generalized linear models. *Journal of the American Statistical Association, 96*, 1022–1030.
Catoni, O. (2012). Challenging the empirical mean and empirical variance: A deviation study. *Annales de l'Institut Henri Poincaré, Probabilités et Statistiques, 48*, 1148–1185.
Chen, J., & Chen, Z. (2008). Extended Bayesian information criteria for model selection with large model spaces. *Biometrika, 95*, 759–771.
Chen, J., & Chen, Z. (2012). Extended BIC for small-n-large-p sparse GLM. *Statistica Sinica, 22*, 555–574.
Chernozhukov, V., Chetverikov, D., Demirer, M., Duflo, E., Hansen, C., & Newey, W. (2017). Double/debiased/neyman machine learning of treatment effects. *American Economic Review, 107*(5), 261–265.
Devroye, L., Lerasle, M., Lugosi, G., & Oliveira, R. (2017). Sub-Gaussian mean estimators. *The Annals of Statistics, 44*, 2695–2725.
Efron, B., Hastie, T. J., Johnstone, I. M., & Tibshirani, R. J. (2004). Least angle regression. *The Annals of Statistics, 32*, 407–499.
Fama, E. F., & French, K. R. (1993). Common risk factors in the returns on stocks and bonds. *Journal of Financial Economics, 33*(1), 3–56.
Fan, J., Feng, Y., & Song, R. (2011). Nonparametric independence screening in sparse ultra-high-dimensional additive models. *Journal of the American Statistical Association, 106*(494), 544–557.
Fan, J., Furger, A., & Xiu, D. (2016). Incorporating global industrial classification standard into portfolio allocation: A simple factor-based large covariance matrix estimator with high-frequency data. *Journal of Business & Economic Statistics, 34*(4), 489–503.
Fan, J., & Kim, D. (2018). Robust high-dimensional volatility matrix estimation for high-frequency factor model. *Journal of the American Statistical Association, 113*(523), 1268–1283.
Fan, J., & Li, R. (2001). Variable selection via nonconcave penalized likelihood and its oracle properties. *Journal of the American Statistical Association, 96*, 1348–1360.
Fan, J., Liao, Y., & Mincheva, M. (2013). Large covariance estimation by thresholding principal orthogonal complements. *Journal of the Royal Statistical Society: Series B, 75*, 603–680.
Fan, J., Liu, H., & Wang, W. (2015). Large covariance estimation through elliptical factor models. *The Annals of Statistics, 46*(4), 1383.
Fan, J., & Lv, J. (2008). Sure independence screening for ultrahigh dimensional feature space. *Journal of the Royal Statistical Society: Series B, 70*(5), 849–911.
Fan, J., & Lv, J. (2011). Nonconcave penalized likelihood with NP-dimensionality. *IEEE Transactions on Information Theory, 57*, 5467–5484.

Fan, J., Ma, Y., & Dai, W. (2014). Nonparametric independence screening in sparse ultra-high-dimensional varying coefficient models. *Journal of the American Statistical Association, 109*(507), 1270–1284.

Fan, J., & Song, R. (2010). Sure independence screening in generalized linear models with np-dimensionality. *The Annals of Statistics, 38*(6), 3567–3604.

Fan, J., Wang, W., & Zhong, Y. (2019). Robust covariance estimation for approximate factor models. *Journal of Econometrics, 208*(1), 5–22.

Fan, Y., & Tang, C. (2013). Tuning parameter selection in high dimensional penalized likelihood. *Journal of the Royal Statistical Society: Series B, 75*, 531–552.

Frank, L. E., & Friedman, J. H. (1993). A statistical view of some chemometrics regression tools. *Technometrics, 35*, 109–135.

Friedman, J., Hastie, T., & Tibshirani, R. (2008). Sparse inverse covariance estimation with the graphical lasso. *Biostatistics, 9*(3), 432–441.

Friedman, J. H., Hastie, T. J., & Tibshirani, R. J. (2010). Regularization paths for generalized linear models via coordinate descent. *Journal of Statistical Software, 33*, 1–22.

Golub, T. R., Slonim, D. K., Tamayo, P., Huard, C., Gaasenbeek, M., Mesirov, J. P., ... Bloomfield, C. D. (1999). Molecular classification of cancer: Class discovery and class prediction by gene expression monitoring. *Science, 286*(5439), 531–537.

Hampel, F. R. (1971). A general qualitative definition of robustness. *Annals of Mathematical Statistics, 42*, 1887–1896.

Hampel, F. R. (1974). The influence curve and its role in robust estimation. *Journal of the American Statistical Association, 69*, 383–393.

Hampel, F. R., Ronchetti, E. M., Rousseeuw, P. J., & Stahel, W. A. (1986). *Robust statistics: The approach based on influence functions*. New York: Wiley.

Heritier, S., Cantoni, E., Copt, S., & Victoria-Feser, M. (2009). *Robust methods in biostatistics*. Chichester: Wiley.

Huber, P. J. (1964). Robust estimation of a location parameter. *Annals of Mathematical Statistics, 35*, 73–101.

Huber, P. J. (1967). The behavior of maximum likelihood estimates under nonstandard conditions. In *Proceedings of the Fifth Berkeley Symposium on Mathematical Statistics and Probability* (pp. 221–233).

Huber, P. J. (1981). *Robust statistics*. New York: Wiley.

Huber, P. J., & Ronchetti, E. M. (1981). *Robust statistics* (2nd edition). New York: Wiley.

Javanmard, A., & Montanari, A. (2014). Confidence intervals and hypothesis testing for high-dimensional regression. *Journal of Machine Learning Research, 15*, 2869–2909.

Knight, K., & Fu, W. (2000). Asymptotics for lasso-type estimators. *The Annals of Statistics, 28*, 1356–1378.

Lambert-Lacroix, S., & Zwald, L. (2011). Robust regression through the Huber's criterion and adaptive lasso penalty. *Electronic Journal of Statistics, 5*, 1015–1053.

Lee, J. D., Sun, D. L., Sun, Y., & Taylor, J. E. (2016). Exact post-selection inference, with application to the lasso. *The Annals of Statistics, 44*, 907–927.

Leeb, H., & Pötscher, B. M. (2005). Model selection and inference: Facts and fiction. *Econometric Theory, 21*, 21–59.

Leeb, H., & Pötscher, B. M. (2008). Sparse estimators and the oracle property, or the return of Hodges' estimator. *Econometric Theory, 142*, 201–211.

Leeb, H., & Pötscher, B. M. (2009). On the distribution of penalized maximum likelihood estimators: The LASSO, SCAD, and thresholding. *Journal of Multivariate Analysis, 100*, 2065–2082.

Li, G., Peng, H., & Zhu, L. (2011). Nonconcave penalized M-estimation with a diverging number of parameters. *Statistica Sinica, 21*, 391–419.

Loh, P. L., & Tan, X. L. (2018). High-dimensional robust precision matrix estimation: Cellwise corruption under ϵ-contamination. *Electronic Journal of Statistics 12*, 1429–1467.

Loh, P.-L., & Wainwright, M. (2015). Regularized M-estimators with nonconvexity: Statistical and algorithmic theory for local optima. *Journal of Machine Learning Research, 16*, 559–616.

Ma, Y., & Genton, M. G. (2001). Highly robust estimation of dispersion matrices. *Journal of Multivariate Analysis, 78*(1), 11–36.
Machado, J. (1993). Robust model selection and M-estimation. *Econometric Theory, 9*, 478–493.
Maronna, R., Martin, R. D., & Yohai, V. J. (2006). *Robust statistics: Theory and methods*. New York: Wiley.
McCullagh, P., & Nelder, J. A. (1989). *Generalized linear models* (2nd edition). London: Chapman & HAll/CRC.
Meinshausen, N., & Bühlmann, P. (2006). High-dimensional graphs and variable selection with the lasso. *The Annals of Statistics, 34*, 1436–1462.
Negahban, S. N., Ravikumar, P., Wainwright, M. J., & Yu, B. (2012). A unified framework for high-dimensional analysis of M-estimators with decomposable regularizers. *Statistical Science, 27*(4), 538–557.
Perou, C. M., Sørlie, T., Eisen, M. B., Van De Rijn, M., Jeffrey, S. S., Rees, C. A., ... Fluge, Ø. (2000). Molecular portraits of human breast tumours. *Nature, 406*(6797), 747.
Rothman, A., Bickel, P., Levina, E., & Zhu, J. (2008). Sparse permutation invariant covariance estimation. *Electronic Journal of Statistics, 2*, 494–515.
Rothman, A., Levina, E., & Zhu, J. (2009). Generalized thresholding of large covariance matrices. *Journal of the American Statistical Association, 104*, 177–186.
Schwartz, L. (1959). *Théorie des distributions. Publications de l'Institut de Mathématique de l'Université de Strasbourg, 2(9–10)*. Paris: Hermann.
Tibshirani, R. J. (1996). Regression shrinkage and selection via the lasso. *Journal of the Royal Statistical Society: Series B, 58*, 267–288.
Tibshirani, R. J. (2011). Regression shrinkage and selection via the lasso: A retrospective. *Journal of the Royal Statistical Society: Series B, 73*, 273–282.
von Mises, R. (1947). On the asymptotic distribution of differentiable statistical functions. *The Annals of Mathematical Statistics, 18*, 309–348.
Wedderburn, R. (1974). Quasi-likelihood functions, generalized linear models, and the Gauss-Newton method. *Biometrika, 61*, 439–447.
Yuan, M. (2010). High dimensional inverse covariance matrix estimation via linear programming. *Journal of Machine Learning Research, 11*, 2261–2286.
Yuan, M., & Lin, Y. (2006). Model selection and estimation in regression with grouped variables. *Journal of the Royal Statistical Society: Series B, 68*, 49–67.
Yuan, M., & Lin, Y. (2007). Model selection and estimation in the Gaussian graphical model. *Biometrika, 94*(1), 19–35.
Zhang, C., Guo, X., Cheng, C., & Zhang, Z. (2014). Robuts-BD estimation and inference for varying dimensional general linear models. *Statistica Sinica, 24*, 515–532.
Zhang, C. H., & Zhang, S. S. (2014). Confidence intervals for low dimensional parameters in high dimensional linear models. *Journal of the Royal Statistical Society: Series B, 76*, 217–242.
Zhao, P., & Yu, B. (2006). On model selection consistency of lasso. *Journal of Machine Learning Research, 7*, 2541–2563.
Zou, H. (2006). The adaptive lasso and its oracle properties. *Journal of the American Statistical Association, 101*, 1418–1429.
Zou, H., & Hastie, T. J. (2005). Regularization and variable selection via the elastic net. *Journal of the Royal Statistical Society: Series B, 76*, 301–320.

Chapter 20
Frequency Domain

Felix Chan and Marco Reale

20.1 Introduction

Frequency domain has always played a central role in analysing and forecasting time series. Yet, the popularity of frequency domain techniques in the empirical literature remains relatively low due to their seemingly high requirement in mathematics. Indeed, much of the theoretical foundation in frequency domain analysis was built on integral transform theory. As such, many of the more practical techniques from frequency domain are hidden behind the somewhat intimidating mathematical details.

The purpose of this chapter is to provide an overview of the recent techniques from frequency or time-frequency domains to analyse time series. The main objective is to provide readers an appreciation on some of the central ideas behind frequency domain and to demonstrate these ideas through some empirical examples. The cost of this approach is the sacrifice of mathematical rigour which is unfortunately inevitable. This is akin to the spirit in Granger and Hatanaka (1964) where the main objective is to highlight the central idea of spectral analysis rather than presenting a rigours mathematical treatment.

This does not, however, diminish the importance of mathematical rigour. It is the hope of the authors that the chapter generates sufficient interest through this 'layman approach' and encourages readers to seek a more precise introduction of the subject. Several excellent references are provided within the chapter and readers

F. Chan (✉)
School of Economics, Finance and Property, Curtin University, Perth, WA, Australia
e-mail: F.Chan@curtin.edu.au

M. Reale
School of Mathematics and Statistics, University of Canterbury, Christchurch, New Zealand
e-mail: marco.reale@canterbury.ac.nz

P. Fuleky (ed.), *Macroeconomic Forecasting in the Era of Big Data*,
Advanced Studies in Theoretical and Applied Econometrics 52,
https://doi.org/10.1007/978-3-030-31150-6_20

are strongly encouraged to consult with these references to fill the theoretical gaps in this chapter.

This chapter focuses on three specific applications of time series analysis in frequency domain, namely Granger causality, forecasts based on wavelet decomposition and the ZVAR model using the generalised shift operator. The organisation of the chapter is as follows: Sect. 20.2 provides some technical background in spectral analysis. Section 20.3 discusses Granger causality from the frequency domain perspective. The concept of wavelets will be introduced in Sect. 20.4 as a generalisation of Fourier analysis. An empirical example on forecasting asset returns using wavelet decomposition will also be presented. Section 20.5 introduces the Generalised Shift Operator and the ZVAR model. An application of the ZVAR model to generate forecasts at a frequency that is different from the sampling frequency of the data will also be discussed. Its forecasting performance will be examined by two Monte Carlo experiments. Finally, Sect. 20.6 will contain some concluding remarks.

20.2 Background

This section provides some basic definitions and concepts that will be used for the rest of the chapter. The main idea here is to establish a connection between time and frequency domains. The section first considers a class of deterministic functions which can be written in terms of both time and frequencies. It then introduces a convenient transform to switch between the two representations. This is followed by extending the analysis to stochastic functions and establishes the analogue results, which lead to the well-known *Spectral Representation Theorem.*

Let $x(t)$ be an analytic function in t with the following representation

$$x(t) = \sum_{k=-\infty}^{\infty} a_k \phi_k(t), \tag{20.1}$$

where $\phi_k(t) : X \to Y$, $k \in \mathbb{Z}$ forms a set of basis functions. The domain, X, and the range, Y, can both be the subsets of $\mathbb{R}$ or $\mathbb{C}$. In this chapter, t bears the interpretation of time, so $X \subseteq \mathbb{R}^+$. Assume there exists a set of orthogonal basis functions, $\psi_k(t) : X \to Y$ such that

$$\int_X \phi_k(t)\overline{\psi_l(t)}dt = \delta_{kl} \tag{20.2}$$

where $\overline{\psi_l(t)}$ denotes the complex conjugate of $\psi_l(t)$ and δ_{kl} denotes the delta function with

$$\delta_{kl} = \begin{cases} 1 & k = l \\ 0 & k \neq l. \end{cases}$$

If $\psi_k(t)$ is a real-valued function then $\overline{\psi_k(t)} = \psi_k(t)$. As it will become clear later, $\psi_l(t) = \phi_l(t)$ in some cases. In those cases, $\{\phi_k(t)\}_{k=-\infty}^{\infty}$ forms the set of orthogonal basis functions.

The basic idea of Eqs. (20.1) and (20.2) is to provide a starting point to analyse a class of functions that can be expressed as a linear combination of a set of functions with special features. To an extent, this can be viewed as the deterministic analogue of a moving average process where the time series can be expressed as a linear combination of a set of independently and identically distributed random variables.

If $x(t)$ is known and conditional on the set of orthogonal basis functions, the coefficient a_k can be calculated as

$$a_k = \int_X x(t)\overline{\psi_k(t)}dt \qquad \forall k \in \mathbb{Z}. \tag{20.3}$$

To see this, consider

$$\begin{aligned}
x(t)\overline{\psi_k(t)} &= \sum_{l=-\infty}^{\infty} a_l\phi_l(t)\overline{\psi_k(t)} \\
\int_X x(t)\overline{\psi_k(t)}dt &= \sum_{l=-\infty}^{\infty} a_l \int_X \phi_l(t)\overline{\psi_k(t)}dt \\
&= \sum_{l=-\infty}^{\infty} a_l\delta_{lk} \\
&= a_k.
\end{aligned}$$

This is a generalisation of the well-known *Fourier transform*. When

$$\phi_l(t) = \exp\left(2\pi i \frac{l}{T} t\right) \text{ and } \psi_k(t) = \frac{1}{T}\phi_k(t)$$

where i denotes the complex number, $i = \sqrt{-1}$, then Eq. (20.1) is the Fourier series of the function $x(t)$ and the coefficients,

$$\begin{aligned}
a_k &= \frac{1}{T}\int_{-T/2}^{T/2} x(t)\exp\left(-2\pi i \frac{k}{T} t\right) dt \\
&= \frac{1}{T}\hat{x}(k)
\end{aligned}$$

where $\hat{x}(k)$ denotes the Fourier Transforms of $x(t)$. The specification of $\phi_l(t)$ introduces the concept of frequency into this analysis. Since $\exp(iwt) = \cos(wt) + i\sin(wt)$, the coefficients, a_k, are the *Fourier Transforms* of $x(t)$ at particular frequencies k/T scaled by $1/T$. Note that in this case, $X = [-T/2, T/2]$ and the scaling factor, $1/T$, leads to another representation of $x(t)$. To derive this, substitute the expression above into Eq. (20.1) which gives

$$\begin{aligned} x(t) &= \sum_{k=-\infty}^{\infty} \frac{1}{T}\hat{x}(k) \exp\left(2\pi i \frac{k}{T} t\right) \\ &= \sum_{k=-\infty}^{\infty} \hat{x}(s) \exp\left(2\pi i s t\right) \Delta s \end{aligned}$$

where $s = k/T$ and $\Delta s = \frac{k}{T} - \frac{k-1}{T} = \frac{1}{T}$. The last line in the expression above is a Riemann sum and since T can be set arbitrary, as $T \to \infty$, the last line becomes

$$x(t) = \int_{-\infty}^{\infty} \hat{x}(s) \exp\left(2\pi i s t\right) ds \tag{20.4}$$

which is the inverse Fourier transform. Thus $(x(t), \hat{x}(k))$ denotes the Fourier transform pair, where $x(t)$ admits a Fourier series representation when $\phi_k(t) = \exp\left(2\pi i \frac{k}{T} t\right)$ and $\psi_k(t) = \frac{1}{T}\phi_k(t)$.

Perhaps more importantly, Eq. (20.4) provides a connection between time and frequencies. Specifically, it expresses the function $x(t)$ at each time point as a sum of functions over all frequencies. The coefficients a_k, $k \in \mathbb{R}$, express the characteristic of the function $x(t)$ in each frequency in terms of a sum of functions over time domain. Together, they establish a mathematical connection between time and frequency domains for the function $x(t)$.

The discussion so far assumed $x(t)$ is a deterministic function. Many of the concepts extend to the case where $x(t)$ is a stochastic process in a seemingly natural way. It must be stressed here that while some of the results carry over to stochastic processes in an expected way, the mathematics required to demonstrate these is substantial. A common starting point is to consider the coefficients a_k in Eq. (20.1) to be independent random variables such that $a_k \sim D\left(0, \sigma_k^2\right)$ with the additional condition that σ_k^2 are absolute summable. That is,

$$\sum_{k=\infty}^{\infty} |\sigma_k| < \infty.$$

In other words, the coefficients a_k, $k \in \mathbb{Z}$, represent the random components in $x(t)$. Since $x(t)$ is still assumed to follow Eq. (20.1), the variance of $x(t)$ will therefore be the sum of variances of a_k. The condition above is a sufficient condition to ensure the existence of the second moment of $x(t)$ under this setting.

Since a_k are functions of frequencies and not time, $x(t)$ is likely to be autocorrelated, that is, $\mathbb{E}\left[x(t)x(t-h)\right] \neq 0$ for some $h > 0$. One way to analyse this is to define the *Weiner auto-covariance* function as

$$C_w(h) = \lim_{T\to\infty} \frac{1}{2T} \int_{-T}^{T} x(t)\overline{x(t-h)}dt. \tag{20.5}$$

Note that this is a continuous time version of the auto-covariance estimator for discrete time series data. See Koopmans (1995) for further exposition. At this point of discussion, the following simplifying assumptions will be useful. Let $\phi_k(t-h) = \phi_k(t)\phi_k(-h)$ with $\phi_k(0) = 1$ and $\psi_k(t) = \frac{1}{T}\phi_k(t)$. Now substitute Eq. (20.1) into $C_w(h)$ gives

$$\begin{aligned}
C_w(h) &= \lim_{T\to\infty} \sum_{k=-\infty}^{\infty} a_k^2 \frac{1}{2T} \int_{-T}^{T} \phi_k(t)\overline{\phi_k(t-h)}dt \\
&\quad + \sum_{k=-\infty}^{\infty} \sum_{\substack{l=-\infty \\ k\neq l}}^{\infty} a_k a_l \frac{1}{2T} \int_{-T}^{T} \phi_k(t)\overline{\phi_l(t-h)}dt \\
&= \lim_{T\to\infty} \sum_{k=-\infty}^{\infty} a_k^2 \frac{1}{2} \int_{-T}^{T} \phi_k(t)\overline{\psi_k(t-h)}dt + \sum_{k=-\infty}^{\infty} \sum_{\substack{l=-\infty \\ k\neq l}}^{\infty} a_k a_l \frac{1}{2} \int_{-T}^{T} \phi_k(t)\overline{\psi_l(t-h)}dt \\
&= \sum_{k=-\infty}^{\infty} a_k^2 \phi_k(h).
\end{aligned}$$

The last line follows from Eq. (20.2). Let $C(h) = \mathbb{E}\left[x(t)\overline{x(t-h)}\right] = \mathbb{E}\left[C_w(h)\right]$, this implies

$$C(h) = \sum_{k=-\infty}^{\infty} \sigma_k^2 \phi_k(h). \tag{20.6}$$

The expression on the right-hand side gives an expression for the auto-covariance structure from the frequency domain perspective. When $h = 0$,

$$\mathbb{E}\left[x(t), x(t)\right] = \sum_{k=-\infty}^{\infty} \sigma_k^2$$

which bears the interpretation that the variance of $x(t)$ is the sum of the variances associated with each $\phi_k(t)$. In the special case that

$$\phi_k(t) = \exp\left(2\pi i \frac{k}{T} t\right) = \cos\left(2\pi \frac{k}{T} t\right) + i \sin\left(2\pi \frac{k}{T} t\right),$$

σ_k^2 is the variance associated with the frequency k/T.

An important feature about the auto-covariance function as defined in Eq. (20.6) is that it is also the *spectral representation* of the auto-covariance function. In fact, the *spectral representation theorem* gives

$$C(h) = \int_{-\infty}^{\infty} \phi_k(h) dF(k) \tag{20.7}$$

and

$$x(t) = \int_{-\infty}^{\infty} \phi_k(t) da(k) \tag{20.8}$$

where $F(k)$ and $dF(k)$ are the *spectral distribution* and *spectral density* of the time series $x(t)$, respectively. Compare to Eq. (20.6), it can be shown that $F(k) = \sum_k \mathbb{E}|a_k|^2$ and thus Eq. (20.7) gives the *continuous frequency* version of Eq. (20.6). In order words, it extends the values of k from being an integer (discrete time series) to a real number (continuous time series).

The importance of this result cannot be overstated. It establishes a rigorous mathematical relation between the variance-covariance structures of discrete time series and its continuous counterpart. This is also reflected by Eq. (20.8) which can be viewed as a random continuous version of Eq. (20.1) where $da(k)$ plays the role of the coefficient, a_k. In other words $x(t)$ can be expressed as a continuous sum of a (infinitely uncountable) set of independently distributed random variables, i.e., $da(k)$. An implication is that $x(t)$ can therefore be approximated with a discrete sum of independently distributed random variables under suitable regularity conditions. In other words, it allows one to interpret a discrete time series as a continuous time series sampled at a regular interval. For a rigorous treatments of these, see Koopmans (1995).

The discussion in this section relates several concepts together. Under certain conditions, it is possible to express both discrete and continuous time series in terms of time or in terms of frequencies. More importantly, it is possible to switch between these representations by leveraging the orthogonality property of the basis functions. It is also possible to interpret a discrete time series as a continuous time series sampled at a regular interval. Big data, especially in the form of tall and huge data, represents a higher sampling frequency, which leads to better approximation of the underlying continuous time series using discrete data. The mathematical results presented above provide a mean to leverage the information in tall and huge data.

20.3 Granger Causality

Apart from understanding the relations between different economic and finance variables, one aspect of Granger Causality is its ability to examine the contribution of additional variables in forecasting. In fact, earlier definition of causality is related to the reduction of forecast variance. That is, if the forecast variance of x_t is reduced when y_t and its past values are included in the forecast model, then y_t is said to 'cause' x_t. Despite its age, Zellner (1979) still contains valuable philosophical discussion on the different definitions of causality and their testing in econometrics.

The empirical literature over the past decades has investigated Granger causality in the following fashion. Assume $\mathbf{x}_t = \left(x_{1t}, \ldots, x_{pt}\right)'$ is a $p \times 1$ vector of variables which admits the following Autoregressive (AR) representation,

$$\mathbf{\Phi}(B)\mathbf{x}_t = \boldsymbol{\varepsilon}_t$$

where $\mathbf{\Phi}(B) = \mathbf{I}_p - \sum_{i=1}^{r} \mathbf{\Phi}_i B^i$ with B being the backward (lag) operator such that $Bx_t = x_{t-1}$. The random vector $\boldsymbol{\varepsilon}_t \sim D\left(\mathbf{0}_p, \mathbf{\Sigma}\right)$ is assumed to be a vector of random variables such that $\mathbb{E}\left(\boldsymbol{\varepsilon}_t \boldsymbol{\varepsilon}_\tau\right) = \mathbf{0}_{p \times p}$ for $t \neq \tau$. The variance-covariance matrix, $\mathbf{\Sigma}$ is typically assumed to be positive definite with an inverse that can be decomposed as $\mathbf{\Sigma}^{-1} = \mathbf{G}\mathbf{G}'$ so that there exist a $p \times 1$ random vector, $\boldsymbol{\eta}_t \sim D\left(\mathbf{0}_p, \mathbf{I}_p\right)$ and $\boldsymbol{\varepsilon}_t = \mathbf{G}\boldsymbol{\eta}_t$.

Thus, if any of the off-diagonal elements in $\mathbf{\Phi}_i$ are non-zero, then there exists Granger causality. For example, if the $(2, 1)$ elements in $\mathbf{\Phi}_1$ is non-zero, then x_{1t} *Granger causes* x_{2t}. Similarly, if the $(1, 2)$ element in $\mathbf{\Phi}_1$ is non-zero then x_{2t} *Granger causes* x_{1t}.

Properties of various estimators for $\mathbf{\Phi}_i$ are well established in the literature, so test of statistical significance in the off-diagonal elements in $\mathbf{\Phi}_i$ can be typically conducted using standard software packages. For details and examples, see Hamilton (1994) and Lütkepohl (2005).

Interestingly, tests for Granger causality using frequency domain techniques can be traced back to the seminal work of Granger (1963) and Granger (1969). Some of the more recent advances include Breitung and Candelon (2006), Geweke (1982), Granger and Lin (1995), Hosoya (1991). An interesting trend is that most of the latest developments considered both stationary and co-integrated series, see for example, Granger and Lin (1995) and Breitung and Candelon (2006).

These techniques are becoming quite popular among empirical researchers in both economics and finance. For examples, see Bahmani-Oskooee, Chang, and Ranjbar (2016), Benhmad (2012), Bouri, Roubaud, Jammazi, and Assaf (2017), Croux and Reusens (2013), Joseph, Sisodia, and Tiwari (2014), Tiwari and Albulescu (2016), Tiwari, Mutascu, Albulescu, and Kyophilavong (2015) for their applications in finance and see Aydin (2018), Bahmani-Oskooee et al. (2016), Bozoklu and Yilanci (2013), Li, Chang, Miller, Balcilar, and Gupta (2015) , Shahbaz, Tiwari, and Tahir (2012), Sun, Chen, Wang, and Li (2018), Tiwari (2012) for their applications in economics.

These empirical studies highlighted two interesting developments. First, some of these studies, such as Croux and Reusens (2013) and Tiwari et al. (2015), obtained empirical evidence on the link between macroeconomic variables and financial variables. Previous empirical studies on establishing connection between financial markets and macroeconomic performance have not been overwhelmingly successful. These recent studies highlighted the importance of examining Granger causality in frequency domain as a complementary technique to the conventional time domain techniques. Recent studies also seem to move towards time-frequency domain analysis, with a specific focus on wavelets which will be the main focus in the next section.

The following exposition of Granger causality follows closely Granger (1963, 1969) which forms the foundation of Granger causality test in frequency domain. First, let us consider a bivariate Vector Autoregressive system

$$x_t = \boldsymbol{\phi}_{11}(B)x_t + \boldsymbol{\phi}_{12}(B)y_t + \varepsilon_{xt} \quad (20.9)$$

$$y_t = \boldsymbol{\phi}_{21}(B)x_t + \boldsymbol{\phi}_{22}(B)y_t + \varepsilon_{yt} \quad (20.10)$$

where $\boldsymbol{\phi}_{ij}(B) = \sum_{k=1}^{p_{ij}} \phi_{ij,k} B^k$ denotes a polynomial in the backward operator B for all $i, j = 1, 2$. These polynomials do not have to share the same order and the orders are not particularly important at this stage. It is assumed, however, that all roots of $1 - \boldsymbol{\phi}_{ii}(x)$ lie outside of the unit circle, i.e., both y_t and x_t are strictly stationary. The case when one of the roots lies on the unit circle is still an area of active research, see for examples Granger and Lin (1995), Breitung and Candelon (2006) and Breitung and Schreiber (2018).

As discussed previously, the central idea of testing Granger causality in time domain is to consider the following hypotheses[1]

$$H_0 : \boldsymbol{\phi}_{12} = \mathbf{0} \quad (20.11)$$

$$H_1 : \boldsymbol{\phi}_{12} \neq \mathbf{0} \quad (20.12)$$

for testing y Granger causes x and

$$H_0 : \boldsymbol{\phi}_{21} = \mathbf{0} \quad (20.13)$$

$$H_1 : \boldsymbol{\phi}_{21} \neq \mathbf{0} \quad (20.14)$$

for testing x Granger causes y. The testing procedure often follows the log-ratio principle. Let $\hat{\boldsymbol{\Sigma}}_x^R$ be the estimated variance-covariance matrix of ε_{xt} with the restriction that $\boldsymbol{\phi}_{12}(B) = 0$ and $\hat{\boldsymbol{\Sigma}}_x$ be the estimated variance-covariance matrix of ε_{xt} in Eq. (20.9). Testing the null in Eq. (20.11) can be done via the test statistic

$$M_{y \to x}^L = \log\left(\frac{|\boldsymbol{\Sigma}_x^R|}{|\boldsymbol{\Sigma}_x|}\right) \sim \chi^2(r) \quad (20.15)$$

where $|\boldsymbol{\Sigma}|$ denotes the determinant of the matrix $\boldsymbol{\Sigma}$ and r is the number of parameter in $\boldsymbol{\phi}_{12}$. The testing of the null in Eq. (20.13) can be done in a similar fashion.

[1]Note that $\boldsymbol{\phi}_{ij}(B)$ and $\boldsymbol{\phi}_{ij}$ denote two different objects. The former denotes the lag polynomial whereas the latter denotes the vector of coefficients.

The analogue in frequency domain can be derived as follows: Recall the spectral representation of a time series as defined in Eq. (20.8) which allows one to convert the time domain representation of a time series into its frequency domain representation. To do this, first set the index k to a frequency parameter, $w \in (0, \pi)$, let $\phi_w = \exp(iwt)$ and $da(K) = dX(w)$ and substitute these into Eq. (20.8) gives

$$x_t = \int_{-\pi}^{\pi} \exp(iwt)\, dX(w)$$

which is the frequency domain representation of x_t. Similar approach can be constructed for y_t, ε_{xt} and ε_{yt}. Substitute their frequency domain representations to Eqs. (20.9) and (20.10) gives

$$\int_{-\pi}^{\pi} \exp(iwt) \left\{ \left[1 - \boldsymbol{\phi}_{11}(\exp(iw))\right] dX(w) - \boldsymbol{\phi}_{12}(\exp(iw))\, dY(w) - d\varepsilon_x(w) \right\} = 0 \tag{20.16}$$

$$\int_{-\pi}^{\pi} \exp(iwt) \left\{ -\boldsymbol{\phi}_{21}(\exp(iw))\, dX(w) + \left[1 - \boldsymbol{\phi}_{22}(\exp(iw))\right] dY(w) - d\varepsilon_y(w) \right\} = 0. \tag{20.17}$$

These imply

$$\begin{bmatrix} a_{11} & a_{12} \\ a_{21} & a_{22} \end{bmatrix} \begin{bmatrix} dX(w) \\ dY(w) \end{bmatrix} = \begin{bmatrix} d\varepsilon_x(w) \\ d\varepsilon_y(w) \end{bmatrix} \tag{20.18}$$

where

$$\begin{aligned} a_{11} &= 1 - \boldsymbol{\phi}_{11}(\exp(iw)) \\ a_{12} &= -\boldsymbol{\phi}_{12}(\exp(iw)) \\ a_{21} &= -\boldsymbol{\phi}_{21}(\exp(iw)) \\ a_{22} &= 1 - \boldsymbol{\phi}_{22}(\exp(iw)). \end{aligned}$$

Let $\mathbf{A} = \{a_{ij}\}$ and if $\mathbf{A}^{-1}$ exists then $dX(w)$ and $dY(w)$ can be expressed in terms of the independent components $d\varepsilon_x(w)$ and $d\varepsilon_y(w)$, respectively. That is,

$$\begin{bmatrix} dX(w) \\ dY(w) \end{bmatrix} = \frac{1}{|\mathbf{A}|} \begin{bmatrix} a_{22} & -a_{12} \\ -a_{21} & a_{11} \end{bmatrix} \begin{bmatrix} d\varepsilon(w) \\ d\varepsilon_y(w) \end{bmatrix} \tag{20.19}$$

with $|\mathbf{A}| = a_{11}a_{22} - a_{12}a_{21}$. For ease of exposition, let us assume $d\varepsilon_x(w)$ and $d\varepsilon_y(w)$ are independent. The case where they are not independent will be considered later in this section. Under independence

$$\mathbb{E}\left(\begin{bmatrix} d\varepsilon_x(w) \\ d\varepsilon_y(w) \end{bmatrix} \begin{bmatrix} \overline{d\varepsilon_x}(w) \ \overline{d\varepsilon_y}(w) \end{bmatrix}\right) = \begin{bmatrix} \sigma_x^2 & 0 \\ 0 & \sigma_y^2 \end{bmatrix}.$$

Using this property, the power and cross spectrum of x_t and y_t can be derived as

$$\mathbb{E}\left(\begin{bmatrix} dX(w) \\ dY(w) \end{bmatrix}\right) = \frac{1}{|\mathbf{A}|^2}\begin{bmatrix} a_{22}^2\sigma_x^2 - a_{12}^2\sigma_y^2 & -a_{21}a_{22}\sigma_x^2 - a_{11}a_{12}\sigma_y^2 \\ -a_{22}a_{21}\sigma_x^2 - a_{11}a_{12}\sigma_y^2 & a_{21}^2\sigma_x^2 + a_{11}^2\sigma_y^2 \end{bmatrix} \tag{20.20}$$

where the power spectra for x_t and y_t are the elements on the main diagonal namely,

$$f_x(w) = \frac{a_{22}^2\sigma_x^2 - a_{12}^2\sigma_y^2}{|\mathbf{A}|^2} \tag{20.21}$$

$$f_y(w) = \frac{a_{21}^2\sigma_x^2 + a_{11}^2\sigma_y^2}{|\mathbf{A}|^2} \tag{20.22}$$

respectively. Note the matrix is symmetric by construction, the cross spectrum is the off-diagonal element, that is

$$Cr(w) = \frac{-a_{21}a_{22}\sigma_x^2 - a_{11}a_{12}\sigma_y^2}{|\mathbf{A}|^2}. \tag{20.23}$$

The cross spectrum has two main components, one contains information about Granger causality from x_t to y_t, while the other contains information about Granger causality from y_t to x_t. Specifically, if y_t does not Granger cause x_t, then $a_{12} = 0$ and if $a_{21} = 0$ then there is no evidence that x_t Granger causes y_t. A unique aspect of this approach is that the cross spectrum is a function of frequency w and so are a_{12} and a_{21}. This means that a_{12} may be 0 at some frequencies but not others. Therefore, the cross spectrum gives a possible mean to test Granger causality at different frequencies.

This idea was further developed by Geweke (1982) and Hosoya (1991). Consider once again the bivariate system of Eqs. (20.9) and (20.10). Let $\boldsymbol{\varepsilon}_t = \left(\varepsilon_{xt}, \varepsilon_{yt}\right)'$ and $\mathbb{E}\left(\boldsymbol{\varepsilon}_t\boldsymbol{\varepsilon}_t'\right) = \mathbf{GG}'$ for some non-singular matrix $\mathbf{G}$, then $\boldsymbol{\varepsilon}_t = \mathbf{G}\boldsymbol{\eta}_t$, where $\boldsymbol{\eta}_t = \left(\eta_{xt}, \eta_{yt}\right)'$ with $\mathbb{E}(\boldsymbol{\eta}_t\boldsymbol{\eta}_t') = \mathbf{I}$. Let $\boldsymbol{\Psi}(B) = \boldsymbol{\Phi}(B)\mathbf{G}^{-1}$ with

$$\boldsymbol{\Phi}(B) = \begin{bmatrix} 1 - \phi_{11}(B) & -\phi_{12}(B) \\ -\phi_{21}(B) & 1 - \phi_{22}(B) \end{bmatrix} \tag{20.24}$$

and

$$\boldsymbol{\Psi}(B) = \begin{bmatrix} \psi_{11}(B) & \psi_{12}(B) \\ \psi_{21}(B) & \psi_{22}(B) \end{bmatrix}. \tag{20.25}$$

This allows one to write the bivariate system in terms of its Wold decomposition with uncorrelated errors, that is,

$$\begin{bmatrix} x_t \\ y_t \end{bmatrix} = \begin{bmatrix} \psi_{11}(B) & \psi_{12}(B) \\ \psi_{21}(B) & \psi_{22}(B) \end{bmatrix} \begin{bmatrix} \eta_{xt} \\ \eta_{yt} \end{bmatrix}. \tag{20.26}$$

Follow a similar argument as before in deriving the power spectra and the cross spectrum, it can be shown that

$$f_x(w) = \left\{ \left| \psi_{11} \left[\exp(iw) \right] \right|^2 + \left| \psi_{12} \left[\exp(iw) \right] \right|^2 \right\}. \tag{20.27}$$

Similar to the previous case, y_t Granger causes x_t if $\psi_{12}(1) \neq 0$ and the power spectra of x_t as stated above provides a natural way to test this. Hosoya (1991) proposed the following measure of causality

$$\begin{aligned} M_{y \to x}(w) &= \log \left(\frac{|f_x(w)|}{\left| \psi_{11} \left[\exp(iw) \right] \right|^2} \right) \\ &= \log \left(1 + \frac{\left| \psi_{12} \left[\exp(iw) \right] \right|^2}{\left| \psi_{11} \left[\exp(iw) \right] \right|^2} \right). \end{aligned} \tag{20.28}$$

$M_{y \to x} \geq 0$ and equals to 0 if and only if $\left| \psi_{12} \left[\exp(iw) \right] \right| = 0$ and therefore a measure of overall causality can be constructed by summing $M_{y \to x}(w)$ over all frequencies and that is

$$M_{y \to x} = \int_{-\pi}^{\pi} M_{y \to x}(w) dw. \tag{20.29}$$

Note that both $\left| \psi_{11} \left[\exp(iw) \right] \right|^2$ and $\left| \psi_{12} \left[\exp(iw) \right] \right|^2$ can be estimated from the estimated parameters in Eqs. (20.9) and (20.10). Therefore, $M_{y \to x}(w)$ can also be estimated empirically. Let $\hat{M}_{y \to x}$ be a consistent estimator of $M_{y \to x}$ based on T observations, by assuming the appropriate regularity conditions, Hosoya (1991) proposed the following test statistics

$$W(w) = T \hat{M}_{y \to x}^2(w) V^{-1}(\hat{\boldsymbol{\Theta}}) \overset{d}{\sim} \chi^2(1) \tag{20.30}$$

where $\hat{\boldsymbol{\Theta}}$ denotes a consistent estimator for the parameter vector in Eqs. (20.9) and (20.10) and

$$V(\boldsymbol{\Theta}) = \left.\frac{\partial \hat{M}_{y\to x}}{\partial \boldsymbol{\Theta}'}\right|_{\hat{\boldsymbol{\Theta}}} \boldsymbol{\Lambda}(\boldsymbol{\Theta}) \left.\frac{\partial \hat{M}_{y\to x}}{\partial \boldsymbol{\Theta}}\right|_{\hat{\boldsymbol{\Theta}}} \tag{20.31}$$

with $\boldsymbol{\Lambda}(\boldsymbol{\Theta})$ being the variance-covariance matrix of $\hat{\boldsymbol{\Theta}}$.

As one may suspect, the test statistics as defined in Eq. (20.30) is difficult to compute due to the presence of various derivatives which are nonlinear functions of the parameter estimates. Interestingly, these numerical challenges can be largely avoided. Breitung and Candelon (2006) shows that the null hypothesis

$$H_0 : M_{y\to x}(w) = 0 \tag{20.32}$$

is equivalent to

$$H_0 : R(w)\boldsymbol{\phi}_{12} = 0. \tag{20.33}$$

$$R(w) = \begin{bmatrix} \cos(w) & \cos(2w) & \dots & \cos(pw) \\ \sin(w) & \sin(2w) & \dots & \sin(pw) \end{bmatrix}. \tag{20.34}$$

Therefore, a standard F-test for the hypothesis in Eq. (20.33), which is approximately $F(2, T - 2p_{12})$ distributed, can be used to test hypothesis (20.32). In other words, the problem has been reduced into testing a set of linear restrictions, which is much simpler to implement. The procedure is summarised in Procedure 1.

Procedure 1 Granger causality test in frequency domain

1: Set w such that $w \in (0, \pi)$.
2: Construct $R(w)$ as defined in Eq. (20.34).
3: Obtain a consistent estimate of $\boldsymbol{\phi}_{12}$ in Eq. (20.9) and calculate the residual sum of squares, RSS_{UR}.
4: Construct the restricted model by imposing the restrictions as implied by the hypothesis in Equation (20.33) into Equation (20.9).
5: Obtain an estimate of the coefficients in the restricted model and calculate the residual sum of squares, RSS_R.
6: Construct the test statistic:

$$F = \frac{(RSS_R - RSS_{UR})/2}{RSS_{UR}/T - 2p_{12}}.$$

7: F is distributed approximately $F(2, T - 2p_{12})$ and thus the testing of Hypothesis (20.33) can be carried out in the usual fashion.

Step 4 in Procedure 1 imposes the restrictions into the VAR system. One way to do this is to partition $R(w)$ and $\boldsymbol{\phi}_{12}$ such that

$$R(w)\boldsymbol{\phi}_{12} = \left[R_1(w)|R_2(w)\right]\begin{bmatrix}\boldsymbol{\phi}_{12}^1\\ \boldsymbol{\phi}_{12}^2\end{bmatrix}$$
$$= R_1(w)\boldsymbol{\phi}_{12}^1 + R_2(w)\boldsymbol{\phi}_{12}^2$$

where $R_1(w)$ is a non-singular matrix and the restrictions can then be imposed by setting $\boldsymbol{\phi}_{12}^1 = R_1^{-1}(w)R_2(w)\boldsymbol{\phi}_{12}^2$. The same procedure with the suitable modification can also be used to test $H_0 : M_{x\rightarrow y} = 0$.

Breitung and Candelon (2006) also extend this approach for bivariate co-integrated system with a focus on long-run causality in a Vector Error Correction (VECM) framework. The overall result is that, if x_t and y_t are both $I(1)$ and co-integrated, then most of the results in Breitung and Candelon (2006) hold. The analysis becomes much more complicated if the two time series are of different orders of integration.

20.4 Wavelet

Recall Eq. (20.1), where the function $x(t)$ is written in terms of a linear combination of a set of basis functions, $\phi_k(t)$. When $\phi_k(t) = \exp(i\omega_k t)$, the expression gives a trigonometry series representation of $x(t)$. Each basis function, $\phi_k(t)$ depends on one parameter namely, ω_k, which represents a specific frequency. A natural extension is to allow $\phi_k(t)$ to depend on more than one parameter. The motivation of such extension is that when $\phi_k(t)$ only depends on the frequency, ω_k, it assumes that the contribution of that frequency is fixed over time. In other words, the coefficient a_k is assumed to be a fixed constant when $x(t)$ is a deterministic function and a_k is a random variable with finite and constant second moment over time when $x(t)$ is a stochastic process. This is restrictive for macroeconomic forecasts because the contributions of different frequencies may vary over time due to factors such as policy changes and therefore, the ability to capture the contribution of a particular frequency at a particular time point would be desirable. One way to approach this is to allow the basis functions to include two parameters where one would bear the interpretation of frequency while the other would represent the location in time. Recall Eq. (20.3), the coefficient can be calculated as

$$a_k = \int_X x(t)\overline{\psi_{k,\tau}(t)}dt. \tag{20.35}$$

Therefore, if the dual function, $\psi_{k,\tau}(t)$, contains two parameters, k and τ, then the coefficient will also be a function of the two parameters.

One approach to incorporate the two parameters is to replace the variable t with $\frac{t-\tau}{\eta_k}$. The frequency parameter in this case is $\omega_k = \eta_k^{-1}$ with $\eta_k > 0\ \forall k$. The parameter τ can then be interpreted as the control on the location of time, t, and thus serves the purpose of shifting the location of time for a particular frequency. The appropriate basis functions, $\phi_{k,\tau}(t)$, are called *wavelet functions* and because it has the ability to analyse the contribution of a particular frequency at a particular time point, wavelet analysis is often referred to as a *time-frequency domain* technique.

The introduction of the additional parameter introduces an extra layer of complexity in terms of the orthogonal nature of the dual functions. So far it is assumed that $\phi_k(t)$ is orthonormal for $k \in \mathbb{Z}$ following Eq. (20.2). The definition of orthogonality must be modified when there are two parameters. One way to proceed is to let $\Phi = \left\{\phi_{k,\tau}(t) : k \in \mathbb{Z}^+, \tau \in \mathbb{Z}\right\}$ be a set of basis functions for $L^2(\mathbb{R})$ such that

$$\int_X \phi_{k,\tau_1}(t)\phi_{k,\tau_2}(t)dt = 0 \qquad \tau_1 \neq \tau_2.$$

The fact that $\phi_{k,\tau}(t) \in L^2(\mathbb{R})$ means that $\int |\phi_{k,\tau}(t)|^2 dt < \infty$ which ensures the wavelet functions possess certain desirable properties. Moreover, the condition above means that $\phi_{k,\tau}(t)$ is orthogonal to its integer translation. In other words, it is orthogonal to its own time location shifts but no assumption is being made about the inner product with its dilation, i.e., different values of k.

Now if one is willing to impose an additional structure on Φ namely, let V_k be the subspace spanned by $\phi_{k,\tau}(t)$ for $\tau \in \mathbb{Z}$ such that $V_k \subseteq L^2(\mathbb{R})$ and $V_0 \subset V_1 \ldots \subset V_k$. That is, the spaces spanned by the wavelet functions with lower k values are nested within those spanned by wavelet functions with higher k values. In such case, one can write

$$V_1 = V_0 \oplus W_0$$

where W_0 denotes the orthogonal complement of V_0 with respect to the space V_1. In other words, V_1 is being decompose into V_0 and its orthogonal complements, W_0. Repeated applications of this expression gives

$$V_k = V_0 \oplus W_0 \oplus \ldots \oplus W_{k-1}.$$

Figure 20.1 provides a graphical representation of this decomposition. Consider the space V_0 which is spanned by $\phi_{0,\tau}(t)$ and note that all $\phi_{0,\tau}(t)$ are orthogonal to

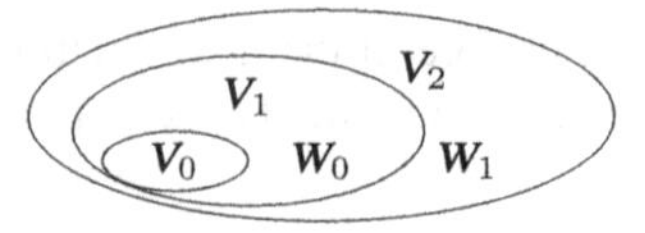

Fig. 20.1 Wavelet decomposition

each other by construction over all values of τ. Next, consider V_1. By construction and as shown in Fig. 20.1, V_1 consists of V_0 and its complement W_0. Since V_1 is spanned by $\phi_{1,\tau}(t)$ which are orthogonal to each other over all values of τ and since V_0 is a subspace of V_1, W_0 must be an orthogonal complement to V_0. Repeat this argument, one can see that the $L^2(\mathbb{R})$ space can be decomposed into V_0 and a set of orthogonal complements, $\{W_k\}_{k=0}^{\infty}$.

Therefore, if there exists a set of orthonormal basis for W_k, $\psi_{k,\tau}$, that can be derived from $\phi_{k,\tau}$, then any continuous function in $f(t) \in V_k$ can be expressed as a linear combination of $\phi_{0,\tau}$ and $\psi_{j,\tau}$, $j = 0, \ldots, k$. Since Φ is a set of basis function of $L^2(\mathbb{R})$, this implies all continuous function in $L^2(\mathbb{R})$ can be expressed as a linear combination of $\phi_{0,\tau}$ and $\psi_{j,\tau}$, $j = 1, \ldots, k$ as $k \to \infty$.
Two technical questions arise naturally. Do functions $\psi_{k,\tau}(t)$ actually exist? If so, under what conditions can they be derived from $\phi_{k,\tau}(t)$? The question concerning existence can be ensured by the following conditions

1.
$$\int_X \phi_{1,0}(t)dt = 0.$$

2.
$$\int_X \phi_{1,0}(t)\overline{\phi_{1,0}(t)}dt = 1.$$

3.
$$\int_X \phi_{k,\tau_1}(t)\overline{\phi_{k,\tau_2}(t)}dt = 0 \qquad \tau_1 \neq \tau_2.$$

4. Let $f(t) \in V_{k,\tau}$ then there exits $\{a_{k,\tau}\}$ $\tau \in \mathbb{R}$, such that
$$f(t) = \sum_{\tau=-\infty}^{\infty} a_{k,\tau}\phi_{k,\tau}(t)$$

5. Let V_k be the space spanned by $\phi_{k,\tau}(t)$, then $V_k \subset V_{k+1}$.
6. $f(t) = 0$ is the only function in $\bigcap_{k=0}^{\infty} V_k$.

Conditions 1 to 4 ensure $\phi_{k,\tau}$ forms an orthonormal basis for V_k for each k. Condition 5 ensures V_k is a sequence of increasing subspaces and Condition 6 ensures separability of each V_k, $k \in \mathbb{Z}^+$. Under Conditions 1–6, it can be shown that there exists a set of orthonormal basis functions, $\{\psi_{k,\tau}(t) : k \in \mathbb{Z}^+, \tau \in \mathbb{Z}\}$ that span $W_0 \oplus W_1 \oplus \ldots W_k$. For details see Bachman, Narici, and Beckenstein (2000).

Under existence, the generation of $\psi_{k,\tau}(t)$ from $\phi_{k,\tau}(t)$ can be achieved by the familiar Gram–Schmidt process. For notational convenient, define the inner product between $g(t) : X \to Y$ and $f(f) : X \to Y$ as

$$< f(t), g(t) >= \int_X f(t)\overline{g(t)}dt.$$

Consider the following iterative procedures

$$\begin{aligned}
\psi^*_{0,\tau}(t) =&\phi_{0,\tau}(t) \\
\psi^*_{1,\tau}(t) =&\phi_{1,\tau}(t) - \frac{< \psi^*_{0,\tau}(t), \phi_{1,\tau}(t) >}{< \psi^*_{0,\tau}(t), \psi^*_{0,\tau}(t) >}\psi^*_{0,\tau}(t) \\
\psi^*_{2,\tau}(t) =&\phi_{2,\tau}(t) - \frac{< \psi^*_{0,\tau}(t), \phi_{2,\tau}(t) >}{< \psi^*_{0,\tau}(t), \psi^*_{0,\tau}(t) >}\psi^*_{0,\tau}(t) - \frac{< \psi^*_{1,\tau}(t), \phi_{2,\tau}(t) >}{< \psi^*_{1,\tau}(t), \psi^*_{1,\tau}(t) >}\psi^*_{1,\tau}(t) \\
&\vdots \\
\psi^*_{k,\tau}(t) =&\phi_{k,\tau}(t) - \sum_{j=0}^{k} \frac{< \psi^*_{j,\tau}(t)\phi_{k,\tau}(t) >}{< \psi^*_{j,\tau}(t), \psi^*_{j,\tau}(t) >}\psi^*_{j,\tau}(t) \\
&\vdots
\end{aligned}$$

and let

$$\psi_{k,\tau}(t) = \frac{\psi^*_{k,\tau}(t)}{\sqrt{< \psi^*_{k,\tau}(t), \psi^*_{k,\tau}(t) >}} \qquad k \in \mathbb{Z}.$$

It is straightforward to check that $\Psi = \{\psi_{k,\tau}(t) : k \in \mathbb{Z}^+, \tau \in \mathbb{Z}\}$ form a set of orthonormal functions. Under the Conditions 1 to 6, it can also be shown that Ψ form sets of basis functions for W_k, $k \in \mathbb{Z}^+$. For details, see Bachman et al. (2000).

An implication of these theoretical foundations is that a time series $\{x(t)\}_{t\in\mathbb{Z}}$, under certain conditions, can be decomposed as a linear combination of wavelets and scaling functions. That is,

$$x(t) = \frac{1}{\sqrt{T}} \sum_{\tau=\infty}^{\infty} a_{k_0,\tau}\phi_{k_0,\tau}(t) + \frac{1}{\sqrt{T}} \sum_{k=k_0}^{\infty} \sum_{\tau=-\infty}^{\infty} b_{k,\tau}\psi_{k,\tau}(t) \tag{20.36}$$

where

$$a_{k_0,\tau} = \frac{1}{\sqrt{T}} \sum_{t=1}^{T} x(t)\phi_{k_0,\tau}(t) \tag{20.37}$$

$$b_{k,\tau} = \frac{1}{\sqrt{T}} \sum_{t=1}^{T} x(t)\psi_{k,\tau}(t). \tag{20.38}$$

Define

$$u(t) = \frac{1}{\sqrt{T}} \sum_{\tau=\infty}^{\infty} a_{k_0,\tau} \phi_{k_0,\tau}(t) \quad (20.39)$$

$$v_k(t) = \frac{1}{\sqrt{T}} \sum_{k=k_0}^{\infty} \sum_{\tau=-\infty}^{\infty} b_{k,\tau} \psi_{k,\tau}(t). \quad (20.40)$$

The series $u(t)$ is often called the approximation series, where $v_k(t)$ are called the detail series. One interpretation of this decomposition is that the approximation series contains the "smoothed" version of the original series where the detail series capture the additional components.

Intuitively the approximation series contains most of the major characteristics of the time series, or characteristics that persist over time. The detail series capture the more irregular aspects of the time series. Note that the approximation series is a single sum whereas the detail series are double sums. This suggests that the level of details, that is, the amount of irregular characteristics of the time series, is controlled by the k parameter. The higher is k, the more 'details' can be captured by this decomposition. Obviously, over-fitting may appear to be a concern with high k but the availability of big data, especially 'tall' data, may lead to an opportunity to increase the number of detail series without the risk of over-fitting and allows potentially more interesting dynamics to be revealed. This would be an interesting area for future research.

From a forecasting perspective, one approach is to forecast the approximation and detail series independently. This is feasible since these series are orthogonal by construction. The aggregation of these independent forecasts will then form the forecast of the original series. This is one of the most popular approaches in forecasting economic and financial time series using wavelets. See for examples, Conejo, Plazas, Espínola, and Molina (2005), Kriechbaumer, Angus, Parsons, and Rivas Casado (2014) and Berger (2016).

Before providing a demonstration of forecasting using wavelets, it would be useful to introduce some wavelet functions. Since the original proposal of wavelet analysis, there are now over 100 different wavelet functions. The choice of the wavelet functions is application dependent and it is mostly empirically driven. Interestingly, but perhaps not surprisingly, the forecast performance is closely related to the choice of wavelets as demonstrated in the next subsection.

Among the different choice of wavelets, the Haar family remains one of the most popular. The Haar function is defined to be

$$\phi(t) = \mathbf{1}_{[0,1)}(2t) - \mathbf{1}_{[0,1)}(2t-1) \quad (20.41)$$

where

$$\mathbf{1}_{[0,1)}(t) = \begin{cases} 0 & t \notin [0, 1) \\ 1 & t \in [0, 1). \end{cases}$$

This leads to the wavelet family by defining

$$\phi_{k,\tau} = k\phi\,(kt - \tau) \qquad k = 2^j, \quad j, \tau \in \mathbb{Z}. \tag{20.42}$$

The popularity of the Haar family is due to its simplicity. As implies by Eq. (20.41), the Haar family is essentially a set of rectangular waves with different width and height. The superposition of these rectangles is able to produce a wide range of shape to approximate any given signal.
Another popular choice is the Daubechies family which extends the Haar family by imposing varying number of vanishing moments, rather than defining them via the scaling and dilation of a father wavelet. As such, it has no closed form expression. For more details see Daubechies (1992). The appropriate choice of wavelet functions is still an area of active research. To the best of the author knowledge, there is no known criteria to select the best wavelet function. See Bachman et al. (2000) for some useful discussion.

20.4.1 Wavelet Forecasting

This subsection provides a demonstration of wavelet decomposition and its potential application for forecasting. Let p_t be the logarithmic transform of an asset price at time t and consider its returns calculated as

$$r_t = 100 \log \left(\frac{p_t}{p_{t-1}} \right).$$

In general, r_t forms a stationary process and assume its underlying data generating process is square-integrable, then returns can be represented as Eq. (20.36).

To consolidate the idea, let p_t be the daily closing price for Intel Incorporate (ticket: Intc) from 2nd August, 2004 to 28nd February 2018. This gives a total of 3416 observations and Fig. 20.2 shows the closing prices and the corresponding returns.

The ability of predicting the dynamics of asset returns has been one of the most popular topics in financial econometrics. It is generally acknowledged that if one considers r_t a random variable, then the prediction of its first moment is difficult. This can be demonstrated through the *autocorrelations* and the *partial autocorrelations* as shown in Figs. 20.3 and 20.4. As can be seen in the figures, the autocorrelation structure in the return data is complicated. A closer look to the graph reveals that a set of selected lags namely, 1, 4 and 24, are statistically significant.

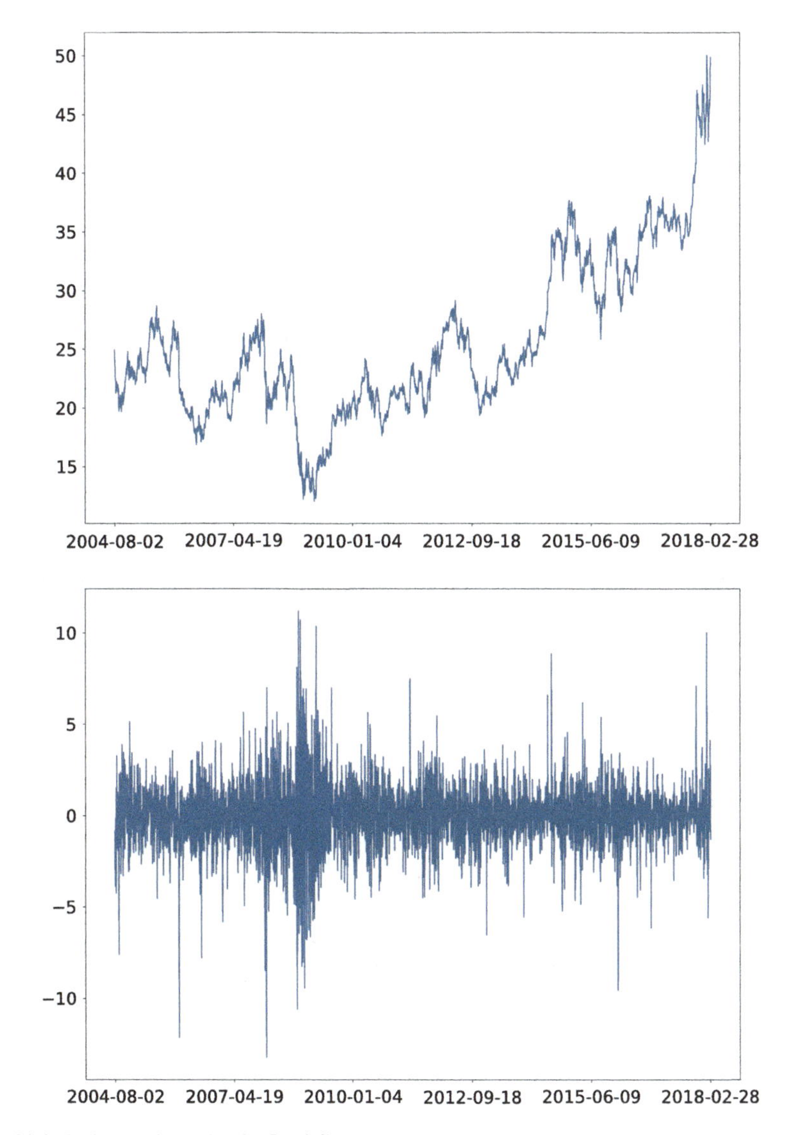

Fig. 20.2 Daily closing price for Intel Core

Similar structure can also be found in the partial autocorrelation graph. Given the returns series is second order stationary, the Wold decomposition asserts that the series can be approximated arbitrarily well by a moving average process. In other words, one should be able to develop an ARMA(p, q) model for this series and use it for forecasting purposes.

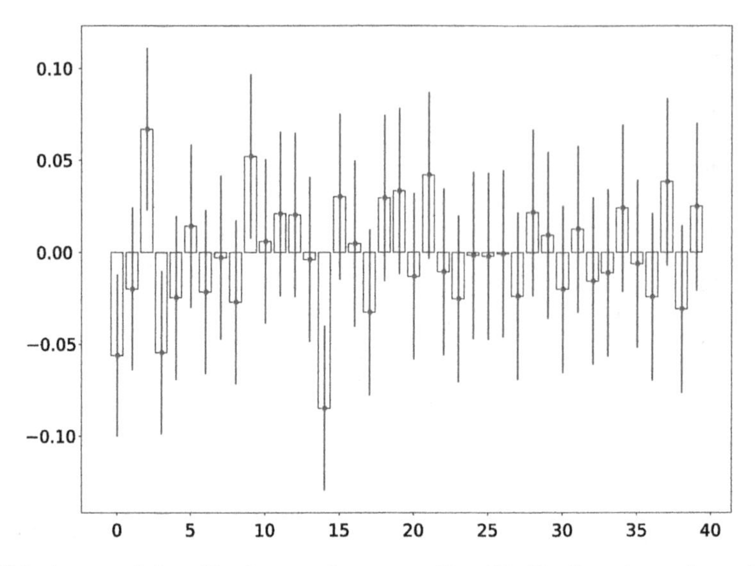

Fig. 20.3 Autocorrelation of Intel corporation returns. *Note:* The blue lines denote the confidence intervals and the red dots denote the actual estimates

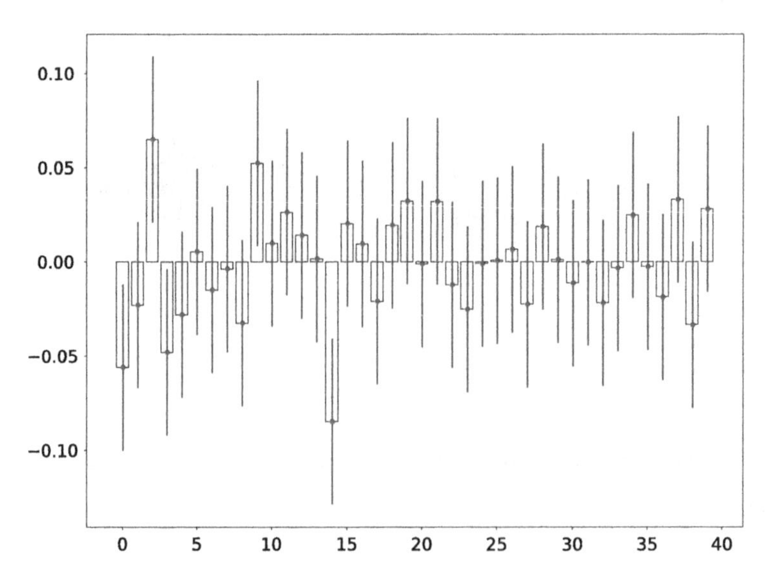

Fig. 20.4 Partial autocorrelation of Intel corporation returns. *Note:* The blue lines denote the confidence intervals and the red dots denote the actual estimates

The question is therefore the determination of the lag orders, p and q. While there exist powerful techniques to determine the optimal lag for fitting purposes, the forecast performance based on these techniques is not always clear. An alternative is to decompose the series using wavelets and develop a forecasting model for

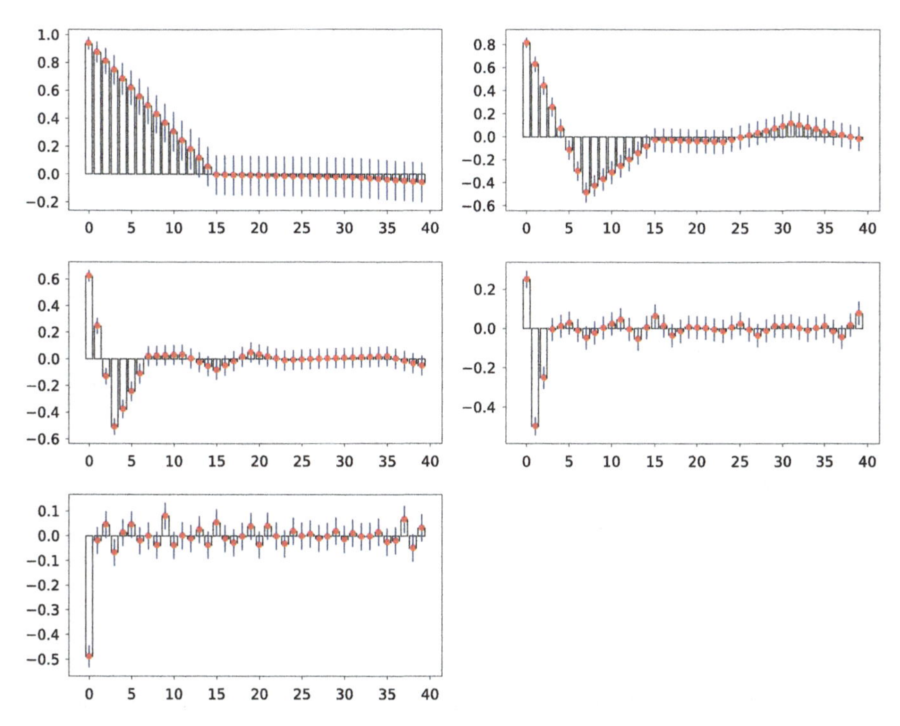

Fig. 20.5 Autocorrelations of wavelet decompositions. *Note:* The blue lines denote the confidence intervals and the red dots denote the actual estimates. The panel on the top right-hand corner contains the autocorrelation plot of the approximation series whereas the remaining panels contain the autocorrelation plots of the four detail series

each of the approximation and detail series. To justify this approach, consider the autocorrelation and partial autocorrelation graphs of the approximation and detail series as shown in Figs. 20.5 and 20.6.

Interestingly, the autocorrelation and the partial autocorrelation functions reveal a much clearer picture on the autocorrelation structure of each wavelet series. In principle, it should be relatively simpler to establish a forecast model for each of these series than the original series. The sum of the forecasts from each of the wavelet series can be used as a forecast of the original series.

For purpose of demonstration, this section specifies four different models to forecast the Intel returns. For each model, the first 3000 observations are used as the training set to estimate the parameters. The four models will then each produce a 1-day ahead forecasts for the Intel returns. The next observation in the sample will then be included in the training set to estimate the models again to produce the next 1-day ahead forecast. This procedure will be repeated until the training set contains the full sample.

The four models are (1) ARMA model, (2) Haar wavelet at level 5, (3) Haar wavelet at level 6 and (4) Daubechics wavelets at level 3. The lag order of the ARMA model as well as the ARMA models of the decomposed series is determined

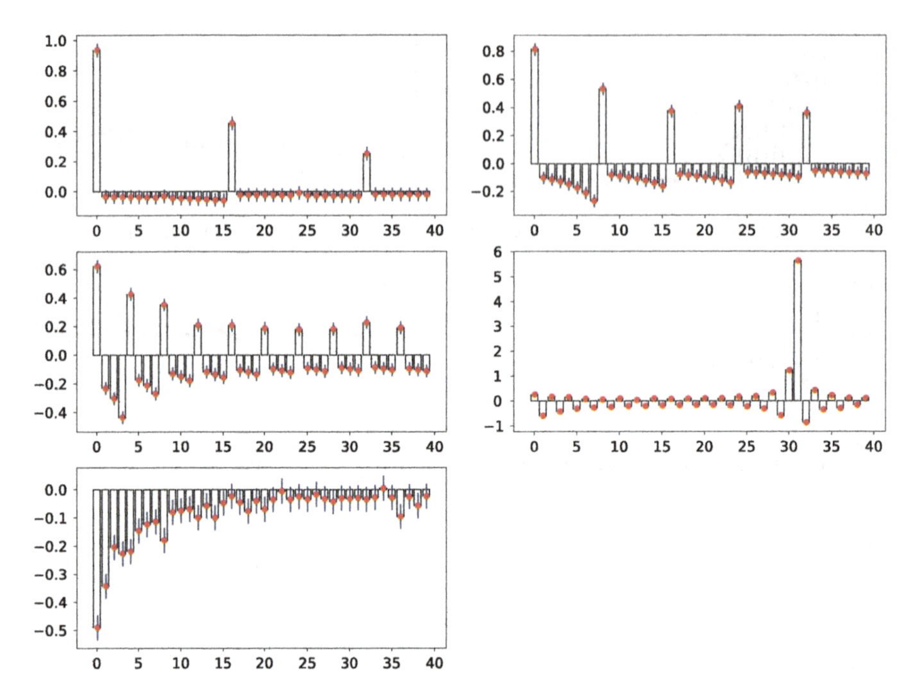

Fig. 20.6 Partial autocorrelations of wavelet decompositions. *Note:* The blue lines denote the confidence intervals and the red dots denote the actual estimates. The panel on the top right-hand corner contains the partial autocorrelation plot of the approximation series whereas the remaining panels contain the partial autocorrelation plots of the four detail series

by maximal AIC with the upper bound set at 12 lags for both AR and MA terms. The procedure of constructing wavelet forecasts can be found in Procedure 2. The computation is carried out using the Statsmodels (see Seabold & Perktold, 2010) and Pywavelet (see Lee et al., 2006) modules in Python 3. The latter module handles Steps 5 to 17 in Procedure 2. The forecasts of the approximation and detail series are constructed using the same procedure as in (1). The forecast performances are measured by both Mean Squared Forecast Error (MSFE) and Mean Absolute Deviation (MAD) and the results can be found in Table 20.1. The plot of the out-of-sample forecasts against the original series can be found in Fig. 20.7.

As shown in Table 20.1, the forecast performances of wavelet functions are generally better than the traditional ARMA model in this case. However, the choice of wavelet functions is clearly of some importance here. While the Haar wavelets generally did perform well, the Daubechies wavelet did worse than the ARMA model in terms of MSFE but slightly better than the ARMA model in MAD. This may be due to a small number of outliers and indeed, the decomposition of series in the presence of outliers and extreme observations would appear to be an important area of future research in wavelet analysis.

Procedure 2 Wavelet forecast

1: Choose a wavelet function $\phi_{k,\tau}(t)$.
2: Set $k_0 = 0$.
3: Set an upper bound of k, $\bar{k} \in \mathbb{Z}^+$.
4: Set an upper bound for τ, $\bar{\tau} \in \mathbb{Z}^+$.
5: Given the time series $x(t)$, $t = 1, \ldots, T$:
6: **for** $k = k_0 \ldots \bar{k}$ **do**
7: **for** $\tau = -\bar{\tau} \ldots \bar{\tau}$ **do**
8: Construct $\psi_{k,\tau}(t)$ using the Gram-Schmidt as presented above, replacing the inner product by the dot product. That is

$$< \psi^*_{k,\tau}(t), \phi_{k,\tau}(t) >= \sum_{t=1}^{T} \psi^*_{k,\tau}(t)\phi_{k,\tau}(t).$$

9: Construct the coefficients $a_{k,\tau}$ and $b_{k,\tau}$ using Equations (20.37) and (20.38), respectively.
10: **end for**
11: **end for**
12: Construct the approximation series $u(t)$ using Equation (20.39).
13: Construct out-of-sample forecasts for $u(t)$. That is, construct $\hat{u}(t)$, $t = T+1, \ldots, T+h$.
14: **for** $k = k_0 \ldots \bar{k}$ **do**
15: Construct $v_k(t)$ using Equation (20.40).
16: Construct out-of-sample forecasts for $v_k(t)$. That is, construct $\hat{v}_k(t)$, $t = T+1, \ldots, T+h$.
17: **end for**
18: $\hat{x}(t) = \hat{u}(t) + \hat{v}_k(t)$ gives the forecast of the original series $x(t)$ for $t = T+1, \ldots, T+h$.

Table 20.1 Forecast performance comparison

	ARMA	Harr 4	Harr 6	DB 3
MSFE	1.806393	1.237853	1.233023	1.894574
MAD	0.903027	0.716970	0.715708	0.826051

Figure 20.7 also reveals that the wavelet functions seem to be able to produce forecasts with much higher degree of variation that matches the original series. While the forecasts produced by ARMA model captured the mean of the time series reasonably well, it does not share the same degree of variability as the wavelet forecasts.

Before finishing the discussion of wavelets, one natural question is whether it is possible to test Granger causality using wavelet functions rather than trigonometry functions as discussed in Sect. 20.3. The motivation of such approach hinges on the fact that macroeconomic time series often exhibited structural breaks due to extreme events and policy changes. Therefore, it is not difficult to imagine that Granger causality may change over different frequencies and different time locations. Wavelets analysis seems to provide such flexibility.

While research in this area is still ongoing, preliminary work seems to be promising. Olayeni (2016) and the references within provide some of the most recent research in this area.

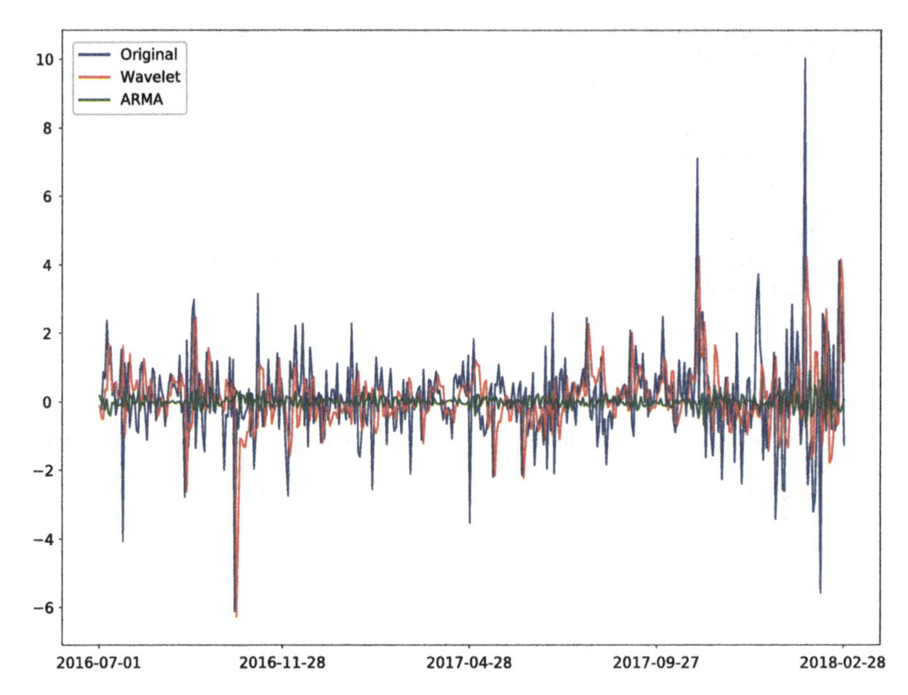

Fig. 20.7 Wavelet and ARMA forecasts. *Note:* This contains the plot of the original series (in blue), the forecast series by Haar family (in red) and the forecast series by the ARMA model (in green)

20.5 ZVAR and Generalised Shift Operator

This section introduces the Generalised Shift Operator and its application to the development of the ZVAR model. The exposition follows closely to Wilson, Reale, and Haywood (2016) whose authors are often credited to be the inventors of the ZVAR model. Its acronym was justified by the use of the letter Z for the generalised shift operator. Wilson et al. (2016), however, light-heartedly used the letter Z to mean an ultimate model in the autoregressive family. Roughly speaking, the generalised shift operator generalises the familiar lag (backward shift) operator to allow continuous shifting, rather than shifting discretely. Consequently, it has the ability to produce forecast at a higher frequency based on data sampled at lower frequency.

An important point to note here is that this approach cannot create additional information on the dynamics between each discrete time point. All it does is utilising information from existing observations and provide a process that connects each discrete time point. This is similar to a polynomial spline in spirit which connects each discrete time point based on some approximation function. Unlike traditional spline approach, however, ZVAR can produce forecasts which cannot be achieved with traditional spline approach. This is due to the fact that most spline techniques are originated as an attempt to construct a smooth function that would pass through a certain set of points. This would require a minimum a starting point and an end point. However, splines do not generally produce values past the end point which makes it difficult to produce out-of-sample forecasts. ZVAR also provides a description of the dynamics of the variables similarly to a typical ARMA type models, which allows one to understand the dependence structure of the time series.

Another interesting feature of the ZVAR model is that it has the ability to carry past information in an extremely parsimonious way. This facilitates greatly with the modelling of tall data, where potentially many lags are required to produce sensible forecasts.

20.5.1 Generalised Shift Operator

The generalised shift operator Z is defined as

$$Z_\theta = \frac{B-\theta}{1-\theta B} \tag{20.43}$$

where B denotes the standard backward shift (lag) operator such that $B\mathbf{x}_t = \mathbf{x}_{t-1}$ where $\mathbf{x}_t$ is a $k \times 1$ vector of time series. Also worth noting is that the inverse of the Z operator is

$$Z^{-1} = \frac{1-\theta B}{B-\theta} = \frac{B^{-1}-\theta}{1-\theta B^{-1}} \tag{20.44}$$

and thus when $\theta = 0$, Z^{-1} defines the forward shift (lead) operator such that $Z^{-1}\mathbf{x}_t = \mathbf{x}_{t+1}$. A special feature of this operator is that the amount of "shifting" is governed by the parameter, $\theta \in (-1, 1)$, and more specifically, $Z_\theta \mathbf{x}_t \approx \mathbf{x}_{t-l}$ where $l = (1+\theta)(1-\theta)^{-1}$ when $\mathbf{x}_t$ is a slowly varying time series. See Wilson et al. (2016) for further discussion. Now define $\mathbf{s}_t^{(k)} = Z_\theta^k \mathbf{x}_t$, then it is straightforward to show that the state variable, $\mathbf{s}_t^{(k+1)}$ satisfies the following recursive relation

$$\mathbf{s}_t^{(k+1)} = \mathbf{s}_{t-1}^{(k)} - \theta \mathbf{s}_t^{(k)} + \theta \mathbf{s}_{t-1}^{(k+1)}. \tag{20.45}$$

To see this, consider

$$\begin{aligned}
\mathbf{s}_t^{(k+1)} &= Z_\theta \mathbf{s}_t^{(k)} \\
&= \frac{B-\theta}{1-\theta B}\mathbf{s}_t^{(k)} \\
(1-\theta B)\,\mathbf{s}_t^{(k+1)} &= \mathbf{s}_{t-1}^{(k)} - \theta \mathbf{s}_{t-1}^{(k)} \\
\mathbf{s}_t^{(k+1)} &= \mathbf{s}_{t-1}^{(k)} - \theta \mathbf{s}_t^{(k)} + \theta \mathbf{s}_{t-1}^{(k+1)}.
\end{aligned}$$

On surface, $\mathbf{s}_t^k$ appears to represent $\mathbf{x}_{t-kl}$, that is, it shifts the time series back by the value of kl. However, an important feature about $\mathbf{s}_t^k$ is that it does not just contain the information of $\mathbf{x}_{t-kl}$ but rather, it carries the information from the past values of $\mathbf{x}_{t-kl}$. To see this, consider

$$\begin{aligned} Z_\theta \mathbf{x}_t =& (B-\theta)(1-\theta B)^{-1}\mathbf{x}_t \\ =& (1-\theta B)^{-1}(\mathbf{x}_{t-1}-\theta \mathbf{x}_t) \\ =& \left(1-\theta^2\right)\sum_{i=0}^{\infty}\theta^i B^{i+1}\mathbf{x}_t - \theta \mathbf{x}_t, \end{aligned}$$

The last line follows by replacing $(1-\theta B)^{-1}$ with its Taylor expansion. The first term in the last line suggests that $\mathbf{s}_t^1$ contains a weighted sum of all past values of $\mathbf{x}_t$. This would be useful in modelling tall data with many time series observations as the number of required lags may be reduced due to this representation. The expression above is also useful in understanding the impact of the generalised shift operator on time series. Let $\mathbf{\Gamma}(h)$ denotes the auto-covariance function of $\mathbf{x}_t$ and consider the variance-covariance matrix of $\mathbf{s}_t$ and $\mathbf{x}_t$, that is

$$\begin{aligned} \mathbb{E}\left(\mathbf{s}_t\mathbf{x}_t'\right) =& \mathbb{E}\left[\left(1-\theta^2\right)\sum_{i=0}^{\infty}\theta^i \mathbf{x}_{t-i-1}\mathbf{x}_t - \theta \mathbf{x}_t\mathbf{x}_t'\right] \\ =& \left(1-\theta^2\right)\sum_{i=0}^{\infty}\theta^i \mathbf{\Gamma}(i+1) - \theta\, \mathbf{\Gamma}(0). \end{aligned}$$

Now assume that $\mathbf{\Gamma}(l) = \phi\mathbf{\Gamma}(l-1) = \phi^l\mathbf{\Gamma}(0)$ where ϕ measures the persistence of $\mathbf{x}_t$ and is assumed to be less than 1.

$$\begin{aligned} \mathbb{E}\left(\mathbf{s}_t\mathbf{x}_t'\right) =& \left[\phi(1-\theta^2)\sum_{i=0}^{\infty}(\theta\phi)^i - \theta\right]\mathbf{\Gamma}(0) \\ =& \left[\phi\frac{1-\theta^2}{1-\phi\theta} - \theta\right]\mathbf{\Gamma}(0). \end{aligned}$$

Therefore the amount of "shifting" implied by the Z operator can be calculated by solving

$$\phi^l = \phi\frac{1-\theta^2}{1-\phi\theta} - \theta$$

which gives

$$l = \frac{\log\left|\phi\frac{1-\theta^2}{1-\phi\theta} - \theta\right|}{\log|\phi|} \tag{20.46}$$

and a straightforward application of L'Hôpital's rule shows that

$$\lim_{\phi \to 1} l = \frac{1+\theta}{1-\theta}. \tag{20.47}$$

Hence, for a reasonably persistent time series, $Z\mathbf{x}_t \approx \mathbf{x}_{t-l}$ with $l = (1+\theta)(1-\theta)^{-1}$. However, the approximation will behave poorly when $\phi = \theta$ because Eq. (20.46) suggests that $l \to -\infty$ as $\phi \to \theta$. Figure 20.8 contains two plots that demonstrate the performance of the approximation. The first plot shows the difference between l and $(1+\theta)(1-\theta)^{-1}$ over different levels of persistence with $\theta = 0.5$ while the surface plot shows the difference between l and $(1+\theta)(1-\theta)^{-1}$ over different levels of persistence and θ values.

As expected, the approximation does reasonably well as long as $\phi \neq \theta$ but it does tend to deviate further as θ approaches ϕ.

Given the relation between the lag order and the parameter in the generalised shift operator, it is then possible to 'shift' the time series to any $l \in (0, 1)$ as long as

$$\theta = \frac{l-1}{1+l} \tag{20.48}$$

lies in $(-1, 1)$, which is clearly the case for $l \in [0, 1]$. Therefore it is possible to forecast using low frequency data at higher frequency. For example, assume $\mathbf{x}_t$ was observed at quarterly frequency and let $\theta = -0.5$, then $Z\mathbf{x}_t \approx \mathbf{x}_{t+1/3}$ which is the value of the first month in the next quarter.

20.5.2 ZVAR Model

Given the relation between the lag order and the generalised shift operator, the ZVAR model can then be used to forecast high frequency observations using lower frequency data. Following Wilson et al. (2016), define the general form of the ZVAR model as

$$L\left(Z_\rho \mathbf{x}_t | \mathbf{F}_{t-1}\right) = \sum_{i=1}^{p} \xi_i \mathbf{s}_t^{(i-1)} \tag{20.49}$$

where L denotes the linear projection of the first argument onto the second with $\mathbf{F}_{t-1}$ containing all past values of $\mathbf{x}_t$ up to $t-1$.

For quarterly data, clearly if $\rho = -0.5$, then $Z_\rho \mathbf{x}_t = \mathbf{x}_{t+1/3}$, in that case, Eq. (20.49) becomes a predictive model for the first month of the next quarter based on past values of $\mathbf{x}_t$. Similarly, set $\rho = -0.2$ then $Z_\rho \mathbf{x}_t \approx \mathbf{x}_{t+2/3}$ which turns Eq. (20.49) into a predictive model for the second month of the next quarter. Finally, when $\rho = 0$, Eq. (20.49) reverts to a standard predictive model for the next time

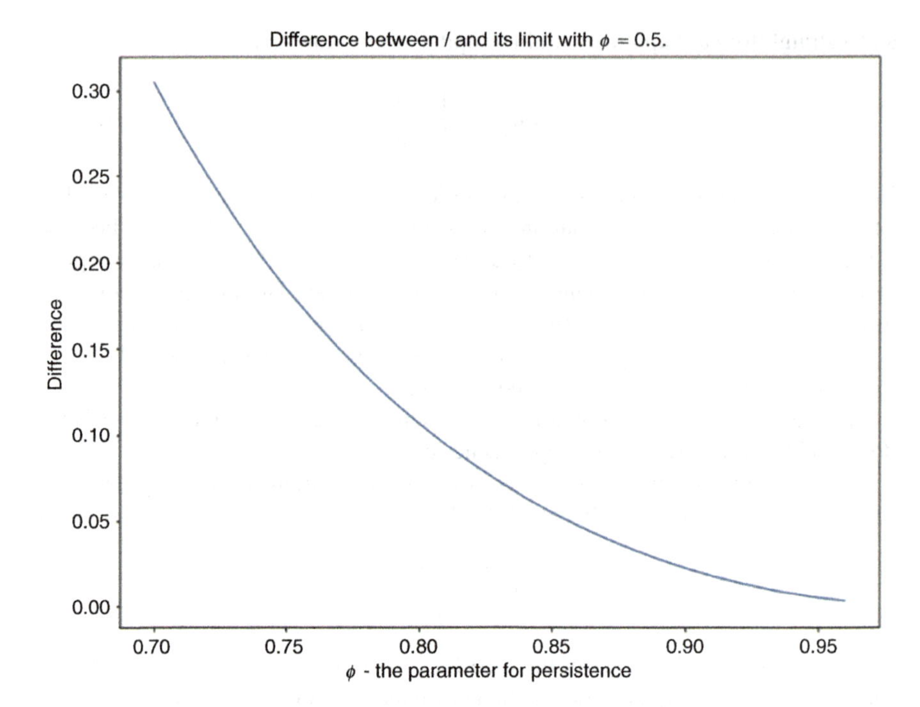

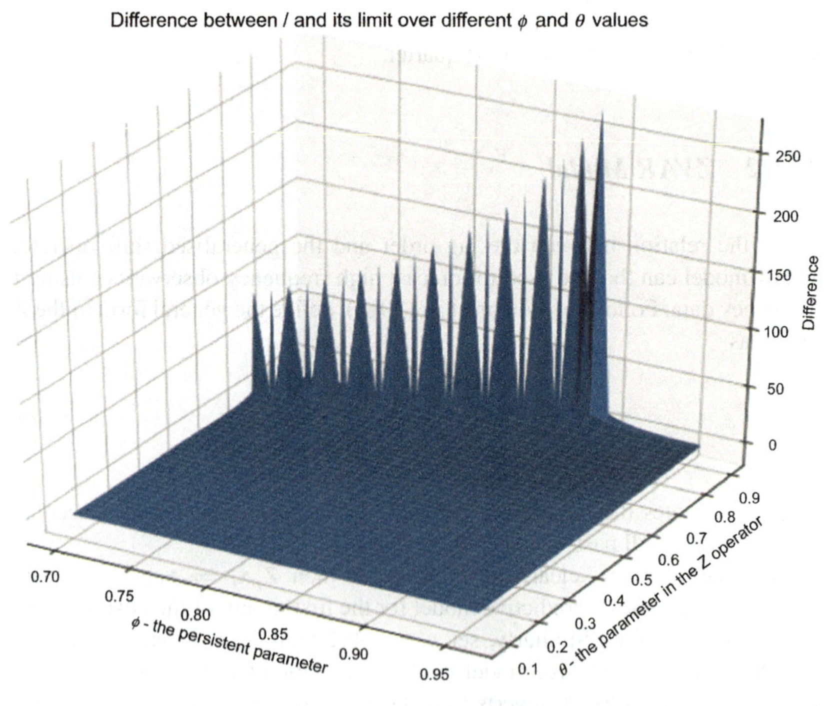

Fig. 20.8 Shifting with the Z operator *Note:* The approximation performs poorly and erratically when $\phi \to \theta$

period. This can be summarised as follows:

$$L\left(Z_{-0.5}\mathbf{x}_t|\mathbf{F}_{t-1}\right)=\sum_{i=1}^{p}\xi_i^1\mathbf{s}_t^{(i-1)} \tag{20.50}$$

$$L\left(Z_{-0.2}\mathbf{x}_t|\mathbf{F}_{t-1}\right)=\sum_{i=1}^{p}\xi_i^2\mathbf{s}_t^{(i-1)} \tag{20.51}$$

$$L\left(Z_{0}\mathbf{x}_t|\mathbf{F}_{t-1}\right)=\sum_{i=1}^{p}\xi_i^0\mathbf{s}_t^{(i-1)}. \tag{20.52}$$

The coefficient vectors, $\boldsymbol{\xi}^i$ for $i=0,1,2$ where $\boldsymbol{\xi}^i=\left(\xi_1^i,\ldots,\xi_p^i\right)'$ can be obtained via *least squares* type estimators. Specifically, let $\mathbf{X}_i=\left(Z_\rho^{-1}\mathbf{s}^0,\ldots,Z_\rho^{-1}\mathbf{s}^{(p-1)}\right)$ where $\mathbf{s}^{(k)}=\left(\mathbf{s}_1^{(k)},\ldots,\mathbf{s}_T^{(k)}\right)'$ with $\rho=-0.5,-0.2,0$ for $i=1/3,2/3,1$, respectively, then

$$\hat{\boldsymbol{\xi}}^i=\left(\mathbf{X}_i'\mathbf{X}_i\right)^{-1}\mathbf{X}_i'Z_\rho^{-1}\mathbf{x}. \tag{20.53}$$

Upon obtaining the coefficient estimates, the forecasts of each month in a quarter can be obtained in a straight forward manner by using Eqs. (20.50)–(20.52).

20.5.3 Monte Carlo Evidence

This subsection conducts some Monte Carlo experiments to examine the performance of the proposed method. The basic idea is to simulate data at the monthly frequency but only utilises quarterly data for purposes of estimating the ZVAR model as defined in Eqs. (20.50)–(20.52). These models will then be used to produce monthly forecasts and their forecast performance will be compared to the true model with full set of observations, i.e., true model using monthly observations. The forecast performance will also be compared with the following misspecified VAR(1) model.

$$x_{it}=\phi_1 x_{it-1}+\varepsilon_{it},\qquad i=1,\ldots,3 \tag{20.54}$$

The idea here is to examine the performance of a ZVAR model using quarterly data (less information) with a misspecified time series model using monthly data (full information).

Two different data generating processes are considered in this subsection. The first DGP follows:

$$\begin{pmatrix} x_{1t} \\ x_{2t} \\ x_{3t} \end{pmatrix} = \begin{pmatrix} 0.7 & 0 & 0 \\ 0 & 0.7 & 0 \\ 0 & 0 & 0.7 \end{pmatrix} \begin{pmatrix} x_{1t-1} \\ x_{2t-1} \\ x_{3t-1} \end{pmatrix} + \begin{pmatrix} 0.4 & 0 & 0 \\ 0 & 0.4 & 0 \\ 0 & 0 & 0.4 \end{pmatrix} \begin{pmatrix} \varepsilon_{1t-1} \\ \varepsilon_{2t-1} \\ \varepsilon_{3t-1} \end{pmatrix} + \begin{pmatrix} \varepsilon_{1t} \\ \varepsilon_{2t} \\ \varepsilon_{3t} \end{pmatrix} \tag{20.55}$$

and the second DGP follows:

$$\begin{pmatrix} x_{1t} \\ x_{2t} \\ x_{3t} \end{pmatrix} = \begin{pmatrix} 0.5 & -0.1 & 0.2 \\ -0.3 & 0.3 & 0.1 \\ -0.1 & 0.2 & 0.3 \end{pmatrix} \begin{pmatrix} x_{1t-1} \\ x_{2t-1} \\ x_{3t-1} \end{pmatrix} + \begin{pmatrix} 0.4 & 0 & 0 \\ 0 & 0.3 & 0 \\ 0 & 0 & -0.1 \end{pmatrix} \begin{pmatrix} \varepsilon_{1t-1} \\ \varepsilon_{2t-1} \\ \varepsilon_{3t-1} \end{pmatrix} + \begin{pmatrix} \varepsilon_{1t} \\ \varepsilon_{2t} \\ \varepsilon_{3t} \end{pmatrix}. \tag{20.56}$$

In both cases, the variance-covariance matrix of $(\varepsilon_{1t}, \varepsilon_{2t}, \varepsilon_{3t})'$, denoted by $\boldsymbol{\Omega}$ is

$$\boldsymbol{\Omega} = \begin{pmatrix} 0.02 & -0.03 & -0.01 \\ -0.03 & 0.06 & 0.025 \\ -0.01 & 0.025 & 0.0125 \end{pmatrix}. \tag{20.57}$$

The number of replications is 1000 for both DGPs. For each replication, 40 years of monthly data will be simulated by the DGP as stated in Eqs. (20.55) and (20.57), which results in 480 observations in total. The first 30 years of data will be used for estimation purposes with the last 10 years reserved for one-step ahead dynamic forecasts. In other words, the first 360 observations will be used to estimate the parameters in the model with the last 120 observations to be used for out-of-sample forecasts.

The forecast performance as measured by Mean Squared Error (MSFE) can be found in Tables 20.2 and 20.3 for DGPs as stated in Eqs. (20.55) and (20.56),

Table 20.2 Forecast performances from DGP following Eq. (20.55)

	True DGP	Estimated true DGP	ZAR monthly	ZAR quarterly	VAR(1)
x_{1t}	0.020078	0.048147	0.048249	0.055375	0.054211
x_{2t}	0.059929	0.142581	0.143016	0.148263	0.147091
x_{3t}	0.012472	0.029940	0.030420	0.040075	0.040065

Table 20.3 Forecast performances from DGP following Eq. (20.56)

	True DGP	Estimated true DGP	ZAR monthly	ZAR quarterly	VAR(1)
x_{1t}	0.019941	0.022553	0.022581	0.041691	0.043512
x_{2t}	0.059887	0.039426	0.040137	0.103939	0.114036
x_{3t}	0.012487	0.049332	0.088825	0.106217	0.121554

respectively. In each case, the forecast performance of ZVAR using quarterly data is compared to other four cases namely, forecasts produced by the true model (no estimation), forecasts produced by estimated true model (correct ARMA specification with estimated parameters using the first 360 observations), estimated ZVAR model using monthly data and an VAR(1) model.

The results broadly follow expectation. That is, the true model performs better than the estimated true model, the ZVAR model based on monthly data performs closely to the estimated true model, while ZVAR based on quarterly data performs worse than the other three cases. This is to be expected because the ZVAR model based on monthly data provided a close approximation to the true model in terms of model specification but the performance would be worse than the true model or the estimated true model due to finite sample errors in estimation. The performance of the ZVAR model using quarterly data would be worse than the other three models because it does not utilise the full data set and therefore, it does not have access to full information. Interestingly, ZVAR using quarterly data performs marginally better than the misspecified VAR(1) model under Eq. (20.56) but not under Eq. (20.55). The reason for that may be due to the fact that Eq. (20.54) does not allow any information from the other time series and this appears to be more important than having less observations in this case.

Obviously, all information should be utilised if available. The main contribution here is that, in the absence of full information, ZVAR still seems to be able to produce forecasts that out-perform a misspecified model with full observations. It may therefore prove to be a useful tool in the presence of big data when the sampling frequency of each variable may not be consistent across the whole dataset.

20.6 Conclusion

This chapter provided an overview on three topics in time series analysis which share a common foundation in frequency domain namely, series decomposition based on Fourier Series and wavelets as well as the generalised shift operator. One common theme of these techniques is their ability to express complex structures as compositions of simpler ones. This allows the analysis of complex time series to be divided as a sequence of much simpler processes and thus contributes to the analysis of big and possibly complex data.

Wavelet analysis can be interpreted as a generalisation of Fourier analysis. The Spectral Representation Theorem provided a powerful justification of approximating a second order stationary process as a linear combination of complex exponentials. This result extends to wavelet analysis in the sense that any continuous function in L^2 can be expressed as a linear combination of wavelets. However, in the case when the continuous function is a stochastic process, there is yet no known result similar to that of the spectral representation theorem. Thus, deriving the conditions in which a stochastic process admits a wavelet representation would appear to be an important area of future research, especially given the potential contribution of wavelets to forecast complex processes as demonstrated in this chapter.

As data becomes more widely available in terms of its frequency and accessibility, models with the ability to capture long memory with parsimonious representations would provide a convenient tool for forecasters. The ZVAR model based on the generalised shift operator provides one of these tools. In addition, the ZVAR model has the ability to 'shift' along the data set continuously, even when the data must be sampled discretely. This potentially allows forecasters to produce forecasts at the frequency that is most relevant, rather than being restricted to the same sampling frequency as the data.

References

Aydin, M. (2018). Natural gas consumption and economic growth nexus for top 10 natural GaseConsuming countries: A granger causality analysis in the frequency domain. *Energy, 165*, 179–186. https://doi.org/10.1016/j.energy.2018.09.149

Bachman, G., Narici, L., & Beckenstein, E. (2000). *Fourier and wavelet analysis*. New York: Springer.

Bahmani-Oskooee, M., Chang, T., & Ranjbar, O. (2016). Asymmetric causality using frequency domain and time-frequency domain (wavelet) approaches. *Economic Modelling, 56*, 66–78. https://doi.org/10.1016/j.econmod.2016.03.002

Benhmad, F. (2012). Modeling nonlinear Granger causality between the oil price and U.S. dollar: A wavelet based approach. *Economic Modelling, 29*, 1505–1514. https://doi.org/10.1016/j.econmod.2012.01.003

Berger, T. (2016). Forecasting based on decomposed financial return series: a wavelet analysis. *Journal of Forecasting, 35*, 419–433. https://doi.org/10.1002/for.2384

Bouri, E., Roubaud, D., Jammazi, R., & Assaf, A. (2017). Uncovering frequency domain causality between gold and the stock markets of China and India: Evidence from implied volatility indices. *Finance Research Letters, 23*, 23–30. https://doi.org/10.1016/j.frl.2017.06.010

Bozoklu, S., & Yilanci, V. (2013). Energy consumption and economic growth for selected OECD countries: Further evidence from the Granger causality test in the frequency domain. *Energy Policy, 63*, 877–881. https://doi.org/10.1016/j.enpol.2013.09.037

Breitung, J., & Candelon, B. (2006). Testing for short- and long-run causality: A frequency-domain approach. *Journal of Econometrics, 132*, 363–378. https://doi.org/10.1016/j.jeconom.2005.02.004

Breitung, J., & Schreiber, S. (2018). Assessing causality and delay within a frequency band. *Econometrics and Statistics, 6*, 57–73. https://doi.org/10.1016/j.ecosta.2017.04.005. arXiv:1011.1669v3

Conejo, A. J., Plazas, M. A., Espínola, R., & Molina, A. B. (2005). Day-ahead electricity price forecasting using the wavelet transform and ARIMA models. *IEEE Transactions on Power Systems, 20*, 1035–1042.

Croux, C., & Reusens, P. (2013). Do stock prices contain predictive power for the future economic activity? A Granger causality analysis in the frequency domain. *Journal of Macroeconomics, 35*, 93–103. https://doi.org/10.1016/j.jmacro.2012.10.001

Daubechies, I. (1992). *Ten lectures on wavelets*. SIAM. Retrieved from http://bookstore.siam.org/cb61/

Geweke, J. (1982). Measurement of linear dependence and feedback between multiple time series measurement of linear dependence and feedback betwveen multiple time series. *Journal of the American Statistical Association, 77*(378), 304–313.

Granger, C. W. J. (1963). Economic processes involving feedback. *Information and Control, 6*(1), 28–48. https://doi.org/10.1016/S0019-9958(63)90092-5
Granger, C. W. J. (1969). Investigating causal relations by econometric models and cross-spectral methods. *Econometrica, 37*, 424–438. https://doi.org/10.2307/1912791
Granger, C. W. J., & Hatanaka, M. (1964). *Spectral analysis of economic time series*. Princeton: Princeton University Press.
Granger, C. W. J., & Lin, J.-L. (1995). Causality in the long run. *Econometric Theory, 11*(2), 530–536.
Hamilton, J. (1994). *Time series analysis*. Princeton: Princeton University Press.
Hosoya, Y. (1991). The decomposition and measurement of the interdependency between second-order stationary processes. *Probability Theory and Related Fields, 88*, 429–444. https://doi.org/10.1007/BF01192551
Joseph, A., Sisodia, G., & Tiwari, A. K. (2014). A frequency domain causality investigation between futures and spot prices of Indian commodity markets. *Economic Modelling, 40*, 250–258. https://doi.org/10.1016/j.econmod.2014.04.019
Koopmans, L. H. (1995). *The spectral analysis of time series*. San Diego: Academic Press.
Kriechbaumer, T., Angus, A., Parsons, D., & Rivas Casado, M. (2014). An improved waveletARIMA approach for forecasting metal prices. *Resources Policy, 39*, 32–41.
Lee, G., Gommers, R., Wasilewski, F., Wohlfahrt, K., O'Leary, A., Nahrstaedt, H., & Contributors. (2006). *Pywavelets – wavelet transforms in python*. https://githubcom/PyWavelets/pywt
Li, X. L., Chang, T., Miller, S. M., Balcilar, M., & Gupta, R. (2015). The co-movement and causality between the U.S. housing and stock markets in the time and frequency domains. *International Review of Economics and Finance, 38*, 220–233. https://doi.org/10.1016/j.iref.2015.02.028
Lütkepohl, H. (2005). *New introduction to multiple time series analysis*. Berlin: Springer.
Olayeni, O. R. (2016). Causality in continuous wavelet transform without spectral matrix factorization: Theory and application. *Computational Economics, 47*(3), 321–340. https://doi.org/10.1007/s10614-015-9489-4
Seabold, S., & Perktold, J. (2010). Statsmodels: Econometric and statistical modeling with python. In *9th Python in Science Conference*.
Shahbaz, M., Tiwari, A. K., & Tahir, M. I. (2012). Does CPI Granger-cause WPI? New extensions from frequency domain approach in Pakistan. *Economic Modelling, 29*, 1592–1597. https://doi.org/10.1016/j.econmod.2012.05.016
Sun, X., Chen, X., Wang, J., & Li, J. (2018). Multi-scale interactions between economic policy uncertainty and oil prices in time-frequency domains. *North American Journal of Economics and Finance*, in press. https://doi.org/10.1016/j.najef.2018.10.002
Tiwari, A. K. (2012). An empirical investigation of causality between producers' price and consumers' price indices in Australia in frequency domain. *Economic Modelling, 29*, 1571–1578. https://doi.org/10.1016/j.econmod.2012.05.010
Tiwari, A. K., & Albulescu, C. T. (2016). Oil price and exchange rate in India: Fresh evidence from continuous wavelet approach and asymmetric, multi-horizon Granger-causality tests. *Applied Energy, 179*, 272–283. https://doi.org/10.1016/j.apenergy.2016.06.139
Tiwari, A. K., Mutascu, M. I., Albulescu, C. T., & Kyophilavong, P. (2015). Frequency domain causality analysis of stock market and economic activity in India. *International Review of Economics and Finance, 39*, 224–238. https://doi.org/10.1016/j.iref.2015.04.007
Wilson, G. T., Reale, M., & Haywood, J. (2016). *Models for dependent time series*. Boca Raton: CRC Press.
Zellner, A. (1979). Causality and econometrics. *Carnegie-Rochester Conference Series on Public Policy, 10*, 9–54.

Chapter 21
Hierarchical Forecasting

George Athanasopoulos, Puwasala Gamakumara, Anastasios Panagiotelis, Rob J. Hyndman, and Mohamed Affan

21.1 Introduction

Accurate forecasting of key macroeconomic variables such as Gross Domestic Product (GDP), inflation, and industrial production, has been at the forefront of economic research over many decades. Early approaches involved univariate models or at best low dimensional multivariate systems. The era of big data has led to the use of regularisation and shrinkage methods such as dynamic factor models, Lasso, LARS, and Bayesian VARs, in an effort to exploit the plethora of potentially useful predictors now available. These predictors commonly also include the components of the variables of interest. For instance, GDP is formed as an aggregate of consumption, government expenditure, investment, and net exports, with each of these components also formed as aggregates of other economic variables. While the macroeconomic forecasting literature regularly uses such sub-indices as predictors, it does so in ways that fail to exploit accounting identities that describe known deterministic relationships between macroeconomic variables.

In this chapter we take a different approach. Over the past decade there has been a growing literature on forecasting collections of time series that follow aggregation

G. Athanasopoulos (✉) · A. Panagiotelis
Department of Econometrics and Business Statistics, Monash University, Caulfield, VIC, Australia
e-mail: George.Athanasopoulos@monash.edu; Anastasios.Panagiotelis@monash.edu

P. Gamakumara · R. J. Hyndman
Department of Econometrics and Business Statistics, Monash University, Clayton, VIC, Australia
e-mail: Puwasala.Gamakumara@monash.edu; Rob.Hyndman@monash.edu

M. Affan
Maldives Monetary Authority (MMA), Malé, Republic of Maldives
e-mail: mohamed.affan@mma.gov.mv

P. Fuleky (ed.), *Macroeconomic Forecasting in the Era of Big Data*, Advanced Studies in Theoretical and Applied Econometrics 52,
https://doi.org/10.1007/978-3-030-31150-6_21

constraints, known as hierarchical time series. Initially the aim of this literature was to ensure that forecasts adhered to aggregation constraints thus ensuring aligned decision-making. However, in many empirical settings the forecast reconciliation methods designed to deal with this problem have also been shown to improve forecast accuracy. Examples include forecasting accidents and emergency admissions (Athanasopoulos, Hyndman, Kourentzes, & Petropoulos, 2017), mortality rates (Shang & Hyndman, 2017), prison populations (Athanasopoulos, Steel, & Weatherburn, 2019), retail sales (Villegas & Pedregal, 2018), solar energy (Yagli, Yang, & Srinivasan, 2019; Yang, Quan, Disfani, & Liu, 2017), tourism demand (Athanasopoulos, Ahmed, & Hyndman, 2009; Hyndman, Ahmed, Athanasopoulos, & Shang, 2011; Wickramasuriya, Athanasopoulos, & Hyndman, 2018), and wind power generation (Zhang & Dong, 2019). Since both aligned decision-making and forecast accuracy are key concerns for economic agents and policy makers we propose the application of state-of-the-art forecast reconciliation methods to macroeconomic forecasting. To the best of our knowledge the only application of forecast reconciliation methods to macroeconomics focuses on point forecasting for inflation (Capistrán, Constandse, & Ramos-Francia, 2010; Weiss, 2018).

The remainder of this chapter is set out as follows: Section 21.2 introduces the concept of hierarchical time series, i.e., collections of time series with known linear constraints, with a particular emphasis on macroeconomic examples. Section 21.3 describes state-of-the-art forecast reconciliation techniques for point forecasts, while Sect. 21.4 describes the more recent extension of these techniques to probabilistic forecasting. Section 21.5 describes the data used in our empirical case study, namely Australian GDP data, that is represented using two alternative hierarchical structures. Section 21.6 provides details on the setup of our empirical study including metrics used for the evaluation of both point and probabilistic forecasts. Section 21.7 presents results and Sect. 21.8 concludes providing future avenues for research that are of particular relevance to the empirical macroeconomist.

21.2 Hierarchical Time Series

To simplify the introduction of some notation we use the simple two-level hierarchical structure shown in Fig. 21.1. Denote as $y_{\text{Tot},t}$ the value observed at time t for the most aggregate (Total) series corresponding to level 0 of the hierarchy. Below level 0, denote as $y_{i,t}$ the value of the series corresponding to node i, observed at time t. For example, $y_{\text{A},t}$ denotes the tth observation of the series corresponding to node A at level 1, $y_{\text{AB},t}$ denotes the tth observation of the series corresponding to node AB at level 2, and so on.

Let $\boldsymbol{y}_t = (y_{\text{Tot},t}, y_{\text{A},t}, y_{\text{B},t}, y_{\text{AA},t}, y_{\text{AB},t}, y_{\text{BA},t}, y_{\text{BB},t}, y_{\text{BC},t})'$ denote a vector containing observations across all series of the hierarchy at time t. Similarly denote as $\boldsymbol{b}_t = (y_{\text{AA},t}, y_{\text{AB},t}, y_{\text{BA},t}, y_{\text{BB},t}, y_{\text{BC},t})'$ a vector containing observations only for the bottom-level series. In general, $\boldsymbol{y}_t \in \mathbb{R}^n$ and $\boldsymbol{b}_t \in \mathbb{R}^m$ where n denotes the number

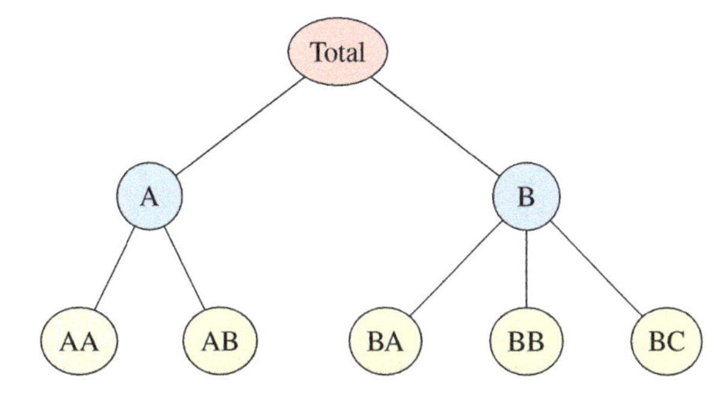

Fig. 21.1 A simple two-level hierarchical structure

of total series in the structure, m the number of series at the bottom level, and $n > m$ always. In the simple example of Fig. 21.1, $n = 8$ and $m = 5$.

Aggregation constraints dictate that $y_{\text{Tot}} = y_{\text{A},t} + y_{\text{B},t} = y_{\text{AA},t} + y_{\text{AB},t} + y_{\text{BA},t} + y_{\text{BB},t} + y_{\text{BC},t}$, $y_{\text{A},t} = y_{\text{AA},t} + y_{\text{AB},t}$ and $y_{\text{B}} = y_{\text{BA},t} + y_{\text{BB},t} + y_{\text{BC},t}$. Hence we can write

$$\boldsymbol{y}_t = \boldsymbol{S}\boldsymbol{b}_t, \tag{21.1}$$

where

$$\boldsymbol{S} = \begin{pmatrix} 1\;1\;1\;1\;1 \\ 1\;1\;0\;0\;0 \\ 0\;0\;1\;1\;1 \\ \boldsymbol{I}_5 \end{pmatrix}$$

is an $n \times m$ matrix referred to as the *summing matrix* and $\boldsymbol{I}_m$ is an m-dimensional identity matrix. $\boldsymbol{S}$ reflects the linear aggregation constraints and in particular how the bottom-level series aggregate to levels above. Thus, columns of $\boldsymbol{S}$ span the linear subspace of $\mathbb{R}^n$ for which the aggregation constraints hold. We refer to this as the *coherent subspace* and denote it by $\mathfrak{s}$. Notice that pre-multiplying a vector in $\mathbb{R}^m$ by $\boldsymbol{S}$ will result in an n-dimensional vector that lies in $\mathfrak{s}$.

Property 21.1 A hierarchical time series has observations that are *coherent*, i.e., $\boldsymbol{y}_t \in \mathfrak{s}$ for all t. We use the term coherent to describe not just $\boldsymbol{y}_t$ but any vector in $\mathfrak{s}$.

Structures similar to the one shown in Fig. 21.1 can be found in macroeconomics. For instance, in Sect. 21.5 we consider two alternative hierarchical structures for the case of GDP and its components. However, while this motivating example involves aggregation constraints, the mathematical framework we use can be applied for any general linear constraints, examples of which are ubiquitous in macroeconomics. For instance, the trade balance is computed as exports minus imports, while the consumer price index is computed as a weighted average of sub-indices, which are in turn weighted averages of sub-sub-indices, and so on. These structures can also be captured by an appropriately designed $\boldsymbol{S}$ matrix.

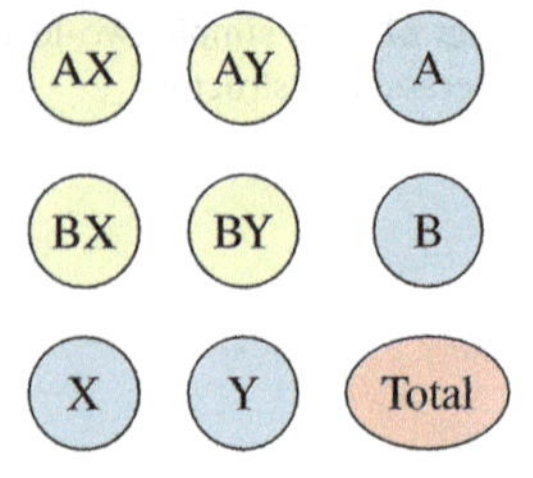

Fig. 21.2 A simple two-level grouped structure

An important alternative aggregation structure, also commonly found in macroeconomics, is one for which the most aggregate series is disaggregated by attributes of interest that are crossed, as distinct to nested which is the case for hierarchical time series. For example, industrial production may be disaggregated along the lines of geography or sector or both. We refer to this as a *grouped* structure. Figure 21.2 shows a simple example of such a structure. The Total series disaggregates into $y_{A,t}$ and $y_{B,t}$, but also into $y_{X,t}$ and $y_{Y,t}$, at level 1, and then into the bottom-level series, $\boldsymbol{b}_t = (y_{AX}, y_{AY}, y_{BX}, y_{BY})'$. Hence, in contrast to hierarchical structures, grouped time series do not naturally disaggregate in a unique manner.

An important implementation of aggregation structures are *temporal hierarchies* introduced by Athanasopoulos et al. (2017). In this case the aggregation structure spans the time dimension and dictates how higher frequency data (e.g., monthly) are aggregated to lower frequencies (e.g., quarterly, annual). There is a vast literature that studies the effects of temporal aggregation, going back to the seminal work of Amemiya and Wu (1972), Brewer (1973), Tiao (1972), Zellner and Montmarquette (1971) and others, including Hotta and Cardoso Neto (1993), Hotta (1993), Marcellino (1999), Silvestrini, Salto, Moulin, and Veredas (2008). The main aim of this work is to find the single best level of aggregation for modelling and forecasting time series. In this literature, the analyses, results (whether theoretical or empirical), and inferences, are extremely heterogeneous, making it very challenging to reach a consensus or to draw firm conclusions. For example, Rossana and Seater (1995) who study the effect of aggregation on several key macroeconomic variables state:

> Quarterly data do not seem to suffer badly from temporal aggregation distortion, nor are they subject to the construction problems affecting monthly data. They therefore may be the optimal data for econometric analysis.

A similar conclusion is reached by Nijman and Palm (1990). Silvestrini et al. (2008) consider forecasting French cash state deficit and provide empirical evidence of forecast accuracy gains from forecasting with the aggregate model rather than aggregating forecasts from the disaggregate model.

The vast majority of this literature concentrates on a single level of temporal aggregation (although there are some notable exceptions such as Andrawis, Atiya, and El-Shishiny (2011), Kourentzes, Petropoulos, and Trapero (2014)). Athanasopoulos et al. (2017) show that considering multiple levels of aggregation via temporal hierarchies and implementing forecast reconciliation approaches rather than single-level approaches results in substantial gains in forecast accuracy across all levels of temporal aggregation.

21.3 Point Forecasting

A requirement when forecasting hierarchical time series is that the forecasts adhere to the same aggregation constraints as the observed data; i.e., they are coherent.

Definition 21.1 A set of h-step-ahead forecasts $\tilde{\boldsymbol{y}}_{T+h|T}$, stacked in the same order as $\boldsymbol{y}_t$ and generated using information up to and including time T, are said to be *coherent* if $\tilde{\boldsymbol{y}}_{T+h|T} \in \mathfrak{s}$.

Hence, coherent forecasts of lower level series aggregate to their corresponding upper level series and vice versa.

Let us consider the smallest possible hierarchy with two bottom-level series, depicted in Fig. 21.3, where $y_{\text{Tot}} = y_{\text{A}} + y_{\text{B}}$. While base forecasts could lie anywhere in $\mathbb{R}^3$, the realisations and coherent forecasts lie in a two dimensional subspace $\mathfrak{s} \subset \mathbb{R}^3$.

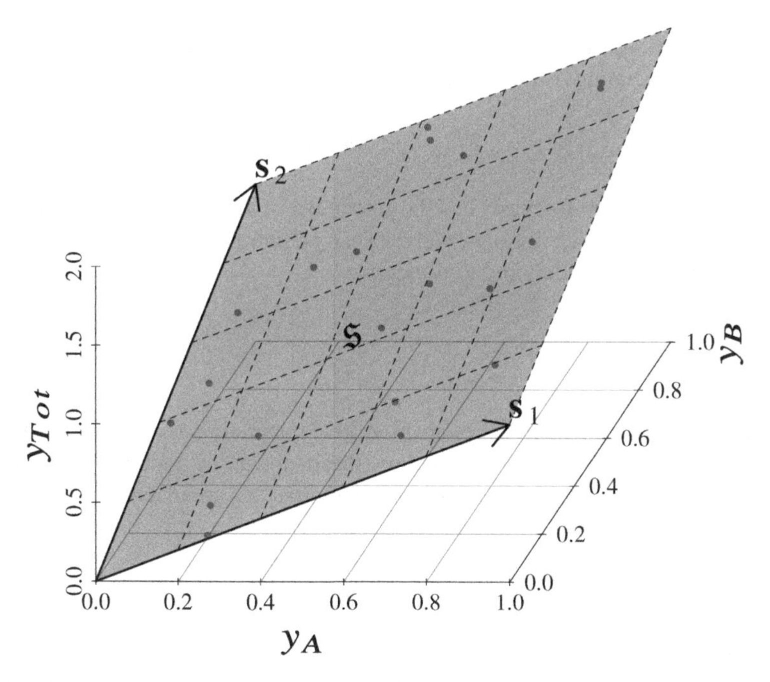

Fig. 21.3 Representation of a coherent subspace in a three dimensional hierarchy where $y_{\text{Tot}} = y_{\text{A}} + y_{\text{B}}$. The coherent subspace is depicted as a grey two dimensional plane labelled $\mathfrak{s}$. Note that the columns of $\mathbf{s}_1 = (1, 1, 0)'$ and $\mathbf{s}_2 = (1, 0, 1)'$ form a basis for $\mathfrak{s}$. The red points lying on $\mathfrak{s}$ can be either realisations or coherent forecasts

21.3.1 Single-Level Approaches

A common theme across all traditional approaches for forecasting hierarchical time series is that a single level of aggregation is first selected and forecasts for that level are generated. These are then linearly combined to generate a set of coherent forecasts for the rest of the structure.

Bottom-Up

In the *bottom-up* approach, forecasts for the most disaggregate level are first generated. These are then aggregated to obtain forecasts for all other series of the hierarchy (Dunn, Williams, & Dechaine, 1976). In general, this consists of first generating $\hat{\boldsymbol{b}}_{T+h|T} \in \mathbb{R}^m$, a set of h-step-ahead forecasts for the bottom-level series. For the simple hierarchical structure of Fig. 21.1, $\hat{\boldsymbol{b}}_{T+h|T} = (\hat{y}_{\text{AA},T+h|T}, \hat{y}_{\text{AB},T+h|T}, \hat{y}_{\text{BA},T+h|T}, \hat{y}_{\text{BB},T+h|T}, \hat{y}_{\text{BC},T+h|T})$, where $\hat{y}_{i,T+h|T}$ is the h-step-ahead forecast of the series corresponding to node i. A set of coherent forecasts for the whole hierarchy is then given by

$$\tilde{\boldsymbol{y}}^{\text{BU}}_{T+h|T} = \boldsymbol{S}\hat{\boldsymbol{b}}_{T+h|T}.$$

Generating bottom-up forecasts has the advantage of no information being lost due to aggregation. However, bottom-level data can potentially be highly volatile or very noisy and therefore challenging to forecast.

Top-Down

In contrast, *top-down* approaches involve first generating forecasts for the most aggregate level and then disaggregating these down the hierarchy. In general, coherent forecasts generated from top-down approaches are given by

$$\tilde{\boldsymbol{y}}^{\text{TD}}_{T+h|T} = \boldsymbol{S}\boldsymbol{p}\hat{y}_{\text{Tot},T+h|T},$$

where $\boldsymbol{p} = (p_1, \ldots, p_m)'$ is an m-dimensional vector consisting of a set of proportions which disaggregate the top-level forecast $\hat{y}_{\text{Tot},T+h|T}$ to forecasts for the bottom-level series; hence $\boldsymbol{p}\hat{y}_{\text{Tot},T+h|T} = \hat{\boldsymbol{b}}_{T+h|T}$. These are then aggregated by the summing matrix $\boldsymbol{S}$.

Traditionally, proportions have been calculated based on the observed historical data. Gross and Sohl (1990) present and evaluate twenty-one alternative approaches. The most convenient attribute of these approaches is their simplicity. Generating a set of coherent forecasts involves only modelling and generating forecasts for the most aggregate top-level series. In general, such top-down approaches seem to produce quite reliable forecasts for the aggregate levels and they are useful with low count data. However, a significant disadvantage is the loss of information due to aggregation. A limitation of such top-down approaches is that characteristics of lower level series cannot be captured. To overcome this, Athanasopoulos et al. (2009) introduced a new top-down approach which disaggregates the top-level based on proportions of forecasts rather than the historical data and showed that this method outperforms the conventional top-down approaches. However, a limitation of all top-down approaches is that they introduce bias to the forecasts even when the top-level forecast itself is unbiased. We discuss this in detail in Sect. 21.3.2.

Middle-Out

A compromise between bottom-up and top-down approaches is the middle-out approach. It entails first forecasting the series of a selected middle level. For series above the middle level, coherent forecasts are generated using the bottom-up approach by aggregating the middle-level forecasts. For series below the middle level, coherent forecasts are generated using a top-down approach by disaggregating the middle-level forecasts. Similarly to the top-down approach it is useful for when bottom-level data is low count. Since the middle-out approach involves generating top-down forecasts, it also introduces bias to the forecasts.

21.3.2 Point Forecast Reconciliation

All approaches discussed so far are limited to only using information from a single level of aggregation. Furthermore, these ignore any correlations across levels of a hierarchy. An alternative framework that overcomes these limitations is one that involves forecast *reconciliation*. In a first step. forecasts for all the series across all levels of the hierarchy are computed, ignoring any aggregation constraints. We refer to these as *base* forecasts and denote them by $\hat{\boldsymbol{y}}_{T+h|T}$. In general, base forecasts will not be coherent, unless a very simple method has been used to compute them such as for naïve forecasts. In this case, forecasts are simply equal to a previous realisation of the data and they inherit the property of coherence.

The second step is an adjustment that reconciles base forecasts so that they become coherent. In general, this is achieved by mapping the base forecasts $\hat{\boldsymbol{y}}_{T+h|T}$ onto the coherent subspace $\mathfrak{s}$ via a matrix $\boldsymbol{S}\boldsymbol{G}$, resulting in a set of coherent forecasts $\tilde{\boldsymbol{y}}_{T+h|T}$. Specifically,

$$\tilde{\boldsymbol{y}}_{T+h|T} = \boldsymbol{S}\boldsymbol{G}\hat{\boldsymbol{y}}_{T+h|T}, \tag{21.2}$$

where $\boldsymbol{G}$ is an $m \times n$ matrix that maps $\hat{\boldsymbol{y}}_{T+h|T}$ to $\mathbb{R}^m$, producing new forecasts for the bottom level, which are in turn mapped to the coherent subspace by the summing matrix $\boldsymbol{S}$. We restrict our attention to projections on $\mathfrak{s}$ in which case $\boldsymbol{S}\boldsymbol{G}\boldsymbol{S} = \boldsymbol{S}$.

This ensures that unbiasedness is preserved, i.e., for a set of unbiased base forecasts reconciled forecasts will also be unbiased.

Note that all single-level approaches discussed so far can also be represented by (21.2) using appropriately designed $\boldsymbol{G}$ matrices, however, not all of these will be projections. For example, for the bottom-up approach, $\boldsymbol{G} = \big(\boldsymbol{0}_{(m\times n-m)}\ \boldsymbol{I}_m\big)$ in which case $\boldsymbol{SGS} = \boldsymbol{S}$. For any top-down approach $\boldsymbol{G} = \big(\boldsymbol{p}\ \boldsymbol{0}_{(m\times n-1)}\big)$, for which $\boldsymbol{SGS} \neq \boldsymbol{S}$.

Optimal MinT Reconciliation

Wickramasuriya et al. (2018) build a unifying framework for much of the previous literature on forecast reconciliation. We present here a detailed outline of this approach and in turn relate it to previous significant contributions in forecast reconciliation.

Assume that $\hat{\boldsymbol{y}}_{T+h|T}$ is a set of unbiased base forecasts, i.e., $\mathrm{E}_{1:T}(\hat{\boldsymbol{y}}_{T+h|T}) = \mathrm{E}_{1:T}[\boldsymbol{y}_{T+h} \mid \boldsymbol{y}_1, \ldots, \boldsymbol{y}_T]$, the true mean with the expectation taken over the observed sample up to time T. Let

$$\hat{\boldsymbol{e}}_{T+h|T} = \boldsymbol{y}_{T+h|T} - \hat{\boldsymbol{y}}_{T+h|T} \tag{21.3}$$

denote a set of base forecast errors with $\mathrm{Var}(\hat{\boldsymbol{e}}_{T+h|T}) = \boldsymbol{W}_h$, and

$$\tilde{\boldsymbol{e}}_{T+h|T} = \boldsymbol{y}_{T+h|T} - \tilde{\boldsymbol{y}}_{T+h|T}$$

denote a set of coherent forecast errors. Lemma 1 in Wickramasuriya et al. (2018) shows that for any matrix $\boldsymbol{G}$ such that $\boldsymbol{SGS} = \boldsymbol{S}$, $\mathrm{Var}(\tilde{\boldsymbol{e}}_{T+h|T}) = \boldsymbol{SGW}_h\boldsymbol{S}'\boldsymbol{G}'$. Furthermore Theorem 1 shows that

$$\boldsymbol{G} = (\boldsymbol{S}'\boldsymbol{W}_h^{-1}\boldsymbol{S})^{-1}\boldsymbol{S}'\boldsymbol{W}_h^{-1} \tag{21.4}$$

is the unique solution that minimises the trace of $\boldsymbol{SGW}_h\boldsymbol{S}'\boldsymbol{G}'$ subject to $\boldsymbol{SGS} = \boldsymbol{S}$. MinT is optimal in the sense that given a set of unbiased base forecasts, it returns a set of best linear unbiased reconciled forecasts, using as $\boldsymbol{G}$ the unique solution that minimises the trace (hence MinT) of the variance of the forecast error of the reconciled forecasts.

A significant advantage of the MinT reconciliation solution is that it is the first to incorporate the full correlation structure of the hierarchy via $\boldsymbol{W}_h$. However, estimating $\boldsymbol{W}_h$ is challenging, especially for $h > 1$. Wickramasuriya et al. (2018) present possible alternative estimators for $\boldsymbol{W}_h$ and show that these lead to different $\boldsymbol{G}$ matrices. We summarise these below.

- Set $\boldsymbol{W}_h = k_h\boldsymbol{I}_n$ for all h, where $k_h > 0$ is a proportionality constant. This simple assumption returns $\boldsymbol{G} = (\boldsymbol{S}'\boldsymbol{S})^{-1}\boldsymbol{S}'$ so that the base forecasts are orthogonally projected onto the coherent subspace $\mathfrak{s}$ minimising the Euclidean distance between $\hat{\boldsymbol{y}}_{T+h|T}$ and $\tilde{\boldsymbol{y}}_{T+h|T}$. Hyndman et al. (2011) come to the same solution, however, from the perspective of the following regression model

$$\hat{\boldsymbol{y}}_{T+h|T} = \boldsymbol{S}\boldsymbol{\beta}_{T+h|T} + \boldsymbol{\varepsilon}_{T+h|T},$$

where $\boldsymbol{\beta}_{T+h|T} = \mathrm{E}[\boldsymbol{b}_{T+h} \mid \boldsymbol{b}_1, \ldots, \boldsymbol{b}_T]$ is the unknown conditional mean of the bottom-level series and $\boldsymbol{\varepsilon}_{T+h|T}$ is the coherence or reconciliation error with mean zero and variance $\boldsymbol{V}$. The OLS solution leads to the same projection matrix $\boldsymbol{S}(\boldsymbol{S}'\boldsymbol{S})^{-1}\boldsymbol{S}'$, and due to this interpretation we continue to refer to this reconciliation method as OLS. A disadvantage of the OLS solution is that the homoscedastic diagonal entries do not account for the scale differences between the levels of the hierarchy due to aggregation. Furthermore, OLS does not account for the correlations across series.

- Set $\boldsymbol{W}_h = k_h \mathrm{diag}(\hat{\boldsymbol{W}}_1)$ for all h ($k_h > 0$), where

$$\hat{\boldsymbol{W}}_1 = \frac{1}{T}\sum_{T=1}^{T} \hat{\boldsymbol{e}}_t \hat{\boldsymbol{e}}_t'$$

is the unbiased sample estimator of the in-sample one-step-ahead base forecast errors as defined in (21.3). Hence this estimator scales the base forecasts using the variance of the in-sample residuals and is therefore described and referred to as a weighted least squares (WLS) estimator applying variance scaling. A similar estimator was proposed by Hyndman et al. (2019).

An alternative WLS estimator is proposed by Athanasopoulos et al. (2017) in the context of temporal hierarchies. Here $\boldsymbol{W}_h$ is proportional to diag($\boldsymbol{S}\mathbf{1}$) where $\mathbf{1}$ is a unit column vector of dimension n. Hence the weights are proportional to the number of bottom-level variables required to form an aggregate. For example, in the hierarchy of Fig. 21.1, the weights corresponding to the Total, series A and series B are proportional to 5, 2 and 3 respectively. This weighting scheme depends only on the aggregation structure and is referred to as structural scaling. Its advantage over OLS is that it assumes equivariant forecast errors only at the bottom level of the structure and not across all levels. It is particularly useful in cases where forecast errors are not available; for example, in cases where the base forecasts are generated by judgemental forecasting.

- Set $\boldsymbol{W}_h = k_h \hat{\boldsymbol{W}}_1$ for all h ($k_h > 0$) to be proportional to the unrestricted sample covariance estimator for $h = 1$. Although this is relatively simple to obtain and provides a good solution for small hierarchies, it does not provide reliable results as m grows compared to T. This is referred to this as the MinT(Sample) estimator.

- Set $\boldsymbol{W}_h = k_h \hat{\boldsymbol{W}}_1^D$ for all h ($k_h > 0$), where $\hat{\boldsymbol{W}}_1^D = \lambda_D \text{diag}(\hat{\boldsymbol{W}}_1) + (1 - \lambda_D)\hat{\boldsymbol{W}}_1$ is a shrinkage estimator with diagonal target and shrinkage intensity parameter

$$\hat{\lambda}_D = \frac{\sum_{i \neq j} \hat{\text{Var}}(\hat{r}_{ij})}{\sum_{i \neq j} \hat{r}_{ij}^2},$$

where $\hat{r}_{ij}$ is the (i, j)th element of $\hat{\boldsymbol{R}}_1$, the one-step-ahead sample correlation matrix as proposed by Schäfer and Strimmer (2005). Hence, off-diagonal elements of $\hat{\boldsymbol{W}}_1$ are shrunk towards zero while diagonal elements (variances) remain unchanged. This is referred to as the MinT(Shrink) estimator.

21.4 Hierarchical Probabilistic Forecasting

A limitation of point forecasts is that they provide no indication of uncertainty around the forecast. A richer description of forecast uncertainty can be obtained by providing a probabilistic forecast, also commonly referred to as a density forecast. For a review of probabilistic forecasts, and *scoring rules* for evaluating such forecasts, see Gneiting and Katzfuss (2014). This chapter and Chapter 16 respectively provide comprehensive summaries of methods for constructing density forecasts and predictive accuracy tests for both point and density forecasts. In recent years, the use of probabilistic forecasts and their evaluation via scoring rules has become pervasive in macroeconomic forecasting, some notable (but non-exhaustive) examples are Geweke and Amisano (2010), Billio, Casarin, Ravazzolo, and Van Dijk (2013), Carriero, Clark, and Marcellino (2015) and Clark and Ravazzolo (2015).

The literature on hierarchical probabilistic forecasting is still an emerging area of interest. To the best of our knowledge the first attempt to even define coherence in the setting of probabilistic forecasting is provided by Taieb, Taylor, and Hyndman (2017) who define a coherent forecast in terms of a convolution. An equivalent definition due to Gamakumara, Panagiotelis, Athanasopoulos, and Hyndman (2018) defines a coherent probabilistic forecast as a probability measure on the coherent subspace $\mathfrak{s}$. Gamakumara et al. (2018) also generalise the concept of forecast reconciliation to the probabilistic setting.

Definition 21.2 Let $\mathcal{A}$ be a subset[1] of $\mathfrak{s}$ and let $\mathcal{B}$ be all points in $\mathbb{R}^n$ that are mapped onto $\mathcal{A}$ after premultiplication by $\boldsymbol{SG}$. Letting $\hat{\nu}$ be a base probabilistic forecast for the full hierarchy, the coherent measure $\tilde{\nu}$ reconciles $\hat{\nu}$ if $\tilde{\nu}(\mathcal{A}) = \hat{\nu}(\mathcal{B})$ for all $\mathcal{A}$.

[1]Strictly speaking $\mathcal{A}$ is a Borel set.

In practice this definition leads to two approaches. For some parametric distributions, for instance the multivariate normal, a reconciled probabilistic forecast can be derived analytically. However, in macroeconomic forecasting, non-standard distributions such as bimodal distributions are often required to take different policy regimes into account. In such cases a non-parametric approach based on bootstrapping in-sample errors proposed Gamakumara et al. (2018) can be used. These scenarios are now covered in detail.

21.4.1 Probabilistic Forecast Reconciliation in the Gaussian Framework

In the case where the base forecasts are probabilistic forecasts characterised by elliptical distributions, Gamakumara et al. (2018) show that reconciled probabilistic forecasts will also be elliptical. This is particularly straightforward for the Gaussian distribution which is completely characterised by two moments. Letting the base probabilistic forecasts be $\mathcal{N}(\hat{\boldsymbol{y}}_{T+h|T}, \hat{\boldsymbol{\Sigma}}_{T+h|T})$, then the reconciled probabilistic forecasts will be $\mathcal{N}(\tilde{\boldsymbol{y}}_{T+h|T}, \tilde{\boldsymbol{\Sigma}}_{T+h|T})$, where

$$\tilde{\boldsymbol{y}}_{T+h|T} = \boldsymbol{S}\boldsymbol{G}\hat{\boldsymbol{y}}_{T+h|T} \tag{21.5}$$

$$\text{and} \qquad \tilde{\boldsymbol{\Sigma}}_{T+h|T} = \boldsymbol{S}\boldsymbol{G}\hat{\boldsymbol{\Sigma}}_{T+h|T}\boldsymbol{G}'\boldsymbol{S}'. \tag{21.6}$$

There are several options for obtaining the base probabilistic forecasts and in particular the variance covariance matrix $\hat{\boldsymbol{\Sigma}}$. One option is to fit multivariate models either level by level or for the hierarchy as a whole leading respectively to a $\hat{\boldsymbol{\Sigma}}$ that is block diagonal or dense. Another option is to fit univariate models for each individual series in which case $\hat{\boldsymbol{\Sigma}}$ is a diagonal matrix. A third option that we employ here is to obtain $\hat{\boldsymbol{\Sigma}}$ using in-sample forecast errors, in a similar vein to how $\hat{\boldsymbol{W}}_1$ is estimated in the MinT method. Here the same shrinkage estimator described in Sect. 21.3.2 is used. The reconciled probabilistic forecast will ultimately depend on the choice of $\boldsymbol{G}$; the same choices of $\boldsymbol{G}$ matrices used in Sect. 21.3 can be used.

21.4.2 Probabilistic Forecast Reconciliation in the Non-parametric Framework

In many applications, including macroeconomic forecasting, it may not be reasonable to assume Gaussian predictive distributions. Therefore, non-parametric approaches have been widely used for probabilistic forecasts in different disciplines. For example, ensemble forecasting in weather applications (Gneiting, 2005; Gneiting & Katzfuss, 2014; Gneiting, Stanberry, Grimit, Held, & Johnson, 2008), and bootstrap-based approaches (Manzan & Zerom, 2008; Vilar & Vilar, 2013). In macroeconomics, Cogley, Morozov, and Sargent (2005) discuss the importance of allowing for skewness in density forecasts and more recently Smith and Vahey (2016) discuss this issue in detail.

Due to these concerns, we employ the bootstrap method proposed by Gamakumara et al. (2018) that does not make parametric assumptions about the predictive distribution. An important result exploited by this method is that applying point forecast reconciliation to the draws from an incoherent base predictive distribution, results in a sample from the reconciled predictive distribution. We summarise this process below:

1. Fit univariate models to each series in the hierarchy over a training set from $t = 1, \ldots, T$. Let these models denote $M_1, \ldots, M_n$.
2. Compute one-step-ahead in-sample forecast errors. Collect these into an $n \times T$ matrix $\hat{\boldsymbol{E}} = (\hat{\boldsymbol{e}}_1, \hat{\boldsymbol{e}}_2, \ldots, \hat{\boldsymbol{e}}_T)$, where the n-vector $\hat{\boldsymbol{e}}_t = \boldsymbol{y}_t - \hat{\boldsymbol{y}}_{t|t-1}$. Here, $\hat{\boldsymbol{y}}_{t|t-1}$ is a vector of forecasts made for time t using information up to and including time $t-1$. These are called in-sample forecasts since while they depend only on past values, information from the entire training sample is used to estimate the parameters for the models on which the forecasts are based.
3. Block bootstrap from $\hat{\boldsymbol{E}}$; that is, choose H consecutive columns of $\hat{\boldsymbol{E}}$ at random, repeating this process B times. Denote the $n \times H$ matrix obtained at iteration b as $\hat{\boldsymbol{E}}^b$ for $b = 1, \ldots, B$.
4. For all b, compute $\hat{\boldsymbol{\Upsilon}}^b = \{\hat{\boldsymbol{y}}_1^b, \ldots, \hat{\boldsymbol{y}}_n^b\}' \in \mathbb{R}^{n \times H} : \hat{y}_{i,h}^b = f(M_i, \hat{e}_{i,h}^b)$ where, $f(.)$ is a function of fitted univariate model in step 1 and associated error. That is, $\hat{y}_{i,h}$ is a sample path simulated from fitted model M_i for ith series and error approximated by the corresponding block bootstrapped sample error $\hat{e}_{i,h}^b$ which is the (i, h)th element of $\hat{\boldsymbol{E}}^b$. Each row of $\hat{\boldsymbol{\Upsilon}}^b$ is a sample path of h forecasts for a single series. Each column of $\hat{\boldsymbol{\Upsilon}}^b$ is a realisation from the joint predictive distribution at a particular horizon.
5. For each $b = 1, \ldots, B$, select the hth column of $\hat{\boldsymbol{\Upsilon}}^b$ and stack these to form an $n \times B$ matrix $\hat{\boldsymbol{\Upsilon}}_{T+h|T}$.
6. For a given $\boldsymbol{G}$ matrix and for each $h = 1, \ldots, H$, compute $\tilde{\boldsymbol{\Upsilon}}_{T+h|T} = \boldsymbol{S}\boldsymbol{G}\hat{\boldsymbol{\Upsilon}}_{T+h|T}$. Each column of $\tilde{\boldsymbol{\Upsilon}}_{T+h|T}$ is a realisation from the joint h-step-ahead reconciled predictive distribution.

21.5 Australian GDP

In our empirical application we consider Gross Domestic Product (GDP) of Australia with quarterly data spanning the period 1984:Q4–2018:Q3. The Australian Bureau of Statistics (ABS) measures GDP using three main approaches namely Production, Income, and Expenditure. The final GDP figure is obtained as an average of these three figures. Each of these measures is aggregates of economic variables which are also targets of interests for the macroeconomic forecaster. This suggests a hierarchical approach to forecasting could be used to improve forecasts of all series in the hierarchy including the headline GDP.

We concentrate on the Income and Expenditure approaches as nominal data are available only for these two. We restrict our attention to nominal data due to the fact that real data are constructed via a chain price index approach with different price deflators used for each series. As a result, real GDP data are not coherent—the aggregate series is not a linear combination of the disaggregate series. For similar reasons we do not use seasonally adjusted data; the process of seasonal adjustment results in data that are not coherent. Finally, although there is a small statistical discrepancy between each series and the headline GDP figure, we simply treat this statistical discrepancy, which is also published by the ABS, as a time series in its own right. For further of the details on the data please refer to Australian Bureau of Statistics (2018).

21.5.1 Income Approach

Using the income approach, GDP is calculated by aggregating all income flows. In particular, GDP at purchaser's price is the sum of all factor incomes and taxes, minus subsidies on production and imports (Australian Bureau of Statistics, 2015):

$$\begin{aligned} GDP = &\ \textit{Gross operating surplus} + \textit{Gross mixed income}\\ &+ \textit{Compensation of employees}\\ &+ \textit{Taxes less subsidies on production and imports}\\ &+ \textit{Statistical discrepancy (I)}. \end{aligned}$$

Figure 21.4 shows the full hierarchical structure capturing all components aggregated to form GDP using the income approach. The hierarchy has two levels of aggregation below the top-level, with a total of $n = 16$ series across the whole structure and $m = 10$ series at the bottom level.

21.5.2 Expenditure Approach

In the expenditure approach, GDP is calculated as the aggregation of final consumption expenditure, gross fixed capital formation (GFCF), changes in inventories of finished goods, work-in-progress, and raw materials and the value of exports

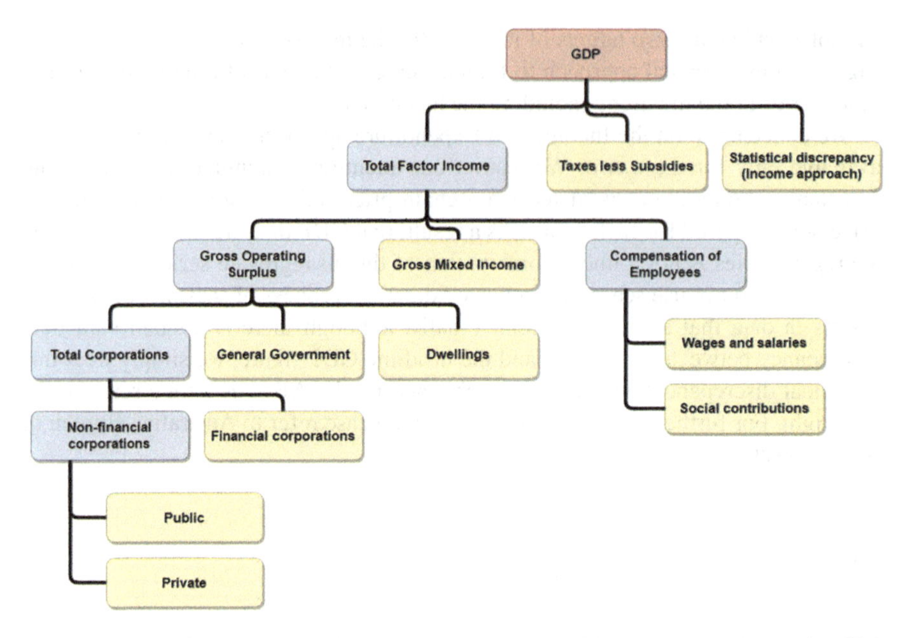

Fig. 21.4 Hierarchical structure of the income approach for GDP. The pink cell contains GDP the most aggregate series. The blue cells contain intermediate-level series and the yellow cells correspond to the most disaggregate bottom-level series

less imports of the goods and services (Australian Bureau of Statistics, 2015). The underlying equation is:

$$\begin{aligned} GDP &= \textit{Final consumption expenditure} + \textit{Gross fixed capital formation} \\ &\quad + \textit{Changes in inventories} + \textit{Trade balance} + \textit{Statistical discrepancy (E)}. \end{aligned}$$

Figures 21.5, 21.6, and 21.7 show the full hierarchical structure capturing all components aggregated to form GDP using the expenditure approach. The hierarchy has three levels of aggregation below the top-level, with a total of $n = 80$ series across the whole structure and $m = 53$ series at the bottom level. Descriptions of each series in these hierarchies along with the series ID assigned by the ABS are given in the Tables 21.1, 21.2, 21.3, and 21.4 in the Appendix.

Figure 21.8 displays time series from the income and expenditure approaches. The top panel shows the most aggregate GDP series. The panels below show series from lower levels for the income hierarchy (left panel) and the expenditure hierarchy (right panel). The plots show the diverse features of the time series with some displaying positive and others negative trending behaviour, some showing no trends but possibly a cycle, and some having a strong seasonal component. These highlight the need to account for and model all information and diverse signals from each series in the hierarchy, which can only be achieved through a forecast reconciliation approach.

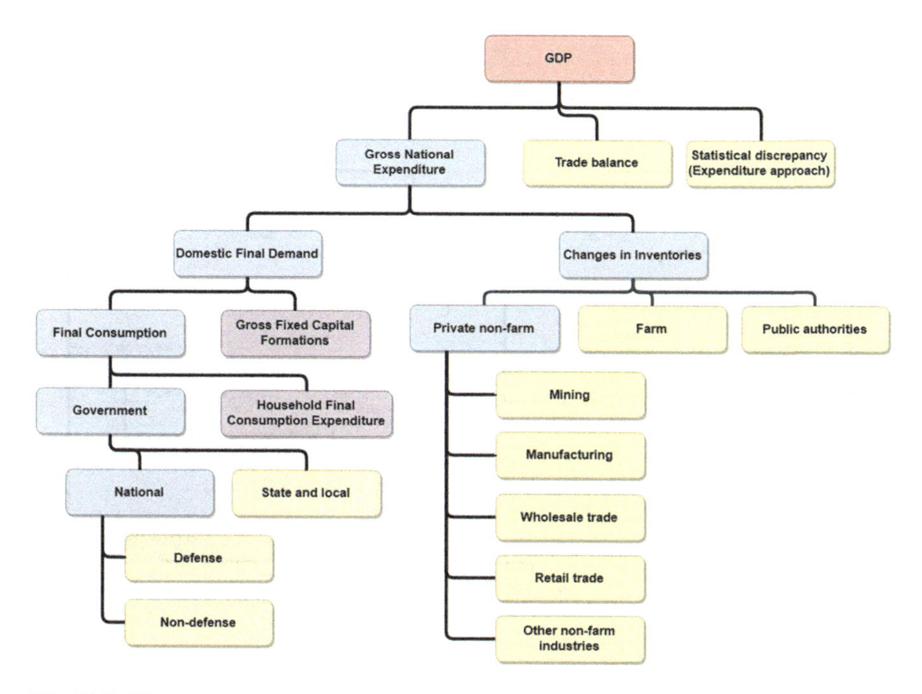

Fig. 21.5 Hierarchical structure of the expenditure approach for GDP. The pink cell contains GDP, the most aggregate series. The blue and purple cells contain intermediate-level series with the series in the purple cells further disaggregated in Figs. 21.6 and 21.7. The yellow cells contain the most disaggregate bottom-level series

21.6 Empirical Application Methodology

We now demonstrate the potential for reconciliation methods to improve forecast accuracy for Australian GDP. We consider forecasts from $h = 1$ quarter ahead up to $h = 4$ quarters ahead using an *expanding* window. First, the training sample is set from 1984:Q4 to 1994:Q3 and forecasts are produced for 1994:Q4 to 1995:Q3. Then the training window is expanded by one quarter at a time, i.e., from 1984:Q4 to 2017:Q4 with the final forecasts produced for the last available observation in 2018:Q1. This leads to 94 1-step-ahead, 93 2-steps-ahead, 92 3-steps-ahead, and 91 4-steps-ahead forecasts available for evaluation.

21.6.1 Models

The first task in forecast reconciliation is to obtain base forecasts for all series in the hierarchy. In the case of the income approach, this necessitates forecasting $n = 16$ separate time series while in the case of the expenditure approach, forecasts

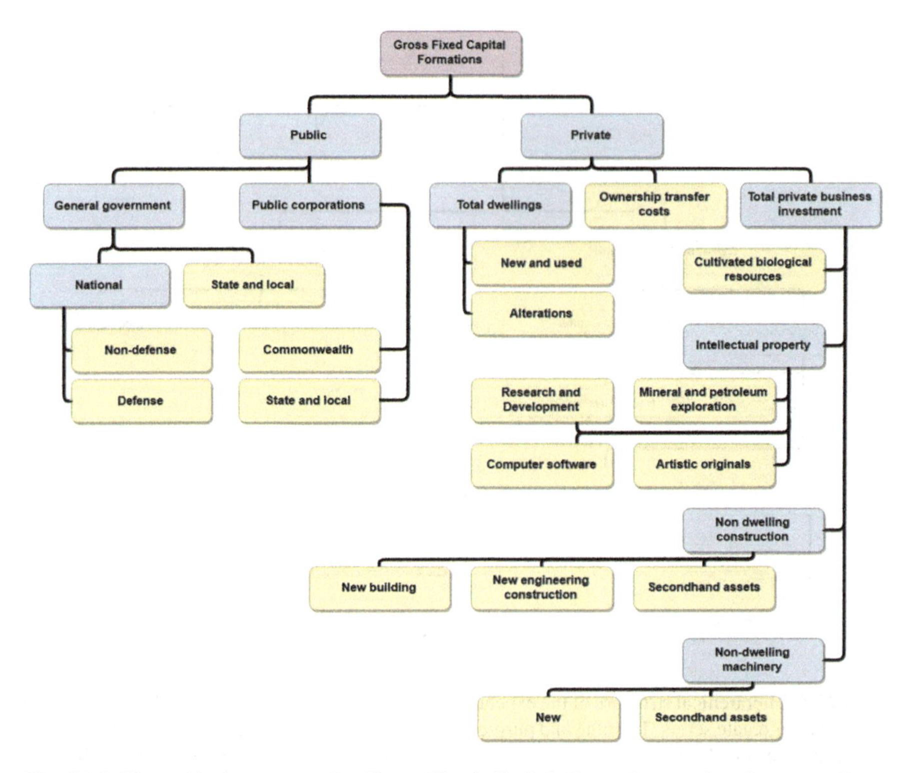

Fig. 21.6 Hierarchical structure for Gross Fixed Capital Formations under the expenditure approach for GDP, continued from Fig. 21.5. Blue cells contain intermediate-level series and the yellow cells correspond to the most disaggregate bottom-level series

for $n = 80$ separate time series must be obtained. Given the diversity in these time series discussed in Sect. 21.5, we focus on an approach that is fast but also flexible. We consider simple univariate ARIMA models, where model order is selected via a combination of unit root testing and the AIC using an algorithm developed by Hyndman, Koehler, Ord, and Snyder (2008) and implemented in the `auto.arima()` function in Hyndman, Lee, and Wang (2019). A similar approach was also undertaken using the ETS framework to produce base forecasts (Hyndman & Khandakar, 2008). Using ETS models to generate base forecasts had minimal impact on our conclusions with respect to forecast reconciliation methods and in most cases ARIMA forecasts were found to be more accurate than ETS forecasts. Consequently for brevity, we have excluded presenting the results for ETS models. However, these are available from github[2] and are discussed in detail in Gamakumara (2019). We note that a number of more complicated approaches could have been used to obtain base forecasts including multivariate models such

[2]The relevant github repository is https://github.com/PuwasalaG/Hierarchical-Book-Chapter.

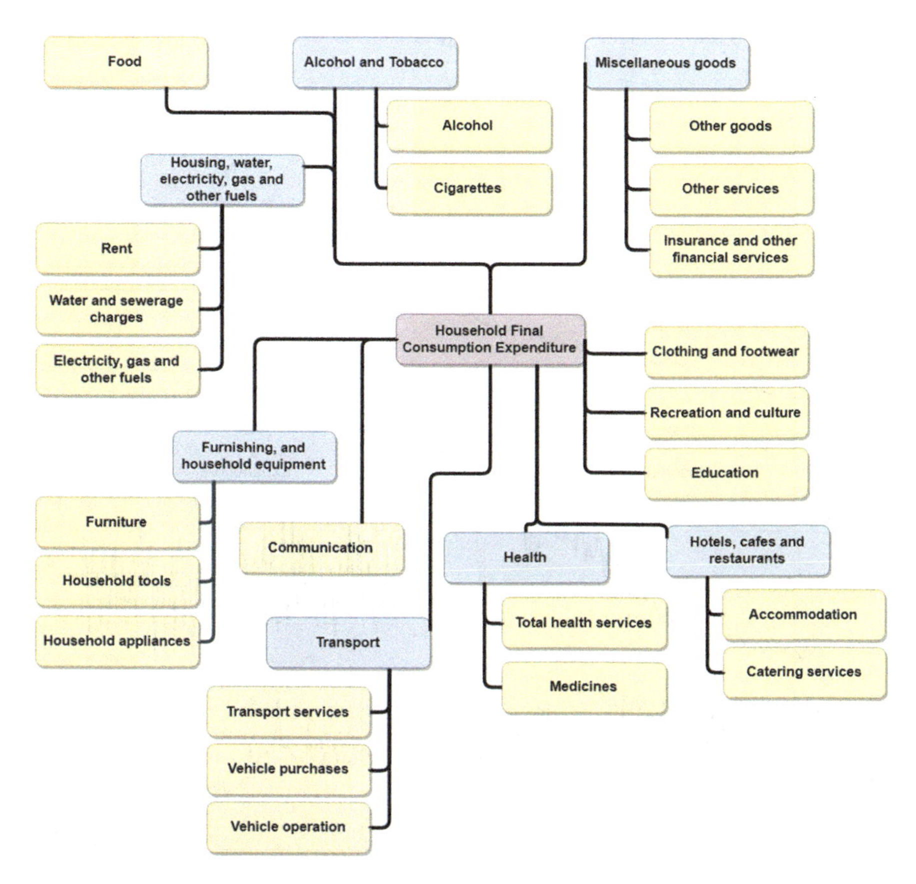

Fig. 21.7 Hierarchical structure for Household Final Consumption Expenditure under the expenditure approach for GDP, continued from Fig. 21.5. Blue cells contain intermediate-level series and the yellow cells correspond to the most disaggregate bottom-level series

as vector autoregressions, and models and methods that handle a large number of predictors such as factor models or least angle regression. However, Panagiotelis, Athanasopoulos, Hyndman, Jiang, and Vahid (2019) show that univariate ARIMA models are highly competitive for forecasting Australian GDP even compared to these methods, and in any case our primary motivation is to demonstrate the potential of forecast reconciliation.

The hierarchical forecasting approaches we consider are bottom-up, OLS, WLS with variance scaling and the MinT(Shrink) approach. The MinT(Sample) approach was also used but due to the size of the hierarchy, forecasts reconciled via this approach were less stable. Finally, all forecasts (both base and coherent) are compared to a seasonal naïve benchmark (Hyndman & Athanasopoulos, 2018); i.e., the forecast for GDP (or one of its components) is the realised GDP in the same quarter of the previous year. The naïve forecasts are by construction coherent and therefore do not need to be reconciled.

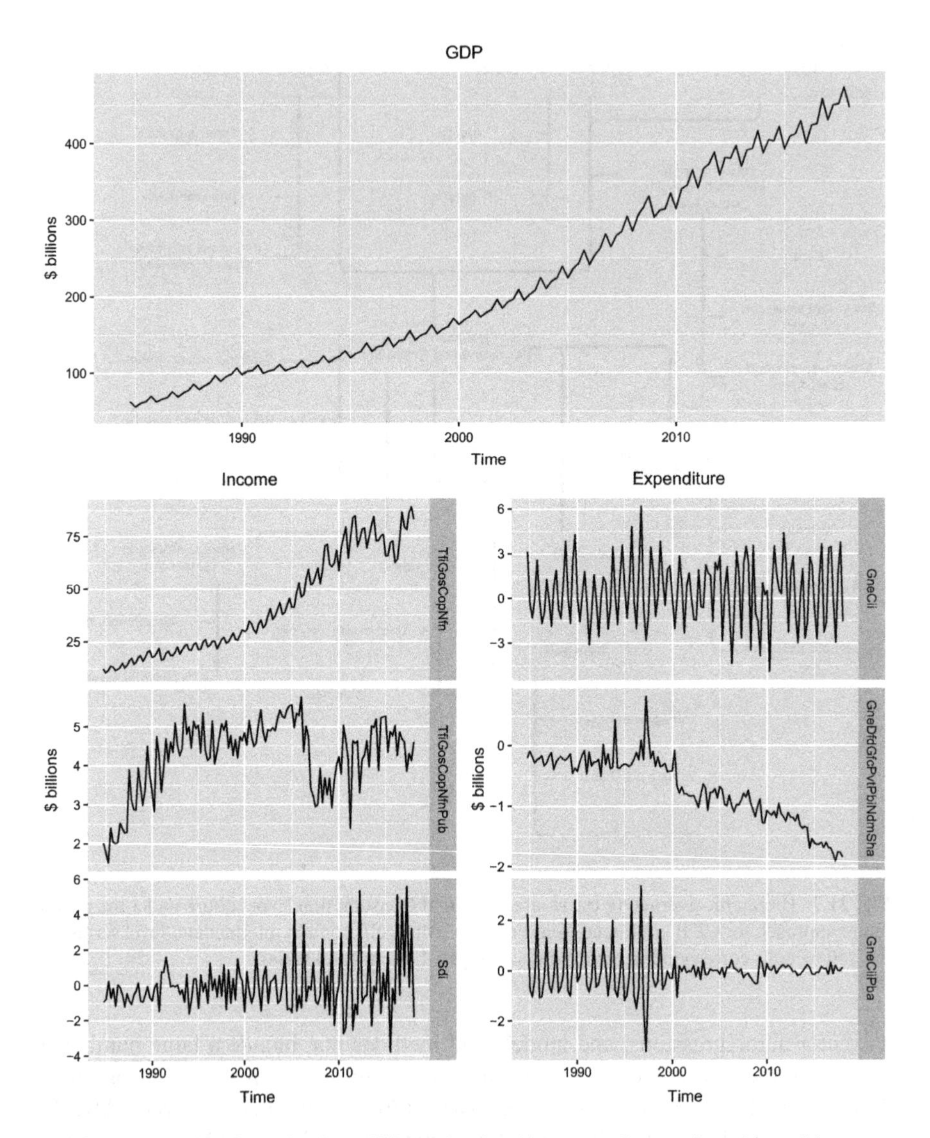

Fig. 21.8 Time plots for series from different levels of income and expenditure hierarchies

21.6.2 Evaluation

For evaluating point forecasts we consider two metrics, the Mean Squared Error (MSE) and the Mean Absolute Scaled Error (MASE) calculated over the expanding window. The absolute scaled error is defined as

$$q_{T+h} = \frac{|\breve{e}_{T+h|T}|}{(T-4)^{-1}\sum_{t=5}^{T}|y_t - y_{t-4}|},$$

where $\breve{e}_{t+h}$ is the difference between any forecast and the realisation,[3] and 4 is used due to the quarterly nature of the data. An advantage of using MASE is that it is a scale independent measure. This is particularly relevant for hierarchical time series, since aggregate series by their very nature are on a larger scale than disaggregate series. Consequently, scale dependent metrics may unfairly favour methods that perform well for the aggregate series but poorly for disaggregate series. For more details on different point forecast accuracy measures, refer to Chapter 3 of Hyndman and Athanasopoulos (2018).

Forecast accuracy of probabilistic forecasts can be evaluated using scoring rules (Gneiting & Katzfuss, 2014). Let $\breve{F}$ be a probabilistic forecast and let $\breve{\boldsymbol{y}} \sim \breve{F}$ where a breve is again used to denote that either base forecasts or coherent forecasts can be evaluated. The accuracy of multivariate probabilistic forecasts will be measured by the energy score given by

$$eS(\breve{F}_{T+h|T}, \boldsymbol{y}_{T+h}) = \mathrm{E}_{\breve{F}} \|\breve{\boldsymbol{y}}_{T+h} - \boldsymbol{y}_{T+h}\|^{\alpha} - \frac{1}{2}\mathrm{E}_{\breve{F}} \|\breve{\boldsymbol{y}}_{T+h} - \breve{\boldsymbol{y}}^{*}_{T+h}\|^{\alpha},$$

where $\boldsymbol{y}_{T+h}$ is the realisation at time $T+h$, and $\alpha \in (0, 2]$. We set $\alpha = 1$, noting that other values of α give similar results. The expectations can be evaluated numerically as long as a sample from $\breve{F}$ is available, which is the case for all methods we employ. An advantage of using energy scores is that in the univariate case it simplifies to the commonly used cumulative rank probability score (CRPS) given by

$$\mathrm{CRPS}(\breve{F}_i, y_{i,T+h}) = \mathrm{E}_{\breve{F}_i} |\breve{y}_{i,T+h} - y_{i,T+h}| - \frac{1}{2}\mathrm{E}_{\breve{F}_i} |\breve{y}_{i,T+h} - \breve{y}^{*}_{i,T+h}|,$$

where the subscript i is used to denote that CRPS measures forecast accuracy for a single variable in the hierarchy.

Alternatives to the energy score were also considered, namely log scores and variogram scores. The log score was disregarded since Gamakumara et al. (2018) prove that the log score is improper with respect to the class of incoherent probabilistic forecasts when the true DGP is coherent. The variogram score gave similar results to the energy score; these results are omitted for brevity but are available from github and are discussed in detail in Gamakumara (2019).

21.7 Results

21.7.1 Base Forecasts

Due to the different features in each time series, a variety of ARIMA and seasonal ARIMA models were selected for generating base forecasts. For example, in the

[3]Breve is used instead of a hat or tilde to denote that this can be the error for either a base or reconciled forecast.

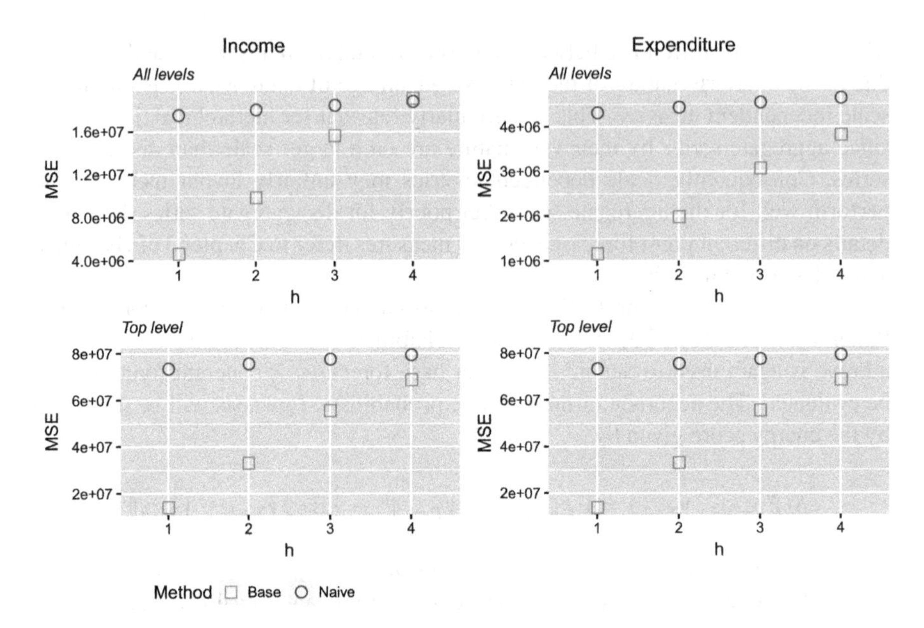

Fig. 21.9 Mean squared errors for naïve and ARIMA base forecasts. Top panels refer to results summarised over all series while bottom panels refer to results for the top-level GDP series. Left panels refer to the income hierarchy and right panels to the expenditure hierarchy

income hierarchy, some series require seasonal differencing while other did not. Furthermore the AR orders vary from 0 to 3, the MA orders from 0 to 2, and their seasonal counterparts SAR from 0 to 2 and SMA from 0 to 1. Figure 21.9 compares the accuracy of the ARIMA base forecasts to the seasonal naïve forecasts over different forecast horizons. The panels on the left show results for the Income hierarchy while the panels on the right show the results for the Expenditure hierarchy. The top panels summarise results over all series in the hierarchy, i.e., we calculate the MSE for each series and then average over all series. The bottom panels show the results for the aggregate level GDP.

The clear result is that base forecasts are more accurate than the naïve forecasts, however, as the forecasting horizon increases, the differences become smaller. This is to be expected since the naïve model here is a seasonal random walk, and for horizons $h < 4$, forecasts from an ARIMA model are based on more recent information. Similar results are obtained when MASE is used as the metric for evaluating forecast accuracy.

One disadvantage of the base forecasts relative to the naïve forecasts is that base forecasts are not coherent. As such we now turn our attention to investigating whether reconciliation approaches can lead to further improvements in forecast accuracy relative to the base forecasts.

21.7.2 Point Forecast Reconciliation

We now turn our attention to evaluating the accuracy of point forecasts obtained using the different reconciliation approaches as well as the single-level bottom-up approach. All results in subsequent figures are presented as the percentage changes in a forecasting metric relative to base forecasts, a measure known in the forecasting literature as *skill scores*. Skill scores are computed such that positive values represent an improvement in forecasting accuracy over the base forecasts while negative values represent a deterioration.

Figures 21.10 and 21.11 show skill scores using MSE and MASE respectively. The top row of each figure shows skill scores based on averages over all series. We conclude that reconciliation methods generally improve forecast accuracy relative to base forecasts regardless of the hierarchy used, the forecasting horizon, the forecast error measure or the reconciliation method employed. We do, however, note that while all reconciliation methods improve forecast performance, MinT(Shrink) is the best forecasting method in most cases.

To further investigate the results we break down the skill scores by different levels of each hierarchy. The second row of Figs. 21.10 and 21.11 shows the skill scores for a single series, namely GDP which represents the top-level of both hierarchies. The third row shows results for all series excluding those of the bottom level, while the final row shows results for the bottom-level series only. Here, we see two general features. The first is that OLS reconciliation performs poorly on the bottom-level series, and the second is that bottom-up performs relatively poorly on aggregate series. The two features are particularly exacerbated for the larger expenditure hierarchy. These results are consistent with other findings in the forecast reconciliation literature (see for instance Athanasopoulos et al., 2017; Wickramasuriya et al., 2018).

21.7.3 Probabilistic Forecast Reconciliation

We now turn our attention towards results for probabilistic forecasts. Figure 21.12 shows results for the energy score which as a multivariate score summarises forecast accuracy over the entire hierarchy. Once again all results are presented

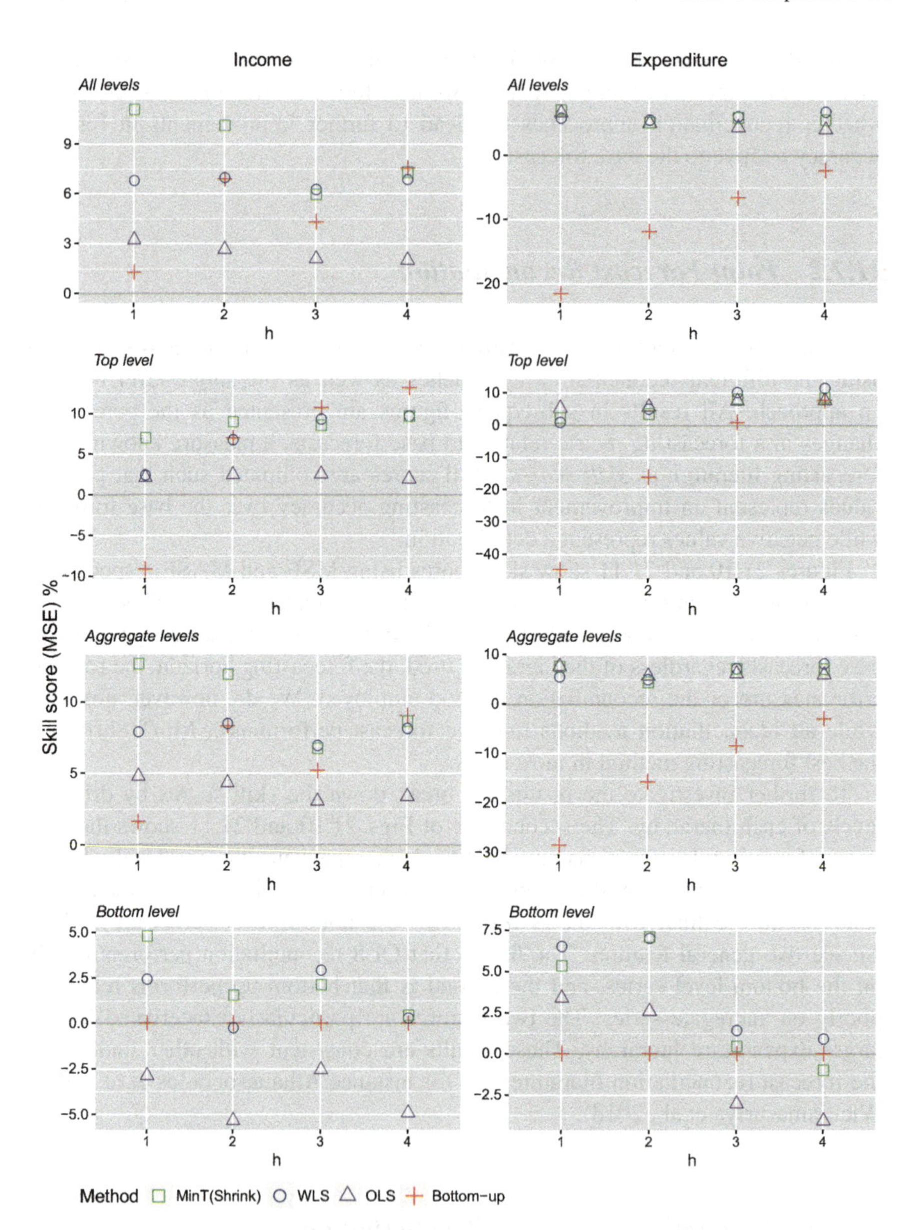

Fig. 21.10 Skill scores for point forecasts from alternative methods (with reference to base forecasts) using MSE. The left panels refer to the income hierarchy while the right panels refer to the expenditure hierarchy. The first row refers to results summarised over all series, the second row to top-level GDP series, the third row to aggregate levels, and the last row to the bottom level

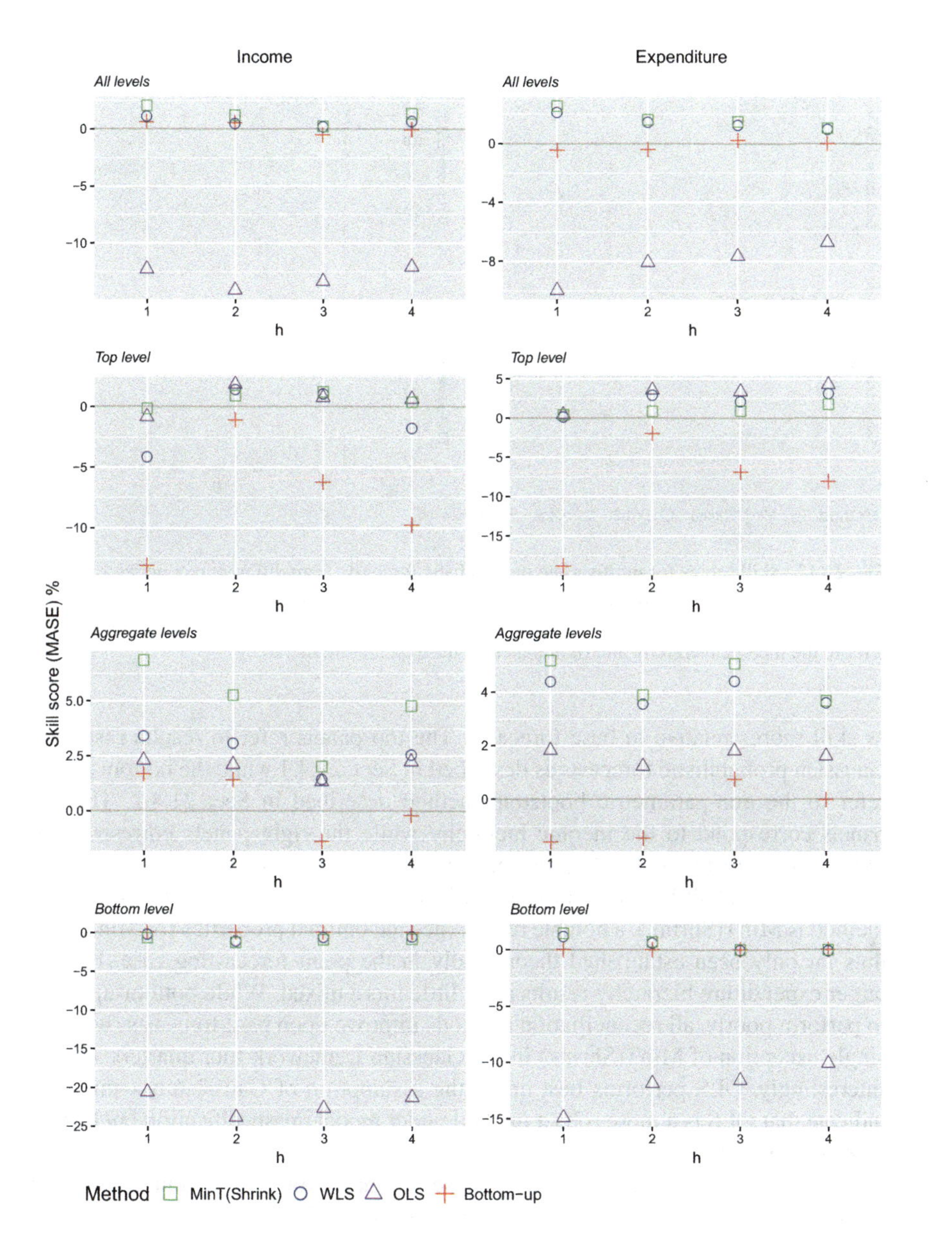

Fig. 21.11 Skill scores for point forecasts from different reconciliation methods (with reference to base forecasts) using MASE. The left two panels refer to the income hierarchy and the right two panels to the expenditure hierarchy. The first row refers to results summarised over all series, the second row to top-level GDP series, the third row to aggregate levels, and the last row to the bottom level

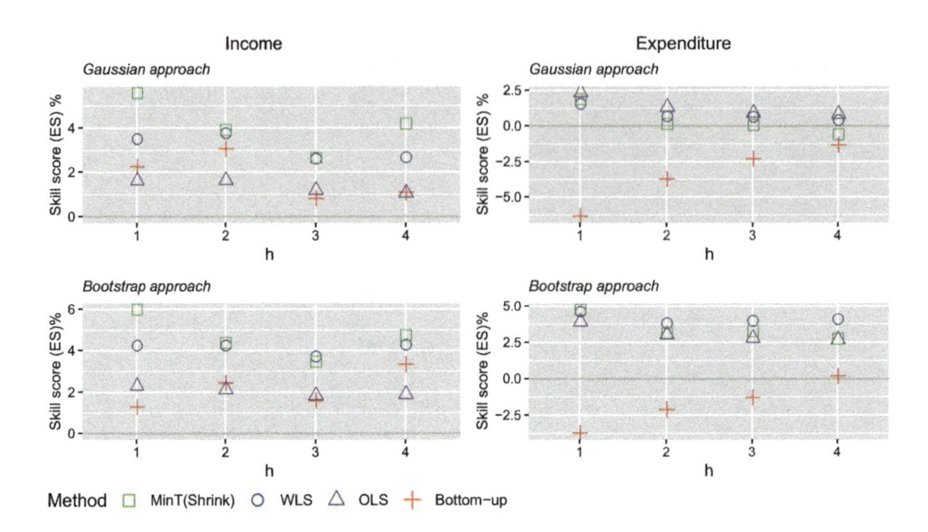

Fig. 21.12 Skill scores for multivariate probabilistic forecasts from different reconciliation methods (with reference to base forecasts) using energy scores. The top panels refer to the results for the Gaussian approach and the bottom panels to the non-parametric bootstrap approach. Left panels refer to the income hierarchy and right panels to the expenditure hierarchy

as skill scores relative to base forecasts. The top panels refer to results assuming Gaussian probabilistic forecasts as described in Sect. 21.4.1 while the bottom panels refer to the non-parametric bootstrap method described in Sect. 21.4.2. The left panels correspond to the income hierarchy while the right panels correspond to the expenditure hierarchy. For the income hierarchy, all methods improve upon base forecasts at all horizons. In nearly all cases the best performing reconciliation method is MinT(Shrink), a notable result since the optimal properties for MinT have thus far only been established theoretically in the point forecasting case. For the larger expenditure hierarchy results are a little more mixed. While bottom-up tends to perform poorly, all reconciliation methods improve upon base forecasts (with the single exception of MinT(Shrink) in the Gaussian framework four quarters ahead). Interestingly, OLS performs best under the assumption of Gaussianity—this may indicate that OLS is a more robust method under model misspecification but further investigation is required.

Finally, Fig. 21.13 displays the skill scores based on the cumulative ranked probability score for a single series, namely top-level GDP. The cause of the poor performance of bottom-up reconciliation as a failure to accurately forecast aggregate series is apparent here.

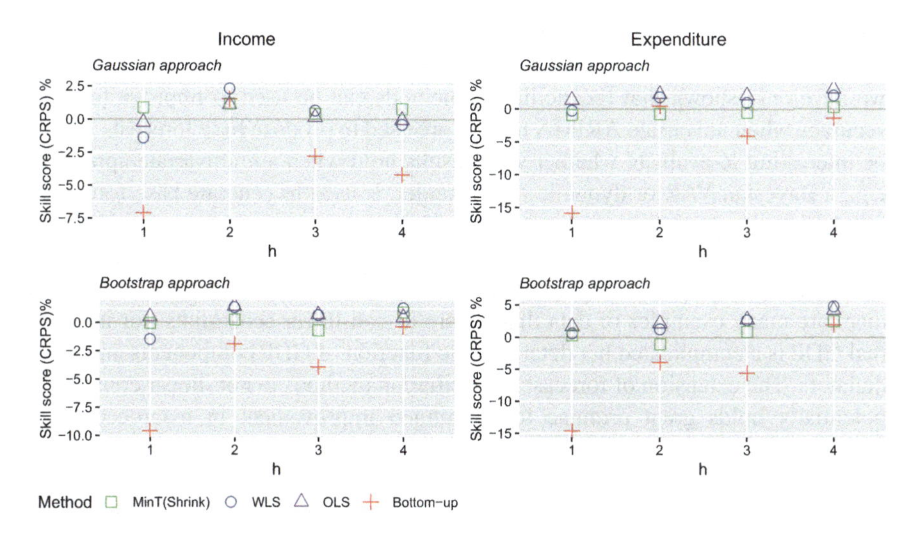

Fig. 21.13 Skill scores for probabilistic forecasts of top-level GDP from different reconciliation methods (with reference to base forecasts) using CRPS. Top panels refer to the results for Gaussian approach and bottom panels refer to the non-parametric bootstrap approach. The left panel refers to the income hierarchy and the right panel to the expenditure hierarchy

21.8 Conclusions

In the macroeconomic setting, we have demonstrated the potential for forecast reconciliation methods to not only provide coherent forecasts, but to also improve overall forecast accuracy. This result holds for both point forecasts and probabilistic forecasts, for the two different hierarchies we consider and over different forecasting horizons. Even where the objective is to only forecast a single series, for instance top-level GDP, the application of forecast reconciliation methods improves forecast accuracy.

By comparing results from different forecast reconciliation techniques we draw a number of conclusions. Despite its simplicity, the single-level bottom-up approach can perform poorly at more aggregated levels of the hierarchy. Meanwhile, when forecast accuracy at the bottom level is evaluated, OLS tends to break down in some instances. Overall, the WLS and MinT(Shrink) methods (and particularly the latter) tend to yield the highest improvements in forecast accuracy. Similar results can be found in both simulations and the empirical studies of Athanasopoulos et al. (2017) and Wickramasuriya et al. (2018).

There are a number of open avenues for research in the literature on forecast reconciliation, some of which are particularly relevant to macroeconomic applications. First there is scope to consider more complex aggregation structures, for instance in addition to the hierarchies we have already considered, data on GDP and GDP components disaggregated along geographical lines are also available. This leads to a grouped aggregation structure. Also, given the substantial literature on the optimal frequency at which to analyse macroeconomic data, a study on forecasting GDP or other variables as a temporal hierarchy may be of interest. In this chapter we have only shown that reconciliation methods can be used to improve forecast accuracy when univariate ARIMA models are used to produce base forecasts. It will be interesting to evaluate whether such results hold when a multivariate approach, e.g., a Bayesian VAR or dynamic factor model, is used to generate base forecasts, or whether the gains from forecast reconciliation would be more modest. Finally, a current limitation of the forecast reconciliation literature is that it only applies to collections of time series that adhere to linear constraints. In macroeconomics there are many examples of data that adhere to non-linear constraints, for instance real GDP is a complicated but deterministic function of GDP components and price deflators. The extension of forecast reconciliation methods to non-linear constraints potentially holds great promise for continued improvement in macroeconomic forecasting.

Appendix

See Tables 21.1, 21.2, 21.3, 21.4.

Table 21.1 Variables, series IDs and their descriptions for the income approach

Variable	Series ID	Description
Gdpi	A2302467A	GDP(I)
Sdi	A2302413V	Statistical discrepancy (I)
Tsi	A2302412T	Taxes less subsidies (I)
TfiCoeWns	A2302399K	Compensation of employees; Wages and salaries
TfiCoeEsc	A2302400J	Compensation of employees; Employers' social contributions
TfiCoe	A2302401K	Compensation of employees
TfiGosCopNfnPvt	A2323369L	Private non-financial corporations; Gross operating surplus
TfiGosCopNfnPub	A2302403R	Public non-financial corporations; Gross operating surplus
TfiGosCopNfn	A2302404T	Non-financial corporations; Gross operating surplus
TfiGosCopFin	A2302405V	Financial corporations; Gross operating surplus
TfiGosCop	A2302406W	Total corporations; Gross operating surplus
TfiGosGvt	A2298711F	General government; Gross operating surplus
TfiGosDwl	A2302408A	Dwellings owned by persons; Gross operating surplus
TfiGos	A2302409C	All sectors; Gross operating surplus
TfiGmi	A2302410L	Gross mixed income
Tfi	A2302411R	Total factor income

Table 21.2 Variables, series IDs and their descriptions for expenditure approach

Variable	Series ID	Description
Gdpe	A2302467A	GDP(E)
Sde	A2302566J	Statistical discrepancy(E)
Exp	A2302564C	Exports of goods and services
Imp	A2302565F	Imports of goods and services
Gne	A2302563A	Gross national exp.
GneDfdFceGvtNatDef	A2302523J	Gen. gov.—National; Final consumption exp.—Defence
GneDfdFceGvtNatNdf	A2302524K	Gen. gov.—National; Final consumption exp.—Non-defence
GneDfdFceGvtNat	A2302525L	Gen. gov.—National; Final consumption exp.
GneDfdFceGvtSnl	A2302526R	Gen. gov.—State and local; Final consumption exp,
GneDfdFceGvt	A2302527T	Gen. gov.; Final consumption exp.
GneDfdFce	A2302529W	All sectors; Final consumption exp.
GneDfdGfcPvtTdwNnu	A2302543T	Pvt.; Gross fixed capital formation (GFCF)
GneDfdGfcPvtTdwAna	A2302544V	Pvt.; GFCF—Dwellings—Alterations and additions
GneDfdGfcPvtTdw	A2302545W	Pvt.; GFCF—Dwellings—Total
GneDfdGfcPvtOtc	A2302546X	Pvt.; GFCF—Ownership transfer costs
GneDfdGfcPvtPbiNdcNbd	A2302533L	Pvt. GFCF—Non-dwelling construction—New building
GneDfdGfcPvtPbiNdcNec	A2302534R	Pvt.; GFCF—Non-dwelling construction—New engineering construction
GneDfdGfcPvtPbiNdcSha	A2302535T	Pvt.; GFCF—Non-dwelling construction—Net purchase of second hand assets
GneDfdGfcPvtPbiNdc	A2302536V	Pvt.; GFCF—Non-dwelling construction—Total
GneDfdGfcPvtPbiNdmNew	A2302530F	Pvt.; GFCF—Machinery and equipment—New
GneDfdGfcPvtPbiNdmSha	A2302531J	Pvt.; GFCF—Machinery and equipment—Net purchase of second hand assets
GneDfdGfcPvtPbiNdm	A2302532K	Pvt.; GFCF—Machinery and equipment—Total
GneDfdGfcPvtPbiCbr	A2716219R	Pvt.; GFCF—Cultivated biological resources
GneDfdGfcPvtPbiIprRnd	A2716221A	Pvt.; GFCF—Intellectual property products—Research and development
GneDfdGfcPvtPbiIprMnp	A2302539A	Pvt.; GFCF—Intellectual property products—Mineral and petroleum exploration
GneDfdGfcPvtPbiIprCom	A2302538X	Pvt.; GFCF—Intellectual property products—Computer software
GneDfdGfcPvtPbiIprArt	A2302540K	Pvt.; GFCF—Intellectual property products—Artistic originals
GneDfdGfcPvtPbiIpr	A2716220X	Pvt.; GFCF—Intellectual property products Total
GneDfdGfcPvtPbi	A2302542R	Pvt.; GFCF—Total private business investment
GneDfdGfcPvt	A2302547A	Pvt.; GFCF
GneDfdGfcPubPcpCmw	A2302548C	Plc. corporations—Commonwealth; GFCF

(continued)

Table 21.2 (continued)

Variable	Series ID	Description
GneDfdGfcPubPcpSnl	A2302549F	Plc. corporations—State and local; GFCF
GneDfdGfcPubPcp	A2302550R	Plc. corporations; GFCF Total
GneDfdGfcPubGvtNatDef	A2302551T	Gen. gov.—National; GFCF—Defence
GneDfdGfcPubGvtNatNdf	A2302552V	Gen. gov.—National; GFCF—Non-defence
GneDfdGfcPubGvtNat	A2302553W	Gen. gov.—National; GFCF Total
GneDfdGfcPubGvtSnl	A2302554X	Gen. gov.—State and local; GFCF
GneDfdGfcPubGvt	A2302555A	Gen. gov.; GFCF
GneDfdGfcPub	A2302556C	Plc.; GFCF
GneDfdGfc	A2302557F	All sectors; GFCF

Table 21.3 Variables, series IDs and their descriptions for changes in inventories—expenditure approach

Variable	Series ID	Description
GneCii	A2302562X	Changes in Inventories
GneCiiPfm	A2302560V	Farm
GneCiiPba	A2302561W	Public authorities
GneCiiPnf	A2302559K	Private; Non-farm Total
GneCiiPnfMin	A83722619L	Private; Mining (B)
GneCiiPnfMan	A3348511X	Private; Manufacturing (C)
GneCiiPnfWht	A3348512A	Private; Wholesale trade (F)
GneCiiPnfRet	A3348513C	Private; Retail trade (G)
GneCiiPnfOnf	A2302273C	Private; Non-farm; Other non-farm industries

Table 21.4 Variables, series IDs and their descriptions for household final consumption—expenditure approach

Variable	Series ID	Description
GneDfdHfc	A2302254W	Household Final Consumption Expenditure
GneDfdFceHfcFud	A2302237V	Food
GneDfdFceHfcAbt	A3605816F	Alcoholic beverages and tobacco
GneDfdFceHfcAbtCig	A2302238W	Cigarettes and tobacco
GneDfdFceHfcAbtAlc	A2302239X	Alcoholic beverages
GneDfdFceHfcCnf	A2302240J	Clothing and footwear
GneDfdFceHfcHwe	A3605680F	Housing, water, electricity, gas and other fuels
GneDfdFceHfcHweRnt	A3605681J	Actual and imputed rent for housing
GneDfdFceHfcHweWsc	A3605682K	Water and sewerage charges
GneDfdFceHfcHweEgf	A2302242L	Electricity, gas and other fuel
GneDfdFceHfcFhe	A2302243R	Furnishings and household equipment
GneDfdFceHfcFheFnt	A3605683L	Furniture, floor coverings and household goods
GneDfdFceHfcFheApp	A3605684R	Household appliances

(continued)

Table 21.4 (continued)

Variable	Series ID	Description
GneDfdFceHfcFheTls	A3605685T	Household tools
GneDfdFceHfcHlt	A2302244T	Health
GneDfdFceHfcHltMed	A3605686V	Medicines, medical aids and therapeutic appliances
GneDfdFceHfcHltHsv	A3605687W	Total health services
GneDfdFceHfcTpt	A3605688X	Transport
GneDfdFceHfcTptPvh	A2302245V	Purchase of vehicles
GneDfdFceHfcTptOvh	A2302246W	Operation of vehicles
GneDfdFceHfcTptTsv	A2302247X	Transport services
GneDfdFceHfcCom	A2302248A	Communications
GneDfdFceHfcRnc	A2302249C	Recreation and culture
GneDfdFceHfcEdc	A2302250L	Education services
GneDfdFceHfcHcr	A2302251R	Hotels, cafes and restaurants
GneDfdFceHfcHcrCsv	A3605694V	Catering services
GneDfdFceHfcHcrAsv	A3605695W	Accommodation services
GneDfdFceHfcMis	A3605696X	Miscellaneous goods and services
GneDfdFceHfcMisOgd	A3605697A	Other goods
GneDfdFceHfcMisIfs	A2302252T	Insurance and other financial services
GneDfdFceHfcMisOsv	A3606485T	Other services

References

Amemiya, T., & Wu, R. Y. (1972). The effect of aggregation on prediction in the autoregressive model. *Journal of the American Statistical Association, 67*(339), 628–632. arXiv: 0026.

Andrawis, R. R., Atiya, A. F., & El-Shishiny, H. (2011). Combination of long term and short term forecasts, with application to tourism demand forecasting. *International Journal of Forecasting, 27*(3), 870–886. https://doi.org/10.1016/j.ijforecast.2010.05.019

Athanasopoulos, G., Ahmed, R. A., & Hyndman, R. J. (2009). Hierarchical forecasts for Australian domestic tourism. *International Journal of Forecasting, 25*(1), 146–166. https://doi.org/10.1016/j.ijforecast.2008.07.004

Athanasopoulos, G., Hyndman, R. J., Kourentzes, N., & Petropoulos, F. (2017). Forecasting with temporal hierarchies. *European Journal of Operational Research, 262*, 60–74.

Athanasopoulos, G., Steel, T., & Weatherburn, D. (2019). *Forecasting prison numbers: A grouped time series approach*. Melbourne: Monash University.

Australian Bureau of Statistics. (2015). Australian system of national accounts: Concepts, sources and methods. Cat 5216.0.

Australian Bureau of Statistics. (2018). Australian national accounts: National income, expenditure and product. Cat 5206.0.

Billio, M., Casarin, R., Ravazzolo, F., & Van Dijk, H. K. (2013). Time-varying combinations of predictive densities using nonlinear filtering. *Journal of Econometrics, 177*(2), 213–232.

Brewer, K. (1973). Some consequences of temporal aggregation and systematic sampling for ARMA and ARMAX models. *Journal of Econometrics, 1*(2), 133–154. https://doi.org/10.1016/03044076(73)900158

Capistrán, C., Constandse, C., & Ramos-Francia, M. (2010). Multi-horizon inflation forecasts using disaggregated data. *Economic Modelling, 27*(3), 666–677.

Carriero, A., Clark, T. E., & Marcellino, M. (2015). Realtime nowcasting with a Bayesian mixed frequency model with stochastic volatility. *Journal of the Royal Statistical Society: Series A (Statistics in Society), 178*(4), 837–862.

Clark, T. E., & Ravazzolo, F. (2015). Macroeconomic forecasting performance under alternative specifications of time-varying volatility. *Journal of Applied Econometrics, 30*(4), 551–575.

Cogley, T., Morozov, S., & Sargent, T. J. (2005). Bayesian fan charts for UK inflation: Forecasting and sources of uncertainty in an evolving monetary system. *Journal of Economic Dynamics and Control, 29*(11), 1893–1925.

Dunn, D. M., Williams, W. H., & Dechaine, T. L. (1976). Aggregate versus subaggregate models in local area forecasting. *Journal of American Statistical Association, 71*(353), 68–71. https://doi.org/10.2307/2285732

Gamakumara, P. (2019). *Probabilistic forecasts in hierarchical time series* (Doctoral dissertation, Monash University).

Gamakumara, P., Panagiotelis, A., Athanasopoulos, G., & Hyndman, R. J. (2018). *Probabilistic forecasts in hierarchical time series* (Working paper No. 11/18). Monash University Econometrics & Business Statistics.

Geweke, J., & Amisano, G. (2010). Comparing and evaluating Bayesian predictive distributions of asset returns. *International Journal of Forecasting, 26*(2), 216–230.

Gneiting, T. (2005). Weather forecasting with ensemble methods. *Science, 310*(5746), 248–249. https://doi.org/10.1126/science.1115255

Gneiting, T., & Katzfuss, M. (2014). Probabilistic forecasting. *Annual Review of Statistics and Its Application, 1*, 125–151. https://doi.org/10.1146/annurev-statistics-062713-085831

Gneiting, T., Stanberry, L. I., Grimit, E. P., Held, L., & Johnson, N. A. (2008). Assessing probabilistic forecasts of multivariate quantities, with an application to ensemble predictions of surface winds. *Test, 17*(2), 211–235. https://doi.org/10.1007/s11749-008-0114-x

Gross, C. W., & Sohl, J. E. (1990). Disaggregation methods to expedite product line forecasting. *Journal of Forecasting, 9*(3), 233–254. https://doi.org/10.1002/for.3980090304

Hotta, L. K. (1993). The effect of additive outliers on the estimates from aggregated and disaggregated ARIMA models. *International Journal of Forecasting, 9*(1), 85–93. https://doi.org/10.1016/0169-2070(93)90056-S

Hotta, L. K., & Cardoso Neto, J. (1993). The effect of aggregation on prediction in autoregressive integrated moving-average models. *Journal of Time Series Analysis, 14*(3), 261–269.

Hyndman, R. J., Ahmed, R. A., Athanasopoulos, G., & Shang, H. L. (2011). Optimal combination forecasts for hierarchical time series. *Computational Statistics and Data Analysis, 55*(9), 2579–2589. https://doi.org/10.1016/j.csda.2011.03.006

Hyndman, R. J., & Athanasopoulos, G. (2018). *Forecasting: Principles and practice*. OTexts. Retrieved from https://OTexts.com/fpp2

Hyndman, R. J., Athanasopoulos, G., Bergmeir, C., Caceres, G., Chhay, L., O'Hara-Wild, M., ... Zhou, Z. (2019). *Forecast: Forecasting functions for time series and linear models*. Version 8.5. Retrieved from https://CRAN.R-project.org/package=forecast

Hyndman, R. J., & Khandakar, Y. (2008). Automatic time series forecasting: The forecast package for R. *Journal of Statistical Software, 26*(3), 1–22.

Hyndman, R. J., Koehler, A. B., Ord, J. K., & Snyder, R. D. (2008). *Forecasting with exponential smoothing: The state space approach*. Berlin: Springer.

Hyndman, R. J., Lee, A. J., & Wang, E. (2016). Fast computation of reconciled forecasts for hierarchical and grouped time series. *Computational Statistics and Data Analysis, 97*, 16–32. https://doi.org/10.1016/j.csda.2015.11.007

Kourentzes, N., Petropoulos, F., & Trapero, J. R. (2014). Improving forecasting by estimating time series structural components across multiple frequencies. *International Journal of Forecasting, 30*(2), 291–302. https://doi.org/10.1016/j.ijforecast.2013.09.006

Manzan, S., & Zerom, D. (2008). A bootstrap-based non-parametric forecast density. *International Journal of Forecasting, 24*(3), 535–550. https://doi.org/10.1016/j.ijforecast.2007.12.004

Marcellino, M. (1999). Some consequences of temporal aggregation in empirical analysis. *Journal of Business & Economic Statistics, 17*(1), 129–136.

Nijman, T. E., & Palm, F. C. (1990). Disaggregate sampling in predictive models. *Journal of Business & Economic Statistics, 8*(4), 405–415.

Panagiotelis, A., Athanasopoulos, G., Hyndman, R. J., Jiang, B., & Vahid, F. (2019). Macroeconomic forecasting for Australia using a large number of predictors. *International Journal of Forecasting, 35*(2), 616–633.

Rossana, R., & Seater, J. (1995). Temporal aggregation and economic times series. *Journal of Business & Economic Statistics, 13*(4), 441–451.

Schäfer, J., & Strimmer, K. (2005). A shrinkage approach to large-scale covariance matrix estimation and implications for functional genomics. *Statistical Applications in Genetics and Molecular Biology, 4*(1), 1–30.

Shang, H. L., & Hyndman, R. J. (2017). Grouped functional time series forecasting: An application to age-specific mortality rates. *Journal of Computational and Graphical Statistics, 26*(2), 330–343. https://doi.org/10.1080/106018600.2016.1237877. arXiv: 1609.04222.

Silvestrini, A., Salto, M., Moulin, L., & Veredas, D. (2008). Monitoring and forecasting annual public deficit every month: The case of France. *Empirical Economics, 34*(3), 493–524. https://doi.org/10.1007/s00181-007-0132-7. arXiv: 0016.

Smith, M. S., & Vahey, S. P. (2016). Asymmetric forecast densities for us macroeconomic variables from a Gaussian copula model of cross-sectional and serial dependence. *Journal of Business & Economic Statistics, 34*(3), 416–434.

Ben Taieb, S., Taylor, J. W., & Hyndman, R. J. (2017). *Hierarchical probabilistic forecasting of electricity demand with smart meter data* (Working paper).

Tiao, G. C. (1972). Asymptotic behaviour of temporal aggregates of time series. *Biometrika, 59*(3), 525–531. arXiv: 0027.

Vilar, J. A., & Vilar, J. A. (2013). Time series clustering based on nonparametric multidimensional forecast densities. *Electronic Journal of Statistics, 7*(1), 1019–1046. https://doi.org/10.1214/13-EJS800

Villegas, M. A., & Pedregal, D. J. (2018). Supply chain decision support systems based on a novel hierarchical forecasting approach. *Decision Support Systems, 114*, 29–36. https://doi.org/10.1016/j.dss.2018.08.003

Weiss, C. (2018). *Essays in hierarchical time series forecasting and forecast combination* (Doctoral dissertation, University of Cambridge).

Wickramasuriya, S. L., Athanasopoulos, G., & Hyndman, R. J. (2019). Optimal forecast reconciliation for hierarchical and grouped time series through trace minimization. *Journal of the American Statistical Association, 114*(526), 804–819. https://doi.org/10.1080/01621459.2018.1448825

Yagli, G. M., Yang, D., & Srinivasan, D. (2019). Reconciling solar forecasts: Sequential reconciliation. *Solar Energy, 179*, 391–397. https://doi.org/10.1016/j.solener.2018.12.075

Yang, D., Quan, H., Disfani, V. R., & Liu, L. (2017). Reconciling solar forecasts: Geographical hierarchy. *Solar Energy, 146*, 276–286. https://doi.org/10.1016/j.solener.2017.02.010

Zellner, A., & Montmarquette, C. (1971). A study of some aspects of temporal aggregation problems in econometric analyses. *The Review of Economics and Statistics, 53*(4), 335–342.

Zhang, Y., & Dong, J. (2019). Least squares-based optimal reconciliation method for hierarchical forecasts of wind power generation. *IEEE Transactions on Power Systems* (forthcoming). https://doi.org/10.1109/TPWRS.2018.2868175

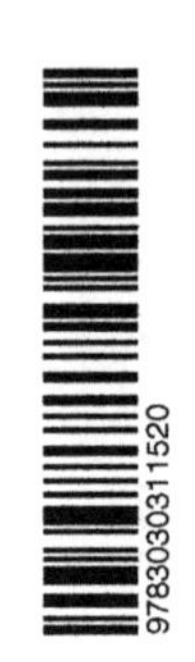